Improving organic crop cultivation

It is widely recognised that agriculture is a significant contributor to global warming and climate change. Agriculture needs to reduce its environmental impact and adapt to current climate change whilst still feeding a growing population, i.e. become more 'climate-smart'. Burleigh Dodds Science Publishing is playing its part in achieving this by bringing together key research on making the production of the world's most important crops and livestock products more sustainable.

Based on extensive research, our publications specifically target the challenge of climate-smart agriculture. In this way we are using 'smart publishing' to help achieve climate-smart agriculture.

Burleigh Dodds Science Publishing is an independent and innovative publisher delivering high quality customer-focused agricultural science content in both print and online formats for the academic and research communities. Our aim is to build a foundation of knowledge on which researchers can build to meet the challenge of climate-smart agriculture.

For more information about Burleigh Dodds Science Publishing simply call us on +44 (0) 1223 839365, email info@bdspublishing.com or alternatively please visit our website at www.bdspublishing.com.

Related titles:

Managing soil health for sustainable agriculture Volume 1: Fundamentals
Print (ISBN 978-1-78676-188-0) Online (ISBN 978-1-78676-190-3; 978-1-78676-191-0)

Managing soil health for sustainable agriculture Volume 2: Monitoring and management
Print (ISBN 978-1-78676-192-7) Online (ISBN 978-1-78676-194-1; 978-1-78676-195-8)

Integrated weed management for sustainable agriculture
Print (ISBN 978-1-78676-164-4) Online (ISBN 978-1-78676-166-8; 978-1-78676-167-5)

Improving organic animal farming
Print (ISBN 978-1-78676-180-4) Online (ISBN 978-1-78676-182-8; 978-1-78676-183-5)

Chapters are available individually from our online bookshop: https://shop.bdspublishing.com

climate•SMART•publishing
IN AGRICULTURAL SCIENCE

BURLEIGH DODDS SERIES IN AGRICULTURAL SCIENCE

NUMBER 47

Improving organic crop cultivation

Edited by Prof. Ulrich Köpke, University of Bonn, Germany

burleigh dodds
SCIENCE PUBLISHING

Published by Burleigh Dodds Science Publishing Limited
82 High Street, Sawston, Cambridge CB22 3HJ, UK
www.bdspublishing.com

Burleigh Dodds Science Publishing, 1518 Walnut Street, Suite 900, Philadelphia, PA 19102-3406, USA

First published 2019 by Burleigh Dodds Science Publishing Limited

Library of Congress Control Number: 2018962571

British Library Cataloguing in Publication Data
A catalogue record for this book is available from the British Library

ISBN 978-1-78676-184-2 (Print)
ISBN 978-1-78676-187-3 (PDF)
ISBN 978-1-78676-186-6 (ePub)
ISSN 2059-6936 (print)
ISSN 2059-6944 (online)

DOI 10.19103/AS.2017.0029

Typeset by Deanta Global Publishing Services, Chennai, India
Printed by Lightning Source

Contents

Series list xi

Introduction xvi

Part 1 Crop breeding and cultivation

1 Key issues in breeding and trialling robust cereal cultivars for organic farming 3
H. Spieß, B. Schmehe and C. Vollenweider, LBS Dottenfelderhof, Germany
1 Introduction 3
2 The origins and aims of organic crop breeding 6
3 Key issues for new cultivars in organic farming systems 6
4 Breeding targets 13
5 Plant health 16
6 Quality requirements 22
7 Conclusions 24
8 Future trends 25
9 Where to look for further information 26
10 References 27

2 Organic seed production, certification and availability 33
Steven P. C. Groot, Wageningen University & Research, The Netherlands; and Maaike H. J. E. Raaijmakers, Bionext, The Netherlands
1 Introduction 33
2 Organic seed quality control issues: seed maturity 35
3 Seed disease and pathogen control 36
4 Application of biologicals to seed 41
5 Seed priming, pelleting and storage 42
6 Seed certification and testing 45
7 Organic seed regulations 47
8 Organic seed production: contamination issues and availability 49
9 Summary 55
10 Future trends 56
11 Where to look for further information 56
12 References 57

3 Maintaining soil fertility and health in organic crop cultivation 61
C. Watson, Scotland's Rural College (SRUC), UK and Swedish University of Agricultural Sciences, Sweden; and E. A. Stockdale, Head of Farming Systems Research, NIAB, UK
1 Introduction 61
2 Some key themes in soil fertility and soil health 63
3 Case study: interactions between soil management and the delivery of soil functions/ecosystem services 69
4 Conclusion and future trends 76
5 Acknowledgements 77

6 Where to look for further information 77
7 References 79

4 Cover crops in organic crop cultivation 87
Peter von Fragstein und Niemsdorff, Universität Kassel, Germany
1 Introduction 87
2 Soil structure and erosion control 89
3 Nutrient cycling 89
4 Weed, pest and disease control 90
5 Promoting biodiversity 92
6 Carbon sequestration and greenhouse gas emissions 94
7 Integration of cover crops in no-till cultivation 95
8 Conclusions and future trends 96
9 Where to look for further information 96
10 References 97

5 The role of crop rotations in organic farming 105
Bernhard Freyer, University of Natural Resources and Life Sciences (BOKU), Austria
1 Introduction 105
2 Principles of crop rotation 106
3 Precrop effects in crop rotations 107
4 Nutrient effects of legumes and other rotation crops 109
5 The role of rotation crops in suppressing weeds, diseases and pests 117
6 Rotations and crop yields 120
7 Designing a crop rotation 122
8 Measuring and modelling crop rotations, nutrient and humus balances 126
9 Crop rotations on stockless farms 129
10 Conclusion and future trends 132
11 Where to look for further information 133
12 References and further reading 133

6 Conservation tillage in organic farming 153
Maike Krauss and Paul Mäder, Research Institute of Organic Agriculture (FiBL), Switzerland; Joséphine Peigné, ISARA-Lyon, France; and Julia Cooper, Newcastle University, UK
1 Introduction 153
2 Main benefits of conservation tillage 155
3 Main challenges of conservation tillage 159
4 Future trends 164
5 Conclusion 170
6 Where to look for further information 170
7 References 171

7 Manure management in organic farming 179
Peter Sørensen, Department of Agroecology, Aarhus University, Denmark; Luca Bechini, University of Milan, Italy; and Lars Stoumann Jensen, University of Copenhagen, Denmark
1 Introduction 179
2 Manure composition and properties in different livestock systems 181
3 Manure storage and treatment 184

4 Manure field application methods 189
5 Turnover and availability of manure N in soil 193
6 Utilization of P, K and S in manures 198
7 Plant-based manures (green manures) 200
8 Future trends and conclusion 201
9 Where to look for further information 202
10 References 203

8 Organic fertilizers and biofertilizers 211
Lidia Sas Paszt and Slawomir Gluszek, Research Institute of Horticulture, Poland
1 Introduction 211
2 Biofertilizers 212
3 Consortia of microorganisms to improve the effectiveness of organic fertilization 216
4 Animal excrement: manures, slurry and guano 217
5 Products and by-products of animal origin 218
6 Products and by-products of plant origin for fertilizers 220
7 Composts 226
8 Untreated minerals and by-products of selected industrial processes 227
9 Biochar 227
10 Conclusion 228
11 Where to look for further information 228
12 References 229

9 Improving water management in organic crop cultivation **243**
Michael J. Goss, University of Guelph, Canada; Adrian Unc, Memorial University of Newfoundland, Canada; and Wilfried Ehlers, Georg-August University, Germany
1 Introduction 243
2 Key aspects of organic farming affecting availability and use of water 255
3 Developments in water management in organic agriculture 261
4 Conclusion 263
5 Where to look for further information 263
6 References 264

Part 2 Crop pests and diseases

10 Disease and pest management in organic farming: a case for applied agroecology **271**
M. R. Finckh, S. M. Junge, J. H. Schmidt and O. D. Weedon, Universität Kassel, Germany
1 Introduction 271
2 General principles of plant protection in organic agriculture 272
3 Case study: soil regeneration and effects on potato health 276
4 Integrating diversity through evolutionary breeding 283
5 Requirements for agroecology-based ecological plant protection 289
6 Conclusion and future trends 292
7 Acknowledgements 293
8 Where to look for further information 294
9 References 294

11 Direct plant protection in organic farming 303
Bernhard Speiser, Hans-Jakob Schärer and Lucius Tamm, Research Institute of Organic Agriculture (FiBL), Switzerland
1 Introduction 303
2 Current practices in organic plant protection 305
3 Currently authorised materials 307
4 Dynamics of authorisation of materials in the EU 310
5 Efforts to replace copper fungicides 316
6 Future trends and conclusion 318
7 Where to look for further information 319
8 References 320

12 Integrated weed management in organic cropping systems 323
Paolo Bàrberi, Scuola Superiore Sant'Anna, Italy
1 Introduction 323
2 Key weed management challenges: intra-row weed control 326
3 Key weed management challenges: problematic weeds and weed management in reduced- and no-till organic systems 330
4 Integrated weed management and sustainable organic crop production 333
5 Future trends 335
6 Where to look for further information 336
7 References 337

Part 3 Standards and monitoring

13 Setting and reviewing standards for organic farming 345
Susanne Padel, The Organic Research Centre, UK
1 Introduction 345
2 Historic development of organic standards and control systems 346
3 Principles of organic production 348
4 Regulation of organic production 349
5 The future of organic standards development 356
6 Future trends and conclusion 359
7 Where to look for further information 361
8 References 361

14 Measuring and improving the environmental performance of organic farming 365
A. Muller, Research Institute of Organic Agriculture (FiBL) and Institute for Environmental Decisions, Department of Environmental Systems Science, Federal Institute of Technology Zurich, Switzerland; M. Meier and C. Schader, Research Institute of Organic Agriculture (FiBL), Switzerland; A. Gattinger, Justus-Liebig University Giessen, Germany; and M. Steffens, Research Institute of Organic Agriculture (FiBL), Switzerland
1 Introduction 365
2 Notions of environmental performance 369
3 Efficiency, consistency and sufficiency 370
4 Methods for assessing environmental performance 371
5 Improving performance 377
6 Conclusions and future trends 378

7 Acknowledgements 379
8 Where to look for further information 380
9 References 380

15 Eco-functional intensification of organic farming 383
Thomas F. Döring, University of Bonn, Germany
1 Introduction 383
2 Material and energy flows 385
3 Ecological regulation 388
4 Technical developments 389
5 Challenges 392
6 Conclusion 394
7 Where to look for further information 395
8 References 395

16 Biodiversity as a prerequisite of sustainable organic farming 401
Lukas Pfiffner and Laura Armengot, Research Institute of Organic Agriculture (FiBL), Switzerland
1 Introduction 401
2 Biodiversity and land-use intensity 402
3 Impact of organic farming on biodiversity 406
4 Biodiversity at different spatial scales 416
5 Impact of organic farming on selected functional groups 418
6 Future trends and conclusion 421
7 References 422

17 The impact of organic agriculture on diet and health 435
Kirsten Brandt, Newcastle University, UK
1 Introduction 435
2 Associations between organic consumption and diet choices 437
3 Organic foods and health 440
4 Combined impact on health of organic foods in diet 450
5 Future trends 452
6 Conclusion 453
7 References 454

Part 4 Organic crop cultivation in the developing world

18 Supporting smallholders in organic crop cultivation: the case of East Africa 463
R. Onwonga, University of Nairobi, Kenya; K. P. Sibuga, Sokoine University of Agriculture, Tanzania; H. Nduku, University of Nairobi, Kenya; L. Sigsgaard, University of Copenhagen, Denmark; A. Saria and L. Shechambo, Sokoine University of Agriculture, Tanzania; M. Montoro, University of Copenhagen, Denmark; C. Chepkoech, Q. Genga and R. Wahome, University of Nairobi, Kenya; N. Halberg, Aarhus University, Denmark; and H. Høgh-Jensen, Technical University of Denmark, Denmark
1 Introduction 463
2 The field study approach 464
3 Challenges and innovations in pest, weed and soil fertility management 466

4 Conclusion 473
5 Future trends 473
6 Where to look for further information 474
7 References 474

19 Improving organic agriculture in the developing world: Africa 477
Victor Olowe, Institute of Food Security, Environmental Resources and Agricultural Research (IFSERAR), Federal University of Agriculture, Abeokuta (FUNAAB), Nigeria
1 Introduction 477
2 Status of organic agriculture in Africa 479
3 EOA-I in Africa 481
4 Steps to improving organic agriculture in Africa 481
5 Conclusion 488
6 Where to look for further information 488
7 References 489

20 Improving organic agriculture in the developing world: Asia 493
Sang Mok Sohn, Dankook University, South Korea
1 Introduction 493
2 Key crops grown in Asia 493
3 Organic production in Asia 494
4 Rice production systems 495
5 Types of organic rice farming 497
6 Organic vegetable production 498
7 Improving organic agriculture in Asia 502
8 Conclusions and future trends 504
9 References 505

21 Organic agriculture and agroecology in Latin America 509
Roberto Ugás, Universidad Nacional Agraria La Molina, Peru
1 Introduction 509
2 Principles 512
3 Organisations and movements 513
4 Production and business 516
5 Institutions and policies 525
6 Performance of organic systems 526
7 Summary and future trends 531
8 Acknowledgements 532
9 References 532

Index 539

Series list

Title	Series number
Achieving sustainable cultivation of maize - Vol 1 From improved varieties to local applications Edited by: Dr Dave Watson, CGIAR Maize Research Program Manager, CIMMYT, Mexico	001
Achieving sustainable cultivation of maize - Vol 2 Cultivation techniques, pest and disease control Edited by: Dr Dave Watson, CGIAR Maize Research Program Manager, CIMMYT, Mexico	002
Achieving sustainable cultivation of rice - Vol 1 Breeding for higher yield and quality Edited by: Prof. Takuji Sasaki, Tokyo University of Agriculture, Japan	003
Achieving sustainable cultivation of rice - Vol 2 Cultivation, pest and disease management Edited by: Prof. Takuji Sasaki, Tokyo University of Agriculture, Japan	004
Achieving sustainable cultivation of wheat - Vol 1 Breeding, quality traits, pests and diseases Edited by: Prof. Peter Langridge, The University of Adelaide, Australia	005
Achieving sustainable cultivation of wheat - Vol 2 Cultivation techniques Edited by: Prof. Peter Langridge, The University of Adelaide, Australia	006
Achieving sustainable cultivation of tomatoes Edited by: Dr Autar Mattoo, USDA-ARS, USA & Prof. Avtar Handa, Purdue University, USA	007
Achieving sustainable production of milk - Vol 1 Milk composition, genetics and breeding Edited by: Dr Nico van Belzen, International Dairy Federation (IDF), Belgium	008
Achieving sustainable production of milk - Vol 2 Safety, quality and sustainability Edited by: Dr Nico van Belzen, International Dairy Federation (IDF), Belgium	009
Achieving sustainable production of milk - Vol 3 Dairy herd management and welfare Edited by: Prof. John Webster, University of Bristol, UK	010
Ensuring safety and quality in the production of beef - Vol 1 Safety Edited by: Prof. Gary Acuff, Texas A&M University, USA & Prof. James Dickson, Iowa State University, USA	011
Ensuring safety and quality in the production of beef - Vol 2 Quality Edited by: Prof. Michael Dikeman, Kansas State University, USA	012
Achieving sustainable production of poultry meat - Vol 1 Safety, quality and sustainability Edited by: Prof. Steven C. Ricke, University of Arkansas, USA	013
Achieving sustainable production of poultry meat - Vol 2 Breeding and nutrition Edited by: Prof. Todd Applegate, University of Georgia, USA	014
Achieving sustainable production of poultry meat - Vol 3 Health and welfare Edited by: Prof. Todd Applegate, University of Georgia, USA	015

Achieving sustainable production of eggs - Vol 1 016
Safety and quality
Edited by: Prof. Julie Roberts, University of New England, Australia

Achieving sustainable production of eggs - Vol 2 017
Animal welfare and sustainability
Edited by: Prof. Julie Roberts, University of New England, Australia

Achieving sustainable cultivation of apples 018
Edited by: Dr Kate Evans, Washington State University, USA

Integrated disease management of wheat and barley 019
Edited by: Prof. Richard Oliver, Curtin University, Australia

Achieving sustainable cultivation of cassava - Vol 1 020
Cultivation techniques
Edited by: Dr Clair Hershey, formerly International Center for Tropical Agriculture (CIAT), Colombia

Achieving sustainable cultivation of cassava - Vol 2 021
Genetics, breeding, pests and diseases
Edited by: Dr Clair Hershey, formerly International Center for Tropical Agriculture (CIAT), Colombia

Achieving sustainable production of sheep 022
Edited by: Prof. Johan Greyling, University of the Free State, South Africa

Achieving sustainable production of pig meat - Vol 1 023
Safety, quality and sustainability
Edited by: Prof. Alan Mathew, Purdue University, USA

Achieving sustainable production of pig meat - Vol 2 024
Animal breeding and nutrition
Edited by: Prof. Julian Wiseman, University of Nottingham, UK

Achieving sustainable production of pig meat - Vol 3 025
Animal health and welfare
Edited by: Prof. Julian Wiseman, University of Nottingham, UK

Achieving sustainable cultivation of potatoes - Vol 1 026
Breeding improved varieties
Edited by: Prof. Gefu Wang-Pruski, Dalhousie University, Canada

Achieving sustainable cultivation of oil palm - Vol 1 027
Introduction, breeding and cultivation techniques
Edited by: Prof. Alain Rival, Center for International Cooperation in Agricultural Research for Development (CIRAD), France

Achieving sustainable cultivation of oil palm - Vol 2 028
Diseases, pests, quality and sustainability
Edited by: Prof. Alain Rival, Center for International Cooperation in Agricultural Research for Development (CIRAD), France

Achieving sustainable cultivation of soybeans - Vol 1 029
Breeding and cultivation techniques
Edited by: Prof. Henry T. Nguyen, University of Missouri, USA

Achieving sustainable cultivation of soybeans - Vol 2 030
Diseases, pests, food and non-food uses
Edited by: Prof. Henry T. Nguyen, University of Missouri, USA

Achieving sustainable cultivation of sorghum - Vol 1 031
Genetics, breeding and production techniques
Edited by: Prof. William Rooney, Texas A&M University, USA

Achieving sustainable cultivation of sorghum - Vol 2 032
Sorghum utilization around the world
Edited by: Prof. William Rooney, Texas A&M University, USA

Achieving sustainable cultivation of potatoes - Vol 2 033
Production, storage and crop protection
Edited by: Dr Stuart Wale, Potato Dynamics Ltd., UK

Achieving sustainable cultivation of mangoes 034
Edited by: Professor Víctor Galán Saúco, Instituto Canario de Investigaciones Agrarias (ICIA), Spain & Dr Ping Lu, Charles Darwin University, Australia

Achieving sustainable cultivation of grain legumes - Vol 1 035
Advances in breeding and cultivation techniques
Edited by: Dr Shoba Sivasankar et al., formerly International Crops Research Institute for the Semi-Arid Tropics (ICRISAT), India

Achieving sustainable cultivation of grain legumes - Vol 2 036
Improving cultivation of particular grain legumes
Edited by: Dr Shoba Sivasankar et al., formerly International Crops Research Institute for the Semi-Arid Tropics (ICRISAT), India

Achieving sustainable cultivation of sugarcane - Vol 1 037
Cultivation techniques, quality and sustainability
Edited by: Prof. Philippe Rott, University of Florida, USA

Achieving sustainable cultivation of sugarcane - Vol 2 038
Breeding, pests and diseases
Edited by: Prof. Philippe Rott, University of Florida, USA

Achieving sustainable cultivation of coffee 039
Edited by: Dr Philippe Lashermes, Institut de Recherche pour le Développement (IRD), France

Achieving sustainable cultivation of bananas - Vol 1 040
Cultivation techniques
Edited by Prof. Gert H. J. Kema, Wageningen University and Research, The Netherlands; and Prof. André Drenth, University of Queensland, Australia

Global Tea Science 041
Current status and future needs
Edited by: Dr V. S. Sharma, formerly UPASI Tea Research Institute, India & Dr M. T. Kumudini Gunasekare, Coordinating Secretariat for Science Technology and Innovation (COSTI), Sri Lanka

Integrated weed management 042
Edited by: Emeritus Prof. Rob Zimdahl, Colorado State University, USA

Achieving sustainable cultivation of cocoa 043
Edited by: Prof. Pathmanathan Umaharan, Cocoa Research Centre – The University of the West Indies, Trinidad and Tobago

Robotics and automation for improving agriculture 044
Edited by: Prof. John Billingsley, University of Southern Queensland, Australia

Water management for sustainable agriculture 045
Edited by: Prof. Theib Oweis, ICARDA, Jordan

Improving organic animal farming 046
Edited by: Dr Mette Vaarst, Aarhus University, Denmark & Dr Stephen Roderick, Duchy College, Cornwall, UK

Improving organic crop cultivation 047
Edited by: Prof. Ulrich Köpke, University of Bonn, Germany

Managing soil health for sustainable agriculture - Vol 1 048
Fundamentals
Edited by: Dr Don Reicosky, Soil Scientist Emeritus USDA-ARS and University of Minnesota, USA

Managing soil health for sustainable agriculture - Vol 2 049
Monitoring and management
Edited by: Dr Don Reicosky, Soil Scientist Emeritus USDA-ARS and University of Minnesota, USA

Rice insect pests and their management 050
E. A. Heinrichs, Francis E. Nwilene, Michael J. Stout, Buyung A. R. Hadi & Thais Freitas

Improving grassland and pasture management in temperate agriculture 051
Edited by: Prof. Athole Marshall and Dr Rosemary Collins, IBERS, Aberystwyth University, UK

Precision agriculture for sustainability 052
Edited by: Dr John Stafford, Silsoe Solutions, UK

Achieving sustainable cultivation of temperate zone tree fruit and berries – Vol 1 053
Physiology, genetics and cultivation
Edited by: Prof. Gregory Lang, Michigan State University, USA

Achieving sustainable cultivation of temperate zone tree fruit and berries – Vol 2 054
Case studies
Edited by: Prof. Gregory Lang, Michigan State University, USA

Agroforestry for sustainable agriculture 055
Edited by: Prof. María Mosquera-Losada, University of Santiago de Compostela, Spain & Dr Ravi Prabhu, World Agroforestry Centre (ICRAF), Kenya

Achieving sustainable cultivation of tree nuts 056
Edited by: Prof. Ümit Serdar, Ondokuz Mayis University, Turkey & Emeritus Prof. Dennis Fulbright, Michigan State University, USA

Assessing the environmental impact of agriculture 057
Edited by: Prof. Bo P. Weidema, Aalborg University/2.-0 LCA Consultants, Denmark

Critical issues in plant health: 50 years of research in African agriculture 058
Edited by: Dr Peter Neuenschwander and Dr Manuele Tamò, IITA, Benin

Achieving sustainable cultivation of vegetables 059
Edited by: Emeritus Prof. George Hochmuth, University of Florida, USA

Advances in breeding techniques for cereal crops 060
Edited by: Prof. Frank Ordon, Julius Kuhn Institute (JKI), Germany & Prof. Wolfgang Friedt, Justus-Liebig University of Giessen, Germany

Advances in Conservation Agriculture – Vol 1 061
Systems and science
Edited by: Prof. Amir Kassam, University of Reading, UK

Advances in Conservation Agriculture – Vol 2 062
Practice and benefits
Edited by: Prof. Amir Kassam, University of Reading, UK

Achieving sustainable greenhouse cultivation 063
Edited by: Prof. Leo Marcelis and Dr Ep Heuvelink, Wageningen University, The Netherlands

Achieving carbon-negative bioenergy systems from plant materials 064
Edited by: Dr Chris Saffron, Michigan State University, USA

Achieving sustainable cultivation of tropical fruits 065
Edited by: Prof. Elhadi Yahia, Universidad Autónoma de Querétaro, Mexico

Advances in postharvest management of horticultural produce 066
Edited by: Prof. Chris Watkins, Cornell University, USA

Pesticides and agriculture 067
Profit, politics and policy
Dave Watson

Integrated management of diseases and insect pests of tree fruit 068
Edited by: Prof. Xiangming Xu and Dr Michelle Fountain, NIAB-EMR, UK

Integrated management of insect pests: Current and future developments 069
Edited by: Emeritus Prof. Marcos Kogan, Oregon State University, USA & Prof. Leon Higley, University of Nebraska-Lincoln, USA

Preventing food losses and waste to achieve food security and sustainability 070
Edited by: Prof. Elhadi M. Yahia, Universidad Autónoma de Querétaro, Mexico

Achieving sustainable management of boreal and temperate forests 071
Edited by: Dr John Stanturf, Estonian University of Life Sciences (formerly US Forest Service), USA

Advances in breeding of dairy cattle 072
Edited by: Prof. Julius van der Werf, University of New England, Australia & Dr Jennie Pryce, DEDJTR-Victoria/La Trobe University, Australia

Improving gut health in poultry 073
Edited by: Prof. Steven C. Ricke, University of Arkansas, USA

Achieving sustainable cultivation of barley 074
Edited by: Dr Glen Fox, University of Queensland, Australia & Prof. Chengdao Li, Murdoch University, Australia

Advances in crop modelling for a sustainable agriculture 075
Edited by: Emeritus Prof. Ken Boote, University of Florida, USA

Achieving sustainable crop nutrition 076
Edited by: Prof. Zed Rengel, University of Western Australia

Achieving sustainable urban agriculture 077
Edited by: Prof Han Wiskerke, Wageningen University, The Netherlands

Introduction

Global sales of organic products have grown significantly, yet organic farming remains a small percentage of overall agricultural production with lower yields than conventional methods. Organic crop cultivation thus faces a range of challenges if it is to grow significantly. This volume reviews the wealth of research addressing these challenges. The volume surveys the key advances and best practice in cultivation techniques across the value chain of organic farming, as well as discussing ways of monitoring and improving the environmental impact of organic crop production. The book has a particular focus on ways of supporting organic farming in the developing world.

Part 1 Crop breeding and cultivation

The first part of the volume reviews developments in improving cultivation across the value chain, from breeding more robust, low input varieties to ways of maintaining soil health and improving crop nutrition. The theme of Chapter 1 is the key issues in breeding and trialling robust cereal cultivars for organic farming. The chapter examines desirable traits for new cultivars in organic farming systems, including seed saving for use in future crops, nutritional quality, plant architecture and health, the competiveness of crops against weeds, and the importance of maintaining genetic diversity. It then reviews how these aims can be achieved in breeding new varieties of organic wheat, oats and winter barley. Finally, the chapter looks at ways of balancing nutritional and other quality attributes.

Moving on from organic breeding to focus on seeds, Chapter 2 deals with organic seed production, certification and availability. Through evolution, plants produce seeds with variation in speed of germination and dormancy in order to spread the risk of germinating in the wrong season. Although domestication and breeding have created crops with more rapid germinating seeds (compared with their wild relatives), the variation in performance for many crops is such that seed treatments and other measures are needed to create starting material for commercial crop production. Further measures are also needed during cultivation, harvesting, cleaning and drying. The chapter looks at the various factors that influence organic seed quality and availability, methods for seed sorting, upgrading and storage, methods of seed sanitation that are applicable for organic seeds and seed certification systems. The chapter identifies research that has resulted in commercially-available methods for upgrading seed quality.

The subject of Chapter 3 is the challenge of maintaining soil fertility and health in organic crop cultivation. The chapter explores key themes in soil fertility and soil health, and describes ways of optimizing organic matter. The chapter considers the challenge of designing appropriate rotations, describes sustainable sources of nutrients in organic farming and looks at the importance of matching supply and demand for nitrogen in organic farming. The chapter includes a detailed case study of interactions between soil management and the delivery of soil functions/ecosystem services.

Building on the foundation of the previous chapter, Chapter 4 moves on to consider the use of cover crops in organic crop cultivation. Cover crops are close-growing crops providing protection for soil and seeding between periods of normal crop production. They are positioned in rotation between two main or cash crops, and not primarily grown for harvest. Although traditionally cover crops have been terminated and incorporated into the soil, the emergence of Conservation Agriculture and no-till systems has led to greater emphasis on retaining cover crop residues as a pre-crop mulch for the following crop. The chapter considers recent research on the benefits of cover crops in areas such as erosion and soil structure, nutrient cycling, and control of weeds, disease and pests. The chapter then looks at the broader role of cover crops in promoting biodiversity, and their role in carbon sequestration and reducing greenhouse gases. Finally, the chapter discusses some of the challenges in using cover crops to optimise crop yields, and presents a number of key questions for the future of this practice.

Following on from the previous chapter's focus on cover crops, Chapter 5 considers the role of crop rotations in organic farming. Crop rotations are defined in terms of a number of crops organized in a specific order over a defined period of time in an individual field, and with specific distribution in a field and landscape, in a way that ensures the best growing conditions for all crops for the benefit of the whole farming system. The chapter examines the principles of crop rotation, precrop effects in crop rotations, and the nutrient effects of legumes and other key rotation crops. The chapter also considers the role of rotation crops in suppressing weeds, diseases and pests, as well as research on rotations and crop yields. The chapter also looks at measuring and modelling crop rotations, nutrient and humus balances, crop rotations on stockless farms, intercropping and crop rotations in tropical and sub-tropical conditions.

Chapter 6 examines the use of conservation tillage in organic farming. Conservation tillage combines the principles of organic farming with the benefits of soil erosion control achieved by the conversion from ploughing to either reduced tillage or no-tillage/direct-seeding. The chapter reviews more than two decades of on-station and on-farm research. The chapter outlines the main benefits of conservation tillage before considering its main challenges. The chapter also contains a number of detailed case studies from the field.

Moving from conservation tillage to manure management, Chapter 7 looks at the use of animal manure in organic farming. Animal manure is the most important source of nutrients applied in most organic farming systems. Nitrogen (N) is often the dominant yield-limiting factor in non-leguminous crops and lack of N may cause reduced yield and low protein content in many crops. In addition, N is easily lost, causing both reduced yields and serious environmental impacts from leaching. The chapter reviews the literature related to manure management and nutrient utilization, including ways to improve manure management in animal housing, during storage and after application in the field. The chapter considers anaerobic versus aerobic manure storage (composting), manure application methods and their effects on N utilization and soil carbon storage. The chapter explains how phosphorus and potassium applied in manure also contribute to crop nutrient supply, especially when nutrient availability in soil is low, as well as maintaining the soil fertility.

The subject of Chapter 8 is organic fertilizers and biofertilizers. An alternative to chemical fertilizers are products of natural origin, organic fertilizers, biofertilizers or natural growth bio-stimulants which increase plant growth, yields, resistance to pathogens and pests, and improve soil fertility, productivity and biodiversity. The chapter examines the role of biofertilizers to improve the effectiveness of organic fertilization, before moving on to consider the use of animal manures, slurry and guano. The chapter also considers the contributions of composts and biochar to crop and soil health.

The final chapter in the section, Chapter 9, looks at improving water management in organic crop cultivation. Managing water for crop production requires capture and storage of water in the soil, which must also be protected from damage by excess water. This protection is provided by increased soil organic matter content, harvest residues, living plants (e.g. cover crops) or a mulch of dead shoot material. The chapter describes actions to deal with excess water and offset water shortage through irrigation, and addresses issues concerning the water cycle, field water balances, soil evaporation, crop transpiration, and crop water use efficiency. The chapter considers new practices that allow soil organisms to protect crops from water shortage.

Part 2 Crop pests and diseases

The second part of the volume discusses the key topic of pests and diseases, with reviews of integrated pest and weed management as well as organic plant protection products. Chapter 10 makes the case for applied agroecology in disease and pest management in organic farming. After outlining the general principles of plant protection in organic agriculture, the chapter presents a detailed case study of soil regeneration and its effects on potato health. The chapter shows how diversity can be integrated through evolutionary breeding and outlines the requirements for agroecology-based ecological plant protection.

Staying with the theme of plant protection, Chapter 11 discusses direct plant protection in organic farming. In organic farming, plant health is maintained primarily through preventive management practices. Nevertheless, there are crops and situations where such measures alone do not provide sufficient protection, and direct plant protection is therefore necessary. The chapter focuses on substances which may be used for plant protection, as well as underlining the importance of indirect plant protection measures for organic farming systems. The chapter describes current practices in organic plant protection, identifies currently authorized materials and discusses the process of authorization of materials in the EU. The chapter examines efforts to find replacements for copper fungicides.

Concluding the section, Chapter 12 looks at integrated weed management in organic crop cultivation. Weed management is a key challenge for organic production. Integrated weed management (IWM), the combination of different tactics within a long-term management strategy, can keep weed abundance low. However, successful IWM relies on maximum diversification of the cropping system whichcan be hard to achieve in practice. The chapter presents examples of successful integration between preventive, cultural and direct tactics in an IWM strategy. It reviews IWM approaches for difficult cases such as intra-row weed control, the presence of problematic weeds, and reduced-tillage systems. Finally, the chapter discusses future advances which could promote adoption of IWM in organic systems.

Part 3 Standards and monitoring

The third part of the volume covers ways of monitoring organic farming, particularly its environmental impact. The theme of Chapter 13 is the setting and reviewing of clear standards for organic farming. There is concern that the current certification system does not fully safeguard the objectives and principles of the movement and encourage further improvements in sustainability. The chapter provides a brief history of organic standard development, and introduces the values and principles of organic farming that have guided this process. The chapter discusses advantages and disadvantages of the different types of rules currently used to achieve common objectives, such as prohibitions or obligations to use specified inputs and practices. The chapter explores whether trends in sustainability assessment could help to close the gap between ambition and rules, concluding that research can improve the evidence base, evaluate the impact of standards and produce a better understanding of the barriers that prevent new producers converting to organic farming.

Complementing the previous chapter's focus on organic standards, Chapter 14 concentrates on measuring and improving the environmental performance of organic farming. There is a large and growing body of literature comparing the environmental impact of organic and conventional farming in specific settings, either on farm or at the product level. However, the question of how 'environmental performance' should be assessed is complex and multi-faceted, and simplistic indicators tend to neglect key aspects of production systems. The chapter evaluates a number of multidimensional methods of assessment, including Life Cycle Assessment, Productivity Analysis, multi-criteria assessment approaches and food systems modelling. The chapter concludes that the assessment of environmental performance must go beyond the level of production, and address consistency and sufficiency aspects that emerge at the consumption stage.

The subject of Chapter 15 is eco-functional intensification of organic farming. Organic farming aims to increase productivity while maintaining a positive impact on biodiversity and using minimal resources. An urgent question of global importance is how organic agricultural systems can be improved with regard to the various functions of producing food, saving biodiversity and reducing resource use – in short, how organic farming can increase its multifunctional performance. The chapter discusses opportunities for and measures of eco-functional intensification (EFI) in organic farming, including material and energy flows, ecological regulation and technical developments. The chapter also summarizes general challenges associated with EFI.

Moving on to the subject of biodiversity, Chapter 16 considers biodiversity as a prerequisite of sustainable organic farming. Over the last century the intensification of agriculture with high inputs of synthetic pesticides and fertilisers, combined with monocrop specialization, has been detrimental to the diversity of genetic resources of crop varieties and livestock breeds, to wild flora and fauna species and to the diversity of ecosystems. Higher levels of biodiversity can strengthen farming systems and their resilience under changing environmental conditions. The chapter gives an overview of the impact of organic farming on biodiversity. The chapter considers land-use intensity, the impact of organic farming on biodiversity, biodiversity at different spatial scales and the impact of organic farming on selected functional groups. Finally, the chapter discusses some examples of the application of functional agro-biodiversity to boost key ecosystem services on organic farms.

The final chapter of the section, Chapter 17, looks at the impact of organic agriculture on diet and health. Consumers of organic vegetables tend to consume them in higher amounts and with more concentrated phytochemicals, so any benefits of phytochemicals for human health are doubled for a typical consumer of organic food. The chapter explores the associations between organic consumption and diet choices, addressing the claimed and perceived links between organic foods and health and assessing the combined nutritional impact of a diet based on organic food.

Part 4 Organic crop cultivation in the developing world

The fourth and final part of the book discusses ways of supporting organic cultivation in the developing world. The subject of Chapter 18 is supporting smallholders in organic crop cultivation, with a particular focus on East Africa. The chapter reports the outcomes of selected ProGrOV (Productivity and Growth in Organic Value Chains) case studies in East Africa involving farmers, students and scientists, together with advisors from organic organizations, in developing and testing agroecological practices for the intensification of kale, maize, sweet pepper and tomatoes for organic markets. After outlining the methodology of the field study approach, the chapter examines challenges and innovations in pest, weed and soil fertility management that can help smallholders.

As Chapter 19 points out, Africa accounts for 25 percent of the world's arable land and yet produces just 10 percent of global agricultural output. Against the backdrop of the challenges to agriculture in Africa, the African Heads of States took a decision in 2011 to integrate organic agriculture into the mainstream agricultural systems of all member states by the year 2020. The chapter describes the current status of organic agriculture in Africa and sets out steps that need to be followed to improve organic agriculture in the continent, from advocacy to improved agricultural techniques and greater government investment.

Moving from Africa to Asia, Chapter 20 considers the improvement of organic agriculture on that continent. Traditional Asian farming methods are often seen as being close to organic production, but there is a need for considerable research and farmer education if the sector is to develop in accordance with scientifically-led organic knowledge and fulfil its potential. The chapter describes key crops grown in Asia and the current state of organic production in Asia, focussing particularly on rice production systems. The chapter examines types of diversified organic rice farming and organic vegetable production. Finally, the chapter deals with the challenge of improving organic agriculture in Asia.

The final chapter of the volume, Chapter 21, examines organic agriculture and agroecology in Latin America. The chapter provides an account of the development and current status of agroecology and organic agriculture in Latin America and the Caribbean (LAC). It highlights the main contributions made by LAC farmers, practitioners, researchers and value chain actors to the worldwide growth of agroecology and organic agriculture. It provides an overview of certified and non-certified organic agriculture in the region, the guarantee systems in place and the markets for such products. The chapter describes national standards and regulations for organic agriculture, as well as programmes for organic agriculture and agroecology which have been developed over the last twenty years. The chapter discusses numerous factors, including favourable policies, capacity building, provision of good quality public services and infrastructure development. All of these are key to promoting the transition of smallholders into more sustainable farming systems. The chapter concludes with a review of the performance of organic systems in the LAC region.

Part 1

Crop breeding and cultivation

Chapter 1

Key issues in breeding and trialling robust cereal cultivars for organic farming

H. Spieß, B. Schmehe and C. Vollenweider, LBS Dottenfelderhof, Germany

1 Introduction
2 The origins and aims of organic crop breeding
3 Key issues for new cultivars in organic farming systems
4 Breeding targets
5 Plant health
6 Quality requirements
7 Conclusions
8 Future trends
9 Where to look for further information
10 References

1 Introduction

Our global food system faces major challenges. A significant increase in crop yields is required in order to secure sufficient food supply for a growing world population. In the past 15 years, world population has grown from 6 billion to 7.5 billion, and according to the United Nations (2015) it is likely to increase up to about ten billion in 2050. The total yield output of our three major food crops has to double within the next 35 years in order to meet the food demand of such a population. According to a study of the research unit of Deutsche Bank, such an increase in production can only be realized when effective cultivation methods are introduced, more crop land is made available and high-yielding cultivars are developed (Schaffnit-Chatterjee et al., 2010). Three methods are discussed to reach these goals: First, the 'biotechnology approach' that relies on methods such as cell culture technologies as well as SMART breeding (selection with markers and advanced reproductive technologies), which is discussed controversially among organic breeders (Lammerts van Bueren et al., 2010). Second, the broader genetic engineering approach that uses different techniques which directly impact the DNA. In recent years the potential and risks of this approach have been vastly extended by the so-called 'new breeding techniques' entailing also new genetic techniques such as the CRISPR/CAS technology (Jinek et al., 2012). Finally, the 'integrated-organic' approach which promises to have the biggest potential to increase average crop

http://dx.doi.org/10.19103/AS.2017.0029.01

Box 1 New technologies in genetic engineering

In recent years, a suite of novel genetic engineering technologies – the so-called 'new breeding techniques' (NBTs) – have emerged in plant breeding. Considerable potential is ascribed in particular to the gene editing technique CRISPR/CAS (Jinek et al., 2012), which seems to be not only much more precise and reliable but also cheaper to use than previous technologies of genetic engineering.

Since some NBTs can be used to induce genetic modifications, which cannot be found in end products (or reliably distinguished from naturally occurring mutations), there is an ongoing political debate whether NBTs should be considered genetic engineering technologies. The organic sector, however, has a clear position in this issue (Nuijten et al., 2017): From the standpoint of the organic sector, a technology or process should always be evaluated in a process rather than a product-oriented framework. Organic produce, for instance, is certified based on the farming process rather than specific characteristics of end products. Therefore, the organic sector holds the view that crop cultivars derived with NBT should be defined as genetically engineered and subject to the mandatory risk assessments.

It remains a question whether the application of NBT should be allowed in organic crop breeding. In a comprehensive analysis, Nuijten et al. (2017) have summarized guiding principles of the organic sector for the evaluation of breeding techniques. In this study, it has been concluded (see also Wickson et al., 2016) that the application of NBT does not comply with the principles and values of organic farming as they are currently incorporated in position papers of IFOAM (2014, 2016) and the European Consortium for Organic Plant Breeding (ECO-PB, 2015). The value system of the organic sector described in these documents is based on a holistic worldview; new technologies have to be evaluated against a set of complexly interrelated agroecological, socio-economic and ethical principles (Verhoog, 2007; Nuijten et al., 2017). Genetic engineering techniques and NBT do not comply with these principles at all the three levels. In particular, NBTs are not compatible with the ethical principle of organic agriculture, which protects the basic integrity of living organisms (Lammerts van Bueren et al., 2003). In organic agriculture an intrinsic value is assigned to all living entities based on their autonomy, level of self-organization, wholeness and dignity. Therefore, breeding techniques that interfere at the subcellular level are not considered suitable for organic crop breeding as the cell can be seen as the lowest level of self-organized life.

Another ethical principle of organic agriculture, which NBTs do not comply with, is the principle of care or precautionary principle (Nuijten et al., 2017). As a consequence of its holistic worldview, the organic sector has a different risk perception than other sectors of society. Unfavourable side effects, including environmental and health risks, are considered inherent to reductionist technological approaches such as gene editing techniques interfering at the level of single nucleotide pairs.

In addition, the application of NBT is incompatible with agroecological and socio-economic principles of organic agriculture (Verhoog, 2007; Nuijten et al., 2017). One of the main concerns of the organic sector is the restriction of the breeder's privilege and farmer's rights to save their own seeds due to intellectual property rights and patents which seem to be inherent to genetic engineering technologies. Additional concerns include the problem of contamination with genetically modified organisms and limited options for breeders, farmers, processors and consumers. Alternative perspectives and development pathways are urgently needed in organic crop breeding (Nuijten et al., 2017; Ceccarelli, 2014).

yields in an environmentally sound approach. Methods that intervene at the subcellular or genomic level are not allowed for use in organic crop breeding (IFOAM, 2014, see also Box 1).

Within the European Union, a number of conventional research projects to breed F1-hybrids have been proposed to reach the goal of nourishing a growing world population. For instance, the German Federal Ministry of Food and Agriculture has allocated a sum of 12 million Euros to eleven innovative and collaborative projects between 2014 and 2019 (BLE, 2014). In the case of bread wheat, the aim is to increase crop yields using F1 hybrids, intensify breeding for resistance to abiotic environmental stresses such as drought and heat stress and implement biotechnological methods. This approach only partially complies with the principles of organic agriculture (Gelinsky and von Frieling, 2016). In contrast to mainstream funding, financial resources for organic breeding are very limited (Kotschi and Wirz, 2015).

In addition to the nutritional gap, another major challenge for agriculture will be climate change. Its impacts range from changing precipitation patterns and more extreme weather events to different and higher disease and pest pressure (Lamichhane et al., 2015). Again, great hope rests on crop breeding to adapt plants to those changing climatic conditions. For example, many scientists attributed the rapid spread of yellow wheat rust (*Puccinia striiformis* f. sp. *tritici*, Fig. 1) that caused major yield losses in Western Europe within the past few years, especially under organic conditions, partially to milder temperatures in winter (Hovmøller et al., 2015). More resistant cultivars have to be developed while keeping pace with rapidly evolving rust strains.

The above-mentioned approaches to face these challenges share the problem of a single-directional perspective. Simple technological solutions to complex problems are proposed, often with no regard to the larger socio-economic, political and legal context. Insufficient food supplies in many developing countries, for instance, often originate from socio-economic problems, such as income insecurity (FAO, 2015). Other factors, in addition to new breeding approaches, like proper food waste management and a reduction of world meat consumption, have been inadequately taken into consideration, as nearly one-third of the food production in developed countries is lost (Gustavsson et al., 2011) and about one-third of the total agricultural cropland is meanwhile used for the production of animal feed (Alexandratos and Bruinsma, 2012). Further, the privatization of water, soil and seeds restricts the availability of these factors to many people and increases global inequity (Rulli et al., 2013).

Figure 1 Yellow stripe rust (*Puccinia striiformis*) on winter wheat.

In order to meet future challenges of food security, food safety and food sovereignty the World Agriculture Report, initiated by the World Bank and the United Nations, demands a systemic transformation of the agricultural sector (Kiers et al., 2008). According to the report, agricultural policy proposals should focus on the availability of resources and infrastructure for local food supply chains rather than maximum output. Small-scale and peasant farming systems are seen as the best choice for national and regional food security and sovereignty. The report highlights the importance of these systems for social and ecological services and welfare and as a basis for a more sustainable food system. Plant breeding and the development of publicly available cultivars suitable for local farming systems are considered important for the development of local, small-scale agriculture.

Despite these findings, multinational seed and agrochemical corporations propose genetic engineering as a panacea for solving the problems of world hunger and agriculture. Consequently, a prominent focus of these companies has been on the development of herbicide-resistant crops, which allow farmers to kill the weeds in their fields while leaving the crop unharmed. Herbicide-resistant crops have the benefit of saving labour costs but encourage the use of herbicides and are therefore associated with a wide range of environmental and health problems. In general, genetic engineering poses systemic risks for natural and agricultural ecosystems that are difficult to estimate. For this reason this technique does not comply with the principles of organic farming (Jacobsen et al., 2013).

2 The origins and aims of organic crop breeding

Aims quite similar to those of the World Agriculture Report served as the motivation for the first organic crop breeding programmes. The origins of organic crop breeding are deeply intertwined with the emergence of biodynamic agriculture in the first half of the twentieth century. It was Rudolf Steiner, the founder of biodynamic agriculture, who proposed in 1922 in a conversation with the farmer Ernst Stegemann that new cultivars should be developed and breeders may resort to wild plants, and in particular wild grasses, in order to achieve this (Mos and Heyden, 2006). Ever since, biodynamic farmers and breeders have been engaged in organic crop breeding and seed conservation (Steiner, 2004; Vogt, 2007).

The aim of organic crop breeding is to develop cultivars suited for the conditions in organic farming systems (Kristiansen et al., 2006; Lammerts van Bueren et al., 2011; Lammerts van Bueren and Myers, 2012). Organic cultivars should possess traits that cannot be found in conventional cultivars (Wolfe et al., 2008). These include the ability to withstand weed competition and mechanical weed control, plant health and natural resistance to seed-borne diseases, as well as excellent nutritional quality. Organically bred cultivars should be open-pollinated, enabling farmers to produce their own seeds. Moreover, organic cultivars should contribute to a higher genetic and biological diversity in agriculture.

3 Key issues for new cultivars in organic farming systems

3.1 Seed saving

Closed nutrient cycles are a key element of organic agriculture. According to Steiner (2004), farms may be considered as self-contained, individual organisms. It seems plausible to extend the concept of 'closed cycles' by including the production and conservation of seeds. Ideally, in organic agriculture, seeds should be produced on-farm. Thus, it is an important aim of organic crop breeding to develop cultivars which allow farmers to save their own seeds. In addition, it is an ideal of many organic crop breeding initiatives that the practical breeding work takes place on-farm and that farmers participate in the breeding process. The advantages of on-farm breeding include improved cooperation with farmers as well as better adaptation to site-specific environmental and socio-economic conditions (Lammerts van Bueren et al., 2011). Strong evidence for the value of closed (nutrient) cycles at the farm level has been put forward by Rashid et al. (2013). This group found that soil organisms on a specific site benefit from using manure from cows whose fodder has been grown on the same site rather than on some other farm. This 'home field advantage' effect resulted in a conversion rate of manure and nitrogen 20% and 14% higher, respectively, compared with the purchased manure from outside the farm site. In other studies, it has been suggested that selection under organic conditions may generally lead to healthier and more robust plants (Burger et al., 2008; Messmer et al., 2009). The Mandaamin Institute in Wisconsin, USA, is currently developing N-efficient/N_2 fixing maize cultivars from landraces (Goldstein, 2015, 2016), and even conventional breeders now acknowledge the advantage of selecting N-efficient plants under organic and low input conditions (Presterl et al., 2002).

3.2 Nutritional quality

Organic produce should deliver excellent nutritional quality and contribute to the mental and physical well-being of consumers. Thus, the improvement of nutritional qualities of cultivars is one of the main goals of organic crop breeding (Zdravkovic et al., 2014). However, increases in crop yields and improvements in nutritional quality are often conflicting goals. Indeed, while crop yields have increased dramatically during the past 100 years due to an intensification of agricultural practices and intensive fertilization, this process frequently went along with a lower concentration of major and minor nutrients, proteins and secondary metabolites in agricultural produce, which is called the 'dilution effect' (Simmonds, 1995). A standard example is given by the negative correlation between yield and gluten content in bread wheat grains (Fig. 2). Nevertheless, there are some lines/strains that combine a (relatively) high gluten content and a high yield potential, possibly as a result of a more efficient nitrogen transfer from vegetative to generative parts of the crop.

3.3 Plant health

Natural morphological and physiological traits, which lead to resistance or tolerance against a wide range of pests and diseases, can be found in all plants. However, there are numerous specific requirements for crop resistance, which are of particular importance under specific environmental and agronomical conditions. As a general principle, it is essential that organic farmers implement preventive steps within their agronomical practice before they resort to resistance breeding programmes. This is especially important in the case of crop diseases which either are difficult to treat or involve the formation of toxins. Examples of cereal crop diseases that involve toxin formation include the well-known ergot (*Claviceps purpurea*), smuts and bunts (*Ustilago* spp., *Tilletia* spp.) and *Fusarium* (*Fusarium* spp.).

While leaf and spike diseases are of minor concern, the development of cultivars resistant to seed-borne diseases is considered important in organic seed breeding (Lorenz et al., 2006). The application of synthetic seed treatment compounds is prohibited in organic

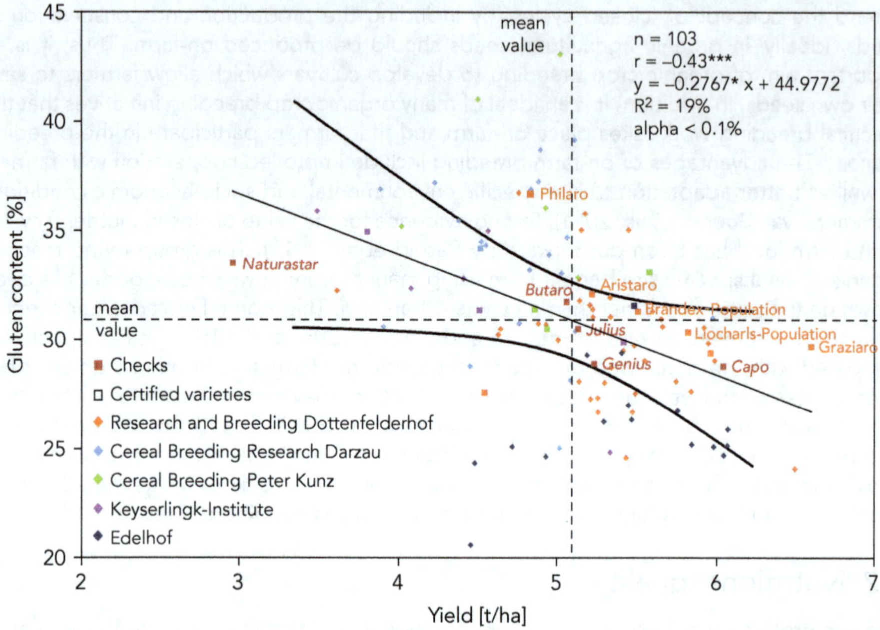

Figure 2 Winter wheat: correlation between total grain yield and gluten content. Source: Winter wheat cultivar trials, Dottenfelderhof 2016.

agriculture and, therefore, there is a lack of efficient control measures against diseases such as dwarf bunt (*Tilletia controversa*), loose smut of wheat, barley and oats (*Ustilago tritici, U. nuda and U. avenae*), barley leaf stripe (*Pyrenophora graminea),* leaf blotch of oats *(Drechslera avenae*) and covered smut of barley (*Ustilago hordei*). It is important to note that the cultivars in many national performance trials, for example, those conducted by the German Federal Plant Variety Office (Bundessortenamt), are not routinely screened for these diseases. It is only through the pioneering work of biodynamic crop breeders in Germany that we have knowledge about resistant cultivars (Müller, 2004; Schmehe and Spieß, 2011, 2015). This knowledge is undoubtedly an important basis for successful future organic crop breeding programmes.

Another important difference between leaf and spike diseases and seed-borne diseases is that in case of the former category, crops are usually endowed with quantitative, horizontal and polygenic resistance, while resistance against seed-borne diseases seem to require qualitative, vertical and mono- or oligogenic resistance mechanisms. Mono- and oligogenic resistance are usually fully effective. However, it is relatively easy for pathogens to overcome the resistance (Robinson, 1996). For this reason, breeders try to combine many different monogenic resistances in a single plant.

3.4 Competitiveness against weeds

Compared with mainstream agriculture, less efficient weed control is one of the main limitations in organic agriculture, resulting in yield losses and higher weed abundance, which may, on the other hand, increase the associated biodiversity. However, competition for light is a key element for weed suppression. Thus, an effective strategy for weed control in organic agriculture is the restriction of light penetration through crop shading. Rademacher (1950) investigated the species- and cultivar-specific weed suppression ability through light absorption in crops. Ammon and Kunz (1982) used the same methodology to compare wheat cultivars and showed that by selection for special plant morphology, especially plant height and planophile instead of erectophile leaf position, the weed suppression was increased. The use of modern short wheat cultivars from conventional breeding programmes under organic conditions can lead to increasing problems, for example, with couch grass (*Elymus repens*). In organic agriculture low crop density and wide row spacing are typical for cereal crops. Thus, shading ability has to be increased by morphological features adapted to these conditions (Eisele and Köpke, 1997a). Important traits are early crop vigour, ground cover and light interception (Drews et al., 2009), high season ground crop cover (Hoad et al., 2012), high tillering ability, prostrate growth, plant height (taller types), high leaf area index (LAI) as well as leaf posture (Eisele and Köpke, 1997b; Hoad et al., 2012). Cultivars with planophile leaf inclination can reduce weed growth significantly, especially at wide row spacings that may be necessary for allowing additional harrowing when pressure of perennial and climbing weeds is high (Eisele and Köpke, 1997b; Drews et al., 2009). Hoad et al. (2012) suggested a method for the assessment of leaf angles and growth habit ideotypes. In practice, measuring the LAI and other plant features is a time-consuming process. However, as these traits are closely correlated with shading ability, that is, crop ground cover, it is sufficient for practical breeder's selection processes if a visual assessment of the crop ground cover at different stages of plant development is made to fulfil the aim of high competitiveness against weeds (see Fig. 6). It was interesting to learn that commercial breeding for nitrogen efficiency resulted in a winter wheat cultivar 'Pegassos' (breeder: Spanakakis, Strube, Germany) that displayed the morphological features which enhance weed competitiveness as outlined above (Neuhoff et al., 2005).

3.5 The importance of biodiversity

A high level of genetic diversity constitutes the foundation of crop breeding. Of particular importance for organic breeders is a diverse gene pool of varieties and lines endowed with various morphological features, disease and pest resistance as well as specific quality traits.

Organic farmers and breeders strive to preserve and enhance the diversity of natural and agricultural ecosystems. There are several ways organic breeders contribute to an improved, planned and associated biodiversity. An integral part of most breeding projects is the *in situ* conservation, which breeders use in order to examine and preserve varieties and breeding lines on their research stations (Fig. 3 and Box 2). Moreover, in recent years, efforts have been made to supplement breeding nurseries and gene banks with more decentralized conservation strategies. In 2008 the European Commission passed new legislation which facilitates the approval of landraces and varieties, which are naturally adapted to local and regional conditions and threatened by genetic erosion. The European Directive 2008/62/EC allows farmers and breeders to grow and market seeds of these 'conservation varieties', which would otherwise not comply with the requirements for cultivar approval (Commission Directive, 2008). An important improvement of the new seed legislation has been to lower the costs and administrative efforts for the approval of these cultivars. In Germany, for instance, the costs to approve

Figure 3 Genetic diversity in wheat. Field trials at Forschung und Züchtung Dottenfelderhof, Bad Vilbel, Germany.

conservation varieties amounts to only 30 Euros/year. Thanks to this regulation, it has been possible to popularize a number of German landraces and heirloom varieties on the seed market.

Another important contribution to biodiversity is expected from the introduction of so-called Composite Cross Populations (CCPs) in organic agriculture, a concept which was originally introduced by Harlan and Martini (1929) and Suneson (1956). CCPs are formed by mating groups of parents of self-pollinating plants. Subsequent generations are then grown and harvested as bulk populations. The key idea of CCPs is to create genetically diverse populations, which are then improved by means of natural selection and supervised selection by breeders. CCPs may offer several potential advantages for farmers and ecosystems, including better site adaptation, higher yield stability and increased and more durable disease and pest resistance (Finckh and Wolfe, 2015; Murphy et al., 2004; Paillard et al., 2000a,b; Rhoné et al., 2007), as well as an ability to adapt to changing and increasingly unstable environmental conditions. The latter trait is of particular importance with regard to climate change. In developing countries CCPs may offer great potential to increase biodiversity and improve food security and food sovereignty, especially when there is a strong focus on farmers' participation.

Until recently, it has not been possible to introduce CCPs on the European seed market since cultivars have to be distinct, uniform and stable as a prerequisite for official registration (UPOV, 1991). However, new regulations were passed in 2014 with the specific purpose to allow the introduction of heterogeneous plant material on the market (European Commission, 2014). Currently, the permission is restricted to four species: wheat, barley, oats and maize. It should be noted that the legislation includes

Box 2 Examples of organic breeding initiatives and networks

Onsite breeding: On the biodynamic farm 'Dottenfelderhof', the organic research and breeding of cereal and vegetable cultivars (Forschung und Züchtung Dottenfelderhof, 'FZD') has been integrated in the farm's crop rotation for the last 40 years, resulting in cultivars that are adapted to the conditions and practices of organic agriculture.

The 'Keyserlingk Institute' at Lake Constance has built a regional network of breeders, farmers and bakers. Locally adapted cultivars are selected by breeders from landraces, grown by farmers and used by bakers to produce bread and other products. Bakers and farmers return some of their surplus to finance organic breeding programmes.

In the Netherlands, a cooperation between 'FZD', the 'Louis Bolk Institute' and local farmers resulted in the release of two spring wheat CCPs with good yield potential and improved baking quality. In this project breeders worked closely with farmers and used on-farm selection methods to improve the populations for local conditions in the Netherlands.

Farmers also participate in organic breeding activities. Organic farmer Wolfgang Kampmann from Germany, for instance, developed a winter wheat cultivar from own crosses within 10 years and released a conservation variety in 2015 ('Jagsttaler').

In Bavaria, Germany, the organic growers association 'Tagwerk' with more than 700 members produces oats for the company 'Barnhouse'. The association is growing a loose smut-resistant oat cultivar developed at Dottenfelderhof. The 'Barnhouse' company is financially supporting organic oat breeding.

Organic vegetable breeders in Germany from the 'Kultursaat' association developed an extra-occupational trainee course for people interested in organic breeding. The cereal breeding company 'Peter Kunz' (GZPK) initiated a vocational training for active breeders where breeders of different crop species share experience and knowledge.

Even large-scale breeding companies such as KWS show interest in promoting organic breeding. On the 'Klostergut Wiebrechtshausen' estate, which is managed according to the certified organic label 'Naturland', field trials for winter wheat are conducted in cooperation with 'FZD'.

In 2017 the first national organic field day was held in Frankenhausen, Germany, which was a great opportunity to bring all stakeholders of the organic sector together. Farmers could see and compare the cultivars bred for organic agriculture.

one outcrossing species with maize. Both CCPs and random mating maize populations are referred to as 'populations' in these regulations.

A number of European countries, including Denmark, France, Germany, the United Kingdom, Italy and the Netherlands, have implemented the new legislation at the national level. In Germany the first populations of wheat, barley and maize have been approved in 2016 (Fig. 4; Spieß et al., 2017). Because of its temporal status, the new legislation

Figure 4 Winter wheat CCPs developed by Forschung und Züchtung Dottenfelderhof, Bad Vilbel, Germany.

will expire at the end of 2018. Since populations are promising to be particularly suited for the conditions of organic farming, organic growers and breeders should demand an extension of this legislation.

3.6 Social aspects

Within the last 40 years the commercial seed industry has undergone profound changes, which resulted in a concentration of the global seed market on a few global players (Howard, 2009). For the great part of human history, however, crop breeding and seed conservation were seen as tasks for society as a whole. Ever since the advent of agriculture about 10 000 years ago, crop varieties have had the status of a common good and seeds were devolved freely from one generation of farmers to the next.

It is important that farmers and breeders continue to have unrestricted access to seeds and crop varieties. As part of global and transgenerational justice, seeds should be preserved in their diversity and distributed freely. Patents and other intellectual property rights that prevent free access to genetic resources for breeders and farmers are therefore not permitted in organic plant breeding. Ensuring transparency is an important aim of organic crop breeding. Organic breeders must provide all the relevant information for the development of new cultivars. Ideally, the breeding targets are set in consultation with farmers, processors, traders and consumers. The inclusion of all the stakeholders in the breeding process is the foundation for the trust of farmers and consumers alike (IFOAM, 2014, see also Box 2).

4 Breeding targets

The aim of this section is to illustrate particular breeding targets in cereals and, in particular, bread wheat. It is important to keep in mind that the ranking of breeding targets depends on various factors including species and variety of the cultivar, growing region, crop management system and marketing requirements. In addition, breeding objectives may be strongly influenced by government policy and seed regulations.

In the case of the organic bread wheat ideotype, the relevant breeding targets have been summarized by Köpke (2006). In total, the targets include 34 traits (see Fig. 5). For the sake of argument, let us assume that we have a plant which already incorporates 'optimal trait values' for 17 of the 34 traits. If we cross this plant with a second plant which incorporates 'optimal trait values' for the remaining 17 traits, there is only one plant in about 17.2 billion plants in the segregating generations with 'optimal trait values' for all traits. Or, to put it differently, there is only one 'optimal' plant in an area of about 4300 ha. This demonstrates that it is very difficult to breed 'ideal' cultivars which incorporate many favourable characteristics. A realistic aim of a breeder will always be to approach 'the ideal plant' as closely as possible.

The traits which influence the very early stages of crop growth include health of the seed grains, germination ability and germination power. The influence of breeders on these traits is limited since different crop management practices and storage conditions tend

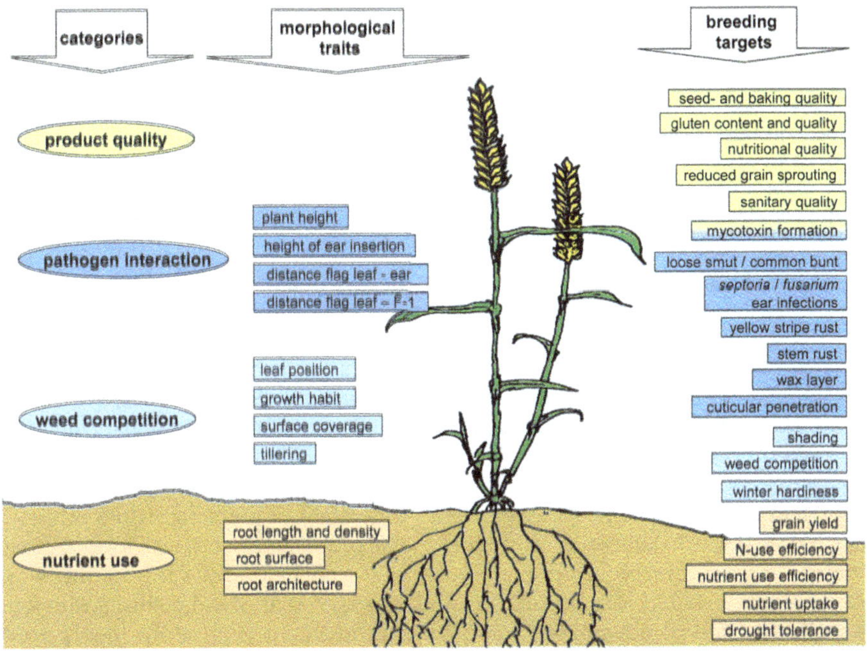

Figure 5 Requirements for the organic crop ideotype in winter wheat. Source: Köpke, 2006 with modifications from H. Spieß.

to mask genetic differences between cultivars. Yet, there are traits with a high heritability (e.g. a high thousand-kernel mass) which support germination and a vigorous early plant growth (Korkut et al., 2001). Breeders may also influence seed quality by endowing the plants with resistance against seed-borne diseases.

In the early stages of crop development, it is important that breeders pay adequate attention to traits such as robustness and winter hardiness. It is plausible that mild winters have led many breeders to neglect these traits in recent years. This may be part of an explanation for the high rates of winter-kill in the growing season 2011/2012 in Germany and other parts of Western Europe. Little is known about phenotypic properties of wheat plants which influence winter hardiness. According to some ancient literature the trait might be linked to the number of low-lying shoots at the tillering stage (Fowler et al., 1981). Breeders' varieties with this property performed well in field trials in 2011/2012.

For vigorous young plant growth, it is important for the seedlings to develop an efficient root system. Embryonic roots (and crown roots at later growth stages) take up water and nutrients and anchor the plant in the soil. A strong root system is of particular importance in organic farming since the plants have to withstand mechanical weed control in the three-to-five leaf stage (Rasmussen, 2004). Resilience to mechanical weed control measures is considered an important objective in organic plant breeding.

Due to a generally lower nutrient availability, high nutrient uptake capacities and nutrient use efficiencies of crop cultivars are particularly relevant in organic agriculture. An effective and vigorous root system is a determining factor for these traits. Nevertheless, assimilates bound in the root systems compete with shoot assimilates as well as a high harvest index. Thus, the ideal root architecture entails a dense fine root system with a maximized root surface, instead of binding assimilates in thick roots and a high total root biomass (Athmann et al., 2013). Field screening for root architecture and root growth might therefore still be a promising strategy (Lammerts van Bueren et al., 2014).

The ability of seedlings to withstand weed competition is another important consideration during early stages of plant growth. Improving weed suppression ability is one of the main objectives of organic crop breeding. Accordingly, breeders prefer cultivars with high soil surface coverage (Fig. 6) and prostrate growth. At the stage of tillering, shoots should cover some ground surface by growing horizontally along the ground.

At later stages of plant growth, organic breeders tend to favour plants with leaves stretching in a more horizontal rather than vertical direction (Eisele and Köpke, 1997a,b). Figure 7 shows the organic cultivar 'Jularo' directly next to a conventional cultivar with poor weed suppression.

Figure 8 depicts the relationship between the traits 'ground coverage at the stage of bolting' and the breeder's visual assessment of the state of the crop before winter. The results clearly show that varieties suited for organic production score well for both traits.

Another important factor, which contributes to the ability of crops to suppress weeds, is plant height. It should also be kept in mind, however, that wheat cultivars with longer stalks offer additional advantages for organic production: High straw yields provide bedding material for livestock farms and help organic farmers to maintain soil organic matter.

Figure 6 Early ground coverage results in a high weed suppression ability and protects the soil. There are considerable varietal differences (from left to right: the biodynamically bred cultivar 'Jularo' and two conventional cultivars often used in organic agriculture).

Figure 7 High weed suppression ability due to a planophile leaf position of the biodynamic cultivar 'Jularo' (on the right) relative to a check variety (left) in the example of common poppy (*Papaver rhoeas* L.).

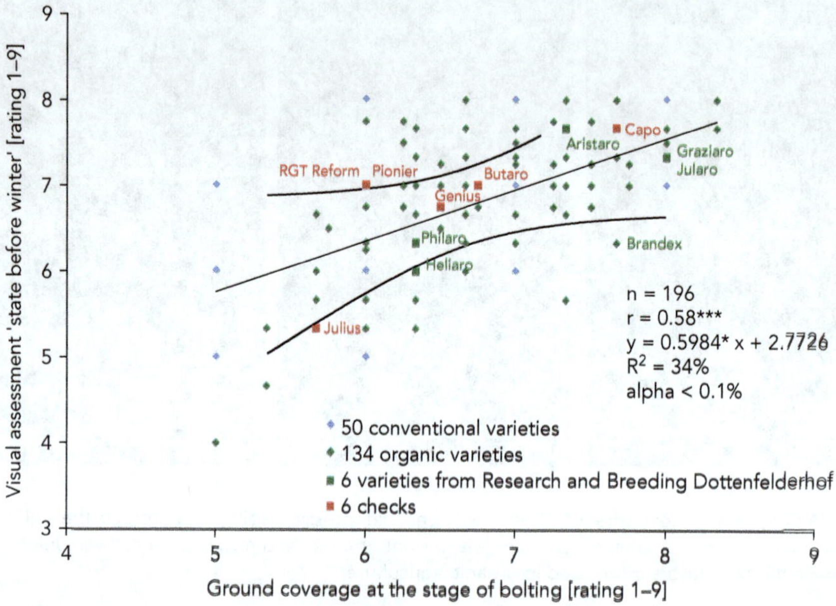

Figure 8 Correlation between the traits 'ground coverage at the stage of bolting' and the breeder's 'visual assessment of the state of the crop before winter'. Trait levels are rated on a scale from 1 to 9, 1 = no expression of the respective trait level, 9 = maximum expression. Source: Research trials Forschung und Züchtung Dottenfelderhof 2016.

5 Plant health

5.1 Wheat

It is important to note that plants with longer stalks benefit from better plant (ear) health in a more favourable microclimate. In fact, dwarf varieties suffer from fungi in the humid milieu closer to the soil surface, even under otherwise favourable growing conditions. Figure 9 shows the negative correlation between *Fusarium* infection and plant length of different winter wheat varieties. Moreover, longer upper internodes tend to reduce the risk of fungal infections such as *Fusarium* head blight (*Fusarium* spp.) or septoria disease (*Septoria nodorum*). The transport of spores via rain drops from the upper leaves to the ear is hindered by longer distances (Schauder et al., 1995). In this context, it should also be mentioned that many breeders prefer ears which are less compact since they dry off more quickly, thus offering better protection from fungal ear and grain infections.

To summarize, cultivars with less susceptibility to *Fusarium* disease are characterized by the following morphological features:

1 tall types with long culms,
2 long distances between flag leaf and ear,
3 awnless with not too dense ears and
4 closed flowering and short flowering phase (Köpke et al., 2007).

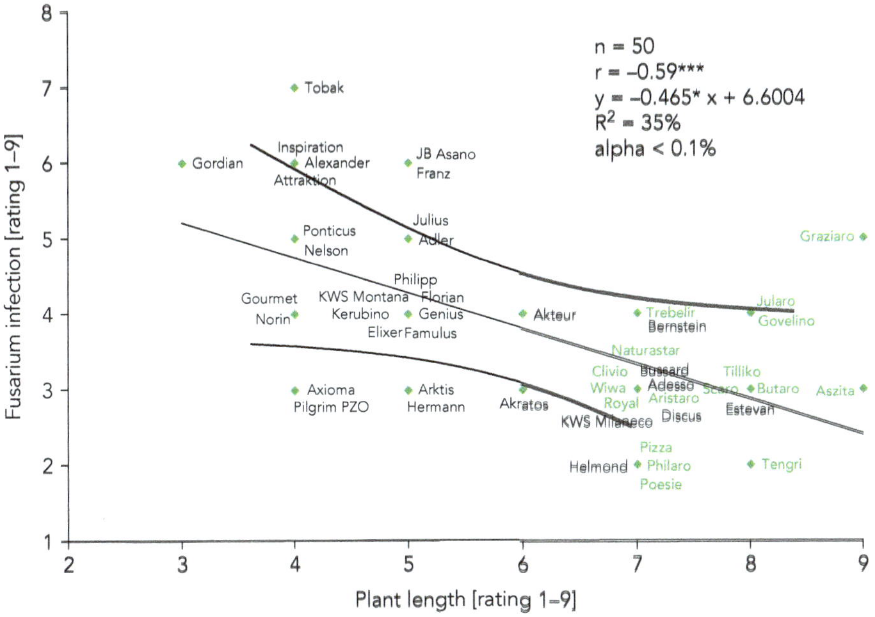

Figure 9 Correlation between culm length and infection with *Fusarium* spp. (on a scale from 1 to 9, where 1 corresponds to a minimal plant length/no infection and 9 to maximum plant length/infection levels). Organic cultivars (green): 'Aristaro', 'Aszita', 'Butaro', 'Clivio', 'Govelino', 'Graziaro', 'Jularo', 'Naturastar', 'Philaro', 'Pizza', 'Poesie', 'Royal', 'Scaro', 'Tengri', 'Tilliko', 'Trebelir', 'Wiwa' can be found at the bottom right. Source: Bundessortenamt (2016) and Agroscope Posieux (2014, 2015).

Indeed, due to potentially significant mycotoxin risks, organic breeders screen their breeding material systematically for lines susceptible to *Fusarium*. For that purpose wheat plants may even be artificially inoculated with spores of *Fusarium culmorum* or *F. graminearum*. Nevertheless, critical values of *Fusarium* toxins (*Deoxynivalenol, Zearalenone*) have been rarely measured under normal cultivation conditions in organic farming (Rembiałkowska, 2007) presumably also as a function of microclimate and low soil nitrogen availability (Köpke et al., 2007).

In recent years, rust diseases such as wheat leaf rust (*Puccinia recondita*) and stripe rust (*P. striiformis*) started to play a more prominent role. Due to warmer winters, it is expected that stem rust (*P. graminis*) will become more relevant in the intermediate term (Junk et al., 2016). In the past, the rule of thumb has been that stripe rust occurs only once in a decade. However, new aggressive strains (Hovmøller et al., 2015) have caused massive yield losses in three consecutive years, 2014–16, throughout Western Europe. Fungicides in mainstream farming are (partially) effective at controlling stripe rust, but in organic agriculture, hopes rest on developing resistant cultivars. Due to rapidly evolving strains, resistance breeding is a challenge. Fortunately, organic and biodynamic breeders have been working on rust resistance for some time and are therefore well prepared. In recent field trials organically bred cultivars have shown breeding success under moderate disease pressure (Fig. 10).

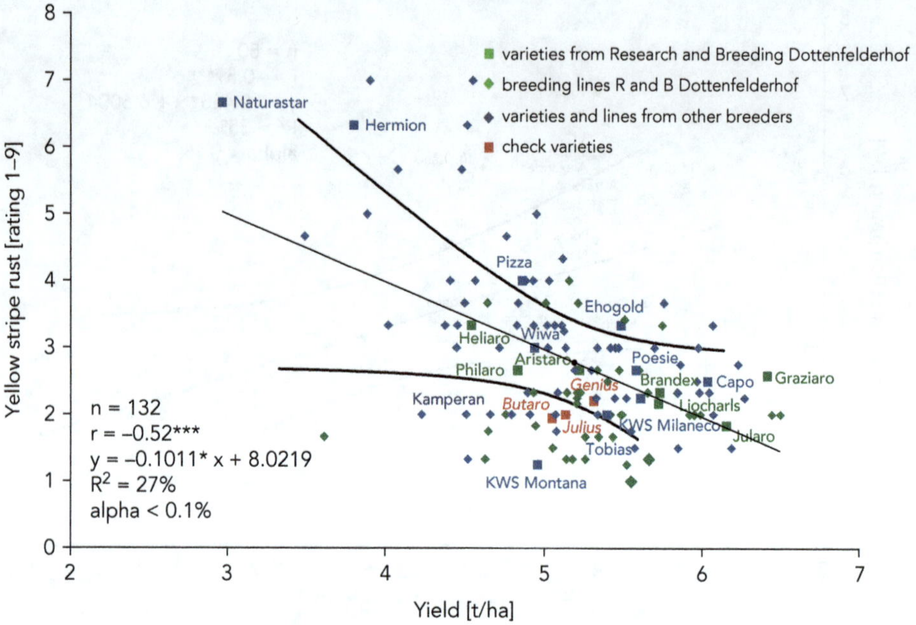

Figure 10 Correlation between total grain yield and infection levels with yellow stripe rust (*Puccinia striiformis*) in a winter wheat trial. Infection levels are rated on a scale from 1 to 9, where 1 indicates no infection and 9 maximum infection levels. Source: trials at Forschung und Züchtung Dottenfelderhof 2016.

As mentioned earlier, seed-borne diseases such as common bunt of wheat play a more important role in organic agriculture than leaf diseases. This is primarily due to the toxic effects of many seed-borne diseases, the problem of seed contamination, limited disease control measures and the lack of implementation of commercial resistance breeding programmes during the past decades (Hagenguth, 2016; Matanguihan et al., 2011).

In mainstream agriculture, the problem of seed-borne diseases has been solved by applying synthetic seed treatment compounds. In contrast, the options in organic agriculture are limited to prophylactic measures. Organic seed treatment preparations, such as Tillecur® or Cerall®, or mechanical seed treatments show varying degrees of effectiveness and increase production costs to the edge of acceptability. In organic production, it would therefore be a significant crop health and economic advantage to develop cultivars with resistance to seed-borne diseases.

One of the most damaging seed-borne diseases occurring in organic cereal cultivation is common bunt of wheat (*Tilletia caries*) (Kristensen and Borgen, 2001; Matanguihan et al., 2011; Spieß, 2014). Common bunt infects all wheat types, including einkorn and spelt wheat, as well as triticale, and can cause yield losses up to 90%. Another seed-borne disease, called dwarf bunt (*Tilletia controversa*), seems to have spread recently to areas of lower altitude in Central Europe (Spieß, 2014). Farmers are particularly concerned about this disease due to the possibility of soil infections and the fact that dwarf bunt spores can survive for about 10 years in the soil. A high contamination with common or dwarf bunt spores may result in a total loss of the crop. Infested grains can be processed into neither flour nor animal feed due to the presence of the toxin *Trimethylamine* and the typical fishy smell. In Germany, for example, wheat cannot be brought onto the market as certified seed if there are more than five infested plants on an area of 150 m² and more than 20 spores/grain. Waldow and Jahn (2007) suggested a threshold of one spore/seed for susceptible wheat cultivars and 20 spores/seed for less susceptible cultivars.

As a first step towards the goal of establishing a successful resistance breeding programme, organic breeders have screened a wide range of wheat cultivars and breeding lines for their susceptibility to common bunt. The results from one of these long-term field trials are depicted in Fig. 11. In a four-year trial, only three out of 177 registered wheat cultivars were visibly free from infestation. Nine cultivars have shown low susceptibility, while the vast majority of the testing material (about 85%) proved to be susceptible to common bunt. Therefore, breeders found it necessary to supplement their breeding material with landraces of international origin as an additional source of resistance. However, this meant that the development of a bunt resistant cultivar suitable for the agricultural and climatic condition of Western Europe would require years of backcrossing. For example the introduction of landraces (of foreign origin) delays the development process by at least five years to about 15 years. Nevertheless, first successes

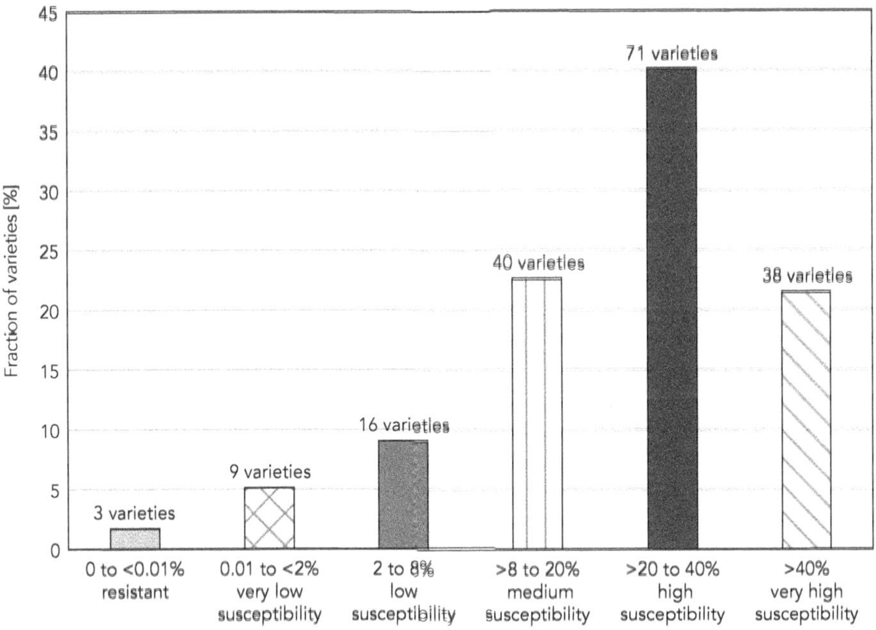

Figure 11 Susceptibility to common bunt of 177 officially registered winter wheat varieties from Germany, Austria and Switzerland. The seeds were inoculated with a dose of 0.2 g common bunt teli-ospores per 100 g of seeds. Source: Spieß (2014).

have already been achieved in recent years. Officially registered bunt resistant cultivars include the winter wheat cultivars 'Aristaro', 'Butaro', 'Graziaro', 'Trebelir' and 'Tilliko', as well as the dwarf bunt tolerant winter wheat cultivars 'Wiwa', 'Aristaro', 'Graziaro' and 'Philaro'.

Organic breeders have also implemented breeding programmes for loose smut of wheat (*Ustilago tritici*), which are just as ambitious as those for common bunt. These programmes have already resulted in the introduction of several loose smut resistant cultivars on the market, which include the winter wheat cultivars 'Jularo', 'Govelino', 'Tilliko' and the spring wheat 'Heliaro'. Similar resistance breeding projects are carried out for loose smut-resistant cultivars of oats (*Avena sativa*) in Central Europe.

In contrast to the aforementioned ear diseases, under present climatic conditions, the resistances incorporated in modern wheat cultivars seem to be sufficient to control leaf diseases such as mildew, *Septoria* or *Drechslera* spp. (Przystalski et al., 2008). However, there are other leaf diseases such as yellow stripe and leaf rust (*Puccinia striiformis, P. recondita*), net blotch (*Pyrenophora teres*) or ramularia leaf spot of barley (*Ramularia collo-cygni*) which seem to be more difficult to control due to their extremely versatile nature and are therefore (potential) targets of resistance breeding programmes.

5.2 Oats

Although the nutritional value of oats is high in comparison to other cereals, oats is considered to be a minor crop due to its shrinking cultivation area. Consequently, the number of oat breeders worldwide is relatively small and the organic oat breeding programmes with the goal of releasing registered cultivars for commercial production are currently limited, to one in Canada and two in Germany, only. In Canada in 2014 the organically bred spring oat 'AAC Oravena' was released (Mitchell Fetch et al., 2014). In Germany a naked oat cultivar 'Talkunar' was released in 2017 and two spring oat varieties 'Kaspero' and 'Sinaba' were released in 2018.

Loose smut (*Ustilago avenae*) resistance plays a major role in organic breeding (Schmehe et al., 2016). The latest comprehensive overview about testing for resistance to smut diseases was given by Menzies et al. (2009). In most countries loose smut does not play a role in the process of cultivar registration, but it can be a severe problem for seed certification. In Canada the Prairie Recommending Committee for Oat and Barley (PRCOB) collects data on the reactions to smuts (*Ustilago* spp.), oat crown rust (*Puccinia coronata*) and oat stem rust (*Puccinia graminis* sp. *avenae*). Entries need to meet a minimum level of resistance to each disease over a period of three years in order for the PRCOB to support the cultivar for commercial production or sale in Canada.

Figure 12 shows the results of an evaluation of oat accessions for resistance to loose smut after artificial inoculation in Germany from 2009 to 2016. Most of the entries with a good level of resistance are varieties with poor agronomic traits that need backcrossings in order to develop cultivars with suitable agronomic traits combined with loose smut resistance. On the other hand, in the group of highly to very highly susceptible entries, a relatively high percentage of cultivars with good agronomic performance can be found. This may result from the fact that under mainstream conditions, loose smut can be controlled with chemical seed treatments and resistance breeding is therefore considered unnecessary.

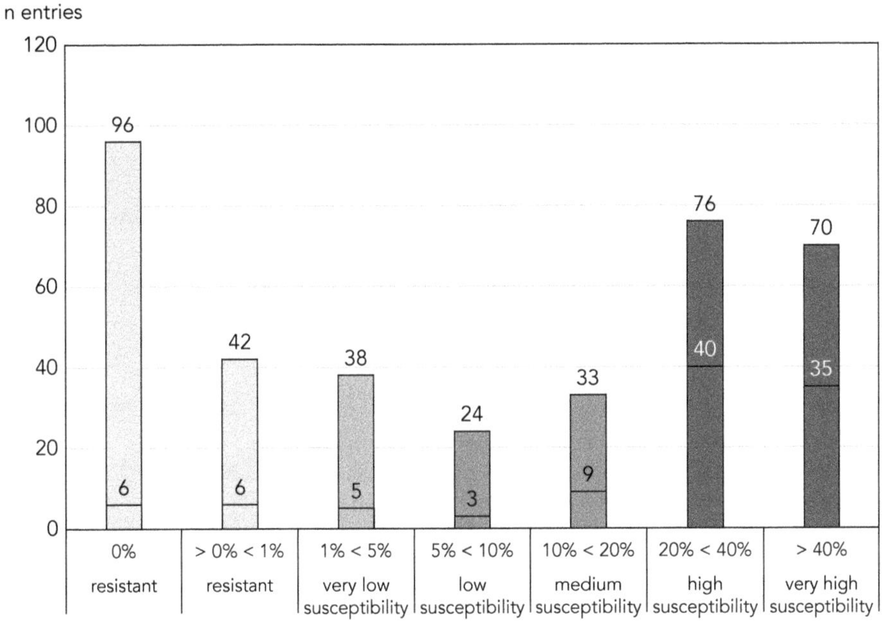

Figure 12 Classes of resistance levels to loose smut (*Ustilago avenae*) as percentage of smutted panicles – lower values represent the fraction of varieties with good agronomic traits. Evaluation from 2009 to 2016, entries tested total n = 379.

5.3 Winter barley

Barley leaf stripe (*Pyrenophora graminea*) does not play a role in VCU tests (value for cultivation and use, Rey et al., 2008) in Europe but can lead to significant yield losses. The disease is strictly seed-borne (Platenkamp, 1976; Delogu et al., 1989). The percentage of ears with symptoms correlates with yield loss in a ratio of 1:0.5–1 (Porta-Puglia et al., 1986; Skou et al., 1992). At present, there is no effective treatment in organic agriculture against stripe disease; thus, resistant varieties are urgently needed in order to avoid severe yield losses.

Artificial inoculation in the laboratory becomes difficult for a large number of entries. Consequently, spreader varieties with a high infection level were used for field infection (cultivars 'Alpaca', 'Etrusco' and 'Landi'). The infection level can then be determined in the following season in the field. Figure 13 shows the results of an evaluation of 192 winter barley entries for resistance to leaf stripe. The number of entries with good agronomic traits (registered cultivars) and low susceptibility seems to be sufficient; however, because of the high yield loss, the tolerable level of infection is much lower than that for loose smut.

In the beginning of the evaluation, winter barley varieties mainly of East Asian origin and those listed in the GENRES final report (Enneking, 2002) with highest levels of resistance were discarded from the test due to poor winter hardiness and weak agronomic traits in Central Europe (Schmehe and Spieß, 2015). Nonetheless, they may perform better in other climates.

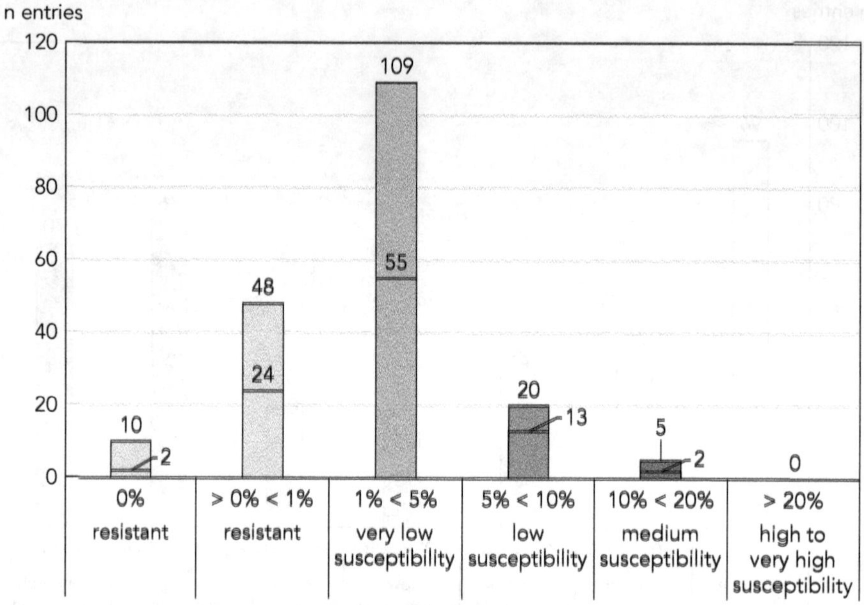

Figure 13 Barley leaf stripe (*Pyrenophora graminea*): classes of resistance levels as percentage of ears with symptoms after field infection – the ordinate shows the number of tested barley entries and lower values indicate the fraction of registered cultivars. Evaluation from 2013 to 2015, entries tested total n = 192.

6 Quality requirements

Organic breeders usually place great emphasis on the physical appearance of kernels. Their bias can be partly explained by a tight correlation between the size of seed kernels and their germination ability and power. In addition, it is assumed that there is a correlation between certain characteristics of their shape and texture and good nutritional and baking qualities (Knott and Talukdar, 1971). Breeders usually select for large, round-oval-shaped, smooth-shelled, fleckless and vitreous kernels. Additional selection criteria for organic breeders include a narrow crease as well as short seed hairs so that fungus spores (e.g. *Tilletia caries*) cannot adhere.

In the case of bread wheat, in mainstream agriculture markets nutritional quality traits are still considered less important than technological baking properties. In conventional as well as organic farming systems, wheat prices are determined by the total (wet) gluten content of the produce, which is supposed to be correlated with a good baking quality. However, new research shows that rather than total gluten content, for bread baking, it is essential that the milled grain contains *specific* amounts of high-quality gluten with an optimal ratio of glutenin and gliadin. In fact, a well-balanced protein composition tends to lead to a high baking volume due to the formation of cross-link molecules (Wieser, 2007).

These findings suggest that it might still be possible to combine (to some extent) a high yield potential and a good baking quality in a single wheat cultivar despite the negative correlation between yield potential and wet gluten content. In organic breeding programmes, it has been demonstrated that by using the so-called 'sedimentation value' based on the SDS method as a selection criterion, it is possible to breed cultivars with a relatively high yield potential, high baking volume and low total gluten content (see Fig. 2 and 14).

In addition, it should be mentioned that superfine and whole wheat flours display different baking characteristics. Whereas a high wet gluten content of >27% is a basic requirement for achieving a good baking quality in the case of superfine flour, the same is not true for whole wheat flour. This difference is particularly relevant under organic conditions since sufficient wet gluten contents can often not be achieved under organic conditions due to limited N availability.

Large annual fluctuations make it important that all technological baking properties, including total gluten content, gluten-index, sedimentation value based on the Zeleny- (Zeleny, 1947) or the SDS-method (McDonald, 1985), or falling number, are determined as early as possible in the selection process. This is also true for alternative methods that determine baking quality properties, which include a direct measurement of baking volumes using a standardized baking test (Linnemann, 2011), the measurement of the content of so-called 'glutenin macropolymers' or GMP (Marti et al., 2015), which seems to be positively correlated to a high baking volume, or a direct evaluation of the taste and consistency of the dough and crumb of bread samples.

The nutritional value of wheat has been recently critically assessed in popular publications with rather sensational titles such as *Wheat Belly* (Davis, 2014) and *Grain Brain* (Perlmutter,

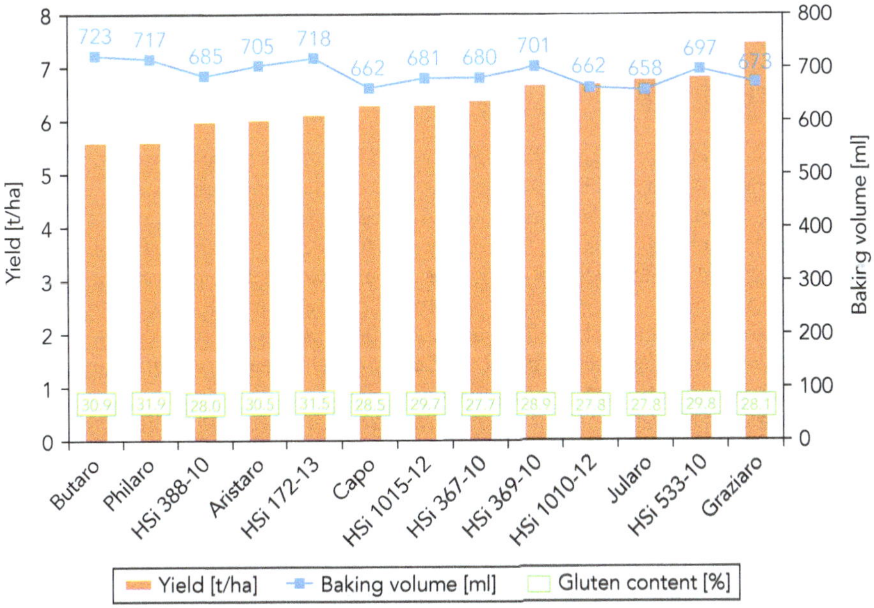

Figure 14 Total grain yield, baking volume and gluten content of selected organically bred cultivars and breeding lines. The cultivar 'Graziaro' combines a high yield potential and a high baking volume. A lower total gluten content is compensated by a high gluten quality. Trials at Forschung und Züchtung Dottenfelderhof 2015 and 2016.

2013). It has been proposed that wheat may cause diabetes, heart diseases, dementia and other illnesses. Yet, there is no scientific evidence for any of these claims. Of greater importance to breeders is the so-called 'non-celiac gluten sensitivity' (NCGS). NCGS is among the most common gluten-related disorders with a prevalence rate between 5 and 10% in the general population. It has been suggested that this disease is caused by specific proteins called amylase trypsin inhibitors (ATIs) (Fasano et al., 2015). Another possible activator for the irritable bowel syndrome could be the so-called 'FODMAPs'. These are fermentable oligo-, di- and monosaccharides and polyols, that is, sugars containing 1–14 sugar molecules, which cannot be fully digested in the small intestine. Ziegler et al. (2016) showed that the content of FODMAPs can be reduced significantly with a longer preparation time of the dough. From a broader perspective, it seems plausible that gluten-related disorders are related to unbalanced dietary habits based on highly processed foods prevalent in most industrialized nations. And indeed, it has been demonstrated that longer preparation times that are necessary for sourdough fermentation can decrease celiac-disease-inducing effects (M'hir et al., 2012). Hence, gluten-related disorders are probably a problem more related to the processing of wheat and not wheat itself.

Today, more than 200 food additives are permitted in commercial baking. Industrially processed dough may contain more than 20 of these compounds including various synthetic enzymes. Moreover, gluten is used as a food supplement to many sweets, desserts and ready-made meals. Thus, the omnipresence of gluten in modern diets might well be part of an explanation for gluten-related disorders including NCGS (Lionetti et al., 2015). Also the abundant use of glyphosate-based herbicides in conventional agriculture is discussed to induce gluten sensitivity (Samsel and Seneff, 2013).

7 Conclusions

Organic crop breeding and organic crop breeding research have entered scientific discourse on IFOAM, ECO-PB and other conferences. Organic breeding is chosen as a central theme for discussion and new breeding methods and ideas as well as breeding successes are presented at these conferences (IFOAM, 2009; Pedersen, 2015).

The general public also increasingly perceives and values the importance of organic crop breeding. This can be seen, for instance, in the example of growing financial support for the 'Seed Fund' of the 'Future Foundation for Agriculture' in Germany, where organic breeders can apply for financial support for their breeding projects (http://www.zukunftsstiftung-landwirtschaft.de/saatgutfonds/).

European seed registration authorities meanwhile offer value for cultivation and use (VCU) tests specifically for cereal crops under organic conditions (Loeschenberger et al., 2008). These allowed private organic breeding initiatives to introduce a number of new cultivars on the European seed market in recent years. Biodynamic crop breeding initiatives from Germany and Switzerland play a leading role in these developments. Among them are cereal breeding initiatives [*Cereal Breeding Research Darzau, www.darzau.de, Forschung und Züchtung* (*Research and Breeding*) *Dottenfelderhof, www.forschung-dottenfelderhof.de, Keyserlingk-Institute, www.saatgut-forschung.de* and *Cereal Breeding Peter Kunz, www.getreidezuechtung.ch*] as well as vegetable breeding initiatives (*Kultursaat, www.kultursaat.org, Dreschflegel, www.dreschflegel-saatgut.de and saat:gut, www.saat-gut.org*).

Due to the limited scale of the organic seed sector, all of these initiatives depend upon financial support from charitable organizations as well as publicly or privately funded research programmes. In order to preserve and expand the organic crop breeding sector, it is necessary to significantly increase awareness among the whole value chain from organic farmers, processors, retailers to end-consumers.

8 Future trends

At the beginning of this chapter, we outlined key future challenges for agriculture including (1) the improvement of global food security and food sovereignty, (2) conservation and enrichment of biodiversity in crop plants and (3) development of strategies for adaptation to the adverse impacts of climate change. We shall complement our introductory remarks with some thoughts and ideas pointing to possible future perspectives for organic plant breeding in these areas:

1 Organic plant breeding seems perfectly capable of making a significant contribution to feeding a growing world population by developing high-yielding crop cultivars without having to rely on (novel techniques of) genetic engineering or F1-hybrids. However, in order to make a lasting contribution to the improvement of global food security and food sovereignty, it is essential that organic breeding programmes take the larger socio-economic context of agriculture and the needs of all stakeholders into account (see e.g. Löwenstein, 2011). Moreover, the enhancement of the nutritional quality of food crops should become an essential consideration in organic plant breeding.
2 The largest symbiotic association on earth is the interlinked, mutually supportive community of crop plants, domestic animals and human beings. This association led to an abundant diversity of crop plants and domestic animals; yet, large portions of this diversity were lost during the past 150 years. The continued existence of crop plants critically depends on the caring hand of mankind. Thus, it is the duty of modern crop breeders to act as a custodian of the symbiotic relationship of crop plants, domestic animals and man and to foster and develop this important foundation of civilization. Food security and food sovereignty cannot be achieved on a global scale unless biodiversity is accessible at the local level. Therefore, regional and local breeding initiatives should be strengthened and locally adapted cultivars made available. Efforts by organic breeding programmes are already contributing to an enrichment of local seed diversity by developing locally adapted conservation-, amateur- and niche-varieties.
3 Global food security for both animals and human beings is threatened by the impacts of climate change ranging from changing precipitation patterns and extreme weather events to higher disease and pest pressures. In September 2017 one of the strongest ever recorded hurricanes, 'Irma', left a trail of destruction in the Caribbean, which affected both people and agriculture. Climatologists expect extreme weather events to become much more frequent in the future, which will have severe negative consequences for agriculture as modern crop plants lack the ability to adapt to rapidly changing conditions. Again, organic breeding might offer a candidate for a viable adaptation strategy with the introduction of so-called 'CCPs'. Due to their

high degree of genetic diversity, it is expected that CCPs can adapt simultaneously to multiple environmental stressors including extreme weather events and higher levels of disease and pest pressure (Ceccarelli et al., 2010; Spieß and Vollenweider, 2017). CCPs should be tested extensively in field trials and practical agriculture.

The enormity of the challenges facing agriculture and the vast scope of possible areas of activity make it clear that sufficient private and public funding for organic breeding programmes should be made available. Only a range of diverse approaches and efforts for continuous improvements will guarantee that organic crop breeding continues to thrive in the near and long-term future, meeting the needs of plants, animals, human beings and nature as a whole.

Contemporary organic crop breeding initiatives are built upon the ancient heritage of human breeding efforts and natural biodiversity. Varieties and their seeds are a cultural treasure from which we develop the plants we are consuming. This cultural asset has to be preserved by everyone who benefits from its yields. Thus, organic crop breeding is a task for society as a whole.

9 Where to look for further information

Organic plant breeding initiatives throughout Europe are working on the improvement and release of organic crop cultivars. The following initiatives are mentioned in this chapter:

- Cereal Breeding Peter Kunz, www.getreidezuechtung.ch,
- Cereal Breeding Research Darzau, www.darzau.de,
- Research and Breeding Dottenfelderhof, www.forschung-dottenfelderhof.de,
- Keyserlingk-Institute, www.saatgut-forschung.de,
- Kultursaat association of vegetable breeders, www.kultursaat.org,
- saat:gut, www.saat-gut.org,
- Louis Bolk Institute, http://www.louisbolk.org/sustainable-agriculture/plant-breeding,
- https://culinaris-saatgut.de/en/
- https://www.seedcooperative.org.uk/
- http://organic-plant-breeding.org/

These initiatives have set a wide range of different (and often complementary) breeding aims and research objectives. General guidelines and principles for organic plant breeding have been developed at the level of IFOAM and by the European Consortium for Organic Plant Breeding (ECO-PB):

- https://www.ifoam.bio/en/sector-platforms/ifoam-seeds-platform
- http://www.eco-pb.org/home.html

Organically bred seeds are supplied by:

- Bingenheimer Saatgut AG, https://www.bingenheimersaatgut.de,
- Sativa Rheinau AG, http://www.sativa-rheinau.ch/
- Dreschflegel, www.dreschflegel-saatgut.de.

10 References

Agroscope Posieux (2014), Cahier Blé d'automne, Plantes Agroscope Transfer Nr. 6/2014.

Agroscope Posieux (2015), Cahier Blé d'automne, Plantes Agroscope Transfer Nr. 76/2015.

Alexandratos, N. and Bruinsma, J. (2012), World agriculture towards 2030/2050: The 2012 revision. ESA Working paper No. 12-03. Rome: FAO, p. 4.

Ammon, H. U. and Kunz, P. (1982), Der Einfluss der Konkurrenzkraft kurzhalmiger Getreidesorten auf die Entwicklung einiger Unkräuter. *Mitt. Schweiz. Landw.* 30, 12–17.

Athmann, M., Kautz, T. and Köpke, U. (2013), Charakterisierung des Wurzelwachstums in Bioporen mit in situ Endoskopie. In: Neuhoff, D., Stumm, C. and Ziegler, S. (Eds), *Ideal und Wirklichkeit - Perspektiven Ökologischer Landbewirtschaftung.* Beiträge zur 12. Wissenschaftstagung Ökologischer Landbau, Bonn. Berlin: Verlag Dr. Köster.

BLE (2014), Getreideforschung: Elf Projekte sollen ertragreiche und umweltstabile Weizensorten liefern. http://www.ble.de/SharedDocs/Downloads/.

Bundessortenamt (2016), *Beschreibende Sortenliste Getreide, Mais, Öl- und Faserpflanzen, Leguminosen, Rüben, Zwischenfrüchte.* Hannover: Bundessortenamt.

Burger, H., Schloen, M., Schmidt, W. and Geiger, H. H. (2008), Quantitative genetic studies on breeding maize for adaptation to organic farming. *Euphytica*, 163(3), 501–10.

Ceccarelli, S. (2014), GM crops, organic agriculture and breeding for sustainability. *Sustainability*, 6(7), 4273–86. doi:10.3390/su6074273.

Ceccarelli, S., Grando, S., Maatougui, M., Michael, M., Slash, M., Haghparast, R., Rahmanian, M., Taheri, A., Al-Yassin, A., Benbelkacem, A., Labdi, M., Mimoun, H. and Nachit, M. (2010), Plant breeding and climate changes. *Journal of Agricultural Science*, 148, 627–37.

Commission Directive (2008), Commission Directive 2008/62/EC of 20 June 2008 providing for certain derogations for acceptance of agricultural landraces and varieties which are naturally adapted to the local and regional conditions and threatened by genetic erosion and for marketing of seed and seed potatoes of those landraces and varieties.

Davis, W. (2014), *Wheat Belly: Lose the Wheat, Lose the Weight, and Find your Path Back to Health*, Rodale, New York, NY, pp. 320.

Delogu, G., Porta-Puglia, A. and Vannacci, G. (1989), Resistance of winter barley varieties subjected to natural inoculum of *Pyrenophora graminea*. *Journal of Genetics and Breeding*, 43(2), 61–6.

Drews, S, Neuhoff, D. and Köpke, U. (2009), Weed suppression ability of three winter wheat varieties of different row spacing under organic farming conditions. *Weed Research* 49, 526–33.

ECO-PB (2015), New Plant Breeding Techniques. Position Paper. Available online: http://www.ifoam-eu.org/sites/default/files/ifoameu_policy_npbts_position_final_20151210.pdf (accessed on 14 June 2017). ECO-PB: European Consortium for Organic Plant Breeding (http://www.eco-pb.org).

Eisele, J. and Köpke, U. (1997a), Choice of cultivars in organic farming new criteria for winter wheat ideotypes. 1: Light conditions in stands of winter wheat affected by morphological features of different varieties. *Pflanzenbauwissenschaften*, 1, 19–24.

Eisele, J. and Köpke, U. (1997b), Choice of cultivars in organic farming: New criteria for winter wheat ideotypes. Vol II: Weed competitiveness of morphologically different cultivars. *Pflanzenbauwissenschaften*, 1(2), 84–9.

Enneking, D. (Ed.). (2002), *GENRES CT-98-104 - Evaluation and Conservation of Barley Genetic Resources to Improve Their Accessibility to Breeders in Europe - Final Report.*

European Commission. (2014), 2014/150/EU: Commission Implementing Decision of 18 March 2014 on the organisation of a temporary experiment providing for certain derogations for the marketing of populations of the plant species wheat, barley, oats and maize pursuant to Council Directive 66/402/EEC (notified under document C(2014) 1681), Text with EEA relevance.

FAO. (2015), *The State of Food Insecurity in the World.* Rome: FAO.

Fasano, A., Sapone, A., Zevallos, V. and Schuppan, D. (2015), Nonceliac gluten sensitivity. *Gastroenterology*, 148(6), 1195–204.

Finckh, M. and Wolfe, M. S. (2015), Biodiversity enhancement. In: Finckh, M., Van Bruggen, H. C. and Tamm, L. (Eds), *Plant Diseases and Their Management in Organic Agriculture*. St. Paul, MN: The American Phytopathological Society (APS Press), pp. 153–74.

Fowler, D. B., Gusta, L. V. and Tyler, N. J. (1981), Selection for winterhardiness in wheat, III. Screening methods. *Crop Science*, 21(6), 896–901.

Gelinsky, E. and von Frieling, H.-D. (2016), Hybridweizen: Vor dem Durchbruch?. *Gen-ethischer Informationsdienst*, 239, 8–11.

Goldstein, W. (2015), Breeding corn for organic farmers with improved N efficiency/N fixation, and protein quality. In: *Proceedings of the Organic Agriculture Research Symposium*, LaCrosse, WI, 25–26 February 2015, pp. 1–4.

Goldstein, W. (2016), Partnership between maize and bacteria for nitrogen efficiency and nitrogen fixation. Bulletin 1. Elkhorn, Wisconsin: Mandaamin Institute, pp. 1–47.

Gustavsson, J., Cederberg, C., Sonesson, U., Van Otterdijk, R. and Meybeck, A. (2011), *Global Food Losses and Food Waste*. Rome: FAO.

Hagenguth, J. F. (2016), Breeding for organic agriculture: Evaluation of common bunt resistance in three winter wheat populations derived from M822123, PI178383 and PI560841-bcl. Diplomarbeit/Masterarbeit - Institut für Biotechnologie in der Pflanzenproduktion, BOKU-Universität für Bodenkultur, pp 135.

Harlan, H. V. and Martini, M. L. (1929), A composite hybrid mixture. *Agronomy Journal*, 21(4), 487.

Hoad, S., Ove Bertholdsson, N., Neuhoff, D. and Köpke, U. (2012), Approaches for improved weed suppression in organically grown cereals. In: Lammerts van Bueren, E. and Myer, J. R. (Eds), *Organic Crop Breeding*. Chichester, UK: Wiley-Blackwell, pp. 61–76.

Hovmøller, M. S., Walter, S., Bayles, R. A., Hubbard, A., Flath, K., Sommerfeldt, N., Leconte, M., Czembor P., Rodriguez-Algaba, J., Thach, T., Hansen J. G., Lassen, P., Justesen, A. F., Ali, S. and de Vallavieille-Pope, C. (2015), Replacement of the European wheat yellow rust population by new races from the centre of diversity in the near-Himalayan region. *Plant Pathology*, 65(3), 402–11.

Howard, P. H. (2009), Visualizing consolidation in the global seed industry: 1996–2008. *Sustainability*, 1(4), 1266–87.

IFOAM. (2009), *Collected papers of the first IFOAM International Conference on Organic Animal and Plant Breeding*. Santa Fe, New Mexico/USA: IFOAM, pp. 247–51.

IFOAM. (2014), *The IFOAM NORMS for Organic Production and Processing Norms 2014*. Bonn, Germany: IFOAM.

IFOAM. (2016), *Principles of Organic Agriculture Preamble*. Bonn, Germany: IFOAM Organics International. Available online: http://www.ifoam.bio/sites/default/files/poa_english_web.pdf (Accessed on 14 June 2017).

Jacobsen, S.-E., Sørensen, M., Pedersen, S. M. and Weiner, J. (2013), Feeding the world. *Agronomy for Sustainable Development*, 4, 651–62.

Jinek, M., Chylinski, K., Fonfara, I., Hauer, M., Doudna, J. A., and Charpentier, E. (2012), A programmable dual-RNA–guided DNA endonuclease in adaptive bacterial immunity. *Science*, 337(6096), 816–21.

Junk, J., Kouadio, L., Delfosse, P. and El Jarroudi, M. (2016), Effects of regional climate change on brown rust disease in winter wheat. *Climatic Change*, 135(3–4), 439-51.

Kiers, E. T., Leakey, R. R. B., Izec, A. M., Heinemann, J. A., Rosenthal, E., Nathan, D. and Jiggins, J. (2008), Agriculture at a crossroads. *Science*, 320, 320–1.

Knott, D. R. and Talukdar, B. (1971), Increasing seed weight in wheat and its effect on yield, yield components, and quality. *Crop Science*, 11(2), 280–3.

Köpke, U. (2006), Welchen Beitrag könnte die Pflanzenzüchtung für die Ertragssicherheit im Ökologischen Landbau leisten? *Vortrag: DFG-Senatskommission für Stoffe und Ressourcen in der Landwirtschaft*, Rundgespräch Ökologischer Landbau, 11 September 2006. Berlin: Wissenschaftsforum.

Köpke, U., Thiel, B. and Elmholt, S. (2007), Strategies to reduce mycotoxin and fungal alkaloid contamination in organic and conventional cereal production systems. In: Cooper, J., Niggli, U. and Leifert, C. (Eds), *Handbook of Organic Food Safety and Quality*. Cambridge, England: Woodhead Publishing Ltd., pp. 353–91.

Korkut, K. Z., Başer, I., and Bilgin, O. (2001), Genotypic and phenotypic variability, heritability and phenotypic correlation for yield and yield components in bread wheat varieties. *Acta Agronomica Hungarica*, 49(3), 237–42.

Kotschi, J. and Wirz, J. (2015), Who pays for seeds? Thoughts on financing organic plant breeding. Working Paper. Marburg and Dornach: AGRECOL and Section for Agriculture. http://www.agrecol.de/files/Kotschi_&_Wirz%20Engl_12_05_15.pdf.

Kristensen, L. and Borgen, A. (2001), Reducing spread of spores of common bunt disease (*Tilletia tritici*) via combining equipment. *Biological Agriculture & Horticulture*, 19(1), 9–18.

Kristiansen, P., Taji, A. and Reganold, J. P. (2006), *Organic Agriculture: A Global Perspective*. Ithaca, NY: CSIRO Publishing.

Lamichhane, J. R., Barzman, M., Booij, K., Boonekamp, P., Desneux, N., Huber, L. and Sarah, J. L. (2015), Robust cropping systems to tackle pests under climate change. A review. *Agronomy for Sustainable Development*, 35(2), 443–59.

Lammerts van Bueren, E. T., and Myers, E. (2012), *Organic Crop Breeding*. Wiley-Blackwell.

Lammerts van Bueren, E. T., Struik, P. C., Tiemens-Hulscher, M. and Jacobsen, E. (2003), Concepts of intrinsic value and integrity of plants in organic plant breeding and propagation. *Crop Science*, 43(6), 1922–9. doi: 10.2135/cropsci2003.1922.

Lammerts van Bueren, E. T., Backes, G., De Vriend, H. and Østergård, H. (2010), The role of molecular markers and marker assisted selection in breeding for organic agriculture. *Euphytica*, 175(1), 51–64.

Lammerts van Bueren, E. T., Jones, S. S., Tamm, L., Murphy, K. M., Myers, J. R., Leifert, C. and Messmer, M. M. (2011), The need to breed crop varieties suitable for organic farming, using wheat, tomato and broccoli as examples: A review. *NJAS-Wageningen Journal of Life Sciences*, 58(3), 193–205.

Lammerts van Bueren, E. T., Thorup-Kristensen, K., Leifert, C., Cooper, J. M. and Becker, H. C. (2014), Breeding for nitrogen efficiency: Concepts, methods, and case studies. *Euphytica*, 199(1–2), 1–2.

Linnemann, L. (2011), Optimierter Backtest für praxisrelevante Mehlbeurteilung. – Am Beispiel von Ökoweizen. *Mühle + Mischfutter*, 148(17), 577–80.

Lionetti, E., Leonardi, S., Franzonello, C., Mancardi, M., Ruggieri, M. and Catassi, C. (2015), Gluten psychosis: Confirmation of a new clinical entity. *Nutrients*, 7(7), 5532–9.

Loeschenberger, F., Fleck, A., Grausgruber, H., Hetzendorfer, H., Hof, G., Lafferty, J. and Birschitzky, J. (2008), Breeding for organic agriculture: The example of winter wheat in Austria. *Euphytica*, 163(3), 469.

Lorenz, N., Klause, S., Müller, K.-J. and Spieß, H. (2006), Screening of winter barley varieties (*Hordeum vulgare*) for resistance against loose smut (*Ustilago nuda*) and covered smut (*U. hordei*) in Germany. *Czech Journal of Genetics and Plant Breeding*, 42(Special Issue), 20–5.

Löwenstein, F. zu (2011), *Food crash - Wir werden uns ökologisch ernähren oder gar nicht mehr.* Pattloch eBook.

M'hir, S., Ziadi, M., Chammem, N. and Hamdi, M. (2012), Gluten proteolysis as alternative therapy for celiac patients: A mini-review, *African Journal of Biotechnology*, 11(29), 7323–30.

Marti, A., Augst, E., Cox, S. and Koehler, P. (2015), Correlations between gluten aggregation properties defined by the GlutoPeak test and content of quality-related protein fractions of winter wheat flour. *Journal of Cereal Science*, 66, 89–95.

Matanguihan, J. B., Murphy, K. M. and Jones, S. S. (2011), Control of common bunt in organic wheat. *Plant Disease*, 2, 92–103.

McDonald, C. E. (1985), Sodium dodecyl sulfate sedimentation test for durum wheat. *Cereals Food World*, 30, 674–7.

Menzies, J. G., Turkington, T. G. and Knox, R. E. (2009), Testing for resistance to smut diseases of barley, oats and wheat in western Canada. *Canadian Journal of Plant Pathology*, 31, 265–79.

Messmer, M., Burger, H., Schmidt, W. and Geiger, H. H. (2009), Importance of appropriate selection environments for breeding maize adapted to organic farming systems. In: Tagungsband der 60. Jahrestagung der Vereinigung der Pflanzenzüchter und Saatgutkaufleute Österreichs.

Züchtung und Genressourcen gegen abiotische Stressfaktoren - Markergestützte Selektion in der Praxis. A-Irdning: Höhere Bundeslehr- und Forschungsanstalt für Landwirtschaft Raumberg-Gumpenstein, pp. 49–51.

Mitchell Fetch, J., Entz, M. and Spaner, D. (2014), OSCII Activity A.2 Organic Oat Breeding. https://www.dal.ca/faculty/agriculture/oacc/en-home/organic-science-cluster/OSCII/theme-a/activity-a2.html.

Mos, U. and Heyden, B. (2006), *Die Wildgrasveredlung, Rudolf Steiners Impuls in der Pflanzenzucht*. Dornach: Verlag am Goetheanum-CH.

Müller, K.-J. (2004), Prüfung der Anfälligkeit aktuell verfügbarer Sommergersten gegenüber Hartbrand (*Ustilago hordei*). In: *LWK Niedersachsen, Ref. Öko-Landbau (Hg.): Versuchsergebnisse im Ökol. Ackerbau in Niedersachsen 2002–2003*, pp. 15–32

Murphy, K., Lammer, D., Lyon, S., Brady, C. and Jones, S. S. (2004), Breeding for organic and low-input farming systems: An evolutionary-participatory breeding method for inbred cereal grains. *Renewable Agriculture and Food Systems*, 20, 48–55

Neuhoff, D., Hoad, S., Davies, K., Gawronski, S., Gawronskaja, H., Drews, S., Juroszek, P., de Lucas de Bueno, C. and R. Zanoli (2005), Strategies of weed control in organic farming (WECOF). Final report of the FP 5 European Combined Project, online publication. Accessed 23 March 2017. http://www.wecof.uni-bonn.de

Nuijten, E., Messmer, M. and Lammerts van Bueren, E. T. (2017), Concepts and strategies of organic plant breeding in light of novel breeding techniques. *Sustainability*, 9(1), 18. doi:10.3390/su9010018.

Paillard, S., Goldringer, I., Enjalbert, J., Doussinault, G., de Vallavieille-Pope, C. and Brabant, P. (2000a), Evolution of resistance against powdery mildew in winter wheat populations conducted under dynamic management. I - Is specific resistance selected? *Theoretical and Applied Genetics*, 101, 449–56.

Paillard, S., Goldringer, I., Enjalbert, J., Doussinault, G., de Vallavieille-Pope, C. and Brabant, P. (2000b), Evolution of resistance against powdery mildew in winter wheat populations conducted under dynamic management. II - Adult plant resistance. *Theoretical and Applied Genetics*, 101, 457–62.

Pedersen, T. M. (Ed.) (2015), COBRA - Coordinating Organic Plant Breeding Activities - for diversity. In: *Proceedings of The Cobra Final Conference*, Denmark, 24–25 November 2015.

Perlmutter, D. (2013), *Grain Brain: The Surprising Truth about Wheat, Carbs, and Sugar – Your Brain's Silent Killers*. London, UK: Hachette UK.

Platenkamp, R. (1976), Investigations on the infection pathway of Drechslera graminea in germinating barley. *Royal Veterinary and Agricultural University Yearbook*, 49–64.

Porta-Puglia, A., Delogu, G. and Vannacci, G. (1986), *Pyrenophora graminea* on winter barley seed: Effect on disease incidence and yield losses. *Journal of Phytopathology*, 117(1), 26–33.

Presterl, T., Groh, S., Landbeck, M., Seitz, G., Schmidt, W. and Geiger, H. H. (2002), Nitrogen uptake and utilization efficiency of European maize hybrids developed under conditions of low and high nitrogen input. *Plant Breeding*, 121(6), 480-6.

Przystalski, M., Osman, A., Thiemt, E. M., Rolland, B., Ericson, L., Østergård, H., Levy, L., Wolfe, M., Büchse, A., Piepho, H.-P. and Krajewski, P. (2008), Comparing the performance of cereal varieties in organic and non-organic cropping systems in different European countries. *Euphytica*, 3, 417–33.

Rademacher, B. (1950), Über die Lichtverhältnisse in Kulturpflanzenbeständen im Hinblick auf den Unkrautwuchs. *Z. Acker- und Pflanzenbau*, 92, 129–65.

Rashid, M. I., de Goede, R. G. M., Brussaard, L. and Lantinga, E. A. (2013), Home field advantage of cattle manure decomposition affects the apparent nitrogen recovery in production grasslands. *Soil Biology and Biochemistry*, 57, 320–6.

Rasmussen, I. A. (2004), The effect of sowing date, stale seedbed, row width and mechanical weed control on weeds and yields of organic winter wheat. *Weed Research*, 44(1), 12–20.

Rembiałkowska, E. (2007), Quality of plant products from organic agriculture. *Journal of the Science of Food and Agriculture*, 15, 2757–62.

Rey, F., Fontaine, L., Osman, A. and Van Waes, J. (Eds). (2008), Value for cultivation and use testing of organic cereal varieties: What are the key issues. In: *Proceedings of the COST ACTION*, pp. 28–9.

Rhoné, B., Raquin, A. L. and Goldringer, I. (2007), Strong linkage disequilibrium near the selected Yr17 resistance gene in a wheat experimental population. *Theoretical and Applied Genetics*, 114(5), 787–802.

Robinson, R. A. (1996), *Return to Resistance*. Ag Access, Davis, CA.

Rulli, M. C., Saviori, A. and D'Odorico, P. (2013), Global land and water grabbing. *Proceedings of the National Academy of Sciences*, 110(3), 892–7.

Samsel, A. and Seneff, S. (2013), Glyphosate, pathways to modern diseases II: Celiac sprue and gluten intolerance. *Interdisciplinary Toxicology* 6 (4), 159–84. doi:10.2478/intox-2013-0026.

Schaffnit-Chatterjee, C. (2010), *Risk Management in Agriculture*. Frankfurt am Main: Deutsche Bank Research.

Schauder, A., Dornbusch, Ch. and Köpke, U. (1995), Produktionsziel Qualität – Qualität bei Saatgut und Speisegetreide im Organischen Landbau, Volume I. In: *SÖL und Fachgebiet Ökologischer Landbau der Universität Kiel (Ed.): Tagungsband 3.* Wissenschaftliche Fachtagung zum Ökologischen Landbau, 21–23 February 1995. Gießen: Wissenschaftlicher Fachverlag, pp. 197–20.

Schmehe, B. and Spieß, H. (2011), Ergebnisse der Prüfung aktuell zugelassener deutscher und österreichischer Hafersorten auf Flugbrandresistenz sowie Evaluierung von Inokulationsverfahren. In: Leithold, G., Becker, K., Brock, C., Fischinger, S., Spiegel, A.-K., Spory, K., Wilbois, K.-P. und Williges, U. (Hrsg.), *Beitr. 11. Wiss.tag. Öko-Landbau, Gießen, Bd. 1.* Berlin: Verlag Dr. Köster, pp. 344–7.

Schmehe, B. and Spieß, H. (2015), Evaluierung von Wintergerste-Sorten und -Zuchtstämmen auf Resistenz gegenüber Streifenkrankheit (*Drechslera graminea*). In: Häring A. M., Hörning, B. and Hoffmann-Bahnsen, R. (Hrsg.), *Am Mut hängt der Erfolg. Beitr. 13. Wissenschaftstagung Ökol. Landbau.* Berlin: Verlag Dr. Köster, pp. 148–9

Schmehe B., Spieß H., Martis S., Szabo L., Klause S., Pregitzer A. and Perez L. (2016), On farm organic oat breeding. In: *Oral Presentation on the 10th International Oat Conference*, 11–15 July 2016, St. Petersburg, Russia.

Simmonds, N. W. (1995), The relation between yield and protein in cereal grain. *Journal of the Science of Food and Agriculture*, 67, 309–15.

Skou, J. P., Nielsen, B. J. and Haahr, V. (1992), The effectivity of Vada resistance against leaf stripe in barley varieties. *Nordisk Jordbruksforskning*, 74(3), 34.

Spieß, H. (2014), Strategie zur Bekämpfung des Steinbrandes (*Tilletia caries*) im Bio-Landbau, 65. Tag. Vereinigung Pflanzenzüchter u. Saatgutkaufleute Österreichs, HBLFA Gumpenstein, pp. 117–21.

Spieß, H. and Vollenweider, C. (2017), Populationen: Vielfalt als Chance. Eine Antwort auf globale Herausforderungen in der Landwirtschaft? *Lebendige Erde*, 5, 39–40.

Spieß, H., Vollenweider, C., Finckh, M. R., Weedon, O. D., Eder, B., Siegmeier, T, Bülow, L. and Frese, L. (2017), *Züchtung von Populationen für den Öko-Landbau – Ein Beitrag zur Steigerung der biologischen Vielfalt und zur Anpassung an die Folgen des Klimawandels.* Freising: Wissenschaftstagung Ökologischer Landbau, in press.

Steiner, R. (2004), *Agriculture Course: The Birth of the Biodynamic Method.* Glasgow, UK: Rudolf Steiner Press. (First published 1924 in German).

Suneson, C. A. (1956), An evolutionary plant breeding method. *Agronomy Journal*, 48(4), 188–91.

United Nations. (2015), World Population Prospects: The 2015 Revision, Key Findings and Advance Tables, Department of Economic and Social Affairs, Population Division. Working Paper No. ESA/P/WP.241. https://esa.un.org/unpd/wpp/publications/files/key_findings_wpp_2015.pdf.

UPOV. (1991), Act of 1991 International convention for the protection of new varieties of plants of 2 December 1961 as Revised at Geneva on 10 November 1972, on 23 October 1978, and on 19 March 1991.

Verhoog, H. (2007), Organic agriculture versus genetic engineering. *NJAS - Wageningen Journal of Life Sciences* 54 (4), 387–400. doi:10.1016/S1573-5214(07)80011-X.

Vogt, G. (2007), The origins of organic farming. In: Lockeretz, W. (Ed.), *Organic Farming: An International History*. Wallingford: CABI, pp. 9–29.

Waldow, F. and Jahn, M. (2007), Investigations in the regulation of common bunt (*Tilletia tritici*) of winter wheat with regard to threshold values, cultivar susceptibility and non-chemical protection measures. *Journal of Plant Diseases and Protection*, 114(6), 269–75.

Wickson, F., Binimelis, R. and Herrero, A. (2016), Should organic agriculture maintain its opposition to GM? New techniques writing the same old story, Sustainability, 8(12), 1105.

Wieser, H. (2007), Chemistry of gluten proteins. *Food Microbiology*, 24(2), 115–19.

Wolfe, M. S., Baresel, J. P., Desclaux, D., Goldringer, I., Hoad, S., Kovacs, G., Löschenberger, F., Miedander, T., Østergård, H. and Lammerts van Bueren, E. T. (2008), Developments in breeding cereals for organic agriculture. *Euphytica*, 163(3), 323.

Zdravkovic, J., Pavlovic, N., Girek, Z., Zdravkovic, M., and Cvikic, D. (2014), Characteristics important for organic breeding of vegetable crops. In: Etingoff, K. (Ed.), *Organic Agricultural Practices: Alternatives to Conventional Agricultural Systems*. Boca Raton, FL: CRC Press, p. 209.

Zeleny, I. (1947), A simple sedimentation test estimating the breadbaking and gluten qualities of wheat flour. *Cereal Chemistry*, 24, 465–75.

Ziegler, J. U., Steiner, D., Longin, C. F. H., Würschum, T., Schweiggert, R. M. and Carle, R. (2016), Wheat and the irritable bowel syndrome – FODMAP levels of modern and ancient species and their retention during bread making. *Journal of Functional Foods*, 257–66.

Chapter 2

Organic seed production, certification and availability

Steven P. C. Groot, Wageningen University & Research, The Netherlands; and Maaike H. J. E. Raaijmakers, Bionext, The Netherlands

1 Introduction
2 Organic seed quality control issues: seed maturity
3 Seed disease and pathogen control
4 Application of biologicals to seed
5 Seed priming, pelleting and storage
6 Seed certification and testing
7 Organic seed regulations
8 Organic seed production: contamination issues and availability
9 Summary
10 Future trends
11 Where to look for further information
12 References

1 Introduction

After preparing the land, organic farmers start crop production by sowing seeds or planting vegetative propagation materials. The quality of organic plant propagation materials has a significant influence on the development of the crop, the final yield and quality. Rapid establishment of seedlings increases their competitiveness with weeds, an important issue for organic farming. A fast-growing root system eases the uptake of nutrients, which is particularly relevant when using organic fertilisers. It is also important to start with seeds free from seed-borne pathogens to reduce the risk of an outbreak of plant disease in the early crop.

Often neither organic nor conventional seed production result in 100% healthy and rapidly germinating seeds. Indeed, through evolution, plants produce seeds which vary in speed of germination and dormancy to spread the risk of germinating in the wrong season. Although domestication and breeding have created crops with more rapidly germinating

http://dx.doi.org/10.19103/AS.2017.0029.02

seeds than their wild relatives, for many crops the variation in performance is such that seed treatments are needed to create appropriate starting materials for commercial crop production. Measures are also needed during seed production, harvesting, cleaning and drying. To increase seed lot quality, sorting and upgrading techniques are employed, from simple sieving through to seed sorting by density or using image analysis to select seeds with the highest quality. Seeds can be primed to improve field emergence and pelleted to improve mechanical sowing.

Many plant pathogens such as viruses, bacterial or fungal pathogens or nematodes have evolved to infect seeds and survive desiccation. As so-called seed-borne diseases, they can be transmitted to the next generation (Fig. 1). Seed production measures are needed to detect potential plant pathogens and to prevent or reduce seed infection. When seeds are infected with seed-borne diseases, sanitation treatments are required to eliminate the pathogens.

Most seed sorting and upgrading techniques can in principle also be applied to organic seeds. Processes where chemicals are employed, such as liquid density sorting and certain priming recipes, are not suited for organic seeds. In the case of organic seeds, disinfection largely relies on physical treatments or the use of natural products with antifungal or antibacterial activity. With the increasing awareness of environmental risks from chemical pesticides and fungicides, physical treatments are also becoming an important alternative for conventionally produced seeds.

In this chapter we will review various factors that influence organic seed quality and availability; methods for seed sorting, upgrading and storage; methods of seed sanitation that are applicable for organic seeds; and seed certification systems. We will assess recent research that has resulted in commercially available methods for improving seed quality and on promising new research lines.

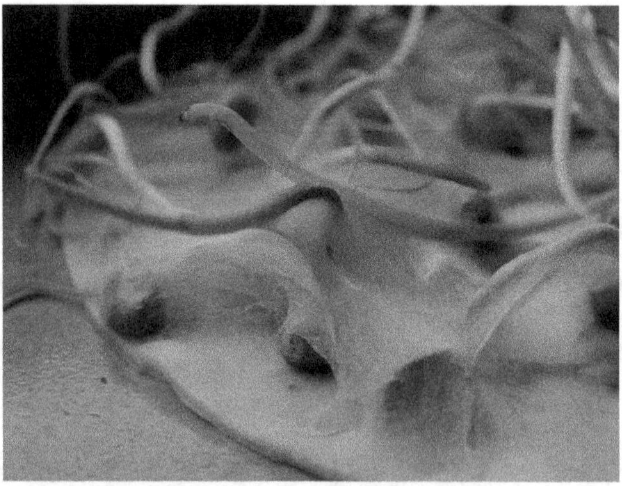

Figure 1 Infected wheat seedlings with a *Fusarium* fungus. The fungus infected the seeds during seed production, survived the drying and storage and infected the emerging seedlings as seed-borne pathogen (Photo: Wageningen UR, NL).

2 Organic seed quality control issues: seed maturity

An important difference between crop production and seeds for propagation is the optimal harvest moment. Seed development has three phases: embryogenesis, seed filling and maturation. Agronomists refer to cereal and legume seeds as 'physiologically mature' when the seeds have reached their maximum dry weight (TeKrony et al., 1979). That moment is the end of the seed filling phase when the production of starch, oil and storage proteins has ceased. Waiting longer to use these seeds results in drier seeds but does not result in higher yield on a dry weight basis. However, in the physiology of seed development this is the moment when the seeds have initiated their maturation process to acquire tolerance for survival after shedding. Finalising this maturation phase is very important for seed quality in the case of propagation material (Zanakis et al., 1994; Bewley et al., 2013). Processes during maturation include the production of soluble sugars, such as sucrose, raffinose and stachyose, as well as the so-called late embryogenesis-abundant proteins. These sugars and proteins are assumed to play a role in the protection of membranes, structural proteins, RNA and the DNA during storage of the seeds under dry conditions. Acquiring true seed maturity by establishing these protection mechanisms is therefore essential to reduce damage occurring during seed drying and storage and physical sanitation treatments.

Most crops flower over an extended period of time and as a consequence the plant simultaneously produces seeds at different developmental stages. With crops such as tomato the seeds are harvested by picking individual mature red fruits, most often coinciding with true seed maturity. But with crops such as cabbages (*Brassica oleracea*) it is not possible to harvest individual fruits. In these cases the oldest fruits (siliques) are already shattering, while the plant is still flowering at the top of its inflorescence. The seed producer has to pick a time to start harvesting the seeds. When the crop is dried fast, seed development is fixed in the maturation status at the moment of cutting. After threshing the really immature seeds can easily be removed by sieving. However, the seeds that have just finished the phase of seed filling, but have not acquired full stress tolerance, are full sized and cannot be separated from the true mature seeds with the highest quality.

Many seeds are green during seed development, because they contain chlorophyll. One of the processes initiated at the end of seed filling is the degradation of chlorophyll, a process that continues during seed maturation. The content of chlorophyll is therefore inversely correlated with seed maturity. The amount of chlorophyll can very sensitively be measured using its fluorescence properties. With this principle, a non-destructive method has been developed to analyse the relative amount of chlorophyll in individual seeds (Jalink et al., 1998). Analysis of a *Brassica oleracea* seed lot sorted in maturity fractions using chlorophyll fluorescence demonstrated the importance of seed maturation (Fig. 2). The initial seed lot had a germination frequency of 80%, too low for a commercial use. The two most mature fractions germinated at 100% or 95%, while the two least mature seed fractions germinated in only about 15% or 50% of cases. Removing the worst performing seeds, with higher chlorophyll levels, will result in upgrading the seed lot from animal feed to one with commercial quality. Upgrading of seed lots using chlorophyll fluorescence works for seeds from many crops and the invention has resulted in commercial seed chlorophyll analysers and related sorter technology.

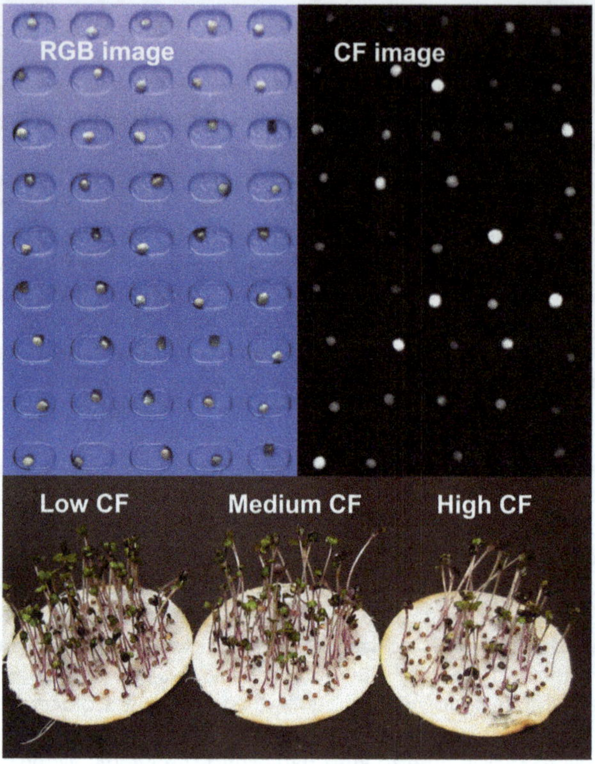

Figure 2 Seed maturity analysis of cabbage seeds using chlorophyll fluorescence. Top left: RGB image; top right: fluorescence image; bottom: germination performance of seeds with a low, intermediate or high level of chlorophyll fluorescence. The top pictures were made with the 'CF Mobile' instrument (www.seqso.nl) (Photo: Wageningen UR, NL).

3 Seed disease and pathogen control

Plant diseases during seed production can reduce the yield and quality of the seeds to be harvested, but also provide a risk for the next crop when the pathogen infects the seeds to be used for propagation. For some pathogens, even low levels of seed infection can result in severe crop losses (Pellegrino et al., 2010). An example of diseases that can be seed-borne is *Alternaria radicina*, a fungal pathogen that can cause black rot in carrots (Fig. 3). To reduce the risk of seed lot contamination with infected seeds, seed producers can take several measures. The first is the choice of production location, taking environmental conditions as well as humidity and temperature into consideration. Next is the use of healthy basic seeds at the start. Thirdly, during seed production the risks of contamination can be reduced as much as possible, and as a last measure, sanitation treatments are used when infection occurs.

It is important to determine the critical control points in preventing pathogen contamination. This requires knowing how the pathogen is transmitted and how it survives in the environment. For example, *Xanthomonas campestris* pv. *campestris* (the causal

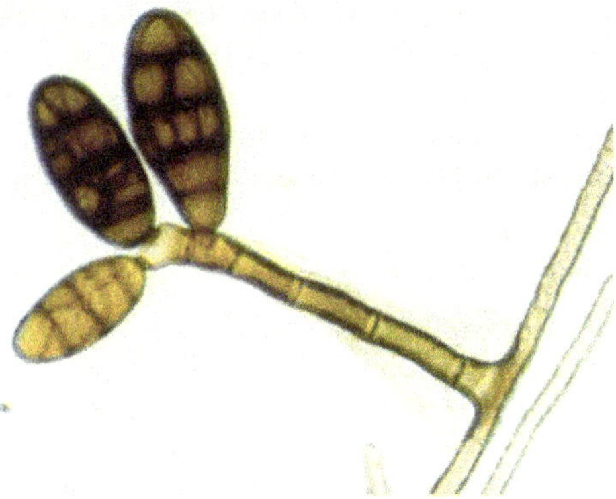

Figure 3 Spore-forming structure from the *Alternaria radicina*, a fungal pathogen that can cause black rot in carrots (Photo: Wageningen UR, NL).

agent of the bacterial disease black rot) can survive up to two years in plant debris in the soil (Alvarez, 2000), and seeds from *Brassica* crops can acquire infection via pollinating flies carrying the bacterium on their feet (van der Wolf and van der Zouwen, 2010). Because the occurrence of a highly infectious black rot disease in a *Brassica* crop can cause such large economic losses, seed health tests are performed using samples of 10 000 seeds to reduce the risk of selling seeds with even very low frequencies of infection. Frequent field inspections during seed production are essential to take measures in case infections occur.

3.1 Control of pathogens

It is not always possible to produce organic seeds without pathogens and it is therefore essential to test new seed harvests for the presence of pathogens. When pathogens are detected, disinfestation, disinfection or protection treatments are needed. Seed disinfestation involves the control of spores and other forms of disease organisms on the seed surface (Mancini and Romanazzi, 2014). Seed disinfection is the elimination of a pathogen that has penetrated into the living cells of the seed, infected it and become established. Seed protection is the application of a treatment to protect the seed from diseases caused by seed-borne or soil-borne organisms.

For conventional farming the use of chemical fungicides is still the main method to combat fungal pathogens, but an increasing awareness of the potential negative effects of chemical fungicides has created a demand for non-chemical alternatives. Moreover, fungicides do not act against bacteria and viruses. As a result, older physical seed sanitation treatments have been revived in recent decades, often combined with new technologies for process control, as well as use of novel approaches (Mancini and Romanazzi, 2014). Research has resulted in a multitude of non-chemical methods. These include physical methods, seed treatment or coating with plant or animal extracts, or addition of living microorganisms to the seeds. Those non-chemical treatments are most often applicable for use in organic farming (Koch et al., 2010; Koch and Groot, 2015).

It should be noted that physical sanitation treatments and treatments with antimicrobial compounds are not selective. They have the disadvantage that non-pathogenic microorganisms are also removed from the seeds. Naturally occurring microorganisms in and on seeds create a kind of buffer against some pathogens in the soil, by occupying the space in and on the seed. When these are killed, there is far less competition against microorganisms from the soil, including pathogens. This can make the seedling more vulnerable to soil-borne pathogens. A potential solution might be to apply beneficial microorganisms to the seeds after the sanitation treatment.

3.2 Physical seed sanitation

There are many physical seed sanitation methods. Most of them are based on heat, with or without moisture, and can be used in organic seed production. For several decades commercial cucumber seed lots have received a standard thermotherapy treatment under dry conditions at 72°C for three days to eradicate the cucumber green mottle mosaic virus (Kim et al., 2003). The efficiency of the treatment is debated and its effectiveness in deactivation of the virus in seeds needs to be confirmed through a bioassay (Reingold et al., 2015).

Treating seeds by dipping them for a short period in hot water can effectively eradicate several pathogens in the seed and is commercially practised on seeds and bulbs of several crops. An example being *Brassica* crops, where it is used to reduce risks of seed infection with the *Xanthomonas* bacterium (causal agent of black rot disease). As with many other sanitation treatments, it is important to find a treatment window that eradicates the pathogen, but does not cause too much damage to the germination ability of the seeds. Ideally, for each seed lot it is important to test variations needed in the water temperature and treatment duration, as seed lots may differ in their response. In the EU-financed research project 'STOVE' (Seed Treatments for Organic Vegetable Production, see Schmitt et al., 2006), potential causes for this variation in sensitivity were studied. It was shown, for instance, that less mature cabbage seeds are more sensitive to a hot water treatment, compared with mature seeds (Groot et al., 2006), while seeds that have started their germination processes before harvest (Fig. 4) are also more sensitive to hot water (Groot et al., 2008). A disadvantage of hot water treatment is the need to dry the seeds afterwards, making it unsuitable for high volume seed lots such as cereals.

Another combination of using moisture and high temperature is the use of steam. In a treatment known as the 'ThermoSeed™' or aerated steam treatment, the seeds are treated for a relatively short duration with a mixture of steam and heated air, which results in moistening the outer layers of the seed. This aerated steam treatment has the advantage that the moisture hardly penetrates inside the seeds, omitting the need for re-drying. This makes it suitable for the treatment of large seed lots as with cereals. The temperature of the air mixture, its moisture level and the duration are the key variables in this treatment. The method was developed in Sweden at the request of conventional farmers who asked for an alternative to fungicide treatment of cereal seeds (Forsberg, 2004). It is efficient in eradicating most fungal pathogens on cereal seeds and is applied on a commercial scale for the treatment of cereal and vegetable seeds (Fig. 5). With this treatment it is essential to determine the optimal treatment parameters before processing the entire seed lot, as the physiological condition of the seed influences its sensitivity (Forsberg, 2004; Groot et al., 2006).

A very different physical seed sanitation method is 'e-ventus' technology, also called e-dressing or electron seed treatment, developed at the German Fraunhofer Institut für Elektronenstrahl- und Plasmatechnik. With this method the seeds pass through a band of low-energy, accelerated electrons that penetrate the seed coat. The ionising effect of the

Figure 4 Kohlrabi seed lot containing seeds that have started their germination process before harvesting. Some seeds show root protrusion, but also other seeds may have started their germination processes and will have lost part of their stress tolerance (Photo: Wageningen UR, NL).

electrons kills the microorganisms on the outside of the seed and in the seed coat. A key variable in the treatment is the acceleration voltage which needs to be adjusted so that the electrons only penetrate the seed coat and not the living endosperm or embryonic cells in the seed. The second key variable is the dose. The technology has been used commercially in Germany since 1995, including the use of a mobile unit, for the treatment of mainly cereal seeds (Röder et al., 2009).

3.3 Natural products with biocidal activity

There are many plant- or animal-based products available that have a biocidal effect. The use of vinegar for food preservation is an example. In collaboration with other partners in the 'STOVE' project, Wageningen University & Research has tested various essential oils, organic acids, Biosept, Tillecur and extracts of stinging nettle for disinfection of vegetable seeds (van der Wolf et al., 2008). Biosept is an extract of grapefruit containing essential oils and organic acids, and Tillicur is based on mustard powder. In *in vitro* assays, thyme oil, oregano oil, cinnamon oil, clove oil and Biosept had the highest activity against the bacterial pathogens *X. campestris* pv. *campestris* and *Clavibacter michiganensis* subsp. *michiganensis* and the fungi *Alternaria dauci* and *Botrytis aclada*. Low antimicrobial activity was found for the organic acids against fungal pathogens. Seed treatment with essential oils or Biosept in concentrations between 0.1 and 1% reduced the percentage of seeds contaminated with fungi in blotter tests from 70% to less than 10% and eliminated >99% of total bacteria associated with cabbage seed. Thyme oil was the most potent tested compound in reducing seed-borne

Figure 5 Experimental unit used by the Agricultural University in Uppsala (Sweden) to develop the thermo seed treatment. The cereal seeds are moving for a defined period in a mixture of steam and air (Photo: Steven Groot, Wageningen UR, NL).

pathogens (Tinivella et al., 2009; Koch et al., 2010; Amein et al., 2011; Schmitt et al., 2011). Antifungal activity of thyme oil has also been observed against *Fusarium oxysporum* f. sp. *basilici* infection of basil seeds and *Alternaria radicina* infection of carrot seeds, though the latter was combined with a hot water treatment (Lopez-Reyes et al., 2016a,b).

4 Application of biologicals to seed

In recent decades, we have become aware that many microorganisms are often beneficial in the ecosystem and important for animal and plant health. Research in the area of microbiology has grown fast and provides very interesting and useful information that can be exploited to understand and improve plant health. In response to the need to find alternatives for the use of chemical pesticides and synthetic fertilisers, application of beneficial microorganisms, also called biologicals, is a rapidly growing field of research and commercial exploitation (Bashan, 1998; Burges, 2012; O'Callaghan, 2016; McQuilken et al., 1998). Adding microorganisms on seeds is often more cost-efficient compared to application in the field (O'Callaghan, 2016) and several seed companies are investing significantly in this area (BASF, 2016; Syngenta, 2016).

A well-known example of biologicals applied to seeds is the common practice of adding *Rhizobia* sp. bacteria to legume seeds. The result is a faster colonisation of the rhizosphere with these bacteria, improved nodulation and atmospheric nitrogen fixation, which gives a higher crop yield and a reduced need for nitrogen fertiliser. This practice was undertaken at the end of the nineteenth century in the United States, when seeds were inoculated with the *Rhizobium* sp. bacteria by mixing them with 'naturally inoculated' soil (Bashan, 1998; Vessey, 2003). This biological is also used in organic farming (Grossman, 2015). A more recent example of a microorganism applied as biofertiliser is coating seeds with the fungus *Penicillium bilaiae*, which may aid in increasing the uptake of phosphorus from the soil (Monsanto BioAg, 2016). The results seem variable as, in some experiments, the effect on yield is even negative (Karamanos et al., 2010).

In recent years the positive effects of inoculation of seeds with microorganisms have been shown for a large number of combinations (O'Callaghan, 2016). Among these are treatments with *Pseudomonas* spp. or *Bacillus* spp. bacteria, which can help the plant to resist soil-borne pathogens such as *Pythium* or *Rhizoctonia*, either by antagonistic activity or by triggering the seed's defence system (O'Callaghan, 2016). The effectiveness of biologicals in suppressing plant diseases has been shown in many laboratory studies, but the number of field studies is rather limited. At this point only a few microbial seed inoculant products are available for commercial seed treatments.

Pseudomonas chlororaphis is the active agent in Cedomon®, Cedress® and Cerall® (Fig. 6), which has been used for commercial treatments of cereal seeds since 1997 (BioAgri, 2017b). It is one of the few seed treatment with biologicals approved in the EU and in some countries also for organic farming (Lötjönen and Torniainen, 2005). The *Pseudomonas chlororaphis* bacterium can stimulate root growth and plant establishment and the treatment seems to be effective against several seed-borne diseases such as *Tilletia caries*, *Septoria nodorum* and *Fusarium* spp. Recently the same company released Amase®, based on the bacteria *Pseudomonas azotoformans*, as a biological seed treatment for cucumber, lettuce, tomato, peppers, eggplant, cabbage and broccoli (BioAgri, 2017a).

Major hurdles in large-scale application are the costs for production of the microorganisms, the product shelf life and the registration process. The latter is comparable in complexity and cost to that of chemical products. Commercial parties need a return on their investment, making it essential that a company gets intellectual property (IP) rights to a product through patenting. The large investments in biologicals made by several seed companies suggest that, in the future, more seed treatments with beneficial microorganisms will become available for farmers to offset the reduction in use

Figure 6 Wheat seeds treated with the biological Cerall®. (Photo: Wageningen UR, NL).

of chemical plant protectants and fertilisers. Whether they can be used for organic farming will also depend on the methods for production of the microbes and the materials used for stabilisation and shelf life extension, and whether these are consistent with organic regulations. This includes the potential use of genetic modification.

5 Seed priming, pelleting and storage

To fasten emergence and increase uniformity, seed companies can apply a technique called priming, basically a wet pre-treatment that initiates germination processes under controlled conditions. The principle of priming is adding a limited amount of water to the seed, allowing germination processes to start, but not enough for root extension (Bewley et al., 2013). This maintains seed desiccation tolerance, so the seed can be dried, stored and shipped to the farmer. This makes most priming recipes compatible with organic farming principles and, as a result, several seed companies sell primed seeds for the organic market (Fig. 7).

An example is celery (*Apium graveolens*). When the seeds are shed from the mother plant the embryo is very small. When the seeds are sown and absorb water from the soil, the embryo starts growing. Only when the embryo is fully grown can the root protrude to initiate the establishment of a seedling. This time needed for embryo growth varies between individual celery seeds from one to several weeks. As a result seedling establishment is far from uniform, a disadvantage for commercial production of a celery crop. Seed technologists have developed a seed priming treatment to obtain celery seed lots with a uniform germination (Brocklehurst and Dearman, 1983). During the priming the embryo grows, but root protrusion is inhibited, so the seeds can still be dried. Comparable priming techniques are also performed for other crops to increase uniformity or, as for lettuce, to enable germination at high field temperatures.

Mechanical sowing of seeds is difficult with small or irregular shaped seeds, such as those of onion or lettuce. To enable mechanical sowing of seed, companies can add a

Figure 7 Organic onion production. Plants on the right were produced from primed seeds, the plants on the left from non-primed seeds (Photo: Wageningen UR, NL).

pellet around the seed to increase its size and weight, as well as creating a spherical shape (Fig. 8). During the pelleting process, the seeds tumble in a rotating pan while an inert powder, such as diatomaceous clay, a binder combining starch or cellulose and water, is added. This slowly builds layer upon layer around the seed. If no chemicals are added, this pelleting process is compatible with organic farming.

5.1 Seed storage

Most crop seeds can be dried and stored. The period up to which the seeds remain alive depends on the storage conditions, the crop and the physiological condition of the seeds (Bewley et al., 2013). In general seeds should be stored under dry and preferably cool conditions. Vegetable seed companies store their seeds most often at 30% relative humidity (RH) and 15–20°C, whereas 'high volume–low cost' seeds, like those of cereals, are most often only dried and stored in warehouses under ambient conditions. The gradual deterioration of seeds during storage is related to oxidation of the membranes, DNA, RNA and proteins. For that reason dry seeds survive longer under low oxygen conditions (Groot et al., 2015). Low oxygen conditions can aid in the tropics, where ambient temperatures are high and cooling is costly. Rice or maize seeds, stored by farmers in a bag that is impermeable to both moisture and oxygen, survive longer (Guenha et al., 2014; Williams et al., 2017). Moreover, low oxygen conditions also eliminate accidental insect infestation of the seed lot in an environmentally friendly way, avoiding the use of insecticides or fumigation.

It is equally important that farm-saved seeds, or leftovers from commercial seed packages, are also stored under optimal conditions, in order to maintain good quality seeds for the next sowing season. The primary concern is to store seeds in dry conditions at an equilibrium RH below 50%, preferably between 20 and 40% RH. Open storage in a refrigerator is not a good idea, as the RH in most refrigerators is close to 100% and seeds will rapidly absorb moisture due to their hygroscopic character. Wageningen University & Research developed a simple storage system for farm-saved or leftover seeds (Fig. 9). The principle is a box, for instance a food container, with a transparent lid that

Figure 8 Celery seedlings germinated from primed and pelleted seeds. Some seeds do not germinate since the seed lot has been stored too long (Photo: Wageningen UR, NL).

Figure 9 Storage box developed at Wageningen UR for farm-saved seeds or leftover seeds. The hermetically sealed box contains silica gel and a hygrometer. Small amounts of seeds can also be dried in the box. Once the RH increases above 40% the silica gel needs to be re-dried in an oven (Photo: Wageningen UR).

can be hermetically sealed. In the box, the dry seeds are stored with a bag containing desiccant (such as silica gel) and a hygrometer on top, that can be read through the lid. The box should be stored preferably in a cool place and the RH regularly checked. If the RH increases above 40%, the desiccant can be dried in an oven (silica gel at 105°C) and placed in the container to reduce the RH.

6 Seed certification and testing

The sales and commercial use of organic plant propagation materials have to comply with international and national regulations. To market organic seeds or other types of crop propagation material, such as trees, tubers, bulbs and cuttings, two types of certification are needed. One is the quality certificate that is obliged for all propagation material and the second is the certification for production under organic conditions.

All propagation materials marketed in the EU and many other countries require a minimal level of quality regarding germination performance, health and purity. This regulation was established in the last century because of the importance of seed quality for agricultural production and food security. The certification of minimal quality is performed by accredited organisations, which can be public or commercial laboratories.

The International Seed Testing Association (ISTA, http://www.seedtest.org) was established in 1924 to create accuracy and uniformity in seed quality testing and optimise tests. At present the ISTA has accredited labs in over 75 countries, which can provide certificates regarding quality of seed lots, using standardised procedures for sampling and testing. In the United States and Canada, the Association of Official Seed Analysts (AOSA, http://www.aosaseed.com) also provides certificates regarding the quality of marketed seed lots. AOSA was founded in 1908 when regulation of seed trade started in the United States. The ISTA procedures for representative sampling and testing are laid down in the 'International Rules for Seed Testing' (ISTA, 2016). These ISTA rules have 19 sections describing standardised methods for sampling, testing seed lot quality and reporting results. Seed purity is an important aspect of quality control, as seed lots may contain seeds from other species (weeds), broken seeds, other plant material or stones. As water contributes to the weight of a seed lot, two standard methods have been developed to determine seed moisture content (ISTA, 2016).

6.1 Seed germination tests

The optimal conditions for seed germination differ between crops, as some are adapted to warm climates (e.g. rice), while others prefer lower temperatures for germination (e.g. spinach). Moreover, some seeds will germinate better using alternating day/night temperatures, while some seeds require light for germination and others prefer darkness. With some crops, or varieties, seeds can be dormant, which means the seeds do not germinate despite optimal temperature conditions. Dormancy is a natural mechanism to prevent germination in the wrong season. In agriculture, this trait can be used to prevent pre-harvest sprouting in the ear or fruit during moist weather conditions. As dormancy can interfere with testing the maximum germination capacity, it is sometimes necessary to break this dormancy during seed testing. Sometimes a period of cold incubation of the wet seeds is enough whilst, in other cases, adding nitrate or the plant hormone gibberellin to the germination medium may help. The ISTA rules (ISTA, 2016) describe germination testing conditions for a large number of crops. Germination tests are most often performed on top or between paper blotters, which makes it possible to detect protrusion of the embryonic root through the covering seed structures. Germination tests on filter paper are usually undertaken in optimal conditions compared to the field, where seeds may experience biotic or abiotic stress. As a result field emergence is often lower compared to germination on blotters in the seed testing lab. It is for this reason that soil tests are also often performed (Fig. 10). Farmers can use the information provided by ISTA on the optimal germination conditions for a crop to get an idea of how they can check germination of farm-saved seeds, using soil, paper or cotton blotters.

6.2 Seed health tests

Seed health is another important issue for seed certification, as many pathogens have evolved to travel with the seeds to infect the next generation. ISTA has developed dedicated tests for seed-borne pathogens (ISTA, 2016; Mathur and Kongsdal, 2003). Some of these, such as many fungi, are traditionally detected and identified using grow-out tests after killing the moist seed by freezing (Fig. 11). More recently techniques based on DNA sequencing have become available for the detection of seed-borne fungi (see Mancini et al., 2016, for a review). Although DNA-based techniques can be fast and highly specific, their disadvantage over grow-out tests is that they may also detect the pathogen when it is non-viable. Serological tests with specific antibodies are used for the detection of viruses.

Figure 10 Soil test for the analysis of germination quality with bean seeds (Photo, Wageningen UR, NL).

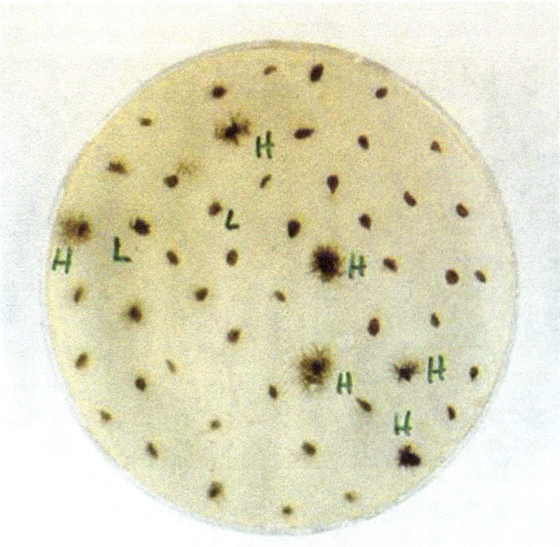

Figure 11 Assay for the detection of carrot seed infection with *Alternaria radicina*, using semi-selective media (Photo: Wageningen UR, NL).

Some diseases can be very harmful and highly infectious and it is essential to detect these at even extreme low incidence of infection. For example, the assay for detection of *X. campestris* pv. *campestris*, the causal agent of black rot on *Brassica* spp. requires sample sizes between 10 000 and 30 000 seeds (Astma et al., 2014). Similar requirements are needed for the detection of *C. michiganensis* subsp. *michiganensis*, the causal agent of bacterial canker in tomato (European and Mediterranean Plant Protection Organization, 2013).

7 Organic seed regulations

Certified organic seed is defined by the International Federation for Organic Agriculture Movements (IFOAM) as seed from cultivars that may be derived from conventional breeding programmes (excluding genetic engineering) and that are produced under organic farming conditions for one growing season for annual crop species, and two growing seasons for perennial and biannual crop species (Fig. 12). Organic regulations in the EU and the United States use almost the same definitions.

In the European Union, inputs in organic farming, including plant propagation material, are currently regulated according to Council Regulation *(EC) Nr. 834/2007* (The Commission of the European communities, 2007) and Commission Regulation (EC) No 889/2008 (The Commission of the European communities, 2008). New legislation is under discussion.

For organic crop production in the EU organic seeds or vegetative propagating materials should be used when available. EU member states may authorise the use of non-organic (conventional) seed or vegetative propagating material if it is not available from organic production, if the available organic seeds are not appropriate for the user or if they cannot

Figure 12 Organic production of leek seeds needs two years under organic growing conditions (Photo: Vitalis, NL).

be delivered before sowing time. In addition, this authorisation can be granted for use in research and small-scale field trials or for variety of conservation purposes.

In the EU, when conventional seed or seed potatoes are used, they must not be chemically treated after harvest. However, chemical seed treatment may be applied when it is required by the competent authority for phytosanitary purposes. An example is organic tomato seeds in Europe, which need to be treated with hydrochloric acid against Pepino mosaic virus since 2003.

The rules for organic plant production apply for the production of organic propagating material. However, no specific rules are defined for the treatment of organic seeds after harvest within EU organic regulations. Consequently, it is not always clear for seed companies (or certifiers) which methods and inputs are allowed for the cleaning, priming and coating of organic seeds. One could argue that since the seeds are organic, only fertilisers and soil conditioners from Annex 1 of the The Commission of the European Communities (2008) and plant protection products from Annex 2 are allowed. This means, for instance, that thyme oil could be used for sanitation and mineral clay for pelleting of seeds. In the EU biological, mechanical and physical methods are allowed for the processing of organic food and feed in article 6 (The Commission of the European Communities, 2007), which suggests that such treatment is also allowed for organic seeds. This includes the treatment of seeds with hot water to remove pathogens, for example. In practice EU member states and certifiers can differ in their interpretation of the EU regulations (The Commission of the European Communities, 2007, 2008), which could be solved by adding a chapter on seed treatment to the (new) organic regulation currently under discussion.

The US formalised its national organic standard in 2002 with the National Organic Program (NOP) (USDA-AMS, 2002). The rules on seed treatment are specified in the guidance on organic seeds from the NOP (United States Department of Agriculture – Agricultural Marketing Service – National Organic Program, 2013). The USDA organic regulations require that organic producers use organic seeds, annual seedlings and planting stocks. These regulations allow producers to utilise non-organic seeds and planting stock when 'equivalent' organic varieties are not commercially available. Relevant quality considerations are germination rate, presence of weeds, shelf life and stability of the seed as well as resistance against diseases. The (higher) price of organic seed is not accepted as an argument to use conventional seed (United States Department of Agriculture – Agricultural Marketing Service – National Organic Program, 2013). Seeds that are used for the production of edible sprouts always have to be organic, both in Europe and the United States.

8 Organic seed production: contamination issues and availability

Organic seeds can be produced by seed companies or by organic farmers. Some seed companies specialise in breeding and seed production for organic agriculture, while other seed companies mainly deliver to the conventional market but also produce organically propagated seeds for some varieties. Globally, the number of seed companies has been decreasing rapidly during the last decades, due to market concentration and the chemical industry buying seed companies. IP practices (e.g. patents on seed) are one of the main causes for this consolidation (Howard, 2015). For organic farmers who want to use organic seeds, this is especially alarming since some of the major agrochemical seed companies are not investing in the organic market. These companies may also be using breeding techniques that are not compatible with the principles of organic farming.

Organic farmers as seed producers may not only save, re-use and exchange seeds from their own harvest (so-called farm-saved seeds). Some of them also produce seeds for sale and select the best plants in order to develop varieties that are adapted to their local conditions. In many developing countries farmers are still the main seed producers. Research in six African countries covering 40 crops showed that farmers access 90% of their seed from informal systems with 51-% of that deriving from local markets (McGuire and Sperling, 2016).

In Western countries farmers' seed networks also exist, often involving organic farmers. The aim of those seed networks is to maintain and improve choice and offer an alternative to commercially produced seed that is mostly bred for conventional, high input farming systems. Examples are Réseau Semences Paysannes in France (www.semencespaysannes.org), and the organic seed alliance in the United States (http://seedalliance.org/).

Production of high-quality seeds is a skill and, in the case of organic seeds, this is performed by (organic) farmers specialised in seed production or by seed companies themselves. With cereals and most legumes, where the seeds serve both as propagation materials and for growing, the basic cultivation conditions for a crop or propagation material may be largely similar. However, the quality requirements for propagation material are much higher, as seeds for propagation material must all be intact and healthy.

For several crops, such as potato and tomato, organic seed production is no greater a challenge than the production of non-organic seeds. However, for biennial vegetable crops as cabbages, carrot and onion, it is difficult to produce high-quality organic seeds, as these crops require two growing seasons for seed production (Fig. 12). The mother plants must stay on the field for a long time, including over winter, which increases the risk of pathogen infection. This usually results in lower seed yields compared with conventional production which can make use of chemical disease control. As a consequence, organic seed production for these crops is much more expensive compared to that for non-organic seeds, resulting in higher seed prices. Another consequence of this challenge is that, from certain conventional varieties, no organic seeds are currently available on the market.

8.1 Accidental chemical contamination of organic seed

As organic seeds are produced without application of chemical plant protection products, most seeds are free from chemical residues. However, a few years ago, a German retailer tested a number of organic seed samples and found that part of the sample contained chemical residues above the so-called baby food threshold (the safety threshold for infants with a higher susceptibility to contamination risks). In response, Wageningen University & Research and the Dutch organic seed companies started a study to find potential causes and sources of contamination. It was concluded that most of the tested organic seeds were free from chemical residues. In cases of contamination, the levels were, with one exception, so low that fraud could be excluded. So, what could be the source of contamination? Since the contaminating molecules were mainly crop protectants presently used in conventional seed production, it was concluded that most of the contamination was due to evaporation from chemically treated seeds and condensation on organic seeds. This happens, for instance, when conventional and organic seeds are processed at the same location. It was therefore recommended that organic seeds should not be dried with air that is in contact with conventional seeds. The study also suggested that hermetically sealed storage or separate warehouses should be considered. Some organic seed lots have been found to contain trace levels of old type of persistent crop protection products. It is known that some of those chemicals can persist in the soil or groundwater for decades. The source for this type of contamination is likely the environment, for instance, the soil in which the seeds are produced. Overall it was concluded that the low levels of contamination found do not indicate that the seeds were not produced according to organic standards. It was found that these low levels of contamination more or less mirror the wider presence of chemical crop protection molecules in the environment.

8.2 Contamination with genetically modified seeds

The use of genetically modified (GM) organisms (GMOs) is prohibited in organic farming. Since seeds are at the beginning of the production chain, access to GMO-free seeds is essential for farmers to be able to produce GMO-free organic food. In countries where GM crops are grown, there is a risk that a seed lot can become contaminated with GM seeds during seed production or handling. The highest risks comes from incidental fertilisation of the seed-producing plants by GM pollen, transferred by wind or pollinating insects. For that reason seed production needs to have a minimal distance to flowering plants from the same crop or wild plants from that species. The distance needed depends on whether the crop is strict self-pollinating or cross-pollinating and whether it is pollinated by wind or insects. For cross- and wind-pollinated crops such as corn and sugar beet, this distance is usually larger than for self- and bee-pollinated crops such as soybean, although bees can transfer pollen over large distances. The effectiveness of the isolation distances between GM and non-GM production also depends on field sizes, prevailing winds and other climatic and geographic conditions. To minimise the risk of cross-pollination, seed producers preferably locate their seed production in a region or country where no GM production of the same crop takes place. But in countries where GM production is widespread, like the United States and Canada, this is becoming increasingly problematic (see Box 1).

Box 1 Some hints for seed production of arable crops

In choosing the site of production should of course be in the right climate region for crop growth, flowering and full seed development. Preferably, it is a site with low humidity and rainfall as many diseases thrive under moist conditions. Diseases will reduce the yield or even worse many can travel with the seed to the next generation. Inoculum pressure should be reduced by avoiding neighbouring crops, pre-crops or weeds that can serve as host for the diseases. The pre-crop can also have a large influence on the occurrence of weeds in the seed production crop. A high weed incidence can result in the contamination of the harvested seeds (lower purity).

To produce a high quality, the seeds should be able to acquire full maturation on the mother plant before weather conditions get unfavourable. Especially cold and wet conditions at the end of seed development will stimulate the spread and development of pathogens as for instance *Fusarium* species. To reduce disease incidence, already during crop production watering through irrigation is preferred above rain, although the latter often cannot be avoided. Moist conditions prior to harvest can also result in pre-harvest sprouting, especially with cereals. Both to reduce seed infection and pre-harvest sprouting, it is important that wetting of the seeds is avoided as much as possible and if rainfall cannot be avoided, it is important the crop can dry quickly. It is better not to include seeds from spots with lodging, as those seeds have a high risk of reduced quality.

After harvesting, drying and threshing seeds can be sorted on size and small seeds can better be discarded as they may derive from infected ears or pods. Good drying of the seeds and storage under dry conditions is even more important than cool storage. Contamination with insects can be cured by temporary storage at sub-zero temperatures or under anoxia conditions, although this can be expensive.

For further reading we can refer to textbook or chapters: McDonald and Copeland, 2012; Köpke et al., 2007; Piorr, 1992.

To keep organic seeds GM-free it is important that routine (instead of incidental) contamination of seeds with GMOs is not accepted by government authorities. For this reason IFOAM is advocating for a legal threshold for all non-GM seed at the lowest detectable level (IFOAM EU Group, 2003). As yet, politicians in Europe have not reached an agreement on a threshold for GMOs in seeds that is both feasible and acceptable. Consequently in the EU a zero tolerance applies; all seeds in which traces of GM are detected need to be labelled as GM seeds.

8.3 Availability of organic seeds

A growing number of seed companies in Europe are producing organic seeds. The agricultural and rural development website from the European Commission (http://ec.europa.eu/agriculture/organic/eu-policy/eu-rules-on-production/seeds-database_en) provides links to all the national databases in EU member states, listed over 300 organic seed suppliers (producers and traders) throughout the EU in 2016 (Renaud et al., 2016). The current availability of organic seeds is far from sufficient to provide all European farmers with organic seeds from all the crops they grow and from the varieties they prefer. In some EU countries there is hardly any organic seed in the market at all. Because of uncertainties in the market (for farmers it is often easy to get derogation to use cheaper conventional seeds), as well as technical difficulties, the propagation of organic seed is not yet economically attractive for many varieties. For some crops or varieties, the organic market is simply too small to interest seed companies in organic seed production. On the other hand, the organic market is rapidly growing and the upscaling of seed production takes time. As a consequence, for some favourable varieties like mildew-resistant onion varieties or *Phytophthora*-resistant seed potatoes, the demand for organic propagation material currently exceeds the supply.

Each EU member state is obliged to keep a national electronically accessible seed database which shows the availability of organic seed and seed potatoes on their territory. These online national databases are used in many European countries as a central tool in managing the use and regulation of organic seed (European Commission, 2017). Seed companies can enter their organic seed products in the database and customers (farmers) can then check the availability of organic seed for numerous varieties. The national databases can also be used to manage derogation requests by producers and generate statistical information such as changes over time in derogations asked for a particular crop (Döring et al., 2012). The permissible reasons for authorising the use of non-organic seed (derogation) are clearly defined (see section above) and EU member states must report annually to the EC about the amount of authorisations granted.

The quality of the national databases and the amount of seed varieties offered differ greatly between EU member states. For instance in Bulgaria the database is nearly empty, while the German database lists a wide range of varieties in vegetables, arable crops and vegetative propagating material (fruit trees and strawberries) from over 80 different suppliers. There seems to be a relation to the availability of organic seeds and the development of the conventional seed sector. In countries where a national seed sector is missing, there is also less availability of organic seeds. In the Netherlands, the market leader in vegetable seed production, a large amount of vegetable seed varieties is offered on the seed database, but there is often a shortage of cereals which have to be bought from the international market.

For international operating seed companies and traders it is not always easy to use foreign databases since they are often only in the national language. As a result they might sell organic seeds through their conventional dealers without offering it on the relevant database. To solve this problem Belgium, Germany, Ireland, Luxemburg, Sweden, Switzerland and the United Kingdom have collaborated to share information and make it easier for seed companies to list new varieties on national databases (http://www.organicxseeds.com/) (Fig. 13).

Since not all organic seeds are registered, the database is not always a good indication for the availability of organic seed. In many southern and eastern European countries, farm-saved seed is still widely used. When this seed is produced on an organic farm, it is accepted as organic seed in most countries. But since farm-saved seed is not certified or registered, it is not included in official data on organic seed.

Over 100 US seed production companies are active in the US organic seed market. However, the US organic seed sector remains challenged by the lack of reliable

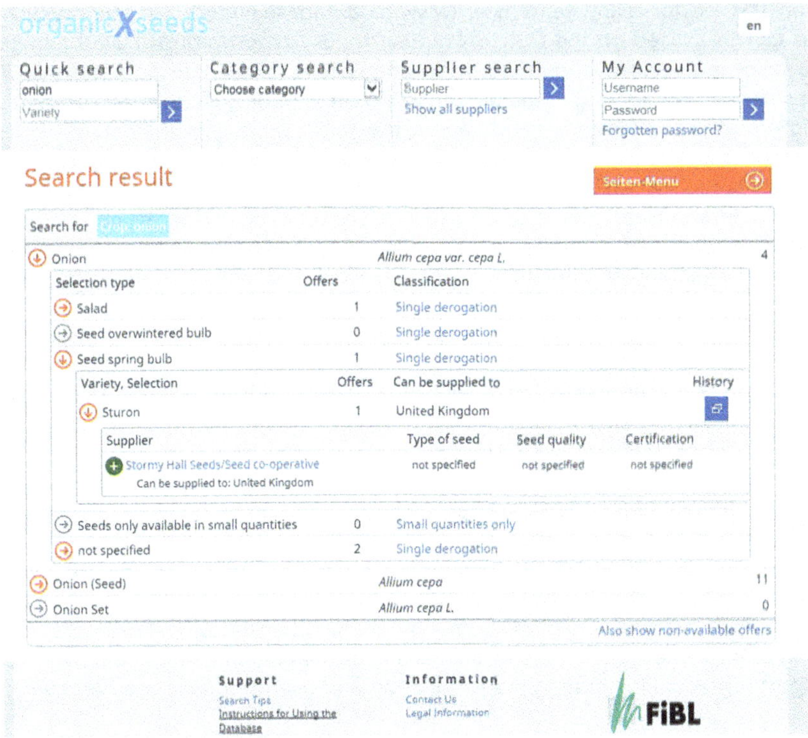

Figure 13 The organicXseeds database (http://www.organicxseeds.com/) provides information on the availability of organic seeds for specific crops within several EU countries. The screenshot was made on 23 January 2017.

information about the availability of crop types and varieties (Renaud et al., 2016). Several organisations have created databases to ensure transparency in the claims made about organic seed availability. However, until the end of 2015, none of the databases has achieved comprehensive coverage and none is officially endorsed by the NOP. This lack of a comprehensive national database clearly discourages seed producers from increasing organic seed production and meeting demand.

8.4 Organic seed policy

In addition to the national databases there is a provision in the EU regulation for an European Annex – a list of 'species for which it is established that organically produced seed or seed potatoes are available in sufficient quantities and for a significant number of varieties in all parts of the Community'. Thus far (early 2017), however, this list has remained empty.

Given the very strong seed sector in the country, in 2003 the Dutch government developed their own National Annex or Category 1. This is a list of species and sub-species for which sufficient organic seed is available in the Netherlands and for which no derogations are granted. Crops such as potatoes, onions, lettuce and cucumber have been on this list for many years now (www.biodatabase.nl/en/). To define which (sub) species can be placed on the Dutch National Annex, expert groups have been created for the following sectors: open field horticulture crops, glass house crops, arable crops and vegetative propagating material. These expert groups consist of representatives of organic farmers, seed producers and traders, transplant producers, the database manager and advisors/researchers. The role of the expert groups is to advise the government (the competent authority) which then makes the final decision. The expert groups have a clear mandate and clear criteria to decide whether a new (sub)species can be added to the National Annex. An important precondition is that the organic range is sufficient (in quantity and variation) for different soil types, seasons and markets. The expert group can also advise the removal of a certain (sub) species from the National Annex if the range of varieties is no longer sufficient. To maintain support from farmers it is necessary to make the system flexible and adaptable to changes in the market. To improve the effectiveness of the system, some supplementary rules have been introduced. As an example, (Fig. 14), some farmers attempted to circumvent the use of organic seed potatoes by delaying their order until organic stock had run out. To prevent this from happening the 'February rule' was introduced. This means organic potato farmers in the Netherlands have to order their seed potatoes before February 1st. After this date no authorisations for the use of conventional seed potatoes are possible.

As things stand in early 2017, National Annexes have also been implemented in France, Germany, Sweden and the Flemish part of Belgium, all with adaptations in procedures related to their national situation. Some other member states, for example, Italy, are planning to implement similar systems. This makes the National Annex a proven concept to support a gradual increase in the production and use of organic seed and vegetative

Figure 14 Potato variety trial for organic farming in the Netherlands (Photo: Louis Bolk Institute, NL).

propagation material. IFOAM EU is lobbying to introduce an article on the National Annex in the new EU organic regulation which is in advanced stage of negotiations.

It has to be noted, however, that using a non-derogation list (National Annex) is not a prerequisite for high usage of organic seed. Some countries perform well in terms of the percentage of organic seed usage due to a strict derogation regime. Countries like Denmark and Switzerland, for instance, consider all derogation requests on a case-by-case basis, using publicly available variety equivalence lists for each species and (sub) species. If sufficiently recommended organic varieties of a certain crop are available, the farmer needs a very good, farm-specific reason to get a derogation. The use of this kind of strict derogation regime requires the support of crop specialists.

In the United States the primary responsibility for enforcement of organic seed requirements is assigned to organic certifiers. Certifiers must check whether organic growers have done their best to find organic seed so that they increase their organic seed usage year-on-year. Growers for their part must demonstrate clearly the steps they take to source organic seed. If growers use conventional instead of organic seed, they must show in their 'Organic Systems Plan' that there are no suitable organic seeds available, for example, the desired crop traits and characteristics are not available in organic form (Renaud et al., 2014). Although these requirements can help to raise awareness and increase the use of organic seeds, the lack of a common agenda and a recognised national seed database in the United States is hindering progress and the development of a level playing field at the national level (Renaud et al., 2016).

It is clear that there is still considerable variability in organic seed regulation systems within the EU and the United States. Although flexibility is necessary, the current absence of a level playing field for the use of organic seeds leads to unfair competition and is undermining support from farmers. For many cultivars a larger guaranteed market is necessary to make the production of organic seeds commercially attractive for seed companies. Further growth in the use and production of organic seeds requires more international cooperation and coordinated regulatory action.

9 Summary

The quality of organic seeds and vegetative propagating materials is of utmost importance to the success of the crop. Its quality is often an important determinant of final yield and quality and thus farmers' income. The seeds or tubers need to be healthy and germinate rapidly and uniformly as well as result in the establishment of a vigorously growing seedling. Many countries have imposed rules for minimal quality of commercial seeds. In organic farming seedling quality is even more important to strengthen competition with weeds and ensure a fast-growing root system able to fully capture nutrients released from organic fertilisers. This chapter provides information on methods to analyse and improve quality, such as sorting based on seed maturity or the use of sanitation treatments. The use of organically produced seeds or vegetative propagation material is required in organic farming. Seed has to be produced for at least one generation under organic conditions. Only when organic seed is not available can conventional non-chemically treated propagation material be used. The current situation regarding the use and availability of organic seeds is described, as well as government initiatives to stimulate the use of organic seeds.

10 Future trends

One of the greatest challenges in the near future is to increase both the availability and the use of organic seeds. Research is needed to support organic seed companies in reducing the costs for seed production for some difficult crops, especially biannual crops. Research is also needed to stimulate the production of organic seeds and vegetative propagation materials from crops for which no suitable varieties are yet available. Seed companies need to see a clear long-term market before they invest in providing organic seeds for such varieties. The organic sector as a whole should realise that seeds are an integral part of organic production systems and that, at present, higher prices for organic seeds cannot always be avoided. However, these costs should be shared by all partners in the supply chain. Research is needed to find out the best policy for different crops and different supply chains to stimulate the production and use of organic seeds and bridge the gap between need and supply. In addition organic breeding should be supported in order to develop varieties that are better adapted to organic growing conditions. If varieties have an added value to organic farmers, the price of the seed becomes less relevant.

Seed health remains an important issue for seed quality. Novel methods, with the use of DNA and RNA sequencing techniques, will aid in early detection of pathogens during seed production and will determine the critical control points to prevent infection or spread of disease as well as to produce healthy seeds. New methods for non-chemical seed sanitation are in development. An example is the application of pulsed electric field treatment technology that is used in the food industry (Evrendilek and Tanasov, 2017).

Use of biologicals with seeds will increase in the near future. Research will be needed to address what advantages these developments bring for organic farming systems, which are already characterised by rich microbial biodiversity in the soil. At present it is not known whether organically produced seeds differ in their microbial biodiversity compared to non-chemically treated conventionally produced seeds. The rapid advancements and cost reductions in DNA sequencing techniques will be of great aid in such studies (Galimberti et al., 2015; Das and Deb, 2015; Thomsen and Willerslev, 2015). Such techniques can also aid in understanding the dynamics of microorganisms, including pathogens, in the organic farming ecosystem and their resilience.

Another important area of research is the interaction between seed vigour and seed health. It is well known that stressed plants and animals are more susceptible to pathogens. Producing seeds with a higher vigour, either through seed production and sorting techniques or through breeding, will likely aid in seeds and seedlings that are more tolerant against pathogens.

11 Where to look for further information

- The IFOAM position paper on the use of organic seed and plant propagation material in Organic Agriculture, published in 2011, http://www.ifoam.bio/sites/default/files/position_seeds.pdf.
- A good textbook on seed physiology: Bewley, J.D., Bradford, K.J., Hilhorst, H.W.M., Nonogaki, H., 2013. *Seeds: Physiology of Development, Germination and Dormancy*, 3rd Edition ed. Springer, New York.

- A good book on plant health management in organic farming, including seed-borne diseases, is Finckh, M.R., van Bruggen, A.H., Tamm, L., 2015. *Plant Diseases and their Management in Organic Agriculture.* APS Press St. Paul, Minnesota.
- A review on the status of organic seed regulations in the United States, Mexico and the EU: Renaud, E.N., Lammerts van Bueren, E.T., Jiggins, J., 2016. The meta-governance of organic seed regulation in the USA, European Union and Mexico. *International Journal of Agricultural Resources, Governance and Ecology* 12, 262–91.
- For a review on the latest efforts by breeders to develop improved varieties for organic production: van Bueren, E. T. L. and J. R. Myers (2011). *Organic Crop Breeding*, John Wiley & Sons.
- For information on organic plant breeding and the regulation on organic seeds in Europe visit for instance the website from the European Consortium for Organic plant breeding (ECO-PB), founded in 2001, with the aim of improving the performance of organic agriculture by boosting organic seed and plant breeding efforts across Europe. The website address is www.eco-pb.org.
- In June 2017 the European project LIVESEED 'Improve performance of organic agriculture by boosting organic seed and plant breeding efforts across Europe' started with the aim to increase the availability and use of organic produced seeds and adapted varieties. The project is financially supported by the European Commission and involves the participation of 35 partners and 17 countries. More information can be obtained through their website: www.liveseed.eu.

12 References

Alvarez, A. M. (2000) Black rot of crucifers. In Slusarenko, A., Fraser, R. and van Loon, L. (Eds) *Mechanisms of Resistance to Plant Diseases.* Dordrecht, The Netherlands: Kluwer Academic Publishers, pp. 21–52.

Amein, I., Wright, S. A., Wikström, M. and et al. (2011) Evaluation of non-chemical seed treatment methods for control of *Alternaria brassicicola* on cabbage seeds. *Journal of Plant Diseases and Protection* 118: 214–21.

Astma, M., Koenraadt, H. M. S. and Politikou, A. (2014) Detection of *Xanthomonas campestris* pv. *campestris* on *Brassica* spp. disinfested/disinfected seed with grinding. Annexe to International Rules for Seed Testing, Chapter 7: Seed Health Testing Methods.

BASF. (2016) Trade News, Apr 21, 2016. BASF opens new R&D Center for biological crop protection and seed solutions. https://www.basf.com/en/company/news-and-media/news-releases/2016/04/p-16-188.html. Accessed 17 January 2017.

Bashan, Y. (1998) Inoculants of plant growth-promoting bacteria for use in agriculture. *Biotechnology Advances* 16: 729–70.

Bewley, J. D., Bradford, K. J., Hilhorst, H. W. M. and et al. (2013) *Seeds: Physiology of Development, Germination and Dormancy*, New York: Springer.

BioAgri. (2017a) Amase®. http://www.bioagri.se/?p=30931&m=4377&topic=amase. Accessed on 17 January 2017.

BioAgri. (2017b) Products. http://www.bioagri.se/?p=30933&m=4353&topic=products. Accessed on 17 January 2017.

Brocklehurst, P. and Dearman, J. (1983) Interactions between seed priming treatments and nine seed lots of carrot, celery and onion. I. Laboratory germination. *Annals of Applied Biology* 102: 577–84.

Burges, H. D. (2012) *Formulation of Microbial Biopesticides: Beneficial Microorganisms, nematodes and seed treatments*, Dordrecht: Springer Science & Business Media.

Das, S. and Deb, B. (2015) DNA barcoding of fungi using Ribosomal ITS Marker for genetic diversity analysis: A Review. *International Journal of Pure and Applied Bioscience* 3: 160–7.

Döring, T. F., Bocci, R. and Hitchings, R. and et al. (2012) The organic seed regulations framework in Europe – current status and recommendations for future development. *Organic agriculture* 2: 173–83.

European and Mediterranean Plant Protection Organization. (2013) PM 7/42 (2) *Clavibacter michiganensis* subsp. *michiganensis*. *EPPO Bulletin* 43: 46–67.

European Commission. (2017) Seed databases. http://ec.europa.eu/agriculture/organic/eu-policy/seed-databases_en. accessed on 24 January 2017.

Evrendilek, G. A. and Tanasov, I. (2017) Configuring pulsed electric fields to treat seeds: An innovative method of seed disinfection. *Seed Science and Technology* 45: 72–80.

Forsberg, G. (2004) Control of cereal seed-borne diseases by hot humid air seed treatment. *Acta Universitatis agriculturae Suecia. Agraria* 443: 1–49.

Galimberti, A., Bruno, A., Mezzasalma, V. and et al. (2015) Emerging DNA-based technologies to characterize food ecosystems. *Food Research International* 69: 424–33.

Groot, S. P. C., Birnbaum, Y., Kromphardt, C. and et al. (2008) Effect of the activation of germination processes on the sensitivity of seeds towards physical sanitation treatments. *Seed Science and Technology* 36: 609–20.

Groot, S. P. C., Birnbaum, Y., Rop, N. and et al. (2006) Effect of seed maturity on sensitivity of seeds towards physical sanitation treatments. *Seed Science and Technology* 34: 403–13.

Groot, S. P. C., de Groot, L., Kodde, J. and et al. (2015) Prolonging the longevity of *ex situ* conserved seeds by storage under anoxia. *Plant Genetic Resources* 13: 18–26.

Grossman, J. (2015) Legume inoculation for organic farming systems. http://articles.extension.org/pages/64401/legume-inoculation-for-organic-farming-systems.

Guenha, R., Salvador, B. d. V., Rickman, J. and et al. (2014) Hermetic storage with plastic sealing to reduce insect infestation and secure paddy seed quality: A powerful strategy for rice farmers in Mozambique. *Journal of Stored Products Research* 59: 275–81.

Howard, P. H. (2015) Intellectual property and consolidation in the seed industry. *Crop Science* 55: 2489–95.

IFOAM EU Group. (2003) Co-existence between GM and non-GM crops – Necessary anti-contamination and liability measures. http://www.ifoam-eu.org/sites/default/files/page/files/ifoameu_policy_gm_coexistence_position_200310.pdf.

ISTA. (2016) *International rules for seed testing*, Basserdorf, Switzerland: International Seed Testing Association.

Jalink, H., van der Schoor, R., Frandas, A. and et al. (1998) Chlorophyll fluorescence of *Brassica oleracea* seeds as a non-destructive marker for seed maturity and seed performance. *Seed Science Research* 8: 437–43.

Karamanos, R. E., Flore, N. A. and Harapiak, J. T. (2010) Re-visiting use of *Penicillium bilaii* with phosphorus fertilization of hard red spring wheat. *Canadian Journal of Plant Science* 90: 265–77.

Kim, S. M., Nam, S. H., Lee, J. M. and et al. (2003) Destruction of Cucumber green mottle mosaic virus by heat treatment and rapid detection of virus inactivation by RT-PCR. *Molecules and Cells* 16: 338–42.

Koch, E. and Groot, S. P. C. (2015) Health management for seeds and other organic propagation material. In Finckh, M. R., Ariena H. C. van Bruggen, A. H. C. and Tamm, L. (Ed.) *Plant Diseaes and Their Management in Organic Agriculture*. The American Phytopathological Society, pp. 189–203.

Koch, E., Schmitt, A., Stephan, D. and et al. (2010) Evaluation of non-chemical seed treatment methods for the control of *Alternaria dauci* and *A. radicina* on carrot seeds. *European Journal of Plant Pathology* 127: 99–112.

Köpke, U., Thiel, B. and Elmholt, S. (2007) Strategies to reduce mycotoxin and fungal alkaloid contamination in organic and conventional cereal production systems. In Cooper, J., Leifert, C. and Niggli, U. (Eds.) *Handbook of Organic Food Safety and Quality*. Woodhead Publishing Ltd, pp. 353–91.

Lopez-Reyes, J. G., Gilardi, G., Garibaldi, A. and et al. (2016a) In vivo evaluation of essential oils and biocontrol agents combined with heat treatments on basil cv Genovese Gigante seeds against *Fusarium oxysporum* f. sp. *basilici*. *Phytoparasitica* 44: 35–45.

Lopez-Reyes, J. G., Gilardi, G., Garibaldi, A. and et al. (2016b) In vivo evaluation of essential oils and biocontrol agents combined with hot water treatments on carrot seeds against alternaria radicina. *Journal of Phytopathology* 164: 131–5.

Lötjönen, T. and Torniainen, M. (2005) Does Cedomon work? *Forskningsnytt om økologisk landbruk i Norden* (1): 26–7.

Mancini, V., Murolo, S. and Romanazzi, G. (2016) Diagnostic methods for detecting fungal pathogens on vegetable seeds. *Plant Pathology* 65: 691–703.

Mancini, V. and Romanazzi, G. (2014) Seed treatments to control seedborne fungal pathogens of vegetable crops. *Pest Management Science* 70: 860–8.

Mathur, S. B. and Kongsdal, O. (2003) *Common Laboratory Seed Health Testing Methods for Detecting Fungi*, Bassersdorf, Switzerland: International Seed Testing Association.

McDonald, M. and Copeland, L. O. (2012) *Seed Production: Principles and Practices*. Dordrecht: Springer Science + Business Media.

McGuire, S. and Sperling, L. (2016) Seed systems smallholder farmers use. *Food Security* 8: 179–95.

McQuilken, M. P., Halmer, P. and Rhodes, D. J. (1998) Application of microorganisms to seeds. In Burges, H. D. (Ed.) *Formulation of Microbial Biopesticides: Beneficial Microorganisms, Nematodes and Seed Treatments*. Dordrecht: Springer Science + Business Media, pp. 255–85.

Monsanto BioAg. (2016) JumpStart®. http://www.monsantobioag.com/global/ca/Products/Pages/jumpstart.aspx. Accessed on 17 January 2017.

O'Callaghan, M. (2016) Microbial inoculation of seed for improved crop performance: Issues and opportunities. *Applied Microbiology and Biotechnology* 100: 5729–46.

Pellegrino, C., Gilardi, G., Gullino, M. L. and et al. (2010) Detection of *Phoma valerianellae* in lamb's lettuce seeds. *Phytoparasitica* 38: 159–65.

Piorr, H. (1992) Phytopathological advantages and risks of organic farming systems: Future perspectives to improve organic cropping systems. In Altman, J. (Ed.) *Pesticide Interactions in Crop Production: Beneficial and Deleterious Effects*. Cleveland: CRC Press, pp. 461–73.

Reingold, V., Lachman, O., Blaosov, E. and et al. (2015) Seed disinfection treatments do not sufficiently eliminate the infectivity of Cucumber green mottle mosaic virus (CGMMV) on cucurbit seeds. *Plant Pathology* 64: 245–55.

Renaud, E. N., Lammerts van Bueren, E. T. and Jiggins, J. (2016) The meta-governance of organic seed regulation in the USA, European Union and Mexico. *International Journal of Agricultural Resources, Governance and Ecology* 12: 262–91.

Renaud, E. N., Lammerts van Bueren, E. T. L. and Jiggins, J. (2014) The development and implementation of organic seed regulation in the USA. *Organic agriculture* 4: 25–42.

Röder, O., Jahn, M., Schröder, T. and et al. (2009) E-ventus technology-an innovative treatment method for sustainable reduction in the use of pesticides with recommendation for organic seed. *Journal für Verbraucherschutz und Lebensmittelsicherheit* 4: 107–17.

Schmitt, A., Jahn, M., Kromphardt, C. and et al. (2006) STOVE: Seed treatments for organic vegetable production. *European Joint Organic Congress*, 30–31 May 2016. Odense, Danmark.

Schmitt, A., Koch, E., Stephan, D. and et al. (2011) Evaluation of non-chemical seed treatment methods for the control of *Phoma valerianellae* on lambs lettuce seeds. *Journal for Plant Diseases and Plant Protection (JPDP)* 5: 200–7.

Syngenta. (2016) Syngenta launches new biological seed treatment. http://www4.syngenta.com/media/media-releases/yr-2016/12-05-2016. Accessed 17 January 2017.

TeKrony, D. M., Egli, D. B., Balles, J. and et al. (1979) Physiological Maturity in Soybean. *Agronomy Journal* 71: 771–5.

The Commission of the European communities. (2007) Council Regulation (EC) No 834/2007. *Official Journal of the European Union*, http://eur-lex.europa.eu/LexUriServ/LexUriServ.do?uri=OJ:L:2007:189:0001:0023:EN:PDF. (accessed 22 January 2017).

The Commission of the European communities. (2008) Commission Regulation (EC) No 889/2008. *Official Journal of the European Union*, http://eur-lex.europa.eu/LexUriServ/LexUriServ.do?uri=OJ:L:2008:250:0001:0084:en:PDF.

Thomsen, P. F. and Willerslev, E. (2015) Environmental DNA – An emerging tool in conservation for monitoring past and present biodiversity. *Biological Conservation* 183: 4–18.

Tinivella, F., Hirata, L., Celan, M. and et al. (2009) Control of seed-borne pathogens on legumes by microbial and other alternative seed treatments. *European Journal of Plant Pathology* 123: 139–51.

United States Department of Agriculture – Agricultural Marketing Service – National Organic Program. (2013) Guidance – Seeds, annual seedlings, and planting stock in organic crop production. https://www.ams.usda.gov/sites/default/files/media/5029.pdf.

USDA-AMS. (2002) *7 CFR Part 205 National Organic Program Standards*. Available at: www.ams.usda.gov/nop/standards/.

van der Wolf, J., Birnbaum, Y., Van Der Zouwen, P. and et al. (2008) Disinfection of vegetable seed by treatment with essential oils, organic acids and plant extracts. *Seed Science and Technology* 36: 76–88.

van der Wolf, J. M. and van der Zouwen, P. S. (2010) Colonization of Cauliflower Blossom (*Brassica oleracea*) by *Xanthomonas campestris* pv. *campestris*, via Flies (*Calliphora vomitoria*) Can Result in Seed Infestation. *Journal of Phytopathology* 158: 726–32.

Vessey, J. K. (2003) Plant growth promoting rhizobacteria as biofertilizers. *Plant and Soil* 255: 571–86.

Williams, S. B., Murdock, L. L. and Baributsa, D. (2017) Storage of maize in purdue improved crop storage (PICS) bags. *PLoS One* 12: e0168624.

Zanakis, G., Ellis, R. and Summerfield, R. (1994) Seed quality in relation to seed development and maturation in three genotypes of soyabean (*Glycine max*). *Experimental Agriculture* 30: 139–56.

Chapter 3

Maintaining soil fertility and health in organic crop cultivation

C. Watson, Scotland's Rural College (SRUC), UK and Swedish University of Agricultural Sciences, Sweden; and E. A. Stockdale, Head of Farming Systems Research, NIAB, UK

1 Introduction

2 Some key themes in soil fertility and soil health

3 Case study: interactions between soil management and the delivery of soil functions/ecosystem services

4 Conclusion and future trends

5 Acknowledgements

6 Where to look for further information

7 References

1 Introduction

The concept of soil health is deeply embedded in organic agriculture as reflected in Lady Eve Balfour's statement 'The health of soil, plant, animal and man is one and indivisible' (Balfour, 1943). But what do we really mean by soil health? Here we use the integrative concept of soil health, closely related to soil quality, which reflects the capacity of a soil to respond to agricultural management practices and maintain agricultural production amongst a range of other ecosystem services (Kibblewhite et al., 2008). This definition of soil health can be seen as an agricultural interpretation of the overarching FAO definition of soil health 'the capacity of soil to function as a living system' (FAO, 2011) compared with soil quality which has been defined as its fitness for a given use (Larson and Pierce, 1994). Soils in cropping systems both give and receive 'supporting' and 'regulating' ecosystem services while providing 'provisioning' services (Fig. 1). Ecosystem services are normally regarded as positive (+); however, agricultural soils also create some negative (−) ecosystem services or 'disservices' which can be viewed as a negative feedback loop to 'regulating' and 'provisioning' services. In managing soil within organic farming systems, our aim must be to minimise these disservices while providing the healthy soils needed to deliver adequate 'provisioning' services. The capacity of a soil to maintain agricultural production or soil fertility (Cooke, 1967) encompasses not only the provision of crop nutrition but also plant disease prevention/plant health promotion and is thus a component of soil health. Soil fertility can therefore be seen as an aggregate of soil biology, chemistry and physics as illustrated in Fig. 2, which shows how the incorporation of crop residues can influence soil

http://dx.doi.org/10.19103/AS.2017.0029.03

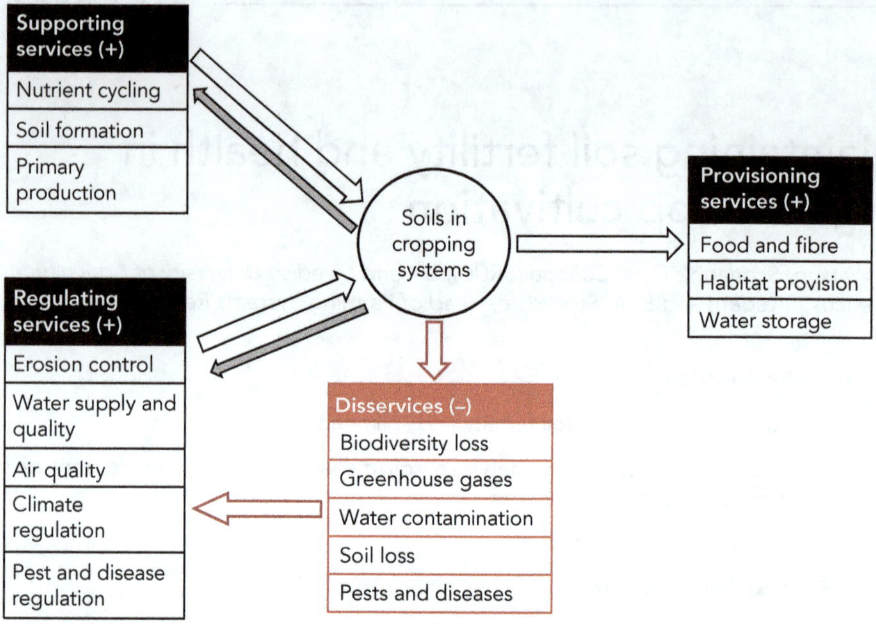

Figure 1 Soils in cropping systems and their relationship to ecosystem services and disservices (based on MEA, 2005; Zhang et al., 2007; Kibblewhite et al., 2008).

fertility through effects on biology, chemistry and physics (Stockdale and Watson, 2012). The extent to which soil fertility can be manipulated and the rate at which changes occur will depend on fundamental properties such as soil texture and parent material, and also on climatic conditions. Many possible indicators and indexes of soil health are described in the literature (see, e.g. Kelly et al., 2009; Idowu et al., 2008), and there is currently much interest in the development of improved tools to communicate the complexity of the impacts of soil management and to allow farmers to develop locally adapted approaches to maintain soil health (Stockdale et al., 2017).

As Fig. 1 illustrates, soils are multifunctional, and thus there is a very large body of research focused on soils and ecosystem services. It is further complicated by the interactions of soil parent material and climate to provide a vast diversity of soil textures, with differing chemical and physical characteristics interacting with changing climates. This chapter focuses on agricultural management and thus provisioning services rather than environmental impact or soil as a biodiversity reservoir although these soil functions are an important part of the case study in Section 3. We focus on organic agricultural systems including temporary grass although we acknowledge that protected cropping systems are a significant challenge for soil health and fertility in organic systems (see Gamliel and van Bruggen, 2016). Where possible we take examples from research in organic systems but use some examples from conventional farming to illustrate basic principles of soil fertility and health.

The 'Ecology' principle in organic farming states that 'organic agriculture should be based on living ecological systems and cycles, work with them, emulate them and help sustain them' (IFOAM, 2005). The EU Council Regulation 834/2007 (EC, 2007) states

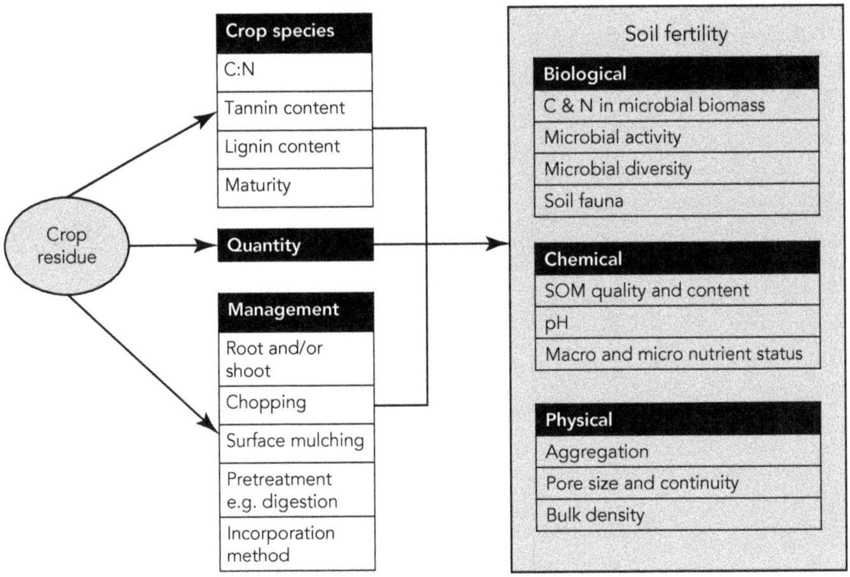

Figure 2 Impact of crop residue incorporation on soil fertility in a given soil type and climate and some factors that affect it.

'the maintenance and enhancement of soil life and natural soil fertility, soil stability and soil biodiversity preventing and combating soil compaction and soil erosion, and the nourishing of plants primarily through the soil ecosystem'. 'Organic plant production should contribute to maintaining and enhancing soil fertility as well as to preventing soil erosion. Plants should preferably be fed through the soil eco-system and not through soluble fertilisers added to the soil'. The ecology principle is thus enacted in terms of soil fertility by relying on the use of multi-annual rotations to maintain or improve soil fertility, together with careful use of species and varieties, appropriate cultivation methods and recycled organic materials. Where necessary, a limited number of specified additional fertilisers and soil conditioners can be brought into the farm system, some of which require permission from certification bodies. In Europe these materials are listed in Annex 1 of the EC Commission Regulation 889/2008 (EC, 2008). The EC regulation is subject to different implementing rules in different member states; examples of the differences in soil and nutrient management that can arise from this can be found in Løes et al. (2017).

2 Some key themes in soil fertility and soil health

2.1 Organic matter maintenance

Soil organic matter is fundamental in terms of its role in providing a substrate for soil biota and thus nutrient cycling in soils. Long-term experiments suggest that soil microbial communities are more influenced by organic matter inputs and soil type than by inorganic fertilisers (e.g. Williams et al., 2013). Organic matter also plays multiple

Table 1 Mean and range of soil organic carbon content (%, in <2 mm air-dried soil) under arable, ley and permanent grass in different soil texture groups (from Loveland and Webb, 2003)

	Arable		Ley		Permanent managed grass	
	Mean (range)	Observations	Mean (range)	Observations	Mean (range)	Observations
Clay	3.38 (1.2–9.9)	290	4.3 (1.3–10.9)	72	5.45 (1.3–10.9)	208
Coarse loamy	1.97 (0.5–8.1)	430	3.05 (0.0–9.0)	125	4.06 (1.1–11.9)	281
Fine loamy	2.6 (0.5–10.9)	651	3.51 (1.0–12.0)	306	4.34 (1.1–11.5)	631
Coarse silty	2.85 (1.1–5.9)	22	3.17 (0.8–6.6)	10	3.86 (1.5–10.6)	18
Fine silty	2.72 (0.8–10.1)	186	3.44 (1.0–9.6)	88	4.36 (1.0–11.0)	166
Sandy	1.75 (0.4–9.1)	75	2.29 (0.7–4.5)	7	2.59 (0.6–5.4)	28

other roles in soil including improved erosion control, water infiltration rates and water holding capacity through its influence on soil structure (Allison, 1973). Brock et al. (2011) demonstrated the importance of maintaining soil organic matter levels for yield in organic farming systems. Increasing soil organic matter inputs in crop residues, manures, slurries or other approved sources generally increases the soil organic matter content of soils although this is site dependent as a result of the influence of pedo-climatic conditions. Soil organic matter levels are generally higher under cool, wet conditions but texture is also important, so the higher the clay content the greater the soil carbon or organic matter content is, under given management and climatic conditions (Table 1). As Table 1 indicates, including temporary grassland within rotations is well known to have a positive impact on soil organic matter content. Working in an organic ley/arable system, Taylor et al. (2006) demonstrated the ability of these systems to maintain high levels of soil organic matter. It is more challenging to retain or build soil organic matter in organic arable systems, particularly those which do not import organic livestock manure (Schulz et al., 2014). Exchanges of organic manures between organic farms to balance crop and livestock production are discussed further in Section 2.4. Harvests remove between one-third and one half of the net primary production of plants (Bolinder et al., 2007), and thus the lower biomass of arable crops compared with perennial crops means that arable systems have lower C inputs to the soil than grassland systems. Shoot to root ratios of different annual and perennial crops also vary widely (Fig. 3) and well-designed organic rotations should contain crops with different rooting depths and volumes in order to make best use of the available soil profile. This may be through time or space that is in intercrops and mixtures as well as through rotations (see Section 2.2 for discussion of rotation design). Frequent tillage has also been demonstrated to result in increased N mineralisation rates and loss of soil C and N (Balesdent et al., 2000) as a result of physical and biological changes.

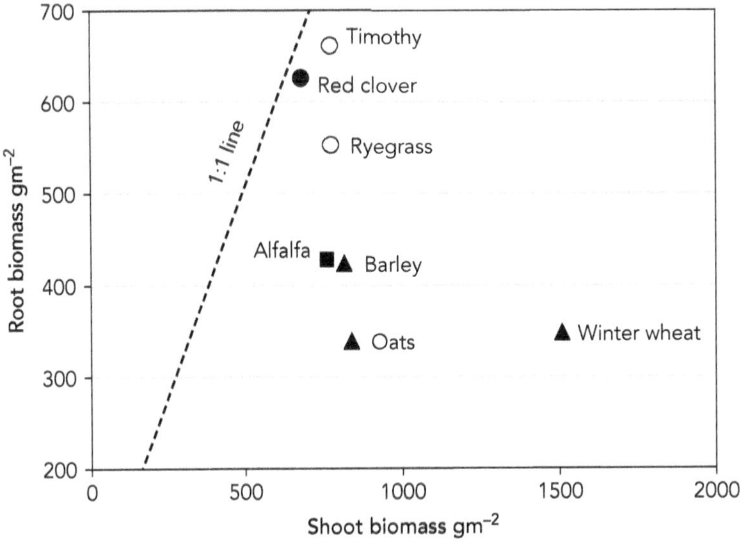

Figure 3 Root (0–30 cm depth) and shoot biomass of some grasses (open circle), legumes (closed circle) and cereals (triangle) from Bolinder et al. (1997, 2002).

2.2 Designing appropriate rotations

Crop rotations play several key roles in organic farming, and these have been described by many authors (e.g. Watson et al., 2002a; Lampkin, 1990); here we highlight some important contemporary issues. In summary, rotations maintain soil nitrogen supply as they contain both nitrogen building phases and exploitative cropping phases. Secondly, the choice of crops influences soil organic matter and thus soil structure. Thirdly, they have a key role in regulating weed, pest and disease populations. Both crop diversity and the sequence affect soil fertility and health, together with overall productivity and environmental impact. The use of crop species and families with different disease susceptibilities is important. For example, average pea yields in France have decreased in recent years, and this has been attributed to increased frequency of peas in rotations, leading to a build-up of soil-borne diseases and a consequent shift of production to less suitable soils and associated lower yields (Magrini et al., 2016). The inclusion of forage and/or grain legumes which fix atmospheric nitrogen through their association with *Rhizobium* which live in the nodules of the legume root system is the key to building soil nitrogen. The issue of synchronising supply and demand for nitrogen within the rotation is so fundamental to soil fertility in organic farming that it is discussed separately in the next section. The amount of N fixed by legumes depends on a large number of management and environmental variables (Peoples et al., 2009; Ledgard and Steele, 1992). However, as regards rotation design, the frequency of legumes in the rotation affects N fixation, with continuous pea cropping reducing the proportion of N derived from the atmosphere to 15% compared with an average of 55% in peas grown every second, third or fourth year (Knight, 2012). The reasons for this remain unclear but may be related to disease as discussed above.

Nitrogen supply is not the only factor that determines the growth of the following crop as the incorporation of the pre-crop residues also influences phosphorus availability (Talgre et al., 2014; Eichler-Löbermann et al., 2009). Residue incorporation also induces changes in soil microbiology (Ratnadass et al., 2012), and can result in increased populations of beneficial soil microorganisms (van Bruggen et al., 2016; Dias et al., 2015), promoting root health of the following crops. Understanding the pre-crop effect on soil fertility and health not only for a single year but also for year-on-year effects through the rotation is a significant challenge, but modelling approaches to this are currently being developed (Reckling et al., 2016). Understanding these effects can also help in planning the need for manures or supplementary nutrients in the following year, thus reducing costs and maximising profit (Preissel et al., 2015).

The choice of crops in rotation also influences soil structure in both the topsoil and the subsoil, which can influence establishment and yield of following crops (Ball et al., 2005). Crop roots have a fundamental effect on subsoil structure and thus water and nutrient use through the formation of biopores. When plant roots die and decompose, biopores remain; these are destroyed by cultivation but survive below plough depth (Ehlers et al., 1983). Biopores are also formed by burrowing earthworms, and thus reduced tillage practices that favour earthworms can result in increased density of biopores (Kautz, 2014). Such biopores allow the roots of following crops to quickly and easily access deeper soil layers and the water and nutrients they hold (Han et al., 2015b; Kautz et al., 2013), potentially influencing the ability of the system to function under drought conditions. Deposits of carbon and nutrients from previous roots and detritus from earthworms create nutrient-rich hotspots in these biopores (Kautz et al., 2013). Including more dicotyledonous crops in the rotation results in a higher density of biopores (Kautz, 2014) as do tap-rooted crops compared with fibrous rooted crops (Han et al., 2015a).

2.3 Sustainable sources of nutrients in organic farming

The long-term sustainability of organic farming through the production of economically viable yields requires acceptable sources of major and micronutrients to be available on-farm in plant available forms. The issues of synchrony between supply and demand for nitrogen are discussed in Section 2.4, and here we address absolute amounts of nutrients. Maintenance of current yields requires that nutrients sold off the farm in produce are replaced. Any overall improvement in organic crop yields is likely to require increased nutrient availability combined with plant breeding and other agronomic approaches. Co-products from processing of organic food are not generally returned to organic farming systems, potentially resulting in nutrient depletion of soils. Sewage sludge which ultimately 'closes the nutrient loop' is not permitted within the EU organic regulations. Oelofse et al. (2013) reviewed the pros and cons of use of sewage sludge and some controversial industrial co-products in organic farming from technical, legislative and social perspectives.

Nutrient budgets are now commonly used to assess the sustainability of farming systems as well as their potential environmental impact. They operate on a mass balance principle using the inputs and outputs of nutrients over a defined area for a defined period of time. Farm-gate budgets for organic farms often show negative balances for phosphorus and potassium, suggesting that these systems are unsustainable in the long term unless they import nutrients (Watson et al., 2002b). Depending on soil type, soil reserves may be able to sustain production in the short term. In general, organic mixed and livestock farming systems show more positive farm-gate balances than arable systems (Berry et al., 2003; Watson et al., 2002b). However, within organic farming, as in conventional farming, there is increasing specialisation, resulting in decoupling of crop and livestock production (David et al., 2012), and this has implications for maintaining soil nutrient levels. Using more detailed soil surface budgets at the field scale can provide useful information to guide manure and fertiliser management (Watson et al., 2002b) and can also be used to understand the origin of nutrients with respect to farm sustainability (Foissy et al., 2013). In a study of 28 farms (including arable, mixed and livestock) in 3 French counties, Foissy et al. (2013) showed that between 35% and 64% on N inputs to crop production came from biological N fixation on average per farm type and region (Table 2). The livestock farms in the study showed minimal dependence on imported N, P and K, whereas the arable farms imported up to 100% of P and K.

Creative solutions are thus needed to maintain soil nutrient and organic matter levels on specialised organic arable farms or horticultural enterprises. The use of 'cut and carry' or 'mobile' green manures is one approach to allowing fertility to be managed across the farm system by moving harvested shoot material to areas of a farm or a field low in organic matter or nutrients (Sorensen and Thorup-Kristensen, 2011). The need for organic matter in organic arable systems together with the need to meet environmental legislation has led to the development of collaborative partnerships between organic farmers to either exchange manure for straw/fodder or exchange manure for money (Asai and Langer, 2014). They note that farmers are rarely paying a price related to the fertiliser or organic matter value of the manure. This kind of approach leads to the question as to whether assessing mixed farms at the regional level may be more appropriate than individual mixed farms when addressing nutrient balances (Moraine et al., 2017; Nesme et al., 2012). This view might facilitate the appropriate uses of locally produced co-products as soil amendments. In the study of Asai and Langer

Table 2 Soil surface N, P, K balances for different farm types on 28 farms in 3 counties of France showing dependence on N fixation and off-farm nutrient sources (based on Foissy et al., 2013)

	Arable Lorraine	Arable plain of Valence	Mixed Lorraine	Livestock Lorraine	Livestock plain of Forez	All
	n = 4	*n* = 4	*n* = 4	*n* = 10	*n* = 6	*n* = 28
Nitrogen						
% input from fixation	64.4	41.6	46.5	46.2	35.1	43.5
% input from off-farm sources	14.3	58.4	5.2	0.0	0.6	9.4
Balance (kg N ha^{-1} year^{-1})	−30.2 (±14.4)	−34.9 (±33.1)	−5.6 (±13.1)	−1.0 (±12.4)	20.8 (±28.0)	−6.0 (±27.2)
Phosphorus						
% input from off-farm sources	47.7	100.0	13.5	0.0	1.6	38.4
Balance (kg P ha^{-1} year^{-1})	−8.5 (±3.6)	16.2 (±20.1)	−4.6 (±2.1)	−5.2 (±1.5)	0.2 (±2.9)	−1.4 (±10.5)
Potassium						
% input from off-farm sources	43.3	100.0	10.6	0.00	1.1	14.5
Balance (kg K ha^{-1} year^{-1})	−37.2 (±28.7)	−2.5 (±40.6)	−18.9 (±11.6)	−28.1 (±16.9)	15.7 (±23.8)	−15.0 (±29.3)

(2014) in Denmark, 77% of exporters of manure were conventional farms and this clearly has implications for the long-term sustainability of organic production. Oelofse et al. (2013) previously reported that in Denmark in 2011, approximately 25% of manure nutrient inflows to organic crops came from conventional farming. Such arrangements can only function while the organic regulations allow for the import of manures and other materials from conventional sources. In some countries, for example UK, there is already a requirement for a written agreement between organic poultry producers and organic landholders for the judicious use of the organic poultry manure on certified land. However, while bringing manures, composts and other organic materials on to the farm as a soil conditioner or fertiliser is positive in terms of building soil organic matter and soil structure, these materials can also bring in undesirable contaminants. For example, manure can become contaminated from the use of veterinary products, for example copper in manures due to contamination with waste from footbaths using copper salts (e.g. Jokela et al., 2010) or zinc from metal equipment or building materials in livestock housing (Öborn et al., 2005).

The use of globally traded commodities and how much organic farming relies on this type of commodity are a source of continuing controversy. Using rock phosphate as an example, there are questions in relation to the philosophy of use of non-renewable resources in organic farming, logistics of global trade and the practicalities of using a relatively insoluble source of phosphorus which can take years to become plant available in soils (see e.g. Paulsen et al., 2016).

2.4 Matching supply and demand for nitrogen

When crop residues (whether leguminous or non-leguminous) are incorporated into soil, nitrogen is released into the soil. Ensuring that this nitrogen is released in synchrony with the demand of the following crop is important both agronomically and environmentally. If the available nitrogen exceeds crop demand, then it may be subject to loss in leaching or through conversion to gaseous forms. On the other hand, delayed mineralisation can limit N uptake and yield of the following crop. The challenge here is the multiple agronomic and environmental factors that influence both N release and crop uptake. The timing of nitrogen mineralisation and crop demand are mediated by environmental factors as well as soil management, the chemical characteristics of the incorporated material and the growth pattern of the following crop (Petersen et al., 2013; Jarvis et al., 1996).

Management factors influencing N release from crop residues include timing of incorporation (Ball et al., 2007; Thorup-Kristensen and Dresbøll, 2010) and crop management prior to incorporation of the residue, for example cutting, mulching or grazing of perennial crops (Hatch et al., 2014; Ball et al., 2007). Soil factors such as organic matter content (Brock et al., 2011) and pH also influence N mineralisation. Carbon:nitrogen ratios vary with plant species as well as stage of maturity of the harvested plant material. For example, in a range of horticultural green manure species, Sorensen and Thorup-Kristensen (2011) showed C:N ratios varying from 9 (buckwheat) to 26 (sorrel). There is also evidence that some plant species, for example ribgrass (*Plantago lanceolata* L.) can have an inhibitory effect on soil N mineralisation due to the production of allelopathic substances (Dietz et al., 2013). This suggests the potential to produce mixtures of crop residues with manageable nitrogen release patterns.

3 Case study: interactions between soil management and the delivery of soil functions/ecosystem services

There are a growing number of published studies which have measured soil biodiversity in organic farming. However, relatively few link management practices to soil biodiversity in terms of outcome. This results in a lack of on-farm understanding of how to manage soil biology on organic farms in order to promote soil fertility and health. In 2011, we worked with groups of organic farmers in England to identify the practices they thought would promote soil biodiversity. We used the list of practices as the basis for a literature review on links between management, biodiversity and the delivery of soil ecosystem services (Stockdale and Watson, 2012). Based on a report to Defra (2010), we divided the practices into (1) systems-oriented approaches that provide energy-containing substrates and/or seek to optimise soil habitat such as amount and quality of organic inputs, tillage practices and the use of diverse cropping and (2) those that target specific often monotonic aspects of the soil biota or their environment (intervention through product application at a specific point in time) such as biocontrol or inoculation with specific species (arbuscular mycorrhizal or ectomycorrhizal fungi, rhizobia) or mixed species cultures (plant-growth-promoting rhizobacteria, plant-growth-promoting fungi). We have summarised and updated that literature review in Table 3 in order to show some of the most important management practices for promoting soil fertility and health in organic farming. Other soil management

Table 3 Expected impacts of key land management practices used in organic farming on soil biota and soil function in relation to provisioning and other ecosystem services derived from expert judgement and a *literature review based on, and updated from*, Stockdale and Watson (2012)

Land management practice	Direct impacts on soil biota	Indirect impacts on soil biota	Likely impacts on soil function for agriculture and other ecosystem services	More information relevant to the use of this practice in the field on soil biota and/or soil functions in organic farming
Managing amount and quality of organic matter inputs				
• Use of green waste compost, mushroom compost, paper waste and coffee grounds, that is, application of (local) waste organic matter Repeated applications	• Provides energy/nutrient source for soil food web • Supports increased biomass • May increase species richness and evenness depending on 'quality' – C/N ratio and so forth	• Stimulates structural formation processes after disturbance • Improves structural stability in many soils • Improves drainage in poorly drained soils • Improves water holding capacity in sandy soils • Fertiliser effects of nutrients supplied stimulate plant growth and C inputs via roots and residues	• Improves nutrient supply for plant growth • Improves soil structure and its stability – reducing sediment loss • Increases soil C content – C sequestration • Improves water balance, regulates water flows • Increases greenhouse gas production – increase soil respiration – CO_2 production; if soils become waterlogged, increase N_2O production	• Heinze et al. (2011) • Henneron et al. (2015) • Sradnick et al. (2013)
• Application of biochar	• Limited evidence of direct effects	• May increase protected habitat space for small organisms – limited evidence of role of pores within biochar in supporting soil organisms • Increased water retention	• Increases soil C content in very stable forms – C sequestration • Other benefits claimed but limited evidence in temperate soils	• Ulyett et al. (2014)

• Increased use of composting rather than slurry application	• Provides energy/nutrient source for soil food web • Reduces direct toxic effects of soluble NH_4 • Reduces indirect toxic effects caused by osmotic shock	• Reduced fertiliser effects following nutrient stabilisation by composting; may reduce plant growth and C inputs via roots and residues compared with slurry use • Stimulates structural formation processes after disturbance • Improves structural stability in many soils • Usually has small liming effect – increasing soil pH	• Improves soil structure and its stability – reducing sediment loss • Increases soil C content – C sequestration • Improves water balance, regulates water flows • Reduces NH_3 volatilisation losses • Reduces greenhouse gas production after soil application – reduced N_2O production. Less difference between manure management methods when accounting for CO_2 production during composting	• Fliessbach et al. (2007)
Modified tillage practice				
• Minimum intensity tillage	• All tillage operations kill soil macrofauna – largest impacts on earthworms and beetles; reduced numbers of tillage operations lead to significant increases in earthworm populations	• All tillage operations that mix soil reduce connectivity of transmission pores to depth • Changes pore size distribution, disrupts pore connectivity • Mixes organic matter (OM) inputs throughout tilled soil	• Improves soil structure – reducing sediment loss • Improves water balance, regulate water flows • Reduces energy requirements of cropping	• Lewis et al. (2011) • Sun et al. (2016) • Săle et al. (2015)

(Continued)

Table 3 *(Continued)*

Land management practice	Direct impacts on soil biota	Indirect impacts on soil biota	Likely impacts on soil function for agriculture and other ecosystem services	More information relevant to the use of this practice in the field on soil biota and/or soil functions in organic farming
• No till includes non-inversion tillage compared with minimum tillage	• Allows development of anecic earthworm populations towards site carrying capacity	• Increases connectivity of transmission pores from surface to depth • Increased profile stratification; higher OM contents in surface soils • Surface mulch of residues provides more suitable end of season habitat for surface dwelling arthropods	• Improves soil structure and its stability – reducing sediment loss • Improves water balance, regulates water flows • Increasing evidence that reduced tillage intensity does not always increase soil organic matter although it often changes its distribution with soil depth • Diversifies farmed landscapes overwinter; provides feeding habitats for seed-eating birds • Reduces energy requirements of cropping • Can enhance N fixation	• Carr et al. (2013) • Cooper et al. (2016) • Mäder and Berner (2012) • Kuntz et al. (2013) • Zotarelli et al. (2012) • Schulz et al. (2014)
• Drilling crops directly into clover swards compared with cultivation and establishment	• No direct effects expected	• Permanent ground cover of clover (usually white clover) provides food resources at soil surface (for earthworms) • Permanent perennial root system providing inputs below ground through exudation and root turnover • Increases connectivity of transmission pores from surface to depth	• Improves nutrient supply for plant growth as a result of N fixation and residue turnover • Improves soil structure and its stability; permanent soil cover – reducing sediment loss • Improves water balance, regulates water flows • May increase soil C content – C sequestration	• Pelosi et al. (2015)

Diversifying cropping systems				
• Locally adapted rotations with grass/clover leys compared with monoculture or minimal break crops	• Diversifies amount and quality of residue inputs, modifying energy/nutrient sources for the soil food web	• More variety in timing and type of cultivation practices • Reduced disturbance compared with arable can promote AMF • Increased variety in rooting patterns • Increased diversity of hosts to support persistence of plant-associating organisms	• Diversifies farmed landscapes • Increases soil C content – C sequestration • Improves soil structure – reducing sediment loss	• Verbruggen et al. (2012)
• Integration of green manures and cover crops into crop rotations	• Provides additional OM inputs as energy/nutrient source for soil food web • Supports increased biomass • May change species richness and evenness depending on 'quality' – C/N ratio and so forth • Depending on crop species provides hosts for mutualistic soil organisms	• Increased variety in root biomass, rooting patterns, amount and quality of root exudates, and amount and quality of residue inputs • Increased duration of ground cover	• Increased duration of soil cover – reducing sediment loss • Increases soil C content – C sequestration • Improves soil structure and its stability • Diversifies farmed landscapes	• Schipanski et al. (2014) • Benitez et al. (2016) • McDaniel et al. (2014)

(Continued)

Table 3 *(Continued)*

Land management practice	Direct impacts on soil biota	Indirect impacts on soil biota	Likely impacts on soil function for agriculture and other ecosystem services	More information relevant to the use of this practice in the field on soil biota and/or soil functions in organic farming
• Introduction of annual intercrops with legume/ non-legumes	• Supports increased biomass • May change species richness and evenness depending on 'quality' – C/N ratio and so forth • Depending on crop species provides hosts for mutualistic soil organisms	• Increased variety in root biomass, rooting patterns, amount and quality of root exudates, and amount and quality of residue inputs	• Improves nutrient use efficiency, may reduce nutrient leaching risk • Can increase losses of N in following year compared with non-legume monocrop • Diversifies farmed landscapes	• Tang et al. (2014)
• Introduction of diverse seed mixes for example deep rooting species and herbs	• Diversifies amount and quality of root exudation, modifying energy/nutrient sources for the soil food web • May change species richness and evenness depending on 'quality' – C/N ratio and so forth • Depending on species mix provides hosts for mutualistic soil organisms	• Increased variety in root biomass, rooting patterns and amount and quality of root exudates	• Improves nutrient use efficiency, may reduce nutrient leaching risk • May improve soil structure and its stability	• Malézieux et al. (2009) • Van Eekeren et al. (2009)

Specific targeted interventions				
• Green manure crops for example mustard incorporated to provide soil fumigation effects • Additional to 'Use of green manures' above	• Root exudates and decomposition products with allelopathic effects on soil biota	• Root exudates and decomposition products with allelopathic effects on seed germination	• May allow reduced use of pesticides and/or sterilisation	• van Bruggen et al. (2016)
• Inoculation of legumes through seed treatments	• No evidence of direct effects on indigenous soil biota	• None expected • If successful in promoting legume growth may increase variety in root biomass, rooting patterns and amount and quality of root exudates	• Range of impacts result from successful cultivation of leguminous crops supported by seed inoculation	• Van Kessel and Hartley (2000)
• Use of mycorrhizal fungal inoculants to support tree/crop establishment	• No evidence of direct effects on indigenous soil biota in the field • May provide added AM diversity through provision of mycorrhizal spores for association with subsequent crops	• None expected • If successful in promoting plant growth may increase variety in root biomass, rooting patterns, and amount and quality of root exudates	• Range of impacts result from successful establishment of inoculated crop	• Gosling et al. (2006) • Verbruggen et al. (2013)
• Application of plant-growth-promoting microorganisms	• No evidence of direct effects on indigenous soil biota • Carrier substrates may have impacts • May provide added bacterial/fungal diversity but numbers added are very small compared to indigenous population	• None expected • If successful in promoting plant growth may increase variety in root biomass, rooting patterns, and amount and quality of root exudates	• Few impacts expected	• Dias et al. (2015)

practices such as liming which are primarily employed to improve soil fertility from a chemical perspective will also affect soil biology; see for example, Sradnick et al. (2013).

Systems approaches such as the use of conservation agriculture in organic farming employ more than one of the practices discussed in Table 3. A useful review of the impacts of conservation agriculture in relation to soil fertility can be found in Peigné et al. (2007) and Henneron et al. (2015).

4 Conclusion and future trends

Organic farming has taken on board the ideas of precision farming in relation to the use of robots for weeding, but less is written about opportunities for precision management of soil fertility. The application of lime or allowable mineral fertilisers based on precision approaches is within the organic standards. The future holds possibilities for organic materials to be formulated in such a way that allows precision applications based on stratified field measurements although this requires technological innovation. Mixtures of crop residues and/or organic manures and/or co-products from industry provide the possibility for developing organic fertilisers which can supply known ratios of nutrients, thus potentially improving plant uptake efficiency and reducing losses. There is a continued interest in developing reduced tillage systems for organic farming (Stockdale and Watson, 2009), and current thinking suggests that precision approaches could be part of this. For example, the development of seeding techniques and weeding techniques which allow more precise placement of crops and intercrops provides new opportunities for designing organic farming systems which maintain soil fertility, control weeds and reduce the need for soil cultivation (e.g. Campiglia et al. 2014).

Climate change provides a huge challenge to on-farm management. There is a need to develop adaptive systems for working with natural resources under current climatic conditions in order to increase the ability to adapt to climate change and other factors, such as changes in market demand, in the future. Consequently developing farmers' skills to understand and manage their farming systems in relation to climatic variation now will enhance resilience to climate change through knowledge-based rather than technologically reliant innovation. One of the biggest advantages organic farmers may have is not only their built resilience into their soils through increasing organic matter levels, but also their adaptive way of working, which encourages considered strategic responses rather than quick-fix solutions. Preventing land degradation, supporting biodiversity and the supply of a range of ecological goods and services may be as important as maintaining yield in the light of climate change; a focus on short-term protection of productivity may lead to medium- to long-term yield loss and misleading conclusions in assessments of the sustainability of farming systems (Gomiero et al., 2008).

There is a need for a holistic view of food systems, beyond productivity, to include the integrated evaluation of environmental, social and health impacts. Only when food systems are viewed in this way might we be able to move forward towards meeting the food-related sustainable development goals and the importance of soils in meeting these goals (Keesstra et al., 2016). Much more remains to be determined about the links between soil fertility, health and nutrition for both humans and animals. There is currently a lack of methodologies available which work across the whole food system and link production systems to product quality and onwards into livestock and human health and well-being

(Watson et al., 2008). Science continues to advance, and it is important that we continue to evaluate the importance of soil management strategies together with environmental factors in controlling the composition of our diet.

Policy support for organic farming has in many EU countries had its origin in the benefits that organic farming brings to biodiversity, largely in relation to aboveground biodiversity. There remains a need to value the ecosystem services that soil provides as a part of future policy; this requires us to better understand links between soil biological processes, management and the factors that determine the economic value of resources, in other words, an interdisciplinary approach which crosses the boundaries of natural and socio-economic sciences (Pascual et al., 2015).

A major challenge is the gap between rapid advances in biological research on soil–plant systems and the practical implementation of the resulting knowledge. Progress in soil microbiology now enables biological indicators akin to the commonly used physical and chemical measures (Lemanceau et al., 2014). Many persistent pests and diseases can be detected through the use of specific primers and quantitative polymerase chain reaction and the consideration of molecular measures for inclusion in an on-farm assessment of soil health and integration in metrics of soil health is beginning.

5 Acknowledgements

The authors acknowledge the financial support for the FertilCrop project provided by transnational funding bodies, being partners of the FP7 ERA-net project, CORE Organic Plus and the co-fund from the European Commission. The text in this report is the sole responsibility of the authors and does not necessarily reflect the views of the national funding bodies having financed this project. The authors also acknowledge support from the Scottish Government RESAS Strategic Research Programme and Agriculture and Horticulture Development Board (AHDB).

6 Where to look for further information

There is a wide gradient of intensity of organic farming and an overlap of management practices which makes comparing ecosystem services on real farms very difficult (Chongtham et al., 2017; Williams and Hedlund, 2013). As a result of this, some studies find positive impacts of organic farming on soil-related ecosystem services (e.g. Mäder et al., 2002) and some do not (Gosling and Shepherd, 2005). However, comparisons between organic and conventional systems are methodologically difficult, and our aim here is not to look at the value and design of comparative studies as many other have debated this (e.g. Kirchmann et al., 2016). However, in Table 4, we list a selection of long-term experiments in Europe some of which compare systems within organic production (e.g. Taylor et al., 2006) and some of which compare organic with low-input or conventional systems. As long as the constraints of these trials are taken into account, they have much to offer in understanding soil fertility and health and links to other ecosystem services. Constraints differ between trials, but for example, some compare different organic and mineral inputs to the same rotation (e.g. Mäder et al., 2002), while others study the whole cropping/farming system (e.g. Eltun, 1994). Although many of the

Table 4 Examples of well documented, long-term experiments on organic farming in Europe

Name and (responsible organisation)	Site	Start date	Systems compared (O,B,L,C[a])	Rotations	References
DOC (FiBL)	Frick, Switzerland	1978	O,B,C	Ley/arable + manure	Mäder et al. (2002)
Long-term fertilisation trial (Institute for Biodynamic Research)	Darmstadt, Germany	1980	O,B,C	Arable + manure	Raupp (2001)
Apelsvoll cropping systems experiment (Bioforsk)	Norway	1990	O,L,C	Ley/arable and arable	Eltun (1994)
Tulloch Rotations Trial (SRUC)	Aberdeen, UK	1991	O	Ley/arable, grazed by sheep compared with arable rotation	Watson et al. (2015) and Taylor et al. (2006)
Montepaldi Long-Term Experiment (MOLTE) (University of Florence)	Florence, Italy	1991	O,C	Arable + organic fertilisers	Migliorini et al. (2014)
Wiesengut (University of Bonn)	Bonn, Germany	1993	O,B	Arable + manure	Zaller and Köpke (2004)
Grabow Trial (IUNG-PIB)	Osiny, Poland	1994	O,L,C	Arable	Stalenga (2007)
La Cage (INRA)	Versailles, France	1997	O,C,CA	Arable	Henneron et al. (2015)
CROPSYS (Aarhus University)	Denmark, initially 3 sites, but now reduced to 1	1997	O,C (previously organic only)	Arable + manure	Knudsen et al. (2014) and Olesen et al. (2000)
Gladbacherhof (University of Giessen)	Vilmar, Germany	1998	O	Ley/arable ± manure, arable	Schulz et al. (2014) and Schmidt et al. (2006)
University of Tuscia	Viterbo, Italy	2001	O,C	Arable/horticulture + guano	Mancinelli et al. (2010)
MASCOT (Scuola Superiore Sant'Anna)	Pisa, Italy	2001	O,C	Arable	Mazzoncini et al. (2010)
Nafferton Factorial System Comparison trial (Newcastle University)	Stocksfield, UK	2001	O,C	Ley/arable, arable with compost	Eyre et al. (2011)
Frick Tillage Trial (FIBL)	Frick, Switzerland	2002	O	Ley/arable with tillage and manure treatments	Krauss et al. (2017)

[a] O = Organic, B = Biodynamic, L = Low input or Integrated, C = Conventional, CA = conservation agriculture.

trials simulate mixed farming systems, very few actually include grazing animals. There are also some very useful larger-scale studies comparing systems such as Logården in Sweden (SLU, Stenberg et al., 2012), at Taastrup in Denmark (University of Copenhagen, Fan et al., 2016) and Scheyern in Germany (Technical University of Munich, Lin et al., 2016). New facilities continue to be developed, for example, the Swedish University of Agricultural Sciences is establishing organic farming systems trials at Lönnstorp in Southern Sweden (https://www.slu.se/en/departments/biosystems-technology/research-facilites/lonnstorp/). The organisations listed here and in Table 4 are just a few of the universities and research institutes providing both research and research translation relevant to soil fertility and health in organic farming.

7 References

Allison, F. E. (1973), *Soil Organic Matter and its Role in Crop Production*, Elsevier, New York, NY.

Asai, M. and Langer, V. (2014), 'Collaborative partnerships between organic farmers in livestock-intensive areas of Denmark', *Org. Agric.*, 4(1), 63–77.

Balesdent, J., Chenu, C. and Balabane, M. (2000), 'Relationship of soil organic matter dynamics to physical protection and tillage', *Soil Till. Res.*, 53(3), 215–30.

Balfour, E. B. (1943), *The Living Soil*, Faber & Faber, London.

Ball, B. C., Bingham, I., Rees, R. M., Watson C. A. and Litterick, A. (2005), 'The role of crop rotations in determining soil structure and rooting conditions', *Can. J. Soil Sci.*, 85(5), 557–77.

Ball, B. C., Watson, C. A. and Crichton, I. (2007), 'Nitrous oxide emissions, cereal growth, N recovery and soil nitrogen status after ploughing organically managed grass/clover swards', *Soil Use Manage.*, 23(2), 145–55.

Benitez, M. S., Taheri, W. I. and Lehman, R. M. (2016), 'Selection of fungi by candidate cover crops', *Appl. Soil Ecol.*, 103, 72–82.

Berry, P. M., Stockdale, E. A., Sylvester-Bradley, R., Philipps, L., Smith, K. A., Lord, E. I., Watson, C. A. and Fortune, S. (2003), 'N, P and K budgets for crop rotations on nine organic farms in the UK', *Soil Use Manage.*, 19(2), 112–18.

Bolinder, M. A., Angers, D. A. and Dubuc, J. P. (1997), 'Estimating shoot to root ratios and annual carbon inputs in soils for cereal crops', *Agric. Ecosyst. Environ.*, 63(1), 61–6.

Bolinder, M. A., Angers, D. A., Bélanger, G., Michaud, R. and Laverdière, M. R. (2002), 'Root biomass and shoot to root ratios of perennial forage crops in eastern Canada', *Can. J. Plant Sci.*, 82(4), 731–7.

Bolinder, M. A., Janzen, H. H., Gregorich, E. G., Angers, D. A. and Vandenbygaart, A. J. (2007), 'An approach for estimating net primary productivity and annual carbon inputs to soil for common agricultural crops in Canada', *Agric. Ecosyst. Environ.*, 118(1), 29–42.

Brock, C., Fließbach, A., Oberholzer, H. R., Schulz, F., Wiesinger, K., Reinicke, F., Koch, W., Pallutt, B., Dittman, B., Zimmer, J. and Hülsbergen, K. J. (2011), 'Relation between soil organic matter and yield levels of nonlegume crops in organic and conventional farming systems', *J. Plant Nutr. Soil Sci.*, 174(4), 568–75.

Campiglia, E., Mancinelli, R., Radicetti, E. and Baresel, J. P. (2014), 'Evaluating spatial arrangement for durum wheat (*Triticum durum* Desf.) and subclover (*Trifolium subterraneum* L.) intercropping systems', *Field Crop Res.*, 169, 49–57.

Carr, P. M., Gramig, G. G. and Liebig, M. A. (2013), 'Impacts of organic zero tillage systems on crops, weeds, and soil quality', *Sustainability*, 5(7), 3172–201.

Chongtham, I. R., Bergkvist, G., Watson, C., Sandström, E., Bengtsson, J. and Öborn, I. (2017), 'Factors influencing crop rotation strategies on organic farms with different time periods since conversion to organic production', *Biol. Agric. Hortic.*, 33(1), 14–27.

Cooke, G. W. (1967), *The Control of Soil Fertility*, Crosby Lockwood and Son, London.

Cooper, J., Baranski, M., Stewart, G., Nobel-de Lange, M., Bàrberi, P., Fließbach, A., Peigné, J., Berner, A., Brock, C., Casagrande, M. and Crowley, O. (2016), 'Shallow non-inversion tillage in organic farming maintains crop yields and increases soil C stocks: A meta-analysis', *Agron. Sustain. Dev.*, 36(1), 1–20.

David, C., Abecassis, J., Carcea, M., Celette, F., Friedel, J. K., Hellou, G., Hiltbrunner, J., Messmer, M., Narducci, V., Peigné, J., Samson, M. F., Schweinzer, A., Thomsen, I. K. and Thommen, A. (2012), 'Organic bread wheat production and market in Europe', *in* E. Lichtfouse (Ed.), *Sustainable Agriculture Reviews*, Springer, the Netherlands, pp. 43–62.

Defra. (2010). Sub Project A: The role of soil biota in soil fertility and quality and approaches to influencing soil communities to enhance delivery of these functions. Report to Defra SP1601, SKM Enviros, http://randd.defra.gov.uk/Default.aspx?Menu=Menu&Module=More&Location=None&Completed=0&ProjectID=16925 (accessed 30 April 2017).

Dias, T., Dukes, A. and Antunes, P. M. (2015), 'Accounting for soil biotic effects on soil health and crop productivity in the design of crop rotations', *J. Sci. Food Agric.*, 95(3), 447–54.

Dietz, M., Machill, S., Hoffmann, H. C. and Schmidtke, K. (2013), 'Inhibitory effects of *Plantago lanceolata* L. on soil N mineralization', *Plant Soil*, 368(1–2), 445–58.

EC. (2007), http://eur-lex.europa.eu/legal-content/EN/ALL/?uri=CELEX:32007R0834 (accessed 27 April 2017).

EC. (2008), http://eur-lex.europa.eu/legal-content/EN/TXT/PDF/?uri=CELEX:02008R0889-20160507&qid=1470303317688&from=EN (accessed 27 April 2017).

Ehlers, W., Köpke, U., Hesse, F. and Böhm, W. (1983), 'Penetration resistance and root growth of oats in tilled and untilled loess soil', *Soil Till. Res.*, 3, 261–75.

Eichler-Löbermann, B., Gaj, R. and Schnug, E. (2009), 'Improvement of soil phosphorus availability by green fertilization with catch crops', *Commun. Soil Sci. Plant Anal.*, 40(1–6), 70–81.

Eltun, R. (1994), 'The Apelsvoll cropping system experiment. I. Background, objectives and methods', *Norweg. J. Agr. Sci.*', 8, 301–15.

Eyre, M. D., Critchley, C. N. R., Leifert, C. and Wilcockson, S. J. (2011), 'Crop sequence, crop protection and fertility management effects on weed cover in an organic/conventional farm management trial', *Eur. J. Agron.*, 34(3), 153–62.

Fan, F., Henriksen, C. B. and Porter, J. (2016), 'Valuation of ecosystem services in organic cereal crop production systems with different management practices in relation to organic matter input', *Ecosyst. Serv.*, 22(Part A), 117–27.

FAO. (2011), *Soil Health: Technologies that Save and Grow*, FAO, Rome.

Fliessbach, A., Oberholzer, H. R., Gunst, L. and Mäder, P. (2007), 'Soil organic matter and biological soil quality indicators after 21 years of organic and conventional farming', *Agric. Ecosyst. Environ.*, 118(1), 273–84.

Foissy, D., Vian, J. F. and David, C. (2013), 'Managing nutrient in organic farming system: Reliance on livestock production for nutrient management of arable farmland', *Org. Agric.*, 3(3–4), 183–99.

Gamliel, A. and van Bruggen, A. H. C. (2016), 'Maintaining soil health for crop production in organic greenhouses', *Sci. Hortic.*, 208, 120–30.

Gomiero, T., Paoletti, M. G. and Pimentel, D. (2008), 'Energy and environmental issues in organic and conventional agriculture', *Crit. Rev. Plant Sci.*, 27(4), 239–54.

Gosling, P. and Shepherd, M. (2005), 'Long-term changes in soil fertility in organic arable farming systems in England, with particular reference to phosphorus and potassium', *Agric. Ecosyst. Environ.*, 105(1), 425–32.

Gosling, P., Hodge, A., Goodlass, G. and Bending, G. D. (2006), 'Arbuscular mycorrhizal fungi and organic farming', *Agr. Ecosyst. Environ.*, 113(1), 17–35.

Han, E., Kautz, T., Perkons, U., Lüsebrink, M., Pude, R. and Köpke, U. (2015a), 'Quantification of soil biopore density after perennial fodder cropping', *Plant Soil*, 394, 73–85. doi:10.1007/s11104-015-2488-3

Han, E., Kautz, T., Perkons, U., Uteau, D., Peth, S., Huang, N., Horn, R. and Köpke, U. (2015b), 'Root growth dynamics inside and outside of soil biopores as affected by crop sequence determined with the profile wall method', *Biol. Fert. Soils*, 51, 847–56. doi:10.1007/s00374-015-1032-1

Hatch, D., Joynes, A., Roderick, S., Shepherd, M. and Goodlass, G. (2014), 'Effects of cutting, mulching and applications of farmyard manure on the supply of nitrogen from a red clover/grass sward', *Org. Agric.*, 4(1), 15–24.

Heinze, S., Oltmanns, M., Joergensen, R. G. and Raupp, J. (2011), 'Changes in microbial biomass indices after 10 years of farmyard manure and vegetal fertilizer application to a sandy soil under organic management', *Plant Soil*, 343(1–2), 221–34.

Henneron, L., Bernard, L., Hedde, M., Pelosi, C., Villenave, C., Chenu, C., Bertrand, M., Girardin, C. and Blanchart, E. (2015), 'Fourteen years of evidence for positive effects of conservation agriculture and organic farming on soil life', *Agron. Sustain. Dev.*, 35, 169–81.

Idowu, O. J., van Es, H. M., Abawi, G. S., Wolfe, D. W., Ball, J. I., Gugino, B. K., Moebius, B. N., Schindelbeck, R. R. and Biligli, A. V. (2008), 'Farmer-oriented assessment of soil quality using field, laboratory, and VNIR spectroscopy methods', *Plant Soil*, 307, 243–53.

IFOAM. (2005), http://www.ifoam.bio/en/principles-organic-agriculture/principle-ecology (accessed 30 April 2017).

Jarvis, S. C., Stockdale, E. A., Shepherd, M. A. and Powlson, D. S. (1996), 'Nitrogen mineralization in temperate agricultural soils: Processes and measurement', *Adv. Agron.*, 57, 187–235.

Jokela, W. E., Tilley, J. P. and Ross, D. S. (2010), 'Manure nutrient content on Vermont dairy farms: Long-term trends and relationships', *Commun. Soil Sci. Plant Anal.*, 41(5), 623–37.

Kautz, T. (2014), 'Research on subsoil biopores and their functions in organically managed soils: A review', *Renew. Agric. Food Syst.*, 30, 318–27.

Kautz, T., Amelung, W., Ewert, F., Gaiser, T., Horn, R., Jahn, R., Javaux, M., Kemna, A., Kuzyakov, Y., Munch, J.-C., Paetzold, S., Peth, S., Scherer, H. W., Schloter, M., Schneider, H., Vanderborght, J., Vetterlein, D., Walter, A., Wiesenberg, G. L. B. and Köpke, U. (2013), 'Nutrient acquisition from arable subsoils in temperate climates: A review', *Soil Biol. Biochem.*, 57, 1003–22.

Keesstra, S. D., Quinton, J. N., van der Putten, W. H., Bardgett, R. D. and Fresco, L. O. (2016), 'The significance of soils and soil science towards realization of the United Nations Sustainable Development Goals', *Soil*, 2(2), 111.

Kelly, B., Allan, C. and Wilson, B. P. (2009), 'Soil indicators and their use by farmers in the Billabong Catchment, southern New South Wales', *Aus. J. Soil Res.*, 47, 234–42.

Kibblewhite, M. G., Ritz, K. and Swift, M. J. (2008), 'Soil health in agricultural systems', *Philos. Trans. R. Soc. Lond., B, Biol. Sci.*, 363(1492), 685–701.

Kirchmann, H., Kätterer, T., Bergström, L., Börjesson, G. and Bolinder, M. A. (2016), 'Flaws and criteria for design and evaluation of comparative organic and conventional cropping systems', *Field Crop. Res.*, 186, 99–106.

Knight, J. D. (2012), 'Frequency of field pea in rotations impacts biological nitrogen fixation', *Can. J. Plant Sci.*, 92(6), 1005–11.

Knudsen, M. T., Meyer-Aurich, A., Olesen, J. E., Chirinda, N. and Hermansen, J. E. (2014), 'Carbon footprints of crops from organic and conventional arable crop rotations using a life cycle assessment approach', *J. Cleaner Prod.*, 64, 609–18.

Krauss, M., Ruser, R., Müller, T., Hansen, S., Mäder, P. and Gattinger, A. (2017), 'Impact of reduced tillage on greenhouse gas emissions and soil carbon stocks in an organic grass-clover ley-winter wheat cropping sequence', *Agric. Ecosyst. Environ.*, 239, 324–33.

Kuntz, M., Berner, A., Gattinger, A., Scholberg, J. M., Mäder, P. and Pfiffner, L. (2013), 'Influence of reduced tillage on earthworm and microbial communities under organic arable farming', *Pedobiologia*, 56(4), 251–60.

Lampkin, N. (1990), *Organic Farming*, Farming Press, Ipswich, UK.

Larson, W. E. and Pierce, F. J. (1994), 'The dynamics of soil quality as a measure of sustainable management', *in* J. W. Doran, D. C. Coleman, D. F. Bezdicek and B. A. Stewart (Eds), *Defining Soil Quality for a Sustainable Environment*, SSSA Spec. Publ. No. 35, Soil Sci. Soc. Am., Inc. and Am. Soc. Agron., Inc., Madison, WI, pp. 37–51.

Ledgard, S. F. and Steele, K. W. (1992), 'Biological nitrogen fixation in mixed legume/grass pastures', *Plant Soil*, 141(1–2), 137–53.

Lemanceau, P., Maron, P. A., Mazurier, S., Mougel, C., Pivato, B., Plassart, P., Ranjard, L., Revellin, C., Tardy, V. and Wipf, D. (2014), 'Understanding and managing soil biodiversity: A major challenge in agroecology', *Agron. Sustain. Dev.*, 35(1), 67–81.

Lewis, D. B., Kaye, J. P., Jabbour, R. and Barbercheck, M. E. (2011), 'Labile carbon and other soil quality indicators in two tillage systems during transition to organic agriculture', *Renew. Agr. Food Syst.*, 26(04), 342–53.

Lin, H. C., Huber, J. A., Gerl, G. and Hülsbergen, K. J. (2016), 'Nitrogen balances and nitrogen-use efficiency of different organic and conventional farming systems', *Nutr. Cycl. Agroecosys.*, 105(1), 1–23.

Løes, A. K., Bünemann, E. K., Cooper, J., Hörtenhuber, S., Magid, J., Oberson, A. and Möller, K. (2017), 'Nutrient supply to organic agriculture as governed by EU regulations and standards in six European countries', *Org. Agric.*, 7(4), 395–418. doi:10.1007/s13165-016-0165-3

Loveland, P. and Webb, J. (2003), 'Is there a critical level of organic matter in the agricultural soils of temperate regions: A review', *Soil Till. Res.*, 70(1), 1–18.

Mäder, P. and Berner, A. (2012), 'Development of reduced tillage systems in organic farming in Europe', *Renew. Agr. Food Syst.*, 27(1), 7–11.

Mäder, P., Fliessbach, A., Dubois, D., Gunst, L., Fried, P. and Niggli, U. (2002), 'Soil fertility and biodiversity in organic farming', *Science*, 296, 1694–7.

Magrini, M. B., Anton, M., Cholez, C., Duc, G., Hellou, G., Jeuffroy, M. H., Meynard, J. M., Pelzer, E., Voisin, A. S. and Walrand, S. (2016), 'Why are grain-legumes rarely present in cropping systems despite their environmental and nutritional benefits? Analyzing lock-in in the French agrifood system', *Ecol. Econ.*, 126, 152–62.

Malézieux, E., Crozat, Y., Dupraz, C., Laurans, M., Makowski, D., Ozier-Lafontaine, H., Rapidel, B., De Tourdonnet, S. and Valantin-Morison, M. (2009), 'Mixing plant species in cropping systems: concepts, tools and models: A review', *Agron. Sustain. Dev.*, 29, 43–62.

Mancinelli, R., Campiglia, E., Tizio, A. D. and Marinari, S. (2010), 'Soil carbon dioxide emission and carbon content as affected by conventional and organic cropping systems in Mediterranean environment', *Appl. Soil Ecol.*, 46, 64–72.

Mazzoncini, M, Canali, S., Giovannetti, M., Castagnoli, M., Tittarelli, F., Antichi, D., Nannelli, R., Cristani, C. and Bàrberi, P. (2010), 'Comparison of organic and conventional stockless arable systems: A multidisciplinary approach to soil quality evaluation', *Appl. Soil Ecol.*, 44, 124–32.

McDaniel, M. D., Tiemann, L. K. and Grandy, A. S. (2014), 'Does agricultural crop diversity enhance soil microbial biomass and organic matter dynamics? A meta-analysis', *Ecol. Appl.*, 24(3), 560–70.

MEA. (2005), *Millennium Ecosystem Assessment Ecosystems and Human Well-being*, Island Press, Washington DC.

Migliorini, P., Moschini, V., Tittarelli, F., Ciaccia, C., Benedettelli, S., Vazzana, C. and Canali, S. (2014), 'Agronomic performance, carbon storage and nitrogen utilisation of long-term organic and conventional stockless arable systems in Mediterranean area', *Eur. J. Agron.*, 52, 138–45.

Moraine, M., Melac, P., Ryschawy, J., Duru, M. and Therond, O. (2017), 'A participatory method for the design and integrated assessment of crop-livestock systems in farmers' groups', *Ecol. Indic.*, 72, 340–51.

Nesme, T., Toublant, M., Mollier, A., Morel, C. and Pellerin, S. (2012), 'Assessing phosphorus management among organic farming systems: A farm input, output and budget analysis in southwestern France', *Nutr. Cycl. Agroecosys.*, 92(2), 225–36.

Öborn, I., Modin-Edman, A.-K., Bengtsson, H., Gustafson, G. M., Salomon, E., Nilsson, S. I., Holmqvist, J., Jonsson, S. and Sverdrup, H. (2005), 'A systems approach to assess farm scale nutrient and trace element dynamics. A case study at the Öjebyn dairy farm', *Ambio*, 34(4), 298–307.

Oelofse, M., Jensen, L. S. and Magid, J. (2013), 'The implications of phasing out conventional nutrient supply in organic agriculture: Denmark as a case', *Org. Agric.*, 3(1), 41–55.

Olesen, J. E., Askegaard, M. and Rasmussen, I. A. (2000), 'Design of an organic farming crop rotation experiment', *Acta Agric. Scand., Sec. B, Soil Plant Sci.*, 50(1), 13–21.

Pascual, U., Termansen, M., Hedlund, K., Brussaard, L., Faber, J. H., Foudi, S., Lemanceau, P. and Jørgensen, S. L. (2015), 'On the value of soil biodiversity and ecosystem services', *Ecosyst. Serv.*, 15, 11–18.

Paulsen, H. M., Köpke, U., Oberson, A. and Rahmann, G. (2016), 'Phosphorus—the predicament of organic farming', *in* E. Schnug and L. J. De Kok (Eds), *Phosphorus in Agriculture: 100% Zero*, Springer, the Netherlands, pp. 195–213.

Peigné, J., Ball, B. C., Roger-Estrade, J. and David, C. (2007), 'Is conservation tillage suitable for organic farming? A review', *Soil Use Manage.*, 23(2), 129–44.

Pelosi, C., Bertrand, M., Thénard, J. and Mougin, C. (2015), 'Earthworms in a 15 years agricultural trial', *Appl. Soil Ecol.*, 88, 1–8.

Peoples, M. B., Brockwell, J., Herridge, D. F., Rochester, I. J., Alves, B. J. R., Urquiaga, S., Boddey, R. M., Dakora, F. D., Bhattarai, S., Maskey, S. L., Sampet, C., Rerkasem, B., Khan, D. F., Hauggaard-Nielsen, H. and Jensen, E. S. (2009), 'The contributions of nitrogen-fixing crop legumes to the productivity of agricultural systems', *Symbiosis*, 48(1–3), 1–17.

Petersen, S. O., Schjonning, P., Olesen, J. E., Christensen, S. and Christensen, B. T. (2013), 'Sources of nitrogen for winter wheat in organic cropping systems', *Soil Sci. Soc. Am. J.*, 77(1), 155–65.

Preissel, S., Reckling, M., Schläfke, N. and Zander, P. (2015), 'Magnitude and farm-economic value of grain legume precrop benefits in Europe: A review', *Field Crop Res.*, 175, 64–79.

Ratnadass, A., Fernandes, P., Avelino, J. and Habib, R. (2012), 'Plant species diversity for sustainable management of crop pests and diseases in agroecosystems: A review', *Agron. Sustain. Dev.*, 32(1), 273–303.

Raupp, J. (2001), 'Manure fertilization for soil organic matter maintenance and its effects upon crops and the environment, evaluated in a long-term trial', *in* R. M. Rees, B. C. Ball, C. D. Campbell and C. A. Watson (Eds), *Sustainable Management of Soil Organic Matter*, CAB International, Wallingford, UK, pp. 301–8.

Reckling, M., Hecker, J.-M., Bergkvist, G., Watson, C., Zander, P., Stoddard, F., Eory, V., Topp, K., Maire, J. and Bachinger, J. (2016), 'A cropping system assessment framework – evaluating effects of introducing legumes into crop rotations', *Eur. J. Agron.*, 76, 186–97.

Säle, V., Aguilera, P., Laczko, E., Mäder, P., Berner, A., Zihlmann, U., van der Heijden, M. G. A. and Oehl, F. (2015), 'Impact of conservation tillage and organic farming on the diversity of arbuscular mycorrhizal fungi', *Soil Biol. Biochem.*, 84, 38–52.

Schipanski, M. E., Barbercheck, M., Douglas, M. R., Finney, D. M., Haider, K., Kaye, J. P., Kemanian,A. R., Mortensen, D. A., Ryan, M. R., Tooker, J. and White, C. (2014), 'A framework for evaluating ecosystem services provided by cover crops in agroecosystems', *Agric. Syst.*, 125, 12–22.

Schmidt, H., Schulz, F., and Leithold, G. (2006), 'Organic farming trial Gladbacherhof: Effects of different crop rotations and tillage systems', *in* J. Raupp, C. Pekrun, M. Oltmanns and U. Koepke (Eds), *Long-term Field Experiments in Organic Farming*. International Society of Organic Agriculture Research (ISOFAR), Berlin, pp. 165–82.

Schulz, F., Brock, C., Schmidt, H., Franz, K. P. and Leithold, G. (2014), 'Development of soil organic matter stocks under different farm types and tillage systems in the Organic Arable Farming Experiment Gladbacherhof', *Arch. Agron. Soil Sci.*, 60(3), 313–26.

Sorensen, J. N. and Thorup-Kristensen, K. (2011), 'Plant-based fertilizers for organic vegetable production', *J. Plant Nutr. Soil Sci.*, 174(2), 321–32.

Sradnick, A., Murugan, R., Oltmanns, M., Raupp, J. and Joergensen, R. G. (2013), 'Changes in functional diversity of the soil microbial community in a heterogeneous sandy soil after long-term fertilization with cattle manure and mineral fertilizer', *Appl. Soil Ecol.* 63, 23–8.

Stalenga, J. (2007), 'Applicability of different indices to evaluate nutrient status of winter wheat in the organic system', *J. Plant Nutr.*, 30(3), 351–65.

Stenberg, M., Ulén, B., Söderström, M., Roland, B., Delin, K. and Helander, C. A. (2012), 'Tile drain losses of nitrogen and phosphorus from fields under integrated and organic crop rotations. A four- year study on a clay soil in southwest Sweden', *Sci. Total Environ.*, 434, 79–89.

Stockdale, E. A. and Watson, C. A. (2009), 'Biological indicators of soil quality in organic farming systems', *Renew. Agr. Food Syst.*, 24(4), 308–18.

Stockdale, E. A. and Watson, C. A. (2012), 'Managing soil biota to deliver ecosystem services', Natural England Commissioned Reports, Number 100. http://publications.naturalengland.org.uk/file/3582455 (accessed 28 April 2017).

Stockdale, E. A., Griffiths, B. S., Hargreaves, P. R., Bhogal, A., Crotty, F. V. and Watson, C. A. (2017), 'Impacts of management of soil health – site-specific delivery of sustainable agro-ecosystems', *Aspects Appl. Biol.*, 137, 87–92.

Sun, H., Koal, P., Liu, D., Gerl, G., Schroll, R., Gattinger, A., Joergensen, R. G. and Munch, J. C. (2016), 'Soil microbial community and microbial residues respond positively to minimum tillage under organic farming in Southern Germany', *Appl. Soil Ecol.*, 108, 16–24.

Talgre, L., Lauringson E., Roostalu, H. and Makke, A. (2014), 'Phosphorus and potassium release during decomposition of roots and shoots of green manure crops', *Biol. Agric. Hortic.*, 30(4), 264–71.

Tang, X., Bernard, L., Brauman, A., Daufresne, T., Deleporte, P., Desclaux, D., Souche, G., Placella, S. A. and Hinsinger, P. (2014), 'Increase in microbial biomass and phosphorus availability in the rhizosphere of intercropped cereal and legumes under field conditions', *Soil Biol. Biochem.*, 75, 86–93.

Taylor, B. R., Younie, D., Matheson, S., Coutts, M., Mayer, C., Watson, C. A. and Walker, R. L. (2006), 'Output and sustainability of organic ley/arable crop rotations at two sites in Northern Scotland', *J. Agric Sci.* 144(5), 435–47.

Thorup-Kristensen, K. and Dresbøll, D. B. (2010), 'Incorporation time of nitrogen catch crops influences the N effect for the succeeding crop', *Soil Use Manage.*, 26(1), 27–35.

Ulyett, J., Sakrabani, R., Kibblewhite, M. and Hann, M. (2014), 'Impact of biochar addition on water retention, nitrification and carbon dioxide evolution from two sandy loam soils', *Eur. J. Soil Sci.*, 65(1), 96–104.

van Bruggen, A. H., Gamliel, A. and Finckh, M. R. (2016), 'Plant disease management in organic farming systems', *Pest Manag. Sci.*, 72(1), 30–44.

Van Eekeren, N., van Liere, D., de Vries, F., Rutgers, M., de Goede, R. and Brussaard, L. (2009), 'A mixture of grass and clover combines the positive effects of both plant species on selected soil biota', *Appl. Soil Ecol.*, 42(3), 254–63.

Van Kessel, C. and Hartley, C. (2000), 'Agricultural management of grain legumes: Has it led to an increase in nitrogen fixation?' *Field Crop Res.*, 65(2), 165–81.

Verbruggen, E., Van Der Heijden, M. G., Weedon, J. T., Kowalchuk, G. A. and Roeling, W. F. (2012), 'Community assembly, species richness and nestedness of arbuscular mycorrhizal fungi in agricultural soils', *Mol. Ecol.*, 21(10), 2341–53.

Verbruggen, E., Heijden, M. G., Rillig, M. C. and Kiers, E. T. (2013), 'Mycorrhizal fungal establishment in agricultural soils: Factors determining inoculation success', *New Phytol.*, 197(4), 1104–9.

Watson, C. A., Atkinson, D., Gosling, P., Jackson, L. R. and Rayns, F. W. (2002a), 'Managing soil fertility in organic farming systems', *Soil Use Manage.*, 18(S1), 239–47.

Watson, C. A., Bengtsson, H., Løes, A-K., Myrbeck, A., Salomon, E., Schroder, J. and Stockdale, E. A. (2002b), 'A review of farm-scale nutrient budgets for organic farms as a tool for management of soil fertility', *Soil Use Manage.*, 18(S1), 264–73.

Watson, C. A., Walker, R. L. and Stockdale, E. A. (2008), 'Research in organic production systems – past, present and future', *J. Agric. Sci.*, 146(1), 1–19.

Watson, C. A., Topp, C. F. E. and Walker, R. L. (2015), 'Long-term productivity of legume based crop rotations', *Aspects Appl. Biol.*, 128, 189–94.

Williams, A. and Hedlund, K. (2013), 'Indicators of soil ecosystem services in conventional and organic arable fields along a gradient of landscape heterogeneity in southern Sweden', *Appl. Soil Ecol.*, 65, 1–7.

Williams, A., Börjesson, G. and Hedlund, K. (2013), 'The effects of 55 years of different inorganic fertiliser regimes on soil properties and microbial community composition', *Soil Biol. Biochem.*, 67, 41–6.

Zaller, J. G. and Köpke, U. (2004), 'Effects of traditional and biodynamic farmyard manure amendment on yields, soil chemical, biochemical and biological properties in a long-term experiment', *Biol. Fertil. Soils*, 40(4), 222–9.

Zhang, W., Ricketts, T. H., Kremen, C., Carney, K. and Swinton, S. M. (2007), 'Ecosystem services and dis-services to agriculture', *Ecol. Econ.*, 64(2), 253–60.

Zotarelli, L., Zatorre, N. P., Boddey, R. M., Urquiaga, S., Jantalia, C. P., Franchini, J. C. and Alves, B. J. (2012), 'Influence of no-tillage and frequency of a green manure legume in crop rotations for balancing N outputs and preserving soil organic C stocks', *Field Crop Res.*, 132, 185–95.

Chapter 4

Cover crops in organic crop cultivation

Peter von Fragstein und Niemsdorff, Universität Kassel, Germany

1 Introduction
2 Soil structure and erosion control
3 Nutrient cycling
4 Weed, pest and disease control
5 Promoting biodiversity
6 Carbon sequestration and greenhouse gas emissions
7 Integration of cover crops in no-till cultivation
8 Conclusions and future trends
9 Where to look for further information
10 References

1 Introduction

Cover crops (CCs) are defined as 'close-growing crop that provide soil protection, seeding protection and soil improvement between periods of normal crop production, or between trees in orchards and vines in vineyards. When plowed under and incorporated into the soil, CCs may be referred to as green manure crops' (SSSA 2008). As the definition suggests, the main functions of CCs lie in areas such as soil protection and improvement as well as weed, disease and pest control. CCs are rotated between two main or cash crops and are not primarily grown for harvest, although they can be used as fodder for livestock in mixed farming systems when other fodder sources are limited (Franzluebbers and Stuedemann 2015). CCs can also function as companion plants to other crops, serving as undersown or understorey crops, or as a living mulch. This practice can be seen in cultivation of perennial crops such as fruit trees and vines, cereals and other tall growing crops, and even in lower growing vegetables (Kaspar et al. 2008; Kołota and Adamczewska-Sowińska 2013; Pfeiffer et al. 2016; Xie et al. 2016). CC biomass is also used in biogas production (Blanco-Canqui et al. 2015).

Traditionally CCs were terminated and incorporated into the soil by methods such as tillage, undercutting, mowing or roller crimping. With the emergence of 'conservation agriculture' and no-till systems, there is a greater emphasis on retaining CC residues on the undisturbed soil surface as a precrop mulch for the following cash crop. On the other hand, complete removal of the shoot biomass enables mulching of other areas through

http://dx.doi.org/10.19103/AS.2017.0029.04

techniques such as 'cut and carry' (Stumm and Köpke 2015), 'transfer mulch' (Storch 2013) or 'mobile green manure' (Sorensen and Thorup-Kristensen 2011). This can be achieved through sole use of the CC or by mixing the fresh material with other harvest by-products such as straw in order to thicken the mulch layer and to slow decomposition of the mulch material. In addition to immediate use, it is also possible to store the crop biomass for later use as fermented silage (Sorensen and Thorup-Kristensen 2011). Figure 1 shows the different types of mulch systems used in organic farming systems with low or no livestock.

The use of different terms for CCs indicates their different functions within cropping systems. The term *cover crops* highlights their role in covering the soil to protect it against sealing, erosion, moisture loss as well as to hinder emergence and development of weeds. CCs are also referred to as *catch crops* to highlight their role in conserving nutrients in soil and reducing nutrient losses from leaching. The term *green manure* (GM) emphasizes the manuring and fertilizing effect of those crops after their incorporation into the soil. *Living mulch* or *undersown* crops are CCs which grow as *companion plants* alongside main crops, and which have a role in weed suppression, soil temperature and moisture regulation. These terms and the functions they highlight represent only a small range of the potential effects of CCs which depend on factors such as species and specific site conditions.

Over the last 20 years research on CCs has shifted from refining agronomic practices to a broader focus on the ecological contribution of CCs to agroecosystems and their potential role in addressing global issues such as greenhouse gas (GHG) emissions from agriculture. A series of recent publications reflect this new understanding of CCs and their role as part of crop rotation management (Damour et al. 2015; Power 2010; Balvanera et al. 2013; Cavigelli et al. 2013; Blanco-Canqui et al. 2015; Storkey et al. 2015; Finney et al. 2016; Schipanski et al. 2014; Tribouillois et al. 2016; Vrignon-Brenas et al. 2016). These concerns are also reflected in the range of research projects on cover crop projects in the EU and the United States (Table 1). This chapter reviews this new research and what it means for

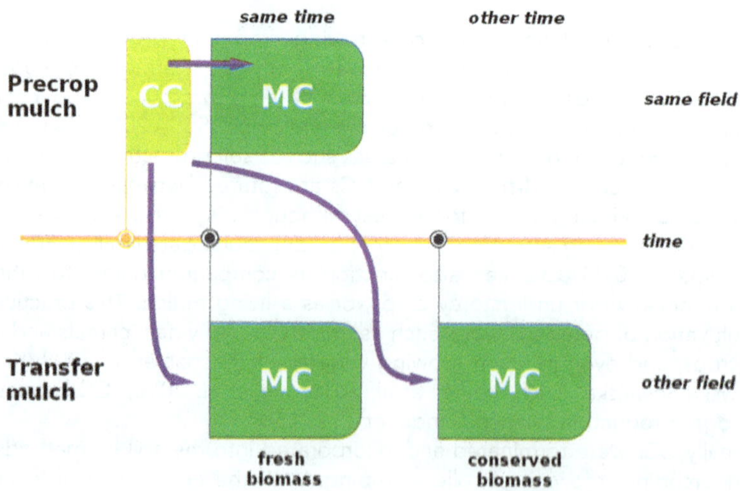

Figure 1 Scheme for the use of cover crops in space and time, as precrop or transfer mulch. Abbreviations: CC = cover crop, MC = main crop.

Table 1 Examples of projects on the evaluation of cover crop systems

Origin	Name	URL
EU	TILMAN-ORG	http://orgprints.org/view/projects/TILMAN-ORG.html
EU	OSCAR	https://web5.wzw.tum.de/oscar/index.php?id=2
USA	ROSE	http://agsci.psu.edu/organic/research-and-extension/rotational-no-till
USA	SASL	https://rodaleinstitute.org/our-work/weed-management-overview/weed-management-current-projects/

the use of CCs in organic farming. The chapter first looks at recent research on the benefits of CCs in areas such as soil structure building and erosion control; nutrient cycling; weed, disease and pest control. It then looks at the newer areas of research such as the broader role of CCs in promoting biodiversity as well as their role in carbon sequestration and reducing GHG emissions. It also looks at some of the challenges in using CCs to optimize crop yields.

2 Soil structure and erosion control

It is well-known that erosion and related nutrient losses from run-off can be reduced by CC and GM. This has been demonstrated, for example, in a study by Maharjan et al. (2016) who introduced rye grass CCs after each harvest in a cabbage–potato–radish–soybean rotation. These CCs helped prevent erosion and run-off in a mountainous landscape in South Korea vulnerable to the effects of the monsoon. Edwards and Burney (2007) found the inclusion of rye as a winter CC more effective than rye as a spring CC in terms of reduced sediment loss and run-off volume, with early sowing more effective than late sowing. Studies of the value of CCs in reducing sediment loss and run-off have been reviewed by Blanco-Canqui et al. (2015).

Deep-rooted CCs can also play a role in improving soil structure and reducing risks from compaction (Chen and Weil 2010; Blanco-Canqui et al. 2012; Steele et al. 2012). CCs can also improve soil aggregate stability, reducing erosion whilst enhancing soil macroporosity, water, nutrient and carbon storage as well as root growth (Liu et al. 2005; Blanco-Canqui et al. 2013a,b). There have been a number of studies on the effects of CCs on water infiltration and retention in soils (Kahimba et al. 2008; Blanco-Canqui et al. 2012; Daigh et al. 2014), though some have highlighted the risk or CC water use reducing water available for following crops (Nielsen and Vigil 2005).

3 Nutrient cycling

A number of studies highlight the role of CCs in improving nutrient cycles in soils with potential benefits for improved and more sustainable yields. CCs play a role in fixing atmospheric nitrogen, scavenging nutrients and reducing nutrient leaching (Dabney et al. 2010; Kaspar et al. 2012; Blanco-Canqui et al. 2011; Xie et al. 2016). Riley et al. (2003) found that the yields of beetroot and white cabbage were augmented by 22% and 27% when chopped grass and/or red clover were applied as a surface mulch, although nutrient use efficiency levels were low. Schipanski and Drinkwater (2011) described a living mulch (or undersown) system – red clover *Trifolium pratense* and a mix of red clover and orchard grass *Dactylis glomerata* – the use of which has a long tradition in European organic crop husbandry. They confirmed the value of including a scavenging graminaceous partner to the legume in order to improve the share of N from fixation.

Lorin et al. (2015) tested living mulch systems in which frost-sensitive legumes were grown as an understorey in a field of oilseed rape. The decayed mulch layer of the legumes served as N source for the cash crop in the second growing period and resulted in a higher N uptake between 20 and 40 kg N ha^{-1}. A living mulch – *Tr. repens* – was tested by Deguchi et al. (2014) as an autumn-seeded CC, terminated by mowing and kept as surface mulch and single N source during cultivation of the following maize (corn) crop. The living mulch treatment produced an N uptake at corn harvest of 62 kg N ha^{-1}, reducing the need for additional N fertilization. The same group (Deguchi et al. 2007) also demonstrated the enhanced transfer of P from a living mulch biomass – *Tr. repens* – to corn, accompanied by higher mycorrhizal infection.

Wortman et al. (2012) focused on the effect of spring-sown CC mixtures, principally legumes and brassicas, as part of a sunflower–soybean–corn rotation. The CCs were terminated in late May using a field disc or a sweep plough undercutter. They provided (1) higher nitrate content in top soil, (2) more soil moisture during the early main crop growth and (3) yield increases of soybean and corn of 0.88 and 1.40 Mg ha^{-1}, respectively. Tosti et al. (2014) confirmed the advantage of CC mixtures using barley *Hordeum vulgare* and hairy vetch *Vicia villosa* which combine N fixing and scavenging properties. Campiglia et al. (2011) found similar results for hairy vetch.

Thomsen and Hansen (2014) tested the effect on N leaching of preharvest and postharvest introduction of white mustard, oil radish and perennial rye grass as a living mulch. Preharvest CCs achieved a maximum N uptake of 24 kg N ha^{-1} when terminated in September. When grown until late autumn the N uptake resulted in 66 kg N ha^{-1} (oil radish) and 35 kg N ha^{-1} (postharvest CCs and ryegrass). Preharvest establishment of brassicas can also be beneficial for reducing N leaching, preferably with extended growing periods. Valkama et al. (2015) conducted a meta-analysis of 35 Scandinavian studies over four decades to investigate the potential of CCs to reduce nitrate leaching. Among the non-legumes ryegrass species were found to be most efficient in reducing N losses through leaching by 50% on average. Italian ryegrass *Lolium multiflorum* reduced leaching by 60%, whilst perennial *L. perenne* and Westerwolds ryegrass *L. multiflorum westerwoldicum* reduced leaching by 25%. White and red clover were inefficient in preventing N losses. The leaching risk was also influenced by management practices (ploughing date and depth) and climatic conditions (such as rainfall). The authors concluded that non-legumes were an 'effective method for reducing N leaching across the varieties of soils and weather conditions in the Nordic countries', though effects on grain yields were not as good as legumes.

4 Weed, pest and disease control

CCs have long been seen as a way of suppressing weeds (Teasdale et al. 2007; Mirsky et al. 2011). Factors such as a short growing period and tolerance to low soil fertility make crops such as buckwheat very suitable CCs for weed suppression in organic vegetable production (Björkman and Shail 2013). Uchino et al. (2012, 2015) found winter rye and hairy vetch particularly effective in weed suppression. Lorin et al. (2015) found living mulch mixes of common vetch, faba bean and berseem clover to be effective, though Pfeiffer et al. (2016) have identified problems with undersown living mulch systems in potentially reducing main crop yields. Wells et al. (2015) tested the suitability of six winter rye cultivars (three early and three late flowering types) at three locations in North Carolina for their use and efficiency as a surface mulch and weed barrier in a no-till soybean stand. In most cases the rye cultivars exceeded the aimed total shoot biomass level of 9 Mg ha^{-1} required to build up an effective mulch layer against weed seed germination and growth. The later the CC precrop was turned into a mulch surface, the better the weed suppression in the following soybean crop. However, higher yields overall were achieved after early termination of the precrop (Teasdale and Mohler 2000). These kinds of trade-off are also highlighted by Mirsky et al. (2012). They identified the conflict between the need for a CC biomass yield of at least 8 Mg ha^{-1} to achieve weed suppression in a no-till crop, on the one hand, and, on the other, the potentially suboptimal conditions this can create for seed and fertilizer placement, including the promotion of seed- and seedling-feeding herbivores.

Research has also highlighted the role of CCs in suppressing crop pests and diseases. Hooks et al. (2010) have, for example, reviewed the use of marigold (*Tagetes* spp.) in nematode suppression. Fourie et al. (2016) have provided a detailed review on the potential of brassica-based management strategies against nematode pests. Other studies found that *Brassica juncea* and *Eruca sativa* GMs suppressed *Globodera pallida* (potato cyst nematode, PCN) (Ngala et al. 2015). Their glucosinolate content was positively correlated with their action against PCN. Curto et al. (2015) confirmed *Crotalaria juncea* has similar properties to those of *Raphanus sativus* ssp. *oleiformis* and *Eruca sativa* in combatting *Meloidogyne incognita* which is a serious problem in tomato production. Similar benefits from the use of *Crotalaria* were reported by Claudius-Cole et al. (2014) in studies of yam-based cropping systems.

MacGuidwin et al. (2012) have investigated the influence of different GM amendments upon *Verticillium dahliae* and *Pratylenchus penetrans* in potato cultivation. Marigold and millet caused significantly less *P. penetrans* population densities, whereas rapeseed and sorgum–sudangrass gave similar results in reducing *V. dahliae*. Keinath et al. (2003) tested the suitability of velvet beans *Mucuna pruriens* L. var. *utilis*, as a disced (i.e. fine turned) GM or mulched CC against *Rhizoctonia solani* infestations in collard *Brassica olearacea* sp. In contrast to mulching, the disced treatments were found effective against *R. solani* infestations. Hansen and Keinath (2013) evaluated the effects of different brassicas on *Rhizoctonia solani*, *Sclerotium rolfsii* and *Pythium* spp. which infect pepper. A key factor is the emission by these plants of isothiocyanates which have pesticidal properties. The highest amounts of isothiocyanates were released by *Br. juncea*, followed by *Br. napus*, reducing the incidence of *R. solani*, but not *S. rolfsii* and *P. spec*. Studies on isothiocyanates have also been undertaken by Trott et al. (2012).

In their review of the suppressive effects of various CCs against potato diseases, Larkin and Griffin (2007) concluded that 'Overall, Indian mustard was most effective for reducing powdery scab *Rhizoctonia solani* and common scab *Streptomyces* sp. diseases, whereas rapeseed and canola were most effective in reducing *Rhizoctonia* diseases'. More recent reviews of Larkin (2013, 2015) also suggest that Indian mustard *Brassica juncea* is particularly effective compared to other brassica species. Other studies by Larkin et al. (2010) on the use of no-till winter rye as a CC in rotations involving potatoes found that the rye reduced black scurf caused by *Rhizoctonia solani* and common scab *Streptomyces* sp. by 12.5% and 7.2% over all rotations tested. In addition, the CC enhanced bacterial populations and microbial activity and slightly improved tuber yield (by about 4%). Larkin et al. (2017) have also investigated the long-term effect of what they defined as a disease-suppressive (DS) system in a rotation of mustard GM/rapeseed–sudangrass GM/rye–potato. The rotation reduced soil-borne diseases such as black scurf and common scab as well as improving tuber yields. The disease suppressive effects of GMs can be observed mainly in plants containing isothiocyanates and glucosinolates (Larkin and Griffin 2007; Larkin et al. 2010; Hansen and Keinath 2013) particularly brassica species (Bangarwa et al. 2011; Smolinska et al. 2003).

5 Promoting biodiversity

Biodiversity can be categorized into the concepts of planned and associated biodiversity, highlighting direct and indirect influences of farming practices on agroecosystem functions. Planned biodiversity can easily be achieved by including one or more cover or intercrop species into a rotational design. The most challenging and limiting factor is the potential competition for water and nutrient resources. On the other hand, under suitable conditions, this enrichment of a rotation provides numerous options for the improvement of ecosystem services. Investigations of two and three crops containing rotations in Michigan (Smith et al. 2008) clearly showed the potential of rotation-based diversity. The integration of one or two more annual (cover) crops (red clover, crimson clover and cereal rye) positively affected various agroecosystem functions, for example crop yield response of corn, soybean and wheat, nitrogen availability and weed suppression.

The case of associated biodiversity is illustrated by the influence of CCs on soil biota. Soil biota play an essential role in maintaining soil fertility, decomposing crop residues at soil surfaces, minimizing ammonium volatilization from degrading residues and improving the physical condition of soil. As living organisms soil biota are critically dependent upon sufficient and continuous sources of nutrition. Any kind of enriched growing system incorporating intercrops (sharing the same space and time with main crops) or CCs (bridging the time between main crops in the same field) can, under suitable conditions, support soil life and enhance ecosystem services.

Among soil biota, earthworm species are key indicators for functioning interactions between CCs and the surrounding agroecosystem. Research by Bogužas et al. (2010) and de Cima et al. (2016), for example, both showed the detrimental impact of intensive tillage and the benefits of organic amendments on the occurrence of earthworms, either from crop-based or animal-based sources. Rotary cultivation plus a mixed CC stand of red clover *Trifolium pratense* and timothy grass *Phleum pratense* increased the density and biomass of earthworms, with shallow tillage and no-till practices achieving similar results (Bogužas et al. 2010). The combination of worm-appropriate tillage (i.e. no-till) plus cover cropping was found to be successful in British cereal–legume intercrop systems (wheat undersown with white clover *Trifolium repens*). The enormous increase in density and biomass of earthworms (around twofold in comparison with a non-CC system) was interpreted as a positive response to a food supply of sufficient quantity, quality and continuity (Schmidt et al. 2003). In a 12-year trial, the cultivation of corn *Zea mays* intercropped with velvet bean *Mucuna pruriens* var. *utilis* in the tropical climate of southern Benin also increased macrofauna density and biomass (earthworms and termites) and affected microfauna, with increased densities of predatory nematodes and decreased densities of obligatory phytophagous nematodes (Blanchart et al. 2006).

A change from corn *Zea mays* [C]-soybean *Glycine max* [S] rotations in the US Upper Midwest to C-cover crop (rye *Secale cereale*)-S rotations achieved 1.2 and 3.2 higher worm densities in autumn and spring measurements compared with a non-CC control, with higher biomass figures (1.4 and 2.5). The significant differences to the control were caused by the high increase in *Aporrectodea* species (Korucu et al. 2018). The use of a CC also substantially reduced losses of broadcast, fall-applied fertilizers from surface run-off. The reduced losses of nutrients and sediments might be caused by the surface-protective action of CCs plus the soil structure-improving actions of earthworms, including higher infiltration rates of surface water.

There have been a number of studies looking at which CCs may provide the best nutrition and habitat for common earthworms, for example, *Lumbricus terrestris.* As an example, Valckx et al. (2011) found a clear preference for Italian ryegrass *Lolium multiflorum* compared to lacy phacelia *Phacelia tanacetifolia*, oats *Avena sativa*, yellow mustard *Sinapis alba* and rapeseed *Brassica napus.* In a three-year trial in Ireland, working with an annual cropping sequence of CC and barley, Roarty et al. (2017) found that pea *Pisum sativum*, followed by oats *Avena sativa* and rye *Secale cereale*, had the most pronounced effect in increasing the abundance and biomass of earthworms. Axelsen and Thorup-Kristensen (2000) tested the effect of winter rye *Secale cereale*, hairy vetch *Vicia villosa* and fodder radish *Raphanus sativus* in a CC-barley sequence on soil microfauna. All CCs induced a positive response of collembola [C] and mites [M] which was highest ([C] 120.000 Ind. m^{-2}, [M] 90.000 Ind. m^{-2}) after fodder radish as a precrop.

As the latter study suggests, research has also focused on the broader role of CCs in promoting the complex microbial communities living in soil and which underpin many of the ecosystem services it delivers. Bàrberi (2015) defined functional agrobiodiversity as a 'subset of total biodiversity identified at the gene, species and habitat level able to deliver a given agroecosystem service'. There is now a wealth of research identifying the importance of the complex insect and microbial community in soils in promoting soil health and the ecosystem services soils deliver. Microorganisms in soil include arbuscular mycorrhizal fungi (AMF) which are particularly important for plant root function. There have been a number of studies on the impact of CCs on soil species biodiversity. Studies of living mulch in cauliflower in a 2-year trial at four different European sites showed a positive impact on arthropods in plant/soil systems (Depalo et al. 2016; Canali et al. 2015). Overall, the living mulch increased the spread and density of carabid beetles and other species. Nair and Ngouajio (2012) also found clear beneficial impacts of CCs (rye and a mix of rye and hairy vetch) and compost amendments on soil microbial biomass, functional microbial diversity and the nematode community. They found the improvement was highest and most sustained in response to the use of composts. In a 3-year trial. Piotrowska and Wilczewski (2012) also confirmed the beneficial influence of both legume and non-legume GMs on soil biological activity.

A Swedish team (Elfstrand et al. 2007) tested the effects of various CC strategies on soil microbial parameters, including a red clover surface mulch, incorporated red clover GM, and composted and fermented red clover as an organic amendment, as compared with a conventionally fertilized control and an untreated control. Mulching induced the highest mycorrhizal colonization at the end of the growing season, whilst green manuring was found to be 'most effective for enhancing and sustaining a high microbial biomass and high rates of enzyme activity in soil'. Studies by Steenwerth and Belina (2008a,b) on the use of cereals as a living mulch in a Californian vineyard concluded that 'cover crops enhanced the soil's capacity for supporting greater microbial biomass N, potential N mineralization, and the microbiological functions of nitrification and denitrification'. Other studies have shown the beneficial effects of hairy vetch as a CC combined with maize in encouraging higher mycorrhizal colonization (Njeru et al. 2014; Campiglia et al. 2010).

Other research suggests, however, that other factors may be more important than CCs in promoting soil biodiversity. As an example, Higo et al. (2014, 2016) did not find that the use of winter CCs, such as winter wheat and red clover, enhanced the diversity of AMF species in soil. Climate or other environmental factors were found to be more important for the proliferation and abundance of mycorrhizal species. In a nine-year field trial of upland rice and soybean cultivation, Ito et al. (2015a,b) investigated the effects on the soil nematode community of using different GMs (fallow, winter rye and hairy vetch), as well as different tillage systems (mouldboard plough/rotary harrow (MP), rotary cultivator and no-tillage (NT)). They found that the health of the nematode population was mainly influenced by the dynamics of soil carbon resources, and that, as a result, NT was of most benefit to the nematode community.

Similarly, Sapkota et al. (2012) demonstrated the benefit of tillage systems and CCs – such as *Vicia villosa* and *Brassica juncea* – on soil microbial indicators as well as arthropod diversity. They found the enhancing effects of long-term NT practices were significantly higher than those of conventional tillage. Even though differences between CCs and the control (cash crop residues only) were less pronounced, biological soil quality indices and richness were still significantly higher than those of the control. An overview of the role of CCs has been provided in a meta-analysis by Bowles et al. (2016) of 54 field studies worldwide. They concluded that, 'less intensive tillage and winter cover cropping similarly increased AMF colonization of summer annual cash crop roots by about 30 per cent. The key variables influencing the change in AMF colonization were the type of cover crop or the type of alternative tillage, suggesting that farmers can optimize combinations of tillage and cover crops that most enhance AM formation, particularly with no-till systems and legume cover crops.' This conclusion confirms the value of using CCs in conjunction with other techniques such as NT.

6 Carbon sequestration and greenhouse gas emissions

Soils were seen as an important sink for reducing CO_2 levels in the atmosphere and thus mitigating global warming. Compared with other soil amendments, like residues of main crops and perennial forage crops, solid manures or composts, the potential of CCs to increase soil organic matter is limited. This limitation is a function of their less lignified shoot mass and low C/N ratio. Nevertheless, growing CCs instead of keeping the field under bare fallow enables at least temporary sequestration of C due to photosynthesis. Abdalla et al. (2014) have highlighted the importance of carbon dioxide uptake by CCs and their influence on soil C dynamics and carbon sequestration, confirmed by the meta-analysis of Poeplau and Don (2015) who used data from 37 studies around the world to demonstrate the degree of carbon sequestration CCs can achieve. Mancinelli et al. (2015) demonstrated the positive effects of mulched CCs on net carbon contribution to the soil in a Mediterranean environment, with hairy vetch most effective, followed by lacy phacelia and white mustard. Hairy vetch was also found to be effective in reducing irrigation requirements whilst achieving comparable yields to more conventional systems. Peregrina et al. (2014) tested the effect on C sequestration of barley *Hordeum vulgare* and Persian clover *Trifolium resupinatum* as a living mulch in a Spanish vineyard over a 3-year period. Both CCs were able to accumulate 1.19 Mg C ha^{-1} $year^{-1}$ corresponding to a sequestration rate of 0.47 compared with conventional tillage. Repullo-Ruiberriz de Torres et al. (2012) analysed carbon sequestration in an olive grove using different types of mowed CCs – *Brachypodium distachyon*, *Eruca vesicaria* and *Sinapis alba* – whilst *B. distachyon* provided the highest level of shoot biomass which included best conditions for soil protection against intense spring and summer rain, *S. alba* produced the highest level of soil organic C in a soil depth of 20 cm in 3 years (7.7 Mg ha^{-1}).

As noted earlier, the fact that CCs mainly provide biomass with low C/N ratios and accelerated rates of decomposition means their potential for C sequestration is limited. As a result, despite high CC inputs (grass/clover leys, manure and catch crops), an 11-year trial by Chirinda et al. (2010) found no noticeable increase of soil C storage in the top soil (0–30 cm). Only deep-rooting species, for example fodder radish *Raphanus sativus* var. *oleiformis* and chicory *Cichorium intybus* (Thorup-Kristensen 2006) provide advantages by accumulating carbon sources in subsoil layers. Chirinda et al. (2010) highlighted the importance of green manuring or cover cropping to improve nutrient cycling, a conclusion also reached in other studies by Nadeem et al. (2012), Brozyna et al. (2013) and Rees et al. (2013).

In studies of the barley, rapeseed and winter vetch, winter vetch resulted in the highest losses of N_2O through denitrification, producing the lowest C/N of all the CCs reviewed (Sanz-Cobena et al. 2014). Basche et al. (2014) conducted a meta-analysis of 26 publications assessing the effects of CC on N_2O emissions which highlighted the superior performance of non-leguminous to leguminous CCs, a conclusion confirmed by Kallenbach et al. (2010) and Pappa et al. (2011). Such studies also emphasize the importance of factors such as precipitation, types of irrigation and decomposition periods for CCs (Petersen et al. 2011; Mitchell et al. 2013).

7 Integration of cover crops in no-till cultivation

A major challenge in the use of CCs is CC regrowth due to incomplete termination which may then compete for water and nutrients with the following cash crop, particularly when no-till techniques are used (Carr et al. 2013; Vincent-Caboud et al. 2017). As an example, Kornecki et al. (2013) found regrowth of hairy vetch had a significant negative impact on the following cereal crop. This problem makes choice and implantation of the appropriate termination method an important issue. Selection of the right tools for mechanical termination of CCs is reviewed in detail by Creamer and Dabney (2002). Techniques include roller crimping, chopping, undercutting and mowing. As an example, Kornecki (2015) tested three roller-crimpers adapted to small-scale farming and their effect on the termination of a rye CC by one, two and three passes and measured the volumetric moisture content after one, two and three weeks. Three passes of roller/crimping achieved higher levels of rye termination at one and two weeks after rolling than one or two passes.

Research comparing the various stages of CC growth in relation to the effectiveness of termination has highlighted the importance of CCs reaching the flowering growth stage if they are to be effectively controlled (Mischler et al. 2010; Mirsky et al. 2012; Peigné et al. 2015; Wells et al. 2015). More broadly, site conditions must be matched with the right CC species before the promotion of ecosystem services can be transferred effectively into practice (Blanco-Canqui et al. 2015).

8 Conclusions and future trends

There is an urgent need to improve cropping systems that include components of cover cropping, undersowing and similar approaches. Key questions include the following:

- Which other plant species meet the requirements for CCs and could increase the number of useful CCs to improve growing systems?
  ▶▶▶ *Screening of herbaceous species in different environments*
- Do we have sufficient cultivars amongst existing CC species suitable for diverse environments, which are flexible enough to adapt to changing site conditions?
  ▶▶▶ *More plant breeding for new CC cultivars*
- *Which mixtures of legumes and non-legumes fit best in different environments?*
  ▶▶▶ *More 'classical' trials with different mixtures, densities, dates of sowing and termination, mechanical tools and so on*
- Which new technological approaches could enable a better management of living (and surface) mulch systems in terms of weed regulation, nutrient supply and different growing systems?
  ▶▶▶ *Improvement of technical tools for the management of CCs*
- How can negative environmental impacts through GHG emissions be minimized through the use of CCs?
  ▶▶▶ *Development of growing systems with minimized GHG emissions*
- What kind of incentives should be offered to the farming community for a more intensive use of the modern mulching or nutrient management concepts on-farm?
  ▶▶▶ *Political measures for increasing adaptation of environmentally friendly farming systems with special emphasis on minimizing negative environmental impacts*

The new scientific interest in green manuring in the context of ecosystem services is built on a new understanding of the complexity of agroecosystems, and highlights the valuable role these important cropping systems may play in delivering a more sustainable agriculture. There needs to be both (1) more basic research on these systems for a better understanding of their functions and their contribution as ecosystem services and (2) improved knowledge transfer for the conversion of scientific concepts into farming practice.

9 Where to look for further information

Online bibliographic databases allow a rapid overview of publications in a particular field. The increasing interest in 'cover crops', 'catch crops' or 'green manure' is easily visible

Table 2 Counts for three key words as part of paper titles

Database keyword	ScienceDirect	Web of Knowledge	Google Scholar	Google Scholar, 1981–2000	Google Scholar, 2001–18
Cover crops	1054	2196	4280	1140	2700
Green manure	515	899	4630	1660	2180
Intercrop	482	3618	1830	677	962

from a literature search using three well-known search engines: Google Scholar, Web of Knowledge and Science Direct (accessed 14 October 2018) (see Table 2).

The first three columns in Table 2 show results up until 2018. These can be compared to Google Scholar Results for the periods 1981–2000 and 2001–18. These clearly show the growing number of publications in recent years. Scientific groups in all continents are active in research on cover crops. Examples of leading research centres include:

- the Research Institute for Organic Agriculture, Frick(CH) (https:\\shop.fibl.org)
- Ryton Organic Gardens, Coventry, Warwickshire, UK (https:\\gardenorganic.org.uk)
- Sustainable Agriculture Research & Education (SARE), College Park, MD 20742, USA (https:\\sare.org)
- National Center for Appropriate Technology, ATTRA NCAT, Butte, MT 59702, USA (https:\\attra.ncat.org)

Knowledge transfer can be found and documented in numerous publications, technical sheets and similar formats in order to share the new knowledge and understanding of ecosystem service approaches with the farming community. As pars pro toto a short list of interesting groups should be mentioned as valuable source for adequate information for practitioners and passionate consumers.

In addition, many countries provide public or private extension services which link farmers with the scientific community.

The research database of the European Union (https:\\cordis.europa.eu\projects\home_en.html) provides very useful information about current and completed projects within the agricultural sector. In the context of data on organic farming, the Danish–Swiss–German database 'orgprints.org' (http:\\orgprints.org) is an excellent source for both scientific research and so-called 'grey literature' (e.g. technical reports) in this area.

10 References

Abdalla, M., Hastings, A., Helmy, M., Prescher, A., Osborne, B., Lanigan, G., Forristal, D., Killi, D., Maratha, P., Williams, M., Rueangritsarakul, K., Smith, P., Nolan, P. and Jones, M. (2014), 'Assessing the combined use of reduced tillage and cover crops for mitigating greenhouse gas emissions from arable ecosystem', *Geoderma* 223, 9–20.

Axelsen, J. A. and Thorup-Kristensen, K. (2000), 'Collembola and mites in plots fertilised with different types of green manure', *Pedobiologia* 44, 556–66.

Balvanera, P., Siddique, I., Dee, L., Paquette, A., Isbell, F., Gonzalez, A., Byrnes, J., O'Connor, M., Hungate, B. and Griffin, J. (2013), 'Linking biodiversity and ecosystem services: Current uncertainties and the necessary next steps', *Bioscience* 64, 49–57.

Bangarwa, S., Norsworthy, J., Mattice, J. and Gbur, E. (2011), 'Glucosinolate and isothiocyanate production from brassicaceae cover crops in a plasticulture production system', *Weed Science* 59, 247–54.

Bàrberi, P. (2015), 'Functional biodiversity in organic systems: The way forward?', *Sustainable Agriculture Research* 4, 26–30.

Basche, A., Miguez, F., Kaspar, T. and Castellano, M. (2014), 'Do cover crops increase or decrease nitrous oxide emissions? A meta-analysis', *Journal of Soil and Water Conservation* 69, 471–82.

Björkman, T. and Shail, J. J. (2013), 'Using a buckwheat cover crop for maximum weed suppression after early vegetables', *HortTechnology* 23, 575–80.

Blanchart, E., Villenave, C., Viallatoux, A., Barthès, B., Girardin, C., Azontonde, A. and Feller, C. (2006), 'Longterm effect of a legume cover crop (*Mucuna pruriens* var. *utilis*) on the communities of soil macrofauna and nematofauna, under maize cultivation, in southern Benin', *European Journal of Soil Biology* 42, 136–44.

Blanco-Canqui, H., Mikha, M., Presley, D. and Claasen, M. (2011), 'Addition of cover crops enhances no-till potential for improving soil properties', *Soil Science Society of America Journal* 75, 1471–82.

Blanco-Canqui, H., Claasen, M. and Presley, D. (2012), 'Summer cover crops fix nitrogen, increase crop yield and improve soil-crop relationships', *Agronomy Journal* 104, 137–47.

Blanco-Canqui, H., Holman, J., Schlegel, A., Takarto, J. and Shaver, T. (2013a), 'Replacing fallow with cover crops in a semiarid soil: Effects on soil properties', *Soil Science Society of America Journal* 77, 1026–34.

Blanco-Canqui, H., Shapiro, C., Wortmann, C. and Drijber, R. (2013b), 'Soil organic carbon: The value to soil properties', *Journal of Soil and Water Conservation* 68, 129–34.

Blanco-Canqui, H., Shaver, T. M., Lindquist, J. L., Shapiro, C., Elmore, R. W., Francis, C. A. and Hergert, G. W. (2015), 'Cover crops and ecosystem services: Insights from studies in temperate soils', *Agronomy Journal* 107, 2449–74.

Bogužas, V., Kairytė, A. and Jodaugienė, D. (2010), 'Soil physical properties and earthworms as affected by soil tillage systems, straw and green manure management', *Žemdirbystė = Agriculture* 97(3), 3–14.

Bowles, T., Jackson, L., Loeher, M. and Cavagnaro, T. (2016), 'Ecological intensification and arbuscular mycorrhizas: A meta-analysis of tillage and cover crop effects', *Journal of Applied Ecology* 54, 1785–93.

Brennan, E. B. (2013), 'Agronomic aspects of strip intercropping lettuce with alyssum for biological control of aphids', *Biological Control* 65, 302–11.

Brennan, E. B. (2016), 'Agronomy of strip intercropping broccoli with alyssum for biological control of aphids', *Biological Control* 97, 109–19.

Brozyna, M., Petersen, S., Chirinda, N. and Olesen, J. (2013), 'Effects of grass-clover management and cover crops on nitrogen cycling and nitrous oxide emissions in a stockless organic crop rotation', *Agriculture Ecosystems and Environment* 181, 115–26.

Campiglia, E., Caporali, F., Radicetti, E. and Mancinelli, R. (2010), 'Hairy vetch (*Vicia villosa* Roth.) cover crop residue management for improving weed control and yield in no-tillage tomato (*Lycopersicon esculentum* Mill.) production', *European Journal of Agronomy* 33, 94–102.

Campiglia, E., Mancinelli, R., Radicetti, E. and Marinari, S. (2011), 'Legume cover crops and mulches: Effects on nitrate leaching and nitrogen input in a pepper crop (*Capsicum annuum* L.)', *Nutrient Cycling in Agroecosystems* 89, 399–412.

Canali, S., Campanelli, G., Ciaccia, C., Tittarelli, F., Bavec, F., von Fragstein, P., Burgio, G., Ortolani, L. and Kristensen, H. L. (2015), 'System assessment of organic living mulch for cauliflower (*Brassica oleracea* L. var. *botrytis*) cropping systems', *Acta Horticulturae* 29, 91–8.

Carr, P., Horsley, R., Gunderson, J., Winch, T. and Martin, G. (2013), 'Weed growth and crop performance following hairy vetch, rye, and wheat cover crops in a cool semiarid region', *Organic Agriculture* 3, 149–61.

Cavigelli, M., Mirsky, S., Teasdale, J., Spargo, J. and Doran, J. (2013), 'Organic grain cropping systems to enhance ecosystem services', *Renewable Agriculture and Food Systems* 28, 145–59.
Chen, G. and Weil, R. (2010), 'Penetration of cover crop roots through compacted soils', *Plant Soil* 331, 31–43.
Chirinda, N., Carter, M. S., Albert, K. R., Ambus, P., Olesen, J. E., Porter, J. R. and Petersen, S. O. (2010), 'Emissions of nitrous oxide from arable organic and conventional cropping systems on two soil types', *Agriculture Ecosystems and Environment* 136, 199–208.
Claudius-Cole, A. O., Fawole, B., Asiedu, R. and Coyne, D. L. (2014), 'Management of *Meloidogyne incognita* in yam-based cropping systems with cover crops', *Crop Protection* 63, 97–102.
Creamer, N. G. and Dabney, S. M. (2002), 'Killing cover crops mechanically: Review of recent literature and assessment of new research results', *American Journal of Alternative Agriculture* 17, 32–40.
Curto, G., Dallavalle, E., Santi, R., Casadei, N., d'Avino, L. and Lazzeri, L. (2015), 'The potential of *Crotalaria juncea* L. as a summer green manure crop in comparison to Brassicaceae catch crops for management of *Meloidogyne incognita* in the Mediterranean area', *European Journal of Plant Pathology* 142, 829–41.
Dabney, S., Delgado, J., Meisinger, J., Schlomberg, H. and Liebig, M. (2010), 'Using cover crops and cropping systems for nitrogen management', in Delgado, J. and Follett, R. (Eds), *Advances in Nitrogen Management for Water Quality*, Soil Water Conservation Society, Ankeny, USA, pp. 231–82.
Daigh, A., Helmers, M., Kladivko, E., Zhou, X. and Goeken, R. (2014), 'Soil water during the drought of 2012 as affected by rye cover crops in Iowa and Indiana', *Journal of Soil and Water Conservation* 69, 564–73.
Damour, G., Garnier, E., Navas, M., Dorel, M. and Risede, J. (2015), 'Using functional traits to assess the services provided by cover plants: A review of potentialities in banana cropping systems', *European Journal of Agronomy* 134, 81–133.
de Cima, D. S., Tein, B., Eremeev, V., Luik, A., Kauer, K., Reintam, E. and Kahu, G. (2016), 'Winter cover crop effects on soil structural stability and microbiological activity in organic farming', *Biological Agriculture and Horticulture* 32, 170–81.
Deguchi, S., Shimazaki, Y., Uozumi, S., Tawaraya, K., Kawamoto, H. and Tanaka, O. (2007), 'White clover living mulch increases the yield of silage corn via arbuscular mycorrhizal fungus colonization', *Plant Soil* 291, 291–9.
Deguchi, S., Uozumi, S., Touno, E., Kaneko, M. and Tawaraya, K. (2014), 'White clover living mulch controlled only by mowing supplies nitrogen to corn', *Soil Science and Plant Nutrition* 60(2), 183–7.
Depalo, L., Burgio, G., von Fragstein, P., Kristensen, H. L., Bavec, M., Robačer, M., Campanelli, G. and Canali, S. (2016), 'Impact of living mulch on arthropod fauna: Analysis of pest and beneficial dynamics on organic cauliflower (*Brassica oleracea* L. var. *botrytis*) in different European scenarios', *Renewable Agriculture and Food Systems* 32, 240–7.
Edwards, L. and Burney, J. (2007), 'Effect of cover crop seeding rate and date on cool period soil erosion control in potato production in PEI, Canada', *Canadian Journal of Plant Science* 87, 851–4.
Elfstrand, S., Båth, B. and Mårtensson, A. (2007), 'Influence of various forms of green manure amendment on soil microbial community composition, enzyme activity and nutrient levels in leek', *Applied Soil Ecology* 36, 70–82.
Finney, D., White, C. M. and Kaye, J. (2016), 'Biomass production and carbon/nitrogen ratio influence ecosystem services from cover crop mixtures', *Agronomy Journal* 108, 39–52.
Fourie, H., Ahuja, P., Lammers, J. and Daneel, M. (2016), 'Brassicacea-based management strategies as an alternative to combat nematode pests: A synopsis', *Crop Protection* 80, 21–41.
Franzluebbers, A. and Stuedemann, J. (2015), 'Does grazing of cover crops impact biologically active soil carbon and nitrogen fractions under inversion or no tillage management?', *Journal of Soil and Water Conservation* 70, 365–73.
Gomes, J., Bayer, C., Costa, F., Piccolo, M., Zanatta, J., Beber Vieira, F. and Six, J. (2009), 'Soil nitrous oxide emissions in long-term cover crops-based rotations under subtropical climate', *Soil & Tillage Research* 106, 36–44.

Hansen, Z. and Keinath, A. (2013), 'Increased pepper yields following incorporation of biofumigation cover crops and the effects on soilborne pathogen populations and pepper diseases', *Applied Soil Ecology* 63, 67–77.

Higo, M., Isobe, K., Drijber, R., Kondo, T., Yamaguchi, M., Takeyama, S., Suzuki, Y., Niijima, D., Matsuda, Y., Ishii, R. and Torigoe, Y. (2014), 'Impact of a 5-year winter cover crop rotational system on the molecular diversity of arbuscular mycorrhizal fungi colonizing roots of subsequent soybean', *Biology and Fertility of Soils* 50, 913–26.

Higo, M., Isobe, K., Miyazawa, Y., Matsuda, Y., Drijber, R. and Torigoe, Y. (2016), 'Molecular diversity and distribution of indigenous arbuscular mycorrhizal communities colonizing roots of two different winter cover crops in response to their root proliferation', *Journal of Microbiology* 54, 86–97.

Hooks, C. R. R., Wang, K. H., Ploeg, A. and McSorley, R. (2010), 'Using marigold (*Tagetes* spp.) as a cover crop to protect crops from plant-parasitic nematodes', *Applied Soil Ecology* 46, 307–20.

Ito, T., Araki, M., Higashi, T., Komatsuzaki, M., Kaneko, N. and Ohta, H. (2015a), 'Responses of soil nematode community structure to soil carbon changes due to different tillage and cover crop management practices over a nine-year period in Kanto, Japan', *Applied Soil Ecology* 89, 50–8.

Ito, T., Araki, M., Komatsuzaki, M., Kaneko, N. and Ohta, H. (2015b), 'Soil nematode community structure affected by tillage systems and cover crop managements in organic soybean production', *Applied Soil Ecology* 86, 137–47.

Kahimba, F., Sri Ranjan, R., Froese, J., Entz M. and Nason, R. (2008), 'Cover crop effects on infiltration, soil temperature and soil moisture distribution in the Canadian prairies', *Applied Engineering in Agriculture*, 24, 321–33.

Kallenbach, C., Rolston, D. and Horwath, W. (2010), 'Cover cropping affects soil N_2O and CO_2 emissions differently depending on type of irrigation', *Agriculture Ecosystems and Environment* 137, 251–60.

Kaspar, T. C., Kladivko, E. J., Singer, J. W., Morse, S. and Mutch, D. R. (2008), 'Potential and limitations of cover crops, living mulches, and perennials to reduce nutrient losses to water sources from agricultural fields in the upper Mississippi river basin', UMRSHNC Final Report, pp. 127–48.

Kaspar, T., Jaynes, D., Parkin, T., Moorman, T. and Singer, J. (2012), 'Effectiveness of oat and rye cover crops in reducing nitrate losses in drainage water', *Agricultural Water Management* 110, 25–33.

Keinath, A., Harrison, H., Marino, P., Jackson, D. and Pullaro, T. (2003), 'Increase in populations of *Rhizoctonia solani* and wirestem of collard with velvet bean cover crop mulch', *Plant Disease* 87, 719–25.

Kołota, E. and Adamczewska-Sowińska (2013), 'Living mulches in vegetable crops production: Perspectives and limitations (a review)', *Acta Scientiarum Polonorum-Hortorum Cultus* 12, 127–42.

Kornecki, T. (2015), 'Rye termination by different rollers/crimpers developed for no-till small-scale farms', *Applied Engineering in Agriculture* 31, 849–56.

Kornecki, T., Arriaga, F., Price, A. and Balkcom, K. (2013), 'Effects of recurrent rolling/crimping operations on cover crop termination, soil moisture, and soil strength for conservation organic systems', *Applied Engineering in Agriculture* 29, 841–50.

Korucu, T., Shiptalo, M. J. and Kaspar, T. C. (2018), 'Rye cover crop increases earthworm populations and reduces losses of broadcast, fall-applied, fertilizers in surface runoff', *Soil & Tillage Research* 180, 99–106.

Kruger, D., Fourie, J. and Malan, A. (2013), 'Cover crops with biofumigation properties for the suppression of plant-parasitic nematodes: A review', *South African Journal of Enology and Viticulture* 34, 287–95.

Larkin, R. P. (2013), 'Green manures and plant disease management', *CAB Reviews* 8, 1–10.

Larkin, R. P. (2015), 'Soil health paradigms and implications for disease management', *Annual Review of Phytopathology* 53, 199–221.

Larkin, R. P. and Griffin, T. S. (2007), 'Control of soilborne potato diseases using *Brassica* green manures', *Crop Protection* 26, 1067–77.

Larkin, R. P., Griffin, T. S. and Honeycutt, C. W. (2010), 'Rotation and cover crop effects on soilborne potato diseases, tuber yield, and soil microbial communities', *Plant Disease* 94, 1491–502.

Larkin, R. P., Honeycutt, C. W., Griffin, T. S., Olanya, O. M., He, Z. and Halloran, J. M. (2017), 'Cumulative and residual effects of different potato cropping system management strategies on soilborne diseases and soil microbial communities over time', *Plant Pathology* 66, 437–49.

Liu, A., Ma, B. and Bomke, A. (2005), 'Effects of cover crops on soil aggregate stability, total organic carbon and polysaccharides', *Soil Science Society of America Journal* 69, 2041–8.

Lorin, M., Jeuffroy, M. H., Butier, A. and Valantin-Morison, M. (2015), 'Undersowing winter oilseed rape with frost-sensitive legume living mulches to improve weed control', *European Journal of Agronomy* 71, 96–105.

MacGuidwin, A., Knuteson, D., Connell, T., Bland, W. and Bartelt, K. (2012), 'Manipulating inoculum densities of *Verticillium dahliae* and *Pratylenchus penetrans* with green manure amendments and solarization influence potato yield', *Phytopathology* 102, 519–27.

Maharjan, G., Ruidisch, M., Shope, C., Choi, K., Huwe, B., Kim, S., Tenhunen, J. and Arnhold, S. (2016), 'Assessing the effectiveness of split fertilization and cover crop cultivation in order to conserve soil and water resources and improve crop productivity', *Agricultural Water Management* 163, 305–18.

Mancinelli, R., Marinari, S., Brunetti, P., Radicetti, E. and Campiglia, E. (2015), 'Organic mulching, irrigation, and fertilization affect soil CO_2 emission and C storage in tomato crop in the Mediterranean environment', *Soil and Tillage Research* 152, 39–51.

Mirsky, S., Curran, W., Mortensen, D., Ryan, M. and Shumway, D. (2011), 'Timing of cover crop management effects on weed suppression in no-tilled planted soybean using a roller crimper', *Weed Science* 59, 380–9.

Mirsky, S., Ryan, M., Curran, W., Teasdale, J., Maul, J., Spargo, J., Moyer, J., Grantham, A., Weber, D., Way, T. and Camargo, G. (2012), 'Conservation tillage issues: Cover crop-based organic rotational no-till grain production in the mid-Atlantic region, USA', *Renewable Agriculture and Food Systems* 27, 31–40.

Mischler, R., Duiker, S., Curran, W. and Wilson, D. (2010), 'Hairy vetch management for no-till organic corn management', *Agronomy Journal* 102, 355–62.

Mitchell, D., Castellano., M., Sawyer, J. and Pantoja, J. (2013), 'Cover crop effects on nitrous oxide emissions', *Soil Science Society of America Journal* 77, 1765–73.

Nadeem, S., Hansen, S., Bleken, M. and Dorsch, P. (2012), 'N_2O emission from organic barley cultivation as affected by green manure management', *Biogeosciences* 9, 2747–59.

Nair, A. and Ngouajio, M. (2012), 'Soil microbial biomass, functional microbial diversity, and nematode community structure as affected by cover crops and compost in an organic vegetable production system', *Applied Soil Ecology* 58, 45–55.

Ngala, B., Haydock, P., Woods, S. and Back, M. (2015), 'Biofumigation with *Brassica juncea*, *Raphanus sativus* and *Eruca sativa* for the management of field population of the potato cyst nematode *Globodera pallida*', *Pest Management Science* 71, 759–69.

Nielsen, D. and Vigil, M. (2005), 'Legume green fallow effect on soil water content in wheat cultivation', *Agronomy Journal* 97, 684–9.

Njeru, E., Avio, L., Sbrana, C., Turrini, A., Bocci, G., Barberi, P. and Giovannetti, M. (2014), 'First evidence for a major cover crop effect on arbuscular mycorrhizal fungi and organic maize growth', *Agronomy for Sustainable Development* 34, 841–8.

Pappa, V., Ress, R., Walker, R., Baddeley, J. and Watson, C. (2011), 'Nitrous oxide emissions and nitrate leaching in an arable rotation resulting from the presence of an intercrop', *Agriculture, Ecosystems and Environment* 141, 153–61.

Peigné, J., Lefevre, V., Viari, J. and Flery, P. (2015), 'Conservation agriculture in organic farming: Experiences, challenges and opportunities in Europe', in Farooq, M. and Siddique, K (Eds), *Conservation Agriculture*, Springer International Publishing, Cham, Switzerland, pp. 559–78.

Peregrina, F., Perez-Alvarez, E. and Garcia-Escudero, E. (2014), 'The short term influence of aboveground biomass cover crops on C sequestration and beta-glucosidase in a vineyard ground under semiarid conditions', *Spanish Journal of Agricultural Research* 12, 1000–7.

Petersen, S., Mutegi, J., Hansen, E. and Munkholm, L. (2011), 'Tillage effects on N_2O emissions as influenced by winter cover crop', *Soil Biology and Biochemistry* 43, 1509–17.

Pfeiffer, A., Silva, E. and Colquhoun, J. (2016), 'Living mulch cover crops for weed control in small-scale applications', *Renewable Agriculture and Food Systems* 31, 309–17.

Piotrowska, A. and Wilczewski, E. (2012), 'Effects of catch crops cultivated for green manure and mineral nitrogen fertilization on soil enzyme activities and chemical properties', *Geoderma* 189, 72–80.

Poeplau, C. and Don, A. (2015), 'Carbon sequestration in agricultural soils via cultivation of cover crops – a meta-analysis', *Agriculture Ecosystems & Environment* 200, 33–41.

Power, A. (2010), 'Ecosystem services and agriculture: Tradeoffs and synergies', *Philosophical Transactions of the Royal Society B* 365, 2959–71.

Rees, R. M., Augustin, J., Alberti, G., Ball, B. C., Boeckx, P., Cantarel, A., Castaldi, S., Chirinda, N., Chojnicki, B., Giebels, M., Gordon, H., Grosz, B., Horvath, L., Juszczak, R., Klemedtsson, A. K., Klemedtsson, L., Medinets, S., Machon, A., Mapanda, F., Nyamangara, J., Olesen, J. E., Reay, D. S., Sanchez, L., Cobena, A. S., Smith, K. A., Sowerby, A., Sommer, M., Soussana, J. F., Stenberg, M., Topp, C. F. E., van Cleemput, O., Vallejo, A., Watson, C. A. and Wuta, M. (2013), 'Nitrous oxide emissions from European agriculture – an analysis of variability and drivers of emissions from field experiments', *Biogeosciences* 10, 2671–82.

Repullo-Ruiberriz de Torres, M. A., Carbonell-Bojollo, R., Alcantara-Brana, C., Rodriguez-Lizana, A. and Ordonez-Fernandez, R. (2012), 'Carbon sequestration potential of residues of different types of cover crops in olive groves under Mediterranean climate', *Spanish Journal of Agricultural Research* 10, 649–61.

Riley, H., Løes, A., Hansen, S. and Dragland, S. (2003), 'Yield responses and nutrient utilization with the use of chopped grass and clover material as surface mulches in an organic vegetable growing system', *Biological Agriculture and Horticulture* 21, 63–90.

Roarty, S., Hackett, R. A. and Schmidt, O. (2017), 'Earthworm populations in twelve cover crop and weed management combinations', *Applied Soil Ecology* 114, 142–51.

Sanz-Cobena, A., Garcia-Marco, S., Quemada, M., Gabriel, J., Almendros, P. and Vallejo, A. (2014), 'Do cover crops enhance N_2O, CO_2 or CH_4 emissions from soil in Mediterranean arable systems?', *Science of the Total Environment* 466, 164–74.

Sapkota, T., Mazzoncini, M., Barberi, P., Antichi, D. and Silvestri, N. (2012), 'Fifteen years of no till increase soil organic matter, microbial biomass and arthropod diversity in cover crop-based arable cropping systems', *Agronomy for Sustainable Development* 32, 853–63.

Schipanski, M. and Drinkwater, L. (2011), 'Nitrogen fixation of red clover interseeded with winter cereals across management', *Nutrient Cycling in Agroecosystems* 90, 105–19.

Schipanski, M. E., Barbercheck, M., Douglas, M. R., Finney, D. M., Haider, K., Kaye, J. P., Kemanian, A. R., Mortensen, D. A., Ryan, M. R., Tooker, J. and White, C. (2014), 'A framework for evaluating ecosystem services provided by cover crops in agroecosystems', *Agricultural Systems* 125, 12–22.

Schmidt, O., Clements, R. O. and Donaldson, G. (2003), 'Why do cereal legume intercrops support large earthworm populations?', *Applied Soil Ecology* 22, 181–90.

Smith, R., Gross, K. and Robertson, G. (2008), 'Effect of crop diversity on agroecosystem function: Crop yield response', *Ecosystem Research* 11, 355–66.

Smolinska, U., Morra, M., Knudsen, G. and James, R. (2003), 'Isothiocyanates produced by Brassicaceae species as inhibitors of *Fusarium oxysporum*', *Plant Disease* 87, 407–12.

Sorensen, J. and Thorup-Kristensen, K. (2011), 'Plant-based fertilizers for organic vegetable production', *Journal for Plant Nutrition and Soil Science* 174, 321–32.

SSSA (2008), *Glossary of Soil Science Terms*, SSSA, Madison, WI.

Steele, M., Coale, F. and Hill, R. (2012), 'Winter annual cover crop impacts on no-till soil physical properties and organic matter', *Soil Science Society of America Journal* 76, 2164–73.

Steenwerth, K. and Belina, K. (2008a), 'Cover crops enhance soil organic matter, carbon dynamics and microbiological function in a vineyard agroecosystem', *Applied Soil Ecology* 40, 359–69.

Steenwerth, K. and Belina, K. (2008b), 'Cover crops and cultivation: Impacts on soil N dynamics and microbiological function in a Mediterranean vineyard agroecosystem', *Applied Soil Ecology* 40, 370–80.

Storch, J. (2013), 'Mulch- und Direktpflanzung im Ökologischen Gemüsebau unter Verwendung der Reihenmulchschneide-Pflanztechnik sowie Aspekte der Betriebswirtschaft', Universität Kassel, Fachbereich Ökologische Agrarwissenschaften, Bachelor thesis.

Storkey, J., Döring, T., Baddeley, J., Collins, R., Roderick, S., Jones, H. and Watson, C. (2015), 'Engineering a plant community to deliver multiple ecosystem services', *Ecological Applications* 25, 1034–43.

Stumm, C. and Köpke, U., 2015, 'Optimierung des Futterleguminosenanbaus im viehlosen Acker- und Gemüsebau', in Häring, A. M., Hörning, B., Hoffmann-Bahnsen, R., Luley, H., Luthardt, V., Pape, J. and Trei, G. (Eds), *'Am Mut hängt der Erfolg – Rückblicke und Ausblicke auf die Ökologische Landbewirtschaftung', Beiträge zur 13. Wissenschaftstagung Ökologischer Landbau, Hochschule für nachhaltige Entwicklung Eberswalde*, 17.-20. März 2015, Verlag Dr Köster, Berlin, pp. 311–14.

Teasdale, J. and Mohler, C. (2000), 'The quantitative relationship between weed emergence and physical properties of mulches', *Weed Science* 48, 385–92.

Teasdale, J., Brandsaeter, L., Calegari, A. and Neto, F. (2007), 'Cover crops and weed management', in Upadhyaya, M. and Blackshaw, R. (Eds), *Non-chemical Weed Management*, CABI, Wallingford, UK, pp. 49–64.

Thomsen, I. and Hansen, E. (2014), 'Cover crop growth and impact on N leaching as affected by pre- and postharvest sowing and time of incorporation', *Soil Use and Management* 30, 48–57.

Thorup-Kristensen, K. (2006), 'Effect of deep and shallow root systems on the dynamics of soil inorganic N during 3-year crop rotations', *Plant Soil* 288, 233–48.

Tosti, G., Benincasa, P., Farneselli, M., Tei, F. and Guiducci, M. (2014), 'Barley-hairy vetch mixture as cover crop for green manuring and the mitigation of N leaching risk', *European Journal of Agronomy* 54, 34–9.

Tribouillois, H., Cohan, J. P. and Justes, E. (2016), 'Cover crop mixtures including legume produce ecosystem services of nitrate capture and green manuring: Assessment combining experimentation and modelling', *Plant and Soil* 401, 347–64.

Trott, D., LePage, J. and Hebert, V. (2012), 'Assessing natural isothiocyanate air emissions after field incorporation of mustard cover crop', *Bulleting of Environmental Contamination and Toxicology* 88, 482–5.

Uchino, H., Iwama, K., Jitsuyama, Y., Ichiyama, K., Sugiura, E., Yudate, T., Nakamura, S. and Gopal, J. (2012), 'Effect of interseeding cover crops and fertilization on weed suppression under an organic and rotational cropping system 1. Stability of weed suppression over years and main crops of potato, maize and soybean', *Field Crops Research* 127, 9–16.

Uchino, H., Iwama, K., Jitsuyama, Y., Yudate, T., Nakamura, S. and Gopal, J. (2015), 'Interseeding a cover crop as a weed management tool is more compatible with soybean than with maize in organic farming systems', *Plant Production Science* 18(2), 187–96.

Valckx, J., Cordón Pina, A., Govers, G., Hermy, M. and Muys, B. (2011), 'Food and habitat preferences of the earthworm *Lumbricus terrestris* L. for cover crops', *Pedobiologia* 545, 139–44.

Valkama, E., Lemola, R., Känkänen, H. and Turtol, E. (2015), 'Meta-analysis of the effects of undersown catch crops on nitrogen leaching loss and grain yields in the Nordic countries ', *Agriculture, Ecosystems and Environment* 203, 93–101.

Vincent-Caboud, L., Peigné, J., Casagrande, M. and Silva, E. (2017), 'Overview of organic cover crop-based no-tillage techniques in Europe: Farmers' practices and research challenges', *Agriculture* 7(42), 1–16.

Vrignon-Brenas, S., Celette, F., Amosse, C. and David, C. (2016), 'Effect of spring fertilization on ecosystem services of organic wheat and clover relay intercrops', *European Journal of Agronomy* 73, 73–82.

Wells, M., Brinton, C. and Reberg-Horton, S. (2015), 'Weed suppression and soybean yield in a no-till cover-crop mulched system as influence by six rye cultivars', *Renewable Agriculture and Food Systems* 31, 429–40.

Wortman, S. E., Francis, C. A., Bernards, M. L., Drijber, R. A., Lindquist, J. L. (2012), 'Optimizing cover crop benefits with diverse mixtures and an alternative termination method', *Agronomy Journal* 104, 1425–34.

Xie, Y., Tittarelli, F., von Fragstein, P., Bavec, M., Canali, S. and Kristensen, H. L. (2016), 'Can living mulches in intercropping systems reduce the potential nitrate leaching? Studies of organic cauliflower (*Brassica oleracea* L. var. *botrytis*) and leek (*Allium porrum* L.) production across European conditions', *Renewable Agriculture and Food Systems* 32, 224–39.

Chapter 5

The role of crop rotations in organic farming

Bernhard Freyer, University of Natural Resources and Life Sciences (BOKU), Austria

1 Introduction
2 Principles of crop rotation
3 Precrop effects in crop rotations
4 Nutrient effects of legumes and other rotation crops
5 The role of rotation crops in suppressing weeds, diseases and pests
6 Rotations and crop yields
7 Designing a crop rotation
8 Measuring and modelling crop rotations, nutrient and humus balances
9 Crop rotations on stockless farms
10 Conclusion and future trends
11 Where to look for further information
12 References and further reading

1 Introduction

Crop rotation has been defined as 'The practice of alternating the species or families of annual and/or biennial crops grown on a specific field in a planned pattern or sequence so as to break weed, pest and disease cycles and to maintain or improve soil fertility and organic matter content' (IFOAM, 2014). As this suggests, crop rotations are defined in terms of a number of crops organized in a specific order over a defined period of time, with specific spatial arrangements and distribution in a field and landscape that ensures the best growing conditions for all crops. Crop rotation is thus a way to use the performance of a single crop in a way that contributes to the whole farming system (Ball et al., 2005; Watson et al., 1999).

Crop rotations have a long history. As an example, the Romans used crop rotations such as wheat followed by a grain legume (lupin, vetch or broad beans) and in the 3rd year by emmer (White, 1970). Crop rotations became more and more widespread in Europe

http://dx.doi.org/10.19103/AS.2017.0029.05

from the sixteenth century (de Molina, 2002; Jones, 1965; Petit and Lambin, 2002; Slicher van Bath, 1963; Stolz, 1932; Vischer, 2005). Scientific work on crop rotations can be traced back to Theodor Brinkmann in Germany during the 1920s (Brinkman, 1917, 1950; Stein and Steinmann, 2018). However, it is important to note that many accepted crop rotation rules are based on practices dating back to the nineteenth century. The first organic crop rotations were implemented in the 1920s within the context of mixed farming systems, in which the function of the crop rotation was to produce both food crops and forage mainly for ruminants, and often also for draft animals. In addition, the majority of the early trials on crop rotations were undertaken without catch crops (Willey, 1979). This reinforces the need for new research to better understand and refine crop rotation practices based on eco-intensification principles so that the yield gap between conventional and organic cultivation can be narrowed or closed (Doré et al., 2011; Tscharntke et al., 2012).

Closing the yield gap is even more important because contemporary organic farmers are confronted with new ecological and economic challenges, and specifically with the question of how to feed a growing population. They are continually challenged to translate research findings about crop functions into solutions suitable to their individual on-farm experiences and changing environmental and economic conditions. Faced with market pressures, over time there has been a tendency towards simplifying crop rotations in conventional farming which undermines their value (Richter, 1967). The same trend has also been noticed in organic farming (Schmidt, 2004). Many farmers tend to overlook the multifunctional benefits of ley and crop rotation for short-term economic benefits and become more reliant on other external inputs (Kirchmann et al., 2009). This failure to follow crop rotation rules has negative consequences for soil quality, weed pressure and crop yields (Chongtham et al., 2017).

2 Principles of crop rotation

Fundamentally crop rotations should meet the guiding principles set down for organic farming by IFOAM (Vereijken, 1997; Luttikholt, 2007). The IFOAM principles are:

- *Principle of Health:* Organic agriculture should sustain and enhance the health of soil, plant, animal and human as one and indivisible.
- *Principle of Ecology:* Organic agriculture should be based on living ecological systems and cycles, work with them, emulate them and help sustain them.
- *Principle of Fairness:* Organic agriculture should build on relationships that ensure fairness with regard to the common environment and life opportunities.
- *Principle of Care:* Organic agriculture should be managed in a precautionary and responsible manner to protect the health and well-being of current and future generations and the environment.

These principles must be met without reliance on herbicides, the majority of mainstream pesticides and with mineral fertilizers limited to calcium carbonate, phosphorus fertilizers of low solubility, potassium, magnesia, sulphur and microelements (Commission Regulation (EC) No 889/2008). As a consequence the organic crop rotation system has to perform several functions, including:

- Soil fertility, water and nutrient management function
- Pest and disease management function
- Weed management function
- Biodiversity function
- Forage function
- Socio-economic function
- Other non-agricultural functions

In general, the better the functions are integrated in the crop rotation, the higher the yield and quality of products.

Starting with soil health, crop rotations need to optimize multiple soil characteristics (Bot and Benites, 2005; Olesen et al., 2000). Rotations help to stabilize soil aggregates against soil erosion; to increase water infiltration and water-holding capacity via biopores (Gaiser et al., 2012; Han et al., 2015); to contribute to an increase of water in subsoil layers (Wang and Smith, 2004) from the subsoil to the topsoil close to the location of highest rooting densities and root lengths (Perkons et al., 2014). Rotations also have a key role in nutrient management. They help to provide nutrients in an available form, specifically nitrogen via N-fixing crops (Köpke, 1996); to increase nutrient availability of fixed nutrients from clay minerals via root exudates and microbiological activity (Köpke et al., 2015); to increase availability of soil phosphorus (Köpke and Nemecek, 2010); and to avoid nutrient loss.

Turning to pest, disease and weed management, crop rotations help to avoid soil-borne diseases (Bockmann, 1954; Freyer, 2003; Raaijmakers et al., 2009) through the increase of the antiphytopathogenic potential of the soil (Watson et al., 2002); to increase species and habitat diversity (Gabriel et al., 2013), which supports predators of pests (e.g. Lichtenberg et al., 2017). Crop rotations also serve to control 'problem' weeds (Massucati et al., 2010; Pallutt, 2000), to limit weed competition with crop for water, light and nutrients, whilst also allowing weeds potentially to serve a positive function, for example as a habitat for insects. As discussed, crop rotations also have a broader role in conserving biodiversity. They protect and maintain endangered plant species (Anderson, 2005; Bengtsson et al., 2005; Gabriel et al., 2006; Hole et al., 2005; Roschewitz et al., 2005).

Rotations have an important forage function. They serve to provide all kinds of forage qualities and amounts to cover the forage demand of farm animals, helping to further the aim of 100% farm self-sufficiency in feed production. More broadly, crop rotations are economically important. They help to optimize and diversify farm income; if well designed and managed, they optimize not only the yield and economic performance of one crop but of all crops in a rotation cycle (Wijnands, 1999).

Finally, rotations have a wider, non-agricultural role. They contribute to a diverse and aesthetically pleasing landscape with diverse options for non-agricultural use. They help to create particular habitats for plants and animals (Borin et al., 2010) and conserve particular types of landscape (Koohafkan and Cruz, 2011; Lütz and Bastian, 2002), which may support agro-tourism and recreational use both for citizens and tourists (Duncan, 2004).

3 Precrop effects in crop rotations

Optimal crop rotations require a fundamental understanding of precrop effects. We can distinguish two types of precrop effect:

- the direct precrop effect, which is the influence of the previous crop (PC) and its agronomic effects (AE) on the yield and quality of the following/subsequent crop,
- the indirect (accumulated) precrop effect, which is the sum of the effects of PCs and accompanying AE over one rotation on the current crop.

These effects may be summarized as follows:

Direct precrop effect (DPE) = PCAE (CSPE; AASE)
Indirect precrop effect (IDPE) = $PCAE_{(n1 + n2 + \ldots + n-1)}$ (CSPE; AASE)
(where CSPE = crop-specific precrop effect; AASE = accompanying agronomic-specific effects; N = Number of precrop year)

Key crop-specific precrop effects (CSPE) are:

- *Crop residues:* shoot and root biomass with a given mineral content and C/N ratio. Root characteristics include (1) biochemical root characteristics: mineralization of nutrients, exudates and allelopathic effects; and (2) biophysical root characteristics: bulk density, volume of biopores, soil air and water retention;
- *Weed effects:* characteristics such as the amount and distribution of weed diaspores and rhizomes in the soil; uptake and storage of (micro-)nutrients;
- *Pests and diseases:* characteristics include whether the precrop is a host plant for specific pests and diseases; the amount and quality of organic matter and the antiphytopathogenic potential of soils, that is, the presence of microorganisms with the capacity to reduce diseases.

Accompanying agronomic-specific precrop effects are:

- *Genotype:* yield potential, potential to support/reduce pest and diseases, weed suppression potential, harvest index, shoot and root biomass, potential for nitrogen fixation; nitrogen allocation between marketable product and shoot and root residue/biomass; root characteristics, that is, exudates;
- *Soil management:* seed bed quality; soil tillage and harvest technology: degree of soil compaction; soil structure; loosening of soils to support root growth; mineralization of convertible organic material;
- *Fertilizer use:* use of organic or mineral fertilizer, liming: amount and quality; mineralization rate (C/N, nutrient content).

A key factor in assessing indirect precrop effects is the share of a crop or crop group (e.g. cereals) in a crop rotation (Freyer, 2003). The higher the share of a crop, and the higher the share of the crop group (i.e. family), the higher the risk of accumulated negative factors diminishing the yield of the following crop (Müller and Ellmer, 1985) and/or the entire crop rotation. A high share of a certain crop in the crop rotation (or other crops with the same characteristics that is of the same plant family) can lead to problems such as nutrient imbalances and soil-borne pests and diseases affecting the following crop(s). This means that a maximum share should not be exceeded for each crop (Table 1).

Table 1 Maximum share of cereal crops and crop groups in crop rotations

Crop, crop groups	High soil quality (%)	Low soil quality (%)
Wheat	33	25
Winter barley, Triticale	40[a]	33
Rye	50[a]	33
Total winter cereals (without fodder barley/oats)	75[b]	67
Oats	25	25
Spring barley	50[a]	33
Spring cereal total (rotations without winter cereals)	50[a]	50

[a] For organic values should not exceed 33%.
[b] For organic values should not exceed 50%.
Source: Baeumer (1992).
Data recommended for mainstream agriculture.

4 Nutrient effects of legumes and other rotation crops

A fundamental role of crop rotation is to optimize the nutrition of the main crops in the rotation. Key factors in assessing crop nutrition are:

- amount and ratio of nutrients exported from the field when the precrop is harvested;
- amount and ratio of nutrients retained in the root and shoot biomass and their potential availability for the following crops after mulching, cutting and/or mixing into the soil (different management methods have different effects);
- amount and ratio of nutrients made available in the soil (unlocked from the solid phase/solubilized nutrients).

These factors then determine any nutrient deficiency and ways this can be met, including:

- the role of catch crop mixtures as an additional measure to make nutrients available for the following crops;
- the need to top up nutrients using additional organic and mineral fertilizers, to ensure the nutrient needs of the main crops are fully met.

A key precrop factor determining the success of the succeeding crop is the organic matter residues left by the precrop. Shoot and root biomass residues are nutrient sources in management of the following crops. Table 2 shows the nutrient content of shoot biomass residues (stubble, straw and leaves) for key rotation crops. As an example, the residues of maize, rape, sunflower and sugar beet contain a high amount of potassium. Sugar beet and sunflower provide a high amount of phosphorus and magnesium. The highest content of sulphur is found in cereals. Maize and maize straw, sugar beet and legume crop residues store a large amount of nitrogen.

Further information on these properties is provided by the harvest index for individual crops. This measures the relationship between the dry matter of marketable products and

Table 2 Harvest index and nutrient content of shoot biomass residues

Crop	Harvest index	Yield (DM t ha^{-1})	N	P	K (kg ha^{-1})	Mg	S	N	P	K (kg 100 kg^{-1})	Mg	S
Field bean	0.5	3	45	4	65	5	12	1.5	0.13	2.16	0.18	0.40
Pea	0.5	2	30	3	43	4	8	1.5	0.13	2.16	0.18	0.40
Soybean	0.5	3	45	5	39	5	12	1.5	0.17	1.29	0.18	0.40
Sweet lupine	0.5	3	45	4	41	5	12	1.5	0.15	1.35	0.18	0.40
Leg. catch crops	0.5	3	87	12	87	1	1	2.90	0.40	2.91	0.03	0.04
Winter wheat	0.6	4	20	5	46	5	6	0.5	0.13	1.16	0.12	0.15
Winter rye	0.5	4	20	5	66	5	6	0.5	0.13	1.66	0.12	0.15
Triticale	0.5	4.5	23	6	63	5	7	0.5	0.13	1.41	0.12	0.15
Winter barley	0.6	2.8	14	4	40	3	4	0.5	0.13	1.41	0.12	0.15
Spring barley	0.6	2.8	14	4	40	3	4	0.5	0.13	1.41	0.12	0.15
Oat	0.5	3	15	4	42	4	5	0.5	0.13	1.41	0.12	0.15
Maize	0.5	10	90	9	166	24	15	0.9	0.09	1.66	0.24	0.15
Oilseed rape	0.4	4.25	30	7	88	5	6	0.7	0.18	2.08	0.11	0.13
Sunflower	0.3	6	60	24	249	11	–	1	0.40	4.15	0.18	–
Potato	0.8	50	10	2	15	2	3	0.2	0.04	0.30	0.05	0.05
Sugarbeet leaf FM	0.6	280	112	14	139	17	14	0.4	0.05	0.50	0.06	0.05

Data are used for calculating nutrient balances.
Source: Fritsch (2012).

the total shoot biomass. Conventional agriculture aims to increase the harvest index, while organic farming focuses also on the value of the shoot biomass as a source of nutrition and other functions. In this respect a redefined harvest index which also included the root biomass would be more relevant in organic farming, particularly as the root biomass is relevant not only for uptake of nutrients but other characteristics such as water use efficiency (http://plantsinaction.science.uq.edu.au/content/641-harvest-index).

The nitrogen content in legume root biomass in particular provides a steady source of nitrogen in a crop rotation over a 2- to 3-year period (Table 3). Forage legumes receive 70–80% of their nitrogen requirement from nitrogen (N) fixation, while, in the case of grain legumes, this share varies, depending on the crop and the variety (Table 4) (Kahnt, 1983). Table 5 provides an overview on the amount of nitrogen in different crop roots potentially available for the following crops. Relevant for the short-term mineralization of root biomass is the percentage of thin roots which are better able to concentrate nitrogen and to mineralize in a short time. Data on degrees of nitrogen fixation in grain legumes are rather diverse (Baddeley et al., 2013). Amongst grain legumes (Table 3), the lupin shows remarkably high amounts of nitrogen content in its root and shoot biomass that are close to the level of several small grain species. In the case of grain legumes the preceding crop of grain legumes influences their performance (Watson et al., 2017). Lizarazo et al. (2015) identified oat, barley and turnip as favourable precrops for lupin in comparison to wheat, rye or other grain legumes.

Legume nitrogen availability depends on several factors. These include the biomass of the precrop legume left in the field; and the N_2 fixation potential and N distribution between root and shoot biomass (Freyer et al., 2005; Pietsch and Freyer, 2004). Data on field bean indicate that the nitrogen content and distribution in the plant is also influenced by factors such as genotype and variety. Yield potential, the potential for weed suppression, nutrients exported and the related biomass residues left in the field, also vary from variety to variety (Drews et al., 2009; Horneburg, 2016; Rengel and Damon, 2008; Clárk, 1983; Nielsen and Barber, 1978; Schenk and Barber, 1979).

Differences among the rhizobium strains in the root nodules also influence the biological nitrogen fixation (BNF) capacity of the crop (Watson et al., 2017). The estimation of legume BNF is still challenging and further research is needed (Liu et al., 2011). It is also important to note that legume root systems are able to uptake nutrients from the lower soil layers through their deep growing root system. This vertical transfer does not affect the nutrient balance since these minerals are sourced from the nutrient stock in the soils, but contribute to a higher supply of available nitrogen and other nutrients for the following crop. Forage legumes, lupin and faba bean are particularly effective in this respect.

A number of soil factors affect nitrogen fixation. The amount of legume-based nitrogen depends on the soil type and the availability of water and soil nutrients. Low availability of sulphur, phosphorus, manganese and potassium limits nitrogen fixation (Chaudhary et al., 2008). The amount of fixed nitrogen of forage legumes also depends on sowing date, month of cultivation, type of tillage, the length of the vegetation period and whether or not the crop is undersown (Table 6).

Drought conditions can drastically limit nitrogen fixation (Farthofer et al., 2002). In dry regions, the nitrogen fixation of lucerne and biomass production slows down after July where rainfall is low (Pietsch et al., 2009). Under these circumstances it is best to terminate the lucerne crop, given its high water consumption, to prevent a lack of water for the

Table 3 Amount of nitrogen in shoots and roots of grain and forage legumes

Crop	Latin name	Shoot	Root and shoot/stubble	Total	Shoot/root and stubble
			(N kg ha^{-1})		Relative
Grain legumes					
White lupine	*Lupinus albus* L.	448	93	541	4.8
Blue lupine	*Lupinus angustifolius* L.	429	73	502	5.9
Yellow lupine	*Lupinus luteus* L.	213	86	299	2.5
Field bean	*V. faba* cv. 'Herra'	320	57	377	5.6
Field bean	*V. faba* cv. 'Minica'	300	52	352	5.8
Field bean	*V. faba* cv. 'Felix'	192	62	254	3.1
Field bean	*V. faba* cv. 'Herz Freya'	168	63	231	2.7
Forage pea	*Pisum sativum* L.	291	40	331	7.3
Flat pea	*Lathyrus tingitanus* L.	273	24	297	11.4
Spring vetch	*Vicia sativa* L.	238	36	274	6.6
Soybean	*Glycine max* (L.) MERR.	178	31	209	5.7
Forage legumes					
Red clover	*Trifolium pratense* L.	381	118	499	3.2
Serradella	*Ornithopus sativus* BROT.	289	77	366	3.8
Egyptian clover	*Trifolium alexandrinum* L.	412	125	537	3.3
Persian clover	*Trifolium resupinatum* L.	402	123	525	3.3
Swedish clover	*Trifolium hybridum* L.	303	173	476	1.8
White clover	*Trifolium repens* L.	322	131	453	2.5
Lucerne (German variety)	*Medicago sativa* L.	469	157	626	3.0
Lucerne (S. Arabian variety)	*Medicago sativa* L.	346	166	512	2.1
Yellow clover	*Melilotus off.* (L.) Pall.	332	132	464	2.5
Kidney vetch	*Anthyllis vulneraria* L.	209	143	352	1.5
Sainfoin	*Onobrychis* Mill.	184	140	324	1.3
Italian clover	*Trifolium incarnatum* L.	210	88	298	2.4

Source: After Kahnt (1986); Hohenheim, temperate climate, South-Germany, 400 m above NN, 8.8°C, 700 mm.

following winter cereal (Raza et al., 2014). There is evidence that lucerne genotypes differ in the efficiency of water use (Moghaddam et al., 2013) and the capacity to cope with diverse soil stress factors (Sadowsky, 2005). This highlights the need for more research on different forage varieties (Ghimire, 2017; Thilakarathna et al., 2017).

Table 4 Nitrogen fixation of legumes and share of total biological nitrogen fixation

Crop	Fixation (kg N ha^{-1} a^{-1})	Average fixation (kg N ha^{-1} a^{-1})	% of Total BFN
Clover	45–670	250	80
Lucerne	90–340	150	70 and above
Pea	50–500	150	40–70
Lupin	140–200	150	50–95
Soybean	60–300	100	50
Field bean	100–450	200	40–75
Bush bean	40–100	70	40

Source: after Hardarson (1993), Hardarson et al. (1993), Kayser et al. (2010) and Stein-Bachinger et al. (2004).
% BFN = Biological nitrogen fixation in relation to the whole plant; average: forage legumes: 70–95%; grain legumes: 60–90% (see Stein-Bachinger et al., 2004); see also (Watson et al., 2017, p. 248).

Table 5 Nitrogen storage in root residues and nitrogen content of a crop in a crop rotation

Crop	Crop yield (DM t ha^{-1})	Nitrogen demand (Total N kg ha^{-1})	Root (DM t ha^{-1})	Root (N kg 100 kg^{-1})	Root (Total N kg ha^{-1})
Lucerne (1 year)[a]	10	–	5	2.2	110
Potato (FM)	200	100	0.5	3.0	45
Rye	4	100	2	1.2	24
Oat	3	60	2	1.2	24
Field bean	3	–	1.8	1.8	32
Winter wheat	5	125	1.5	1.2	18
Spelt	3	60	1.5	1.2	18

[a] Mineralization over 1–3 years; further 130–220 kg N ha^{-1} can be expected by the above ground biomass.
Source: Several authors in Freyer (2003).
Catch crop: mixture of legumes and non-legumes; further 40–80 kg N ha^{-1} can be expected by the above ground biomass.

Given the capacity of precrops such as legumes to fix nitrogen, it is important to avoid nitrogen leaching by ensuring that the succeeding main crop has a high demand for nitrogen (Tables 5 and 7). It is important to select specific soil and climate-adapted leguminous–non-leguminous catch crop mixtures and soil tillage systems to avoid nitrogen leaching (Askegaard et al., 2005, 2011; Conrad and Fohrer, 2009; Drinkwater et al., 1998; Haas et al., 2002; Hansen et al., 2000; Hauggaard-Nielsen et al., 2003; Jensen and Hauggaard-Nielsen, 2003; Justus and Köpke, 1995; Kirchmann and Bergström, 2001; Köpke, 1995; Möller et al., 2008; Peigné et al., 2007; Rauber et al., 2001; Schröder and Köpke, 2012; Stopes et al., 2002; Thorup-Kristensen et al., 2003; Tosti et al., 2014; Watson et al., 2002). As an example, oil radish planted as a catch crop stores significant amounts of nitrogen and thus prevents nitrogen leaching (Table 7). Legume/non-legume catch crop mixtures are often recommended as most efficient both in terms of nitrogen fixation and nutrient storage (Watson et al., 2017, p. 253).

Table 6 Forage legumes performances dependent on month of cultivation

Cultivation of forage legumes from. . .	. . . until	Month of cultivation	Accumulated DM yield (t ha^{-1})	BNF (kg ha^{-1})	Effect on SOM increase	Amount of root biomass	Weed suppression/ reduction
March (DS)	Sept	7	7	105	+	+	+
March (BS)	Dec	10	8	120	+	+	+
March (BS)	Febr (FY)	12	9	135	+	+	+
March (BS)	Sept (FY)	19	18	270	++	++	++
April (US)	Sept (FY)	19 (12)	15	225	++	++	++
Aug (BS)	Sept (FY)	12	12	180	+	+	+
April (US)	Sept (SY)	30 (24)	30	450	+++	+++	+++
April (MS)	Sept (SY)	30 (24)	35	525	+++	+++	+++

Source: Data are estimations; production under ceteris paribus conditions.
+ = low; ++ = medium; +++ = high.
FY = first year after sowing; SY = second year after sowing; DS = direct sowing; MS = mixed sowing with oat, peas; US = undersown; month of cultivation (): month without a cover crop; BNF = biological nitrogen fixation; SOM = soil organic matter.

Table 7 Nitrogen in shoot and root biomass of catch crops

Catch crop	Shoot biomass	Root biomass
	(N kg ha^{-1})	
Oil radish, sinapis (green manure)	135–160	25–65
Autumn turnip, marrow-stem kale (*Brassica oleracea* var. *medullosa*) (green manure)	140–190	n.d.
Winter, spring canola (*Brassica rapa*)	100–140	n.d.
Grain legumes	100–130	50
Winter, spring rape	90–130	45
Phacelia	90–100	30
Annual ryegrass (*Lolium multiflorum*)	90–120	n.d.
Clover varieties	80–100	30
Lentil	52[a]	n.d.

[a] Anglade et al. (2015).
Source: Lütke Entrup and Zerhusen (1992, p. 157); based on average yields.
n.d. = not determined.
It has to be considered that these data are collected in a conventional cropping system.

It also is important, however, to take into account the broader role of precrop root systems, not just as a source of biomass but also in contributing more broadly to soil health. Some of these broader characteristics are summarized in Tables 8–10. Root depth, distribution, density and root biomass contribute to the soil humus content, water-holding capacity and use efficiency and water drainage, as well as aggregate

Table 8 Agronomic characteristics of crops

Crop	Layer with highest concentration of root biomass . . . up to soil depth (cm)[a]	Impact on suppressing weeds	Nitrogen fixation	Root density
Forage legumes				
Clover, lucerne, desmodium	−60	+++	+++	++
Clover, lucerne, desmodium with grass	−60	+++	+++	+++
Legume-non-legume (without grass) mixtures	−40	++	+++	++
Mucuna, vetch	−30	+++	++	+
Grain legumes	−20	+	+–++	+
Non-leguminous crops				
Leaf crops				
Potato, turnip, maize	−20	+[a]	0	+
Cereals				
Winter	−20	+	0	+
Spring	−20	+	0	+

[a] Soil depth where 90–95% of root biomass is concentrated; see also Anderson et al. (2001).
Source: Könnecke (1967), added by the author; 0 = without impact/+ = little, ++ = medium, +++ = high.

Table 9 Main zones with highest density of root biomass

1.5–2 m and more	0.8–1.5 m	<0.8 m
Lupine, lucerne, red clover, sunflower, oil radish	Field bean, pea, yellow clover, spring vetch, sinapis, rape, phacelia	White clover, Italian clover, peluskins, buckwheat

Source: Scheffer (1941).

Table 10 Impact of crops on soil aggregate water stability

Crops	% Water-resistant soil aggregates
Root crops (potato, turnip, maize)	10–15
Cereals	15–20
Clover (Red, Yellow, Persian, Egyptian)	30–35
Rape seed	40–50
Grass species	50–60
2-year clover-grass mixtures	70

Source: Freyer (2003).

stability and the aeration of the soil through an increase of biopores (Athmann et al., 2013; Kautz et al., 2014; Köpke et al., 2015). Crop rotations without investment into root biomass result in soil compaction, decreasing soil fertility and poor growth of crops in the rotation. As a general rule, the longer the soil is covered with green shoot biomass, the higher the root biomass and density, resulting in characteristics such as more pores and stable soil aggregates which contribute to the long-term soil fertility of the soils (Freyer, 2003).

Crops influence soil aggregate development differently. Freyer (2003) has suggested the following hierarchy in the impact of crops on soil aggregates (from low to high): spring cereals, early potato, soybean < winter cereals < maize, sugar beets, grain legumes < late potato < rape < forage legumes and grass. Also catch crops can improve soil aggregation and water infiltration (see Table 10). Earthworms, fed by plant biomass and dead plant roots, are responsible for soil aggregate development (Beare et al., 1994; Sandhu et al., 2008; Six et al., 2004). High shoot biomass cover positively impacts earthworm growth and activity (Ouellet et al., 2008). Other important factors are reduced soil tillage intensity (Lofs-Holmin, 1983; Chan, 2001; Emmerling, 2001).

As the previous discussion suggests, these nutrient and broader soil health functions are fulfilled particularly by legumes (Murphy-Bokern et al., 2017). Forage legumes have long been seen as the drivers of the organic crop rotation system (Böhm, 2013; Divito and Sadras, 2014; Høgh-Jensen et al., 2002; Kouas et al., 2005; Römer and Lehne, 2004). They collect high amounts of nitrogen and carbon while consumption of fossil fuels for their establishment is low. Forage legumes fulfil key functions in organic crop rotations such as nitrogen fixation, carbon sequestration, provision of root biomass, sanitation of diverse soil-borne diseases and weed suppression (Chorley, 1981).

To achieve their potential, a 2-year cultivation of forage legumes is necessary to develop deep rooting, biomass production or weed suppression. As an example, a 2-year forage legume (red clover, desmodium, lucerne and other clover types) with highest root concentrations between 30 and 60 cm, combined with diverse grass species with high root biomass and density in the upper soil layer (−30 cm), optimizes soil health and the success of the whole crop rotation. It is important to note that forage legumes cannot be substituted with grain legumes because grain legumes perform less efficiently in terms of nitrogen fixation, humus production, weed suppression and root biomass and distribution (Brankatschk and Finkbeiner, 2014). Since grain legumes cannot fully substitute forage legumes, arable-only farms have problems in overcoming deficits in nitrogen and humus balances, as well as an increased weed pressure, which together result in a decrease in cash crops yields.

Catch crops (potentially including wild legumes) can also fulfil some of the roles performed by legumes (Baresel et al., 2003). Catch crops influence P as well as N availability, microbiological properties in the rhizosphere, soil structure and weed development (Hartwig and Ammon, 2002; Song et al., 2007). Catch crops can also have allelopathic effects on soil organisms and crops through root exudates (Rovira, 1969; Woods, 1960). Kolbe (2004) compared the effect of 13 catch crops on maize and potato. N content (2–5%) and N quantities (10–90 kg N ha^{-1} a^{-1}), identifying the benefits of crops such as winter vetch, are highlighted in earlier studies (Möller and Reents, 1999). These results underline the benefits of mixtures of legumes and non-legumes to reduce the risk of nitrogen leaching as well as provide a rich root system with positive effects on biopore development and other soil properties.

5 The role of rotation crops in suppressing weeds, diseases and pests

Weeds reduce crop yields by competing with crops for resources. Some also have allelopathic properties which can inhibit crop growth (Marinov-Serafimov, 2010). There are many ways crop rotations can be used to control weeds (Albrecht and Mattheis, 1998; Frieben and Köpke, 1994; Gabriel et al., 2006; Tscharntke et al., 2005; van Elsen, 2000). The best success is given by combining different approaches within a rotation rather than relying on a single measure (Bàrberi, 2002). Planting the same main crop to follow up is the exception and should be 'buffered' with a catch crop, in part because of the adaptation of weeds under similar production conditions. An example of a combined approach is the integration of a 2-year forage legume crop (25% of the crop rotation), the change between winter and spring crops, legumes, cereals and root crops, as well as the implementation of catch crops (cutting/mulching) and intercropping (Liebman and Dyck, 1993). The combination of crop rotations with tillage systems, different types of mechanical weed control, certain cereal varieties for example with long straw and other morphological features that increase the competitiveness of the crop, or appropriate spacing between the crops, reduces the risk of a weed population becoming established (Eisele and Köpke, 1997; Bàrberi, 2002; Blackshaw et al., 2001; Drews et al., 2009; Silva and Delate, 2017).

However, even when farmers follow best practice in organic crop rotations, so-called 'problem' weeds such as *Cirsium arvense* or *Rumex obtusifolius* can still gain a foothold (Graglia et al., 2006; Hodgson, 1958; Böhm and Verschwele, 2004). In such cases it may be necessary to interrupt the crop rotation, and move over to the use of 2-year forage legumes with early cutting or mulching, combined with tillage strategies, which raise the weed roots towards the soil surface (which can then be removed for example with a cultivator or finger weeder).

One option is to exploit the allelopathic properties of some crops which can inhibit the germination and/or growth of weeds (Watson et al., 2002; Wezel et al., 2014). There have been a number of overviews of the allelopathic effects of crops in supressing weeds (Bàrberi, 2002; Peigné et al., 2007; Verma et al., 2016). These effects are known for crops such as rye (Putnam, 1988; Yenish et al., 1995), vetch and triticale (Watson et al., 2002), sorghum (Duke et al., 2000) and rice (Olofsdotter, 2001). Glucosinolates from kale, rocket and mustard demonstrate allelopathic effects against weeds (Jiménez-Osornio and Gleissman, 1987; Angelini et al., 1998). Organic substances from brassicas also indirectly reduce weeds, soil-borne pests and diseases such as fungi, bacteria and nematodes, and have a potential role as bioherbicides (Watson et al., 2002; Batish et al., 2001; Massucati et al., 2009). It has been found that the allelopathic effects of *Brassicaceae* and *Graminaceae* are superior to those *Leguminosae* (Bàrberi, 2002). A noted example of these properties is *Desmodium triflorum* intercropped with maize which is well known for supressing *Striga* spp. as part of the 'push-and-pull system' developed by Khan et al. (2000, 2006). Chemicals in crops can also help to combat pests such as onion flies or nematodes (Wezel et al., 2014; Caamal-Maldonado et al., 2001).

Vereijken (1994) distinguishes two categories of pests and diseases (PD) with different requirements for crop rotations to prevent PD becoming established and damaging crops:

1. *High potential:* (1) PD that are crop-specific and non-mobile for example cyst nematodes or *Rhizoctonia*: low crop frequency; (2) PD that are non-crop-specific and non-mobile for example *Sclerotinia* or root knot nematodes: low crop frequency and management.
2. *Low potential:* (3) PD that are crop-specific and mobile for example *Phytophthora infestans*: specific types of crop rotation; (4) PD that are non-specific and mobile: specific types of crop rotation and other agronomic interventions.

In general, the more diverse crop rotations are, the lower the risk of PD (Rusch et al., 2010). If crops are repeated sequentially in a rotation, the risk from PD increases together with a resulting loss in yield (which can be as high as 50%). Research suggests the following hierarchy in vulnerability to such yield loss (with winter rye least and flax most vulnerable): winter rye < maize < spring barley < winter barley < winter wheat < oats < potato < sugar beet < faba bean < clover, alfalfa < pea < sunflower < flax (Freyer, 2003). However, it is important to note that this does not mean, for example, that maize can be planted with a high density in organic cropping systems, even as part of a rotation.

Soil-borne diseases are diverse and can be initiated through fungi, bacteria, nematodes, larvae of insects living in soils or on roots, leaves and stems. Each soil-borne disease follows its own life cycle and requires certain conditions to survive. The idea of systematically organized crop rotations is to avoid the development and transfer of soil-borne diseases from one crop to another. To interrupt the cycle, and prevent PD from being transferred, certain crops have to be taken out of the rotation (Table 11). Table 12 illustrates examples of crop rotations which avoid enabling the development and transfer of soil-borne diseases. Further research is necessary to better understand the mechanisms behind the 'break-crop effect' involved in disrupting the cycles of soil-borne root diseases (Watson et al., 2017).

It is important to be aware that catch crops between the main crops can serve as infection bridges in transferring infection onto the following crop (Finckh et al., 2015).

Table 11 Cropping distances in organic crop rotations in number of years

Crop	Years
Rye	0–1
Millet, hemp	1–2
Maize, soybeans	2–3
Barley, spelt, triticale, wheat	2–4
Yellow, Persian and Egyptian clover, crowtoes (*Lotus corniculatus*)	3–4
Sugar beets, forage beets, cabbages, potato	3–6
Oat, rape seed	4–5
Lucerne, red-, swedish- (*Trifolium hybridum*), Italian clover (*Trifolium incarnatum*)	4–8
Field bean (*Vicia faba*), peas, lupines	4–14[a]
Sainfoin (*Onobrychis vicifolia*)	5–6
Sunflower, flax	7–8

[a] In case the soil is already infected by fusariosis (see Kolbe, 2006).
Source: after Freyer (2003), Kahnt (1986), Könnecke (1967) and others, adapted.

Table 12 Crop rotation types (T) and examples (E)

T/E	Year 1	Year 2	Year 3	Year 4	Year 5	Year 6	Year 7	Year 8
T1	Forage legume	Winter cereal	Spring cereal CC	Grain legume	Winter cereal	Spring cereal		
E1	Clover-grass/ lucerne-grass	Wheat CC	Barley CC	Pea CC	Rye/triti-cale CC	Barley UC		
E2	Clover-grass/ lucerne-grass	Wheat UC	Oat UC	Field bean (*Vicia faba*)	Wheat CC	Barley UC		
T2	Forage legume	Winter cereal	Root crop	Grain legume	Winter cereal	Spring cereal		
E1	Clover-grass/ lucerne	Wheat UC	Potato/ maize	Field bean (*Vicia faba*)	Rye/triti-cale UL	Barley UC		
T3	Forage legume	Forage legume	Winter cereal UC	Cereal/root crop	Grain legume	Winter cereal UC	Root crop	Spring cereal
E1	Clover-grass/ lucerne-grass	Clover-grass/ lucerne-grass	Wheat CC	Barley CC/Maize UC/potato UC	Peas UC/ soybean	Rye/ spelt UC	Potato/ vegetables	Barley/ UC

Remark: Catch crops are possible before each spring cereal, potato, maize, rapeseed, grain legumes; mostly excluded after soybean, late potatoes, maize (corn production); catch crop with different amount of legumes (CC); Undersown crop (UC); / = alternative.

Another challenge is so-called legume fatigue which arises from an increased share of legumes in rotations, mainly through diverse clover types or lucerne.

Crop rotation techniques for managing PD include:

- Increasing the period of excluding a crop or one of the same crop family from the crop rotation (see Table 11)
- Use of plants which inhibit disease (e.g. *Tagetes* spp.)
- Resistant or tolerant crops
- Replacement with a crop from another crop family that has similar biological characteristics and economic potential, but is immune to the plant pest or disease

Non-crop rotation strategies include:

- Soil tillage strategies to intensify the decomposition of crop residues
- Application of mechanical techniques to collect or destroy pests, use of organic pesticides or seed treatment with compost extracts, effective microorganisms or fungi such as mycorrhiza

As noted, crop rotations can be combined with other agronomic practices such as the application of compost (Ponti et al., 2007). The higher microbiological activity induced by the compost can inhibit infection potential (Chen and Nelson, 2008; Noble and Coventry, 2005). It can also improve crop growth and health, which enhances resistance to disease, though the full effects of manures and other types of organic fertilizer are often not observed until the 2nd or 3rd year (Raviv, 2005; Suárez-Estrella et al., 2007; Bonanomi et al., 2007; Bruns, 2016). Rules for the use of compost in inhibiting PD depend on factors such as compost processing and composition (Bonanomi et al., 2010; Tilston et al., 2002). Soil tillage also influences the risk of infection (Bailey and Lazarovits, 2003). Research is divided in viewing reduced soil tillage as increasing soil health but potentially allowing weeds (esp. grasses) to develop, compared to ploughing which can reduce weed pressure but potentially transfer infectious material to lower soil levels as well as potentially damage overall soil health (Abawi and Widmer, 2000; Craven and Nel, 2016).

In the case of seriously infected soils, there is often only one radical solution to fully exclude a crop for a longer period. It may be necessary to collaborate with another farm to exchange fields to enable the production of the excluded crop in a separate crop rotation. This strategy allows continuing production of a specific crop in another system, which can be particularly important in vegetable production. The need for collaboration between farms highlights the need also for a broader landscape approach to pest and disease management. Research suggests, for example, that lower incidence of pests is associated with landscape heterogeneity (Aviron et al., 2017; Dye and Begg, 2017; Letourneau et al., 2011; Maisonhaute et al., 2017; Tscharntke et al., 2008; van Rijn et al., 2013).

6 Rotations and crop yields

The amount of available nitrogen made available by a leguminous precrop is a key factor in the yield potential of the following crop. As an example, a yield of 4 t ha^{-1} winter wheat grain with 12% protein (Table 13) can be achieved using a leguminous catch crop as precrop, while a yield of 6 t ha^{-1} requires a 18–24 month grass–clover mixture precrop (Tables 7 and 14). A combination of leguminous catch crop/main crop/follow-up catch crop has also been shown to optimize the protein content of a following winter wheat crop (Pietsch, 2006) (Table 14).

There have been a range of studies assessing the impact of legume precrops and catch crop mixtures using legumes on cereal and potato yields (Snapp et al., 1998). Since trials have been done in different environments with different crop sequences and varieties, results may be different at other sites and are not generally valid under all kinds of agro-ecological conditions (Knight, 2012).

Gruber and Thamm (2007) studied the precrop effects of blue lupin and pea, pure stands of grass–clover sown in spring and phacelia/Persian clover, on the following winter wheat yields. There was no significant impact on yields of one legume crop, but the protein content after a pea precrop was significantly higher than after lupin. An explanation for this

Table 13 The relation between wheat yield, raw protein (RP) content of wheat grain and total nitrogen content of wheat

Yield (DM t ha^{-1})	% RP	RP (t ha^{-1})	N (kg ha^{-1})
4	12	0.5	84
4	14	0.6	98
6	12	0.7	126
6	14	0.8	147

Source: Own data; DM yield × % RP/5.7 = Nitrogen; RP = raw protein.

Table 14 Potential nitrogen reserves in the shoot biomass in a combination of legume catch crops mulch and pea straw

Crop sequence	Crop	Biomass type	Yield (DM t^{-1} ha^{-1})	N content (kg t^{-1} ha^{-1})	Total (kg N ha ha^{-1})
Catch crop	Legume-non-legume mix	Mulch	2.5	29	73
Main crop	Winter/Spring peas	Straw	2	15	30
Catch crop	Pea (seeds from main crop)	Mulch	2.5	29	73
Total N accumulated in above ground biomass					176

Source: After Fritsch (2012).

difference might be the faster mineralization of pea residues and their earlier harvest in comparison to the other legumes. In contrast, Wichmann (2003) found that pea increased the yield of the following winter wheat while blue lupin produced the highest wheat protein content. Trials by Dreymann (2005) on different grass–clover management strategies did not result in significantly higher wheat yields or quality.

Whilst it had no effect on the first winter wheat crop, a 2-year alfalfa grass rotation increased rye grain yield, protein content and shoot N uptake in the second following crop (de Kruijff et al., 2008). In a trial in the same region, the effect of a 2-year mulched alfalfa in comparison to a cutting system showed a 13% higher wheat yield and improved protein content (Surböck et al., 2015). In both these trials, soil water and rainfall distribution had a significant impact on the effects of precrops on following crops. Other trials, however, suggest that mulching may be less effective than cutting in precrop management (Hatch et al., 2007). Trials by Stopes et al. (1996) gave higher winter wheat yields after a 1-year green manure using red clover compared with white clover, trefoil and ryegrass control. A second year of green manure showed no significant effect on wheat yield.

There have been similarly mixed results from research groups studying the impact of legume precrops on potato performance. Different studies have highlighted the superior performance of faba bean over a 2-year grass-red clover mixture (Paffrath, 2005); 1-year grass–clover over 2-year pea (Möller, 2001); and pea with an intercropped oil radish over 1-year grass–clover (Haase et al., 2007). In a trial by Dresow et al. (2013) clover grass (cut/mulch) did not significantly improve potato yields or tuber quality parameters. Another study also found that precrops of lucerne, spring pea and spring barley did not improve potato yields or tuber starch content (Nyamdavaa and Friedel, 2016). However, a study by Paffrath (2003) found that a mulched grass–clover, and a bush bean (*Phaseolus vulgaris*) precrop combined with a phacelia intercrop, led to significantly higher yields in the following crop compared with faba bean or pea precrops. Schneider et al. (2012) compared trials with crop rotations known from mixed farms with those from stockless farms. They found a 2-year clover-based crop rotation was the most successful in yield of the following crop in comparison to 1-year clover or with grain legumes only. These diverse results demonstrate that site- and farm-specific conditions, and the diversity of growth and soil–water conditions, makes it hard to generalize results. However, it has been found that, on average, forage legume-based organic cropping systems in the long run perform better than other types of rotation in terms of cash crop yield and quality (Nkurunziza et al., 2017).

Mayer and Mäder (2016) determined average long-term organic yields relative to conventional crop yields: wheat (70%), maize (82%), potato (75%), soybean (92%), grass–clover and lucerne (87%). Meta-analyses by other authors broadly confirm these results (De Ponti et al., 2012; Ponisio et al., 2015; Seufert et al., 2012). Nitrogen limitation, that is, the amount and availability of nitrogen when needed by crops, is the main factor in the yield gap between organic and non-organic systems (Berry et al., 2002). Other reasons for low yields in organic systems in practice include the level of expertise, management and resources needed to manage a number of crops in a rotation, though many organic farmers focus on long-term sustainability rather than maximum yield per hectare (Freyer, 2016). One option to improve yields is biogas slurry (Grass, 2016; Möller et al., 2008; Terhoeven-Urselmans et al., 2009), which can also compensate for nitrogen deficits as a result of the low nitrogen fixation rates of leguminous precrops (Kirchmann and Witter, 1992; Blagodatskaya and Kuzyakov, 2008; Kuzyakov et al., 2000).

7 Designing a crop rotation

Planning crop rotations requires both long- and short-term approaches. In the long term, it is more practical to plan first around a crop group (crop families or crops with similar characteristics). This provides a foundation for fine-tuning annual crop rotations in the light of production and market conditions (Freyer, 2003; Kolbe, 2006) (see Table 15). The same approach holds for catch crops. Long-term planning involves identifying the general opportunities for catch crop integration, intercropping and crop selection. Short-term planning involves decisions such as whether to undersow, whether the crop should be sown following the harvest of the main crop, the selection of cultivars, soil tillage system and fertilization.

Table 15 Recommended precrop-crop follow-ups

Precrop (examples)	Crop (examples)	Remarks
Clover, lucerne, desmodium, pigeon pea (*Cajanus cajan*), lablab (*Dolichos lablab*), velvet bean (*Mucuna* spp.), *Sesbania sesban*	Wheat, maize, sorghum, potatoes	High availability of nitrogen
Peas, soybeans (partly intercropped with maize), groundnut (*Arachis hypogaea*), lupine; under biannual rainfall conditions (tropical and subtropical climates): lablab or cow pea	Wheat, rye, barley, teff, finger millet, maize, sorghum	Added with legume-non-legume mixed catch crop in temperate zones
Field beans (*Vicia faba*) (in (semi-)humid regions often also intercropped with maize), cow pea	Diverse winter cereals, maize	High availability of nitrogen
Potatoes	Wheat, rye, triticale, vegetables	Crumb structure in upper soil layer, well aggregated; high N concentration in potato roots[a]
Winter cereals	Spring cereals, oil crops, potato	With legume-non-legume mixed catch crops in between

[a] The total N of root biomass is relatively low, however the C/N is tight and mineralization of the roots is fast; together with harvest measures, temperatures above the freezing mark and rainfall, there is risk of nitrogen leaching in temperate climates (Askegaard et al., 2011).
Source: own data; Snapp et al. (1998).

Crop rotations can be defined in terms of:

- the length of crop rotations
- temporal distribution of crops in one field (time)
- the distribution of crops in one field (area)
- the spatial arrangement of crops, shrubs and trees (space) in a landscape.

This section first discusses the conversion period and then reviews these dimensions of crop rotation.

7.1 The conversion period

When farmers convert to organic, the crop rotation is the means to reactivate the functionality of ecosystems to support sustainable crop production. The conversion period to organic farming is therefore a specific phase in crop rotation development. Farmers have to reset their practices and to reorganize the distribution of crops over time (rotation in and over years), space (planting approach) and fields (total farm area, landscape), adding new crops while excluding others. While a considerable body of knowledge is available to guide the planning process, it is critical to take account of farm-specific conditions in planning and implementing crop rotations.

The first step is the analysis of the previous 4–6 years of field-specific rotations. This allows an assessment of current and expected soil fertility (humus and nutrient status), related risks of soil-borne diseases, and the potential weed pressure in the conversion period. The crop rotation in the conversion period has to react to these conditions. The conversion period involves three crop rotations over a 3-year period to build up the share of forage legumes to 25% (Table 16). This leads to a positive humus and nitrogen balance in the soil in rebuilding soil fertility. It is possible that, in the first years, strategies I and II could be more profitable than III, due to the fact that more land can be cultivated with cash crops. However, strategy III, involving a high investment in forage legumes in the first 3 years, will have a positive long-term impact on the development of the whole crop rotation. The overall plan for rotations can be followed up with more detailed management activities (e.g. a 1-year mixture of specific deep-rooting legumes and non-legumes combined with a mechanical loosening of the soil in different layers), with the aim of repairing the accumulated problems from former conventional farming (e.g. soil compaction, low humus content, risks from soil-borne diseases or weeds due

Table 16 Forage legume strategies (S) for entering into the organic crop rotation aiming for a share of 25% forage legumes in the 3rd year

	Year			Nitrogen supply by legumes	Soil organism activity	Weed control	Profitability in the first 3 years	Profitability after first 3 years
	1	2	3					
S	% Forage legumes							
I	25	25	25	++	++	++	++	++
II	10	15	25	+	+	+	+++	+
III	50	35	25	+++	+++	+++	+	+++

Source: own data; S = strategy; +=low; ++= medium; +++= high.

to narrow crop sequences of the same plant family over successive years). In addition to the main crop rotation, the share of legume-based catch crops ('catch crop rotation') is also important for the regeneration of soil fertility. Catch crops are also of specific relevance in mixed farms with animals to cover the forage demand at the end of the season.

7.2 Length of crop rotations

The total length of most crop rotations is between 3 and 12 years, though the latter is exceptional (Sattler and von Wistinghausen, 1985) (Table 12). In practice, a crop sequence of three or four crops starts with a legume followed by cash crops, including one or two catch crops. Without a 2-year clover/lucerne crop in a maximum 8-year rotation (25% in a rotation), soil fertility remains weak, the humus content decreases (Kasper et al., 2015; Surböck et al., 2013), nitrogen content is low, the risk of weed development increases (Nkurunziza et al., 2017) and biophysical soil functions, such as water-holding capacity and water drainage, are poor.

In practice, many farms use less than 15% forage legumes in a crop rotation, and instead use a higher proportion of grain legumes. However, in the long run, a grain legume cannot substitute the performance of forage legumes in a crop rotation. The same holds for compost, which also cannot replicate the functions of forage legumes rooting system. The less forage legumes are part of the rotation, and the longer the rotation, the more legume-based catch crops between winter and spring crops have to be added, but these can only partly compensate the lack of forage legumes.

7.3 Sequence of crop rotations (time)

There are some general rules for sequences of crops in a rotation (Table 12). These are:

- Nitrogen-demanding non-legume crops follow nitrogen-fixing crops
- Humus-demanding follow humus-producing crops
- Spring crops follow a sequence of winter crops and catch crop
- Weed-sensitive crops or crops supporting weed growth follow weed-reducing crops
- Crops that are potentially vulnerable to soil-borne diseases follow those that are resistant to soil-borne disease

These rules should be followed to ensure fertile soils, nitrogen supply, healthy crops, low weed, pest and disease pressure and, finally, sustainable yields. Temporary limited exceptions are possible without negative consequences for crop performance and soil quality, if the crop is supported in other ways, for example through tillage, additional farmyard manure or a catch crop.

7.4 Distribution of crops in one field (area)

There are several strategies to distribute crops in one field at the same time or in sequence (Willey, 1979). Typical arrangements for main crops are:

- *Arable production:* intercropping of two main crops in rows: for example maize with beans; oat with peas

- *Vegetable production:* inter- or mixed cropping, multi-species or double cropping (polyculture) of different vegetables
- *Grassland:* Permanent grassland or the combination of forage legumes with different grasses in arable land (ley)

Arrangements for catch crops include:

- Undersown crop (inclusive relay cropping): A crop undersown in a main crop (e.g. clover in cereals); *Sinapis* sp. under field beans (*Vicia faba*); field beans under potatoes (relay cropping); lablab under maize
- Stubble catch crop: Mainly a mixture of crops sown after harvesting the main crop in a sequence: non-legumes/legumes with non-legumes/legumes

Other forms include:

- Sequential cropping, which is growing two or more crops in a sequence in 1 year (double cropping). This is mostly practised under irrigation, with rice or in systems including vegetables, and is combined with minimum tillage. In these systems nitrogen-fixing and humus-enriching crops are combined with nutrient-demanding crops
- Relay cropping, which is a form of intercropping involving growing two or more crops in a sequence. This involves planting the succeeding crop after flowering but before harvesting of the preceding crop
- Ratoon cropping, which is cultivation of an additional crop from the re-growth of stubbles of a main crop after its harvest, thereby avoiding replanting; as in sugar cane, sorghum and rice cultivation
- Mixed cropping where two or more crops are grown simultaneously in the same field without rows

The aims of these diverse crop arrangements are to optimize the use of resources (e.g. water and nutrients), to reduce the risk of erosion, to increase nutrient uptake and to reduce leaching of nitrogen during the winter period (Reents and Möller, 1999), to suppress and reduce weeds, to reduce pest and disease pressure, to offer habitats for pest predators, to support crops that lodge (e.g. peas, lentils) and to increase the anti-phytopathogenic potential of the soil. Their management involves decisions about factors such as sowing date, length of vegetation period, mulching or mixing organic matter into the soils. This requires complex management (decisions) and is dependent on the residues of the former crop, the soil status, the needs and demands of the following crop, and the weather conditions.

7.5 Spatial arrangements of crop fields within landscapes (space)

Crop rotations have a broader role/function within landscapes (Malézieux et al., 2009):

- Pest and disease control: This includes managing fields to interrupt the ability of PD to move from field to field; as well as distribution of hedges and species-rich grass/herb strips in between arable fields to increase beneficials
- Leguminous agroforestry/alley cropping systems in and in between arable land: These have a multifunctional role in such areas as erosion control and prevention of contamination via pesticides from neighbouring farms

- Recreation and aesthetics: Rotations may contribute to the aesthetic quality of the landscape and help to conserve local characteristics which for example promote tourism or other types of recreation

Fulfilling these functions is often not a decision of just one farm. A landscape approach integrating all farms is necessary that allows planning beyond a single field.

8 Measuring and modelling crop rotations, nutrient and humus balances

As the previous sections suggest, rotations are complex to plan, manage and monitor. As a result there have been attempts to develop models to help farmers making decisions in such areas as calculating nutrient and humus balances which then help to decide on additional fertilizer requirements (Bachinger and Zander, 2007; Dogliotti et al., 2003; Doltra et al., 2011; Janssen et al., 2010; Müller et al., 2006). They have also been used to assess the environmental impact and sustainability of crop rotations (Sadok et al., 2009; Van Ittersum et al., 2008).

Several software-based models have been developed to help in designing and assessing crop rotations (Dury et al., 2012; Olesen et al., 1999a, 2002; Rossing et al., 1997; Stöckle et al., 2003; Vereijken et al., 1995; Wijnands, 1999). Some of these models are limited in terms of the range of crops or the consistency of their assumptions with organic principles (e.g. in terms of permitted inputs) (Castellazzi et al., 2008; Keating et al., 2003; McCown et al., 1996; Salmon-Monviola et al., 2012; Schönhart et al., 2011). A recent model which addresses these problems is that of Reckling et al. (2016b). Their model makes it possible to assess and systematically compare cropping systems, including the legume family, showing environmental impacts and economic impacts. They highlight that one of the challenges is the need for large amounts of data and databanks that are sensitive to the diversity of organic farms.

One potential application of models is calculating nutrient balances. In organic farming the nutrient cycle is nearly closed which means that losses are minimized and nitrogen is provided mainly through nitrogen fixation, while over the years the stock of macro-elements gradually decreases. The cycle starts with the uptake of nutrients through crops from the site-specific nutrient stock of soils. Mineralization processes of the rocks, the organic matter, that is, crop residues, nitrogen fixation by legumes and free-living soil microorganisms, and some nutrients from airborne sources together represent the reservoir of nutrient sources.

Nutrients leave this cycle through gaseous forms (C, N), leaching, cash crops and animal products. Stockless farms, where forage legumes may sometimes be sold, export a significant amount of nitrogen and other nutrients. As this suggests, the export of nutrients depends on the farm type. The nutrient export is increasing following the suggested hierarchy: cattle permanent grassland systems < mixed farming systems < arable farming systems < vegetable farming systems. In mixed arable-livestock farms for example, approximately 50–70% of nitrogen, and 80–95% of all other macro- and micronutrients of forage crops will be recycled through different kinds of animal manures (Freyer and Pericin, 1996).

Organic farms need to compensate for nutrient export to avoid nutrient deficits. Nitrogen export can be balanced via nitrogen-fixing crops. A crop rotation with a 25% forage legume component with additional leguminous catch crops will meet this need. However, there is also a need to ensure a balance in other nutrients that crops require. As an example, phosphorus and potassium may need to be supplemented by mineral sources from outside the farm, preferably through organic matter (e.g. all kind of bio-composts that are accepted by organic guidelines).

Calculating nutrient balances helps to assess the nutrient status of the farm as a whole and the overall effectiveness of crop rotations in particular (Bachinger and Zander, 2007; Freyer and Pericin, 1996; Watson et al., 2002). Field balances can be used to assess the nutrient input–output of each field, including the estimation of nitrogen fixation, and can help in detailed planning and monitoring of the cropping system and fertilizer management. Since data may be limited and assessing balances an estimate, it is essential to combine field and farm-gate nutrient balance calculations.

Estimations of the need for fertilizers in mainstream agriculture are based on the export of nutrients from the field. An additional amount depends on the availability of nutrients in soils, identified via chemical soil analysis (availability of nutrients), the soil and climatic conditions, and the expected yield of the following crops. In the organic system the calculation of fertilizer requires three different steps:

- Nutrients made available by precrops through their root activity, the decomposition of crop residues, soil biological activity, soil chemical (e.g. pH, cation exchange capacity of the soil etc.) and soil physical parameters (e.g. soil texture, soil structure) and the overall nutrient stock of the soil accessible by roots
- Availability of the farm's own animal manure, the form (compost or liquid) and the availability of nutrients to meet crop requirements
- Additional fertilizers purchased from outside the farm

Soil-specific data and investigations can also be used to assess the fertility of the soil, entailing for example the spade diagnosis (soil aggregates, biopores and earthworms), root growth and nodulation of legume roots via rhizobia (Ball and Douglas, 2003; Munkholm, 2000), or pH analysis.

Yield measurement of the PCs (total biomass), crop health status, weed development, intensity of soil tillage, manures and fertilizers given to the PCs (one rotation) provide information for a nutrient balance calculation and the long-term nutrient status at specific field level, which can be used to assess the crop rotation system. Together with assessment of the humus balance, a decision can be made, whether additional nutrients should be provided via organic matter or mineral fertilizers purchased from outside the farm.

Reasons for increases in negative nutrient balances on organic farms include:

- nutrient losses via erosion as a result of a low share of forage legumes, limited soil cover and/or lack of catch crop
- nitrogen leaching through poor forage legume management (Korsaeth and Eltun, 2000; Stopes et al., 2002) or after potato (Haas et al., 2002)
- significant export of nutrients including nitrogen via intense cash cropping and, in particular, through intensive vegetable cultivation (Goulding et al., 2008)

Phosphorus, potassium, sulphur and other macro- and micronutrient deficiencies and poor soil pH (Vargas and Graham, 1988) can significantly reduce yields and, in the case of legumes, to low nitrogen fixation rates (Fischinger et al., 2011). While nitrogen deficits can be addressed by a higher share of legumes in a crop rotation, other nutrient deficits may require importing external inputs (Divito and Sadras, 2014; Foissy et al., 2013; Schulz et al., 2014). In assessing the use of external inputs, it is important to take account of the following:

- the extent to which imported fertilizers can balance the nutrient deficit
- the way different nutrients react to one another (nutrient sensitivity)
- the extent to which importing nutrients contributes to balance a carbon deficit
- the overall environmental impact

Six options for coping with nutrient deficits are discussed in Table 17. Strategy (1) is only sustainable if the supplying farms show a positive nutrient and carbon balance. However, this is rarely the case, except for nitrogen and carbon. Strategies (2), (3) and (4) balance phosphorus and potassium as well as offering additional nitrogen and other nutrients. Long-term research trials have shown that the additional input of certified bio-compost is able to balance a nutrient deficit and can lead to higher yields of cash crops (Stinner et al., 2008). However, the limited nitrogen availability of these composts results in low wheat protein contents (Surböck et al., 2017). Low nutrient sensitivity in organic fertilizers (in the relationship of P and K) is a problem in organic vegetable production. Most organic fertilizers are not adapted to the specific nutrient demands of vegetable crops (Tittatelli et al., 2016; Zikeli et al., 2016).

Strategies (3) to (5) are promising because they close the nutrient cycle but could be criticized since nutrients may come from less environmentally friendly sources. Mineral fertilizers, for example sulphur-based potassium fertilizer or rock phosphate, are an alternative if organic matter is not available. However, rock phosphate production can also be criticized as less environmentally friendly in the way it is produced and as a limited resource in the long run. Animal residues may be problematic where animal welfare is a concern or because of potential health risks.

One option is to mine nutrients from deeper soils via deep-rooting systems. Though valuable, the nutrient amount, however, is limited (e.g. 0–2 kg ha^{-1} P, 1–8 kg ha^{-1} K) (Zhang and DePaolo, 2017). Leguminous alley cropping can also collect nutrients with their deep-rooting system and contribute to nutrient accumulation via composting tree and shrub branches (Fageria, 2007). In the long-term, one option might be 'bioregions', where most farms and food consumption is organic, allowing recirculation of organic matter to help fill the gap created by nutrient export from the farms and to close the nutrient cycle.

A central goal of organic crop rotations and organic matter management is also to achieve positive humus balances to maintain and increase soil fertility, optimal crop growth and quality (Brock et al., 2012; Deike et al., 2008; Hodges, 1991). Besides long-term soil analysis to measure soil organic matter development, humus balancing models can help to assess humus requirements as a result of crop rotations and organic matter supply. A range of models, varying in terms of data demand and use in the field, are discussed by Brock (2016). Results highlight the superior performance of organic over conventional farms, due to a higher share of forage legumes, catch crops and use of farmyard manure (Kasper et al., 2015). Calculations done on 29 farms found that those

Table 17 Strategies for nutrient management

Strategy	Description	(a) Environmental impact	(b) Compensation of nutrient deficit/nutrient sensitivity	(c) Compensation of carbon deficit
1	Import of farmyard manure and slurry from other organic farms	Transfer of surplus from other organic farms only to avoid a deficit there	✓/no	✓
2	Import of bio-compost from villages and urban garden areas	Closing the cycle; can negatively affect the communal and urban garden nutrient and carbon cycle	✓/no	✓
3	Import of regional industrial produced crop-based residues from food industries (e.g. residues from breweries, vinasse, residues of potato starch production . . .)	Closing the cycle; residues should be re-transferred to barley and hop producing farms, potato producing farms; mostly too expensive for arable farms, only viable for vegetable producers	✓/partly	✓
4	Import of industrial animal-based organic fertilizers/residues (e.g. horn and blood meal)	Closing the cycle; should be re-transferred to animal farms; mostly too expensive for arable farms, only viable for vegetable producers	✓/partly but often insufficient	✓
5	Import of urban kitchen waste	Closing the rural-urban nutrient cycle; critical because most of the organic matter is a mixture of food residues from mainly mainstream farming	✓/no	✓
6	Import of mineral fertilizers	Energy intensive, environmentally critical connected to mining activities	✓/no	no

Source: Own data.

combining forage production with grassland had the highest positive humus balances (219–890 kg C ha^{-1} year^{-1}), suggesting that cropping systems with more than 20% forage legumes, use of catch crops and farmyard manure guarantee a positive humus balance (Haas et al., 2001). Other systems with positive humus balances include organic rice cultivation (Sihi et al., 2017) and agroforestry systems (Fassbender, 1998; Abbas et al., 2017).

9 Crop rotations on stockless farms

A key issue in organic farming is the development of more and more stockless farms (Cormack, 1999). Not including animals raises the issue of how to reorganize crop rotations without the need for forage but with the requirement to compensate for the absence of on-farm animal manure (Schmidt, 2004). Two options to address this issue are:

- legume-based crop rotation strategies with different shares of forage, grain legumes and leguminous catch crops
- ways to use forage legumes shoot residues efficiently as a nutrient source

9.1 Legume-based crop rotation strategies

There are various potential crop combinations, depending on the share of forage legumes and percentage of grain legumes and catch crops.

1 Forage legume-based crop rotations:
  a Forage legumes only
  b Forage legumes with catch crops
  c Forage legumes with grain legumes
  d Forage legumes with grain legumes and catch crops
2 Crop rotations without forage legumes as the main crop:
  e Grain legumes
  f Leguminous catch crops
  g Grain legumes and leguminous catch crops

Currently there is no systematic research identifying which percentage of forage legumes in combination with grain legumes and catch crops is most beneficial for stockless farms. Some trials have suggested that crop rotation systems with less than 15% forage legumes, pure grain legume or catch crop systems often lead to negative humus balances and nitrogen deficits (Schulz et al., 2014). Høgh-Jensen (1999) suggests that a threshold of about 11% forage legumes is critical, because of the long-term decrease of soil organic matter. All forage legume-based cropping systems profit from the multipurpose functions of clover or lucerne in combination with grasses and a longer soil rest. A share of 25% in the crop rotation (Kasper et al., 2015), with a 2-year forage legume, will meet humus and nitrogen demand (a). Below this threshold, catch crops and grain legumes need to be added ((b)–(d)). Chongtham et al. (2017) has identified three different crop rotation strategies, from a 'strict' approach which adheres to a fixed rotation, to 'flexible', which adapts from season to season within broad leys and crop rotation rules, and, finally, a 'liberal' approach which may depart from standard crop rotation principles. Farmers' trade-offs in stockless farms are between perennial forage legumes to provide nitrogen for the crop rotation as well as weed control, on the one hand, and, on the other, the use of more inputs such as purchased nutrients and mechanical weed control in cash crop production. In livestock farms, farmers need to balance between forage production, weed control, the cost of re-seeding leys and decisions about whether to grow crops to feed animals or to sell.

Research suggests that short-term leguminous green manure catch crops have the potential to cover the crop rotation requirement for nitrogen (Bergkvist, 2003; Mueller and Thorup-Kristensen, 2001; Poutala et al., 1994; Thorsted et al., 2006). However, these studies do not show whether these crops alone can meet humus and nitrogen requirements in the long run (Kasper et al., 2015) or replicate the functions of deep-rooting crops in accessing nutrients from deeper soil layers (Köpke, 2016). An additional challenge is dealing with weed pressure (Nkurunziza et al., 2017; Schulz and Leithold, 2014). Trials suggest that, on average, grain legume-based crop rotations lead to considerably lower yields of cash crops in comparison to legume-based crop rotations (Schmidt et al., 2006, 2012). Comparisons of stockless farms with those keeping livestock demonstrate the higher productivity of mixed systems (Reckling et al., 2016a).

9.2 Management of forage legume biomass in stockless farms

As noted earlier, a key issue for stockless farms is how to optimize use of forage legume biomass if there is no forage demand by animals. Options include:

1 Field distribution of organic matter
   a Mulching of biomass *in situ*
   b Mulching and incorporating biomass *in situ*
   c Cutting and transferring the biomass to other fields as a green manure (,'*cut and carry*' system strategy A)
2 Farm internal distribution and recirculation of organic matter
   d Cutting and transferring the biomass to a farm's own biogas plant and recirculating the slurry and compost ('*cut and carry*' system strategy B)
   e Cutting and transferring the biomass for producing silage and recirculating to another field ('*cut and carry*' system strategy B)
   f Cutting and producing pellets as a fertilizer (e.g. for vegetable production) ('*cut and carry*' system strategy B)
   g Cutting the second growth for legume seed production while mulching the biomass
3 Farm-to-farm collaboration and recirculation of organic matter
   h Cutting and transferring the biomass as forage (also partly with straw) or pellets to partner farms with animals, recycling slurries and farmyard manures
4 Export of organic matter
   i Cutting and selling the biomass as forage (hay, pellets) or fertilizer (pellets) without recycling the organic matter
   j Cutting the second growth for legume seed production for own purpose or selling

Strategy (a) and (b) run the risk of nitrogen leaching (Schmidt, 2004) and increased greenhouse gas emissions from mulched material (Kulozik-Erben, 2007). Strategies (a), (b) and (c) have similar effects on the nutrient and humus balances at farm level. In strategy (c), other fields profit from the organic matter, but not from the multifunctional qualities of the forage legume growth that is, the rooting system (Stumm and Köpke, 2015; Weller, 2011). This strategy also reduces the risk of leaching. Strategy (c) is specifically of interest for vegetable-producing farms (Stumm and Köpke, 2015).

Strategies (d), (e) and (f) allow the farmer more flexibility in timing fertilization, compared to strategy (c) or (g), and reduce weed seeding. The use of biogas slurry (strategy d) can lead to higher yields (Brozyna et al., 2013; Serdjuk et al., 2017). Strategy (g) is nearly similar to strategies of subgroup (1), but contributes additional product for farm use.

Strategy (h) simulates the farmyard manure of a mixed farm. Applying slurry generates a significant increase in cash crop yields (Stumm and Köpke, 2015). Strategy (i) initially appears to be of economic interest; however, it can lead to serious losses of nutrients and carbon, that is, negative balances of nutrients, and finally lower yields (Brozyna et al., 2013). In case of strategy (j) both the biological and the economic advantages of forage legume cultivation are combined. In this context Råberg et al. (2017) compared the redistribution of ensiled and digested green manure via biogas with *in situ* incorporation on crop yields. The results do not favour either approach but show that strategic biomass redistribution has the potential to enhance biomass productivity while maintaining food crop yields, thereby enhancing whole system productivity and reducing risk of nutrient loss.

The different strategies have broader ecological side effects that should be part of the assessment of forage-based organic matter strategies. In trials by Larsson et al. (1998), they estimated losses of N_2O-N ha^{-1} emissions in a 3-month period were 6 kg ha^{-1} for grass, 13 kg ha^{-1} for alfalfa and 0.2 kg ha^{-1} for bare soil. Rochette et al. (2004) measured average N_2O emissions from legumes, with 1.0 kg N ha^{-1} for annual crops, 1.8 kg N ha^{-1} for pure forage crops and 0.4 kg N ha^{-1} for grass legume mixes. In another trial emissions associated with alfalfa (0.67–1.45 kg N ha^{-1}) and soybean (0.46–3.08 kg N ha^{-1}) production were smaller than those predicted using the emission coefficient proposed for the national inventory of greenhouse gases (alfalfa = 1.60–5.21 kg N ha^{-1}; soybean = 2.76–4.97 kg N ha^{-1}) (Rochette and Janzen, 2005). Trials of Möller and Stinner (2009), comparing a mulching system with the transformation of clover grass via the biogas plant, resulted in reduced N_2O emissions for the latter. Migliorati et al. (2015) identified 50% lower N_2O emissions in a rotation of sorghum after legumes, compared with sorghum after a grass pasture, and suggest that C supplied by crop residues can play a major role in stimulating denitrification.

It can be concluded that there are manifold management strategies for forage legumes in stockless farms, with currently no clear advantage of the one over the others, due to many factors that influence nutrient efficiency environmental impact. More research is necessary for identifying the best site and farm-specific approaches, both in terms of productivity and environmental impact (Köpke, 2016).

10 Conclusion and future trends

Over the past decades crop rotation research has made significant progress. Nevertheless, research on specific crop rotation systems under organic conditions is still rare. Whilst general rules governing crop rotation have been established, we are still some way from understanding the full complexity of interactions between crops and agronomic interventions. Many questions remain unresolved, for example about the impact of crops on soil organic matter, microorganism communities, gaseous loss of nitrogen, mineralization of crop residues, the availability of minerals from catch crops to the following crops, the impact of different crop residues on water-holding capacity of the soils and the impact of agronomic interventions on mineralization processes, as well as the dynamics of soil-borne diseases.

Other questions include the impact of root exudates on soil-borne diseases, that is, the antiphytopathogenic potential of different crops (Bailey and Lazarovits, 2003; Bruns, 2016).Knowledge about root biomass and root distribution, the relation between shoot and root biomass, specifically of forage legumes, and the potential to influence root characteristics through crop breeding remains limited and more emphasis needs to be given to root systems in general ('root rotations') (Watson et al., 2017).

Furthermore, we still do not have sufficient trial data about direct and indirect precrop effects (Mayer and Mäder, 2016; Paulsen et al., 2016). Research on these effects gives inconsistent and sometimes conflicting results (Casagrande et al., 2009). These results underline the need for precise and comprehensive trial documentation, a detailed protocol to take account of the many agronomic factors, as well as information about the genotype-specific harvest index and root-specific characteristics, in order to allow comparison between trials (Paulsen et al., 2016). Only with this information can crop rotation rules be properly identified, verified, applied and, where necessary, adapted to local conditions. Long-term on-farm and on-station trials are needed as well as a holistic approach of studying farms as a whole.

As well as legumes, non-legume precrop effects should be studied in detail in order to optimize crop rotations, including use of catch crops, intercropping and organic matter management. Different strategies for use of mulch material versus silage, slurry and compost of biogas plants need further investigation. There is also need for more research on combined systems of crop rotations and agroforestry/alley systems, specifically in poor soil conditions and more extreme weather conditions, where drought, flood and erosion events are increasing. Diversified and healthy diets and diversification of crop rotations, or 'edible' landscapes are other topics that need more attention.

Finally, a less discussed issue of organic crop rotations is the aesthetic dimension that might contribute to the IFOAM Principle of Health (Hauck et al., 2014). There is currently little discussion of crop rotations within the context of landscapes aesthetics, but this could become an issue within the context of biodiversity and agro-tourism, or as an ethical dimension in the overall discourse on sustainability and multifunctionality of land use systems.

11 Where to look for further information

Further literature on crop rotations can be studied with the book from Mohler and Johnson (2009) and the proceedings of a workshop on cropping systems held in 1999 in Denmark (Olesen et al., 1999b). Other sources are a book summarizing crop rotations in conventional, integrated and organic farming systems (Freyer, 2003) and a review from Mayer and Mäder (2016) with focus on yield development in long-term trials, both in German. The latter topic is further presented in a comparison of yields between organic and conventional farming (Schrama et al., 2018) and critically commented by van de Ven et al. (2018). Latest publication from Delate et al. (2017) informs about organic long-term trials in the United States. Furthermore there are efforts at the European level to initiate a COST-Action on long-term trials in organic farming initiated by ITAB.[1]

12 References and further reading

Abawi, G. and Widmer, T. 2000. Impact of soil health management practices on soilborne pathogens, nematodes and root diseases of vegetable crops. *Applied Soil Ecology*, 15(1), 37–47.

1. http://www.acta.asso.fr/le-reseau/acces-par-institut/detail-institut/i/institut/detail/itab/print.html

Abbas, F., Hammad, H. M., Fahad, S., Cerdà, A., Rizwan, M., Farhad, W., . . . Bakhat, H. F. 2017. Agroforestry: A sustainable environmental practice for carbon sequestration under the climate change scenarios – a review. *Environmental Science and Pollution Research*, 24(12), 11177–91.

Adamtey, N., Musyoka, M. W., Zundel, C., Cobo, J. G., Karanja, E., Fiaboe, K. K., . . . Berset, E. 2016. Productivity, profitability and partial nutrient balance in maize-based conventional and organic farming systems in Kenya. *Agriculture, Ecosystems & Environment*, 235, 61–79.

Akinnifesi, F., Chirwa, P., Ajayi, O., Sileshi, G., Matakala, P., Kwesiga, F., . . . Makumba, W. 2008. Contributions of agroforestry research to livelihood of smallholder farmers in Southern Africa: 1. Taking stock of the adaptation, adoption and impact of fertilizer tree options. *Agricultural Journal*, 3(1), 58–75.

Albrecht, H. and Mattheis, A. 1998. The effects of organic and integrated farming on rare arable weeds on the Forschungsverbund Agrarökosysteme München (FAM) research station in southern Bavaria. *Biological Conservation*, 86(3), 347–56.

Anderson, R. L. 2005. A multi-tactic approach to manage weed population dynamics in crop rotations. *Agronomy Journal*, 97(6), 1579–83.

Anderson, S., Gündel, S., Pound, B. and Triomphe, B. 2001. *Cover Crops in Smallholder Agriculture: Lessons from Latin America*. ITDG Publishing, London.

Andow, D. 1991. Yield loss to arthropods in vegetationally diverse agroecosystems. *Environmental Entomology*, 20(5), 1228–35.

Angelini, L. G., Lazzeri, L., Galletti, S., Cozzani, A. I., Macchia, M. and Palmieri, S. 1998. Antigerminative activity of three glucosinolate-derived products generated by mirosinase hydrolysis. *Seed Science and Technology*, 26(3), 771–9.

Anglade, J., Billen, G. and Garnier, J. 2015. Relationships for estimating N2 fixation in legumes: Incidence for N balance of legume-based cropping systems in Europe. *Ecosphere*, 6(3), 1–24.

Askegaard, M., Olesen, J. E. and Kristensen, K. 2005. Nitrate leaching from organic arable crop rotations: Effects of location, manure and catch crop. *Soil Use and Management*, 21(2), 181–8.

Askegaard, M., Olesen, J. E., Rasmussen, I. A. and Kristensen, K. 2011. Nitrate leaching from organic arable crop rotations is mostly determined by autumn field management. *Agriculture, Ecosystems & Environment*, 142(3), 149–60.

Athmann, M., Kautz, T., Pude, R. and Köpke, U. 2013. Root growth in biopores – evaluation with in situ endoscopy. *Plant and Soil*, 371(1–2), 179–90.

Aviron, S., Monteiro, A., Baudry, J., Djoudi, E. A., Pétillon, J., Plantegenest, M. and Puech, C. 2017. Effects of organic farming at different spatial scales on natural enemies of crop pests and pest predation levels. Dundee (Scotland, UK), 29–31 March 2017, IOBC-WPRS Bulletin 122, pp. 74–8.

Bachinger, J. and Zander, P. 2007. ROTOR, a tool for generating and evaluating crop rotations for organic farming systems. *European Journal of Agronomy*, 26(2), 130–43.

Baddeley, J., Jones, S., Topp, C., Watson, C., Helming, J. and Stoddard, F. 2013. Biological nitrogen fixation (BNF) by legume crops in Europe. Legume Futures Report 1.5.

Baeumer, K. 1992. *Allgemeiner Pflanzenbau*. Stuttgart: Eugen Ulmer.

Bailey, K. and Lazarovits, G. 2003. Suppressing soil-borne diseases with residue management and organic amendments. *Soil and Tillage Research*, 72(2), 169–80.

Ball, B. and Douglas, J. 2003. A simple procedure for assessing soil structural, rooting and surface conditions. *Soil Use and Management*, 19(1), 50–6.

Ball, B., Bingham, I., Rees, R., Watson, C. and Litterick, A. 2005. The role of crop rotations in determining soil structure and crop growth conditions. *Canadian Journal of Soil Science*, 85(5), 557–77.

Bàrberi, P. 2002. Weed management in organic agriculture: Are we addressing the right issues? *Weed Research*, 42(3), 177–93.

Baresel, J., Reents, H. and Schenkel, W. 2003. Screening alternativer Leguminosenarten auf ihre Eignung zur Gründüngung und zum Mischanbau im Ökologischen Landbau. http://org.eprints.org/8353.

Basha, S. J., Sarma, A. S. R. and Ahammed, S. K. 2017. Agronomic manipulations for pests and diseases management in chickpea: A review. *International Journal of Pure & Applied Bioscience*, 5(2), 842–9.

Batish, D. R., Singh, H. and Kaur, S. 2001. Crop allelopathy and its role in ecological agriculture. *Journal of Crop Production*, 4(2), 121–61.

Beare, M., Hendrix, P., Cabrera, M. and Coleman, D. 1994. Aggregate-protected and unprotected organic matter pools in conventional- and no-tillage soils. *Soil Science Society of America Journal*, 58(3), 787–95.

Bengtsson, J., Ahnström, J. and Weibull, A. C. 2005. The effects of organic agriculture on biodiversity and abundance: A meta-analysis. *Journal of Applied Ecology*, 42(2), 261–9.

Bergez, J.-E., Etienne, M. and Balandier, P. 1999. ALWAYS: A plot-based silvopastoral system model. *Ecological Modelling*, 115(1), 1–17.

Bergkvist, G. 2003. Effect of white clover and nitrogen availability on the grain yield of winter wheat in a three-season intercropping system. *Acta Agriculturae Scandinavica, Section B-Plant Soil Science*, 53(3), 97–109.

Berry, P., Sylvester-Bradley, R., Philipps, L., Hatch, D., Cuttle, S., Rayns, F. and Gosling, P. 2002. Is the productivity of organic farms restricted by the supply of available nitrogen? *Soil Use and Management*, 18(suppl. 1), 248–55.

Bharadwaj, V. and Omanwmar, P. 1994. Long term effects of continuous rotational cropping and fertilization on crop yields and soil properties. II: Effects on EC, pH, organic matter and available nutrients of soil. *Journal of the Indian Society of Soil Science*, 42(3), 387–92.

Bhat, N. A., Riar, A., Ramesh, A., Iqbal, S., Sharma, M. P., Sharma, S. K. and Bhullar, G. S. 2017. Soil biological activity contributing to phosphorus availability in vertisols under long-term organic and conventional agricultural management. *Frontiers in Plant Science*, 8, 1523.

Birech, R. and Freyer, B. 2007. Effect of plant biomass and their incorporation depth on organic wheat production in Kenya. In: S. Zikeli, W. Claupein, S. Dabbert, B. Kaufmann, T. Müller und A. Valle Zárate (Eds), 9. *Wissenschaftstagung Ökologischer Landbau. 'Zwischen Tradition und Globalisierung'*. Universität Hohenheim: Köster.

Birech, R. and Freyer, B. 2008. Optimising cropping systems with green manure and tillage techniques in Organic Farming in Kenya. In: U. Köpke (Ed.), *Organic Agriculture in the Tropics and Subtropics* (pp. 69–80). Berlin: Köster.

Birech, R., Freyer, B., Friedel, J., Leonhartsberger, P., Macharia, J. and Obiero, C. 2014. Biomass production in the short rains and its influence on crops in the long rains: A systems approach in organic farming. *American Journal of Experimental Agriculture*, 4(9), 1047.

Blackshaw, R., Larney, F., Lindwall, C., Watson, P. and Derksen, D. 2001. Tillage intensity and crop rotation affect weed community dynamics in a winter wheat cropping system. *Canadian Journal of Plant Science*, 81(4), 805–13.

Blagodatskaya, E. and Kuzyakov, Y. 2008. Mechanisms of real and apparent priming effects and their dependence on soil microbial biomass and community structure: Critical review. *Biology and Fertility of Soils*, 45(2), 115–31.

Bockmann, H. 1954. Die Bedeutung der Pflanzenschutzforschung für die landwirtschaftliche Betriebsorganisation. *Zeitschrift für Pflanzenkrankheiten (Pflanzenpathologie) und Pflanzenschutz*, 61(1), 1–4.

Böhm, H. 2013. Auswirkungen einer Schwefeldüngung auf den Ertrag und die Nährstoffzusamm-ensetzung eines Kleegrasbestandes. *Ideal und Wirklichkeit: Perspektiven ökologischer Landbewirtschaftung, Beiträge zur 12. Wissenschaftstagung Ökologischer Landbau* (pp. 216–19).

Böhm, H. and Verschwele, A. 2004. Ampfer-und Distelbekämpfung im ökologischen Landbau. *Ressortforschung für den ökologischen Landbau* 2004. *Statusseminar 'Ressortforschung für den ökologischen Landbau 2003', BBA, Kleinmachnow, März* 2004 (pp. 39–48).

Bonanomi, G., Antignani, V., Pane, C. and Scala, F. 2007. Suppression of soilborne fungal diseases with organic amendments. *Journal of Plant Pathology*, 311–24.

Bonanomi, G., Antignani, V., Capodilupo, M. and Scala, F. 2010. Identifying the characteristics of organic soil amendments that suppress soilborne plant diseases. *Soil Biology and Biochemistry*, 42(2), 136–44.

Bond, W. and Grundy, A. 2001. Non-chemical weed management in organic farming systems. *Weed Research*, 41(5), 383–405.

Borin, M., Passoni, M., Thiene, M. and Tempesta, T. 2010. Multiple functions of buffer strips in farming areas. *European Journal of Agronomy*, 32(1), 103–11.

Bot, A. and Benites, J. 2005. *The Importance of Soil Organic Matter: Key to Drought-Resistant Soil and Sustained Food Production*. Rome: FAO.

Brankatschk, G. and Finkbeiner, M. 2014. Application of the Cereal Unit in a new allocation procedure for agricultural life cycle assessments. *Journal of Cleaner Production*, 73, 72–9.

Brinkman, T. 1917. Die Oekonomik des landwirtschaftlichen Betriebes. In: *Grundriß der Sozialökonomik. VII. Tübingen* (S. 27–123).

Brinkman, T. 1950. *Das Fruchtfolgebild des deutschen Ackerbaues*. Scheur, Bonn: Bonner Universitäts-Buchdruckerei Gebr.

Brock, C. 2016. Organische Bodensubstanz in Ackerbausystemen. In: B. Freyer (Ed.), *Ökologischer Landbau: Grundlagen, Wissensstand und Herausforderungen* (p. 317). Stuttgart: UTB.

Brock, C., Hoyer, U., Leithold, G. and Hülsbergen, K.-J. 2012. The humus balance model (HU-MOD): A simple tool for the assessment of management change impact on soil organic matter levels in arable soils. *Nutrient Cycling in Agroecosystems*, 92(3), 239–54.

Brozyna, M. A., Petersen, S. O., Chirinda, N. and Olesen, J. E. 2013. Effects of grass-clover management and cover crops on nitrogen cycling and nitrous oxide emissions in a stockless organic crop rotation. *Agriculture, Ecosystems & Environment*, 181, 115–26.

Brundage, A., Taylor, R. and Burton, V. 1979. Relative yields and nutritive values of barley, oats, and peas harvested at four successive dates for forage 1, 2. *Journal of Dairy Science*, 62(5), 740–5.

Bruns, C. 2016. Pflanzengesundheit und Pflanzenschutz. In: B. Freyer (Ed.), *Ökologischer Landbau. Grundlagen, Wissensstand und Herausforderungen* (pp. 502–22). Stuttgart: UTB.

Caamal-Maldonado, J. A., Jiménez-Osornio, J. J., Torres-Barragán, A. and Anaya, A. L. 2001. The use of allelopathic legume cover and mulch species for weed control in cropping systems. *Agronomy Journal*, 93(1), 27–36.

Carr, P. M., Martin, G. B., Caton, J. S. and Poland, W. 1998. Forage and nitrogen yield of barley = pea and oat – pea intercrops. *Agronomy Journal*, 90(1), 79–84.

Casagrande, M., David, C., Valantin-Morison, M., Makowski, D. and Jeuffroy, M.-H. 2009. Factors limiting the grain protein content of organic winter wheat in south-eastern France: A mixed-model approach. *Agronomy for Sustainable Development*, 29(4), 565–74.

Castellazzi, M., Wood, G., Burgess, P. J., Morris, J., Conrad, K. and Perry, J. 2008. A systematic representation of crop rotations. *Agricultural Systems*, 97(1), 26–33.

Chan, K. 2001. An overview of some tillage impacts on earthworm population abundance and diversity – implications for functioning in soils. *Soil and Tillage Research*, 57(4), 179–91.

Chapko, L., Brinkman, M. and Albrecht, K. 1991. Oat, oat-pea, barley, and barley-pea for forage yield, forage quality, and alfalfa establishment. *Journal of Production Agriculture*, 4(4), 486–91.

Chaudhary, M. I., Adu-Gyamfi, J. J., Saneoka, H., Nguyen, N. T., Suwa, R., Kanai, S., . . . Fujita, K. 2008. The effect of phosphorus deficiency on nutrient uptake, nitrogen fixation and photosynthetic rate in mashbean, mungbean and soybean. *Acta Physiologiae Plantarum*, 30(4), 537–44.

Chen, M.-H. and Nelson, E. 2008. Seed-colonizing microbes from municipal biosolids compost suppress Pythium ultimum damping-off on different plant species. *Phytopathology*, 98(9), 1012–18.

Chen, C., Westcott, M., Neill, K., Wichman, D. and Knox, M. 2004. Row configuration and nitrogen application for barley–pea intercropping in Montana. *Agronomy Journal*, 96(6), 1730–8.

Chongtham, I. R., Bergkvist, G., Watson, C. A., Sandström, E., Bengtsson, J. and Öborn, I. 2017. Factors influencing crop rotation strategies on organic farms with different time periods since conversion to organic production. *Biological Agriculture & Horticulture*, 33(1), 14–27.

Chorley, G. P. H. 1981. The agricultural revolution in northern Europe, 1750–1880: Nitrogen, legumes, and crop productivity. *The Economic History Review*, 34(1), 71–93.

Clárk, R. B. 1983. Plant genotype differences in the uptake, translocation, accumulation, and use of mineral elements required for plant growth. *Plant and Soil*, 72(2–3), 175–96.

Clark, M. S., Horwath, W. R., Shennan, C., Scow, K. M., Lantni, W. T. and Ferris, H. 1999. Nitrogen, weeds and water as yield-limiting factors in conventional, low-input, and organic tomato systems. *Agriculture, Ecosystems & Environment*, 73(3), 257–70.

Conrad, Y. and Fohrer, N. 2009. Modelling of nitrogen leaching under a complex winter wheat and red clover crop rotation in a drained agricultural field. *Physics and Chemistry of the Earth, Parts A/B/C*, 34(8), 530–40.

Cormack, W. F. 1999. Testing a stockless arable organic rotation on a fertile soil. In: J. E. Olesen, R. Eltun, M. J. Gooding, E. S. Jensen and U. Köpke (Eds), *Designing and Testing Crop Rotations for Organic Farming: Proceedings from an International Workshop* (pp. 115–23). Denmark: Danish Research Centre for Organic Farming.

Corre-Hellou, G., Faure, M., Launay, M., Brisson, N. and Crozat, Y. 2009. Adaptation of the STICS intercrop model to simulate crop growth and N accumulation in pea–barley intercrops. *Field Crops Research*, 113(1), 72–81.

Craven, M. and Nel, A. A. 2016. Effect of conservation agriculture associated crop rotation systems on root and crown rot severity and respective soil-borne pathogens of maize (*Zea mays* L.) in the Highveld area of South Africa. *South African Journal of Plant and Soil*, 34, 87–95.

Dagar, J. C. 2016. Agroforestry: Four decades of research development. *Indian Journal of Agroforestry*, 18(2), 1–32.

Deike, S., Pallutt, B., Melander, B., Strassemeyer, J. and Christen, O. 2008. Long-term productivity and environmental effects of arable farming as affected by crop rotation, soil tillage intensity and strategy of pesticide use: A case-study of two long-term field experiments in Germany and Denmark. *European Journal of Agronomy*, 29(4), 191–9.

de Kruijff, R., Pietsch, G., Freyer, B. and Friedel, J. K. 2008. Pre-crop effects of alfalfa management systems on inorganic soil nitrogen and cereals in organic farming under pannonian site conditions. *Journal of Plant Nutrition and Soil Science*, 171(4), 576–9.

Delate, K., Cambardella, C., Chase, C. and Turnbull, R. 2017. A review of long-term organic comparison trials in the US. In: K. Etingoff (Ed.), *Sustainable Development of Organic Agriculture: Historical Perspectives*. Oakville, ON and Waretown, NJ: Apple Academic Press, p. 79.

de Molina, M. G. 2002. Environmental constraints on agricultural growth in 19th century Granada (Southern Spain). *Ecological Economics*, 41(2), 257–70.

De Ponti, T., Rijk, B. and Van Ittersum, M. K. 2012. The crop yield gap between organic and conventional agriculture. *Agricultural Systems*, 108, 1–9.

Divito, G. A. and Sadras, V. O. 2014. How do phosphorus, potassium and sulphur affect plant growth and biological nitrogen fixation in crop and pasture legumes? A meta-analysis. *Field Crops Research*, 156, 161–71.

Dogliotti, S., Rossing, W. and Van Ittersum, M. 2003. ROTAT, a tool for systematically generating crop rotations. *European Journal of Agronomy*, 19(2), 239–50.

Doltra, J., Lægdsmand, M. and Olesen, J. E. 2011. Cereal yield and quality as affected by nitrogen availability in organic and conventional arable crop rotations: A combined modeling and experimental approach. *European Journal of Agronomy*, 34(2), 83–95.

Doré, T., Makowski, D., Malézieux, E., Munier-Jolain, N., Tchamitchian, M. and Tittonell, P. 2011. Facing up to the paradigm of ecological intensification in agronomy: Revisiting methods, concepts and knowledge. *European Journal of Agronomy*, 34(4), 197–210.

Dresow, J., Krause, T., Haase, N. U., Loges, R., Heß, J. and Böhm, H. 2013. Effect of different defoliation systems of ryegrass–clover on yield and selected quality parameters of organic potatoes (*Solanum tuberosum* L.) for industrial processing at harvest and after storage. *Potato Research*, 56(3), 179–204.

Drews, S., Neuhoff, D. and Köpke, U. 2009. Weed suppression ability of three winter wheat varieties at different row spacing under organic farming conditions. *Weed Research*, 49(5), 526–33.

Dreymann, S. 2005. Bedeutung einer differenzierten Rotkleegrasbewirtschaftung sowie einer organischen Düngung auf Ertragsbildung, N-Aufnahme und Qualität der Folgefrucht Winterweizen. Dissertation. Kiel, Germany: Christian-Albrechts-Universität zu Kiel - Grünland und Futterbau/Ökologischer Landbau.

Drinkwater, L. E., Wagoner, P. and Sarrantonio, M. 1998. Legume-based cropping systems have reduced carbon and nitrogen losses. *Nature*, 396(6708), 262–5.

Duke, S. O., Romagni, J. G. and Dayan, F. E. 2000. Natural products as sources for new mechanisms of herbicidal action. *Crop Protection*, 19(8–10), 583–9.

Duncan, N. 2004. *Landscapes of Privilege: The Politics of the Aesthetic in an American Suburb*. London: Routledge.

Dury, J., Schaller, N., Garcia, F., Reynaud, A. and Bergez, J. E. 2012. Models to support cropping plan and crop rotation decisions. A review. *Agronomy for Sustainable Development*, 32(2), 567–80.

Dye, R. and Begg, G. S. 2017. Landscape scale management of insect populations. Dundee (Scotland, UK), 29–31 March 2017, 122, pp. 143–6.

Egle, K., Römer, W. and Keller, H. 2003. Exudation of low molecular weight organic acids by *Lupinus albus* L., *Lupinus angustifolius* L. and *Lupinus luteus* L. as affected by phosphorus supply. *Agronomie*, 23(5–6), 511–18.

Ehui, S. K., Kang, B. and Spencer, D. S. 1990. Economic analysis of soil erosion effects in alley cropping, no-till and bush fallow systems in South Western Nigeria. *Agricultural Systems*, 34(4), 349–68.

Eisele, J. and Köpke, U. 1997, Choice of cultivars in organic farming new criteria for winter wheat ideotypes. 1: Light conditions in stands of winter wheat affected by morphological features of different varieties. *Pflanzenbauwissenschaften*, 1, 19–24.

El Kasmioui, O. and Ceulemans, R. 2012. Financial analysis of the cultivation of poplar and willow for bioenergy. *Biomass and Bioenergy*, 43, 52–64.

Emmerling, C. 2001. Response of earthworm communities to different types of soil tillage. *Applied Soil Ecology*, 17(1), 91–6.

Fageria, B. 2007. Green manuring in crop production. *Journal of Plant Nutrition*, 30: 691–719.

Farthofer, R., Friedel, J. K., Pietsch, G. and Freyer, B. 2002. Stickstoff-Auswaschungsverluste und Nachfruchteffekte von Futterleguminosen auf Getreide-Nachfrüchte im Ökologischen Landbau unter pannonischen Standortbedingungen in Österreich 114. *VDLUFA-Kongress, 16–20 September 2002. Generalthema:' Ressourcenschutz und Produktsicherheit-Qualitätsicherung in der Landwirtschaft'* (Vol. 114). Leipzig: Verband Deutscher Landwirtschaftlicher Untersuchungs- und Forschungsanstalten, Bonn.

Fassbender, H. W. 1998. Longterm studies of soil fertility in cacao-shade trees-agroforestry systems: Results of 15 years of organic matter and nutrients research in Costa Rica. In: A. Schulte and D. Ruhiyat (Eds), *Soils of Tropical Forest Ecosystems* (pp. 150–8). Berlin, Heidelberg: Springer.

Fathi, S. 2017. Effect of strip-intercropping of spring canola with clover in improvement of natural biological control of *Plutella xylostella* (L.). *Plant Pest Research*, 7(1), 73–86.

Finckh, M. R., Yli-Mattila, T., Nykänen, A., Kurki, P. and Hannukkala, A. 2015. Organic temperate legume disease management. In: M. R. Finckh, A. H. C. van Bruggen and L. Tamm (Eds), *Plant Diseases and their Management in Organic Agriculture*. St Paul, MN: APS.

Fischinger, S. A., Becker, K. and Leithold, G. 2011. Auswirkungen unterschiedlicher S Versorgungszustände auf den N Flächenertrag eines Luzerne-Kleegrasbestandes. In: G. Leithold, K. Becker, C. Brock, S. Fischinger, A.-K. Spiegel, K. Spory, K.-P. Wilbois und U. Williges (Eds), 11. *Wissenschaftstagung Ökologischer Landbau*, 16–18 *März* 2011, *Gießen*. Berlin: Dr Köster.

Foereid, B., Bro, R., Mogensen, V. O. and Porter, J. R. 2002. Effects of windbreak strips of willow coppice – modelling and field experiment on barley in Denmark. *Agriculture, Ecosystems & Environment*, 93(1), 25–32.

Foissy, D., Vian, J.-F. and David, C. 2013. Managing nutrient in organic farming system: Reliance on livestock production for nutrient management of arable farmland. *Organic Agriculture*, 3(3–4), 183–99.

Forster, D., Christian, A., Verma, R., Zundel, C., Messmer, M. M. and Mäder, P. 2013. Yield and economic performance of organic and conventional cotton-based farming systems – Results from a field trial in India. *PLoS ONE*, 8(12), 1–15.

Freyer, B. 2003. *Fruchtfolgen–Konventionell, Integriert. Biologisch/Crop Rotations–Conventional, Integrated, Organic*. Stuttgart: Eugen Ulmer GmbH & Co.

Freyer, B. 2016. Ernährungssicherung. In: B. Freyer (Ed.), *Ökologischer Landbau-Grundlagen, Wissensstand und Herausforderungen* (pp. 183–91). Stuttgart: UTB.

Freyer, B. and Pericin, C. 1996. Naehrstoffhaushalt in biologisch bewirtschafteten Betrieben. *Agrarforschung*, 3(1), 29–32.

Freyer, B., Pietsch, G., Hrbek, R. and Winter, S. 2005. Futter- und Körnerleguminosen im biologischen Landbau. Agrarverlag, Wien.

Frieben, B. and Köpke, U. 1994. Bedeutung des Organischen Landbaus für den Arten-und Biotopschutz in der Agrarlandschaft. In: Rheinische Friedrich-Wilhelmsuniversität Bonn–Landwirtschaftliche Fakultät (Ed.), *Integrative Extensivierungs- und Naturschutzstrategien* (Vol. 15, pp. 77–88). Bonn: Forschungsberichte Rheinische Friedrich-Wilhelms-Universität.

Fritsch, F. 2012. *Nährstoffgehalte in Düngemitteln Und Im Erntegut; Für die Düngeplanung; Für Nährstoffvergleiche*. Bad Kreuznach: Dienstleistungszentrum Ländlicher Raum Rheinhessen-Nahe-Hunsrück.

Gabriel, D., Roschewitz, I., Tscharntke, T. and Thies, C. 2006. Beta diversity at different spatial scales: Plant communities in organic and conventional agriculture. *Ecological Applications*, 16(5), 2011–21.

Gabriel, D., Sait, S. M., Kunin, W. E. and Benton, T. G. 2013. Food production vs. biodiversity: Comparing organic and conventional agriculture. *Journal of Applied Ecology*, 50(2), 355–64.

Gaiser, T., Perkons, U., Küpper, P. M., Puschmann, D. U., Peth, S., Kautz, T., . . . Köpke, U. 2012. Evidence of improved water uptake from subsoil by spring wheat following lucerne in a temperate humid climate. *Field Crops Research*, 126, 56–62.

Garrity, D. P. 1997. ICRAF's mission in Asia: A test of three-systems hypotheses. PCARRD: Developments in agroforestry research. PCARRD Book Series(160), pp. 8–21.

Ghimire, K. (2017). Identification of Physiological and Morphological Traits Governing High Water Use Efficiency in Alfalfa. Theses and Dissertations. 1211. http://openprairie.sdstate.edu/etd/1211

Giller, K. E. 2001. *Nitrogen Fixation in Tropical Cropping Systems*. Wallingford, UK: CABI Publishing.

Giller, K. E., Ormesher, J. and Awah, F. M. 1991. Nitrogen transfer from Phaseolus bean to intercropped maize measured using 15N-enrichment and 15N-isotope dilution methods. *Soil Biology and Biochemistry*, 23(4), 339–46.

Gollner, G., Gabler, C., Grausgruber-Gröger, S., Friedel, J. K., Grausgruber, H. and Freyer, B. 2010. Körnerleguminosen in Mischkulturen mit Leindotter (*Camelina sativa*) im Ökologischen Landbau unter pannonischen Standortbedingungen. *Journal fur Kulturpflanzen-Journal of Cultivated Plants*, 62(11), 402.

González, E., Salvo, A. and Valladares, G. 2017. Arthropod communities and biological control in soybean fields: Forest cover at landscape scale is more influential than forest proximity. *Agriculture, Ecosystems & Environment*, 239, 359–67.

Goulding, K., Jarvis, S. and Whitmore, A. 2008. Optimizing nutrient management for farm systems. *Philosophical Transactions of the Royal Society of London B: Biological Sciences*, 363(1491), 667–80.

Govaerts, B., Fuentes, M., Mezzalama, M., Nicol, J. M., Deckers, J., Etchevers, J. D., . . . Sayre, K. D. 2007. Infiltration, soil moisture, root rot and nematode populations after 12 years of different tillage, residue and crop rotation managements. *Soil and Tillage Research*, 94(1), 209–19.

Graglia, E., Melander, B. and Jensen, R. K. 2006. Mechanical and cultural strategies to control *Cirsium arvense* in organic arable cropping systems. *Weed Research*, 46(4), 304–12.

Grass, R. 2016. Biogas. In B. Freyer (Ed.), *Ökologischer Landbau. Grundlagen, Wissensstand und Herausforderungen* (pp. 626–39). Stuttgart: UTB.

Gronle, A., Böhm, H. and Heß, J. 2014. Effect of intercropping winter peas of differing leaf type and time of flowering on annual weed infestation in deep and shallow ploughed soils and on pea pests. *Landbauforschung-vTI Agriculture and Forestry Research*, 64(1), 31–44.

Gronle, A., Heß, J. and Böhm, H. 2015a. Effect of intercropping normal-leafed or semi-leafless winter peas and triticale after shallow and deep ploughing on agronomic performance, grain quality and succeeding winter wheat yield. *Field Crops Research*, 180, 80–9.

Gronle, A., Heß, J. and Böhm, H. 2015b. Weed suppressive ability in sole and intercrops of pea and oat and its interaction with ploughing depth and crop interference in organic farming. *Organic Agriculture*, 5(1), 39–51.

Gronle, A., Lux, G., Böhm, H., Schmidtke, K., Wild, M., Demmel, M., . . . Heß, J. 2015c. Effect of ploughing depth and mechanical soil loading on soil physical properties, weed infestation, yield performance and grain quality in sole and intercrops of pea and oat in organic farming. *Soil and Tillage Research*, 148, 59–73.

Gruber, H. and Thamm, U. 2007. Einfluss legumer Vorfrüchte auf Ertrag und Qualität von Winterweizen im ökologischen Landbau. Landesforschungsanstalt für Landwirtschaft und Fischerei Mecklenburg-Vorpommern Institut für Acker- und Pflanzenbau.

Gurr, G. M., Wratten, S. D. and Luna, J. M. 2003. Multi-function agricultural biodiversity: Pest management and other benefits. *Basic and Applied Ecology*, 4(2), 107–16.

Haas, G., Wetterich, F. and Köpke, U. 2001. Comparing intensive, extensified and organic grassland farming in southern Germany by process life cycle assessment. *Agriculture, Ecosystems & Environment*, 83(1), 43–53.

Haas, G., Berg, M. and Köpke, U. 2002. Nitrate leaching: Comparing conventional, integrated and organic agricultural production systems. *International Association of Hydrological Sciences, Publication*, 273, 131–6.

Haase, T., Schüler, C. and Heß, J. 2007. The effect of different N and K sources on tuber nutrient uptake, total and graded yield of potatoes (*Solanum tuberosum* L.) for processing. *European Journal of Agronomy*, 26(3), 187–97.

Han, E., Kautz, T., Perkons, U., Lüsebrink, M., Pude, R. and Köpke, U. 2015. Quantification of soil biopore density after perennial fodder cropping. *Plant and Soil*, 394(1–2), 73–85.

Hansen, B., Kristensen, E. S., Grant, R., Høgh-Jensen, H., Simmelsgaard, S. E. and Olesen, J. E. 2000. Nitrogen leaching from conventional versus organic farming systems – a systems modelling approach. *European Journal of Agronomy*, 13(1), 65–82.

Hardarson, G. 1993. Methods for enhancing symbiotic nitrogen fixation. In: F. A. Bliss and G. Hardarson (Eds), *Enhancement of Biological Nitrogen Fixation of Common Bean in Latin America* (pp. 1–17). Developments in Plant and Soil Sciences, Vol 52. Dordrecht: Springer.

Hardarson, G., Bliss, F. A., Cigales-Rivero, M., Henson, R. A., Kipe-Nolt, J. A., Longeri, L., . . . Sanabria, C. 1993. Genotypic variation in biological nitrogen fixation by common bean. In: F. A. Bliss and G. Hardarson (Eds), *Enhancement of Biological Nitrogen Fixation of Common Bean in Latin America* (pp. 59–70). Developments in Plant and Soil Sciences, Vol 52. Dordrecht: Springer.

Hartwig, N. L. and Ammon, H. U. 2002. Cover crops and living mulches. *Weed Science*, 50(6), 688–99.

Hatch, D., Goodlass, G., Joynes, A. and Shepherd, M. 2007. The effect of cutting, mulching and applications of farmyard manure on nitrogen fixation in a red clover/grass sward. *Bioresource Technology*, 98(17), 3243–8.

Hauck, J., Wolfrum, S. and Hülsbergen, K.-J. 2014. Landscape Aesthetics as an indicator for social sustainability of crop rotations. *Building Organic Bridges*, 2, 527–30.

Hauggaard-Nielsen, H., Ambus, P. and Jensen, E. S. 2001. Interspecific competition, N use and interference with weeds in pea–barley intercropping. *Field Crops Research*, 70(2), 101–9.

Hauggaard-Nielsen, H., Ambus, P. and Jensen, E. S. 2003. The comparison of nitrogen use and leaching in sole cropped versus intercropped pea and barley. *Nutrient Cycling in Agroecosystems*, 65(3), 289–300.

Herzog, F., Oehen, B. and Weibel, F. P. 2016. Agroforstsysteme. In B. Freyer (Ed.), *Ökologischer Landbau. Grundlagen, Wissensstand und Herausforderungen* (pp. 392–405). Stuttgart: UTB.

Hess, J. 1989. Kleegrasumbruch im Organischen Landbau: Stickstoffdynamik im Fruchtfolgeglied Kleegras-Kleegras-Weizen-Roggen. Dissertation. Bonn, Germany: Rheinische Friedrich-Wilhelms-Universität Bonn, Department of Agronomy.

Hodges, R. 1991. Soil organic matter: Its central position in organic farming. In: Wilson, W. S. (Ed.), *Advances in Soil Organic Matter Research: The Impact on Agriculture and the Environment* (pp. 355–64). Cambridge, UK: Royal Society of Chemistry.

Hodgson, J. M. 1958. Canada thistle (*Cirsium arvense* Scop.) control with cultivation, cropping, and chemical sprays. *Weeds*, 1–11.

Høgh-Jensen, H. 1999. The proportion of green fallow in stockless farming systems: Grain N yield, N leaching, and soil organic N. In: J. E. Olesen, R. Eltun, M. J. Gooding, E. S. Jensen and U. Köpke (Eds), *Designing and Testing Crop Rotations for Organic Farming*. DARCOF Report 1 (pp. 223–34). Denmark: Danish Research Centre for Organic Farming.

Høgh-Jensen, H., Schjoerring, J. K. and Soussana, J.-F. 2002. The influence of phosphorus deficiency on growth and nitrogen fixation of white clover plants. *Annals of Botany*, 90(6), 745–53.

Hole, D., Perkins, A., Wilson, J., Alexander, I., Grice, P. and Evans, A. D. 2005. Does organic farming benefit biodiversity? *Biological Conservation*, 122(1), 113–30.

Horneburg, B. 2016. Ökologische Pflanzenzüchtung. In: B. Freyer (Ed.), *Ökologischer Landbau. Grundlagen, Wissensstand, Herausforderungen* (pp. 406–20). Stuttgart: UTB.

IFOAM. 2014. The IFOAM NORMS for Organic Production and Processing. Version 2014. https://www.ifoam.bio/sites/default/files/ifoam_norms_july_2014_t.pdf

Janssen, S., Louhichi, K., Kanellopoulos, A., Zander, P., Flichman, G., Hengsdijk, H., . . . Blanco, M. 2010. A generic bio-economic farm model for environmental and economic assessment of agricultural systems. *Environmental Management*, 46(6), 862–77.

Javanmard, A., Nasab, A. D. M., Javanshir, A., Moghaddam, M. and Janmohammadi, H. 2009. Forage yield and quality in intercropping of maize with different legumes as double-cropped. *Journal of Food, Agriculture and Environment*, 7(1), 163–6.

Jensen, E. S. 1996. Grain yield, symbiotic N2 fixation and interspecific competition for inorganic N in pea-barley intercrops. *Plant and soil*, 182(1), 25–38.

Jensen, E. S. and Hauggaard-Nielsen, H. 2003. How can increased use of biological N2 fixation in agriculture benefit the environment? *Plant and soil*, 252(1), 177–86.

Jensen, E. S., Bedoussac, L., Carlsson, G., Journet, E.-P., Justes, E. and Hauggaard-Nielsen, H. 2015. Enhancing yields in organic crop production by eco-functional intensification. *Sustainable Agriculture Research*, 4(3), 42.

Jiménez-Osornio, J. J. and Gliessman, S. R. 1987. Allelopathic interference in a wild mustard (*Brassica campestris* L.) and broccoli (*Brassica oleracea* L. var. *italica*) intercrop agroecosystem. Paper presented at the *Allelochemicals: Role in Agriculture and Forestry*. ACS Symposium Series.

Jones, E. L. 1965. Agriculture and economic growth in England, 1660–1750: Agricultural change. *The Journal of Economic History*, 25(1), 1–18.

Jones, S., Harris, P. and Gooding, M. 1999. Measurement of nitrogen partitioning within different organic systems incorporating strip intercropping, sheep and crop rotation. Paper presented at the *Designing and Testing Crop Rotations for Organic Farming. Proceedings from an International workshop*. Denmark: Danish Research Centre for Organic Farming.

Jose, S. and Bardhan, S. 2012. Agroforestry for biomass production and carbon sequestration: An overview. *Agroforestry Systems*, 86(2), 105–11.

Jose, S., Gold, M. A. and Garrett, H. E. 2012. The future of temperate agroforestry in the United States. In: P. Nair and D. Garrity (Eds), *Agroforestry-The Future of Global Land Use* (pp. 217–45). Dordrecht: Springer.

Justus, M. and Köpke, U. 1995. Strategies to reduce nitrogen losses via leaching and to increase precrop effects when growing faba beans. *Biological Agriculture & Horticulture*, 11(1–4), 145–55.

Kahnt, G. 1983. *Gründüngung*. Frankfurt/M: DLG Verlag.

Kahnt, G. 1986. *Biologischer Pflanzenbau*. Stuttgart: Eugen Ulmer.

Kang, B. T., Grimme, H. and Lawson, T. L. 1985. Alley cropping sequentially cropped maize and cowpea with Leucaena on a sandy soil in Southern Nigeria. *Plant and Soil*, 85(2), 267–77.

Kasper, M., Freyer, B., Hülsbergen, K. J., Schmid, H. and Friedel, J. K. 2015. Humus balances of different farm production systems in main production areas in Austria. *Journal of Plant Nutrition and Soil Science*, 178(1), 25–34.

Kautz, T., Lüsebrink, M., Pätzold, S., Vetterlein, D., Pude, R., Athmann, M., . . . Köpke, U. 2014. Contribution of anecic earthworms to biopore formation during cultivation of perennial ley crops. *Pedobiologia*, 57(1), 47–52.

Kayser, M., Müller, J. and Isselstein, J. 2010. Nitrogen management in organic farming: Comparison of crop rotation residual effects on yields, N leaching and soil conditions. *Nutrient Cycling in Agroecosystems*, 87(1), 21–31.

Keating, B. A., Carberry, P. S., Hammer, G. L., Probert, M. E., Robertson, M. J., Holzworth, D., . . . Hochman, Z. 2003. An overview of APSIM, a model designed for farming systems simulation. *European Journal of Agronomy*, 18(3), 267–88.

Khan, Z. R., Pickett, J. A., Berg, J.v.d., Wadhams, L. J. and Woodcock, C. M. 2000. Exploiting chemical ecology and species diversity: Stem borer and striga control for maize and sorghum in Africa. *Pest Management Science*, 56(11), 957–62.

Khan, Z. R., Midega, C. A., Hassanali, A., Pickett, J. A., Wadhams, L. J. and Wanjoya, A. 2006. Management of witchweed, Striga hermonthica, and stemborers in sorghum, Sorghum bicolor, through intercropping with greenleaf desmodium, Desmodium intortum. *International Journal of Pest Management*, 52(4), 297–302.

Kirchmann, H. and Bergström, L. 2001. Do organic farming practices reduce nitrate leaching? *Communications in Soil Science and Plant Analysis*, 32(7–8), 997–1028.

Kirchmann, H. and Witter, E. 1992. Composition of fresh, aerobic and anaerobic farm animal dungs. *Bioresource Technology*, 40(2), 137–42.

Kirchmann, H., Bergström, L., Kätterer, T., Mattsson, L. and Gesslein, S. 2007. Comparison of long-term organic and conventional crop–livestock systems on a previously nutrient-depleted soil in Sweden. *Agronomy Journal*, 99(4), 960–72.

Kirchmann, H., Kätterer, T. and Bergström, L. 2009. Nutrient supply in organic agriculture–plant availability, sources and recycling. In: Kirchmann, H. and Bergström, L. (Eds), *Organic Crop Production–Ambitions and Limitations* (pp. 89–116). Dordrecht: Springer.

Knight, J. 2012. Frequency of field pea in rotations impacts biological nitrogen fixation. *Canadian Journal of Plant Science*, 92(6), 1005–11.

Knudsen, M. T., Hauggaard-Nielsen, H., Joernsgaard, B. and Jensen, E. S. 2004. Comparison of interspecific competition and N use in pea–barley, faba bean–barley and lupin–barley intercrops grown at two temperate locations. *The Journal of Agricultural Science*, 142(06), 617–27.

Kolbe, H. 2004. *Zwischenfrüchte im Ökologischen Landbau*. Dresden: Sächsische Landesanstalt für Landwirtschaft.

Kolbe, H. 2006. Fruchtfolgegestaltung im ökologischen und extensiven Landbau: Bewertung von Vorfruchtwirkungen. *Pflanzenbauwissenschaften*, 10(2), 82–9.

Könnecke, G. 1967. *Fruchtfolgen*. Berlin: Deutscher Landwirtschaftsverlag VEB.

Koohafkan, P. and Cruz, M. J. D. 2011. Conservation and adaptive management of globally important agricultural heritage systems (GIAHS). *Journal of Resources and Ecology*, 2(1), 22–8.

Köpke, U. 1995. Nutrient management in organic farming systems: The case of nitrogen. *Biological Agriculture & Horticulture*, 11(1–4), 15–29.

Köpke, U. 1996. Symbiotische Stickstoff-Fixierung und Vorfruchtwirkung von Ackerbohnen (*Vicia faba* L.). Schriftenreihe / Institut für organischen Landbau, 6. Berlin: Köster.

Köpke, U. 2016. Ressourcenschutz und Ökologische Leistungen. In: B. Freyer (Ed.), *Ökologischer Landbau – Grundlagen, Wissensstand und Herausforderungen* (pp. 590–612). Stuttgart: UTB.

Köpke, U. and Nemecek, T. 2010. Ecological services of faba bean. *Field Crops Research*, 115(3), 217–33.

Köpke, U., Athmann, M., Han, E. and Kautz, T. 2015. Optimising cropping techniques for nutrient and environmental management in organic agriculture. *Sustainable Agriculture Research*, 4(3), 15.

Korsaeth, A. and Eltun, R. 2000. Nitrogen mass balances in conventional, integrated and ecological cropping systems and the relationship between balance calculations and nitrogen runoff in an 8-year field experiment in Norway. *Agriculture, Ecosystems & Environment*, 79(2), 199–214.

Kouas, S., Labidi, N., Debez, A. and Abdelly, C. 2005. Effect of P on nodule formation and N fixation in bean. *Agronomy for Sustainable Development*, 25(3), 389–93.

Kruess, A. and Tscharntke, T. 2000. Species richness and parasitism in a fragmented landscape: Experiments and field studies with insects on Vicia sepium. *Oecologia*, 122(1), 129–37.

Kulozik-Erben, A. 2007. Emissionen gasförmiger Stickstoff-Verbindungen (NH3 und N2O) aus Pflanzen in Abhängigkeit von der N-Versorgung und pflanzenartspezifischen Stoffwechseleigenschaften. Dissertation. Giessen: Universitätsbibliothek Giessen.

Kuzyakov, Y., Friedel, J. and Stahr, K. 2000. Review of mechanisms and quantification of priming effects. *Soil Biology and Biochemistry*, 32(11), 1485–98.

Lampkin, N., Pearce, B., Leake, A., Creissen, H., Gerrard, C., Girling, R., . . . Smith, L. 2015. The role of agroecology in sustainable intensification. Report for the Land Use Policy Group, 163.

Larsson, L., Ferm, M., Kasimir-Klemedtsson, A. and Klemedtsson, L. 1998. Ammonia and nitrous oxide emissions from grass and alfalfa mulches. *Nutrient Cycling in Agroecosystems*, 51(1), 41–6.

Lauk, R. and Lauk, E. 2008. Pea-oat intercrops are superior to pea-wheat and pea-barley intercrops. *Acta Agriculturae Scandinavica Section B-Soil and Plant Science*, 58(2), 139–44.

Letourneau, D. K., Armbrecht, I., Rivera, B. S., Lerma, J. M., Carmona, E. J., Daza, M. C., . . . López, S. D. 2011. Does plant diversity benefit agroecosystems? A synthetic review. *Ecological Applications*, 21(1), 9–21.

Lichtenberg, E. M., Kennedy, C. M., Kremen, C., Batáry, P., Berendse, F., Bommarco, R., . . . Williams, N. M. 2017. A global synthesis of the effects of diversified farming systems on arthropod diversity within fields and across agricultural landscapes. *Global Change Biology*, 23(11), 4946–57.

Liebman, M. and Dyck, E. 1993. Crop rotation and intercropping strategies for weed management. *Ecological Applications*, 3(1), 92–122.

Lithourgidis, A., Dordas, C., Damalas, C. and Vlachostergios, D. 2011. Annual intercrops: An alternative pathway for sustainable agriculture. *Australian Journal of Crop Science*, 5(4), 396.

Liu, Y., Wu, L., Baddeley, J. A. and Watson, C. A. 2011. Models of biological nitrogen fixation of legumes. In: E. Lichtfouse, M. Hamelin, M. Navarrete and P. Debaeke (Eds), *Sustainable Agriculture Volume 2* (pp. 883–905). Dordrecht: Springer.

Lizarazo, C. I., Yli-Halla, M. and Stoddard, F. L. 2015. Pre-crop effects on the nutrient composition and utilization efficiency of faba bean (*Vicia faba* L.) and narrow-leafed lupin (*Lupinus angustifolius* L.). *Nutrient Cycling in Agroecosystems*, 103(3), 311–27.

Lofs-Holmin, A. 1983. Earthworm population dynamics in different agricultural rotations. In: J. E. Satchell (Ed.), *Earthworm Ecology* (pp. 151–60). Dordrecht: Springer.

Lütke Entrup, N. and Zerhusen, P. 1992. *Mais und Umwelt. Eine Studie über die Umweltverträglichkeit des Maisanbaus. Probleme und Lösungsansätze*. Hamburg, Germany: Kovac.

Luttikholt, L. W. 2007. Principles of organic agriculture as formulated by the International Federation of Organic Agriculture Movements. *NJAS-Wageningen Journal of Life Sciences*, 54(4), 347–60.

Lütz, M. and Bastian, O. 2002. Implementation of landscape planning and nature conservation in the agricultural landscape – a case study from Saxony. *Agriculture, Ecosystems & Environment*, 92(2), 159–70.

Maisonhaute, J.-É., Labrie, G. and Lucas, E. 2017. Is higher landscape heterogeneity associated with lower variation of abundances of pests and natural enemies? Dundee (Scotland, UK) March 29–31, 2017, 122, pp. 33–7.

Makumba, W., Janssen, B., Oenema, O., Akinnifesi, F. K., Mweta, D. and Kwesiga, F. 2006. The long-term effects of a gliricidia–maize intercropping system in Southern Malawi, on gliricidia and maize yields, and soil properties. *Agriculture, Ecosystems & Environment*, 116(1), 85–92.

Malézieux, E., Crozat, Y., Dupraz, C., Laurans, M., Makowski, D., Ozier-Lafontaine, H., . . . Valantin-Morison, M. 2009. Mixing plant species in cropping systems: Concepts, tools and models: A review. In: E. Lichtfouse, M. Navarrete, P. Debaeke, V. Souchere and C. Alberola *Sustainable Agriculture* (pp. 329–53). Dordrecht: Springer.

Mann, L. and Tolbert, V. 2000. Soil sustainability in renewable biomass plantings. *AMBIO: A Journal of the Human Environment*, 29(8), 492–8.

Marinov-Serafimov, P. 2010. Determination of allelopathic effect of some invasive weed species on germination and initial development of grain legume crops. *Pesticides and Phytomedicine*, 25(3).

Massucati, P., Luiz, F., Windisch, E., Täufer, F. and Köpke, U. 2009. *Kontrolle von Rumex spp. mit Citronella-Öl im Organischen Landbau.* Paper presented at the *10. Wissenschaftstagung*

Ökologischer Landbau: Werte - Wege - Wirkungen: Biolandbau im Spannungsfeld zwischen Ernährungssicherheit, Markt und Klimawandel, Zürich, Band 1: Boden, Pflanzenbau, Agrartechnik, Umwelt- und Naturschutz, Biolandbau international, Wissensmanagement, Zürich.

Massucati, L., Perkons, U. and Köpke, U. 2010. Unkrautregulierung durch Mulchmanagement in Direktsaatverfahren von Ackerbohnen (*Vicia faba* L.) im Ökologischen Landbau. *Mitteilungen der Gesellschaft für Pflanzenbauwissenschaften*, 22, 83–4.

Mayer, J. and Mäder, P. 2016. Langzeitversuche – Eine Analyse der Ertragsentwicklung/Longterm trials – An analysis of yield development. In: B. Freyer (Ed.), *Ökologischer Landbau. Grundlagen, Wissensstand und Herausforderungen/Organic Farming: Basics, State of the Art, Challenges* (pp. 421–45). UTB 4639, Bern: Haupt.

McCown, R., Hammer, G., Hargreaves, J., Holzworth, D. and Freebairn, D. 1996. APSIM: A novel software system for model development, model testing and simulation in agricultural systems research. *Agricultural Systems*, 50(3), 255–71.

Migliorati, M. D. A., Bell, M., Grace, P. R., Scheer, C., Rowlings, D. W. and Liu, S. 2015. Legume pastures can reduce N_2O emissions intensity in subtropical cereal cropping systems. *Agriculture, Ecosystems & Environment*, 204, 27–39.

Moghaddam, A., Raza, A., Vollmann, J., Ardakani, M., Wanek, W., Gollner, G. and Friedel, J. 2013. Carbon isotope discrimination and water use efficiency relationships of alfalfa genotypes under irrigated and rain-fed organic farming. *European Journal of Agronomy*, 50, 82–9.

Mohler, C. L. and Johnson, S. E. 2009. Crop rotation on organic farms: A planning manual. Ithaca, NY: Natural Resource, Agriculture, and Engineering Service (NRAES) Cooperative Extension.

Möller, K. 2001. Einfluss und Wechselwirkung von Krautfäulebefall (*Phytophthora infestans* (Mont.) *de Bary*) und Stickstoffernährung auf Knollenwachstum und Ertrag von Kartoffeln (*Solanum tuberosum* L.) im ökologischen Landbau. Dissertation. Shaker.

Möller, K. and Reents, H. 1999. Einfluss verschiedener Zwischenfrüchte nach Körnererbsen auf die Nitratdynamik im Boden und das Wachstum der Folgefrüchte Kartoffeln und Weizen im ökologischen Landbau. *Mitteilungen der Gesellschaft für Pflanzenbauwissenschaften*, 12, 119–20.

Möller, K. and Stinner, W. 2009. Effects of different manuring systems with and without biogas digestion on soil mineral nitrogen content and on gaseous nitrogen losses (ammonia, nitrous oxides). *European Journal of Agronomy*, 30(1), 1–16.

Möller, K., Stinner, W., Deuker, A. and Leithold, G. 2008. Effects of different manuring systems with and without biogas digestion on nitrogen cycle and crop yield in mixed organic dairy farming systems. *Nutrient Cycling in Agroecosystems*, 82(3), 209–32.

Müller, P. and Ellmer, F. 1985. Fruchtfolge. In Anonym (Ed.), *Grundlagen der Pflanzenproduktion*. 4. *Aufl.* Berlin: VEB Landwirtschaftsverlag.

Müller, T. and Thorup-Kristensen, K. 2001. N-fixation of selected green manure plants in an organic crop rotation. *Biological Agriculture & Horticulture*, 18(4), 345–63.

Müller, T., Thorup-Kristensen, K., Magid, J., Jensen, L. S. and Hansen, S. 2006. Catch crops affect nitrogen dynamics in organic farming systems without livestock husbandry – Simulations with the DAISY model. *Ecological Modelling*, 191(3), 538–44.

Munkholm, L. J. 2000. The spade analysis - a modification of the qualitative spade diagnosis for scientific use. DIAS report, Plant Production (28).

Murphy-Bokern, D., Stoddard, F. and Watson, C. 2017. *Legumes in Cropping Systems*. Wallingford, UK: CABI.

Musyoka, M. W., Adamtey, N., Muriuki, A. W. and Cadisch, G. 2017. Effect of organic and conventional farming systems on nitrogen use efficiency of potato, maize and vegetables in the Central Highlands of Kenya. *European Journal of Agronomy*, 86, 24–36.

Nielsen, N. and Barber, S. 1978. Differences among genotypes of corn in the kinetics of P uptake. *Agronomy Journal*, 70(5), 695–8.

Nkurunziza, L., Chongtham, I. R., Watson, C. A., Marstorp, H., Öborn, I., Bergkvist, G. and Bengtsson, J. 2017. Understanding effects of multiple farm management practices on barley performance. *European Journal of Agronomy*, 90, 43–52.

Noble, R. and Coventry, E. 2005. Suppression of soil-borne plant diseases with composts: A review. *Biocontrol Science and Technology*, 15(1), 3–20.

Nyamdavaa, T. and Friedel, J. K. 2016. Effect of pre-crops, cover crops and manure application on organic potato production in a field experiment in eastern Austria. *Acta Fytotechnica et Zootechnica*, 18(5), 16–18.

Olesen, J. E., Eltun, R., Gooding, M. J., Jensen, E. S. and Köpke, U. (Eds). 1999a. Designing and testing crop rotations for organic farming. *Proceedings from an International Workshop*. DARCOF Report no. 1. Printed from www.foejo.dk. Tjele: Danish Research Centre for Organic Farming. ISSN: 1399-915X.

Olesen, J., Rasmussen, I., Askegaard, M. and Kristensen, K. 1999b. Design of the Danish crop rotation experiment. In: J. E. Olesen, R. Eltun, M. J. Gooding, E. S. Jensen and U. Köpke (Eds), *Designing and Testing Crop Rotations for Organic Farming: Proceedings from an International Workshop* (pp. 49–62). Denmark: Danish Research Centre for Organic Farming.

Olesen, J. E., Askegaard, M. and Rasmussen, I. A. 2000. Design of an organic farming crop-rotation experiment. *Acta Agriculturae Scandinavica, Section B - Plant Soil Science*, 50(1), 13–21.

Olesen, J. E., Rasmussen, I. A., Askegaard, M. and Kristensen, K. 2002. Whole-rotation dry matter and nitrogen grain yields from the first course of an organic farming crop rotation experiment. *The Journal of Agricultural Science*, 139(04), 361–70.

Olofsdotter, M. 2001. Rice—a step toward use of allelopathy. *Agronomy Journal*, *93*(1), 3–8.

Ooro, P. A., Birech, R., Maling'a, J., Germer, J., Asch, F., Freyer, B. and Taiy, R. (2015). *Effect of Pre-Crop on Growth and Yield of Potato in Kenya*. Paper presented at the Tropentag 2015, Management of land use systems for enhanced food security: Conflicts, controversies and resolutions, Berlin, Germany.

Orwa, C., Mutua, A., Kindt, R., Jamnadass, R. and Simons, A. 2009. *Agroforestree Database: A Tree Species Reference and Selection Guide Version 4.0*. Nairobi, KE: World Agroforestry Centre ICRAF.

Ouellet, G., Lapen, D., Topp, E., Sawada, M. and Edwards, M. 2008. A heuristic model to predict earthworm biomass in agroecosystems based on selected management and soil properties. *Applied Soil Ecology*, 39(1), 35–45.

Paffrath, A. 2003. Nachwirkung verschiedener Leguminosen auf die Folgefrüchte unter den Anbaubedingungen des ökologischen Landbaus. *Landwirtschaftskammer Nordrhein-Westfalen, Bonn*. Internet: www. oekolandbau. nrw. de/pdf/projekte-versuche/leitbetriebe-2005/20-fruc htfolge-vorfruchtwirkung. pdf.

Paffrath, A. 2005. Einfluss von Ackerbohnen, Körnererbsen, Buschbohnen und Rotkleegras auf die Folgefrucht Kartoffeln *Ende der Nische, Beiträge zur 8. Wissenschaftstagung Ökologischer Landbau*. Universität Kassel.

Pallutt, B. 2000. Einfluss der Konkurrenzkraft von Getreidebeständen auf das Unkrautwachstum und den Getreideertrag. *Zeitschrift. Pflanzenkrankheiten. Pflanzenschutz, Sonderheft*, 17, 265–74.

Paulsen, H. M., Schochow, M. and Reents, H.-J. 2007. Unkrautvorkommen und Unkrautunterdrückung in Mischfruchtanbausystemen mit Ölpflanzen im ökologischen Landbau. *Landbauforschung Völkenrode SH*, 309, 81–95.

Paulsen, H. M., Böhm, H. and Freyer, B. 2016. Fruchtfolgen/Crop rotations. In: Freyer, B. (Ed.), *Ökologischer Landbau: Grundlagen, Wissensstand und Herausforderungen/Organic Farming: Basics, State of the Art, Challenges* (pp. 336–66). UTB 4639, Bern: Haupt.

Peigné, J., Ball, B., Roger-Estrade, J. and David, C. 2007. Is conservation tillage suitable for organic farming? A review. *Soil Use and Management*, 23(2), 129–44.

Perkons, U., Kautz, T., Uteau, D., Peth, S., Geier, V., Thomas, K., . . . Köpke, U. 2014. Root-length densities of various annual crops following crops with contrasting root systems. *Soil and Tillage Research*, 137, 50–7.

Petit, C. and Lambin, E. 2002. Long-term land-cover changes in the Belgian Ardennes (1775–1929): Model-based reconstruction vs. historical maps. *Global Change Biology*, 8(7), 616–30.

Pietsch, G. 2006. Optimierung von Körner-und Futtererbsenanbau und -verwertung unter den Bedingungen des Ökologischen Landbaus im pannonischen Klimagebiet. BMLFUW Projekt 1290. Online-Fachzeitschrift des BMLFUW.

Pietsch, G. and Freyer, B. 2004. Ökologischer Anbau von Körner- und Grünfuttererbsen im Trockengebiet Österreichs–Ertragsleistung und Vorfruchtwirkung. *Mitt. Ges. Pflanzenbauwiss.*(16), 145–6.

Pietsch, G., Hrbek, R., Laubhann, D. and Friedel, J. 2009. Effect of mulching dates modified for nature conservation on the yield and nitrogen fixation of green manure lucerne crops. *Agronomy for Sustainable Development*, 29(2), 353–62.

Ponisio, L. C., M'Gonigle, L. K., Mace, K. C., Palomino, J., de Valpine, P. and Kremen, C. 2015. Diversification practices reduce organic to conventional yield gap. *Philosophical Transactions of the Royal Society B: Biological Sciences*, 282(1799), 20141396.

Ponti, L., Altieri, M. A. and Gutierrez, A. P. 2007. Effects of crop diversification levels and fertilization regimes on abundance of *Brevicoryne brassicae* (L.) and its parasitization by *Diaeretiella rapae* (M'Intosh) in broccoli. *Agricultural and Forest Entomology*, 9(3), 209–14.

Posner, J. L., Baldock, J. O. and Hedtcke, J. L. 2008. Organic and conventional production systems in the Wisconsin integrated cropping systems trials: I. Productivity 1990–2002. *Agronomy Journal*, 100(2), 253–60.

Poutala, R., Kuoppamäki, O., Korva, J. and Varis, E. 1994. The performance of ecological, integrated and conventional nutrient management systems in cereal cropping in Finland. *Field Crops Research*, 37(1), 3–10.

Putnam, A. R. (1988). Allelochemicals from plants as herbicides. *Weed Technology*, 2(4), 510–18.

Quinkenstein, A., Woellecke, J., Böhm, C., Gruenewald, H., Freese, D., Schneider, B. U. and Huettl, R. F. 2009. Ecological benefits of the alley cropping agroforestry system in sensitive regions of Europe. *Environmental Science & Policy*, 12(8), 1112–21.

Raaijmakers, J. M., Paulitz, T. C., Steinberg, C., Alabouvette, C. and Moënne-Loccoz, Y. 2009. The rhizosphere: A playground and battlefield for soilborne pathogens and beneficial microorganisms. *Plant and Soil*, 321(1–2), 341–61.

Råberg, T. M., Carlsson, G. and Jensen, E. S. 2017. Productivity in an arable and stockless organic cropping system may be enhanced by strategic recycling of biomass. *Renewable Agriculture and Food Systems*, 1–13. doi:10.1017/S1742170517000242.

Rauber, R., Schmidtke, K. and Kimpel-Freund, H. 2001. The performance of pea (*Pisum sativum* L.) and its role in determining yield advantages in mixed stands of pea and oat (*Avena sativa* L.). *Journal of Agronomy and Crop Science*, 187(2), 137–44.

Raviv, M. (2005). The use of compost in growing media as suppressive agent against soil-borne diseases. Paper presented at the International Symposium on Growing Media, KU Leuven.

Raza, A., Bodner, G., Moghaddam, A., Ardakani, M. R., Loiskandl, W., Himmelbauer, M., . . . Friedel, J. K. 2014. Assessing the effect of lucerne utilization systems in the Pannonian region of Austria. *Archives of Agronomy and Soil Science*, 60(3), 297–311.

Reckling, M., Bergkvist, G., Watson, C. A., Stoddard, F. L., Zander, P. M., Walker, R. L., . . . Bachinger, J. 2016a. Trade-offs between economic and environmental impacts of introducing legumes into cropping systems. *Frontiers in Plant Science*, 7, 1–7.

Reckling, M., Hecker, J.-M., Bergkvist, G., Watson, C. A., Zander, P., Schläfke, N., . . . Maire, J. 2016b. A cropping system assessment framework – evaluating effects of introducing legumes into crop rotations. *European Journal of Agronomy*, 76, 186–97.

Reents, H. and Möller, K. 1999. Strategies to avoid nitrate leaching after potato crops by applying different cultivation methods to the following cereals. In: J. E. Olesen, R. Eltun, M. J. Gooding, E. S. Jensen and U. Köpke (Eds), *Designing and Testing Crop Rotations for Organic Farming: Proceedings from an International Workshop* (pp. 213–22). Denmark: Danish Research Centre for Organic Farming.

Rengel, Z. and Damon, P. M. 2008. Crops and genotypes differ in efficiency of potassium uptake and use. *Physiologia Plantarum*, 133(4), 624–36.

Reynolds, L. and Jabbar, M. 1994. The role of alley farming in African livestock production. *Outlook on Agriculture*, 23(2), 105–13.

Richter, H. 1967. Phytopathologische konsequenzen der sich ändernden landwirtschaftlichen anbauverfahren. *European Journal of Plant Pathology*, 73, 81–96.

Rochette, P. and Janzen, H. H. 2005. Towards a revised coefficient for estimating N_2O emissions from legumes. *Nutrient Cycling in Agroecosystems*, 73(2), 171–9.

Rochette, P., Angers, D. A., Bélanger, G., Chantigny, M. H., Prévost, D. and Lévesque, G. 2004. Emissions of NO from alfalfa and soybean crops in eastern Canada. *Soil Science Society of America Journal*, 68(2), 493–506.

Römer, W. and Lehne, P. 2004. Vernachlässigte Phosphor- und Kaliumdüngung im ökologischen Landbau senkt die biologische Stickstofffixierung bei Rotklee und den Kornertrag bei nachfolgendem Hafer. *Journal of Plant Nutrition and Soil Science*, 167(1), 106–13.

Roschewitz, I., Gabriel, D., Tscharntke, T. and Thies, C. 2005. The effects of landscape complexity on arable weed species diversity in organic and conventional farming. *Journal of Applied Ecology*, 42(5), 873–82.

Rossing, W., Meynard, J. and Van Ittersum, M. 1997. Model-based explorations to support development of sustainable farming systems: Case studies from France and the Netherlands. *Developments in Crop Science*, 25, 339–51.

Rovira, A. D. 1969. Plant root exudates. *The Botanical Review*, 35(1), 35–57.

Ruhigwa, B., Gichuru, M., Spencer, D. and Swennen, R. 1994. Economic analysis of cut-and-carry, and alley cropping systems of mulch production for plantains in south-eastern Nigeria. *Agroforestry Systems*, 26(2), 131–8.

Rusch, A., Valantin-Morison, M., Sarthou, J.-P. and Roger-Estrade, J. 2010. Chapter 6: Biological control of insect pests in agroecosystems: Effects of crop management, farming systems, and seminatural habitats at the landscape scale: A review. *Advances in Agronomy*, 109, 219–59.

Sacco, D., Moretti, B., Monaco, S. and Grignani, C. 2015. Six-year transition from conventional to organic farming: Effects on crop production and soil quality. *European Journal of Agronomy*, 69, 10–20.

Sadok, W., Angevin, F., Bergez, J.-É., Bockstaller, C., Colomb, B., Guichard, L., . . . Doré, T. 2009. Ex ante assessment of the sustainability of alternative cropping systems: Implications for using multi-criteria decision-aid methods-a review. In: Lichtfouse, E., Navarrete, M., Debaeke, P., Véronique, S. and Alberola, C. (Eds), *Sustainable Agriculture* (pp. 753–67). Dordrecht: Springer.

Sadowsky, M. J. 2005. Soil stress factors influencing symbiotic nitrogen fixation. In: W. Dietrich and W. E. Newton (Eds), *Nitrogen Fixation in Agriculture, Forestry, Ecology, and the Environment* (Vol. 4, pp. 89–112). Dordrecht, The Netherlands: Springer.

Salmon-Monviola, J., Durand, P., Ferchaud, F., Oehler, F. and Sorel, L. 2012. Modelling spatial dynamics of cropping systems to assess agricultural practices at the catchment scale. *Computers and Electronics in Agriculture*, 81, 1–13.

Sandhu, H. S., Wratten, S. D., Cullen, R. and Case, B. 2008. The future of farming: The value of ecosystem services in conventional and organic arable land. An experimental approach. *Ecological Economics*, 64(4), 835–48.

Sattler, F. and von Wistinghausen, E. 1985. *Der landwirtschaftliche Betrieb: Biologisch-dynamisch*. Suttgart: Eugen Ulmer.

Saucke, H. and Ackermann, K. 2006. Weed suppression in mixed cropped grain peas and false flax (*Camelina sativa*). *Weed Research*, 46(6), 453–61.

Sauerborn, J., Sprich, H. and Mercer-Quarshie, H. 2000. Crop rotation to improve agricultural production in Sub-Saharan Africa. *Journal of Agronomy and Crop Science*, 184(1), 67–72.

Scheffer, F. 1941. *Agrikulturchemie, Teil C: Humus und Humusdüngung*. Stuttgart: Enke.

Schenk, M. and Barber, S. 1979. Root characteristics of corn genotypes as related to P uptake. *Agronomy Journal*, 71(6), 921–4.

Schmidt, H. 2004. *Viehloser Öko-Ackerbau: Beiträge, Beispiele, Kommentare*. Berlin: Verlag Dr Köster.

Schmidt, H., Schulz, F., Leithold, G., Raupp, J., Pekrun, C., Oltmanns, M. and Köpke, U. 2006. Organic Farming Trial Gladbacherhof. Effects of different crop rotations and tillage systems. *Long-term Field Experiments in Organic Farming* (pp. 165–82).

Schneider, R., Heiles, E., Salzeder, G., Wiesinger, K., Schmidt, M. and Urbatzka, P. 2012. Auswirkungen unterschiedlicher Fruchtfolgen im ökologischen Landbau auf den Ertrag und die Produktivität. *Angewandte Forschung und Beratung für den ökologischen Landbau in Bayern* (pp. 87–93).

Schönhart, M., Schmid, E. and Schneider, U. A. 2011. CropRota–A crop rotation model to support integrated land use assessments. *European Journal of Agronomy*, 34(4), 263–77.

Schrama, M., De Haan, J., Kroonen, M., Verstegen, H. and Van der Putten, W. (2018). Crop yield gap and stability in organic and conventional farming systems. *Agriculture, Ecosystems & Environment*, 256, 123–30.

Schröder, D. and Köpke, U. 2012. Faba bean (*Vicia faba* L.) intercropped with oil crops–a strategy to enhance rooting density and to optimize nitrogen use and grain production? *Field Crops Research*, 135, 74–81.

Schulz, F. 1936. Obstpflanzung und Unterkultur in gegenseitiger Beeinflussung. *Landwirtsch. Jb*, 82, 697–776.

Schulz, F. and Leithold, G. 2014. Effekte von Fruchtfolge und Bodenbearbeitung auf die Segetalflora im Ökologischen Landbau. *Julius-Kühn-Archiv*(443), 441.

Schulz, F., Brock, C., Schmidt, H., Franz, K.-P. and Leithold, G. 2014. Development of soil organic matter stocks under different farm types and tillage systems in the Organic Arable Farming Experiment Gladbacherhof. *Archives of Agronomy and Soil Science*, 60(3), 313–26.

Seidel, E. P., dos Reis, W., Mottin, M. C., Fey, E., Schneider, A. P. R. and Sustakowski, M. C. 2017. Evaluation of aggregate distribution and selected soil physical properties under maizejack bean intercropping and gypsum rates. *African Journal of Agricultural Research*, 12(14), 1209–16.

Serdjuk, M., Bodmer, U. and Hülsbergen, K.-J. 2017. Integration of biogas production into organic arable farming systems: Crop yield response and economic effects. *Organic Agriculture* (pp. 1–14).

Seufert, V., Ramankutty, N. and Foley, J. A. 2012. Comparing the yields of organic and conventional agriculture. *Nature*, 485(7397), 229–32.

Shibabaw, A., Alemayehu, G., Adgo, E., Germer, J., Asch, F. and Freyer, B. 2017a. Effects of manuring and crop rotation system on potato in the highlands of Ethiopia (*in press*).

Shibabaw, A., Alemayehu, G., Adgo, E., Germer, J., Asch, F. and Freyer, B. 2017b. The effects of organic treatments and crop rotation systems on the grain yield of wheat (*Triticum aestivum* L.) in the Highlands of Ethiopia (*in press*).

Shibabaw, A., Alemayehu, G., Adgo, E., Germer, J., Asch, F. and Freyer, B. 2017c. Growth and biomass yield response of clover (*Trifolium decorum*) to preceding crop and organic treatment in the highlands of Awi Administrative Zone, Ethiopia. *Ethiopian Journal of Science and Technology*, 10(3), 151–64.

Shults, P. 2017. *Exploring the Benefits of Cover Crops to Agroforestry Tree Plantations: An Analysis of Direct and Indirect Nitrogen Transfer in Alley Cropping Systems*. Michigan State University.

Sihi, D., Dari, B., Sharma, D. K., Pathak, H., Nain, L. and Sharma, O. P. 2017. Evaluation of soil health in organic vs. conventional farming of basmati rice in North India. *Journal of Plant Nutrition and Soil Science*, 180(3), 389–406.

Silva, E. M. and Delate, K. 2017. A decade of progress in organic cover crop-based reduced tillage practices in the Upper Midwestern USA. *Agriculture*, 7(5), 44.

Six, J., Bossuyt, H., Degryze, S. and Denef, K. 2004. A history of research on the link between (micro) aggregates, soil biota, and soil organic matter dynamics. *Soil and Tillage Research*, 79(1), 7–31.

Slicher van Bath, B. H. 1963. The agrarian history of Western Europe, AD 500–1850. *The agrarian history of Western Europe, AD* 500–1850.

Smith, J. 2010. The history of temperate agroforestry. Progressive Farming Trust Limited Trading as The Organic Research Centre, Elm Farm 2010.

Smith, R. G., Gross, K. L. and Robertson, G. P. 2008. Effects of crop diversity on agroecosystem function: Crop yield response. *Ecosystems*, 11(3), 355–66.

Smith, J., Leach, K., Rinne, M. and Kuoppala, K. 2012a. Integrating willow-based bioenergy and organic dairy production–the role of tree fodder for feed supplementation. *Agriculture and Forestry Research* (362), 394–7.

Smith, J., Pearce, B. D. and Wolfe, M. S. 2012b. A European perspective for developing modern multifunctional agroforestry systems for sustainable intensification. *Renewable Agriculture and Food Systems*, 27(4), 323–32.

Smith, J., Pearce, B. D. and Wolfe, M. S. 2013. Reconciling productivity with protection of the environment: Is temperate agroforestry the answer? *Renewable Agriculture and Food Systems*, 28(01), 80–92.

Snapp, S., Mafongoya, P. and Waddington, S. 1998. Organic matter technologies for integrated nutrient management in smallholder cropping systems of Southern Africa. *Agriculture, Ecosystems & Environment*, 71(1), 185–200.

Song, Y., Zhang, F., Marschner, P., Fan, F., Gao, H., Bao, X., . . . Li, L. 2007. Effect of intercropping on crop yield and chemical and microbiological properties in rhizosphere of wheat (*Triticum aestivum* L.), maize (*Zea mays* L.), and faba bean (*Vicia faba* L.). *Biology and Fertility of Soils*, 43(5), 565–74.

Stein, S. and Steinmann, H.-H. 2018. Identifying crop rotation practice by the typification of crop sequence patterns for arable farming systems–A case study from Central Europe. *European Journal of Agronomy*, 92, 30–40.

Stein-Bachinger, K., Bachinger, J. and Schmitt, L. 2004. *Nährstoffmanagement im ökologischen Landbau-Ein Handbuch für Beratung und Praxis mit Anwendungs - CD* (Vol. 1). Münster: Landwirtschaftsverlag.

Stinner, W., Möller, K. and Leithold, G. 2008. Effects of biogas digestion of clover/grass-leys, cover crops and crop residues on nitrogen cycle and crop yield in organic stockless farming systems. *European Journal of Agronomy*, 29(2), 125–34.

Stöckle, C. O., Donatelli, M. and Nelson, R. 2003. CropSyst, a cropping systems simulation model. *European Journal of Agronomy*, 18(3), 289–307.

Stolz, O. 1932. 'Der rechte Kreis', eine dreijährige Fruchtfolge mit Flurzwang im Pustertal nach Belegen des 16. bis 18. Jahrhunderts. *Vierteljahrschrift für Sozial- und Wirtschaftsgeschichte*, 25(H. 4), 353–7.

Stopes, C., Millington, S. and Woodward, L. 1996. Dry matter and nitrogen accumulation by three leguminous green manure species and the yield of a following wheat crop in an organic production system. *Agriculture, Ecosystems & Environment*, 57(2), 189–96.

Stopes, C., Lord, E., Philipps, L. and Woodward, L. 2002. Nitrate leaching from organic farms and conventional farms following best practice. *Soil Use and Management*, 18(s1), 256–63.

Stumm, C. and Köpke, U. 2015. Optimierung des Futterleguminosenanbaus im viehlosen Acker- und Gemüsebau 13. *Wissenschaftstagung Ökologischer Landbau*. Hochschule für nachhaltige Entwicklung Eberswalde.

Suárez-Estrella, F., Vargas-Garcia, C., Lopez, M., Capel, C. and Moreno, J. 2007. Antagonistic activity of bacteria and fungi from horticultural compost against *Fusarium oxysporum* f. sp. *melonis*. *Crop Protection*, 26(1), 46–53.

Surböck, A., Heinzinger, M., Schmid, H., Friedel, J. K. and Freyer, B. 2013. Auswirkungen unterschiedlicher Luzerneanteile in der Fruchtfolge auf Humus- und Stickstoffhaushalt 12. *Wissenschaftstagung Ökologischer Landbau*. D. Neuhoff, C. Stumm, S. Ziegler, G. Rahmann, U. Hamm und U. Köpke (Hrsg.) (2013), *Ideal und Wirklichkeit - Perspektiven Ökologischer Landbewirtschaftung. Beiträge zur* 12. *Wissenschaftstagung Ökologischer Landbau, Bonn*, 5. – 8. *März* 2013. *Verlag Dr Köster, Berlin* (pp. 164–5).

Surböck, A., Heinzinger, M., Schweinzer, A., Friedel, J. K. and Freyer, B. 2015. Auswirkungen unterschiedlicher Luzernenutzung auf Ertrag und Qualität der Nachtfrucht Winterweizen - Ergebnisse aus dem Langzeitversuch MUBIL *Fachtagung Biologische Landwirtschaft* (pp. 29–32). Raumberg-Gumpenstein: Höhere Bundeslehr- und Forschungsanstalt für Landwirtschaft.

Surböck, A., Friedel, J. K., Heinzinger, M., Schweinzer, A. and Freyer, B. 2017. Auswirkungen unterschiedlicher Vorfrüchte und Düngungssysteme auf Ertrag und Qualität von Winterweizen. In S. Wolfrum (Ed.), 14. *Wissenschaftstagung Ökologischer Landbau*. München.

Tanimu, J., Iwuafor, E., Odunze, A. and Tian, G. 2007. Effect of incorporation of leguminous cover crops on yield and yield components of maize. *World Journal of Agricultural Sciences*, 3(2), 243–9.

Terhoeven-Urselmans, T., Scheller, E., Raubuch, M., Ludwig, B. and Joergensen, R. G. 2009. CO2 evolution and N mineralization after biogas slurry application in the field and its yield effects on spring barley. *Applied Soil Ecology*, 42(3), 297–302.

Thilakarathna, M. S., Papadopoulos, Y. A., Grimmett, M. G., Fillmore, S. A., Crouse, M. and Prithiviraj, B. 2017. Red clover varieties with nitrogen fixing advantage during the early stages of seedling development. *Canadian Journal of Plant Science*, 98(3), 517–26.

Thorsted, M. D., Olesen, J. E. and Weiner, J. 2006. Width of clover strips and wheat rows influence grain yield in winter wheat/white clover intercropping. *Field Crops Research*, 95(2), 280–90.

Thorup-Kristensen, K., Magid, J. and Jensen, L. S. 2003. Catch crops and green manures as biological tools in nitrogen management in temperate zones. *Advances in Agronomy*, 79, 227–302.

Tilander, Y. 1993. Effects of mulching with Azadirachta indica and Albizia lebbeck leaves on the yield of sorghum under semi-arid conditions in Burkina Faso. *Agroforestry Systems*, 24(3), 277–93.

Tilston, E., Pitt, D. and Groenhof, A. 2002. Composted recycled organic matter suppresses soil-borne diseases of field crops. *New Phytologist*, 154(3), 731–40.

Tittatelli, F., Bath, B., Ceglie, F. G., Garcia, M. a. d. C., Möller, K., Reents Jr., H., . . . Voogt, W. 2016. Soil fertility management in organic greenhouses in Europe: BioGreenhouse.

Tosti, G., Benincasa, P., Farneselli, M., Tei, F. and Guiducci, M. 2014. Barley–hairy vetch mixture as cover crop for green manuring and the mitigation of N leaching risk. *European Journal of Agronomy*, 54, 34–9.

Tscharntke, T., Klein, A. M., Kruess, A., Steffan-Dewenter, I. and Thies, C. 2005. Landscape perspectives on agricultural intensification and biodiversity–ecosystem service management. *Ecology Letters*, 8(8), 857–74.

Tscharntke, T., Sekercioglu, C. H., Dietsch, T. V., Sodhi, N. S., Hoehn, P. and Tylianakis, J. M. 2008. Landscape constraints on functional diversity of birds and insects in tropical agroecosystems. *Ecology*, 89(4), 944–51.

Tscharntke, T., Clough, Y., Wanger, T. C., Jackson, L., Motzke, I., Perfecto, I., . . . Whitbread, A. 2012. Global food security, biodiversity conservation and the future of agricultural intensification. *Biological Conservation*, 151(1), 53–9.

Ulber, L., Steinmann, H. H., Klimek, S. and Isselstein, J. 2009. An on-farm approach to investigate the impact of diversified crop rotations on weed species richness and composition in winter wheat. *Weed Research*, 49(5), 534–43.

Vakali, C., Zaller, J. G. and Köpke, U. 2011. Reduced tillage effects on soil properties and growth of cereals and associated weeds under organic farming. *Soil and Tillage Research*, 111(2), 133–41.

Van de Ven, G., Schröder, J., Hijbeek, R., Velthof, G. and van Ittersum, M. (2018). Comment on Schrama et al. (2018) 'Crop yield gap and stability in conventional and organic farming systems.' [Agric. Ecosyst. Environ. (256) 123–130]. *Agriculture, Ecosystems & Environment*, 261, 133–5.

Van Elsen, T. 2000. Species diversity as a task for organic agriculture in Europe. *Agriculture, Ecosystems & Environment*, 77(1), 101–9.

Van Ittersum, M. K., Ewert, F., Heckelei, T., Wery, J., Olsson, J. A., Andersen, E., . . . Flichman, G. 2008. Integrated assessment of agricultural systems–A component-based framework for the European Union (SEAMLESS). *Agricultural Systems*, 96(1), 150–65.

Van Rijn, P. C., Kooijman, J. and Wäckers, F. L. 2013. The contribution of floral resources and honeydew to the performance of predatory hoverflies (Diptera: Syrphidae). *Biological Control*, 67(1), 32–8.

Vargas, A. A. and Graham, P. H. 1988. Phaseolus vulgaris cultivar and Rhizobium strain variation in acid-pH tolerance and nodulation under acid conditions. *Field Crops Research*, 19(2), 91–101.

Vereijken, P. 1994. Designing prototypes. *Progress Report 1 of the Research Network on Integrated and Ecological Arable Farming Systems for EU and Associated Countries*. AB-DLO, Wageningen, The Netherlands.

Vereijken, P. 1997. A methodical way of prototyping integrated and ecological arable farming systems (I/EAFS) in interaction with pilot farms. *Developments in Crop Science*, 25, 293–308.

Vereijken, P., Wijnands, F. and Stol, W. 1995. Designing and testing prototypes *Progress Report 1 of the Research Network on Integrated and Ecological Arable Farming Systems for EU and Associated Countries*. AB-DLO, Wageningen, The Netherlands.

Verma, V., Pyare, R., Srivastava, A., Singh, V., Husain, K. and Goyal, G. 2016. Eco-friendly weed management options through allelopathy: A review. *Agriways*, 4(1), 93–106.

Versteeg, M., Amadji, F., Eteka, A., Gogan, A. and Koudokpon, V. 1998. Farmers' adoptability of Mucuna fallowing and agroforestry technologies in the coastal savanna of Benin. *Agricultural Systems*, 56(3), 269–87.

Vischer, D. 2005. Wässerwiesen, Wässermatten – eine irreführende Bezeichnung. *Géomatique Suisse*, 103(10), 572–4.

Vrignon-Brenas, S., Celette, F., Piquet-Pissaloux, A., Jeuffroy, M.-H. and David, C. 2016. Early assessment of ecological services provided by forage legumes in relay intercropping. *European Journal of Agronomy*, 75, 89–98.

Wall, G., Pringle, E. and Sheard, R. 1991. Intercropping red clover with silage corn for soil erosion control. *Canadian Journal of Soil Science*, 71(2), 137–45.

Wang, E. and Smith, C. J. 2004. Modelling the growth and water uptake function of plant root systems: A review. *Crop and Pasture Science*, 55(5), 501–23.

Waters-Bayer, A., Kristjanson, P., Wettasinha, C., Veldhuizen, L., Quiroga, G., Swaans, K. and Douthwaite, B. 2015. Exploring the impact of farmer-led research supported by civil society organisations. *Agriculture & Food Security*, 4(1), 1.

Watson, C., Younie, D. and Armstrong, G. 1999. Designing crop rotations for organic farming: Importance of the ley/arable balance. Designing and testing crop rotations for organic farming. DARCOF report(1), pp. 91–8.

Watson, C., Atkinson, D., Gosling, P., Jackson, L. and Rayns, F. 2002. Managing soil fertility in organic farming systems. *Soil Use and Management*, 18(s1), 239–47.

Watson, C. A., Reckling, M., Preissel, S., Bachinger, J., Bergkvist, G., Kuhlman, T., . . . Vanhatalo, A. 2017. Chapter Four - Grain Legume Production and Use in European Agricultural Systems. *Advances in Agronomy*, 144, 235–303.

Weller, S. 2011. Gründüngung einmal anders. *Bioland*, 2, 16–17.

Wezel, A., Casagrande, M., Celette, F., Vian, J.-F., Ferrer, A. and Peigné, J. 2014. Agroecological practices for sustainable agriculture. A review. *Agronomy for Sustainable Development*, 34(1), 1–20.

White, K. D. 1970. Fallowing, crop rotation, and crop yields in Roman times. *Agricultural History*, 44(3), 281–90.

Wichmann, S. 2003. Vergleich von Körnererbsen in Reinsaat und im Gemenge mit Sommergerste in Hinblick auf Ertrag und Ertragsentwicklung sowie N-Fixierungsleistung, Ernterückstandsmengen und Vorfruchtwirkung. *Beiträge zur 7. Wissenschaftstagung zum Ökologischen Landbau 'Ökologischer Landbau der Zukunft'*, 185–8.

Wijnands, F. 1999. Crop rotation in organic farming: Theory and practice. Paper presented at the *Designing and Testing Crop Rotations for Organic Farming. Proceedings from an International workshop*. Denmark: Danish Research Centre for Organic Farming.

Willey, R. 1979. Intercropping: Its Importance And Research Needs Part 1. Competition And Yield Advantages Vol-32: MPKV; Maharastra.

Wilson, G., Kang, B. and Mulongoy, K. 1986. Alley cropping: Trees as sources of green-manure and mulch in the tropics. *Biological Agriculture & Horticulture*, 3(2–3), 251–67.

Wilson, P. and Wyss, O. 1938. Mixed cropping and the excretion of nitrogen by leguminous plants. *Soil Science Society of America Journal*, 2(C), 289–97.

Woods, F. W. 1960. Biological antagonisms due to phytotoxic root exudates. *The Botanical Review*, 26(4), 546–69.

Yenish, J. P., Worsham, A. D. and Chilton, W. S. 1995. Disappearance of DIBOA-glucoside, DIBOA, and BOA from rye (*Secale cereale* L.) cover crop residue. *Weed Science*, 43(1), 18–20.

Zewdu, T. 2003. Effect of defoliation and intercropping with forage legumes on maize yield and forage production. *Tropical Science*, 43(4), 204–7.

Zhang, S. and DePaolo, D. J. 2017. Rates of CO_2 Mineralization in Geological Carbon Storage. *Accounts of Chemical Research*, 50(9), 2075–84.

Zikeli, S., Gottschlich, H., Fürnkranz, M. and Szin, S. 2016. Gemüsebau. In B. Freyer (Ed.), *Ökologischer Landbau. Grundlagen, Wissensstand, Herausforderungen* (pp. 367–91). Stuttgart: UTB.

Zundel, C., Schneider, M., Musyoka, M., Muriuki, A., Vanlauwe, B., Mucheru-Muna, M., . . . Niggli, U. 2009. Long-term performance of organic crop rotations in the tropics: First results from a high and a medium potential site in sub-humid Central Kenya. African Organic Conference, Kampala, Uganda, 19–22 May 2009. In: Ssekyewa, C. and Neuhoff, D. (Eds), *Fast tracking sustainable development in Africa through harnessing Organic Agriculture and Bio-technology* (pp. 30–1). Kampala, Uganda: Uganda Martyrs University, Faculty of Agriculture.

Chapter 6

Conservation tillage in organic farming

Maike Krauss and Paul Mäder, Research Institute of Organic Agriculture (FiBL), Switzerland; Joséphine Peigné, ISARA-Lyon, France; and Julia Cooper, Newcastle University, UK

1 Introduction
2 Main benefits of conservation tillage
3 Main challenges of conservation tillage
4 Future trends
5 Conclusion
6 Where to look for further information
7 References

1 Introduction

Improved soil erosion control, increasing soil fertility, retaining soil moisture and resource efficiency (including labour and fuel) has driven farmers towards more conservation tillage practices and away from conventional ploughing methods. This development started roughly in the 1930s after the 'Dust Bowl' events in the US prairies where widespread ploughing caused massive wind erosion (Lal et al., 2007). Ploughing inverts the soil layers and entirely buries all crop residues, weeds and fertilisers, leaving a bare soil surface with often less structured and aggregated soil from deeper layers on top. The 'mixed blessings' from ploughing (Lal et al., 2007) encompass risks, like soil erosion, crusting, degradation, loss of soil organic carbon (SOC), nutrient depletion, subsoil compaction and eutrophication of surface waters, and benefits from aeration, nutrient mineralisation to a good weed control, promoting crop growth (Lal et al., 2007). Furthermore, aeration in combination with soil warming in spring, which is achieved by ploughing, is important for early nutrient provision, influencing yields to a great extent in cool humid climates (Soane et al., 2012). In drier and warmer climates, early spring mineralisation is normally not limited, which may explain the widespread adoption of conservation tillage and, especially, no-tillage in semi-arid regions (Derpsch et al., 2010), while farmers in temperate regions of, for example, Europe are more reluctant to adopt conservation tillage techniques (Carr, 2017; Casagrande et al., 2015; Mäder and Berner, 2012; Massucati, 2013; Wilhelm, 2010).

Broadly, conservation tillage addresses the first two elements of the conservation agriculture approach promoted by the Food and Agriculture Organisation of the United Nations (FAO) to meet future demands in a sustainable way: 'continuous minimum

http://dx.doi.org/10.19103/AS.2017.0029.06

mechanical soil disturbance' and 'permanent organic soil cover' (FAO, 2017). The second and third objectives ['diversification of crop species grown in sequences and/or associations' (FAO, 2017)] are met in organic farming systems. Thus, integration of conservation tillage into organic farming might be a promising direction to improve the sustainability of its farming approach.

What exactly is conservation tillage? The Conservation Technology Information Center defines conservation tillage by specifying the outcome of the tillage operation: that plant residue cover of the soil surface should be at least 30% (CTIC, 2017), rather than describing tillage techniques. In the organic farming context, conservation tillage strategies have been described more by practice rather than outcome and include i) reduced tillage systems 'that operate at shallower depths or at lower intensity compared with customary ploughing in a given region' (Mäder and Berner, 2012, p. 8) and ii) cover-crop-based no-tillage systems that evolved mainly in the United States (Silva and Delate, 2017) and Canada (Halde et al., 2017). Reduced tillage can mean less frequent operations, shallow inversion tillage or non-inversion tillage (Zikeli and Gruber, 2017). These are more frequently practised than organic no-tillage strategies, which have only slowly spread in Europe (Vincent-Caboud et al., 2017). These organic no-tillage systems use cover crops that are terminated by a roller crimper instead of herbicides, with subsequent direct drilling of mainly row crops like maize and soybean, but also cereals (Silva and Delate, 2017; Vincent-Caboud et al., 2017). In light of the range of conservation tillage systems used in organic farming, we have tried in Fig. 1 to give an overview of techniques and machines used. Reducing tillage intensity by ploughing to a shallower depth and using on-land ploughs that drive only on non-tilled land are one form of conservation tillage, but the appropriate tillage depth for shallow inversion tillage is still a matter of debate. This view depends much on the common ploughing depth used in a region that varies from 20 cm to 40 cm. Shallow inversion tillage would thus vary between 10 and 20 cm. Chisel ploughs with mixing sweeps can be applied for deep loosening and it is debatable whether this is still a form of reduced tillage. Apart from this, there are many options including different kinds of mouldboard, double-layer or skim ploughs, chisel ploughs, disk harrows and cultivators or power take-off driven machines for superficial mixing. Another type of conservation tillage is strip tillage, where only strips and not the whole soil surface are tilled. And there are also approaches of ridge tillage that were considered reduced (Zikeli and Gruber, 2017). Reducing tillage intensity to an absolute minimum is finally achieved by the so-called 'no-tillage' or 'direct-seeding' systems. In organic farming, crops are thereby directly sown into a mulch cover of a rolled green manure (Silva and Delate, 2017; Vincent-Caboud et al., 2017).

There is no systematic information on the share of organic farmers using conservation tillage globally. Wilhelm (2010) performed a survey among 367 German organic farmers of whom 56% used ploughing solely, mainly to depths <20 cm, and 40% omitted ploughing at least once in a rotation. A survey among organic conservation tillage farmers in Europe revealed that farmers from the southern Mediterranean part apply conservation tillage more than those in northern Europe (Casagrande et al., 2015). In Switzerland, conservation tillage is subsidised since 2014 with additional payments for herbicide-free applications. This attracts more and more farmers. In 2015, about 17% of land under organic management was worked with conservation tillage, with organic no-tillage accounting for <1% (Federal Office for Agriculture, Switzerland, unpublished).

While farmer motivations regarding conservation tillage clearly mirror attitudes towards soil and environmental conservation (Casagrande et al., 2015), the challenges are more

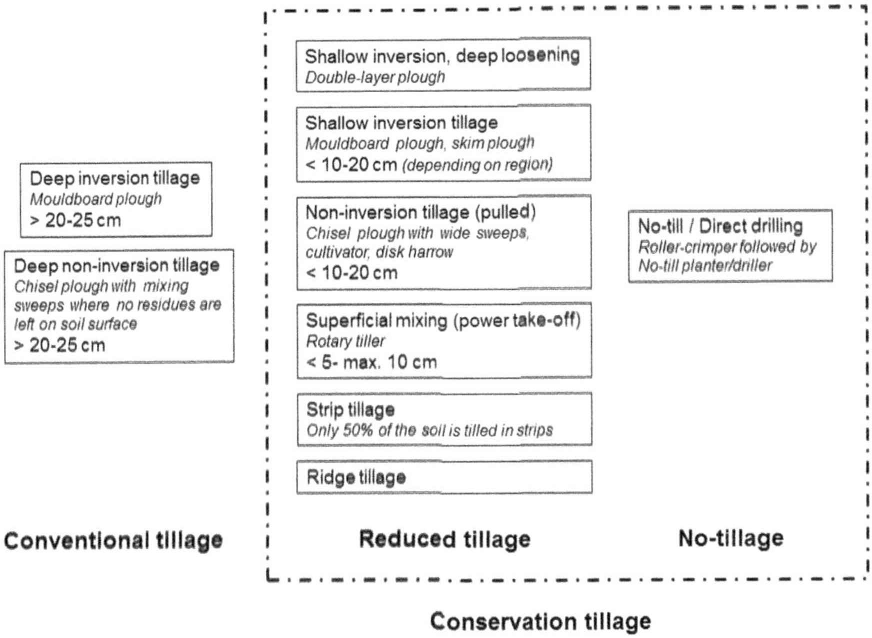

Figure 1 Overview of conservation tillage techniques used in organic farming.

diverse. These encompass concerns about yield variability and crop failure, weed control, timely nutrient supply to crops and technical aspects like lack of machinery and skills along with high costs (Zikeli and Gruber, 2017; Casagrande et al., 2015; Vincent-Caboud et al., 2017), as well as occurrence of pests and diseases (Silva and Delate, 2017). Further development of adjusted rotations and machinery as well as approaches like occasional ploughing or 'strategic tillage', which will be discussed later in this chapter, can provide more flexibility while still benefiting from increased soil fertility.

This chapter reviews the state of the art in research concerning the main benefits and challenges in adopting conservation tillage for organic farmers. Besides scientific articles, conference proceedings, public articles and theses were also included to broaden the database and to address practical farming aspects. Practical experiences from two on-farm case studies in France and Switzerland are presented in text boxes. An outlook on future trends serves to complement ongoing research activities and needs.

2 Main benefits of conservation tillage

Numerous studies have been published, highlighting the benefits of conservation tillage strategies for delivery of key ecosystem services (Holland, 2004; Powlson et al., 2014). In many cases these ecosystem services are already enhanced in organic farming systems (Reganold and Wachter, 2016) and there may be added benefit from implementing conservation tillage within an organic system. Here, we only report results gained from experiments on conservation tillage systems in organic farming. Additive effects can be hard to quantify as this would require design and implementation of complex factorial experiments looking at combined organic/conventional and conservation tillage/ploughing treatments which are seldom available [exception e.g. Wittwer et al. (2017)].

2.1 Soil physical properties

Water infiltration and holding capacity, porosity for aeration and root penetration and finally erodibility and susceptibility to sealing or crusting of the soil surface are dependent on the soil physical properties, especially soil aggregation. Aggregate stability is closely related to SOC levels (Abiven et al., 2009).

SOC concentrations have been widely described to be enriched in the upper soil layer in organic farming (Mäder et al., 2002; Gattinger et al., 2012) and conservation tillage systems (Luo et al., 2010; Soane et al., 2012) and consequently also in organic conservation tillage systems (Cooper et al., 2016). Changing SOC dynamics were already reported soon after conversion from ploughing to conservation tillage management. Light SOC fractions (especially particulate organic matter, POM) increased in topsoils of organic arable fields (Butler et al., 2016; Lewis et al., 2011) but less in intensive vegetable production (Pritchett et al., 2011). More water stable aggregates or aggregate stability in topsoils of organic conservation tillage systems compared with ploughing were reported by Emmerling (2007), Bilalis and Karamanos (2010), Vakali et al. (2011) and Crittenden and de Goede (2016) in various crop rotations and pedoclimatic conditions, while Fontana et al. (2015) found an increased porosity and stability in reduced tilled soils.

There have been, however, no studies measuring soil erodibility in organic conservation tillage systems directly so far. Compared with ploughing, Krauss et al. (2010) observed higher grass–clover yields and clover percentages in the reduced tillage system during dry years, indicating a higher water availability in reduced tillage under drought stress. Crittenden and de Goede (2016) also reported increased plant available water in 0–5 cm in reduced tillage related to ploughing. To further substantiate these soil structure effects on water supply (see Chapter 9) and erosion, more studies are needed.

2.2 Soil fauna

Compared with soil physical aspects, impacts of organic conservation tillage on indicators of soil biological quality such as microbial activity and biomass, as well as diversity, have been shown by many studies. Reducing tillage intensity with its effects of a lower soil disturbance and SOC enrichment seem to promote overall soil life from the macro- to the micro-fauna, especially in topsoils.

The lack of soil mixing in conservation tillage leads to a typical distribution of SOC with concentrations decreasing gradually with increasing soil depth (Krauss et al., 2017c). Enrichment of SOC in topsoils provides food for the soil micro- to macro-fauna, which is mirrored by higher microbial biomass contents in topsoils in reduced tillage systems compared with ploughing (Emmerling, 2007; Krauss et al., 2017c; Lehocka et al., 2009; Sun et al., 2016; Vian et al., 2009). Microbial activity was, however, not always found to be higher in reduced tillage systems. For example, basal respiration was similar in reduced tillage and ploughing systems (Vakali et al., 2011) or higher in reduced tillage (Emmerling, 2007), while increased dehydrogenase or phosphatase activity was found in reduced tilled plots (Gadermaier et al., 2012; Emmerling, 2007).

Microbial community structure and diversity may also be affected by conservation tillage systems, although the impacts of these changes on delivery of ecosystem services are not always clear. Higher ergosterol contents detected by Emmerling (2007) suggested a shift to more fungal-based communities in reduced tillage in arable soil, which was emphasised by phospholipid fatty acid analysis in other long-term experiments (Kuntz et al., 2013; Sun et al., 2016). Pronounced enrichment of SOC and reduced disturbance were suggested as the reasons why fungal communities seem to be favoured in arable reduced tillage systems in comparison to ploughing (Kuntz et al., 2013). In vegetable production, however, such a community shift could not be observed with the change to a low tillage intensity (Pritchett et al., 2011).

Some fungal groups like arbuscular mycorrhizal fungi (AMF) are known to improve water and nutrient uptake from the soil environment (Parniske, 2008). AMF have been suggested to be sensitive to tillage operations that affect hyphae and the distribution of spores. AMF are thus argued to be less affected in no-tillage than ploughed systems (Kabir, 2005). A higher root colonisation (Bilalis and Karamanos, 2010), AMF spore density and species richness (Säle et al., 2015) as well as a higher intraspecific diversity of *Glomus intraradices* (Börstler et al., 2010) confirm this observation for organic reduced tillage systems. Yet, no impacts of a no-tillage system on AMF were also found after 21 years of conversion from ploughing to organic reduced tillage (Sun et al., 2016). Other studies on microbial communities report occasional community shifts, for example, of gram-negative bacteria, protozoa (Kuntz et al., 2013) or nitrifiers (Krauss et al., 2017a) that were more abundant in reduced tilled clayey topsoil compared with ploughing. Gram-positive and -negative bacteria were, however, not more abundant in loamy soil under reduced tillage (Sun et al., 2016). For two out of five species, collembola abundance was greater in the topsoil layer of a reduced tillage system with superficial rototilling and subsoiling compared with ploughing (Petersen, 2002). More studies are needed to substantiate the tillage system effect on micro- and mesofaunal communities and to link their abundance and composition to their functional role towards crop growth and soil fertility.

In contrast, effects on earthworms are well studied. Earthworms profit from a reduced tillage intensity in organic farming, as has been documented also in conventional systems. Crittenden et al. (2014) observed a decrease in earthworm abundance directly after tillage operations, which was more pronounced after ploughing than in the reduced tillage system. They related this to direct effects like physical damage but also to indirect effects like food redistribution. In organic reduced tillage systems, increased earthworm abundances and/or biomass were thus often (Emmerling, 2001; Kainz et al., 2005; Kuntz et al., 2013; Lehocka et al., 2009) though not always observed compared with ploughing (Metzke et al., 2007; Peigné et al., 2009). Decreased earthworm abundance compared with ploughing in a medium term of four years was also reported (Crittenden et al., 2014). On the other hand, organic no-tillage further increased earthworm populations in comparison with both reduced tillage and ploughing in a French study (Peigné et al., 2009) and occasional reduced tillage was also found to already increase earthworm abundances (Moos et al., 2016). In particular, juvenile earthworms (Emmerling, 2001; Kuntz et al., 2013), anecic species (Kainz et al., 2005; Metzke et al., 2007) and also endogeic species (Kuntz et al., 2013) were favoured by reduced tillage systems. Kuntz et al. (2013) concluded that less disturbance and a better food supply in reduced tillage are expressed in lower efforts to rebuild burrows and thus a higher reproduction rate.

2.3 Energy and labour efficiency

The reduction in energy and labour requirements under organic conservation tillage is often claimed but has not yet been systematically assessed. Crowley and Döring (2012) reported that organic reduced tillage saved fuel use and working hours compared with conventional tillage, while Schader et al. (2011) reported negligible savings, as more passes were required. In contrast, organic no-tillage seems to be more efficient than ploughing with regard to fuel use and working hours (Bernstein et al., 2011; Köpke and Schultes, 2008). Due to more unstable and frequent lower yields, gross revenues were, however, lower in organic no-tillage (Bernstein et al., 2011, Halde et al., 2017, Silva and Delate, 2017). In conventional conservation tillage systems, reduced fuel and machine operation costs are often offset by higher energy use due to increased herbicide and fertiliser application (Zentner et al., 2004). This might explain why these systems were comparable to organic plough-based systems from a whole farm perspective (Lynch et al., 2011). As organic farming was also shown to be more energy efficient than conventional farming on an area basis and often but not always on a product basis (Gomiero et al., 2008), overall, organic conservation tillage may be the most efficient system. This, however, has to be underlined with a broader database of studies at farm level.

2.4 Impact on climate change

While addressing effects on climate change, three main aspects have to be considered when comparing tillage systems: sequestration of carbon dioxide (CO_2) from the atmosphere as SOC into the soil, direct emissions of nitrous oxide (N_2O) and potential savings in diesel fuel. Methane (CH_4), a further greenhouse gas, is of minor importance concerning aerated upland soils. While a slight uptake was expected by conversion from ploughing to conservation tillage (Hütsch, 2001) this could not be confirmed for organic farming conditions in a field monitoring (Krauss et al., 2017c).

Enhanced SOC sequestration has often been claimed for conservation tillage systems compared with ploughing, although carbon stock measurements including subsoil did not confirm this (Luo et al., 2010). Yet, sampling bias and different approaches in the calculation of carbon stocks question the quality of studies (Luo et al., 2010; Powlson et al., 2014). Carbon stocks were mainly determined for conventional farming systems and with a lack for organic farming. Taking into account reported SOC stocks from different soil layers and performing a random-effect meta-analysis on the five studies available (Crittenden et al., 2015; Kainz et al., 2005; Krauss et al., 2017c; Schulz et al., 2014; Zikeli et al., 2013), a greater stratification but no overall increase in total SOC stocks was indicated when reduced tillage was used in organic systems (Krauss et al., 2017b) similar to the results of Luo et al. (2010). This preliminary finding will need to be validated with a larger number of studies, as there was a net increase in SOC stocks on some occasions (Krauss et al., 2017c).

Nitrous oxide (N_2O) is a potent greenhouse gas contributing 298 times more to radiative forcing than CO_2 (Myhre et al., 2013). Agriculture contributes to about 60% of anthropogenic N_2O emissions globally with direct emissions from arable fields as a main driving factor (Syakila and Kroeze, 2011). Conservation tillage is considered as an N_2O mitigation strategy. However, indicating higher N_2O emissions in the years after conversion from ploughing, only a few differences between tillage systems in humid climates and lower N_2O emissions in conservation tillage systems in dry climates were measured on average in long-term studies (<10 years) (van Kessel et al., 2013). In organic farming systems, there are relatively few field studies.

In temperate organic arable systems, Krauss et al. (2017c) found no differences between reduced tillage and ploughing in the cropping seasons of grass–clover and winter wheat. N_2O pulse emissions were triggered by management operations only. In this regard, higher N_2O fluxes were detected in the reduced tillage system in the weeks after tillage operations, which was associated with the higher SOC content and in turn higher microbial activity. In organic vegetable production, no overall tillage system effect (Yagioka et al., 2015) or slightly higher N_2O emissions in reduced tilled soils were reported (Kong et al., 2009). There is thus no clear trend, and the variability between sites and years may be larger than that of the tillage system.

Pedoclimatic site conditions consequently seem to largely drive SOC stocks and N_2O emissions under organic farming conditions, which makes it difficult to draw more general conclusions. As mentioned in the section earlier, there might be a potential in saving fuel in organic reduced tillage, even more so in organic no-tillage systems. This needs further verification and means that effects of a change from ploughing to conservation tillage in organic systems is likely to range from no impact on climate change to a small mitigation potential on an area-scale. Considering the lower average yields, the potential may be even smaller on a yield-scale basis.

3 Main challenges of conservation tillage

3.1 Yields

A recent meta-analysis on yields in organic conservation tillage systems including published and unpublished work summarises available data (Cooper et al., 2016). On average, different organic reduced tillage systems reduced yields by 7.6% compared with deep ploughing (873 observations pairs, 21 studies). While shallow ploughing reduced yields by about 5.5% and shallow non-inversion tillage by only 3%, the use of deep non-inversion tillage within organic systems showed the largest gaps, with 11.6% yield reductions compared to organic deep inversion tillage systems. Most data covered temperate climates. Reductions were similar in loamy and sandy soils. Shallow non-inversion, however, seemed to achieve similar yields in loamy soils. Combinable crops such as cereals and maize were more affected by yield reductions than root crops where secondary tillage operations seemed to help in suppressing weeds. There was a huge variation between yield differences in cover crops and leys with no significant difference in the mean. The yield gap between reduced tillage and ploughing was constant throughout the decade where field data were available. However, most yield data were reported from the first five years after conversion. There is thus the need for more data in drier climates and in the long run, but also relating to hybrid systems where strategic tillage is used at critical stages (e.g. ley termination, excessive weed infestation) to elucidate whether a reduced tillage system can be designed in a way where yield stability is assured. Results of Kainz et al. (2003) on reduced ploughing frequency head in that direction and need further substantiation.

Apart from a higher weed infestation in organic reduced tillage systems (Cooper et al., 2016), there are indications that the lack of nitrogen availability is another major driver limiting yields in reduced tillage systems. Shoot nitrogen in barley and barley yield was, for example, lower in a reduced tilled system where lower nitrate contents in soil was measured compared with ploughing (Vakali et al., 2015). This effect was, however, only seen in spring-sown barley and not in autumn-sown rye. Lack of nitrogen in spring may be balanced by cover crops: Higher grass-clover yields and an additional legume catch crop boosted silage maize performance after ley termination with about 34% higher yields in reduced tillage (Krauss et al., 2010). Cover crops decreased the yield gap between reduced tillage and ploughing also in winter wheat and maize at another site. Cover crops had a positive effect on both crop nutrition and weed suppression in this case (Wittwer et al., 2017). An adjusted nutrient management is consequently important for the success of reduced tillage systems and should be the focus of upcoming studies.

3.2 Weeds

In organic reduced tillage systems, an increase in weed abundance and/or biomass in relation to ploughing is commonly described (Armengot et al., 2015; Bilalis et al., 2012; Butler et al., 2016; Crowley and Döring, 2012; Gruber and Claupein, 2009; Massucati and Köpke, 2014; Sans et al., 2011; Schmidt, 2010; Vakali et al., 2002, 2011; Weber et al., 2017; Wittwer et al., 2017; Yagioka et al., 2014) though weed infestation is not always higher throughout the growing season (Crowley and Döring, 2012) or in each crop of a rotation (Armengot et al., 2015; Schulz et al., 2009, 2015). Similar yields between tillage systems despite a higher weed infestation in reduced tillage were observed at some sites (Armengot et al., 2015; Gruber and Claupein, 2009; Schulz et al., 2009), challenging the perception of weeds as a critical competitor in general. Indeed, as tillage operations impact weed abundance and also stimulate nutrient mineralisation, it is difficult to attribute yield decreases in reduced tillage to either a higher weed infestation or a deficiency in nutrient supply (Gruber and Claupein, 2009; Schulz et al., 2009).

Apart from this short-term perspective, weed infestation and also community composition need to be critically monitored in the long run. The soil weed seedbank was found to be 50–60% larger in organically managed chisel-ploughed plots compared with shallow, double layer and deep ploughing after six years (Gruber and Claupein, 2009). The seedbank of Canada thistle (*Cirsium arvense*) was reduced to low levels by ploughing, which was reflected by lower thistle abundance in relation to chisel ploughing. Additional stubble tillage (tillage after harvest under dry conditions in summer for weed control and residue incorporation), but more importantly a grass–clover ley, decreased infestation by *Cirsium arvense* in reduced tillage plots (Gruber and Claupein, 2009). A slight decrease in *Cirsium arvense* was found in shallow compared with deep ploughing [Roda trial in Schmidt (2010)]. A shift of the weed community to more problematic perennial weeds like *Convolvulus arvense* and *Alopecurus myosuroides* (Armengot et al., 2015) or *Apera spica-venti* and *Cirsium arvense* [Güterfelde and Frankenhausen trial in Schmidt (2010)] was reported with decreasing tillage intensity. At a different site, failure of ley termination led to increased occurrence of volunteer grass species in the following years (Kainz et al., 2005). These examples show that the potential of reduced tillage to promote problematic weeds or volunteers and to increase weed seedbanks, particularly in organic systems, requires even more careful rotation management and correct implementation of machinery.

Armengot et al. (2016), who analysed weed communities over a range of pedoclimatic conditions, confirmed that weed communities in organic reduced tillage systems produced more seeds and favoured perennials and also showed that communities had a lower affinity to nutrients and were smaller in size than communities in ploughed systems. Their competitiveness towards cash crops and their functional role is thus not completely clear.

As several reviews in Europe, Northern America and Canada have well summarised (Halde et al., 2017; Silva and Delate, 2017; Vincent-Caboud et al., 2017), weed suppression in organic no-tillage farming is most crucial for its success. This is effective when the cover crop mulch or residues are sufficiently dense for light competition (Bernstein et al., 2011; Bilalis et al., 2012; Halde and Entz, 2014; Massucati and Köpke, 2011). In contrast, cover crop failure leads to a high weed infestation and yield losses (Böhler and Dierauer, 2017; Weber et al., 2017). High seeding densities and cover crop varieties that yield a high biomass are therefore crucial for organic no-tillage. Research on good cover and cash crop combinations is still ongoing. Rye–soybean, oat–faba bean, hairy vetch or winter pea–maize and rye–vegetable sequences are currently most promising combinations in weed suppression (Böhler and Dierauer, 2017; Halde et al., 2017; Massucati and Köpke, 2011; Porter et al., 2005; Silva and Delate, 2017).

3.3 Soil density, microclimate and nutrient supply

In a review on no-tillage in conventional agriculture in Europe, Soane et al. (2012) reported that no-tillage reduces soil compaction below the plough furrow and increases bearing capacity and trafficability. However, according to soil and climate conditions, no-tillage could also increase topsoil density and may be unsuitable for poorly structured sandy soils and poorly drained soil (Soane et al., 2012). Compared with ploughing, penetration resistance measurements (Crittenden et al., 2015; Schulz et al., 2011), bulk density distribution (Krauss et al., 2017c (Supplement); Peigné et al., 2009) and morphological assessment of soil structure (Peigné et al., 2013; Vian et al., 2009) revealed similar densities in topsoils but denser soil layers below the new tillage layer till the former ploughing depth in conservation tillage systems for a range of soil textures. This seems to be a soil density increase due to the lack of tillage-induced loosening rather than compaction caused by heavy machinery that e.g. cause plough pans.

Subsoil non-inversion loosening with intensive secondary tillage (Munkholm et al., 2001) and deep rooting tap roots are approaches to remove plough pans and may help to improve porosity in lower soil layers (Kautz, 2015; Köpke et al., 2015). Applying no-tillage systems on a longer run also showed that plough pans can be remediated (Peigné et al., 2013). According to Soane et al. (2012) or Holland (2004), soil structure improves with no and reduced tillage after a transition period of several years. This is, however, not supported for the whole soil profile and also seems to depend on soil texture. Peigné et al. (2018), for example, showed that the topsoil of a sandy loam soil was still denser after ten years of organic no-tillage and reduced tillage in organic farming compared with ploughing. The natural compaction of sandy soils due to their grain texture seems to be more difficult to manage. And in a clayey soil, continuous tilling to only 10 cm of soil depth led to a dense soil layer between 10 and 20 cm after twelve years (Krauss et al., 2017c, Supplement), despite higher earthworm abundances (Kuntz et al., 2013).

Nutrient supply may also be impacted in both positive and negative ways by reduced tillage intensity (Soane et al., 2012). In years with weather conditions optimal for grass-clover and cereal growth, Krauss et al. (2017c, Supplement) measured higher topsoil nitrate contents in reduced tilled soils, suggesting that these soils were microbially more active and could mineralise soil nitrogen better than ploughed soils. In contrast, insufficient nutrient supply to crops in wet springs in the upcoming years provide evidence that this high bulk density could be problematic for aeration and timely mineralisation, impacting crop growth in unfavourable years. Data available from tracking available nitrogen concentrations during spring and autumn over several years in the study of Schulz et al. (2011) support these observations. They measured lower N concentrations in 0–90 cm of chiselled soils in four out of six spring sampling dates (although a lack of climate data limit potential for interpretation with regard to soil temperatures and moisture). In organic rotational no-tillage systems applied in cool humid climates, lower nitrate contents were measured across three sites on different sampling dates, although this did not affect crop N uptake at two of three sites (Halde and Entz, 2014). Lack of density measurements makes it difficult, however, to link low nitrate concentrations to soil density and thus aeration problems. There are consequently some indications that a high soil density by conversion to conservation tillage may impose problems in aeration and consequently nutrient mineralisation under cool moist conditions. This is a more important challenge in organic farming as plants rely more on nutrient supply from the soil pool for their nutrition (Watson et al., 2002). However, systematic soil structure and nutrient availability measurements to substantiate this relationship are lacking.

3.4 Pests and diseases

Experiences from conventional conservation tillage studies show that residues left on the soil surface may provide shelter not only for beneficial organisms but also for pests and diseases (Page et al., 2013; Roger-Estrade et al., 2010; Sturz et al., 1997). They, however, emphasise that increased above- and below-ground biological activity found in conservation tillage systems may suppress pest and disease organisms by competition at the same time. Page et al. (2013) thus recommend putting emphasis on crop rotations, to use resistant cultivars and to promote antagonists for the regulation of pests and diseases regardless of the tillage system, which is already common practice in organic farming. Practical experiences of cover crop/cash crop combinations with suppressive character that are specifically suitable for conservation tillage systems – systematically documented and spread among farmers – may be sufficient for pest and disease regulation in organic conservation tillage systems.

This may explain why the few studies on pest and disease occurrence in organically managed tillage system comparisons report higher fluctuations between years than tillage systems regarding the abundance of entomopathogenic *Metarhizium anisopliae* (Jabbour and Barbercheck, 2009) or *Fusarium gramineum* and respective deoxynivalenol (DON) concentrations in wheat (Munger et al., 2014). In untreated wheat in the study of Koch et al. (2006), DON concentrations were also similar between tillage systems and more affected by the choice of more or less *Fusarium* resistant cultivars. In potatoes, infestation with wireworms (*Agriotes* spp.) was higher in ploughed soils than soils that were tilled with a chisel plough (Schepl and Paffrath, 2003). In contrast, army worm (*Mythimna unipuncta*) and corn maggot (*Delia platura*) infestation were occasionally problematic in organic cover crop based no-tillage, where cover crops acted as hosts between main crops (Silva and Delate, 2017). Similarly, cabbage pests like cabbage worm (*Pieris rapae*) but not the diamondback moth (*Plutella xylostella*) profited from no-tillage rye mulch, while rye–pumpkin sequences showed no increased incidence on pests and diseases (Halde et al., 2017). Silva and Delate (2017) thus recommended careful planning of cover crops within the rotations to avoid a carryover effect.

3.5 Suitable farm machinery

There has been considerable development in machines suitable for conservation tillage systems. For reduced tillage systems, machinery used is not strictly defined and different approaches are feasible (see Fig. 1). Regarding soil inversion techniques, mouldboard ploughs that are able to work in shallower depths may be suitable, with on-land ploughs being the more soil-friendly alternative. On-land ploughs drive only on untilled land, thus reducing subsoil compaction. In Switzerland, for example, mouldboard ploughing that can regulate tillage depth to a maximum of 10 cm are considered as a reduced tillage technique for the subsidiary scheme (unpublished, source: Federal Office for Agriculture, Switzerland). Double-layer ploughs were introduced to turn the soil shallowly and to loosen the soil to deeper layers simultaneously, thus preserving the naturally stratified soil to a large extent (Zikeli and Gruber, 2017). Overall, all kinds of pulled machines where tillage depths can be adjusted by wheels or rollers – like skim ploughs for shallow inversion, chisel ploughs with mainly flat sweeps that undercut the soil entirely, disc harrows and cultivators for superficial soil mixing – are feasible and used at different soil depths. These machines are the most widely used and several companies offer a great variety in the meanwhile. There are also specifically developed systems that some farmers apply. Ridge systems, like the Turiel or Kemink system, rely on deep loosening and the establishment of ridges (Müller et al., 2009; Zikeli and Gruber, 2017). Power take-off driven machines like rotary tillers and rotary hoes are used to mix the soil superficially to a few centimetres. The Glyphomulch or Geohobel that are specialised rotary tillers were developed to directly incorporate green manures without antecedent mulching. Strip tillage is another approach that turns only parts of the soil, leaving fixed untilled strips in the field that provide good trafficability and erosion control (Morris et al., 2010). While strip tillage is successfully applied in conventional farming with herbicide application, it may not be suitable in organic farming due to difficulties with weed regulation (Dierauer, 2017). There is thus a great variety which means that farmers need to know which machine is most suitable for their conditions and how they are used. Ordering new machines is costly and seen as a constraint in adopting conservation tillage (Casagrande et al., 2015). A good starting point for farmers is thus the possibility to get in touch with companies during field days with machine demonstrations or to borrow new machinery from neighbours. This avoids large costs in the beginning.

Direct seeding or no-tillage requires seeders that are able to cut through a thick mulch layer, to enable good contact between the seeds and the soil and to close the seeding slot. In organic-cover-crop-based no-tillage systems, a roller crimper is additionally used to roll but not cut dense cover crop stands before seeding (Halde et al., 2017; Silva and Delate, 2017; Vincent-Caboud et al., 2017). Afterwards, no additional mechanical weeding is necessary as the thick mulch layer covers the soil, preventing weeds from growing due to light competition. A prerequisite to success is that the cover crop is successfully terminated and enough biomass is available, ideally exceeding 4–5 t ha^{-1} (Massucati and Köpke, 2011; Silva and Delate, 2017). Specific recommendations and experiences concerning cover crop choice and rotations for organic no-tillage can be found in Silva and Delate (2017) and Böhler and Dierauer (2017).

3.6 Farmers' reluctance

According to a survey conducted among 159 European organic farmers in 2012, Casagrande et al. (2015) highlighted the main motivations and problems encountered by organic farmers when they adopt conservation agriculture practices. Several conservation agriculture practices were described, including reduced tillage and no-tillage techniques.

Among farmers using conservation tillage techniques, no-tillage was used only by 27% of the interviewed farmers for an average duration of 5.5 years, whereas reduced tillage was used by 89% of the interviewed farmers for an average duration of 19 years. Moreover, the plough had not been totally abandoned by farmers and was still used for specific crops or in weed-infested fields. Mean frequency of ploughing in an otherwise no-tillage field was 1.2 times in 20 years, whereas in reduced tillage fields it was twice in ten years (Peigné et al., 2015a). Thus, most interviewed farmers adopt a 'hybrid' approach where soil inversion techniques such as ploughing and reduced or no-tillage practices were both used strategically at appropriate stages of the rotation.

No-tillage is rarely practised by organic farmers in Europe (Casagrande et al., 2015), and its adoption is quite recent compared with reduced tillage. Farmers using no-tillage are mainly located in the southern part of Europe. The main problems encountered by farmers using no-tillage in organic systems are in order of importance: weed infestation and thus limited weed control, lack of adequacy of machinery, information and specific technical skills, and lower or unstable crop yields (Casagrande et al., 2015). As ploughing was traditionally also used to control weeds in organic farming, it is not surprising that weeds are the main problems in organic no-tillage systems. Associated with this problem, organic farmers cited the lack of machinery and above all skills and knowledge to face this problem.

Reduced tillage is more common than no-tillage among interviewed European organic farmers. As for no-tillage, the main challenges encountered were weed infestation and control, lack of adequate machinery and technical issues like the termination of green manures or leys without ploughing.

According to Casagrande et al. (2015), there are three different categories of farmer attitudes towards conservation agriculture techniques. The first group of farmers, called 'indifferent farmers', do not present general motivations and problems; they use conservation agriculture techniques such as reduced tillage or no-tillage because of specific farm constraints such as stony fields. The second group, termed 'soil conservationist farmers', are highly motivated to use reduced tillage or no-tillage to improve soil fertility. Those farmers have a positive attitude towards conservation tillage and do not regard challenges as too problematic. In contrast, the third group, termed 'agro-technically challenged farmers', experience a lot of challenges when adopting conservation tillage techniques. They still persist with this practice but could not easily find solutions.

In conclusion, there are several problems, such as weed control and lack of adequate machinery and skills, that limit adoption of conservation tillage techniques in organic farming. However, organic farmers adopting or trying to adopt these techniques are very motivated above all to preserve soil fertility and to work on agronomical and technical solutions.

4 Future trends

Continuous reduced tillage or no-tillage in organic farming is agronomically challenging according to the drawbacks summarised in preceding sections. A hybrid system of reduced tillage and occasional ploughing at key stages, maybe occasional no-tillage, seems to be the practical solution to meet agronomic viability while still preserving soil fertility. Such a hybrid system is also termed as 'strategic tillage' and is already practised by some organic farmers (Peigné et al., 2015b; Zikeli and Gruber, 2017).

In organic farming, there are occasions where ploughing is advantageous. Ley termination without a plough is, for example, difficult as regrowth of perennial grasses and legumes should be avoided. Successful reduced ley termination can, however, be achieved in dry conditions (see Box 1). Also weed-sensitive crops like soybean or root crops that require loose soil are challenging reduced tillage approaches. Mechanical weeding may be disadvantageous in terms of fuel and labour costs in this case. In contrast, seeding of green manures, grass–clover ley mixtures or sowing crops after, for example, potatoes where soils are already sufficiently loose after harvest are good occasions within crop rotations to renounce ploughing. Cereals, grain legume–cereal mixtures like pea–barley or faba bean–oat mixtures and row crops like maize or sunflower are less susceptible and reduced tillage can be applied successfully. They either sufficiently suppress weeds or allow for a good mechanical weed control during susceptible growth stages.

Strategic tillage is also becoming more popular in conventional no-tillage systems though motivations differ. In Australia, for example, increase in herbicide-resistant weeds and diseases as well as soil crusting and unfavourable nutrient accumulation in topsoils unavailable in dry conditions drives discussions about limitations of continuous no-tillage (Dang et al., 2015b).

Independent of the farming system used, there are doubts whether occasional ploughing would counteract the enrichment of SOC in topsoils and even the increase in SOC stocks (if there is any) with conservation tillage. These doubts were raised after a study in temperate climate showed that a single ploughing event depleted all of the SOC sequestered by long-term conservation tillage (Stockfisch et al., 1999). A recent review on one-time tillage of long-term no-tillage systems, however, reported inconsistencies between studies and emphasised the importance of evaluating SOC stocks with the consideration of equivalent soil masses, which was lacking in the study of Stockfisch et al. (1999) (Dang et al., 2015a). The observation that SOC distribution changed by one-time

---

Box 1 Grass-clover ley termination with conservation tillage techniques in Switzerland (FiBL Switzerland)

Organic arable crop rotations include leys with perennial crops that need to be terminated for the cultivation of annuals. This is normally done by mouldboard ploughing. If soils are sufficiently dry, termination by reduced tillage techniques can be successfully applied. Weakening grass–clover plants by successional chiselling with wide sweeps in one or two passes of 3–5-cm depth followed by shallow inversion to 10 cm with a skim plough (Fig. 2) successfully removes leys in summer periods with normal rainfall. In dry summers, successive chiselling from 3–6 cm to max. 10 cm with the aim of complete ley desiccation can superficially terminate leys. In all cases, cutting the ley shortly before termination reduces plant biomass for an efficient incorporation.

Link to video: https://youtu.be/McZU2lTYfyI

Figure 2 Ley termination with a skim plough inverts the soil to only shallow depths. Source: FiBL Switzerland, Maurice Clerc.

ploughing due to soil mixing was, however, clearly shown (Dang et al., 2015a). Ploughing a long-term no-tillage soil affects microorganisms. Wortmann et al. (2008) found reduced microbial biomass contents in the freshly ploughed soils. Yet, most assessed microbial groups recovered within 1–3 years with the exception of AMF. Soil fertility thus seems to be at risk when soils continuously managed with conservation tillage are intensively tilled. As the number of studies is, however, small, more evidence under varying pedoclimatic conditions is needed.

Soil quality considerations of a systems change are finally dependent on the starting point. For current organic farming systems with annual ploughing, strategic tillage has clear advantages. Based on the evidence that reduction of ploughing frequency to three times in a seven-year rotation (excluding the two-year grass–clover ley period) led to minimal impacts on yields, even lower production costs and improvements in soil quality parameters (Kainz et al., 2003) or that occasional reduced tillage favoured earthworms compared with ploughing in the short-term (Moos et al., 2016), reducing ploughing frequency seems to be an agronomically viable and environmentally friendly alternative to annual ploughing and a step forward in the development of organic conservation tillage systems.

Box 2 Shallow on-land ploughing to maintain good soil structure in soils sensitive to compaction (ISARA Lyon)

In the Rhône Alpes region, on-farm trials compared the effect of traditional deep ploughing (about 30 cm soil depth) and reduced tillage on soil fertility and crop performances. After several years, non-inversion techniques such as those with a chisel plough at 5–15 cm depth showed variable results: the technique gave disappointing results with sandy and silty soils, where soils tend to compact. In contrast, results from reduced tillage were encouraging if weeds were successfully controlled and soils presented a good soil structure, especially in loamy soils with shrinking–swelling dynamics. Soil compaction was mainly caused by vehicle passages, for example, by spring crop harvesters. Organic farmers used non-inversion subsoilers to 35–40 cm soil depth to break up this compaction in the reduced tilled fields. The effect lasted about 1–2 years. A promising alternative to this combined strategy for soils sensitive to compaction was found to be shallow on-land ploughing to 18 cm without skim coulters. The tractor is thereby driving on-land which means on non-tilled soil next to the plough avoiding plough pans. Some crop residues were still preserved at soil surface and soil compaction was removed by creating homogeneously distributed mechanical porosity (Fig. 3) while yields and weed regulation were assured.

Figure 3 Vertical soil profile under shallow on-land ploughing assessed according to the French 'profil cultural' measurement (Boizard et al., 2017). Different soil layers are marked with woolly yarn. Soil clods are marked with coloured dots (Peigné et al., 2013). Source: ISARA Lyon, Jean-François Vian.

Box 3 Organic-cover-crop-based no-tillage in Switzerland – Silage maize after a winter pea cover crop (FiBL Switzerland)

Direct seeding of silage maize into a thick mulch cover of a frost-resistant winter pea cover crop (EFB33) that was terminated with a roller crimper, was most promising during the study period (Fig. 4) with average yield reductions of only 7% compared with ploughing. The system works well in dry years (Böhler and Dierauer, 2017). The pea only provides sufficient weed suppression when enough biomass is produced and termination takes place during flowering. Slugs and reduced soil organic matter mineralisation under the mulch cover, impacting nutrient supply, can be challenging. Maize was therefore additionally fertilised with organic commercial fertiliser during seeding at this experimental stage. As the level of weed infestation and yields are dependent on weather conditions and successful cover crop termination, it is still a high-risk system that needs further development.

Figure 4 Silage maize directly sown into a winter pea cover crop mulch. Source: FiBL Switzerland, Thomas Alföldi.

Box 4 Organic-cover-crop-based no-tillage in France – Soybean after rye cover crop (ISARA Lyon)

Apart from the cereal rye cover crop species, early cover crop seeding (e.g. mid-to-end of August in southern France) with a higher seeding rate than typical farmers' practices (e.g. 200 kg ha^{-1} of cereal rye) promotes both higher biomass and rapid cover crop establishment for weed suppression. In spring, successful cover crop termination is achieved when the rye is rolled at least at 50% of the flowering stage. The rolling should be performed at an angle of 90° to the seeding direction and the roller crimper should reach enough weight to crush plant stems. The blades welded on the roller crimper are angled on the cylinder for enhancing cover crop cutting, thereby, limiting cover crop regrowth. High seeding rates of the soybean cash crop and reduced row spacing may encourage rapid soybean canopy closure to suppress weeds (Fig. 5). Regarding the potential lower soil temperature below the mulch cover, direct seeding of soybean can be delayed until the soil is sufficiently warm. As the cash crop emergence depends on the seed–soil contact, the no-tillage seeder should have enough weight to cut through the mulch layer.

Figure 5 Soybean directly sown into a cereal rye cover crop mulch. Source: ISARA LYON, Laura Vincent-Caboud.

5 Conclusion

Active research is ongoing on-station and on-farm experimenting with conservation tillage approaches in organic farming. There are many benefits shown, mostly in terms of increased soil fertility and better topsoil structure, but there is no real breakthrough as yet. Mainly weeds and a timely nutrient supply still pose challenges for continuous conservation tillage especially in applying no-tillage. Yields are on average slightly lower, a fact that is not compensated by premiums as in the case for organic compared with conventional products. Yield stability is additionally lower than with ploughing, a risk that is faced by the farmer. Many organic farmers are therefore reluctant to adopt conservation tillage on their farms.

To make conservation tillage in organic systems effective, many efforts will have to be made to address agronomic challenges like weed regulation and nutrient management as well as socio-economic and monetary barriers. More data related to the economic viability, revenues, working hours and energy savings are needed to evaluate which systems are suitable in the different pedoclimatic conditions. Apart from those practical aspects, more measured data are needed to underline the positive effect of conservation tillage regarding erosion prevention and climate change mitigation and adaptation. This data may affect potential subsidy schemes to incentivise farmers to adopt more risky tillage systems.

Nevertheless, conservation tillage in organic farming can be successfully implemented. It requires more knowledge and attention and there is a need to share experiences within the farmer and researcher community. Practically viable options to start with include hybrid systems where reduced tillage is applied most of the time, and no-tillage on some occasions combined with an occasional strategic use of mouldboard ploughing. These systems need to be tested to assess the long-term benefits and possible drawbacks.

6 Where to look for further information

6.1 Main reviews

Peigné, J., Ball, B. C., Roger-Estrade, J. and David, C. (2007). Is conservation tillage suitable for organic farming? A review. *Soil Use Manage.*, 23, 129–144. doi: 10.1111/j.1475-2743.2006.00082.x.

Cooper, J., Baranski, M., Stewart, G., Nobel-de Lange, M., Bàrberi, P., Fließbach, A., Peigné, J., Berner, A., Brock, C., Casagrande, M., Crowley, O., David, C., De Vliegher, A., Döring, T. F., Dupont, A., Entz, M., Grosse, M., Haase, T., Halde, C., Hammerl, V., Huiting, H., Leithold, G., Messmer, M., Schloter, M., Sukkel, W., van der Heijden, M. G. A., Willekens, K., Wittwer, R. and Mäder, P. (2016). Shallow non-inversion tillage in organic farming maintains crop yields and increases soil C stocks: a meta-analysis. *Agron. Sustain. Develop.*, 36, 22. doi: 10.1007/s13593-016-0354-1.

Special Issue 'Conservation Tillage for Organic Farming' in Agriculture 2017, Volume 7 (http://www.mdpi.com/journal/agriculture/special_issues/conservation_tillage_organic_farming.

6.2 Networks and partner institutes

Core Organic II projects with a good overview on European institutes working on organic conservation tillage: www.tilman-org.net; www.fertilcrop.net

BioBo Project Austria (https://ec.europa.eu/eip/agriculture/en/find-connect/projects/og-biobo-ertragsentwicklung-und-humusaufbau-%C3%BCber)

Rodale Institute (USA) with construction manual of a roller crimper: https://rodaleinstitute.org/our-work/organic-no-tillage/

Running long-term tillage system comparison trials in organic farming in Europe (list may not be complete, status 2017):

CH: Frick trial, Aesch trial (FiBL Switzerland); FAST trial (Agroscope Reckenholz).

D: Gladbacher Hof (University of Giessen); Kleinhohenheim (University of Hohenheim); Puch trial, Neuhof trial (LfL Bavaria).

E: Gallecs trial (University of Barcelona).

F: Thil trial (ISARA Lyon).

NL: BASIS trial (WUR Wageningen).

6.3 Video material

www.tilman-org.net/to-videos

www.bioaktuell.ch/pflanzenbau/ackerbau/bodenbearbeitung/filme-bodenbearbeitung (in German)

7 References

Abiven, S., Menasseri, S. and Chenu, C. 2009. The effects of organic inputs over time on soil aggregate stability - A literature analysis. *Soil Biol. Biochem.*, 41, 1–12.

Armengot, L., Berner, A., Blanco-Moreno, J., Mäder, P. and Sans, F. X. 2015. Long-term feasibility of reduced tillage in organic farming. *Agron. Sustain. Dev.*, 35, 339–46.

Armengot, L., Blanco-Moreno, J. M., Barberic, P., Bocci, G., Carlesi, S., Aendekerk, R., Berner, A., Celette, F., Grosse, M., Huiting, H., Kranzler, A., Luik, A., Mader, P., Peigné, J., Stoll, E., Delfosse, P., Sukkel, W., Surbock, A., Westaway, S. and Sans, F. X. 2016. Tillage as a driver of change in weed communities: A functional perspective. *Agric. Ecosyst. Environ.*, 222, 276–85.

Bernstein, E. R., Posner, J. L., Stoltenberg, D. E. and Hedtcke, J. L. 2011. Organically managed no-tillage rye–soybean systems: Agronomic, economic and environmental assessment. *Agron. J.*, 103, 1169–79.

Bilalis, D. J. and Karamanos, A. J. 2010. Organic maize growth and mycorrhizal root colonization response to tillage and organic fertilization. *J. Sustain. Agr.*, 34, 836–49.

Bilalis, D., Karkanis, A., Pantelia, A., Patsiali, S., Konstantas, A. and Efthimiadou, A. 2012. Weed populations are affected by tillage systems and fertilization practices in organic flax (*Linum usitatissimum* L.) crop. *Aust. J. Crop Sci.*, 6, 157–63.

Böhler, D. and Dierauer, H. 2017. Messerwalze statt Glyphosat. *Landwirtschaft ohne Pflug (LOP)*, 5, 39–43.

Boizard, H., Peigné, J., Sasal, M. C., De Fátima Guimarães, M., Piron, D., Tomis, V., Vian, J.-F., Cadoux, S., Ralisch, R., Tavares Filho, J., Heddadj, D., De Battista, J., Duparque, A., Franchini, J. C. and Roger-Estrade, J. 2017. Developments in the 'profil cultural' method for an improved assessment of soil structure under no-till. *Soil Tillage Res.*, 173, 92–103.

Börstler, B., Thiéry, O., Sykorova, Z., Berner, A. and Redecker, D. 2010. Diversity of mitochondrial large subunit rDNA haplotypes of *Glomus intraradices* in two agricultural field experiments and two semi-natural grasslands. *Mol. Ecol.*, 19, 1497–511.

Butler, D. M., Bates, G. E. and Inwood, S. E. E. 2016. Tillage system and cover crop management impacts on soil quality and vegetable crop performance in organically managed production in Tennessee. *Hortscience*, 51, 1038–44.

Carr, P. 2017. Guest editorial: Conservation tillage for organic farming. *Agriculture*, 7, 19.

Casagrande, M., Peigné, J., Payet, V., Mäder, P., Sans, F. X., Blanco-Moreno, J. M., Antichi, D., Bàrberi, P., Beeckman, A., Bigongiali, F., Cooper, J., Dierauer, H., Gascoyne, K., Grosse, M., Heß, J., Kranzler, A., Luik, A., Peetsmann, E., Surböck, A., Willekens, K. and David, C. 2015. Organic farmers' motivations and challenges for adopting conservation agriculture in Europe. *Org. Agr.*, 6(4), 281–95.

Cooper, J., Baranski, M., Stewart, G., Nobel-De Lange, M., Bàrberi, P., Fließbach, A., Peigné, J., Berner, A., Brock, C., Casagrande, M., Crowley, O., David, C., De Vliegher, A., Döring, T. F., Dupont, A., Entz, M., Grosse, M., Haase, T., Halde, C., Hammerl, V., Huiting, H., Leithold, G., Messmer, M., Schloter, M., Sukkel, W., Van Der Heijden, M. G. A., Willekens, K., Wittwer, R. and Mäder, P. 2016. Shallow non-inversion tillage in organic farming maintains crop yields and increases soil C stocks: a meta-analysis. *Agron. Sustain. Dev.*, 36, 22.

Crittenden, S. J. and De Goede, R. G. M. 2016. Integrating soil physical and biological properties in contrasting tillage systems in organic and conventional farming. *Eur. J. Soil Biol.*, 77, 26–33.

Crittenden, S. J., Eswaramurthy, T., De Goede, R. G. M., Brussaard, L. and Pulleman, M. M. 2014. Effect of tillage on earthworms over short- and medium-term in conventional and organic farming. *Appl. Soil Ecol.*, 83, 140–8.

Crittenden, S. J., Poot, N., Heinen, M., Van Balen, D. J. M. and Pulleman, M. M. 2015. Soil physical quality in contrasting tillage systems in organic and conventional farming. *Soil Till. Res.*, 154, 136–44.

Crowley, O. and Döring, T. F. 2012. Using reduced tillage to improve the efficiency of ecosystem service delivery on organic farms. In: *Agriculture and the Environment IX, Valuing Ecosystems: Policy, Economic and Management Interactions.*

CTIC. 2017. *Tillage Type Definitions* [Online]. Conservation Technology Information Center. Available: www.ctic.purdue.edu/resourcedisplay/322 [Accessed 26 May 2017].

Dang, Y. P., Moody, P. W., Bell, M. J., Seymour, N. P., Dalal, R. C., Freebairn, D. M. and Walker, S. R. 2015a. Strategic tillage in no-till farming systems in Australia's northern grains-growing regions: II. Implications for agronomy, soil and environment. *Soil Till. Res.*, 152, 115–23.

Dang, Y. P., Seymour, N. P., Walker, S. R., Bell, M. J. and Freebairn, D. M. 2015b. Strategic tillage in no-till farming systems in Australia's northern grains-growing regions: I. Drivers and implementation. *Soil Till. Res.*, 152, 104–14.

Derpsch, R., Friedrich, T., Kassam, A. and Li, H. 2010. Current status of adoption of no-till farming in the world and some of its main benefits. *Int. J. Agric. Biol. Eng.*, 3, 1–25.

Emmerling, C. 2001. Response of earthworm communities to different types of soil tillage. *Appl. Soil Ecol.*, 17, 91–6.

Emmerling, C. 2007. Reduced and conservation tillage effects on soil ecological properties in an organic farming system. *Biol. Agric. Hortic.*, 24, 363–77.

FAO. 2017. *Conservation Agriculture* [Online]. Food and Agriculture Organisation of the United Nations. Available: www.fao.org/ag/ca/1a.html [Accessed 31 July 2017].

Fontana, M., Berner, A., Mäder, P., Lamy, F. and Boivin, P. 2015. Soil organic carbon and soil bio-physicochemical properties as co-Influenced by tillage treatment. *Soil Sci. Soc. Am. J.*, 79, 1435–45.

Gadermaier, F., Berner, A., Fließbach, A., Friedel, J. K. and Mäder, P. 2012. Impact of reduced tillage on soil organic carbon and nutrient budgets under organic farming. *Renew. Agr. Food Syst.*, 27, 1–13.

Gattinger, A., Muller, A., Haeni, M., Skinner, C., Fliessbach, A., Buchmann, N., Mäder, P., Stolze, M., Smith, P., Scialabba, N. E.-H. and Niggli, U. 2012. Enhanced top soil carbon stocks under organic farming. *Proc. Natl. Acad. Sci. USA*, 109, 18226–31.

Gomiero, T., Paoletti, M. G. and Pimentel, D. 2008. Energy and environmental issues in organic and conventional agriculture. *Crit. Rev. Plant Sci.*, 27, 239–54.

Gruber, S. and Claupein, W. 2009. Effect of tillage intensity on weed infestation in organic farming. *Soil Till. Res.*, 105, 104–11.

Halde, C. and Entz, M. H. 2014. Flax (*Linum usitatissimum* L.) production system performance under organic rotational no-till and two organic tilled systems in a cool subhumid continental climate. *Soil Till. Res.*, 143, 145–54.

Halde, C., Gagné, S., Charles, A. and Lawley, Y. 2017. Organic no-till systems in eastern Canada: A review. *Agriculture*, 7, 36.

Holland, J. M. 2004. The environmental consequences of adopting conservation tillage in Europe: Reviewing the evidence. *Agr. Ecosyst. Environ.*, 103, 1–25.

Hütsch, B. W. 2001. Methane oxidation in non-flooded soils as affected by crop production - invited paper. *Eur. J. Agron.*, 14, 237–60.

Jabbour, R. and Barbercheck, M. E. 2009. Soil management effects on entomopathogenic fungi during the transition to organic agriculture in a feed grain rotation. *Biol. Control*, 51, 435–43.

Kabir, Z. 2005. Tillage or no-tillage: Impact on mycorrhizae. *Canadian Journal of Plant Science*, 85, 23–9.

Kainz, M., Kimmelmann, S. and Reents, H.-J. 2003. Bodenbearbeitung im Ökolandbau – Ergebnisse und Erfahrungen aus einem langjährigen Feldversuch. *In:* Freyer, B. (Ed.), *7. Wissenschaftstagung zum Ökologischen Landbau*, Ökologischer Landbau der Zukunft, Wien, 24–26 Februar 2003. Universität für Bodenkultur Wien, Institut für Ökologischen Landbau, pp. 33–6.

Kainz, M., Gerl, G., Lemnitzer, B., Bauchenß, J. and Hülsbergen, K.-J. 2005. Wirkungen differenzierter Bodenbearbeitungssysteme im Dauerversuch Scheyern. *In:* Heß, J. and Rahmann, G. (Eds), *8. Wissenschaftstagung Ökologischer Landbau*, Ende der Nische, Kassel, 01–04 March 2005. Kassel University Press GmbH, Kassel.

Kautz, T. 2015. Research on subsoil biopores and their functions in organically managed soils: A review. *Renew. Agr. Food Syst.*, 30, 318–27.

Koch, H. J., Pringas, C. and Maerlaender, B. 2006. Evaluation of environmental and management effects on Fusarium head blight infection and deoxynivalenol concentration in the grain of winter wheat. *Eur. J. Agron.*, 24, 357–66.

Kong, A. Y. Y., Fonte, S. J., Van Kessel, C. and Six, J. 2009. Transitioning from standard to minimum tillage: Trade-offs between soil organic matter stabilization, nitrous oxide emissions, and N availability in irrigated cropping systems. *Soil Till. Res.*, 104, 256–62.

Köpke, U. and Schultes, H. 2008. Direct seeding of faba beans in organic agriculture. *In:* Neuhoff, D., Halberg, N., Alföldi, T., Lockeretz, W., Thommen, A., Rasmussen, I. A., Hermansen, J., Vaarst, M., Lueck, L., Caporali, F., Hogh-Jensen, H., Migliorini, P. and Willer, H. (Eds), *Cultivating the Future Based on Science. Volume 1 - Organic Crop Production*. Proceedings of the Second Scientific Conference of the International Society of Organic Agriculture Research (ISOFAR), held at the 16th IFOAM Organic World Congress in Cooperation with the International Federation of Organic Agriculture Movements (IFOAM) and the Consorzio ModenaBio, 18–20 June 2008 in Modena, Italy, pp. 418–21.

Köpke, U., Athmann, M., Han, E. and Kautz, T. 2015. Optimising cropping techniques for nutrient and environmental management in organic agriculture. *Sustain. Agric. Res.*, 4(3), 11–21.

Krauss, M., Berner, A., Burger, D., Wiemken, A., Niggli, U. and Mäder, P. 2010. Reduced tillage in temperate organic farming: Implications for crop management and forage production. *Soil Use Manage.*, 26, 12–20.

Krauss, M., Krause, H.-M., Spangler, S., Kandeler, E., Behrens, S., Kappler, A., Mäder, P. and Gattinger, A. 2017a. Tillage system affects fertilizer-induced nitrous oxide emissions. *Biol. Fert. Soils*, 53, 49–59.

Krauss, M., Perrochet, F., Lori, M., Ruser, R., Müller, T., Zikeli, S., Gruber, S., Claupein, W., Mäder, P. and Gattinger, A. 2017b. Reduzierte Bodenbearbeitung im Biolandbau - Klimaaspekte. *Agrarforschung Schweiz*, 8(6), 226–31.

Krauss, M., Ruser, R., Müller, T., Hansen, S., Mäder, P. and Gattinger, A. 2017c. Impact of reduced tillage on greenhouse gas emissions and soil carbon stocks in an organic grass-clover ley - winter wheat cropping sequence. *Agr. Ecosyst. Environ.*, 239, 324–33.

Kuntz, M., Berner, A., Gattinger, A., Scholberg, J. M., Mäder, P. and Pfiffner, L. 2013. Influence of reduced tillage on earthworm and microbial communities under organic arable farming. *Pedobiologia*, 56, 251–60.

Lal, R., Reicosky, D. C. and Hanson, J. D. 2007. Evolution of the plow over 10,000 years and the rationale for no-till farming. *Soil Till. Res.*, 93, 1–12.

Lehocka, Z., Klimekova, M., Bielikova, M. and Mendel, L. 2009. The effect of different tillage systems under organic management on soil quality indicators. *Agron. Res.*, 7, 369–73.

Lewis, D. B., Kaye, J. P., Jabbour, R. and Barbercheck, M. E. 2011. Labile carbon and other soil quality indicators in two tillage systems during transition to organic agriculture. *Renew. Agr. Food Syst.*, 26, 342–53.

Luo, Z. K., Wang, E. L. and Sun, O. J. 2010. Can no-tillage stimulate carbon sequestration in agricultural soils? A meta-analysis of paired experiments. *Agr. Ecosyst. Environ.*, 139, 224–31.

Lynch, D., Macrae, R. and Martin, R. 2011. The carbon and global warming potential impacts of organic farming: Does it have a significant role in an energy constrained world? *Sustainability*, 3, 322.

Mäder, P. and Berner, A. 2012. Development of reduced tillage systems in organic farming in Europe. *Renew. Agr. Food Syst.*, 27, 7–11.

Mäder, P., Fließbach, A., Dubois, D., Gunst, L., Fried, P. and Niggli, U. 2002. Soil fertility and biodiversity in organic farming. *Science*, 296, 1694–7.

Massucati, L. and Köpke, U. 2011. Faba bean (*Vicia faba* L.): Response to occasional direct seeding into straw residues under temperate climate conditions. *In:* Neuhoff, D., Mok Sohn, S., Ssekyewa, C., Halberg, N., Rasmussen, I. A. and Hermansen, J. (Eds), *Organic is Life - Knowledge for Tomorrow. Volume 1 - Organic Crop Production*. Proceedings of the Third Scientific Conference of the International Society of Organic Agriculture Research (ISOFAR).

Massucati, L. 2013. *Temporäre Direktsaat von Ackerbohnen (Vicia faba L.) im Ökologischen Landbau*. Verlag Dr. Köster, Berlin.

Massucati, L. and Köpke, U. 2014. Effect of straw mulch residues of previous crop oats on the weed population in direct seeded faba bean in organic farming. *In: 26th German Conference on weed Biology an Weed Control*, 11–13 March 2014, Braunschweig, Germany.

Metzke, M., Potthoff, M., Quintern, M., Heß, J. and Joergensen, R. G. 2007. Effect of reduced tillage systems on earthworm communities in a 6-year organic rotation. *Eur. J. Soil Biol.*, 43(Suppl. 1), S209–15.

Moos, J. H., Schrader, S., Paulsen, H. M. and Rahmann, G. 2016. Occasional reduced tillage in organic farming can promote earthworm performance and resource efficiency. *Appl. Soil Ecol.*, 103, 22–30.

Morris, N. L., Miller, P. C. H., J. H. Orson and Froud-Williams, R. J. 2010. The adoption of non-inversion tillage systems in the United Kingdom and the agronomic impact on soil, crops and the environment—A review. *Soil Till. Res.*, 108, 1–15.

Müller, E., Wildhagen, H., Quintern, M., Heß, J., Wichern, F. and Joergensen, R. G. 2009. Spatial patterns of soil biological and physical properties in a ridge tilled and a ploughed Luvisol. *Soil Till. Res.*, 105, 88–95.

Munger, H., Vanasse, A., Rioux, S. and Legere, A. 2014. Bread wheat performance, fusarium head blight incidence and weed infestation response to low-input conservation tillage systems in eastern Canada. *Can. J. Plant Sci.*, 94, 193–201.

Munkholm, L. J., Schjønning, P. and Rasmussen, K. J. 2001. Non-inversion tillage effects on soil mechanical properties of a humid sandy loam. *Soil Till. Res.*, 62, 1–14.

Myhre, G., D. Shindell, F.-M. Bréon, W. Collins, J. Fuglestvedt, J. Huang, D. Koch, J.-F. Lamarque, D. Lee, B. Mendoza, T. Nakajima, A. Robock, G. Stephens, Takemura, T. and Zhang, H. 2013. Anthropogenic and natural radiative forcing. *In:* Stocker, T. F., D. Qin, G.-K. Plattner, M. Tignor, S. K. Allen, J. Boschung, A. Nauels, Y. Xia, Bex, V. and Midgley, P. M. (Eds), *Climate Change 2013: The Physical Science Basis. Contribution of Working Group I to the Fifth Assessment Report of the Intergovernmental Panel on Climate Change*. Cambridge University Press, Cambridge, UK and New York, NY.

Page, K., Dang, Y. and Dalal, R. 2013. Impacts of conservation tillage on soil quality, including soil-borne crop diseases, with a focus on semi-arid grain cropping systems. *Australas. Plant Path.*, 42, 363–77.

Parniske, M. 2008. Arbuscular mycorrhiza: The mother of plant root endosymbioses. *Nat. Rev. Microbiol.*, 6, 763–75.

Peigné, J., Cannavaciuolo, M., Gautronneau, Y., Aveline, A., Giteau, J. L. and Cluzeau, D. 2009. Earthworm populations under different tillage systems in organic farming. *Soil Till. Res.*, 104, 207–14.

Peigné, J., Vian, J.-F., Cannavacciuolo, M., Lefevre, V., Gautronneau, Y. and Boizard, H. 2013. Assessment of soil structure in the transition layer between topsoil and subsoil using the profil cultural method. *Soil Till. Res.*, 127, 13–25.

Peigné, J., Casagrande, M., Payet, V., David, C., Sans, F. X., Blanco-Moreno, J. M., Cooper, J., Gascoyne, K., Antichi, D., Bàrberi, P., Bigongiali, F., Surböck, A., Kranzler, A., Beeckman, A., Willekens, K., Luik, A., Matt, D., Grosse, M., Heß, J., Clerc, M., Dierauer, H. and Mäder, P. 2015a. How organic farmers practice conservation agriculture in Europe. *Renew. Agr. Food Syst.*, 31, 72–85.

Peigné, J., Lefevre, V., Vian, J. and Fleury, P. 2015b. Conservation agriculture in organic farming: Experiences, challenges and opportunities in Europe. *In:* Farooq, M. and Siddique, K. (Eds), *Conservation Agriculture*. 1 ed. Springer International Publishing, Switzerland.

Peigné, J., Vian, J.-F., Payet, V. and Saby, N. P. A. 2018. Soil fertility after 10 years of conservation tillage in organic farming. *Soil Till. Res.*, 175, 194–204.

Petersen, H. 2002. Effects of non-inverting deep tillage vs. conventional ploughing on collembolan populations in an organic wheat field. *Eur. J. Soil Biol.*, 38, 177–80.

Porter, P., Feyereisen, G., De Bruin, J. and Johnson, G. 2005. No-till organic soybean production following a fall-planted rye cover crop. *In:* Köpke, U., Niggli, U., Neuhoff, D., Cornish, P., Lockeretz, W. and Willer, H. (Eds), *Researching Sustainable Systems*. Proceedings of the First Scientific Conference of the International Society of Organic Agriculture Research (ISOFAR), held in Cooperation with the International Federation of Organic Agriculture Movements (IFOAM) and the National Association for Sustainable Agriculture, Australia (NASAA), 21–23 September 2005 in Adelaide, South Australia.

Powlson, D. S., Stirling, C. M., Jat, M. L., Gerard, B. G., Palm, C. A., Sanchez, P. A. and Cassman, K. G. 2014. Limited potential of no-till agriculture for climate change mitigation. *Nat. Clim. Change*, 4, 678–83.

Pritchett, K., Kennedy, A. C. and Cogger, C. G. 2011. Management effects on soil quality in organic vegetable systems in western Washington. *Soil Sci. Soc. Am. J.*, 75, 605–15.

Reganold, J. P. and Wachter, J. M. 2016. Organic agriculture in the twenty-first century. *Nat. Plant.*, 2, 15221.

Roger-Estrade, J., Anger, C., Bertrand, M. and Richard, G. 2010. Tillage and soil ecology: Partners for sustainable agriculture. *Soil Till. Res.*, 111, 33–40.

Säle, V., Aguilera, P., Laczko, E., Mäder, P., Berner, A., Zihlmann, U., Van Der Heijden, M. G. A. and Oehl, F. 2015. Impact of conservation tillage and organic farming on the diversity of arbuscular mycorrhizal fungi. *Soil Biol. Biochem.*, 84, 38–52.

Sans, F. X., Berner, A. , Armengot, L. and Mäder, P. 2011. Tillage effects on weed communities in an organic winter wheat–sunflower–spelt cropping sequence. *Weed Res.*, 51, 413–21.

Schader, C., Meier, M. S. and Mäder, P. 2011. Sparsamer durch reduzierte Bodenbearbeitung. *Ökologie und Landbau*, 160, 15–18.

Schepl, U. and Paffrath, A. 2003. Entwicklung von Strategien zur Regulierung des Drahtwurmbefalls (*Agriotes* spp. L.) im Ökologischen Kartoffelanbau. *In:* Freyer, B. (Ed.), *7. Wissenschaftstagung zum Ökologischen Landbau*, Universität für Bodenkultur, Wien - Institut für Ökologischen Landbau, pp. 133–6.

Schmidt, H. 2010. *Öko-Ackerbau ohne tiefes Pflügen, Praxisbeispiele und Forschungsergebnisse.* Verlag Dr. Köster, Berlin.

Schulz, F., Brock, C. and Leithold, G. 2009. Effekte unterschiedlicher Systeme der Grundbodenbearbeitung auf Erträge und Beikraut im Dauerfeldversuch Gladbacherhof. *In:* Mayer, J., Alföldi, T., Leiber, F., Dubois, D., Fried, P., Heckendorn, F., Hillmann, E., Klocke, P., Lüscher, A., Riedel, S., Stolze, M., Strasser, F., Van Der Heijden, M. and Willer, H. (Eds), *10. Wissenschaftstagung Ökologischer Landbau, Werte - Wege - Wirkungen: Biolandbau im Spannungsfeld zwischen Ernährungssicherung*, Markt und Klimawandel, ETH Zürich, 11–13 Februar 2009. Verlag Dr. Köster, Berlin.

Schulz, F., Brock, C. and Leithold, G. 2011. Einfluss reduzierter Grundbodenbearbeitung auf Erträge und Bodenparameter im Dauerfeldversuch Gladbacherhof. *In:* Leithold, G., Becker, K., Brock, C., Fischinger, S., Spiegel, A. -K., Spory, K., Wilbois, K. -P. and Williges, U. (Eds), *11. Wissenschaftstagung Ökologischer Landbau, Es geht ums Ganze: Forschen im Dialog von Wissenschaft und Praxis*, Justus-Liebig-Universität Gießen, 15–18 März 2011. Verlag Dr. Köster, Berlin.

Schulz, F., Brock, C., Schmidt, H., Franz, K.-P. and Leithold, G. 2014. Development of soil organic matter stocks under different farm types and tillage systems in the Organic Arable Farming Experiment Gladbacherhof. *Arch. Agron. Soil Sci.*, 60, 313–26.

Schulz, F., Brock, C. and Leithold, G. 2015. Effekte von Betriebssystem und Bodenbearbeitung auf die Segetalflora. *In:* Häring, A. M., Hörning, B., Hoffmann-Bahnsen, R., Luley, H., Luthardt, V., Pape, J. and Trei, G. (Eds), *13. Wissenschaftstagung Ökologischer Landbau, Am Mut hängt der Erfolg - Rückblicke und Ausblicke auf die ökologische Landbewirtschaftung*, Hochschule für nachhaltige Entwicklung Eberswalde, 17–20 März 2015. Verlag Dr. Köster, Berlin.

Silva, E. and Delate, K. 2017. A decade of progress in organic cover crop-based reduced tillage practices in the upper midwestern USA. *Agriculture*, 7, 44.

Soane, B. D., Ball, B. C., Arvidsson, J., Basch, G., Moreno, F. and Roger-Estrade, J. 2012. No-till in northern, western and south-western Europe: A review of problems and opportunities for crop production and the environment. *Soil Till. Res.*, 118, 66–87.

Stockfisch, N., Forstreuter, T. and Ehlers, W. 1999. Ploughing effects on soil organic matter after twenty years of conservation tillage in Lower Saxony, Germany. *Soil Till. Res.*, 52, 91–101.

Sturz, A. V., Carter, M. R. and Johnston, H. W. 1997. A review of plant disease, pathogen interactions and microbial antagonism under conservation tillage in temperate humid agriculture. *Soil Till. Res.*, 41, 169–89.

Sun, H., Koal, P., Liu, D., Gerl, G., Schroll, R., Gattinger, A., Joergensen, R. G. and Munch, J. C. 2016. Soil microbial community and microbial residues respond positively to minimum tillage under organic farming in Southern Germany. *Appl. Soil Ecol.*, 108, 16–24.

Syakila, A. and Kroeze, C. 2011. The global nitrous oxide budget revisited. *Greenhouse Gas Meas. Manage.*, 1, 17–26.

Vakali, C., Köpke, U., Sidiras, N. and Bilalis, D. 2002. Possibilities and limits of reduced primary tillage in organic farming. *In:* Thompson, R. (ed.) *14th IFOAM Organic World Congress 'Cultivating Communities'*. Canadian Organic Growers, Ca – Ottawa, Ontario.

Vakali, C., Zaller, J. G. and Kopke, U. 2011. Reduced tillage effects on soil properties and growth of cereals and associated weeds under organic farming. *Soil Till. Res.*, 111, 133–41.

Vakali, C., Zaller, J. G. and Köpke, U. 2015. Reduced tillage in temperate organic farming: Effects on soil nutrients, nutrient content and yield of barley, rye and associated weeds. *Renew. Agr. Food Syst.*, 30, 270–9.

Van Kessel, C., Venterea, R., Six, J., Adviento-Borbe, M. A., Linquist, B. and Van Groenigen, K. J. 2013. Climate, duration, and N placement determine N_2O emissions in reduced tillage systems: A meta-analysis. *Glob. Change Biol.*, 19, 33–44.
Vian, J. F., Peigné, J., Chaussod, R. and Roger-Estrade, J. 2009. Effects of four tillage systems on soil structure and soil microbial biomass in organic farming. *Soil Use Manage.*, 25, 1–10.
Vincent-Caboud, L., Peigné, J., Casagrande, M. and Silva, E. 2017. Overview of organic cover crop-based no-tillage technique in Europe: Farmers' practices and research challenges. *Agriculture*, 7, 42.
Watson, C. A., Atkinson, D., Gosling, P., Jackson, L. R. and Rayns, F. W. 2002. Managing soil fertility in organic farming systems. *Soil Use Manage.*, 18, 239–47.
Weber, J., Kunz, C., Peteinatos, G., Zikeli, S. and Gerhards, R. 2017. Weed control using conventional tillage, reduced tillage, no-tillage, and cover crops in organic soybean. *Agriculture*, 7, 43.
Wilhelm, B. 2010. Konservierende Bodenbearbeitung im Ökolandbau - Analyse einer Verfahrenstechnik im Kontext der Bodenfruchtbarkeit. Dissertation, Universität Kassel, Witzenhausen, Germany.
Wittwer, R. A., Dorn, B., Jossi, W. and Van Der Heijden, M. G. A. 2017. Cover crops support ecological intensification of arable cropping systems. *Sci. Report.*, 7, 41911.
Wortmann, C. S., Quincke, J. A., Drijber, R. A., Mamo, M. and Franti, T. 2008. Soil microbial community change and recovery after one-time tillage of continuous no-till. *Agron. J.*, 100, 1681–6.
Yagioka, A., Komatsuzaki, M. and Kaneko, N. 2014. The effect of minimum tillage with weed cover mulching on organic daikon (*Raphanus sativus* var. *longipinnatus* cv. Taibyousoufutori) yield and quality and on soil carbon and nitrogen dynamics. *Biol. Agric. Hortic.*, 30, 228–42.
Yagioka, A., Komatsuzaki, M., Kaneko, N. and Ueno, H. 2015. Effect of no-tillage with weed cover mulching versus conventional tillage on global warming potential and nitrate leaching. *Agr. Ecosyst. Environ.*, 200, 42–53.
Zentner, R. P., Lafond, G. P., Derksen, D. A., Nagy, C. N., Wall, D. D. and May, W. E. 2004. Effects of tillage method and crop rotation on non-renewable energy use efficiency for a thin Black Chernozem in the Canadian Prairies. *Soil Till. Res.*, 77, 125–36.
Zikeli, S. and Gruber, S. 2017. Reduced tillage and no-till in organic farming systems, Germany - Status quo, potentials and challenges. *Agriculture*, 7, 35.
Zikeli, S., Gruber, S., Teufel, C.-F., Hartung, K. and Claupein, W. 2013. Effects of reduced tillage on crop yield, plant available nutrients and soil organic matter in a 12-year long-term trial under organic management. *Sustainability*, 5, 3876–94.

Chapter 7

Manure management in organic farming

Peter Sørensen, Department of Agroecology, Aarhus University, Denmark; Luca Bechini, University of Milan, Italy; and Lars Stoumann Jensen, University of Copenhagen, Denmark

1 Introduction
2 Manure composition and properties in different livestock systems
3 Manure storage and treatment
4 Manure field application methods
5 Turnover and availability of manure N in soil
6 Utilization of P, K and S in manures
7 Plant-based manures (green manures)
8 Future trends and conclusion
9 Where to look for further information
10 References

1 Introduction

In all agricultural systems, nutrients removed with harvested crops or lost from the soil need to be replaced by new inputs of nutrients, at least in the long term, in order to maintain and improve soil fertility and sustain crop production. In most organic farming systems, animal manures are the most important source of both macro- and micronutrients. Nitrogen (N) is also supplied via N-fixing crops and a variable amount by atmospheric deposition, but manures are usually an important source of N which is normally the most yield-limiting nutrient in non-N-fixing crops. At the same time, manure N is the most challenging nutrient to manage and utilize efficiently, as it can be lost in a number of processes, both in gaseous forms and by leaching and runoff. Therefore, a first priority in manure management should be to focus on reducing N losses in all steps of the manure handling chain. Secondly, manure management should aim to sustain or improve soil fertility, to secure a long-term supply of other nutrients in each agricultural field on the farm.

An important goal in organic farming is to reduce environmental damages due to agricultural production (Rahmann et al., 2016), and here N losses from manures are central, as manure management affects ammonia emission, emission of the important greenhouse gas N_2O, as well as loss of N_2 and pollution of the aquatic environment through leaching and surface runoff of N and P (Fig. 1).

http://dx.doi.org/10.19103/AS.2017.0029.07

Proper use of manure is also essential for increasing yields in organic production of non-N-fixing crops and is important for securing crops that have a competitive advantage against weeds (Olesen et al., 2007). Placement of liquid manure by injection before sowing may improve the competition of the crop against weeds (Rasmussen, 2002). Injection of manure in growing winter cereals can be difficult without causing significant damage to the crop and surface banding of manure in growing crops may even increase weed biomass (Olesen et al., 2009).

Recommendations on manure management are nearly the same in organic and conventional farming, but proper manure management is even more important in organic farming, where nutrients are more often limiting crop production.

In livestock housing, ammonia losses are often significant; for this reason it is especially important to ensure that animal urine is removed from the floors as fast as possible and the liquid part is stored under conditions with low potential ammonia losses and no risk of leaching/runoff. During subsequent manure storage, the conditions are very important for N losses, which can occur by ammonia volatilization and denitrification. At the field application step, both the timing and application methods are very important, especially to avoid ammonia volatilization, surface runoff of phosphate and organic matter and nitrate leaching from the applied manures.

Manures are also a source of organic matter (Zavattaro et al., 2017; Maillard and Angers, 2014) that has a number of beneficial effects on soil fertility, like increasing soil organic carbon (SOC), soil microbial activity and nutrient turnover (Fliessbach et al., 2007; Mader et al., 2002). However, it should be emphasized that an important contribution of manures to SOC is indirect through increased crop production from improved nutrient supply and thereby increased return of crop residues, including roots and root exudates, to the soil (Hu et al., 2018).

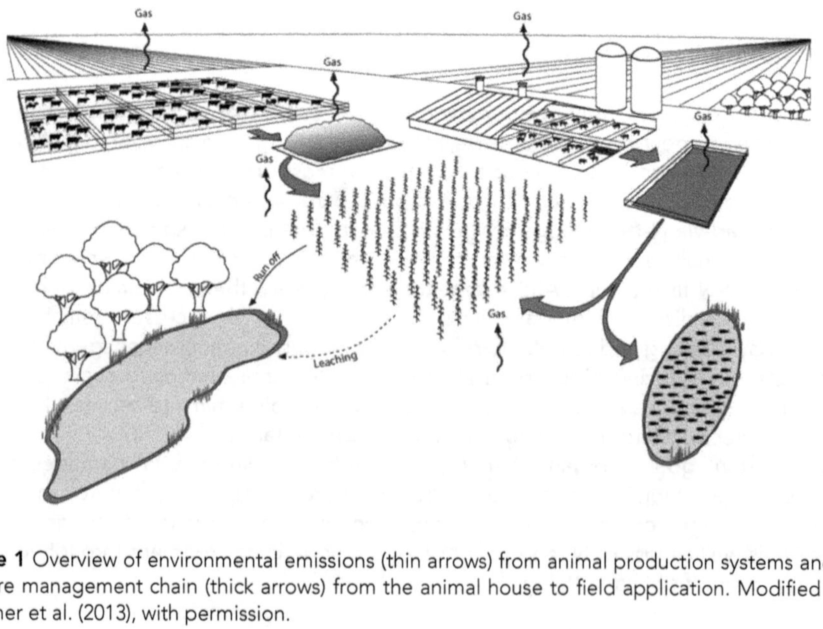

Figure 1 Overview of environmental emissions (thin arrows) from animal production systems and the manure management chain (thick arrows) from the animal house to field application. Modified from Sommer et al. (2013), with permission.

In this chapter, the main focus is on how storage and application of manures can be managed to improve the retention and utilization of manure N in the system and on factors influencing its plant availability. The role of other nutrients in manure is also described, but this is mainly a matter of providing a good distribution of manure between fields on the farm to secure the soil fertility at a level where sufficient amounts of available nutrients are present in soil also in the longer term.

2 Manure composition and properties in different livestock systems

The concentration of different nutrients in manures is very variable depending on various factors, with the most important being livestock species, feeding, housing systems and storage conditions. Therefore, manure composition needs to be chemically analysed when more precise knowledge of the actual manure composition is needed. In Table 1 typical compositions of different manure types and products derived from manure are indicated. Besides classical methods of wet chemistry analysis, rapid methods based on near-infrared spectroscopy are available to analyse slurries (e.g. Sørensen et al., 2007). Alternatively, dry matter content and electrical conductivity of slurries are good proxies for other variables like total C, total N, ammonium-N and P concentrations (e.g. Marino et al., 2008).

Solid manures can be divided into four main categories: deep litter, feedlot manure, solid fraction from slurry separation and farmyard manure (FYM). Deep litters are defined as manures from systems where excreta are allowed to accumulate in the barn over time and litter (e.g. straw) is added regularly in a way that the livestock can reside comfortably on top of the accumulated manure. Such manures also contain all the excreted urine. Feedlot manure is in principle the same but with significantly less litter added; it comes from outdoor feedlots where they have often been exposed to ammonia losses and to leaching in connection with precipitation. Feedlot manure is not further treated here, as we don't find such manure management compatible with organic farming. FYM is derived from buildings where urine and faeces are collected separately, typically by daily scraping and removal of the faeces mixed with bedding litter and feed residues. In systems with FYM, the urine is usually removed by drained floors and stored separately. However, some urine will inevitably end up in the FYM and some faecal particles in the urine fraction. Solid manure fractions derive from mechanical separation of liquid manures containing particulate material (slurry).

The diet composition typically ends up being reflected in the resulting manure C/N ratio, and the manure C/N ratio is very decisive for the short-term N availability (Sørensen et al., 2003b; Sørensen and Fernandez, 2003; Delin et al., 2012). A low C/N ratio is normally also related to a high proportion of ammonium-N in the manure after storage, which is immediately available for plant uptake.

During storage of manures, dry matter is lost during decomposition of organics, releasing CO_2. This means that the C/N ratio will decrease, unless a corresponding amount of N is also lost during storage.

Untreated and stored manures usually have a pH of 7–8 and contain bicarbonate (Sommer and Husted, 1995a). The bicarbonate can both derive from carbonates in the diet, from the animal digestion or from CO_2 released by the microbial decomposition of the manure after excretion. This also means that manures most often have a liming effect

Table 1 Typical properties for a range of manure types: dry matter, total P, K and N content, NH_4-N share of total N, C/N ratio as well as a characterization of the biodegradability of the organic material and typical potential first-year NFRV (nitrogen fertilizer replacement value)

Manure type	Dry matter (%)	P (g kg^{-1} fw)	K (g kg^{-1} fw)	Total N (g kg^{-1} fw)	NH_4-N (% of total N)	C/N	Biodegradability	Potential NFRV (%)
Deep litters	25–30	1–6	5–12	5–10	10–25	20–30	Medium	20–30
FYM[b], pigs	20–25	3.5–5	2–9	5–12	30–45	12–15	Medium	20–50
FYM, cattle	18–20	1.7–2	3–4	4–7	20–30	15–20	Low	15–30
Broiler manure	45–50	7–10	13–19	20–37	10–25[a]	5	High	50–65
Layer manure	50–60	4–16	7–20	15–30	5–35[a]	10	Medium	40–50
Slurry (pig)	4–7	0.25–1.5	1–5	2–6	70–80	4–8	Medium	40–80
Slurry (cattle)	7–10	0.4–1	2–6	2–5	40–60	6–10	Low	35–65
Slurry (poultry)	10–15	1–2	2–3	4–10	60–70	4	Medium	70–85
Biogas digestate[c]	1–5	0.4–1.2	0.5–3	3–10	60–85	3–5	Low	60–90
Sep. slurry liquid	1–2	0.1–1	1.5–6	2–5.5	80–95	1–2	High	80–100
Sep. slurry, solids	9–24	4–30	1–5	5–14	10–50	7–31	Medium	30–45
Urine	2–3	0.1–1	1.6–6	2–5	90	1–2	High	90–100
Compost, cattle manure	30–60	0.9–2	4.5–10	7–12	20–30	15–20	Low	10–25
Compost, org. MSW[d]	40–75	1.9–3.3	4.1–6.6	8–17	10–30	10–15	Low	5–20
Compost, garden waste	50–75	0.7–1.4	1.0–6.0	3–6	0–15	20–30	Low	0–10

[a] Depends on storage time and degree of uric acid hydrolysis.
[b] Farmyard manure, mixture of faeces and straw bedding.
[c] Mixed animal slurry with biowaste co-fermentation.
[d] Source separated organic fraction from municipal solid waste (MSW).
Data from Peters et al. (2011), Carlsbæk (2011), Jensen (2013), Bernal et al. (2017) and own data.

in acidic soils (with pH below 7). During anaerobic storage of manures, volatile fatty acids (VFAs) are formed, and VFA-C may constitute up to 30% of the total C in some manures (Kirchmann and Lundvall, 1993). Bicarbonate, VFAs and ammonium are all buffering compounds that contribute to the buffering property of manures (Sommer and Husted, 1995a). Release of CO_2 from bicarbonates in the slurry after application will drive up pH rapidly, and furthermore VFAs are decomposed within a few days after application to soil (Kirchmann and Lundvall, 1993), resulting in further soil pH increases (Sørensen, 1998a). A rapid, but temporary increase in pH therefore occurs shortly after application, which increases the risk of ammonia losses, but that is later followed by a decrease of pH due to nitrification of ammonium-N and mineralized N. Therefore, it is of utmost importance to minimize the extent of exposure of the manure to the atmosphere after application (e.g. by injection, rapid incorporation or natural infiltration) in order to minimize ammonia losses and maximize the N-fertilizing efficiency of the manure.

The properties of manures and derived products influence which crops they are best suited for. Liquid manures with a low dry matter content like urine and pig slurry may infiltrate into the soil naturally. Nevertheless, surface application in growing crops like grass and cereals is not optimal regarding the risk of ammonia losses. Thus, injection to top-soil layers of soil is recommended and ensures the highest efficiency. Solid manures and cattle slurry with a higher dry matter content are not able to infiltrate to the same extent, which means that high losses of ammonia may occur by surface application (Sommer and Hutchings, 2001). Thus, such manures are therefore better suited in crops where they can be incorporated or injected to the soil before sowing of a new crop. As solid manures contain a higher proportion of organic N that needs to be mineralized before plant uptake, such manures are also most suitable in crops with a long growing season (Jensen, 2013).

The NH_4-N/total N ratio is highly variable among manure types. In general, it is higher for slurries (40–80%), slurry liquid fraction (80–95%) and digestates (60–85%), while it is lower for solid manures (<30%); see Table 1. Nitrate concentration of liquid manures is generally very low, due to anoxic storage conditions. The NH_4-N/total N ratio often gives a good measure of the proportion of manure N that is potentially available in the first year. The manure N availability can also be expressed as the nitrogen fertilizer replacement value (NFRV) (also termed MFE, the mineral fertilizer equivalent) which is calculated as $ANR_{manure}/ANR_{mineral}$, where ANR_{manure} and $ANR_{mineral}$ are the apparent N recovery of treatments receiving manure and mineral fertilizers, respectively (Jensen, 2013).

2.1 Nitrogen losses from housing systems

The most important manure N loss from buildings is via ammonia emission (Webb et al., 2012). Different housing systems exist that also may vary between livestock species. For dairy, varying forms of slurry systems are applied as well as at least two systems with solid manure, namely tied stall and deep litter systems.

Webb et al. (2012) reviewed studies of gaseous emissions from different housing systems with solid manures, and found that a number of manure management factors influence emissions of NH_3 and N_2O: type of litter; amount of litter; amount of area covered by litter; depth of the litter bed; frequency of fresh litter application and litter removal. For the dairy solid manure systems, more NH_3 is emitted from deep litter systems than tied stalls. Peat litter may reduce NH_3 emission compared with straw litter, as peat has a high water-binding capacity, low pH and therefore the ability to chemically bind NH_3. However, peat is also a limited natural resource.

Poulsen et al. (2001) and Sommer et al. (2006b) have estimated NH_3 emission factors from dairy houses to be 6–8% of total N excretion in slurry systems and 6% from deep litter under Danish conditions. In tied stalls, they only found a loss of 3% of total N in NH_3 emission. From pig systems with slatted floor (slurry), they estimated losses of 8–16% of total N excretion with the highest emission from a fully slatted floor, while they estimated 25% loss of total N from pig deep litter indoor systems. Thus, the choice of housing system can have significant implications on N retention in manure, especially on pig farms. Sommer et al. (2006a) estimated NH_3 emission factors from poultry houses with sand or litter on floors to be 25% of total N excretion.

3 Manure storage and treatment

Storage of manures is usually necessary for practical reasons and especially in order to apply the manure to the field at the optimal time for attaining better synchrony with crop demand and thus a high manure N utilization by the crop. A major problem with manure storage is that losses of N, especially ammonia, are inevitable. However, by proper management – especially by coverage of manure storages – ammonia losses can be significantly reduced (Shah et al., 2016).

3.1 Hygienic aspects of manure storage and use

In some parts of the world, it has been of concern that fresh manure or manure stored over a short period may contain animal or human pathogens and thus should not be used directly in the field before it has been matured/stored for a long period. However, the practical experiences from Northern Europe with extensive use of manures in mixtures containing both fresh and stored manures have indicated that this does not appear to be a major problem when the manure is applied at least some weeks before harvest of a crop. For example, it is common practice in Denmark to apply dairy slurry to grass after each grass cut. However, it is recommended not to use fresh slurry, and in particular not let the cattle graze the slurry-fertilized sward until minimum 60 days afterwards.

In crops for direct human consumption, like fresh vegetables, it is important to leave more time between manure application and crop establishment to avoid contamination of vegetables with pathogens, especially for vegetables consumed in a fresh form.

Nicholson et al. (2005) found that pathogens like *Escherichia coli* O157, *Salmonella* and *Listeria* could survive for up to 3 months in a slurry store in England where temperatures were generally <20°C . In contrast, pathogens survived less than 1 month in solid manure heaps where temperatures greater than 55°C were obtained. Nicholson et al. (2005) found that some pathogens survived for up to 1 month after application to soil although the pathogens often could not be detected in soil 4 days after application. More information can be found in Venglovsky et al. (2009) and Köpke et al. (2017).

Anaerobic digestion (AD) of manure for biogas production is not a hygienization treatment *per se*; however, it depends on the mode of operation. To obtain full hygienization effect by AD, the digestion should be performed at temperatures above 50°C for a certain period of time (Vinneras, 2013), and while some biogas plants work at process temperatures around 35–45°C (mesophilic mode), others operate at temperatures around 50–55°C (thermophilic mode), with hydraulic retention times of 1–4 weeks; the latter will surely result in a rather efficient hygienization, though it will typically not be sufficient for elimination of viruses.

3.2 Liquid manure storage

In liquid manure stores, significant losses of ammonia can take place whether the manure consists of mainly urine or it is a slurry, as the total ammoniacal N (TAN, NH_4^+ + NH_3) concentration is always high in such manures. Losses of up to 30% of total N have been estimated under cool temperate conditions (Sommer, 1997). Ammonia losses are significantly influenced by temperature (Sommer, 1997; Misselbrook et al., 2016) and can be expected to be even higher at higher temperatures. Losses of NH_3 from stored liquid manure can be very high under uncovered conditions (Table 2). On the other hand, it is possible to achieve very low losses from liquid manure stores if they are carefully covered or closed; that is losses can be reduced by up to 61–84% (Sommer et al., 1993; Misselbrook et al., 2016). Covers can be concrete roofs, tents and different floating materials like a natural crust, leca pebbles, cut straw or plastic (Misselbrook et al., 2016). Ammonia losses from covered liquid manures stored under cool temperate conditions were assessed to constitute only 1–4% of total N in the manure (Hansen et al., 2008). In some liquid manures like cattle slurry, a naturally formed surface crust can also constitute an effective barrier to ammonia losses.

In liquid manures, anaerobic conditions prevail unless the manure is continuously aerated by forced air, as microbes in the manure use any available oxygen and exchange with the atmosphere is very limited except in the upper layer (Jensen and Sommer, 2013).

Table 2 Ammonia emission factors from manure storage for solid manures (incl. dry matter rich solid fraction from slurry separation and deep litters) and liquid manures (incl. slurry and digestate)

		NH_3-N emission in % of total N		
Animal category	Manure category	No treatment	Compacted	Covered (plastic or tarpaulin)
Fattening pigs and sows	Solid manure	25	13	13
	Deep litter, sep. solids	25	13	13
Dairy cows, beef cattle	Solid manure	5	3	3
	Deep litter, sep. solids	5	3	3
Poultry	Deep litter	10	5	5
		NH_3-N emission in % of TAN		
		No cover	Cover of surface crust, straw etc.	Roof
Dairy cows, beef cattle	Slurry (concrete tank)	10	3	2
Pigs	Slurry (concrete tank)	11	3	1
Cattle + pig	Digestate	27	5	3

Data from Sommer et al. (2006b) and Hansen et al. (2008).

Losses by denitrification are usually negligible from liquid manures due to a lack of nitrate (Sommer, 1997) unless the manure is aerated by forced air.

3.3 Solid manure storage

In solid manure with high moisture content (>70%), oxygen availability is usually low and decomposition relatively limited. Ammonia emission from heaps or stacks of such solid manures therefore occurs exclusively from the outer surface which is quickly depleted of ammonium, limiting NH_3 emission, but regular additions of fresh manure create a new pulse of NH_3 emission every time. However, if the moisture content is lower than 70% or the manure very straw rich, higher oxygen availability will promote rapid decomposition, resulting in self-heating (composting), mineralization of organic N and organic acids and increasing pH, consequently leading to high concentrations of gaseous NH_3 and rapid and substantial emission. Losses of 25–30% of the total N in stored pig manure and cattle deep litter have been recorded, although losses as low as 1–10% also have been observed (Table 2). Ammonia losses can be lowered by up to 50–90% by decreasing the convection of air through the heap with a PVC cover or through compaction of the litter (Table 2).

3.4 Solid manure composting

3.4.1 Definitions: composting vs. anaerobic storage of solid manures

Composting of manure is defined as the microbially mediated decomposition and subsequent stabilization of organic materials under truly aerobic conditions. In a heap of stored solid manure there will always be spots with both aerobic and anaerobic conditions. However, it is possible to manage the balance between aerobic and anaerobic conditions. Under aerobic conditions, for example by addition of extra straw to the manure and by repeated turning of the heap, the temperature in the manure heap can initially rise to about 70°C due to self-heating from microbial activity (Larney et al., 2006; Webb et al., 2012). By compaction and/or sheeting of the manure, anaerobic conditions can be favoured and heap temperature increase can be controlled (Sommer, 2001; Webb et al., 2012).

3.4.2 Ammonia losses

There has for many years been some controversy over whether aerobic or anaerobic storage of manures should be recommended in organic farming. We found no evidence that composting should always be preferable.

A number of studies indicate that large ammonia losses are almost unavoidable by composting during handling and storage amounting to more than 40% of total manure N. On the contrary, it is possible to reduce ammonia losses by manure storage under mostly anaerobic conditions. This is most easily done in systems with liquid manure management, but coverage of solid manures is also possible using, for example plastic sheets or tarpaulins (Sommer, 2001; Shah et al., 2012).

Webb et al. (2012) showed that both heap temperature and ammonia emission are related to manure density. With higher manure density, air exchange with the atmosphere and ammonia emission is reduced.

3.4.3 Other gaseous losses

During storage of solid manures, significant losses of gaseous N may also take place due to coupled nitrification and subsequent denitrification. This can also result in significant emission of the intermediary product N_2O as shown by Yamulki (2006) and Hansen et al. (2006). In most cases, N_2O emissions have been estimated to be between 0.1% and 0.9% of total N, while total denitrification losses can be significantly higher (Sommer, 2001; Webb et al., 2012). Petersen et al. (1998) found unaccounted N losses during storage of solid pig and cattle manures of 23–33% and 13%, respectively, which were attributed to denitrification. Such high denitrification losses require access to oxygen, as the first step is nitrification of ammonium, and may be favoured under conditions with alternating aerobic-anaerobic conditions. However, emission of N_2O from poultry manure tends to be small (Webb et al., 2012).

3.4.4 Crop nitrogen availability

Shah et al. (2012) made a direct comparison of crop N utilization from solid cattle manures managed by composting (repeated turning), uncovered storage (stockpiled) or anaerobic storage by plastic covering (covering). They measured total N losses of 10% during covered storage, 31% by uncovered storage and 46% by composting (Fig. 2). The apparent N recovery in a grass crop was 20% after covered storage and only 7% after composting when related to the original manure N taken from the animal house before storage. In this study, the manure was surface applied on growing grass and ammonia volatilization was probably significant, especially from the covered manure. If the manures had been incorporated, the benefits of manure covering would be expected to be even larger. Sommer (2001) found an even poorer N utilization of less than 2% in a spring barley crop after incorporation of composted cattle deep litter and similarly found that covered storage of the solid manure improved N utilization to about 9%.

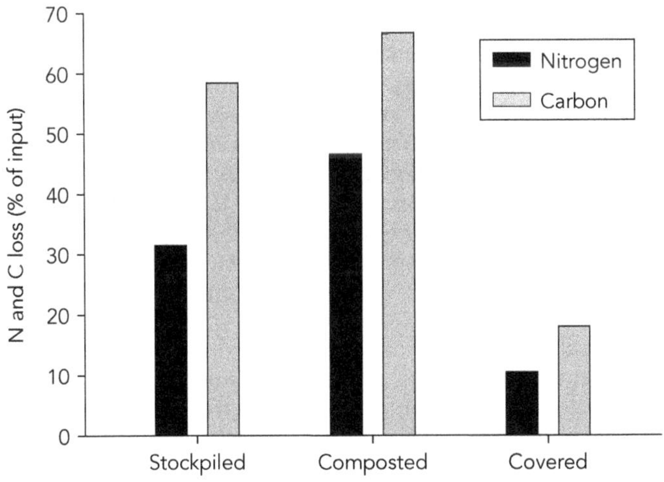

Figure 2 Losses of nitrogen and carbon after 130 days of storage of solid cattle manure using stockpiling, composting or covering by a plastic sheet. Based on data from Shah et al. (2012).

3.4.5 Effects on soil organic carbon

After composting, a certain fraction of the manure organic carbon has been decomposed and respired, but the carbon remaining is more stable (Helgason et al., 2005), and will decompose more slowly once applied to the soil. More than 50% of the initial carbon can be released by composting, whereas less C is lost under anaerobic conditions (Thomsen, 2000; Shah et al., 2012) (Fig. 2). The low losses observed from solid manure covered with plastic (Fig. 2) can probably be explained by a combination of low oxygen level due to low aeration and by low heap temperature (17–25°C). In practice, temperatures may be higher in covered manure, as the plastic covered storage by Shah et al. (2012) was performed under a roof with limited solar radiation. Some studies have suggested that composting should be preferred due to a higher long-term carbon/organic matter retention in soil (Bernal et al., 1998). However, it is important to take into account that the total C input to soil is also significantly lower after composting. In controlled studies where both C losses during storage and decomposition during 266 days in soil were accounted for, a higher C retention in soil was observed after anaerobic storage compared to composting of solid manure (Thomsen and Olesen, 2000).

In the long-term perspective, the amount of C retained in soil can be expected to be nearly the same, whether the same source of C in fresh manure is applied directly to soil or is composted before application (Larney et al., 2006).

3.4.6 Advantages from composting

Advantages from composting of manures are reduced mass, volume and water content, which is most beneficial if manures need to be transported over larger distances (Larney et al., 2006). Other benefits include elimination of pathogens, parasites and weed seeds; reduced odour emissions from land application (Larney et al., 2006) and easier handling and spreading.

Many studies deal with benefits of compost as a disease suppressant (Litterick et al., 2004). Composted manure was demonstrated to suppress *Sclerotinia* which affects a number of agricultural crops (Larney et al., 2006). Different manure composts were found to suppress damping-off caused by *Pythium* and *Rhizoctonia* (Ringer et al., 1997). The magnitude of suppression of some fungal diseases depends on the stability of the compost (Stone et al., 2004) cited from Larney et al. (2006).

Most seeds including weed seeds are killed during composting due to the high process temperatures, but seeds from different species have different sensibility to high temperatures (Larney and Blackshaw, 2003). As heap surfaces do not reach high temperatures, composts cannot be expected to be totally free of viable seeds.

In a 4-year field experiment, De Cauwer et al. (2011) found that the weed seed bank was highest in plots amended with slurry, and lowest where compost was used. According to the authors, this result was not generated by differences in the amount of viable seeds applied with slurry vs. compost, but could have been caused by differences in seed degradation in soil.

A problem with fresh and anaerobically stored manures is odour emission during and after spreading. A benefit of composting is that the compost can be nearly odourless, but the composting process itself is usually not odour free (Larney et al., 2006).

3.5 Anaerobic digestion of manure

AD of organic materials including crop residues and green manures for biogas production provides an interesting opportunity for optimizing N supply in organic farming systems

(Pugesgaard et al., 2014; Rahmann et al., 2016). In AD, organic compounds contained in manure and other biodegradable substrates (like organic residues and wastes and the so-called 'energy crops') are decomposed by microorganisms in the absence of oxygen. The end products are biogas (a mixture of mainly methane and carbon dioxide) and digested manure and biowaste. By AD of manure, part of the organic N in manure is mineralized to TAN and the easily decomposable organic compounds (especially VFAs) are converted to biogas. This implies less immobilization after manure application and a higher first year availability of N in the manure. By AD, pH in the digestate increases compared to untreated manure (Sommer and Husted, 1995b). Thus, there is a risk of increased ammonia emission and to benefit fully from the increased N availability, it is important to store the digestate covered and to use application techniques with low risk of ammonia emission. Typical properties of digestates are shown in Table 1.

Co-digestion of manures with other organic wastes is commonly done in order to increase the biogas yield and economic efficiency of the biogas digester, but is also an attractive way to recycle and reuse nutrients in different organic wastes from society back into the agricultural food chain. This requires that the chemical quality of the waste, like the content of heavy metals and organic contaminants, is acceptable for use on organic farms. Organic wastes with a high C/N ratio can have low or even negative N availability, but after digestion, the N availability is increased.

After AD, manures contain significantly less organic C, but the residual C is significantly more recalcitrant and therefore stable in soil (Moeller, 2015). Thomsen et al. (2013) made a direct comparison of C retention in soil with and without AD and they found that the long-term retention of carbon in soil is very similar when the same amount of C in plant material is applied directly to the soil, fed to a cow and the manure then applied to the soil, or used as feedstock to a biogas plant and the digestate then applied to the soil.

The high temperature that is reached in AD plants can also kill weed seeds. For example, Johansen et al. (2013) found that thermophilic conditions (55°C) killed weed seeds in a few days, while mesophilic conditions (37°C) required more time (11 days) to inactivate the seeds of all species tested. Westerman and Gerowitt (2013) provided a comprehensive review of the effects of AD on weed seed survival.

Moeller (2015) reviewed consequences of manure AD for soil biological activity and fertility, and concluded that the direct long-term effects of AD on N emissions, soil microbial biomass and soil fertility are very similar compared to other ways of manure management.

4 Manure field application methods

Using a proper application technique is important for achieving an even and accurate distribution of manure in the field, which is a prerequisite for reaching an efficient utilization of plant-available nutrients. The application technique and manure exposure time to the atmosphere are especially important for ammonia losses from manures and therefore also important for the first year crop utilization of manure N. Ammonia volatilization starts immediately after manure application, and the volatilization rate is typically highest in the beginning, so the sooner manures can be incorporated, the more can volatilization be reduced.

Runoff loss of manure nutrients is often not so important in agronomic terms, but can cause environmental eutrophication problems even when a low proportion of applied manure nutrients ends up in the water catchments (Sørensen and Jensen, 2013). Surface application of manures on sloping land causes a large risk of runoff losses, whereas injection of manure or rapid incorporation is an effective way to reduce these risks (Eghball and Gilley, 1999). On steep slopes, manure injection is also a problem because the liquid manure can start to run in the slots formed in soil by the injection tines.

4.1 Liquid manure application methods

For application of liquid manures, broad spreading, trailing hoses, trailing shoes and direct injection can be used (Sommer and Hutchings, 2001). Liquid manures usually contain a large proportion of N in the ammonium form and therefore have a large potential for ammonia losses after application. Ammonia losses are favoured by factors impacting on the equilibrium reaction of ammonium ions in the soil solution with gaseous ammonia in soil and air. After surface application, ammonia volatilization is influenced by the liquid infiltration rate in soil, which is most affected by the dry matter content of the manure, but infiltration rate can also be improved by soil cultivation shortly before application (Sommer and Hutchings, 2001). Since broad spreading results in the largest ammonia volatilization due to the large specific surface area of the applied manure (in a thin layer) and therefore high air/manure contact, this technique is no longer permitted in several countries.

Ammonia volatilization rates can be reduced by surface banding, for example using trailing hoses, because the manure-specific surface area is reduced by placing the manure in narrow bands. Incorporation of surface applied slurry by ploughing is more effective for reducing NH_3 loss than by harrowing (Fig. 3). The most effective application method to reduce NH_3 volatilization is direct injection at 10 cm depth or deeper (Fig. 3). However, deeper injection is also more energy demanding and a shallow injection of about 10 cm depth appears sufficient. In growing crops, a reduction in ammonia volatilization of 50% has been observed after shallow injection compared with the use of trailing hoses (Nyord et al., 2012). In growing crops, injection can only be shallow to avoid large damages to the crop. Other important factors affecting ammonia volatilization during field application of manure are soil pH (more acidic soils reduce losses), soil water content and temperature, solar radiation and wind (which favour losses).

Different application methods also influence the initial turnover of manure N and thereby the plant availability. By placement of manure in a more concentrated string in the soil by injection, without extensive mixing with the soil, less N is immobilized and more N is available to the crop (Sørensen and Jensen, 1995). This effect is more significant for cattle slurry than for pig slurry (Table 3) and it is more significant in soils with a higher clay content (Sørensen and Jensen, 1995).

Placement of slurry close to crop rows by injection can be an advantage, especially in crops with a large row distance. For instance, in maize crops, early access to applied nutrients in manure is a clear advantage (Schröder et al., 2015) and placement of cattle slurry close to maize rows can improve early uptake of both N and P in the crop (Pedersen et al., 2017). A faster initial growth of the crop obtained by manure placement also has the potential to improve the crop competition against weeds.

4.2 Solid manure application methods

For solid manures, the normal application technique is a carrier wagon, with a bottom conveyer belt which moves the solid manure continuously towards some rear-mounted

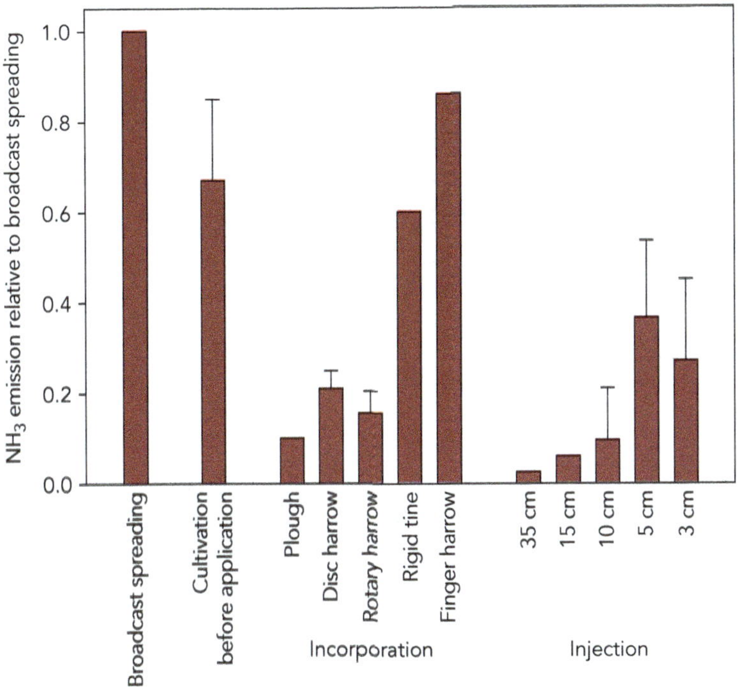

Figure 3 Ammonia emission relative to broadcasting of animal slurry by use of different application techniques onto soil or plant-covered soil. Data summarizing results from a range of experiments performed in different countries and soil types ranging from sandy to loamy soils. Modified from Sommer and Hutchings (2001), with permission.

spreading augers, mounted either vertically, producing a broader spreading pattern or horizontally, producing a narrow, but more homogeneous application strip on the soil surface. Such surface application should be followed by rapid incorporation into the soil by harrowing or ploughing, except by application on a growing crop. Attempts have also been made for direct injection of solid poultry manures (Pote et al., 2011). Webb et al. (2012) reviewed a number of studies of the effects of solid manure incorporation on ammonia losses. They found an average reduction in ammonia emission of 92% and 85% for incorporation within less than 4 h for pig and poultry manures (compared with no incorporation), and a reduction of 20%, 56% and 50% for incorporation within 24 h for cattle, pigs and poultry, respectively. Incorporation of manure by ploughing is more effective to reduce ammonia emission than incorporation by disc or harrow. On the other hand, the working capacity of ploughing is lower, which can mean a longer average incorporation time when using ploughing under practical conditions.

4.3 Timing of manure application

Ammonium in manure is normally nitrified to nitrate within a few days or weeks after application to soil. Both nitrate and sulphate are very vulnerable to leaching. Therefore, manures should generally be applied shortly before the crop needs the manure nutrients in

Table 3 Nitrogen fertilizer replacement value (NFRV) of pig and cattle slurry after surface application + incorporation (by a roto-tiller) or direct injection to a spring barley crop

Fertilizer/manure, method of application	Grain yield (t ha^{-1}, @15% moisture)	N uptake (kg N in grain + straw ha^{-1})	NFRV (% of total N)
Cattle slurry, surface + incorporated	4.26	81	41
Cattle slurry, direct injected	5.14	109	68
Pig slurry, surface + incorporated	4.63	90	63
Pig slurry, direct injected	5.13	103	79
NPK fertilizer	5.19	104	100
LSD ($p < 0.95$)	0.20	9.0	4.8

Average of 3 years of repeated field experiments on a sandy loam soil in Denmark. All slurry treatments received 100 kg NH_4-N ha^{-1}.
Data from Sørensen et al. (2003a) and Webb et al. (2013).

areas with risk of water transport out of the root zone. Under temperate climatic conditions, this means application should be in spring, when most crops have the largest N demand. Perennial crops like grass and a few autumn-sown crops like oil-seed rape (canola) can also take up significant amounts of N in autumn. Proper timing of manure implicates a need for manure storage capacity for almost a year in areas with only one crop per year. Such large storage capacity can be a significant cost to the farmer.

Application of solid manure to bare soil in the early autumn (September) results in extra leaching that is often equal to the ammonium content under humid North European conditions (Thomsen, 2005). By postponing the application to December or March, Thomsen (2005) observed no extra N leaching in the first year compared with no manure application, probably because of delayed nitrification and lower precipitation than needed to transport manure N out of the root zone by the late applications. Some autumn-sown cereals like winter wheat have a relatively small autumn N uptake and are therefore not able to utilize manure N applied in autumn. After application of solid manures just before sowing winter wheat, Sørensen and Rubaek (2012) observed an extra nitrate leaching equivalent to the ammonium content (20–30% of total manure N) under conditions with surplus drainage of 300–400 mm $year^{-1}$. Also on grassland soil, Smith et al. (2002) observed extra nitrate leaching losses equal to the ammonium content of applied solid manures by application to grass in October or later in autumn under cool moist conditions in England.

Application of liquid manures in autumn can cause even higher nitrate losses as they contain a higher proportion of ammonium-N, especially on sandy soils and climates with high precipitation during autumn and winter (Thomsen et al., 1993).

On more heavy soils, application of manures in autumn can result in significant denitrification losses under wet conditions with high rainfall during winter. Thompson et al. (1987) found denitrification losses above 20% of total N in cattle slurry after slurry injection in early winter on a loam soil with grassland. This is probably not an environmental concern (as these losses are mostly N_2), but is a loss of an important N source. On heavy textured soils in areas with wet winter and spring conditions, it can be impossible to apply manure in spring without causing severe soil compaction and damage to soil structure. The soil may only be workable over a very short period, and it can be impossible to incorporate or inject manures. Under such conditions, it is challenging to obtain the full benefits of especially manure N and the best option can be to make a surface application after sowing or in the growing crop.

Even though most manures show a net N immobilization phase after application to soil, it is normally advisable to apply the manure just before sowing the crop or even later when the crop has a high demand for N. By postponing the manure application until sowing time or later, the risk of leaching losses can be minimized in years with high precipitation in spring.

Surface application of manures on non-sloping land can be made during light rain or just before rainfall; this practice can reduce ammonia emission after the manure application due to infiltration of TAN into the soil where it is adsorbed (Misselbrook et al., 2005). Sommer et al. (1991) and Rodhe et al. (1996) found reductions in ammonia emission of 30–50% by irrigation right after solid manure application, resulting in more N available for the crop.

5 Turnover and availability of manure N in soil

Manure contains N in ammoniacal and organic form (Webb et al., 2013). Ammonium is plant-available and thus mostly affects short-term N availability (weeks or months after application), while organic N must be mineralized before it is available to plants. Most of the organic N is slowly mineralized and is only available in the longer term (months to years). Direct uptake of organic N in plants is considered to be possible but negligible.

5.1 Short-term nitrogen availability

After manure application to soil, ammonium usually undergoes one of the following processes that are relatively fast (hours, days, weeks at most): volatilization, nitrification, microbial immobilization, denitrification or fixation in less or non-exchangeable form in clay minerals (Fig. 4).

Nitrification of NH_4^+ occurs in aerobic soils that are not strongly acidic and most of the NH_4^+ is transformed into NO_3^- during a short period (usually 7–14 days) under optimal soil temperature and water content; for example Loria and Sawyer (2005). The nitrification rate is significantly influenced by soil temperature and is relatively slow at temperatures close to 0°C.

Two simultaneous, but opposite processes affect soil mineral N (SMN, the sum of soil nitrate and ammonium): mineralization of organic N and microbial N immobilization (Fig. 4). After application of manure, both processes increase significantly due to the easily decomposable organic matter in manure, but often immobilization is higher than mineralization shortly after manure application, resulting in a negative net mineralization of N (defined as the difference between mineralization and immobilization).

Nitrogen mineralization and immobilization are linked to the respiration of manure C and microbial immobilization occurs when microbes decompose N-poor compounds (Kirchmann and Lundvall, 1993; Sørensen, 1998b). Net N immobilization starts immediately after manure application to soil, and typically reaches a peak 1–4 weeks later (Sørensen, 2004; Bechini and Marino, 2009; Webb et al., 2013). Typical values for net N immobilization correspond to 20–40% of applied organic N (Morvan et al., 2006; Sørensen et al., 2017). Due to microbial N immobilization, not all the ammonium applied is plant-available in the months after manure addition to soil (Sørensen, 2004; Cavalli et al., 2016a). Microbially immobilized N is slowly remineralized, concurrent with a continuous mineralization of N from more recalcitrant manure compounds. Even several months after manure application,

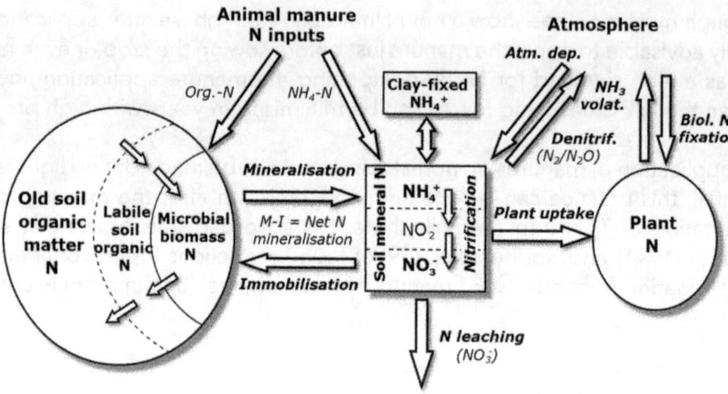

Figure 4 The principal pools and processes of the N cycle in the soil–plant–atmosphere system which can be influenced by animal manure N inputs. Modified from Jensen (2013), with permission.

N mineralization may not fully compensate the initial N immobilization (Thomsen and Olesen, 2000; Cavalli et al., 2016b; Delin et al., 2012). In these cases, therefore, not all ammonium-N applied with manure is plant-available in the year of manure application. Net mineralization of manure N is also influenced by the manure distribution in soil and an uneven distribution of manure in soil, for example by band injection, causes less N immobilization and therefore typically results in a higher N availability for crops (Sørensen and Jensen, 1995) as described in Section 4.1.

Ammonium can also be fixed very quickly (within hours after manure addition to the soil) in non-exchangeable form by 2:1 clay minerals (Nieder et al., 2011; Fig. 4). The amount of ammonium fixed depends on soil characteristics (type and quantity of clay minerals). Release of fixed ammonium is a slow process that can take several weeks. After repeated applications of two slurries, Cavalli et al. (2016b) found that about 20% of applied ammonium was in non-exchangeable form in a clay loam soil, whereas clay fixation has been found to be negligible in more sandy soils (Sørensen and Jensen, 1995).

During the decomposition of the manure organic N in soil, mineralized N can be lost by emission of N_2 and N_2O by denitrification. In addition, N_2 and N_2O emissions occur also during nitrification. In general, the denitrification losses are small (Webb et al., 2013; Velthof et al., 2003). Manures favour denitrification by providing easily decomposable organic matter, which reduces soil oxygen concentration, resulting in more reducing conditions (Charles et al., 2017), but on the other hand, a smaller proportion of manure N is initially transformed to nitrate compared to mineral fertilizer N. Charles et al. (2017) found in a meta-analysis that the average N_2O emission factor from organic manures is lower than that from synthetic N sources (emission related to total N application). However, they also found large differences in emission factors among manures with relatively high emission from liquid manures with a high proportion of inorganic N, lower emission from solid manures and lowest emission from composts.

5.1.1 Factors affecting nitrogen processes in the soil

All biological processes mentioned above are strongly influenced by soil water content and temperature (Griffin et al., 2002). Soil texture can also affect the decomposition rate of organic materials added to soils, due to physicochemical protection exerted by clay minerals (Six et al., 2002). However, experiments comparing the effects of contrasting soil textures with the effects of contrasting manures show that in general, decomposition rate is more affected by manure characteristics than by soil texture, the effect of which is small in most cases (Bechini and Marino, 2009; Thomsen and Olesen, 2000; Thomsen et al., 2003), and if present, influenced by the distribution of manure in soil (Sørensen and Jensen, 1995).

Microbial N immobilization depends on the presence of easily degradable compounds with a high C/N ratio. The critical threshold, that is the C/N ratio corresponding to equal rates of N mineralization and immobilization, is about 15 (Delin et al., 2012; Webb et al., 2013).

Factors affecting the C/N ratio of the manure are animal species, diet (Powell et al., 2006; Sørensen et al., 2003b; Sørensen and Fernandez, 2003), amount and type of bedding used in the stable (Sørensen, 1998b), type and duration of storage, and manure treatments (separation and composting) (Webb et al., 2013; Möller and Müller, 2012). In many studies, the C/N ratio was found to be the best predictor of N mineralization (e.g. Morvan et al., 2006; Fig. 5). While the manure C/N ratio has an impact on the mineralization/immobilization equilibrium, the quantity and the stability of manure organic pools influence the extent and duration of the two processes. In fact, mineralization and immobilization are high for manures with large and easily decomposable pools, while they are less pronounced in more stable manures, for example after composting (Bechini and Marino, 2009; Morvan et al., 2006). Therefore, microbial N immobilization is generally more pronounced in undigested compared to digested slurries (Sørensen et al., 2012), because digestates have a larger fraction of stable organic matter, less VFAs and a lower C/N ratio compared to undigested slurries (Möller and Müller, 2012). These characteristics of digestates, together with a higher NH_4-N/total N ratio, can contribute to a more efficient crop N utilization, particularly in the short term (Schröder et al., 2007; de Boer, 2008; Cavalli et al., 2016a), even if the

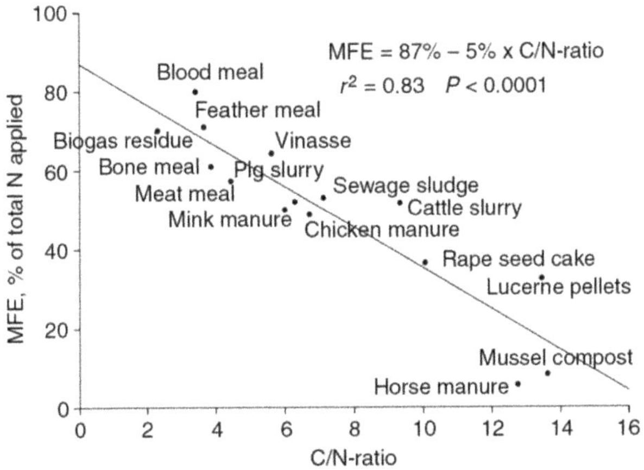

Figure 5 Influence of the total C/N ratio of different organic fertilizer products on the mineral fertilizer equivalent (MFE) value. Modified from Delin et al. (2012), with permission.

higher pH and the higher ammonium content of digestates make them more prone to ammonia volatilization losses.

The previous management of soil has minimal influence on the availability of newly applied manure (Bosshard et al., 2009; Nett et al., 2011). Bosshard et al. (2009) compared the utilization of sheep manure N applied to soil that had been managed either organically with organic manures or conventionally with mineral fertilizers for a number of years, and they could not detect differences in the crop utilization of applied manure N. However, Rashid et al. (2013) observed an effect of previous field management on the release of N after surface application of manure on grassland. They found a faster litter decomposition and N release from manure when applied close to the home farm, and this effect is called 'home field advantage'.

5.1.2 Management criteria

Organic farmers only have limited amounts of manure and need to prioritize where to use it most optimally. Manures typically contain a significant amount of organic matter, which upon application will promote microbial turnover and nutrient immobilization, reducing crop available nutrients through pool substitution. A basic concept in manure management in organic farming should therefore be to maximize the crop utilization of immediately available nutrients in the applied manure, while at the same time promoting the plant utilization of mineralized nutrients in subsequent growing seasons of the manure nutrients inevitably immobilized in the soil upon application. The fraction of manure N that is plant-available in the short term (first crop after application) is largely determined by the manure content of mineral N, because the mineralization rate of manure organic N is low. This is true not only for composts (Sommer, 2001; Eghball and Power, 1999; Thomsen and Olesen, 2000), but also for slurries and digestates if ammonia losses after application are low (de Boer, 2008; Sørensen et al., 2012).

From a practical point of view, first-year N availability can be estimated using tabbed NFRV values which vary depending on the type of manure. They range from 0% to 5% for composts to nearly 100% for some liquid manures (Table 1).

The application date for spring-sown annual crops should be shortly before crop establishment to minimize leaching losses from any winter precipitation. However, in winter-annual crops, manure N should preferentially be applied to the growing crop in spring, especially under conditions with cool humid winters like in Northern Europe (e.g. Sørensen and Rubaek, 2012). Initial immobilization and clay fixation, which reduce available N, should be taken into account to adapt manure application rate accordingly to avoid crop N stress.

5.2 Long-term nitrogen availability

5.2.1 Processes occurring in the long term following manure application

The long-term availability of manure N (i.e. after the year of application) is due to the mineralization of three pools of organic N: the N immobilized by microbes after manure application, the manure organic N that was not decomposed during the first year, and the manure N that has been taken up by the crop, but now has been returned to the soils in residue and root exudates. These residual effects are small after a single manure application, due to the recalcitrance of these pools (Sørensen and Amato, 2002; Muñoz et al., 2003).

When a soil is subjected to repeated manure applications for a series of years, these pools accumulate in the soil and the cumulative effect becomes more and more important (Nevens and Reheul, 2005; Schröder et al., 2007; Cela et al., 2011). In principle, in continuously manured soils, the ANR increases until a steady state is reached, when the cumulative amount of mineralized manure organic N is equal to the annual input of organic N (Schröder et al., 2013), but this may take more than 100 years. Schröder et al. (2013) have summarized residual effects and found average (across different manure types) mineralization values for the second, third and fourth years of 13%, 8% and 6% of the organic N applied in the first year. Similar values are provided by Jensen (2013) and illustrated in Fig. 6. When manuring is halted on a historically manured soil, the annual N mineralization rate decreases similarly slowly over time, until it reaches a baseline (absence of residual effects).

The fraction of manure N that is not mineralized or taken up by plants remains in the soil in organic form. Zavattaro et al. (2017) have estimated, based on an extensive review of European long-term experiments, that soils of manured treatments, in comparison with treatments receiving equal amounts of mineral fertilizers only, contain 15% more organic N after long-term cattle slurry application and 22% more after long-term solid cattle manure application.

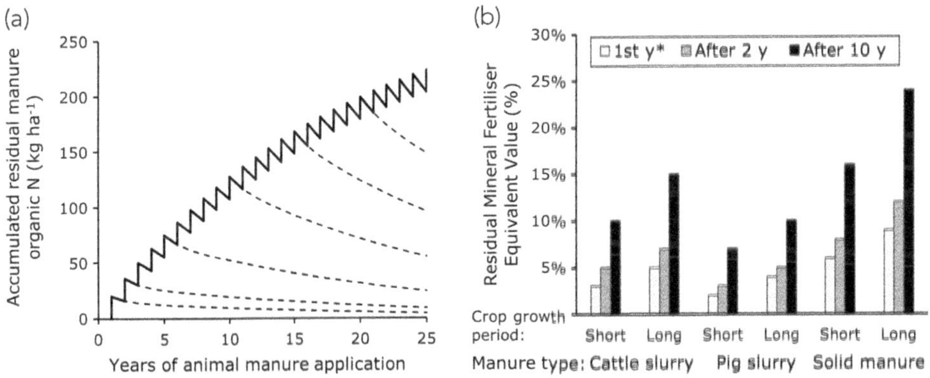

Figure 6 (a) Modelled residual manure organic N accumulated (full line) in soil over 30 years with an annual application of 100 kg total N ha^{-1} in pig slurry (25% org. N/total N, dashed line indicating decay of org. N after first, second, fifth, tenth, fifteenth and twentieth application) and (b) the corresponding equivalent fertilizer replacement value (NFRV) value of 1, 2 or 10 years of repeated applications of different types of animal manure, compared with soil only receiving mineral fertilizers. The NFRV value is given in per cent of the annual manure total N application for crops with either a short (e.g. spring cereals, 50% of mineralized N available for crop) or a long growing period (e.g. maize, beet, 75% of mineralized N available for crop). Solid manure: cattle FYM and deep litter.*: effect in the year after application, the 1st year with residual N effect. Figure based on principles in Sørensen et al. (2002) and reprinted from Jensen (2013) with permission.

5.2.2 Factors affecting the residual effects

The type of manure affects the residual effects by regulating both the size of the microbially immobilized fraction and, more importantly, the size of the recalcitrant fraction: residual effects are more important for manures containing a higher proportion of organic N and for manures having a larger recalcitrant fraction (Webb et al., 2013). Thus, fibrous materials like the solid fraction obtained after physical separation of liquid manures, and composted materials like FYM have larger residual effects compared with more easily decomposable materials like the liquid fraction. AD likely decreases the residual effects (Schröder et al., 2007; Webb et al., 2013) due to lower organic N/total N ratio in digestates compared to undigested manures. Soil texture may also modify residual effects. Sørensen and Amato (2002) found lower remineralization rates in a soil with 19% clay compared to a soil with 9% clay, suggesting that residual organic N might be more stabilized in more clayey soils.

5.2.3 Management criteria

From a practical point of view, residual effects should be managed by: (1) adopting the right values of residual NFRV in the nutrient management plan to take account of the residual N effect, thus avoiding excess N fertilizer rates; and (2) ensuring efficient sinks of crops during the winter to avoid leaching of N mineralized from bare soil, for example by using catch crops. In general, crops with a longer growing season will be able to use more N mineralized from soil, including residual manure N, compared to crops with a shorter growing season, as illustrated in Fig. 6. The use of catch crops or cash crops with a long growing season is particularly important in organic farming, where, in comparison with conventional farming, it is more likely that the soil is rich in labile soil organic matter, and the economic value of the residual N effects is higher than in conventional farming.

6 Utilization of P, K and S in manures

An advantage of manures is that they contain the full range of essential nutrients in combined form so that crops can benefit from having access to all the nutrients needed in relatively high concentrations locally in soil.

6.1 Phosphorus

Soils with sufficient P test levels will normally be capable of supplying most crops, including all cereals, with their demand for available P by mobilization from the labile soil P pool, meaning that crop performance is not dependent on new applications of manure P every year. As P is normally strongly bound in soil, losses of P are low and surplus P application (in excess of crop removal) will accumulate in soil. Surplus P application may result from the fact that manures have a larger N:P ratio compared to crop tissues. Thus, if manure application rates are calibrated on crop N needs, too much P is typically applied for that particular crop. Results from long-term experiments show that surplus P applied with manures and mineral fertilizers contribute similarly to soil available P in the long-term (Huang et al., 2011). Therefore, manures should normally be applied to crops that benefit from the manure N, whereas legumes that don't need an extra N supply should not receive manure, as they can cover their need for P (and other nutrients) from P present in soil, provided that the soil P availability level is sufficient.

However, many organic farms do not have sufficient animal production to have an average surplus or even balance of manure P vs. crop removal, for example arable farms or farms with only limited animal production, and no import of feed. On such farms, soils will gradually reach a low or very low level of available P (low P soil test), and on such soils/farms, it is important to ensure alternative P inputs, for example manure/composts/digestates from other farms or from urban recycling, and to ensure application every year in order to avoid the risk of yield loss.

Some crops like maize and crops with a small root system may benefit from a placement of fertilizer P close to the seed in connection with the sowing, the so-called 'starter-effect'. Such a placement can also be made with liquid manures (Bittman et al., 2012; Schröder et al., 2015). Pedersen et al. (2017) found that the method of manure placement also affects the P availability. Most of the P in manures is normally in water-extractable inorganic form (Sharpley and Moyer, 2000; Kleinman et al., 2005), but manures also contain some insoluble inorganic P and some organic P in both soluble and insoluble form. This implies that a significant part of the manure P is expected to be plant available. In a pot experiment, Delin (2016) found that the 1st-year plant availability was only about half that of superphosphate for a range of manure types. Some studies indicate that manure P is not always easily available to a crop shortly after application. A reason for this is that P is very immobile in soil, meaning that roots need to be close to the manure for uptake of manure P, and the (chemical) environment around newly applied manure may not be suitable to roots (Sawyer and Hoeft, 1990). Another reason can be microbial immobilization of P shortly after the application.

6.2 Potassium

Potassium (K) in manures is in inorganic and mainly soluble form (K^+), meaning that most of the manure K is also immediately available for crops. The high solubility also means that K will mostly follow the water fraction in manure, with risk of leaching. However, K^+ is adsorbed on clay particles in soil and it is therefore not highly mobile in soil. Therefore, leaching losses of K are normally low on more clayey soils, but on sandy soils, K^+ adsorption is less effective and significant K losses by leaching can occur on coarse sandy soils in humid climates (Askegaard et al., 2004). Loamy or clayey soils rich in clays of the illite and montmorillonite type (containing K^+ in the interlayer) will also have a large capacity for K^+ release by weathering (Askegaard et al., 2004).

This means that the K supply for crops on organic farms depends largely on soil type, climate and average manure K balance. If the soil K test level is in the sufficient range, the soil available K pool will be able to supply most crops with their demand for K; however, for highly K-demanding crops like potatoes and grass-clover on very sandy soils with large water surplus, more regular K application may be needed.

As grass-clover crops can cover most of their N needs through symbiotic N fixation, they don't need to be supplied with large amounts of manure N. In order to utilize the manure N efficiently, manures should therefore be used on more N-demanding crops. If a grass-clover crop needs extra supply of K, other sources of K should be considered, for example sources like patentkali consisting of mainly potassium sulphate.

6.3 Sulphur

Sulphur (S) is an essential plant nutrient that is mainly present as organic S in soil and that needs to be mineralized to inorganic sulphate to be taken up by the crop. Sulphate is vulnerable to leaching, similarly to nitrate, meaning that surplus sulphate is easily lost by leaching. Mineralization of soil S cannot always cover the crop needs for S. Previously, S in deposition from air could cover the need for S in most crops throughout mainland Europe, but as fossil fuel exhaust-gas cleaning has significantly improved, atmospheric S deposition has declined dramatically in most parts, and there is a higher risk for S deficiency in many crops. Manures mainly contain S in organic form, as most sulphate is lost as hydrogen sulphide during manure storage (Eriksen et al., 1995). After manure application relatively low S mineralization rates are usually observed, meaning that most animal manures only provide a small amount of available S to the crop which they are applied to (Eriksen et al., 1995) while they supply S to the organic S pool in soil and can help to increase soil S mineralization capacity in the longer term.

7 Plant-based manures (green manures)

Green manure crops are defined as single or mixtures of plant species grown in between main crops (normally in autumn, but also during summer, if water availability allows) with the sole purpose of providing beneficial functions for subsequent crops. However, also full-year crops like grass-clover may serve green manure functions. Green manures can have important functions in organic cropping systems, that is to supply extra N to the system if they contain leguminous species, to reduce nitrate leaching (Tonitto et al., 2006; Thorup-Kristensen et al., 2003), to help with sanitizing weeds and to provide an extra input of organic matter (Poeplau and Don, 2015). However, they do not provide other nutrient inputs to the farm, and they are therefore not suitable as a sole nutrient supply. Most green manure species belong to three botanical families: *Poaceae* (grasses), *Brassicaceae* (brassicas) and *Fabaceae* (legumes). Only legume cover crops can fix atmospheric N: normally they are characterized by higher N uptake compared to other species. Species mixes, for example grass-clover, are used to complement botanical and agronomic characteristics from the different species used.

Dedicated full-year green manure crops like grass-clover can provide significant amounts of N up to 300–400 kg N ha^{-1} to the system by symbiotic N fixation (Rasmussen et al., 2012), but if they are used solely as a green manure the farmer will miss an economic income from this land in the green manure year. Economically, it is therefore more attractive if the shoots of the green manures can be used as cattle feed or for biogas production and the animal manure or digestate then returned where most needed in the crop rotation.

N-fixing cover crops can both (1) fix significant amounts of N symbiotically (Cherr et al., 2006; Li et al., 2015); and (2) at the same time reduce N leaching significantly (Askegaard et al., 2011; De Notaris et al., 2018). However, cover crops only fix significant amounts of N when they are allowed to grow long enough to attain a significant biomass yield. Similarly, non-N-fixing cover crops need a long growing period to take up residual N from the soil. This means that, for fall-planted cover crops, the sowing date should be early enough (in early August in cool temperate climates and early September in warmer European climates) to allow fast growth before temperature and solar radiation decrease during the autumn. In organic crop rotations where N-fixing cover crops were grown every year, De Notaris et al. (2018) found that the N contribution from cover crops to main crop yields could be as high as an average animal manure application of 70 kg total N ha^{-1} yr^{-1}. Cover crops do also have the benefit of conserving other leachable nutrients like sulphur (Thorup-Kristensen et al., 2003).

Green manure growth can be terminated by frost (if so termed 'winter-killed', e.g. white mustard, *Sinapis alba*); if frost-resistant (i.e. winter hardy, e.g. rye, *Secale cereale*) they must be terminated by mechanical operations like ploughing in conventional tillage systems; rolling, crimping or discing in conservation tillage systems.

Green manure crops can also be harvested and applied as mobile green manure to another crop (Sorensen and Thorup-Kristensen, 2011). This is most relevant for N-demanding high-value crops in farms where there is only restricted access to animal manures, as harvest and transport of the green manure is rather costly. The N fertilizer value of green manures is significantly influenced by their C/N ratio (Sorensen and Thorup-Kristensen, 2011). As C/N ratio increases with age of herbage, the N availability is higher in green manure crops harvested at a young stage. Sørensen et al. (2013) measured a NFRV of 33% by incorporation of grass-clover silage (C/N 14.9) and 48% by incorporation of lucerne silage (C/N 12.4) before sowing a spring barley crop.

Sørensen et al. (2013) compared storage of green manures as either compost or as silage. They found significant gaseous losses of N during composting and lower plant availability of N, similarly to what has been determined for composted animal manure. The utilization of N in green manure was about twice as high by ensiling compared to composting. In silage, there is a high concentration of organic acids like acetic and propionic acid and therefore silage should not be used as a fertilizer applied directly on growing crops, as organic acids in high concentrations can have detrimental effects on living plants. However, organic acids are quickly decomposed in soil under aerobic conditions, and the use of silage as fertilizer is not a problem when it is applied before sowing a new crop as the acids are decomposed within a few days, before a new crop is germinating.

Weed control capabilities of cover crops arise from their competitive ability against weeds (for radiation, water and nutrients), from their allelopathic characteristics and from phytotoxic compounds released by cover crop tissue during their decomposition that delay weed germination (Thorup-Kristensen et al., 2003). Competitive ability of cover crops depends on their timely establishment, and rapid initial growth rate. Allelopathic effects contributing to weed control are documented for rye, sorghum, brassicas (rapeseed and mustards), buckwheat and subterranean clover. More details on this issue are provided by Singh et al. (2003) and Haramoto and Gallandt (2004).

8 Future trends and conclusion

The main challenges with manure and other waste-based fertilizers are i) low nutrient concentrations and bulkiness, ii) slow/uncertain nutrient availability, iii) inappropriate nutrient ratios, iv) higher risk of environmental emissions and v) odour, biosecurity and other nuisances for neighbours. These challenges are common for conventional and organic farmers, but some of them pose special challenges for organic agriculture, where supplementation with synthetic fertilizer nutrients cannot be relied on.

Agricultural conditions vary across different geographical settings of organic agriculture, and also when it comes to most optimal strategies for manure and biowaste utilization for crop and soil fertilization, there is a need for adaption of manure management to the local production conditions, including climate, soils and crops. Under some agricultural conditions, composting may be the most appropriate processing technology to apply, while under other agricultural conditions, processing in biogas plants may be the most optimal. We need more research on these issues, with good case illustrations of advantages and drawbacks of the individual solutions under varying conditions. This could include more life-cycle and sustainability (including economic and social aspects) assessments of nutrient supply chains for organic farming products.

Organic farming systems may also have to rely more on the utilization of efficient green manures. These practices are well established, but there is a need for more research on how to increase the efficiency of green manures to grow well under less fertile soil conditions, to fix more N from the atmosphere and to mobilize more soil P, K, S and base cations, including both better plant genotypes and management. Furthermore, we also need more research on how best to integrate green manures with processing of other nutrient sources, for example through co-digestion of harvested green manure biomass in biogas plants.

With organic agricultural products increasing their market share in many affluent countries, the demand for larger quantities of recycled organic fertilizers will be increasing rapidly (Oelofse et al., 2013), and appropriate fertilizer quality (content of nutrients, absence of contaminants) and application methods are needed. However, high quality in itself is not sufficient justification for the application of manure or organic waste derived fertilizers or soil amendments in the organic agricultural sector; the procurement history or origin must also have a sustainable 'footprint'; that is the derived environmental, economic and social impacts have to be acceptable.

In the future, more urban-derived wastes and residues are expected to be recycled back to agriculture (Buckwell and Nadeu, 2016), including to organic farms; this would align well with cyclic philosophy deeply embedded in the concept of organic agriculture. A way to do this may be through co-digestion of such wastes with manures in biogas plants. There is a need for research in this area both in terms of optimizing the utilization of nutrients and investigating how this can be done in an acceptable way without compromising soil and crop quality.

Also the formulation of bio-based fertilizers, which can be certified and approved for organic production (inclusion of organic waste and bio-based materials in the EU Fertilizer Regulation and Organic Certification schemes) are needed. There is a need for research unravelling the consequences of certification schemes for end user (farmer and food product consumers) acceptability of the use of externally derived, bio-based fertilizers, what their perceptions and preferences are, giving perspectives for the market-driven development of more sustainable, bio-based, organic 'green' fertilizers (Case et al., 2017).

9 Where to look for further information

Further information on manure management can be found via informal organizations like the RAMIRAN network: Research Network on Recycling of Agricultural and Industrial

Residues in Agriculture (www.ramiran.uvlf.sk). International conferences are organized regularly by the RAMIRAN network, and a 'Glossary of terms on livestock and manure management' is available (http://ramiran.uvlf.sk/doc11/RAMIRAN%20Glossary_2011.pdf). Recently, the ManuResource conference has been organized regularly with a focus on manure management (http://manuresource2017.org).

A more detailed introduction to manure management is given in the book *Animal Manure Recycling: Treatment and Management* (Sommer et al., 2013).

Regarding the of use cover crops, a very good technical manual is available in Clark (2008), but with a focus on US conditions. The AHDB Nutrient Management Guide (RB209) is a good technical source on nutrient management (https://ahdb.org.uk/projects/RB209.aspx). Their website also features a number of videos and an *app*, with a focus on UK conditions.

10 References

Askegaard, M., Eriksen, J. and Johnston, A. E. (2004). Sustainable management of potassium. In: Schjonning, P., Elmholt, S. and Christensen, B. T. (Eds), *Managing Soil Quality: Challenges in Modern Agriculture*. Wallingford, UK: CABI Publishing, pp. 85–102.

Askegaard, M., Olesen, J. E., Rasmussen, I. A., et al. (2011). Nitrate leaching from organic arable crop rotations is mostly determined by autumn field management. *Agriculture, Ecosystems & Environment* 142: 149–60.

Bechini, L. and Marino, P. (2009). Short-term nitrogen fertilizing value of liquid dairy manures is mainly due to ammonium. *Soil Science Society of America Journal* 73: 2159–69.

Bernal, M. P., Sanchez-Monedero, M. A., Paredes, C., et al. (1998). Carbon mineralization from organic wastes at different composting stages during their incubation with soil. *Agriculture, Ecosystems & Environment* 69: 175–89.

Bernal, M. P., Sommer, S. G., Chadwick, D., et al. (2017). Chapter 3: Current approaches and future trends in compost quality criteria for agronomic, environmental, and human health benefits. In: Sparks, D. L. (Ed.), *Advances in Agronomy*. London, UK: Elsevier Academic Press, pp. 143–233.

Bittman, S., Liu, A., Hunt, D. E., et al. (2012). Precision placement of separated dairy sludge improves early phosphorus nutrition and growth in corn (*Zea mays* L.). *Journal of Environmental Quality* 41: 582–91.

Bosshard, C., Sorensen, P., Frossard, E., et al. (2009). Nitrogen use efficiency of N-15-labelled sheep manure and mineral fertiliser applied to microplots in long-term organic and conventional cropping systems. *Nutrient Cycling in Agroecosystems* 83: 271–87.

Buckwell, A. and Nadeu, E. (2016). Nutrient Recovery and Reuse (NRR) in European agriculture. A review of the issues, opportunities, and actions. Brussels: RISE foundation. http://www.risefoundation.eu/projects/nrr

Carlsbæk, M. (2011). Use of compost in horticulture and landscaping. In: Christensen, T. H. (Ed.), *Solid Waste Technology and Management*. Chichester, UK: John Wiley & Sons, Ltd, pp. 515–32.

Case, S. D. C., Oelofse, M., Hou, Y., et al. (2017). Farmer perceptions and use of organic waste products as fertilisers - A survey study of potential benefits and barriers. *Agricultural Systems* 151: 84–95.

Cavalli, D., Cabassi, G., Borrelli, L., et al. (2016a). Nitrogen fertilizer replacement value of undigested liquid cattle manure and digestates. *European Journal of Agronomy* 73: 34–41.

Cavalli, D., Marino Gallina, P., Sacco, D., et al. (2016b). Soil mineral nitrogen dynamics following repeated application of dairy slurry. *European Journal of Soil Science* 67: 804–15.

Cela, S., Santiveri, F. and Lloveras, J. (2011). Residual effects of pig slurry and mineral nitrogen fertilizer on irrigated wheat. *European Journal of Agronomy* 34: 257–62.

Charles, A., Rochette, P., Whalen, J. K., et al. (2017). Global nitrous oxide emission factors from agricultural soils after addition of organic amendments: A meta-analysis. *Agriculture, Ecosystems & Environment* 236: 88–98.

Cherr, C. M., Scholberg, J. M. S. and McSorley, R. (2006). Green manure approaches to crop production: A synthesis. *Agronomy Journal* 98: 302–19.

Clark, A. (2008). *Managing Cover Crops Profitably*. Beltsville, MD: DIANE Publishing.

de Boer, H. C. (2008). Co-digestion of animal slurry can increase short-term nitrogen recovery by crops. *Journal of Environmental Quality* 37: 1968–73.

De Cauwer, B., D'Hose, T., Cougnon, M., et al. (2011). Impact of the quality of organic amendments on size and composition of the weed seed bank: Effect of fertiliser quality on weed seed bank. *Weed Research* 51: 250–60.

Delin, S. (2016). Fertilizer value of phosphorus in different residues. *Soil Use and Management* 32: 17–26.

Delin, S., Stenberg, B., Nyberg, A., et al. (2012). Potential methods for estimating nitrogen fertilizer value of organic residues. *Soil Use and Management* 28: 283–91.

De Notaris, C., Rasmussen, J., Sørensen, P., et al. (2018). Nitrogen leaching: A crop rotation perspective on the effect of N surplus, field management and use of catch crops. *Agriculture, Ecosystems & Environment* 255, 1–11.

Eghball, B. and Gilley, J. E. (1999). Phosphorus and nitrogen in runoff following beef cattle manure or compost application. *Journal of Environmental Quality* 28: 1201–10.

Eghball, B. and Power, J. F. (1999). Composted and noncomposted manure application to conventional and no-tillage systems: Corn yield and nitrogen uptake. *Agronomy Journal* 91: 819–25.

Eriksen, J., Mortensen, J. V., Kjellerup, V. K., et al. (1995). Forms and plant-availability of sulfur in cattle and pig slurry. *Zeitschrift für Pflanzenernährung und Bodenkunde* 158: 113–16.

Fliessbach, A., Oberholzer, H. R., Gunst, L., et al. (2007). Soil organic matter and biological soil quality indicators after 21 years of organic and conventional farming. *Agriculture, Ecosystems & Environment* 118: 273–84.

Griffin, T., Honeycutt, C. and He, Z. (2002). Effects of temperature, soil water status, and soil type on swine slurry nitrogen transformations. *Biology and Fertility of Soils* 36: 442–6.

Hansen, M. N., Henriksen, K and Sommer, S. G. (2006). Observations of production and emission of greenhouse gases and ammonia during storage of solids separated from pig slurry: Effects of covering. *Atmospheric Environment* 40: 4172–81.

Hansen, M. N., Sommer, S. G., Hutchings, N. J., et al. (2008). Emissionsfaktorer til beregning af ammoniakfordampning ved lagring og udbringning af husdyrgødning [Emission factors for calculation of ammonia volatilization by storage and application of animal manure]. *DJF Rapport Husdyrbrug 84*. Tjele, Denmark: Ministry of Food, Agriculture and Fisheries, Danish Institute of Agricultural Sciences, pp. 1–43.

Haramoto, E. R. and Gallandt, E. R. (2004). Brassica cover cropping for weed management: A review. *Renewable Agriculture and Food Systems* 19: 187–98.

Helgason, B. L., Larney, F. J. and Janzen, H. H. (2005). Estimating carbon retention in soils amended with composted beef cattle manure. *Canadian Journal of Soil Science* 85: 39–46.

Hu, T., Sørensen, P. and Olesen, J. E. (2018). Soil carbon varies between different organic and conventional management schemes in arable agriculture. *European Journal of Agronomy* 94: 79–88.

Huang, S., Ma, Y., Bao, D., et al. (2011). Manures behave similar to superphosphate in phosphorus accumulation in long-term field soils. *International Journal of Plant Production* 5: 135–46.

Jensen, L. S. (2013). Animal manure fertiliser value, crop utilisation and soil quality impacts. In: Sommer, S. G., Christensen, M. L., Schmidt, T., et al. (Eds), *Animal Manure Recycling: Treatment and Management*. Chichester, UK: John Wiley & Sons Ltd, pp. 295–328.

Jensen, L. S. and Sommer, S. G. (2013). Manure organic matter - characteristics and microbial transformations. In: Sommer, S. G., Christensen, M. L., Schmidt, T., et al. (Eds), *Animal Manure Recycling: Treatment and Management*. Chichester, UK: John Wiley & Sons Ltd, pp. 67–90.

Johansen, A., Nielsen, H. B., Hansen, C. M., et al. (2013). Survival of weed seeds and animal parasites as affected by anaerobic digestion at meso- and thermophilic conditions. *Waste Management* 33: 807–12.

Kirchmann, H. and Lundvall, A. (1993). Relationship between N immobilization and volatile fatty acids in soil after application of pig and cattle slurry. *Biology and Fertility of Soils* 15: 161–4.

Kleinman, P. J. A., Wolf, A. M., Sharpley, A. N., et al. (2005). Survey of water-extractable phosphorus in livestock manures. *Soil Science Society of America Journal* 69: 701–8.

Köpke, U., Krämer, J. and Leifert, C. (2017). Pre-harvest strategies to ensure the microbiological safety of fruit and vegetables from manure-based production systems. In: Cooper, J., Niggli, U. and Leifert, C. (Eds), *Handbook of Organic Food Safety and Quality*. Cambridge, UK: Woodhead Publishing Ltd., pp. 413–29.

Larney, F. J. and Blackshaw, R. E. (2003). Weed seed viability in composted beef cattle feedlot manure. *Journal of Environmental Quality* 32: 1105–13.

Larney, F. J., Sullivan, D. M., Buckley, K. E., et al. (2006). The role of composting in recycling manure nutrients. *Canadian Journal of Soil Science* 86: 597–611.

Li, X., Sørensen, P., Li, F., et al. (2015). Quantifying biological nitrogen fixation of different catch crops, and residual effects of roots and tops on nitrogen uptake in barley using in-situ 15N labelling. *Plant and Soil* 395: 273–87.

Litterick, A. M., Harrier, L., Wallace, P., et al. (2004). The role of uncomposted materials, composts, manures, and compost extracts in reducing pest and disease incidence and severity in sustainable temperate agricultural and horticultural crop production - A review. *Critical Reviews in Plant Sciences* 23: 453–79.

Loria, E. R. and Sawyer, J. E. (2005). Extractable soil phosphorus and inorganic nitrogen following application of raw and anaerobically digested swine manure. *Agronomy Journal* 97: 879–85.

Mader, P., Fliessbach, A., Dubois, D., et al. (2002). Soil fertility and biodiversity in organic farming. *Science* 296: 1694–7.

Maillard, E. and Angers, D. A. (2014). Animal manure application and soil organic carbon stocks: A meta-analysis. *Global Change Biology* 20: 666–79.

Marino, P., De Ferrari, G. and Bechini, L. (2008). Description of a sample of liquid dairy manures and relationships between analytical variables. *Biosystems Engineering* 100: 256–65.

Misselbrook, T. H., Nicholson, F. A., Chambers, B. J., et al. (2005). Measuring ammonia emissions from land applied manure: An intercomparison of commonly used samplers and techniques. *Environmental Pollution* 135: 389–97.

Misselbrook, T., Hunt, J., Perazzolo, F., et al. (2016). Greenhouse gas and ammonia emissions from slurry storage: Impacts of temperature and potential mitigation through covering (pig slurry) or acidification (cattle slurry). *Journal of Environmental Quality* 45: 1520–30.

Möller, K. (2015). Effects of anaerobic digestion on soil carbon and nitrogen turnover, N emissions, and soil biological activity. A review. *Agronomy for Sustainable Development* 35: 1021–41.

Möller, K. and Müller, T. (2012). Effects of anaerobic digestion on digestate nutrient availability and crop growth: A review. *Engineering in Life Sciences* 12: 242–57.

Morvan, T., Nicolardot, B. and Péan, L. (2006). Biochemical composition and kinetics of C and N mineralization of animal wastes: A typological approach. *Biology and Fertility of Soils* 42: 513–22.

Muñoz, G. R., Powell, J. M. and Kelling, K. A. (2003). Nitrogen budget and soil N dynamics after multiple applications of unlabeled or 15Nitrogen-enriched dairy manure. *Soil Science Society of America Journal* 67: 817–25.

Nett, L., Ruppel, S., Ruehlmann, J., et al. (2011). Influence of soil amendment history on decomposition of recently applied organic amendments. *Soil Science Society America Journal* 76: 1290–300.

Nevens, F. and Reheul, D. (2005). Agronomical and environmental evaluation of a long-term experiment with cattle slurry and supplemental inorganic N applications in silage maize. *European Journal of Agronomy* 22: 349–61.

Nicholson, F. A., Groves, S. J. and Chambers, B. J. (2005). Pathogen survival during livestock manure storage and following land application. *Bioresource Technology* 96: 135–43.

Nieder, R., Benbi, D. K. and Scherer, H. W. (2011). Fixation and defixation of ammonium in soils: A review. *Biology and Fertility of Soils* 47: 1–14.

Nyord, T., Hansen, M. N. and Birkmose, T. S. (2012). Ammonia volatilisation and crop yield following land application of solid-liquid separated, anaerobically digested, and soil injected animal slurry to winter wheat. *Agriculture, Ecosystems & Environment* 160: 75–81.

Oelofse, M., Jensen, L. S. and Magid, J. (2013). The implications of phasing out conventional nutrient supply in organic agriculture: Denmark as a case. *Organic Agriculture* 3: 41–55.

Olesen, J. E., Hansen, E. M., Askegaard, M., et al. (2007). The value of catch crops and organic manures for spring barley in organic arable farming. *Field Crops Research* 100: 168–78.

Olesen, J. E., Askegaard, M. and Rasmussen, I. A. (2009). Winter cereal yields as affected by animal manure and green manure in organic arable farming. *European Journal of Agronomy* 30: 119–28.

Pedersen, I. F., Rubaek, G. H. and Sorensen, P. (2017). Cattle slurry acidification and application method can improve initial phosphorus availability for maize. *Plant and Soil* 414: 143–58.

Peters, K., Hjorth, M., Jensen, L. S., et al. (2011). Carbon, nitrogen, and phosphorus distribution in particle size-fractionated separated pig and cattle slurry. *Journal of Environmental Quality* 40: 224–32.

Petersen, S. O., Lind, A. M. and Sommer, S. G. (1998). Nitrogen and organic matter losses during storage of cattle and pig manure. *Journal of Agricultural Science* 130: 69–79.

Poeplau, C. and Don, A. (2015). Carbon sequestration in agricultural soils via cultivation of cover crops – A meta-analysis. *Agriculture, Ecosystems & Environment* 200: 33–41.

Pote, D. H., Way, T. R., Kleinman, P. J. A., et al. (2011). Subsurface application of poultry litter in pasture and no-till soils. *Journal of Environmental Quality* 40: 402–11.

Poulsen, H. D., Børsting, C. F., Rom, H. B., et al. (2001). Kvælstof, fosfor og kalium i husdyrgødning – normtal 2000. *DJF Rapport Husdyrbrug 36*. Tjele, Denmark: Danmarks JordbrugsForskning.

Powell, J. M., Wattiaux, M. A., Broderick, G. A., et al. (2006). Dairy diet impacts on fecal chemical properties and nitrogen cycling in soils. *Soil Science Society of America Journal* 70: 786–94.

Pugesgaard, S., Olesen, J. E., Jorgensen, U., et al. (2014). Biogas in organic agriculture- effects on productivity, energy self- sufficiency and greenhouse gas emissions. *Renewable Agriculture and Food Systems* 29: 28–41.

Rahmann, G., Reza Ardakani, M., Bàrberi, P., et al. (2016). Organic Agriculture 3.0 is innovation with research. *Organic Agriculture* 7: 169–97.

Rashid, M. I., de Goede, R. G. M., Brussard, L., et al. (2013). Home field advantage of cattle manure decomposition affects the apparent nitrogen recovery in production grasslands. *Soil Biology and Biochemistry* 57: 320–6.

Rasmussen, K. (2002). Influence of liquid manure application method on weed control in spring cereals. *Weed Research* 42: 287–98.

Rasmussen, J., Soegaard, K., Pirhofer-Walzl, K., et al. (2012). N-2-fixation and residual N effect of four legume species and four companion grass species. *European Journal of Agronomy* 36: 66–74.

Ringer, C. E., Millner, P. D., Teerlinck, L. M., et al. (1997). Suppression of seedling damping-off disease in potting mix containing animal manure composts. *Compost Science & Utilization* 5: 6–14.

Rodhe, L., Salomon, E. and Rammer, C. (1996). Spreading of farmyard manure to ley with different methods - Yield and silage quality. *Swedish Journal of Agricultural Research* 26: 43–51.

Sawyer, J. E. and Hoeft, R. G. (1990). Effect of injected liquid beef manure on soil chemical properties and corn root distribution. *Journal of Production Agriculture* 3: 50–5.

Schröder, J. J., Uenk, D. and Hilhorst, G. (2007). Long-term nitrogen fertilizer replacement value of cattle manures applied to cut grassland. *Plant and Soil* 299: 83–99.

Schröder, J., Bechini, L., Bittman, S., et al. (2013). Residual N effects from livestock manure inputs to soils. *RAMIRAN 2013–15th International Conference*. RAMIRAN - Network on Recycling of Agricultural, Municipal and Industrial Residues in Agriculture.

Schröder, J. J., Vermeulen, G. D., van der Schoot, J. R., et al. (2015). Maize yields benefit from injected manure positioned in bands. *European Journal of Agronomy* 64: 29–36.

Shah, G. M., Groot, J. C. J., Oenema, O., et al. (2012). Covered storage reduces losses and improves crop utilisation of nitrogen from solid cattle manure. *Nutrient Cycling in Agroecosystems* 94: 299–312.

Shah, G. M., Shah, G. A., Groot, J. C. J., et al. (2016). Effect of storage conditions on losses and crop utilization of nitrogen from solid cattle manure. *Journal of Agricultural Science* 154: 58–71.

Sharpley, A. and Moyer, B. (2000). Phosphorus Forms in Manure and Compost and Their Release during Simulated Rainfall. *Journal of Environmental Quality* 29: 1462–9.

Singh, H. P., Batish, D. R. and Kohli, R. K. (2003). Allelopathic interactions and allelochemicals: New possibilities for sustainable weed management. *Critical Reviews in Plant Sciences* 22: 239–311.

Six, J., Conant, R. T., Paul, E. A., et al. (2002). Stabilization mechanisms of soil organic matter: Implications for C-saturation of soils. *Plant and Soil* 241: 155–76.

Smith, K. A., Beckwith, C. P., Chalmers, A. G., et al. (2002). Nitrate leaching following autumn and winter application of animal manures to grassland. *Soil Use and Management* 18: 428–34.

Sommer, S. G. (1997). Ammonia volatilization from farm tanks containing anaerobically digested animal slurry. *Atmospheric Environment* 31: 863–8.

Sommer, S. G. (2001). Effect of composting on nutrient loss and nitrogen availability of cattle deep litter. *European Journal of Agronomy* 14: 123–33.

Sommer, S. G. and Husted, S. (1995a). The chemical buffer system in raw and digested animal slurry. *Journal of Agricultural Science* 124: 45–53.

Sommer, S. G. and Husted, S. (1995b). A simple model of pH in slurry. *The Journal of Agricultural Science* 124: 447–53.

Sommer, S. G. and Hutchings, N. J. (2001). Ammonia emission from field applied manure and its reduction - invited paper. *European Journal of Agronomy* 15: 1–15.

Sommer, S. G., Olesen, J. E. and Christensen, B. T. (1991). Effects of temperature, wind-speed and air humidity on ammonia volatilization from surface applied cattle slurry. *Journal of Agricultural Science* 117: 91–100.

Sommer, S. G., Christensen, B. T., Nielsen, N. E., et al. (1993). Ammonia volatilization during storage of cattle and pig slurry - effect of surface cover. *Journal of Agricultural Science* 121: 63–71.

Sommer, S. G., Jensen, B.-E., Hutchings, N., et al. (2006a). Emissionskoefficienter til brug ved beregning af ammoniakfordampning fra stalde. *DJF Rapport Husdyrbrug 70*. Tjele, Denmark: Ministry of Food, Agriculture and Fisheries and Danish Institute of Agricultural Sciences.

Sommer, S. G., Zhang, G. Q., Bannink, A., et al. (2006b). Algorithms determining ammonia emission from buildings housing cattle and pigs and from manure stores. *Advances in Agronomy* 89: 261–335.

Sommer, S. G., Christensen, M. L., Schmidt, T., et al. (2013). *Animal Manure Recycling: Treatment and Management*. Chichester, UK: John Wiley & Sons Ltd.

Sørensen, J. N. and Thorup-Kristensen, K. (2011). Plant-based fertilizers for organic vegetable production. *Journal of Plant Nutrition and Soil Science* 174: 321–32.

Sørensen, L. K., Sørensen, P. and Birkmose, T. S. (2007). Application of reflectance near infrared spectroscopy for animal slurry analyses. *Soil Science Society of America Journal* 71: 1398–405.

Sørensen, P. (1998a). Carbon mineralization, nitrogen immobilization and pH change in soil after adding volatile fatty acids. *European Journal of Soil Science* 49: 457–62.

Sørensen, P. (1998b). Effects of storage time and straw content of cattle slurry on the mineralization of nitrogen and carbon in soil. *Biology and Fertility of Soils* 27: 85–91.

Sørensen, P. (2004). Immobilisation, remineralisation and residual effects in subsequent crops of dairy cattle slurry nitrogen compared to mineral fertiliser nitrogen. *Plant and Soil* 267: 285–96.

Sørensen, P. and Amato, M. (2002). Remineralisation and residual effects of N after application of pig slurry to soil. *European Journal of Agronomy* 16: 81–95.

Sørensen, P. and Fernandez, J. A. (2003). Dietary effects on the composition of pig slurry and on the plant utilization of pig slurry nitrogen. *Journal of Agricultural Science* 140: 343–55.

Sørensen, P. and Jensen, E. S. (1995). Mineralization-immobilization and plant uptake of nitrogen as influenced by the spatial distribution of cattle slurry in soils of different texture. *Plant and Soil* 173: 283–91.

Sørensen, P. and Jensen, L. S. (2013). Nutrient leaching and runoff from land application of animal manure and measures for reduction. In: Sommer, S. G., Christensen, M. L., Schmidt, T. and Jensen, L. S. (Eds), *Animal Manure Recycling: Treatment and Management*. Chichester, UK: Wiley, pp. 195–210.

Sørensen, P., Khan, A. R., Møller, H. B., et al. (2012). Effects of anaerobic digestion of organic manures on N turnover and N utilization. In: Richards, K. G., Fenton, O and Watson, C. J. (Eds), *The 17th Nitrogen Workshop – Innovations for Sustainable Use of Nitrogen Resources*, 26–29 June 2012. Wexford, Ireland, pp. 80–1.

Sørensen, P., Kristensen, E., Odokonyero, K., et al. (2013). Utilization of nitrogen in legume-based mobile green manures stored as compost or silage. In: Løes A.-K., Askegaard, M., Langer, V., et al. (Eds), Organic farming systems as a driver for change. *NJF Report*, no. 9(3), pp. 157–8.

Sørensen, P. and Rubaek, G. H. (2012). Leaching of nitrate and phosphorus after autumn and spring application of separated solid animal manures to winter wheat. *Soil Use and Management* 28: 1–11.

Sørensen, P., Thomsen, I. K., Jensen, B., et al. (2002). Residual nitrogen effects of animal manure measured by 15N. In: Østergaard, H. S., Fystro, G. and Thomsen, I. K. (Eds), *Optimal Nitrogen Fertilization Tools for Recommendation: Proceedings from NJF Seminar 322*. DIAS Report Plant Production 84. Copenhagen, Denmark: Danish Institute of Agricultural Sciences, pp. 37–41.

Sørensen, P., Thomsen, I. K. and Schroder, J. J. (2017). Empirical model for mineralisation of manure nitrogen in soil. *Soil Research* 55: 500–5.

Sørensen, P., Vinther, F. P., Petersen, S. O., et al. (2003a). Høj udnyttelse af gyllens kvælstof ved direkte nedfældning. *Grøn Viden*. Research Centre Foulum and Danmarks JordbrugsForskning, pp. 1–6.

Sørensen, P., Weisbjerg, M. R. and Lund, P. (2003b). Dietary effects on the composition and plant utilization of nitrogen in dairy cattle manure. *The Journal of Agricultural Science* 141: 79–91.

Stone, A. G., Scheuerell, S. J. and Darby, H. M. (2004). Suppression of soilborne diseases in field agricultural systems: Organic matter management, cover cropping, and other cultural practices. In: Magdoff, F. and Weil, R. R. (Eds), *Soil Organic Matter in Sustainable Agriculture*. Boca Raton, FL: CRC Press, pp. 131–77.

Thompson, R. B., Ryden, J. C. and Lockyer, D. R. (1987). Fate of nitrogen in cattle slurry following surface application or injection to grassland. *Journal of Soil Science* 38: 689–700.

Thomsen, I. K. (2000). C and N transformations in (15)N cross-labelled solid ruminant manure during anaerobic and aerobic storage. *Bioresource Technology* 72: 267–74.

Thomsen, I. K. (2005). Crop N utilization and leaching losses as affected by time and method of application of farmyard manure. *European Journal of Agronomy* 22: 1–9.

Thomsen, I. K. and Olesen, J. E. (2000). C and N mineralization of composted and anaerobically stored ruminant manure in differently textured soils. *Journal of Agricultural Science* 135: 151–9.

Thomsen, I. K., Hansen, J. F., Kjellerup, V., et al. (1993). Effects of cropping system and rates of nitrogen in animal slurry and mineral fertilizer on nitrate leaching from a sandy loam. *Soil Use and Management* 9: 53–8.

Thomsen, I. K., Schjønning, P. and Christensen, B. T. (2003). Mineralisation of 15N-labelled sheep manure in soils of different texture and water contents. *Biology and Fertility of Soils* 37: 295–301.

Thomsen, I. K., Olesen, J. E., Moller, H. B., et al. (2013). Carbon dynamics and retention in soil after anaerobic digestion of dairy cattle feed and faeces. *Soil Biology & Biochemistry* 58: 82–7.

Thorup-Kristensen, K., Magid, J. and Jensen, L. S. (2003). Catch crops and green manures as biological tools in nitrogen management in temperate zones. *Advances in Agronomy* 79: 227–302.

Tonitto, C., David, M. B. and Drinkwater, L. E. (2006). Replacing bare fallows with cover crops in fertilizer-intensive cropping systems: A meta-analysis of crop yield and N dynamics. *Agriculture, Ecosystems & Environment* 112: 58–72.

Velthof, G. L., Kuikman, P. J. and Oenema, O. (2003). Nitrous oxide emission from animal manures applied to soil under controlled conditions. *Biology and Fertility of Soils* 37: 221–30.

Venglovsky, J., Sasakova, N. and Placha, I. (2009). Pathogens and antibiotic residues in animal manures and hygienic and ecological risks related to subsequent land application. *Bioresource Technology* 100: 5386–91.
Vinneras, B. (2013). Sanitation and hygiene in manure management. In: Sommer, S. G., Christensen, M. L., Schmidt, T., et al. (Eds), *Animal Manure Recycling: Treatment and Management*. Chichester, UK: John Wiley and Sons Ltd, pp. 91–104.
Webb, J., Sommer, S. G., Kupper, T., et al. (2012). Emissions of ammonia, nitrous oxide and methane during the management of solid manures. *Agroecology and Strategies for Climate Change* 8: 67–107.
Webb, J., Sørensen, P., Velthof, G., et al. (2013). An assessment of the variation of manure nitrogen efficiency throughout Europe and an appraisal of means to increase manure-N efficiency. *Advances in Agronomy* 119: 371–442.
Westerman, P. R. and Gerowitt, B. (2013). Weed seed survival during anaerobic digestion in biogas plants. *The Botanical Review* 79: 281–316.
Yamulki, S. (2006). Effect of straw addition on nitrous oxide and methane emissions from stored farmyard manures. *Agriculture, Ecosystems & Environment* 112: 140–5.
Zavattaro, L., Bechini, L., Grignani, C., et al. (2017). Agronomic effects of bovine manure: A review of long-term European field experiments. *European Journal of Agronomy* 90: 127–38.

Chapter 8

Organic fertilizers and biofertilizers

Lidia Sas Paszt and Slawomir Gluszek, Research Institute of Horticulture, Poland

1 Introduction
2 Biofertilizers
3 Consortia of microorganisms to improve the effectiveness of organic fertilization
4 Animal excrement: manures, slurry and guano
5 Products and by-products of animal origin
6 Products and by-products of plant origin for fertilizers
7 Composts
8 Untreated minerals and by-products of selected industrial processes
9 Biochar
10 Conclusion
11 Where to look for further information
12 References

1 Introduction

Organic fertilizers contain organic substances, including amino acids, sugars, lipids, vitamins, enzymes, plant hormones, macro- and microelements and other active compounds, which stimulate plant growth and development (Crouch and van Staden, 1993; Stirk and Staden, 1996; Khan et al., 2009; Ji et al., 2017).

Products permitted for use in organic farming in the European Union (EU) are listed in 'Annex I' to the EEC Regulation 2092/1991. At present, there is 'Annex II' – 'Products for use in fertilization and soil conditioning' (Løes et al., 2016). This list covers a wide range of products, from manures and composts to sediments from freshwater bodies. All of them are used as raw materials or, nowadays, as highly processed, modern products in the form of concentrated liquids, enriched powders, granulates or similar products. New bio-based formulations make it possible to save energy, labour input and time needed for efficient field application and plant nutrition (Vaneeckhaute et al., 2013). Biofertilizers contain, among other components, living cells of microorganisms as the biologically active ingredient. The products suitable for organic agriculture that are available on the European market include a large group of products ranging from dried manures to highly processed organic-mineral fertilizers and mycorrhizal and bacterial inocula. Despite numerous studies

http://dx.doi.org/10.19103/AS.2017.0029.08

on the use of preparations of organic origin, there is still a lack of bio-stimulants and protective substances for enriching the habitats of plants microbiologically. Further work is necessary on improving the technologies of cultivation and fertilization of fruit plants.

2 Biofertilizers

There has been, so far, no precise definition for the term 'biofertilizer'. Some authors define it as a substance that contains, as an active ingredient, live cells of microorganisms separately or in combination with other active ingredients that colonize the rhizosphere or spaces inside plant tissues and that help in enhancing the nutrient uptake by plants, and is thus able to enrich the nutrient quality of the soil (Vessey, 2003). This definition limits biofertilizers to microbes that can assimilate free nitrogen or dissolve water-insoluble salts of phosphorus, potassium or other nutrients. Sometimes, the term 'biofertilizer' is used for artificially multiplied microbial inoculants that can improve soil fertility and crop productivity (Roychowdhury et al., 2015).

Others define biofertilizers as biologically active products containing organic compounds such as amino acids, sugars, vitamins and other substances that have a direct nutritional impact on plant growth and yielding.

However, limiting the action only to nutritional effects means that in many cases, there is a problem with a wide group of organic products (both microbial and of organic origin) which also have other effects, for example, protection against diseases or pests or plant growth regulation.

A large group of compounds for agriculture, especially for organic farming, that contain living microbial cells are called biofertilizers. This category covers products that contain live cells of bacteria, cyanobacteria, actinomycetes, yeast cells and spores and/or hyphae of microscopic, filamentous or mycorrhizal fungi. The bacterial and other microbial strains used in the products for organic agriculture belong to various genera and species.

The mechanisms by which plant-growth-promoting rhizobacteria (PGPR) enhance plant growth are not fully understood. However, it is believed that the PGPR promote plant growth and yielding by either direct or indirect mechanisms, where direct mechanisms include the ability to produce phytohormones like indole acetic acid (IAA), gibberellin, cytokinins and ethylene; asymbiotic N_2 fixation; antagonism against phytopathogenic microorganisms by the production of siderophores; and solubilization of mineral phosphates and other nutrients; and indirect mechanisms entail the extracellular production of antibiotics, synthesis of antifungal metabolites, production of fungal cell wall lysin enzymes, depletion of iron from the rhizosphere and competition for sites on roots and induced systemic resistance (Salantur et al., 2006; Egamberdiyeva, 2007; Ahmad et al., 2008; Aznar et al., 2014; Gusain et al., 2015).

2.1 Nitrogen-fixing microorganisms

'Traditional' biofertilizers contain symbiotic bacteria, for example, rhizobia (*Rhizobium, Neorhizobium, Allorhizobium, Pararhizobium, Mesorhizobium, Bradyrhizobium* and *Ensifer*), free-living bacteria, such as *Azotobacter, Azospirillum*, or cyanobacteria like *Nostoc, Anabaena, Tolypothrix* or *Aulosira*, which are able to fix free nitrogen from the air (Souza et al., 2014; Dhar et al., 2015).

2.1.1 Rhizobia

Rhizobia are bacteria that formerly belonged to the family Rhizobiaceae, but at present, some of them are classified into new families: Phyllobacteriaceae into the genus *Mesorhizobium*, Nitrobacteriaceae (Bradyrhizobiaceae) into the genus *Bradyrhizobium* and Burkholderiaceae into the genus *Paraburkholderia* (formerly *Burkholderia*) (Velázquez et al., 2017). The complete list of valid species of rhizobia is constantly updated and recorded as the 'List of Prokaryotic names with Standing in Nomenclature' (http://www.bacterio.cict.fr).

These bacteria are mutualistic symbionts, and fix nitrogen environmentally depended highly variable at about 50–350 kg/ha, with legumes only. They are useful for pulse legumes like chickpea, red-gram, pea, lentil, black gram and so forth; oilseed legumes like soybean and groundnut; and forage legumes like berseem, clovers and lucerne. Successful nodulation of leguminous crops by a *Rhizobium* largely depends on the availability of compatible strains for a particular legume. Rhizobia colonize the roots of specific legumes to form tumour-like growths called root nodules, which act as factories of ammonia production. Rhizobia have the ability to fix atmospheric nitrogen in a symbiotic association with legumes and certain non-legumes, like *Parasponia*. Each legume requires a specific species of *Rhizobium* to form effective nodules. Many legumes may be nodulated by diverse strains of Rhizobia, but growth is enhanced only when nodules are produced by the effective strains of Rhizobia. It is thus extremely important to match microsymbionts prudently for maximum nitrogen fixation, especially because some native strains are not as effective as host-specific rhizobial strains, and there is a possibility of establishing symbiosis with wild 'cheating', non-effective nitrogen-fixing strains (Kiers et al., 2003; Vicario et al., 2016). A strain of *Rhizobium* that nodulates and fixes a large amount of nitrogen in association with one legume species may also do the same in association with certain other legume species. Leguminous plants demonstrate the tendency to respond similarly to particular strains of rhizobia (Wani and Lee, 2002).

The population of *Rhizobium* in the soil depends on the presence of legume crops in the field. In the absence of legumes, its population decreases. Artificial seed inoculation is often needed to restore the population of effective *Rhizobium* strains near the rhizosphere to hasten N_2-fixation. Symbiotic associations of legumes and rhizobia can be a major source of nitrogen in organic and sustainable systems, whereas free-living nitrogen-fixing microorganisms are helpful in non-symbiotic systems (Rashid et al., 2016a).

2.1.2 *Azospirillum*

Azospirillum belongs to the family Spirillaceae and is heterotrophic and associative in nature. It has nitrogen-fixing ability of about 20–40 kg/ha. Besides nitrogen fixation, bacteria from the genus *Azospirillum* are also able to produce phytohormones: auxins, gibberellins, cytokinins and phytohormone-like agents, such as indole derivatives (Cassán et al., 2014). Although there are many species under this genus, such as *Azospirillum amazonense*, *Azospirillum halopraeferens* and *Azospirillum brasilense*, the worldwide distribution and benefits of inoculation have been proved mainly with the *Azospirillum lipoferum* and *Azospirillum brasilense*. *Azospirillum* forms associative symbiosis with many plants, particularly with those having the C4-dicarboxylic pathway of photosynthesis (the Hatch and Slack pathway), because they grow and fix nitrogen on salts of organic acids such as malic and aspartic (Arun, 2007). Thus, they are mainly recommended for maize, sugarcane, sorghum, pearl millet and so forth, especially in environmental stress conditions like drought (Moutia et al., 2010). *Azospirillum* colonizing the roots do not only remain on the root surface, but a sizable proportion of them also penetrate into the root tissues and live in harmony with the plants. They do not, however, produce any visible nodules or outgrowth on root tissue.

2.1.3 *Azotobacter*

Azotobacter belongs to the family Azotobacteraceae, and is aerobic, free living and heterotrophic in nature. *Azotobacter* are present in neutral or alkaline soils, with *Azotobacter chroococcum* being the most commonly occurring species in arable soils. *Azotobacter vinelandii*, *Azotobacter beijerinckii*, *Azotobacter insignis* and *Azotobacter macrocytogenes* are other reported species. The number of *Azotobacter* rarely exceeds 10^4–10^5 g^{-1} of soil due to lack of organic matter and presence of antagonistic microorganisms in the soil. The bacterium produces antifungal antibiotics which inhibit the growth of several pathogenic fungi in the root region, thereby preventing seedling mortality to a certain extent (Nagaraja et al., 2016). An isolated culture of *Azotobacter* fixes about 10 mg nitrogen g^{-1} of carbon source under *in vitro* conditions. *Azotobacter* are also known to synthesize biologically active growth-promoting substances such as vitamins of the B group, IAA and gibberellins. Many strains of *Azotobacter* also exhibit antifungal properties against plant pathogens such as *Fusarium* or *Alternaria* (Chetverikov and Loginov, 2009). They can also be applied in environmental stress management (Di Stasio et al., 2017). The population of *Azotobacter* is generally low in the rhizosphere of crop plants and in uncultivated soils. The occurrence of this organism has been reported in the rhizosphere of a number of crop plants such as rice, maize, sugarcane, vegetables and plantation crops (Arun, 2007).

2.1.4 Blue-green algae (Cyanobacteria)

Blue-green algae belong to eight different families and are phototrophic in nature, producing auxin, IAA and gibberellic acid, fixing about 20–30 kg N/ha in submerged rice fields. Since they are abundantly found in paddies, they are also referred to as 'paddy organisms'. N is the key input required in large quantities for lowland rice production. Soil N and biological nitrogen fixation (BNF) by associated organisms are major sources of N for lowland rice. The 50–60% N requirement is met through the combination of mineralization of soil organic N and BNF by free-living and rice-plant-associated bacteria (Roger and Ladha, 1992). To achieve food security through sustainable agriculture, the requirement for fixed nitrogen must be increasingly met by BNF rather than by industrial nitrogen fixation. Most N_2-fixing BGA are filamentous, consisting of a chain of vegetative cells, including specialized cells called heterocysts which function as micronodules for the synthesis and N_2-fixing process. BGA forms a symbiotic association capable of fixing nitrogen with fungi, liverworts, ferns and flowering plants. The most effective common symbiotic association has been found for the *Azolla* and *Anabaena azollae* (BGA) complex. Besides N_2 fixation, BGA are also capable of producing siderophores which can complex metal ions into stable forms (Singh et al., 2016).

Strains of other free-living and endophytic bacterial species are also capable of fixing nitrogen from the air (Fox et al., 2016). Nitrogen-fixing microorganisms seem to be an economically attractive and ecologically friendly alternative to artificial nitrogen fertilizers, especially in sustainable agricultural production.

2.2 Non-nitrogen-fixing bacteria

Non-nitrogen-fixing bacteria are also suitable for agricultural use both as biofertilizers and plant growth promoters, especially as nutrient (e.g. phosphate) solubilizers, which are able to solubilize inorganic phosphate compounds, such as tricalcium phosphate, dicalcium phosphate, hydroxyapatite and rock phosphate. Among the bacterial genera with this capacity are *Pseudomonas*, *Bacillus*, *Rhizobium*, *Burkholderia*, *Achromobacter*, *Agrobacterium*, *Micrococcus*, *Aerobacter*, *Flavobacterium* and *Erwinia*. There are considerable populations of phosphate-solubilizing bacteria in the soil and the plant rhizosphere. These include both aerobic and anaerobic strains, with a prevalence of aerobic strains in submerged soils. A considerably higher concentration of phosphate-solubilizing bacteria is commonly found in the rhizosphere in comparison with the bulk soil (Roychowdhury et al., 2015). The soil bacteria which belong to the genera *Pseudomonas* and *Bacillus*, and fungi are more common. The major microbiological means by which insoluble P compounds are mobilized is by the production of organic acids, accompanied by acidification of the micro-environment (Raghu and MacRae, 1966). The organic and inorganic acids convert tricalcium phosphate to di- and monobasic phosphates with the net result of an enhanced availability of the element to the plant. The type of organic acids produced and their amounts differ with different organisms (Chen et al., 2006). Tri- and di-carboxylic acids are more effective compared to monobasic and aromatic acids. Aliphatic acids are also found to be more effective in P solubilization compared to phenolic, citric and fumaric acids (Walpola and Yoon, 2012). The analysis of culture filtrates of phosphate-solubilizing microorganisms has shown the presence of a number of organic acids, including citric, fumaric, lactic, 2-ketogluconic, gluconic, glyoxylic and ketobutyric acids (Wani et al., 2015).

Other rhizosphere microorganisms can reduce manganese and iron into water-soluble forms which are available for uptake by plants. In particular, bacteria belonging to the genera *Bacillus*, *Pseudomonas*, *Citrobacter*, *Shewanella*, *Alcaligenes*, *Clostridium* and *Enterobacter* are capable of reducing insoluble salts of iron and manganese (Ebrahiminezhad et al., 2017). The reduction of metal oxides in the soil can be linked to anaerobic oxidation of methane, one of the important greenhouse gases in low-oxygen soil conditions (Oni and Friedrich, 2017). Some beneficial bacteria, like *Pseudomonas*, *Rhizobium*, *Azotobacter* or *Erwinia* sp., can produce chelating agents (siderophores) that can form soluble complexes with soluble ions, preventing them from reprecipitation (Born et al., 2016; Kumar et al., 2017). The production of chelating agents is also one of the mechanisms of pathogen suppression by beneficial microorganisms (Sasirekha and Srividya, 2016).

2.3 Mycorrhiza

Another important group of beneficial microorganisms are mycorrhizal fungi. In agriculture and horticulture, the most common are arbuscular mycorrhizal fungi (AMF) belonging to Glomeromycota. They are obligatory symbionts of plants and form symbiosis with almost 80% of plant species on the earth (Davison et al., 2015; Öpik et al., 2010). These fungi are associated with the majority of agricultural crops, except the crops/plants belonging to the families of Chenopodiaceae, Amaranthaceae, Caryophyllaceae, Polygonaceae,

Brassicaceae, Commelinaceae, Juncaceae and Cyperaceae. AMF can produce a network of extraradical mycelia, which gives access to a greater quantity of water and soil minerals not available to the roots of the host plants. AMF also produce the peptide-like substance – glomalin, which plays a key role in the stabilization of soil structure and quality (Singh et al., 2013; Bedini et al., 2010; Vasconcellos et al., 2016). The fungi receive the necessary carbohydrates from plants for the completion of their life cycle. Arbuscular mycorrhizal associations have been shown to reduce damage caused by soil-borne plant pathogens. It is well documented that the arbuscular mycorrhizal symbiosis can increase plant growth and nutrient uptake, improve fruit quality and alleviate several abiotic stresses such as low temperature stress, drought, salt stress and so forth (Cavagnaro et al., 2015; Rouphael et al., 2015; Taktek et al., 2015; Zhao et al., 2015; Yooyongwech et al., 2016).

Application of microorganisms can increase the bioavailability of nutrients and the stability of soils and can support formation of soil aggregates in poor soils (Rashid et al., 2016b).

3 Consortia of microorganisms to improve the effectiveness of organic fertilization

Application of selected microorganisms may increase the effectiveness of organic fertilizers in plant production. Especially effective is co-application of microorganisms with organic fertilizers.

Microorganisms can be applied alone, as a single strain, but in many cases, better effects are observed after application of microbial consortia. The increased effects are achieved through the synergic relation between different groups of microorganisms. Application of consortia increases the nutritional effect of organic fertilizers, with additional 'value added' contributions, such as the production of phytohormones, antimicrobial agents, siderophores and others. Pseudomonads applied with AMF stimulate flower production and increase the yielding of strawberry, increasing yield parameters and amounts of biologically active substances in strawberry fruits (Bona et al., 2015). Microbial consortia are helpful in better utilization of organic or mineral fertilizers.

Grzyb et al. (2015) have shown that the application of the commercially available microbial consortium 'Mycosat' with different organic preparations and fertilizers increased plant growth in sour cherry 'Debreceni Bötermö' and apple 'Topaz' than the application of fertilizers alone. 'Mycosat' stimulated the growth of plants and promoted better bud-take on inoculated rootstocks. The resistance of bud grafts to freezing was increased. A disadvantage, however, is its relatively low effectiveness when the soil has become excessively dry. A double application of this preparation during the growing season gave better results in the nursery than a one-off treatment.

Common applications of AMF (*Rhizophagus intraradices, Glomus aggregatum*, *Glomus viscosum*, *Claroideoglomus etunicatum* and *Claroideoglomus claroideum*) and plant-growth-promoting *Pseudomonas* strains (C7 and 19Fv1T) positively affected the flowering and fruit production, and also the amounts of vitamin C, sugars and organic acids in tomato plants cv. 'TC 2000' (Bona et al., 2017).

Application of a mixture of beneficial bacteria namely *Pantoea* sp., *Pseudomonas fluorescens*, *Klebsiella oxytoca* and *Rhizobium* sp. on organically or mineral-NPK fertilized fields caused better plant growth and yielding of apple cv. 'Topaz' in comparison with application of single mineral or organic fertilizers alone (Mosa et al., 2016). Preparations with AMF phosphate-solubilizing bacteria and *Azotobacter* combined with vermicompost increased the growth and yielding of strawberry plants in comparison with vermicompost single application (Singh et al., 2015). Similar effects were observed in basil (*Ocimum basilicum*) after co-application of vermicompost, *Dietzia natronolimnaea* and *Glomus intraradices* (Bharti et al., 2016b).

The synergic effect of different microorganisms is taken advantage of in the production of commercial products which contain beneficial bacterial strains and arbuscular or filamentous fungi, sometimes mixed with rock meal as a source of phosphorus, potassium and other minerals (Grzyb et al., 2015).

A group of products called 'Effective Microorganisms' (EM), popular in organic agriculture, is obtained during molasses fermentation (see Section 6.1) using a consortium of many microbial strains. EM, when applied as fertilizer, resulted in positive impacts on plant growth parameters (Kleiber et al., 2014; Salama et al., 2014). For example, strawberry plants cv. 'Honeoye' treated with EM gave similar results to those under mineral fertilization, whereas cv. 'Selva' showed better growth parameters after the application of 'Effective Microorganisms' (Glinicki et al., 2011). Indirect, positive effects in plant growth and yielding are obtained when 'Effective Microorganisms' are used as microbial inoculum for compost preparation. For example, in an eleven-year long-term experiment, wheat fertilized with 'Effective Microorganisms' compost gave better plant growth parameters and yield in comparison to traditional compost (Hu and Qi, 2013). 'Effective Microorganisms' are especially useful in co-application with organic matter like agricultural wastes or with bioorganic fertilizers while increasing the effectiveness of nutrients utilization (Chantal et al., 2013; Shaheen et al., 2017).

4 Animal excrement: manures, slurry and guano

The most common organic fertilizer used in organic farming is farmyard animal manure, which is the basis of organic fertilization of many crops. In recent years, there has been a problem with adequate amounts of this resource for both organic and integrated/conventional agriculture. Manure can be applied in fresh, dried or composted form. However, for safe organic production, to avoid the risk of contamination with pathogenic or opportunistic strains of microorganisms, it is recommended that manure be thermally processed, for example, during thermophilic composting. If raw manure is used, it should be applied 90 days before harvest for crops that do not come into direct contact with the soil, and 120 days prior to harvest for crops that have direct contact with the soil (Köpke et al., 2007). The use of manure originating from factory farming is forbidden.

Long-term application of farmyard manure increases the levels of P, K and S available to plants, and also the amounts of microelements like B, Zn, Fe or Mo in the soil, but can decrease the concentration of Cu and Mn (Rutkowska et al., 2014). Application of manures modifies the amounts of available forms of nutrients in the soil profile. This effect is dependent on the origin and form of the applied manure (Schlegel et al., 2017).

Manures are a source of soluble substances which are easily utilized by soil microorganisms. These substances are sugars, alcohols and volatile fatty acids such as acetic, butyric and valeric acids. Slowly biodegradable organic polymers must be hydrolysed before their utilization by soil microorganisms (Boursier et al., 2005). At present, manures are available on the market in processed form, dried or composted or as dried granules or pellets. Some of these products are supplemented with rock powders, animal blood, bone or feather meals, plant by-products or other materials (Malusà et al., 2014).

Other traditional products of animal origin, widely used as fertilizers, are slurry (liquid animal excrement) used after controlled fermentation and/or appropriate dilution.

Guano, accumulated excrement of seabirds, seals or cave-dwelling bats, is a highly effective fertilizer with high nutrient contents essential for plant growth. However, its mineral content is strictly associated with the feeding regimes of guano-producing animals (Szpak et al., 2012). For example, the guano produced by the sanguivorous bat *Desmodus rotundus* has more than 17% nitrogen, that produced by the frugivorous bat *Pteropus rodricensis* has about 2% and the guano produced by the insectivorous bat *Tadarida brasiliensis* has a nitrogen content of about 10% (Emerson and Roark, 2007). Large deposits of guano are formed in dry climates, where they are protected from nitrogen leaching by rain.

5 Products and by-products of animal origin

5.1 Raw or processed animal residues

A large group of organic fertilizers are raw or processed animal residues, dried and pulverized blood, hoof, horn, bone, degelatinized bone, fish, meat, feather, hair, 'chiquette' meal,'chiquette', wool, fur, hair, dairy products and hydrolysed proteins.

Animal residues are sources of organic matter and minerals N, P, Ca and others. But the usage of these materials, especially meat and bones, for agriculture is restricted by safety rules, particularly those relating to TSE (transmissible spongiform encephalopathy) incidents (Möller, 2015). Substrates allowed for the preparation of organic fertilizers without pressure sterilization are those classified to Category 3 according to the EC Regulation 1069/2009. By-products in this category are obtained from the processing of goods intended for human consumption obtained by slaughtering animals, leftovers from food processing facilities or canteens. This category of by-products, when used for the preparation of organic fertilizers, must be treated by pasteurization at 70°C for more than 1 h. Category 2 covers by-products from animals that died in ways other than slaughtering, animals killed to prevent epizootic diseases or those posing a risk of infection with diseases other than TSE. This category also covers the contents of digestive tracts obtained by slaughtering healthy animals. These residues may be used for the production of fertilizers after pressure sterilization at a temperature of more than 133°C for more than 20 min under a pressure above three bars. Processing of meals into fertilizers may be done by simple granulation to obtain a solid phase. Category 2 by-products can be processed by acid hydrolysis to obtain short peptide chains suitable for the production of solid nitrogen-rich fertilizers. For example, 'Biollsa', prepared by Ilsa Group, Italy, is a granular nitrogen fertilizer produced by acid hydrolysis.

Another group of fertilizers of animal origin, popular in recent years, are liquid hydrolysates rich in amino acids and short peptide chains. The best of these preparations are obtained by enzymatic hydrolysis of purified animal residues such as collagen. For maximum safety, some producers use only Category 3 residues to avoid the risk of spreading disease. Liquid formulation makes them easy to use by spraying, especially in pomological orchards, berries plantations, vineyards and fields. Liquid hydrolysates are also used via irrigation systems, in greenhouse production of high-quality vegetables and fruits. Hydrolysed proteins must not be applied to edible parts of the crop, and in the case of pastures, the feeding of animals must be prohibited for at least three weeks to avoid the potential risk of disease infection. Amino acids are also a source material for the production of complexes with micronutrients, which are eco-friendly plant growth promoters (Ghasemi et al., 2013). The effectiveness of amino acid preparations in plant nutrition and for improving soil quality has been investigated in many trials, which have shown that applications of such preparations increase the yield quantity and quality (Lisiecka et al., 2011; Corte et al., 2014; Colla et al., 2015; Mobini et al., 2014). For example, the Terra-Sorb Foliar amino acid preparation applied on oilseed rape increases the levels of fats and proteins and decreases the amounts of glucosinolates in seeds in comparison with the control (Jakienė, 2013).

Wool, fur and hair are a source of slowly released nitrogen when used in a low-processed form like pellets or meal. They are also used as a feedstock for the production of more processed preparations such as hydrolysates (Nustorova et al., 2006).

Dairy products and by-products, especially whey, have been a point of interest as a form of nitrogenous fertilizer for a long time (Sharratt et al., 1959; Peterson et al., 1979; Marwaha and Kennedy, 1988). Direct application of whey on arable fields could be a method for improving soil fertility and for improving the physical properties of some types of degraded soils (Robbins and Lehrsch, 1992; Jones et al., 1993).

Good results of whey utilization are obtained when it is used as a raw material for organic fertilizer production. Whey-based fertilizers are compositions of different organic amendments like sawdust and straw, and also with minerals like zeolite. An example of such a product is 'Condit' – a pelletized sawdust, straw and zeolite mixture enriched with whey as a source of nitrogen (Cuijpers and Hospers-Brands, 2008). Whey is also a good supplement for bioconversion of some substrates of organic origin into valuable organic fertilizers that have positive effects on plant growth parameters (Stępień et al., 2012). They are also an effective amendment for compost teas, especially in the protection against plant pathogens (Pane et al., 2012).

5.2 Chitin

Chitin fertilizers are obtained from shells of crustaceans. This product is allowed only if obtained from sustainable fisheries, as defined in EC 2371/2002, or organic aquaculture. Chitin and its derivatives are not only sources of nitrogen and carbon, but also active against viruses, bacteria and other pests; they also induce plant defence mechanisms and stimulate the growth and activity of beneficial microorganisms (Schäfer et al., 2012; Cretoiu et al., 2013; Sharp, 2013), and can also increase the effectiveness of beneficial bacteria against pests (Yu et al., 2008). Chitin fragments are known to have eliciting activities, leading to a variety of defence responses in host plants. Based on these and other properties that help to strengthen host plant defences, interest has been growing in using the technique in agricultural systems to reduce the negative impact of diseases on the yield and quality of crops. When applied as a foliar spray, 0.1% (w/v) chitosan solution increases turmeric plant (*Curcuma longa* L.) growth parameters, such as shoot height, number of leaves per plant and plant fresh weight, and also has a positive impact on plant yielding and yield quality, including active substances content (Anusuya and Sathiyabama, 2016). Chitosan is especially effective when used in combination with microelements and humic acids, increasing, for example, dry bean (*Phaseolus vulgaris* L.) plant growth parameters and yield quality (Ibrahim and Ramadan, 2015). Recently, positive results with chitin have been obtained in a study conducted by Mahdavi and Rahimi (2013). They tested the effect of 'Chitosan' on the germination and growth of ajowan (*Trachyspermum ammi*). Stimulation of germination and growth was observed with a decrease in the harmful impact of abiotic stress such as high salinity. Significant benefits have been observed for soybean yield, seed germination and plant growth in a study by Zeng et al. (2012), in which soybeans were coated with chitosan. These properties make chitin and its derivatives a useful material for coating seeds used in organic agriculture.

Chitin is also a raw material for the production of innovative liquid organic fertilizers, which are easy to use and effective when applied to plants and soil (Struszczyk et al., 1989). Innovative formulations, like the preparation 'Apol-Humus' allowed for use in organic agriculture in Poland, are a liquid mixture of chitosan and humic acids for foliar and soil application. This preparation stimulates the growth of plants in two ways: through a direct nutritional effect and disease suppression.

6 Products and by-products of plant origin for fertilizers

The 'products and by-products of plant origin' category covers a wide group of leftovers from the food processing industry, for example, oilseed meals, cocoa husks and malt culms. These by-products are sources of slow releasing nitrogen suitable for use in agriculture and horticulture. The levels of nutrients, especially of nitrogen forms, vary according to the source of the raw material and the extent of processing. For example, the average nitrogen content in sunflower meals is about 5.0%, *Brassica* meal 5.7%, neem (*Azadirachta indica*) cake 5.2%, castor cake 4.25% and cotton seed cake 4.6% (Kumar et al., 2014; Mazzoncini et al., 2015). In the case of meals obtained from seeds of some *Brassica* species, there are high levels of glucosinolates, which have an inhibitory effect on nitrification processes (Mazzoncini et al., 2015). The nutritional effect of plant seed meals can be enhanced due to their additional properties, for example, antifungal or zoocidal activity, especially against root-knot nematodes such as *Meloidogyne* (Duong et al., 2014; Sumbul et al., 2015; Yang et al., 2015a).

For fertilization purposes, processed residues of many plant species, especially legumes, are also used. For example, pelletized meal of red clover (Ekofert K) gave the highest marketable yield of celeriac cv. 'Diamant' in comparison with traditional mineral fertilization and non-fertilized control (Kaniszewski et al., 2013).

6.1 Molasses

Another plant product, molasses, is a raw material for the production of some soil improvers and fertilizers. For example, Vinassa is obtained as a by-product during the production of baker's yeast. Vinassa, when applied as a foliar spray, positively affects plant growth and yielding, but its effects are associated with the plant species, variety and also growing conditions such as the type of soil or field localization. For example, large-fruited cranberry (*Vaccinium macrocarpon* L.) cv. 'Stevens' shows good results in terms of plant growth parameters in comparison to 'Pilgrim' and 'Ben Lear' after applications of

Vinassa (Derkowska et al., 2015b). In nursery stock production, Vinassa, applied at a rate of 5 L/ha, moderately stimulates the growth of maiden trees, increasing their height and the length of lateral shoots, and has a beneficial effect on the development of the root system and flower buds. Apple trees cv. 'Ariva' produced the best quality fruits as a result of treatment with Vinassa (Rozpara et al., 2014). Also soil microbial community parameters, such as the formation of mycorrhizal structures and the population size of free-living bacteria and fungi communities, positively affected after treatments with Vinassa (Derkowska et al., 2015a; Sas Paszt et al., 2015).

6.2 Natural waste fibres

Natural waste fibres are used as raw materials of plant origin from the textile industry. New products based on these materials are also obtained. These materials are used as a slow release fertilizer when mixed with a source of nitrogen, for example, residues of legume plants. For example, they can be used for the production of agro-fleece for organic vegetable production, which protects crops against weeds, and after crop harvest, the fleece is tilled under and acts as a slow release fertilizer and soil structure improver (Babik et al., 2013). In that case, agro-fleece was enriched with clover leaves as a source of nitrogen, and the obtained yield was similar to that obtained using mineral plant nutrition, but with minimal weed growth in the covered fields.

6.3 Plant-derived hydrolysates

Plants are a source of raw material for the production of plant-derived hydrolysates used in agriculture as bio-stimulants. Hydrolysates of plant origin show not only a nutritional effect but also a hormone-like activity, which results in higher biomass, SPAD index and nutrient content in the tissues of the treated plants (Colla et al., 2014).

'BioFeed Amin', a product based on plant hydrolysates, applied in a dose of 5 L/ha, intensifies the growth of maiden trees, markedly improves their branching ability, stimulates the growth of lateral shoots and improves considerably the quality of the root system. Application of plant-derived hydrolysates has a positive impact on soil microbial communities in comparison with mineral fertilization regimes, resulting in higher number of bacteria in the soil and better mycorrhiza formation in the roots (Sas Paszt et al., 2011; Derkowska, et al., 2015a).

6.4 Seaweeds and seaweed products

Seaweed and seaweed-based products have been used in agriculture for a long time, first as plant material directly applied to the soil, later as solid meals used as fertilizers or soil improvers, with their mode of action based on the minerals released into the soil (Crouch and van Staden, 1993). Seaweeds are a rich source of natural compounds such as amino acids, polysaccharides, lipids, trace elements, vitamins and especially plant growth regulators (auxins, cytokinins, gibberellins, betains, etc.), making seaweed suitable for sustainable, especially organic, agriculture. Seaweed products affect plant growth and biotic and abiotic stress resistance when applied directly onto plants and stimulate the rhizosphere microorganisms when applied to the soil (Crouch et al., 1990; Kuwada et al., 2006; Thorsen et al., 2010; Panda et al., 2012). Seaweed preparations directly affect plant growth and yield quality, both in agricultural and horticultural production (Fan et al.,

2011; Hernández-Herrera et al., 2014; Kumar and Sahoo, 2011; Papenfus et al., 2013). In strawberry, for example, the effects of seaweed preparations Kelpak® and Goëmar BM 86® interacted with the plant variety. In the case of 'Elkat', the preparations Kelpak SL and Goëmar BM 86® significantly improved fruit yield, but in the case of 'Salut', there was no effect on yield (Masny et al., 2004). The seaweed preparations Kelpak® and Goëmar BM 86® applied to apple trees, on their own or together in a mixture, stimulated the growth of shoots and leaves and modified fruit coloration, but the effects were also dependent on the cultivar (Basak, 2008). In the case of agricultural crops, applications of the preparation 'Maxicrop' increased shoot and root dry weight of radish in comparison with another seaweed-based product – 'Seasol', and ashes obtained from these products, which were not as effective as Maxicrop (Yusuf et al., 2016). Seaweed extracts can be used to control pests on plants via direct effects or by increasing plant resistance (Hankins and Hockey, 1990; Jayaraj et al., 2008; Jayaraman et al., 2011; Stadnik and Freitas, 2014; Ngala et al., 2016). In the tests conducted in a nursery of the Research Institute of Horticulture in Skierniewice, Poland, the seaweed-based preparation 'BioFeed Quality', applied twice per season at 5 L/ha, intensified the growth of maiden trees, markedly stimulated their branching, caused a significant increase in the length of lateral shoots and had a beneficial effect on the development of the root system of the maiden trees.

In organic agriculture in the EU, only marine seaweed products obtained directly by physical processes including dehydration, freezing, grinding, extraction with water or aqueous acid and/or alkaline solutions, or fermentation can be used. The conditions of extraction have a significant impact on the physical and chemical properties of the final product and its effect on plants under cultivation conditions (Briceño-Domínguez et al., 2014).

6.5 Sawdust and wood chips

Sawdust and wood chips obtained from chemically untreated wood are a soil improver used especially in horticulture. Sawdust, wood chips, wood bark and other similar residues are commonly used as mulch in the cultivation of different horticultural plants, small fruits, vegetables and ornamentals (Haynes and Swift, 1986; Sanderson and Cutcliffe, 1991; Abouziena and Radwan, 2015; Lima et al., 2016). Application of these materials prevents the growth of weeds, but simultaneously 'starves' the cultivated trees of nitrogen, causing their leaves to become yellowish, so that in the following year, the trees grow less vigorously as the nitrogen content in the soil decreases compared to the control (Hoagland et al., 2008).

Another use of sawdust is as a medium or component of media for soilless production of plants, especially in protected (under cover) cultivation, being a promising alternative to the traditional growing media such as rockwool, peat or coconut fibre (Jung et al., 2015; Depardieu et al., 2016). Results of hydroponic production of strawberry fruit on sawdust show that the yield obtained on pure sawdust is lower than that obtained on media based on sawdust mixed with pumice or with coconut fibre (Marinou et al., 2013). Strawberry plants planted in sawdust-based media need additional fertilizer supplementation to satisfy the initial nutrient requirements of bare-root plants (Depardieu et al., 2016).

Sawdust and other wood derivatives are also suitable amendments for manure composting, for the production of composted fertilizers or soil improvers allowed for use in organic agriculture (Cuijpers and Hospers-Brands, 2008; Troy et al., 2012; Khan et al.,

2014). Moreover, sawdust and wood chips are used for biochar production (Spokas et al., 2009; Ghani et al., 2013; Laghari et al., 2016).

6.6 Wood ash as fertilizer and soil improver

Wood ash used in organic agriculture must be obtained from wood that has not been treated with chemicals (preservatives, paints, etc.) after felling because this can negatively affect soil quality and plant growth (Jones and Quilliam, 2014). Wood ash contains all the essential plant nutrients, except nitrogen. Wood ash may be directly used to improve crop production and improve soil pH. Wood ash as a natural mineral fertilizer can induce negative priming and promote soil organic matter from grassland soil (Reed et al., 2017). Wood ash reduces fertilization costs and improves crop yields and soil properties, with residual effects observed for up to seven years or more. Wood ash is a good improver during composting of organic matter. Additions of up to 15% do not affect the composting process, but high doses of ash can increase the input of heavy metals present in the ash to limits that are not acceptable for good-quality composts (Fernández-Delgado Juárez et al., 2015).

6.7 Mushroom culture wastes (spent mushroom substrates – SMS)

The composition of this category is limited to products listed in 'Annex II' – 'Products for use in fertilization and soil conditioning'. A typical mushroom substrate is prepared by using straw, poultry litter or horse manure with the addition of gypsum, fermented and inoculated with fungal mycelia. This method is typical for *Agaricus bisporus* cultures. Other cultivable mushrooms, such as *Pleurotus* sp., are cultivated on wood or on cereal straw (Roy et al., 2015). These two kinds of spent substrates differ in nutrient content and physical properties (Paredes et al., 2016). Mushroom cultivation is one of the methods for the utilization of organic residues. However, large amounts of spent substrates are a serious problem for their utilization (Das and Mukherjee, 2007; Koutrotsios et al., 2014). They can be used as raw material in the field, or can be processed to obtain products with better effectiveness.

Spent mushroom substrates or their derivatives promote the growth and yields of plants and increase crop quality. For example, the growth of pepper plants was strongly induced by *Pleurotus* SMS leachate, in comparison with pure or weathered SMS from *Pleurotus* and *Agaricus* cultures (Roy et al., 2015).

6.8 Fermented products

Other traditional products permitted for use in organic agriculture are fermented household wastes or plant matter. In this group, we can also include biogas-digestate-containing plant residues and animal by-products co-digested with materials of plant origin. Agricultural utilization of digestates is one method of managing these by-products (Dahlin et al., 2017).

Biogas digestates may have a beneficial impact on soil properties due to the presence of macro- and microelements and a high organic matter content (Arthurson, 2009; Möller and Müller, 2012). The digestion process and subsequent processing, such as composting, increase the percentage of nitrogen, phosphorus and potassium in the final product (Macias-Corral et al., 2017). Biogas digestates obtained from plants such as maize affect the mineral uptake by crop plants and have an impact on the composition of soil microbial communities (Hupfauf et al., 2016). Biogas digestates used correctly as fertilizer positively impact the growth of cultivated plants, and yields can be similar to those obtained using traditional mineral fertilizers (Barbosa et al., 2014). On the other hand, utilization of the non-separated or liquid phase of biogas plant digestates in high doses may have toxic effects on soil biota and cultivated plants (Stefaniuk et al., 2015).

6.9 Stillage and stillage extract

Materials obtained during ethanol production are suitable as an effective soil amendment in field and horticultural crops. Stillage is a rich source of residual sugars, organic nitrogen forms and other nutrients whose content is associated with the feedstock material used for fermentation (Wilkie et al., 2000; Alotaibi et al., 2014,).

These properties make stillage and its derivatives a good and promising fertilizer, but utilization of raw stillage is restricted by some legal rules due to its potential negative effect on soil properties (Sajbrt et al., 2010; Fuess and Garcia, 2014). Good results with the use of stillage have been obtained in canola and wheat production (Qian et al., 2011; Alotaibi et al., 2014).

6.9.1 Solid phase residues from biogas production

The solid phase of digestate obtained from biogas production also has beneficial properties. Good-quality feedstock allows the obtaining of a digestate suitable for use in agriculture (Al Seadi et al., 2013). Utilization of the separated solid phase of biogas digestate or stillage is one way of alleviating eco-toxic effects of biogas by-products on plants and soil organisms. Application of the solid phase stimulates root growth, whereas liquid phase digestates show inhibiting effects on the growth of roots in some plants (Stefaniuk et al., 2015). Separated, dried solid phase wastes have a higher C:N ratio in comparison with the liquid or non-separated residues. Also their mineral content is associated with the nature of the biogas digestate.

Residues from biogas production can also be used as feedstock for biochar production, but there is a risk of potential toxicity of biochar for soil organisms associated both with the nature of feedstock material production and pyrolysis temperature. For example, Stefaniuk et al. (2016) showed that biochar obtained from non-separated digestate, pyrolysed at 800°C, is much more toxic for soil organisms in comparison with the separated, solid phase of digestate prepared at 400°C.

6.10 Leonardites

Leonardite is naturally oxidized lignite (brown coal), raw organic sediment rich in humic and fulvic acids, mainly obtained as a by-product of near-surface mining activities. It is used as a raw material as a soil conditioner and for the production of pure humic acid preparations, humates (e.g. potassium or iron salts of humic acids) (Kovács et al., 2013). The production of humic acids is conducted with strong alkali or using physical methods (Canieren et al., 2017).

Humic acids stabilize ion exchange in soils and are very useful in contaminated soils. The effects of humic acids on plant growth and nutrient uptake are associated with physical properties of humic acids, especially their molecular weight (Qian et al., 2015; Sun et al., 2016). Leonardites, or humic acids, accelerate microbial activity in soils, including decomposition of organic pollutants (Tejeda-Agredano et al., 2014). Humic substances extracted from leonardites also have bio-stimulant properties due to the presence of polycyclic hydrocarbons, similar in structure to plant growth regulators (Conselvan et al., 2017).

Similar properties as a raw material are exhibited by brown coal (lignite), which is also a good source of humic substances used as soil improvers. Direct application of lignite to the soil is not recommended due to economic considerations, especially the high cost of long-distance transport, in relation to raw lignite costs and the negative impact of raw lignite on soil properties. Possible negative effects on soil microbial communities or nutrient cycling are visible mainly on a short-time scale. For example, lignite added to a sandy or clay soil decreases microbial activity, measured as respiration, and increases the activity of peroxidase and phenol oxidase enzymes (Kim Thi Tran et al., 2015). A much more effective way is to process lignite into concentrated easy-to-use preparations, rich in humic acids. The methods of extraction of humic acids from brown coal are similar to those used for leonardite, but sometimes with additional treatments for better quality and usability of the final product (Doskočil et al., 2014; Ozkan and Ozkan, 2016). Brown coal and peat, mentioned below, are promising sources of humic acids for the organic fertilizer industry (Liu et al., 2014).

6.11 Peat

Peat is a material of moor origin with use limited to horticulture in organic agriculture in the EU, used as a soil conditioner, a growth substrate for plants, or as soil mulch. Peat is also a source of substances that promote plant growth, for example, humic acids, and phytohormone-like organic substances (Klavins and Purmalis, 2013; Boguta et al., 2016; Szajdak, 2016). Peat is used as a raw material for the production of humic acid preparations and fertilizers for use in agriculture and horticulture (Saito and Seckler, 2014; Agafonova et al., 2015). On the EU market, there are some products obtained from peat, for example, a Latvian liquid preparation called Humate Green OK Universal-Pro and similar products by the same producer, prepared on the basis of peat extracts; it can be applied directly to the soil or on plants to increase the nutritional effect.

6.12 Sapropel and similar sediments

Sapropel refers to sediments rich in organic matter formed under exclusion of oxygen in stagnant water bodies. Apart from humic acids, sapropels contain macro- and microelements, simple organic molecules like amino acids, vitamins, antibiotics and other substances that promote the growth of plants (Blečić et al., 2014; Grantina-Ievina et al., 2014; Szajdak, 2016). The amounts of organic substances are associated with the conditions of sapropel formation and the degree of its transformation (Dmitrieva et al., 2015). A characteristic feature of some sapropels is the number of microorganisms that mineralize the organic substances to form sapropel (Tretjakova et al., 2015). These properties make sapropels a good substrate for the production of new organic fertilizers (Agafonova et al., 2015). For example, the sapropel-based organic fertilizer 'Humin Plus' is composed of sapropel, peat and chicken manure, additionally treated for improved effectiveness (Ostrovskij, 2014). Due to the high content of nitrogen in some sapropels, there is a need to determine this content and use sapropels correctly to avoid over-fertilization with nitrogen according to the 91/676/EEB rules (Klepeckas and Januškaitienė, 2017). For organic agriculture use in the EU, only fresh water sediments are allowed. Experiments have shown that sapropel is a good material positively affecting soil quality, plant growth and yield quality parameters. Applications of sapropel increase the chlorophyll content in cereals, such as wheat and barley (Klepeckas and Januškaitienė, 2017). Sapropel-based fertilizer 'Humin Plus' increased the yields of tested crops by 23–25% in wheat, 18–27% in corn, 11–32% in sugar beet, 32% in potato and 45–55% in canola, depending on the location of the experiment (Ostrovskij, 2014).

7 Composts

Composts are produced from separated domestic residues, but there is a possibility to use other substrates like mushroom culture wastes, vegetable matter or other materials alone or mixed with farmyard manure. Composting is also a good way to stabilize nutrient content and dehydration of animal manures. Nowadays, compost is one of the organic fertilizers allowing utilization of organic wastes during the composting process (Yu et al., 2016). Yearly compost applications promote plant growth and reduce the risk of soil and groundwater contamination with nitrates (Toselli et al., 2013). Long-term compost application can increase organic carbon content and improve soil structure, mainly due to the formation of soil macro-aggregates, which are fundamental as habitats for soil microorganisms (Zhang et al., 2014). Effects on yields do depend on the nutrient contents, their availability as well as the crop and its environment. Applications of compost can result in the increase of the yield of some crops, for example, peas (*Pisum sativum* L.), but can also have no impact on the yields of other crops, such as oats (*Avena sativa* L.) (Jannoura et al., 2014).

As a slow nutrient releasing agent, applications of compost as a fertilizer or a component of growing media can increase the yield and quality of harvested fruits and vegetables. The use of compost resulted in a larger size and higher number of 'extra class' strawberry fruits of the cultivars 'Senga Sengana', Kent' and 'Elsanta' in comparison with the control (Frąc et al., 2009). Strawberry plants cv. 'Marmolada' and 'Maya' planted in a compost–peat mixture gave higher yields than those planted in peat; cv. 'Patty' gave higher yields in the peat control, whereas in the case of cv. 'Irma', the effect was correlated with the compost content in the medium – with 25% compost, the yields were better in peat, and with 50% compost in the medium, the yield was lower than that in peat (Altieri et al., 2014).

Compost application has a big impact, mainly positive, on soil microbiota such as soil bacteria or fungi, but compost can also decrease the colonization rate of AMF (Derkowska et al., 2008; Zhen et al., 2014).

Vermicompost is the result of conversion of organic residues by earthworms into humus-like substances. There are many types of raw material used for vermicompost production: precomposted animal faeces, crop residues, food processing by-products, industrial residues and others (Sim and Wu, 2010; Das et al., 2016; Domínguez et al., 2017). Vermicompost can enhance soil fertility physically, chemically and biologically. In physical terms, a vermicompost-treated soil has better aeration, porosity, bulk density, drainage and water retention. Chemical properties such as pH, electrical conductivity and organic matter content are also improved for better crop yield (Lim et al., 2015). Nutrient composition in vermicompost depends on the raw material used as feedstock and the species of earthworms used for the vermicomposting process (Lim et al., 2015; Joshi et al., 2015). Vermicompost is a better nutritional source than traditional composts due to its increased rate of mineralization and degree of humification by the action of earthworms. Nutrients are present in forms readily available to plants. The beneficial properties of vermicomposts are increased by the presence of fungi, bacteria and actinomycetes, producers of plant growth regulators and other substances which can affect plant growth and yields (Joshi et al., 2015).

When applied on their own or with selected strains of beneficial microorganisms, vermicomposts have a positive effect on plant growth and yield (Yang et al., 2015b), and can be a part of sustainable, integrated or certified organic production (Bharti et al., 2016a; Beck et al., 2016).

Vermicomposts are also a raw material for the production of liquid formulations, which can have a positive impact on plant growth, but are easier to use for field applications, especially in perennial, for example, pomological crops (Singh et al., 2010; Grzyb et al., 2012; Rozpara et al., 2014; Mosa et al., 2016). Additions of other natural compounds can improve the survivorship of beneficial bacterial strains in vermicompost-based formulations (Kalra et al., 2010).

8 Untreated minerals and by-products of selected industrial processes

Untreated minerals are not strictly organic fertilizers, but are allowed for application in organic agriculture as a source of phosphorus, potassium, magnesium, calcium and microelements.

This category covers crude potassium salt or kainite, potassium sulphate (K_2SO_4) possibly containing Mg salts, calcium carbonate, magnesium and calcium carbonate, magnesium sulphate (kieserite), calcium chloride solution, calcium sulphate (gypsum), industrial lime from sugar production, industrial lime from vacuum salt production, elemental sulphur and trace elements as inorganic micronutrients listed in EC Regulation 2003/2003 (Ciesielska et al., 2011). Untreated minerals are also used as raw materials for the production of organic fertilizers.

9 Biochar

Biochar is a new promising material for organic agriculture, which can be used as a fertilizer, soil improver or pollutant binder, alone or in a mixture with other amendments. Currently, it is not present on the list of products allowed for use in organic agriculture in the EU. Biochar is a material made from organic matter, pyrolysed at high temperature in the absence of oxygen. Charred materials include materials of plant origin, such as wood residues, crop residues, microalgae biomass, animal residues (bones or meals), animal manure and mixed materials, for example, food industry wastes or sewage sludge (Chan et al., 2007, 2008; Sohi et al., 2010; Bird et al., 2011; Enders et al., 2012; Angst et al., 2014; Hosseini Bai et al., 2015). The chemical and physical characteristics of biochars vary depending on the feedstock material and conditions of the conversion, including temperature, time of processing and the technology used (Lehmann et al., 2011; Cantrell et al., 2012; Baronti et al., 2014; Akhter et al., 2015). Biochars produced in different plants but from the same biomass and under similar pyrolysis conditions can result in various properties of the final product (Spokas et al., 2012a). The final product is a material that can have residual (or relic) structures of the original feedstock material or no residual structures (Spokas, 2010). The physical structure of biochars affects the organic and inorganic composition: the pH can range from 5.6 to 13.0, the C content from 33.0% to 82.7%, N content from 0.1% to 6.0% and the C: N ratio from 19 to 221 (Jha et al., 2010; Spokas, 2010; Spokas et al., 2012b). Biochar can also contain appreciable quantities of P, K, Ca, Mg and micronutrients (Cu, Zn, Fe, Mn), with ashes accounting for 5–60% of the weight, depending on the source of the biomass and pyrolysis conditions (Cheng et al., 2008; Enders et al., 2012).

The main goal of biochar applications in previous years was carbon sequestration in soil deposits (Jha et al., 2010). Now the biochar utilization is also focused on increasing crop yields (Jeffery et al., 2011). The main focus nowadays is also to increase soil fertility (Atkinson et al., 2010) using biochar-induced specific properties of soil and biochar's impact on soil microbiota (Steiner et al., 2008, 2009; Anderson et al., 2011; Parvage et al., 2013; Vanek et al., 2016). Biochar can also be used as a component of growing media for plant production, especially as an alternative to non-renewable materials like peat (Kern et al., 2017). Biochars of animal origin are a suitable source of phosphorus and other macro-elements (Siebers et al., 2014).

10 Conclusion

Utilization of organic or non-processed mineral fertilizers is obligatory in organic agriculture. The wide range of products allows correct soil management and plant fertilization, according to sustainable agricultural practice rules. Organic agriculture is also an area with scope for innovative product creation like organic fertilizers enriched with strains of beneficial microorganisms, for example soil bacteria, filamentous or mycorrhizal fungi. Utilization of organic fertilizers will contribute to the development of organic and sustainable nutrient management strategies and cultivation measures in agriculture.

Organic agriculture also allows better utilization of neglected resources like agricultural or food industry wastes, meat industry residues or biogas station residues. On the market there are a number of products based on processed organic wastes, but utilization rate of these wastes is not adequate to the amount of produced wastes.

Innovations from organic agriculture will spread to other areas of human activity, like bio-fortification of conventionally managed arable fields, restoration of degraded soils and industrial areas or in creation of new green spaces in urbanised areas.

11 Where to look for further information

The present and future research in the area of organic fertilizers will be focused on the utilization of a wide range of natural resources and food industry processing wastes for plant fertilization and soil improvement. The group of industry processing wastes as raw materials for fertilizer production is especially valuable, because its utilization will help to avoid air, water and soil contamination from decayed organic residues. The better utilization of wastes within one process should deliver a raw material for further processes. For example, wastes from biogas production should be used as raw material for production of organic soil improvers. Other wastes, like animal bones or fur feather and horn wastes, can serve as raw materials for the production of slow release of phosphorus or nitrogen fertilizers.

There is a need to develop innovative biofertilizers, new fertilization techniques and nutrient management strategies, which will be applicable in modern, environmentally friendly, sustainable agriculture.

12 References

Abouziena, H. F. and Radwan, S. M. (2015), Allelopathic effects of sawdust, rice straw, bur- clover weed and cogongrass on weed control and development of onion, *International Journal of ChemTech Research*, 7, pp. 337–45.

Agafonova, L., Alsina, I., Sokolov, G., Kovrik, S., Bambalov, N., Apse, J. and Rak, M. (2015), New kinds of sapropel and peat based fertilizers, in: *Environment. Technology. Resources*. Rezekne, Latvia, Volume 2, pp. 20–6.

Ahmad, F., Ahmad, I. and Khan, M. S. (2008), Screening of free-living rhizospheric bacteria for their multiple plant growth promoting activities, *Microbiological Research*, 163(2), pp. 173–81. DOI:10.1016/j.micres.2006.04.001.

Akhter, A., Hage-Ahmed, K., Soja, G. and Steinkellner, S. (2015), Compost and biochar alter mycorrhization, tomato root exudation, and development of *Fusarium oxysporum* f. sp. *lycopersici*, *Frontiers in Plant Science*, 6, pp. 529. DOI:10.3389/fpls.2015.00529.

Al Seadi, T., Drosg, B., Fuchs, W., Rutz, D. and Janssen, R. (2013), Biogas digestate quality and utilization, in: Murphy, J. and Baxter, D. (Eds), *The Biogas Handbook: Science, Production and Applications*. Sawston/Cambridge, UK: Woodhead Publishing, pp. 267–301.

Alotaibi, K. D., Schoenau, J. J. and Hao, X. (2014), Fertilizer potential of thin stillage from wheat-based ethanol production, *BioEnergy Research*, 7(4), pp. 1421–9. DOI:10.1007/s12155-014-9473-1.

Altieri, R., Esposito, A., Baruzzi, G. and Nair, T. (2014), Corroboration for the successful application of humified olive mill waste compost in soilless cultivation of strawberry, *International Biodeterioration & Biodegradation*, 88, pp. 118–24. DOI:10.1016/j.ibiod.2013.12.006.

Anderson, C. R., Condron, L. M., Clough, T. J., Fiers, M., Stewart, A., Hill, R. A. and Sherlock, R. R. (2011), Biochar induced soil microbial community change: Implications for biogeochemical cycling of carbon, nitrogen and phosphorus, *Pedobiologia*, 54(5–6), pp. 309–20. DOI:10.1016/j.pedobi.2011.07.005.

Angst, T. E., Six, J., Reay, D. S. and Sohi, S. P. (2014), Impact of pine chip biochar on trace greenhouse gas emissions and soil nutrient dynamics in an annual ryegrass system in California, *Environmental Benefits and Risks of Biochar Application to Soil*, 191, pp. 17–26. DOI:10.1016/j.agee.2014.03.009.

Anusuya, S. and Sathiyabama, M. (2016), Effect of chitosan on growth, yield and curcumin content in turmeric under field condition, *Biocatalysis and Agricultural Biotechnology*, 6, pp. 102–6. DOI:10.1016/j.bcab.2016.03.002.

Arthurson, V. (2009), Closing the global energy and nutrient cycles through application of biogas residue to agricultural land–potential benefits and drawback, *Energies*, 2(2), pp. 226–42.

Arun, K. S. (2007), *Bio-fertilizers for Sustainable Agriculture. Mechanism of P-Solubilization*. Sixth Edition, Jodhpur, India: Agribios Publishers, pp. 196–7.

Atkinson, C. J., Fitzgerald, J. D. and Hipps, N. A. (2010), Potential mechanisms for achieving agricultural benefits from biochar application to temperate soils: A review, *Plant and Soil*, 337(1), pp. 1–18. DOI:10.1007/s11104-010-0464-5.

Aznar, A., Chen, N. W. G., Rigault, M., Riache, N., Joseph, D., Desmaële, D., Mouille, G., Boutet, S., Soubigou-Taconnat, L., Renou, J. P., Thomine, S., Expert, D. and Dellagi, A. (2014), Scavenging iron: A novel mechanism of plant immunity activation by microbial siderophores, *Plant Physiology*, 164(4), pp. 2167–83. DOI:10.1104/pp.113.233585.

Babik, J., Babik, I. and Kaniszewski, S. (2013), New biodegradable agro-fleece for soil mulching in organic vegetable production, in: De Neve, S. (Ed.), *NUTRIHORT: Nutrient Management, Innovative Techniques and Nutrient Legislation in Intensive Horticulture for an Improved Water Quality*. Ghent, Netherlands, pp. 343–9.

Barbosa, D. B. P., Nabel, M. and Jablonowski, N. D. (2014), Biogas-digestate as nutrient source for biomass production of *Sida hermaphrodita*, *Zea mays* L. and *Medicago sativa* L., *Energy Procedia*, 59, pp. 120–6. European Geosciences Union General Assembly 2014, EGU Division Energy, Resources & the Environment (ERE). DOI:10.1016/j.egypro.2014.10.357.

Baronti, S., Vaccari, F. P., Miglietta, F., Calzolari, C., Lugato, E., Orlandini, S., Pini, R., Zulian, C. and Genesio, L. (2014), Impact of biochar application on plant water relations in *Vitis vinifera* (L.), *European Journal of Agronomy*, 53, pp. 38–44. DOI:10.1016/j.eja.2013.11.003.

Basak, A. (2008), Effect of preharvest treatment with seaweed products, Kelpak® and Goëmar BM 86®, on fruit quality in apple, *International Journal of Fruit Science*, 8(1–2), pp. 1–14. DOI:10.1080/15538360802365251.

Beck, J. E., Schroeder-Moreno, M. S., Fernandez, G. E., Grossman, J. M. and Creamer, N. G. (2016), Effects of cover crops, compost, and vermicompost on strawberry yields and nitrogen availability in North Carolina, *HortTechnology*, 26(5), pp. 604–13. DOI:10.21273/HORTTECH03447-16.

Bedini, S., Turrini, A., Rigo, C., Argese, E. and Giovannetti, M. (2010), Molecular characterization and glomalin production of arbuscular mycorrhizal fungi colonizing a heavy metal polluted ash disposal island, downtown Venice, *Soil Biology and Biochemistry*, 42(5), pp. 758–65. DOI:10.1016/j.soilbio.2010.01.010.

Bharti, N., Barnawal, D., Shukla, S., Tewari, S. K., Katiyar, R. S. and Kalra, A. (2016a), Integrated application of *Exiguobacterium oxidotolerans*, *Glomus fasciculatum*, and vermicompost improves growth, yield and quality of *Mentha arvensis* in salt-stressed soils, *Industrial Crops and Products*, 83, pp. 717–28. DOI:10.1016/j.indcrop.2015.12.021.

Bharti, N., Barnawal, D., Wasnik, K., Tewari, S. K. and Kalra, A. (2016b), Co-inoculation of *Dietzia natronolimnaea* and *Glomus intraradices* with vermicompost positively influences *Ocimum basilicum* growth and resident microbial community structure in salt affected low fertility soils, *Applied Soil Ecology*, 100, pp. 211–25. DOI:10.1016/j.apsoil.2016.01.003.

Bird, M. I., Wurster, C. M., de Paula Silva, P. H., Bass, A. M. and de Nys, R. (2011), Algal biochar – production and properties, *Bioresource Technology*, 102(2), pp. 1886–91. DOI:10.1016/j.biortech.2010.07.106.

Blečić, A., Railić, B., Dubljević, R., Mitrović, D. and Spalevic, V. (2014), Application of sapropel in agricultural production, *Agriculture and Forestry*, 60(2), pp. 243–50.

Boguta, P., D'Orazio, V., Sokołowska, Z. and Senesi, N. (2016), Effects of selected chemical and physicochemical properties of humic acids from peat soils on their interaction mechanisms with copper ions at various pHs, *Journal of Geochemical Exploration*, 168, pp. 119–26. DOI:10.1016/j.gexplo.2016.06.004.

Bona, E., Lingua, G., Manassero, P., Cantamessa, S., Marsano, F., Todeschini, V., Copetta, A., D'Agostino, G., Massa, N., Avidano, L., Gamalero, E. and Berta, G. (2015), AM fungi and PGP pseudomonads increase flowering, fruit production, and vitamin content in strawberry grown at low nitrogen and phosphorus levels, *Mycorrhiza*, 25(3), pp. 181–93. DOI:10.1007/s00572-014-0599-y.

Bona, E., Cantamessa, S., Massa, N., Manassero, P., Marsano, F., Copetta, A., Lingua, G., D'Agostino, G., Gamalero, E. and Berta, G. (2017), Arbuscular mycorrhizal fungi and plant growth-promoting pseudomonads improve yield, quality and nutritional value of tomato: A field study, *Mycorrhiza*, 27(1), pp. 1–11. DOI:10.1007/s00572-016-0727-y.

Born, Y., Remus-Emsermann, M. N., Bieri, M., Kamber, T., Piel, J. and Pelludat, C. (2016), Fe^{2+} chelator proferrorosamine A: A gene cluster of Erwinia rhapontici P45 involved in its synthesis and its impact on growth of Erwinia amylovora CFBP1430, *Microbiology*, 162(2), pp. 236–45.

Boursier, H., Béline, F. and Paul, E. (2005), Piggery wastewater characterisation for biological nitrogen removal process design, *Bioresource Technology*, 96(3), pp. 351–8. DOI:10.1016/j.biortech.2004.03.007.

Briceño-Domínguez, D., Hernández-Carmona, G., Moyo, M., Stirk, W. and van Staden, J. (2014), Plant growth promoting activity of seaweed liquid extracts produced from Macrocystis pyrifera under different pH and temperature conditions, *Journal of Applied Phycology*, 26(5), pp. 2203–10. DOI:10.1007/s10811-014-0237-2.

Canieren, O., Karaguzel, C. and Aydin, A. (2017), Effect of physical pre-enrichment on humic substance recovery from leonardite, *Physicochemical Problems of Mineral Processing*, 53(1), pp. 502–14.

Cantrell, K. B., Hunt, P. G., Uchimiya, M., Novak, J. M. and Ro, K. S. (2012), Impact of pyrolysis temperature and manure source on physicochemical characteristics of biochar, *Bioresource Technology*, 107, pp. 419–28. DOI:10.1016/j.biortech.2011.11.084.

Cassán, F., Vanderleyden, J. and Spaepen, S. (2014), Physiological and agronomical aspects of phytohormone production by model Plant-Growth-Promoting Rhizobacteria (PGPR) belonging to the genus Azospirillum, *Journal of Plant Growth Regulation*, 33(2), pp. 440–59. DOI:10.1007/s00344-013-9362-4.

Cavagnaro, T. R., Bender, S. F., Asghari, H. R. and Heijden, M. G. A. van der (2015), The role of arbuscular mycorrhizas in reducing soil nutrient loss, *Trends in Plant Science*, 20(5), pp. 283–90. DOI:10.1016/j.tplants.2015.03.004.

Chan, K. Y., Van Zwieten, L., Meszaros, I., Downie, A. and Joseph, S. (2007), Agronomic values of greenwaste biochar as a soil amendment, *Soil Research*, 45(8), pp. 629–34.

Chan, K. Y., Van Zwieten, L., Meszaros, I., Downie, A. and Joseph, S. (2008), Using poultry litter biochars as soil amendments, *Soil Research*, 46(5), pp. 437–44.

Chantal, K., Shao, X., Jing, B., Yuan, Y., Hou, M. and Liao, L. (2013), Effects of effective microorganisms (EM) and bio-organic fertilizers on growth parameters and yield quality of flue-cured tobacco (Nicotiana tabacum), *Journal of Food, Agriculture and Environment*, 11(2), pp. 1212–15.

Chen, Y. P., Rekha, P. D., Arun, A. B., Shen, F. T., Lai, W.-A. and Young, C. C. (2006), Phosphate solubilizing bacteria from subtropical soil and their tricalcium phosphate solubilizing abilities, *Applied Soil Ecology*, 34(1), pp. 33–41. DOI:10.1016/j.apsoil.2005.12.002.

Cheng, C.-H., Lehmann, J., Thies, J. E. and Burton, S. D. (2008), Stability of black carbon in soils across a climatic gradient, *Journal of Geophysical Research: Biogeosciences*, 113(G2), pp. G02027. DOI:10.1029/2007JG000642.

Ciesielska, J., Malusa, E. and Sas-Paszt, L. (2011), Środki ochrony roślin stosowane w rolnictwie ekologicznym, *Komentarz Do Załącznika II Rozporządzenia Komisji (WE) Nr*, 889(2008), pp. 17–19.

Chetverikov, S. P. and Loginov, O. N. (2009), New metabolites of Azotobacter vinelandii exhibiting antifungal activity, *Microbiology*, 78(4), pp. 428–32. DOI:10.1134/S0026261709040055.

Colla, G., Rouphael, Y., Canaguier, R., Svecova, E. and Cardarelli, M. (2014), Biostimulant action of a plant-derived protein hydrolysate produced through enzymatic hydrolysis, *Frontiers in Plant Science*, 5, pp. 448. DOI:10.3389/fpls.2014.00448.

Colla, G., Nardi, S., Cardarelli, M., Ertani, A., Lucini, L., Canaguier, R. and Rouphael, Y. (2015), Protein hydrolysates as biostimulants in horticulture, *Biostimulants in Horticulture*, 196, pp. 28–38. DOI:10.1016/j.scienta.2015.08.037.

Conselvan, G. B., Pizzeghello, D., Francioso, O., Di Foggia, M., Nardi, S. and Carletti, P. (2017), Biostimulant activity of humic substances extracted from leonardites, *Plant Soil*, 420(1–2), pp. 119–34. DOI:10.1007/s11104-017-3373-z.

Corte, L., Dell'Abate, M. T., Magini, A., Migliore, M., Felici, B., Roscini, L., Sardella, R., Tancini, B., Emiliani, C., Cardinali, G. and Benedetti, A. (2014), Assessment of safety and efficiency of nitrogen organic fertilizers from animal-based protein hydrolysates—a laboratory multidisciplinary approach, *Journal of the Science of Food and Agriculture*, 94(2), pp. 235–45. DOI:10.1002/jsfa.6239.

Cretoiu, M. S., Korthals, G. W., Visser, J. H. M. and van Elsas, J. D. (2013), Chitin amendment increases soil suppressiveness toward plant pathogens and modulates the actinobacterial and oxalobacteraceal communities in an experimental agricultural field, *Applied and Environmental Microbiology*, 79(17), pp. 5291–301. DOI:10.1128/AEM.01361-13.

Crouch, I. J. and van Staden, J. (1993), Evidence for the presence of plant growth regulators in commercial seaweed products, *Plant Growth Regulation*, 13(1), pp. 21–9. DOI:10.1007/BF00207588.

Crouch, I. J., Beckett, R. P. and Staden, J. (1990), Effect of seaweed concentrate on the growth and mineral nutrition of nutrient-stressed lettuce, *Journal of Applied Phycology*, 2(3), pp. 269–72. DOI:10.1007/BF02179784.

Cuijpers, W. J. M. and Hospers-Brands, A. (2008), *Hulpmeststoffen: beschikbaarheid en opname van stikstof in de biologische teelt van zomertarwe*. Driebergen, the Netherlands: Louis Bolk Instituut, 33pp.

Dahlin, J., Nelles, M. and Herbes, C. (2017), Biogas digestate management: Evaluating the attitudes and perceptions of German gardeners towards digestate-based soil amendments, *Resources, Conservation and Recycling*, 118, pp. 27–38. DOI:10.1016/j.resconrec.2016.11.020.

Das, N. and Mukherjee, M. (2007), Cultivation of Pleurotus ostreatus on weed plants, *Bioresource Technology*, 98(14), pp. 2723–6. DOI:10.1016/j.biortech.2006.09.061.

Das, V., Satyanarayan, S. and Satyanarayan, S. (2016), Value added product recovery from sludge generated during gum arabic refining process by vermicomposting, *Environmental Monitoring and Assessment*, 188(9), pp. 523. DOI:10.1007/s10661-016-5516-8.

Davison, J., Moora, M., Öpik, M., Adholeya, A., Ainsaar, L., Bâ, A., Burla, S., Diedhiou, A. G., Hiiesalu, I., Jairus, T., Johnson, N. C., Kane, A., Koorem, K., Kochar, M., Ndiaye, C., Pärtel, M., Reier, Ü., Saks, Ü., Singh, R., Vasar, M. and Zobel, M. (2015), Global assessment of arbuscular mycorrhizal fungus diversity reveals very low endemism, *Science*, 349(6251), pp. 970. DOI:10.1126/science.aab1161.

Depardieu, C., Prémont, V., Boily, C. and Caron, J. (2016), Sawdust and bark-based substrates for soilless strawberry production: Irrigation and electrical conductivity management, *PLoS ONE*, 11(4), pp. e0154104 (1–20). DOI:10.1371/journal.pone.0154104.

Derkowska, E., Sas-Paszt, L., Sumorok, B., Szwonek, E. and Gluszek, S. (2008), The influence of mycorrhization and organic mulches on mycorrhizal frequency in apple and strawberry roots, *Journal of Fruit and Ornamental Plant Research*, 16, pp. 227–42.

Derkowska, E., Sas Paszt, L. S., Harbuzov, A. and Sumorok, B. (2015a), Root growth, mycorrhizal frequency and soil microorganisms in strawberry as affected by biopreparations, *Advances in Microbiology*, 5(1), pp. 65–73.

Derkowska, E., Sas Paszt, L. and Szwonek, E. (2015b), Influence of biological products on the growth and development of large-fruited cranberry under greenhouse conditions, *Folia Horticulturae*, 27(1), pp. 71–7. DOI:10.1515/fhort-2015-0016.

Dhar, D. W., Prasanna, R., Pabbi, S. and Vishwakarma, R. (2015), Significance of cyanobacteria as inoculants in agriculture, in: Das, D. (Ed.), *Algal Biorefinery: An Integrated Approach*. Cham, Switzerland: Springer International Publishing, pp. 339–74.

Di Stasio, E., Maggio, A., Ventorino, V., Pepe, O., Raimondi, G. and De Pascale, S. (2017), Free-living (N2)-fixing bacteria as potential enhancers of tomato growth under salt stress. *Acta Horticulturae*, 1164, pp. 151–6. DOI:10.17660/ActaHortic.2017.1164.19

Dmitrieva, E., Efimova, E., Siundiukova, K. and Perelomov, L. (2015), Surface properties of humic acids from peat and sapropel of increasing transformation, *Environmental Chemistry Letters*, 13(2), pp. 197–202. DOI:10.1007/s10311-015-0497-3.

Domínguez, J., Sanchez-Hernandez, J. C. and Lores, M. (2017), Vermicomposting of winemaking by-products, in: Galanakis, C. M. (Ed.), *Handbook of Grape Processing By-Products*. London, UK: Academic Press, Elsevier, pp. 55–78.

Doskočil, L., Grasset, L., Válková, D. and Pekař, M. (2014), Hydrogen peroxide oxidation of humic acids and lignite, *Fuel*, 134, pp. 406–13. DOI:10.1016/j.fuel.2014.06.011.

Duong, D. H., Ngo, X. Q., Do, D. G., Le, T. A. H., Nguyen, V. T. and Nic, S. (2014), Effective control of neem (*Azadirachta indica* A. Juss) cake to plant parasitic nematodes and fungi in black pepper diseases in vitro, *Journal of Vietnamese Environment*, 6 (3), pp. 233–8.

Ebrahiminezhad, A., Manafi, Z., Berenjian, A., Kianpour, S. and Ghasemi, Y. (2017), Iron-reducing bacteria and iron nanostructures, *Journal of Advanced Medical Sciences and Applied Technologies*, 3(1), pp. 9–16.

Egamberdiyeva, D. (2007), The effect of plant growth promoting bacteria on growth and nutrient uptake of maize in two different soils, *Applied Soil Ecology*, 36(2–3), pp. 184–9. DOI:10.1016/j.apsoil.2007.02.005.

Emerson, J. K. and Roark, A. M. (2007), Composition of guano produced by frugivorous, sanguivorous, and insectivorous bats, *Acta Chiropterologica*, 9(1), pp. 261–7. DOI:10.3161/1733-5329(2007)9[261:COGPBF]2.0.CO;2.

Enders, A., Hanley, K., Whitman, T., Joseph, S. and Lehmann, J. (2012), Characterization of biochars to evaluate recalcitrance and agronomic performance, *Bioresource Technology*, 114, pp. 644–53. DOI:10.1016/j.biortech.2012.03.022.

Fan, D., Hodges, D. M., Zhang, J., Kirby, C. W., Ji, X., Locke, S. J., Critchley, A. T. and Prithiviraj, B. (2011), Commercial extract of the brown seaweed *Ascophyllum nodosum* enhances phenolic antioxidant content of spinach (*Spinacia oleracea* L.) which protects *Caenorhabditis elegans* against oxidative and thermal stress, *Food Chemistry*, 124(1), pp. 195–202. DOI:10.1016/j.foodchem.2010.06.008.

Fernández-Delgado Juárez, M., Gómez-Brandón, M. and Insam, H. (2015), Merging two waste streams, wood ash and biowaste, results in improved composting process and end products, *Science of the Total Environment*, 511, pp. 91–100. DOI:10.1016/j.scitotenv.2014.12.037.

Fox, A. R., Soto, G., Valverde, C., Russo, D., Lagares, A., Zorreguieta, Á., Alleva, K., Pascuan, C., Frare, R., Mercado-Blanco, J., Dixon, R. and Ayub, N. D. (2016), Major cereal crops benefit from biological nitrogen fixation when inoculated with the nitrogen-fixing bacterium Pseudomonas protegens Pf- 5 X940, *Environmental Microbiology*, 18(10), pp. 3522–34. DOI:10.1111/1462-2920.13376.

Frąc, M., Michalski, P. and Sas Paszt, L. (2009), The effect of mulch and mycorrhiza on fruit yield and size of three strawberry cultivars, *Journal of Fruit and Ornamental Plant Research*, 17(2), pp. 85–93.

Fuess, L. T. and Garcia, M. L. (2014), Implications of stillage land disposal: A critical review on the impacts of fertigation, *Journal of Environmental Management*, 145, pp. 210–29. DOI:10.1016/j.jenvman.2014.07.003.

Ghani, W. A. W. A. K., Mohd, A., da Silva, G., Bachmann, R. T., Taufiq-Yap, Y. H., Rashid, U. and Al-Muhtaseb, A. H. (2013), Biochar production from waste rubber-wood-sawdust and its potential use in C sequestration: Chemical and physical characterization, *Industrial Crops and Products*, 44, pp. 18–24. DOI:10.1016/j.indcrop.2012.10.017.

Ghasemi, S., Khoshgoftarmanesh, A., Hadadzadeh, H. and Afyuni, M. (2013), Synthesis, characterization, and theoretical and experimental investigations of zinc(II)–amino acid complexes as ecofriendly plant growth promoters and highly bioavailable sources of zinc, *Journal of Plant Growth Regulation*, 32(2), pp. 315–23. DOI:10.1007/s00344-012-9300-x.

Glinicki, R., Sas-Paszt, L. and Jadczuk-Tobjasz, E. (2011), The effect of microbial inoculation with EM-Farming inoculum on the vegetative growth of three strawberry cultivars, *Annals of Warsaw University of Life Sciences- SGGW. Horticulture and Landscape Architecture*, 32, pp. 3–14.

Grantina-Ievina, L., Karlsons, A., Andersone-Ozola, U. and Ievinsh, G. (2014), Effect of freshwater sapropel on plants in respect to its growth-affecting activity and cultivable microorganism content, *Zemdirbyste-Agriculture*, 101(4), pp. 355–66.

Grzyb, Z. S., Piotrowski, W., Bielicki, P., Sas Paszt, L. and Malusà, E. (2012), Effect of different fertilizers and amendments on the growth of apple and sour cherry rootstocks in an organic nursery, *Journal of Fruit and Ornamental Plant Research*, 20(1), pp. 43–53. DOI:10.2478/v10290-012-0004-x.

Grzyb, Z. S., Sas Paszt, L., Piotrowski, W. and Malusa, E. (2015), The influence of mycorrhizal fungi on the growth of apple and sour cherry maidens fertilized with different bioproducts in the organic nursery, *Journal of Life Sciences*, 9, pp. 221–8.

Gusain, Y. S., Kamal, R., Mehta, C. M., Singh, U. S. and Sharma, A. K. (2015), Phosphate solubilizing and indole- 3-acetic acid producing bacteria from the soil of Garhwal Himalaya aimed to improve the growth of rice, *Journal of Environmental Biology*, 36(1), pp. 301.

Hankins, S. D. and Hockey, H. P. (1990), The effect of a liquid seaweed extract from *Ascophyllum nodosum* (Fucales, Phaeophyta) on the two-spotted red spider mite *Tetranychus urticae*, *Hydrobiologia*, 204–205(1), pp. 555–9. DOI:10.1007/BF00040286.

Haynes, R. J. and Swift, R. S. (1986), Effect of soil amendments and sawdust mulching on growth, yield and leaf nutrient content of highbush blueberry plants, *Scientia Horticulturae*, 29(3), pp. 229–38. DOI:10.1016/0304-4238(86)90066-X.

Hernández-Herrera, R. M., Santacruz-Ruvalcaba, F., Ruiz-López, M. A., Norrie, J. and Hernández-Carmona, G. (2014), Effect of liquid seaweed extracts on growth of tomato seedlings (*Solanum lycopersicum* L.), *Journal of Applied Phycology*, 26(1), pp. 619–28. DOI:10.1007/s10811-013-0078-4.

Hoagland, L., Carpenter-Boggs, L., Granatstein, D., Mazzola, M., Smith, J., Peryea, F. and Reganold, J. P. (2008), Orchard floor management effects on nitrogen fertility and soil biological activity in a newly established organic apple orchard, *Biology and Fertility of Soils*, 45(1), pp. 11. DOI:10.1007/s00374-008-0304-4.

Hosseini Bai, S., Xu, C.-Y., Xu, Z., Blumfield, T., Zhao, H., Wallace, H., Reverchon, F. and Van Zwieten, L. (2015), Soil and foliar nutrient and nitrogen isotope composition (δ15N) at 5 years after poultry litter and green waste biochar amendment in a macadamia orchard, *Environmental Science and Pollution Research*, 22(5), pp. 3803–9. DOI:10.1007/s11356-014-3649-2.

Hu, C. and Qi, Y. (2013), Long-term effective microorganisms application promote growth and increase yields and nutrition of wheat in China, *European Journal of Agronomy*, 46, pp. 63–7. DOI:10.1016/j.eja.2012.12.003.

Hupfauf, S., Bachmann, S., Fernández-Delgado Juárez, M., Insam, H. and Eichler-Löbermann, B. (2016), Biogas digestates affect crop P uptake and soil microbial community composition, 542, pp. 1144–54. Special Issue on Sustainable Phosphorus Taking Stock: Phosphorus Supply from Natural and Anthropogenic Pools in the 21st Century. DOI:10.1016/j.scitotenv.2015.09.025.

Ibrahim, E. A. and Ramadan, W. A. (2015), Effect of zinc foliar spray alone and combined with humic acid or/and chitosan on growth, nutrient elements content and yield of dry bean (*Phaseolus vulgaris* L.) plants sown at different dates, *Scientia Horticulturae*, 184, pp. 101–5. DOI:10.1016/j.scienta.2014.11.010.

Jakienė, E. (2013), The effect of the microelement fertilizers and biological preparation Terra Sorb Foliar on spring rape crop, *Žemės Ūkio Mokslai*, 20(2), pp. 75–83.

Jannoura, R., Joergensen, R. G. and Bruns, C. (2014), Organic fertilizer effects on growth, crop yield, and soil microbial biomass indices in sole and intercropped peas and oats under organic farming conditions, *European Journal of Agronomy*, 52(Part B), pp. 259–70. DOI:10.1016/j.eja.2013.09.001.

Jayaraj, J., Wan, A., Rahman, M. and Punja, Z. K. (2008), Seaweed extract reduces foliar fungal diseases on carrot, *Crop Protection*, 27 (10), pp. 1360–6. DOI:10.1016/j.cropro.2008.05.005.

Jayaraman, J., Norrie, J. and Punja, Z. (2011), Commercial extract from the brown seaweed *Ascophyllum nodosum* reduces fungal diseases in greenhouse cucumber, *Journal of Applied Phycology*, 23(3), pp. 353–61. DOI:10.1007/s10811-010-9547-1.

Jeffery, S., Verheijen, F. G. A., van der Velde, M. and Bastos, A. C. (2011), A quantitative review of the effects of biochar application to soils on crop productivity using meta-analysis, *Agriculture, Ecosystems & Environment*, 144(1), pp. 175–87. DOI:10.1016/j.agee.2011.08.015.

Jha, P., Biswas, A. K., Lakaria, B. L. and Rao, A. S. (2010), Biochar in agriculture–prospects and related implications, *Current Science*, 99(9), pp. 1218–25.

Ji, R., Dong, G., Shi, W. and Min, J. (2017), Effects of liquid organic fertilizers on plant growth and rhizosphere soil characteristics of Chrysanthemum, *Sustainability*, 9(5), pp. 841 (1–16). DOI:10.3390/su9050841.

Jones, D. L. and Quilliam, R. S. (2014), Metal contaminated biochar and wood ash negatively affect plant growth and soil quality after land application, *Journal of Hazardous Materials*, 276, pp. 362–70. DOI:10.1016/j.jhazmat.2014.05.053.

Jones, S. B., Robbins, C. W. and Hansen, C. L. (1993), Sodic soil reclamation using cottage cheese (acid) whey, *Arid Soil Research and Rehabilitation*, 7(1), pp. 51–61. DOI:10.1080/15324989309381334.

Joshi, R., Singh, J. and Vig, A. (2015), Vermicompost as an effective organic fertilizer and biocontrol agent: Effect on growth, yield and quality of plants, *Reviews in Environmental Science and Bio/ Technology*, 14(1), pp. 137–59. DOI:10.1007/s11157-014-9347-1.

Jung, J. Y., Kim, J. S., Ha, S. Y., Choi, J. H. and Yang, J.-K. (2015), Suitability of thermal treated sawdust as replacements for peat moss in horticultural media, *Journal of Agriculture and Life Science*, 49(4), pp. 105–15.

Kalra, A., Chandra, M., Awasthi, A., Singh, A. K. and Khanuja, S. P. S. (2010), Natural compounds enhancing growth and survival of rhizobial inoculants in vermicompost-based formulations, *Biology and Fertility of Soils*, 46(5), pp. 521–4. DOI:10.1007/s00374-010-0443-2.

Kaniszewski, S., Babik, I. and Babik, J. (2013), Pelletized legume plants as fertilizer for vegetables in organic farming, in: De Neve, S. (Ed.), *NUTRIHORT: Nutrient Management, Innovative Techniques and Nutrient Legislation in Intensive Horticulture for an Improved Water Quality*. Ghent, Netherlans, pp. 330–42.

Kern, J., Tammeorg, P., Shanskiy, M., Sakrabani, R., Knicker, H., Kammann, C., Tuhkanen, E.-M., Smidt, G., Prasad, M., Tiilikkala, K., Sohi, S., Gascó, G., Steiner, C. and Glaser, B. (2017), Synergistic use of peat and charred material in growing media – an option to reduce the pressure on peatlands? *Journal of Environmental Engineering and Landscape Management*, 25(2), pp. 160–74. DOI:10 .3846/16486897.2017.1284665.

Khan, W., Rayirath, U., Subramanian, S., Jithesh, M., Rayorath, P., Hodges, D., Critchley, A. T., Craigie, J. S., Norrie, J. and Prithiviraj, B. (2009), Seaweed extracts as biostimulants of plant growth and development, *Journal of Plant Growth Regulation*, 28 (4), pp. 386–99.

Khan, N., Clark, I., Sánchez-Monedero, M. A., Shea, S., Meier, S. and Bolan, N. (2014), Maturity indices in co- composting of chicken manure and sawdust with biochar, *Bioresource Technology*, 168, pp. 245–51. DOI:10.1016/j.biortech.2014.02.123.

Kiers, E. T., Rousseau, R. A., West, S. A. and Denison, R. F. (2003), Host sanctions and the legume-rhizobium mutualism, *Nature*, 425(6953), pp. 78–81. DOI:10.1038/nature01931.

Kim Thi Tran, C., Rose, M. T., Cavagnaro, T. R. and Patti, A. F. (2015), Lignite amendment has limited impacts on soil microbial communities and mineral nitrogen availability, *Applied Soil Ecology*, 95, pp. 140–50. DOI:10.1016/j.apsoil.2015.06.020.

Klavins, M. and Purmalis, O. (2013), Properties and structure of raised bog peat humic acids, *Journal of Molecular Structure*, 1050, pp. 103–13. DOI:10.1016/j.molstruc.2013.07.021.

Kleiber, T., Starzyk, J., Górski, R., Sobieralski, K., Siwulski, M., Rempulska, A. and Sobiak, A. (2014), The studies on applying of Effective Microorganisms (EM) and CRF on nutrient contents in leaves and yielding of tomato, *Acta Scientiarum Polonorum – Hortorum Cultus*, 13 (1), pp. 79–90.

Klepeckas, M. and Januškaitienė, I. (2017), Changes in Triticum aestivum and Hordeum vulgare chlorophyll content and fluorescence parameters under impact of various sapropel concentrations, *Biologija*, 62 (4), pp. 216–26.

Köpke, U., Krämer, J. and Leifert, C. (2007), Pre-harvest strategies to ensure the microbiological safety of fruit and vegetables from manure-based production systems, in: Cooper, J., Niggli, U. and Leifert, C. (Eds), *Handbook of Organic Food Safety and Quality*. Cambridge, UK: Woodhead Publishing Ltd., pp. 413–29.

Koutrotsios, G., Mountzouris, K. C., Chatzipavlidis, I. and Zervakis, G. I. (2014), Bioconversion of lignocellulosic residues by *Agrocybe cylindracea* and *Pleurotus ostreatus* mushroom fungi – assessment of their effect on the final product and spent substrate properties, *Food Chemistry*, 161, pp. 127–35. DOI:10.1016/j.foodchem.2014.03.121.

Kovács, K., Czech, V., Fodor, F., Solti, A., Lucena, J. J., Santos-Rosell, S. and Hernández- Apaolaza, L. (2013), Characterization of Fe–Leonardite complexes as novel natural iron fertilizers, *Journal of Agricultural and Food Chemistry*, 61(50), pp. 12200–10. DOI:10.1021/jf404455y.

Kumar, G. and Sahoo, D. (2011), Effect of seaweed liquid extract on growth and yield of Triticum aestivum var. Pusa Gold, *Journal of Applied Phycology*, 23(2), pp. 251–5. DOI:10.1007/s10811-011-9660-9.

Kumar, V., Kumar, S., Jha, S. K. and Jijeesh, C. M. (2014), Influence of de-oiled seed cakes on seedling performance of East Indian Rosewood (Dalbergia latifoila Roxb.), *Soil Environment*, 33(2), pp. 169–74.

Kumar, V., Menon, S., Agarwal, H. and Gopalakrishnan, D. (2017), Characterization and optimization of bacterium isolated from soil samples for the production of siderophores, *Resource-Efficient Technologies*, 3(4), 434–9. DOI:10.1016/j.reffit.2017.04.004.

Kuwada, K., Kuramoto, M., Utamura, M., Matsushita, I. and Ishii, T. (2006), Isolation and structural elucidation of a growth stimulant for arbuscular mycorrhizal fungus from Laminaria japonica Areschoug, *Journal of Applied Phycology*, 18(6), pp. 795–800.

Laghari, M., Hu, Z., Mirjat, M. S., Xiao, B., Tagar, A. A. and Hu, M. (2016), Fast pyrolysis biochar from sawdust improves the quality of desert soils and enhances plant growth, *Journal of the Science of Food and Agriculture*, 96(1), pp. 199–206. DOI:10.1002/jsfa.7082.

Lehmann, J., Rillig, M. C., Thies, J., Masiello, C. A., Hockaday, W. C. and Crowley, D. (2011), Biochar effects on soil biota – A review, *Soil Biology & Biochemistry*, 43(9), pp. 1812–36. 19th International Symposium on Environmental Biogeochemistry. DOI:10.1016/j.soilbio.2011.04.022.

Lim, S. L., Wu, T. Y., Lim, P. N. and Shak, K. P. Y. (2015), The use of vermicompost in organic farming: Overview, effects on soil and economics, *Journal of the Science of Food and Agriculture*, 95(6), pp. 1143–56. DOI:10.1002/jsfa.6849.

Lima, J. D., Zanetti, S., Nomura, E. S., Fuzitani, E. J., Rozane, D. E. and Iori, P. (2016), Growth and yield of anthurium in response to sawdust mulching, *Ciência Rural*, 46, pp. 440–6.

Lisiecka, J., Knaflewski, M., Spizewski, T., Fraszczak, B., Kaluzewicz, A. and Krzesinski, W. (2011), The effect of animal protein hydrolysate on quantity and quality of strawberry daughter plants cv.'Elsanta', *Acta Scientiarum Polonorum, Hortorum Cultus*, 10, pp. 31–40.

Liu, M., Wang, T., Cheng, Y., Fu, Y., Zhang, P., Xie, M., Sun, M. and Yang, D. (2014), Peat and brown coal resources in China and its potential for developing potassium humate fertilizer, *Earth Science Frontiers*, 5, pp. 25.

Løes, A.-K., Bünemann, E. K., Cooper, J., Hörtenhuber, S., Magid, J., Oberson, A. and Möller, K. (2016), Nutrient supply to organic agriculture as governed by EU regulations and standards in six European countries, *Organic Agriculture*, 7, pp. 395–418. DOI:10.1007/s13165-016-0165-3.

Macias-Corral, M. A., Samani, Z. A., Hanson, A. T. and Funk, P. A. (2017), Co-digestion of agricultural and municipal waste to produce energy and soil amendment, *Waste Management & Research*, 35(9), pp. 991–6. DOI:10.1177/0734242X17715097.

Mahdavi, B. and Rahimi, A. (2013), Seed priming with chitosan improves the germination and growth performance of ajowan (*Carum copticum*) under salt stress, *EurAsian Journal of BioSciences*, 7, 69–76.

Malusà, E., Sas Paszt, L., Głuszek, S. and Ciesielska, J. (2014), Organic fertilizers to sustain soil fertility, in: Sinha, S., Pant, K., Bajpai, S., and Govil, J. N. (Eds), *Fertilizers Technology Vol. I: Synthesis*. Houston, TX: Studium Press LLC, pp. 256–81.

Marinou, E., Chrysargyris, A. and Tzortzakis, N. (2013), Use of sawdust, coco soil and pumice in hydroponically grown strawberry, *Plant, Soil and Environment*, 59(10), pp. 452–9.

Marwaha, S. S. and Kennedy, J. F. (1988), Whey—pollution problem and potential utilization, *International Journal of Food Science and Technology*, 23(4), pp. 323–36. DOI:10.1111/j.1365-2621.1988.tb00586.x.

Masny, A., Basak, A. and Żurawicz, E. (2004), Effects of foliar applications of Kelpak SL and Goëmar BM 86®; preparations on yield and fruit quality in two strawberry cultivars, *Journal of Fruit and Ornamental Plant Research*, 12, pp. 23–7.

Mazzoncini, M., Antichi, D., Tavarini, S., Silvestri, N., Lazzeri, L. and D'Avino, L. (2015), Effect of defatted oilseed meals applied as organic fertilizers on vegetable crop production and environmental impact, *Industrial Crops and Products*, 75(Part A), pp. 54–64. DOI:10.1016/j.indcrop.2015.04.061.

Mobini, M., Khoshgoftarmanesh, A. H. and Ghasemi, S. (2014), The effect of partial replacement of nitrate with arginine, histidine, and a mixture of amino acids extracted from blood powder on yield and nitrate accumulation in onion bulb, *Scientia Horticulturae*, 176, pp. 232–7. DOI:10.1016/j.scienta.2014.07.014.

Möller, K. (2015), Assessment of alternative phosphorus fertilizers for organic farming: Meat and bone meal, Improve-P Factsheet, Universität Hohenheim, ETH Zürich, FiBL, Bioforsk, Universität für Bodenkultur Wien, Newcastle University, University of Copenhagen, pp. 1–8.

Möller, K. and Müller, T. (2012), Effects of anaerobic digestion on digestate nutrient availability and crop growth: A review, *Engineering in Life Sciences*, 12(3), pp. 242–57. DOI:10.1002/elsc.201100085.

Mosa, W.-G., Paszt, L. S., Frąc, M., Trzciński, P., Przybył, M., Treder, W. and Klamkowski, K. (2016), The influence of biofertilization on the growth, yield and fruit quality of cv. Topaz apple trees, *Horticultural Science*, 43(3), pp. 105–11.

Moutia, J.-F. Y., Saumtally, S., Spaepen, S. and Vanderleyden, J. (2010), Plant growth promotion by *Azospirillum* sp. in sugarcane is influenced by genotype and drought stress, *Plant and Soil*, 337(1), pp. 233–42. DOI:10.1007/s11104-010-0519-7.

Nagaraja, H., Chennappa, G., Rakesh, S., Naik, M. K., Amaresh, Y. S. and Sreenivasa, M. Y. (2016), Antifungal activity of *Azotobacter nigricans* against trichothecene-producing *Fusarium* species associated with cereals, *Food Science and Biotechnology*, 25(4), pp. 1197–204. DOI:10.1007/s10068-016-0190-8.

Ngala, B. M., Valdes, Y., dos Santos, G., Perry, R. N. and Wesemael, W. M. L. (2016), Seaweed-based products from *Ecklonia maxima* and *Ascophyllum nodosum* as control agents for the root-knot nematodes *Meloidogyne chitwoodi* and *Meloidogyne hapla* on tomato plants, *Journal of Applied Phycology*, 28(3), pp. 2073–82. DOI:10.1007/s10811-015-0684-4.

Nustorova, M., Braikova, D., Gousterova, A., Vasileva-Tonkova, E. and Nedkov, P. (2006), Chemical, microbiological and plant analysis of soil fertilized with alkaline hydrolysate of sheep's wool waste, *World Journal of Microbiology and Biotechnology*, 22(4), pp. 383–90. DOI:10.1007/s11274-005-9045-9.

Oni, O. E. and Friedrich, M. W. (2017), Metal oxide reduction linked to anaerobic methane oxidation, *Trends in Microbiology*, 25(2), pp. 88–90. DOI:10.1016/j.tim.2016.12.001.

Öpik, M., Vanatoa, A., Vanatoa, E., Moora, M., Davison, J., Kalwij, J. M., Reier, Ü. and Zobel, M. (2010), The online database MaarjAM reveals global and ecosystemic distribution patterns in arbuscular mycorrhizal fungi (Glomeromycota), *New Phytologist*, 188(1), pp. 223–41. DOI:10.1111/j.1469-8137.2010.03334.x.

Ostrovskij, M. (2014), Testing HUMIN PLUS microfertilizer, *European Agrophysical Journal*, 1 (2), pp. 79–84.

Ozkan, S. and Ozkan, S. G. (2016), Investigation of humate extraction from lignites, *International Journal of Coal Preparation and Utilization*, 37(6), pp. 285–92. DOI:10.1080/19392699.2016.1171761.

Panda, D., Pramanik, K. B. and Naya, R. (2012), Use of sea weed extracts as plant growth regulators for sustainable agriculture, *International Journal of Stress Management*, 3(3), pp. 404–11.

Pane, C., Celano, G., Villecco, D. and Zaccardelli, M. (2012), Control of *Botrytis cinerea*, *Alternaria alternata* and *Pyrenochaeta lycopersici* on tomato with whey compost-tea applications, *Crop Protection*, 38, pp. 80–6. DOI:10.1016/j.cropro.2012.03.012.

Papenfus, H. B., Kulkarni, M. G., Stirk, W. A., Finnie, J. F. and Van Staden, J. (2013), Effect of a commercial seaweed extract (Kelpak®) and polyamines on nutrient-deprived (N, P and K) okra seedlings, *Scientia Horticulturae*, 151, pp. 142–6. DOI:10.1016/j.scienta.2012.12.022.

Paredes, C., Medina, E., Bustamante, M. A. and Moral, R. (2016), Effects of spent mushroom substrates and inorganic fertilizer on the characteristics of a calcareous clayey-loam soil and lettuce production, *Soil Use and Management*, 32(4), pp. 487–94. DOI:10.1111/sum.12304.

Parvage, M., Ulén, B., Eriksson, J., Strock, J. and Kirchmann, H. (2013), Phosphorus availability in soils amended with wheat residue char, *Biology and Fertility of Soils*, 49(2), pp. 245–50. DOI:10.1007/s00374-012-0746-6.

Peterson, A. E., Walker, W. G. and Watson, K. S. (1979), Effect of whey applications on chemical properties of soils and crops, *Journal of Agricultural and Food Chemistry*, 27(4), pp. 654–8. DOI:10.1021/jf60224a064.

Qian, P., Schoenau, J. and Urton, R. (2011), Effect of soil amendment with thin stillage and glycerol on plant growth and soil properties, *Journal of Plant Nutrition*, 34(14), pp. 2206–21. DOI:10.1080/01904167.2011.618579.

Qian, S., Ding, W., Li, Y., Liu, G., Sun, J. and Ding, Q. (2015), Characterization of humic acids derived from Leonardite using a solid-state NMR spectroscopy and effects of humic acids on growth and nutrient uptake of snap bean, *Chemical Speciation & Bioavailability*, 27 (4), pp. 156–61. DOI:10.1080/09542299.2015.1118361.

Raghu, K. and MacRae, I. C. (1966), Occurrence of phosphate-dissolving micro-organisms in the rhizosphere of rice plants and in submerged soils, *Journal of Applied Bacteriology*, 29(3), pp. 582–6. DOI:10.1111/j.1365-2672.1966.tb03511.x.

Rashid, A., Mir, M. R. and Hakeem, K. R. (2016a), Biofertilizer use for sustainable agricultural production, in: Hakeem, K. R., Akhtar, M. S., and Abdullah, S. N. A. (Eds), *Plant, Soil and Microbes: Volume 1: Implications in Crop Science*. Cham, Switzerland: Springer International Publishing, pp. 163–80.

Rashid, M. I., Mujawar, L. H., Shahzad, T., Almeelbi, T., Ismail, I. M. I. and Oves, M. (2016b), Bacteria and fungi can contribute to nutrients bioavailability and aggregate formation in degraded soils, *Microbiological Research*, 183, pp. 26–41. DOI:10.1016/j.micres.2015.11.007.

Reed, E. Y., Chadwick, D. R., Hill, P. W. and Jones, D. L. (2017), Critical comparison of the impact of biochar and wood ash on soil organic matter cycling and grassland productivity, *Soil Biology and Biochemistry*, 110, pp. 134–42. DOI:10.1016/j.soilbio.2017.03.012.

Robbins, C. W. and Lehrsch, G. A. (1992), Effects of acidic cottage cheese whey on chemical and physical properties of a sodic soil, *Arid Soil Research and Rehabilitation*, 6 (2), pp. 127–34. DOI:10.1080/15324989209381305.

Roger, P. A. and Ladha, J. K. (1992), Biological N2 Fixation in wetland rice fields: Estimation and contribution to nitrogen balance, *Plant and Soil*, 141(1), pp. 41–55. DOI:10.1007/BF00011309.

Rouphael, Y., Franken, P., Schneider, C., Schwarz, D., Giovannetti, M., Agnolucci, M., Pascale, S. D., Bonini, P. and Colla, G. (2015), Arbuscular mycorrhizal fungi act as biostimulants in horticultural crops, *Biostimulants in Horticulture*, 196, pp. 91–108. DOI:10.1016/j.scienta.2015.09.002.

Roy, S., Barman, S., Chakraborty, U. and Chakraborty, B. (2015), Evaluation of spent mushroom substrate as biofertilizer for growth improvement of *Capsicum annuum* L, *Journal of Applied Biology and Biotechnology*, 3(03), pp. 22–7.

Roychowdhury, D., Paul, M. and KumarBanerjee, S. (2015), Isolation identification and characterization of phosphate solubilising bacteria from soil and the production of biofertilizer, *International Journal of Current Microbiology and Applied Sciences*, 4 (11), pp. 808–15.

Rozpara, E., Pąśko, M., Bielicki, P. and Sas Paszt, L. (2014), Influence of various bio-fertilizers on the growth and fruiting of 'Ariwa' apple trees growing in an organic orchard, *Journal of Research and Applications in Agricultural Engineering*, 59(4), pp. 65–8.

Rutkowska, B., Szulc, W., Sosulski, T. and Stepien, W. (2014), Soil micronutrient availability to crops affected by long-term inorganic and organic fertilizer applications, *Plant, Soil and Environment*, 60(5), pp. 198–203.

Saito, B. and Seckler, M. M. (2014), Alkaline extraction of humic substances from peat applied to organic-mineral fertilizer production, *Brazilian Journal of Chemical Engineering*, 31, pp. 675–82.

Sajbrt, V., Rosol, M. and Ditl, P. (2010), A comparison of distillery stillage disposal methods, *Acta Polytechnica*, 50(2), pp. 63–9.

Salama, A. S., El-Sayed, O. M. and El Gammal, O. (2014), Effect of Effective Microorganisms (EM) and potassium sulphate on productivity and fruit quality of 'Hayany' date palm grown under salinity stress, *Journal of Agriculture and Veterinary Sciences*, 7, pp. 90–9.

Salantur, A., Ozturk, A. and Akten, S. (2006), Growth and yield response of spring wheat (*Triticum aestivum* L.) to inoculation with rhizobacteria, *Plant, Soil and Environment*, 52(3), pp. 111–18.

Sanderson, K. R. and Cutcliffe, J. A. (1991), Effect of sawdust mulch on yields of select clones of lowbush blueberry, *Canadian Journal of Plant Science*, 71(4), pp. 1263–6. DOI:10.4141/cjps91-175.

Sas Paszt, L., Sumorok, B., Malusá, E., Głuszek, S. and Derkowska, E. (2011), The influence of bioproducts on root growth and mycorrhizal occurrence in the rhizosphere of strawberry plants 'Elsanta', *Journal of Fruit and Ornamental Plant Research*, 19(1), pp. 13–34.

Sas Paszt, L., Malusá, E., Sumorok, B., Canfora, L., Derkowska, E. and Głuszek, S. (2015), The influence of bioproducts on mycorrhizal occurrence and diversity in the rhizosphere of strawberry plants under controlled conditions, *Advances in Microbiology*, 5(1), pp. 40–53.

Sasirekha, B. and Srividya, S. (2016), Siderophore production by *Pseudomonas aeruginosa* FP6, a biocontrol strain for *Rhizoctonia solani* and *Colletotrichum gloeosporioides* causing diseases in chilli, *Agriculture and Natural Resources*, 50(4), pp. 250–6. DOI:10.1016/j.anres.2016.02.003.

Schäfer, T., Hanke, M.-V., Flachowsky, H., König, S., Peil, A., Kaldorf, M., Polle, A. and Buscot, F. (2012), Chitinase activities, scab resistance, mycorrhization rates and biomass of own-rooted and grafted transgenic apple, *Genetics and Molecular Biology*, 35(2), pp. 466–73. DOI:10.1590/S1415-47572012000300014.

Schlegel, A. J., Assefa, Y., Bond, H. D., Haag, L. A. and Stone, L. R. (2017), Changes in soil nutrients after 10 years of cattle manure and swine effluent application, *Soil and Tillage Research*, 172, pp. 48–58. DOI:10.1016/j.still.2017.05.004.

Shaheen, S., Khan, M., Khan, M. J., Jilani, S., Bibi, Z., Munir, M. and Kiran, M. (2017), Effective Microorganisms (EM) co-applied with organic wastes and NPK stimulate the growth, yield and quality of spinach (*Spinacia oleracea* L.), *Sarhad Journal of Agriculture*, 33 (1), pp. 30–41.

Sharp, G. R. (2013), A review of the applications of chitin and its derivatives in agriculture to modify plant- microbial interactions and improve crop yields, *Agronomy*, 3(4), pp. 778–93. DOI:10.3390/agronomy3040757.

Sharratt, W. J., Peterson, A. E. and Calbert, H. E. (1959), Whey as a source of plant nutrients and its effect on the soil, *Journal of Dairy Science*, 42(7), pp. 1126–31. DOI:10.3168/jds.S0022-0302(59)90705-2.

Siebers, N., Godlinski, F. and Leinweber, P. (2014), Bone char as phosphorus fertilizer involved in cadmium immobilization in lettuce, wheat, and potato cropping, *Journal of Plant Nutrition and Soil Science*, 177(1), pp. 75–83. DOI:10.1002/jpln.201300113.

Sim, E. Y. S. and Wu, T. Y. (2010), The potential reuse of biodegradable municipal solid wastes (MSW) as feedstocks in vermicomposting, *Journal of the Science of Food and Agriculture*, 90(13), pp. 2153–62. DOI:10.1002/jsfa.4127.

Singh, R., Gupta, R. K., Patil, R. T., Sharma, R. R., Asrey, R., Kumar, A. and Jangra, K. K. (2010), Sequential foliar application of vermicompost leachates improves marketable fruit yield and quality of strawberry (*Fragaria × ananassa* Duch.), *Scientia Horticulturae*, 124(1), pp. 34–9. DOI:10.1016/j.scienta.2009.12.002.

Singh, P., Singh, M. and Tripathi, B. (2013), Glomalin: An arbuscular mycorrhizal fungal soil protein, *Protoplasma*, 250(3), pp. 663–9. DOI:10.1007/s00709-012-0453-z.

Singh, A. K., Beer K., and Kumar Pal, A. K. (2015), Effect of vermicompost and biofertilizers on strawberry I: Growth, flowering and yield, *Annals of Plant and Soil Research*, 17, pp. 196–9.

Singh, A., Kaushik, M. S., Srivastava, M., Tiwari, D. N. and Mishra, A. K. (2016), Siderophore mediated attenuation of cadmium toxicity by paddy field cyanobacterium *Anabaena oryzae*, *Algal Research*, 16, pp. 63–8. DOI:10.1016/j.algal.2016.02.030.

Sohi, S. P., Krull, E., Lopez-Capel, E. and Bol, R. (2010), A review of biochar and its use and function in soil, *Advances in Agronomy*, 105, pp. 47–82. DOI:10.1016/S0065-2113(10)05002-9.

Souza, E. M., Chubatsu, L. S., Huergo, L. F., Monteiro, R., Camilios-Neto, D., Wassem, R. and de Oliveira Pedrosa, F. (2014), Use of nitrogen-fixing bacteria to improve agricultural productivity, *BMC Proceedings*, 8(4), pp. O23. DOI:10.1186/1753-6561-8-S4-O23.

Spokas, K. A. (2010), Review of the stability of biochar in soils: Predictability of O:C molar ratios, *Carbon Management*, 1(2), pp. 289–303. DOI:10.4155/cmt.10.32.

Spokas, K. A., Koskinen, W. C., Baker, J. M. and Reicosky, D. C. (2009), Impacts of woodchip biochar additions on greenhouse gas production and sorption/degradation of two herbicides in a Minnesota soil, *Chemosphere*, 77(4), pp. 574–81. DOI:10.1016/j.chemosphere.2009.06.053.

Spokas, K. A., Cantrell, K. B., Novak, J. M., Archer, D. W., Ippolito, J. A., Collins, H. P., Boateng, A. A., Lima, I. M., Lamb, M. C., McAloon, A. J., Lentz, R. D. and Nichols, K. A. (2012a), Biochar: A synthesis of its agronomic impact beyond carbon sequestration, *Journal of Environmental Quality*, 41(4), pp. 973–89.

Spokas, K., Novak, J. and Venterea, R. (2012b), Biochar's role as an alternative N-fertilizer: Ammonia capture, *Plant and Soil*, 350(1–2), pp. 35–42. DOI:10.1007/s11104-011-0930-8.

Stadnik, M. J. and Freitas, M. B. de (2014), Algal polysaccharides as source of plant resistance inducers, *Tropical Plant Pathology*, 39(2), pp. 111–18.

Stefaniuk, M., Bartmiński, P., Różyło, K., Dębicki, R. and Oleszczuk, P. (2015), Ecotoxicological assessment of residues from different biogas production plants used as fertilizer for soil, *Journal of Hazardous Materials*, 298, pp. 195–202. DOI:10.1016/j.jhazmat.2015.05.026.

Stefaniuk, M., Oleszczuk, P. and Bartmiński, P. (2016), Chemical and ecotoxicological evaluation of biochar produced from residues of biogas production, *Journal of Hazardous Materials*, 318, pp. 417–24. DOI:10.1016/j.jhazmat.2016.06.013.

Steiner, C., Das, K. C., Garcia, M., Förster, B. and Zech, W. (2008), Charcoal and smoke extract stimulate the soil microbial community in a highly weathered xanthic Ferralsol, *Pedobiologia*, 51(5–6), pp. 359–66. DOI:10.1016/j.pedobi.2007.08.002.

Steiner, C., Garcia, M. and Zech, W. (2009), Effects of charcoal as slow release nutrient carrier on N-P-K dynamics and soil microbial population: Pot experiments with Ferralsol substrate, in: Woods, W., Teixeira, W., Lehmann, J., Steiner, C., WinklerPrins, A., and Rebellato, L. (Eds), *Amazonian Dark Earths: Wim Sombroek's Vision*. Dordrecht, the Netherlands: Springer Netherlands, pp. 325–38.

Stępień, W., Malusà, E., Paszt, L. S., Renzi, G. and Ciesielska, J. (2012), Effect of brown coal- based composts produced with the use of white rot fungi on the growth and yield of strawberry plants, in: Ecofruit. *15th International Conference on Organic Fruit-Growing. Proceedings for the Conference*. Hohenheim, Germany: Fördergemeinschaft Ökologischer Obstbau eV (FÖKO), pp. 302–6.

Stirk, W. A. and Staden, J. (1996), Comparison of cytokinin- and auxin-like activity in some commercially used seaweed extracts, *Journal of Applied Phycology*, 8(6), pp. 503–8. DOI:10.1007/BF02186328.

Struszczyk, H., Pospieszny H., and Kotliński S. (1989), Some new application of chitosan, in: Skjoak-Bræk G., Anthonsen T., Sandford P. (Eds), *Chitin and Chitosan*, Elsevier Science, London. pp. 733–42.

Sumbul, A., Rizvi, R., Mahmood, I. and Ansari, R. A. (2015), Oil-cake amendments: Useful tools for the management of phytonematodes, *Asian Journal of Plant Pathology*, 9(3), pp. 91–111.

Sun, Q., Ding, W., Yang, Y., Sun, J. and Ding, Q. (2016), Humic acids derived from leonardite-affected growth and nutrient uptake of corn seedlings, *Communications in Soil Science and Plant Analysis*, 47(10), pp. 1275–82. DOI:10.1080/00103624.2016.1178767.

Szajdak, L. W. (2016), Phytohormone in peats, sapropels, and peat substrates, in: Szajdak, L.W. (Ed.), *Bioactive Compounds in Agricultural Soils*. Cham: Springer International Publishing, pp. 247–72.

Szpak, P., Longstaffe, F. J., Millaire, J.-F. and White, C. D. (2012), Stable isotope biogeochemistry of seabird guano fertilization: Results from growth chamber studies with maize (*Zea mays*), *PLoS ONE*, 7 (3), pp. e33741 (1–16). DOI:10.1371/journal.pone.0033741.

Taktek, S., Trépanier, M., Servin, P. M., St-Arnaud, M., Piché, Y., Fortin, J.-A. and Antoun, H. (2015), Trapping of phosphate solubilizing bacteria on hyphae of the arbuscular mycorrhizal fungus Rhizophagus irregularis DAOM 197198, *Soil Biology and Biochemistry*, 90, pp. 1–9. DOI:10.1016/j.soilbio.2015.07.016.

Tejeda-Agredano, M.-C., Mayer, P. and Ortega-Calvo, J.-J. (2014), The effect of humic acids on biodegradation of polycyclic aromatic hydrocarbons depends on the exposure regime, *Environmental Pollution*, 184, pp. 435–42. DOI:10.1016/j.envpol.2013.09.031.

Thorsen, M., Woodward, S. and McKenzie, B. (2010), Kelp (*Laminaria digitata*) increases germination and affects rooting and plant vigour in crops and native plants from an arable grassland in the

Outer Hebrides, Scotland, *Journal of Coastal Conservation*, 14(3), pp. 239–47. DOI:10.1007/s11852-010-0091-6.

Toselli, M., Sorrenti, G., Marangoni, B., Innocenti, A., Baldi, E., Marcolini, G. and Quartieri, M. (2013), Effect of organic fertilization on soil fertility, tree nutritional status and nutrient removal of mature nectarine trees, *Acta Horticulturae*, 1001, pp. 303–10.

Tretjakova, R., Grebeža, J. and Martinovs, A. (2015), Research into biological characteristics of dried sapropel, in: *Environment. Technology, Resources*. Rezekne, Latvia, 1, pp. 223–7.

Troy, S. M., Nolan, T., Kwapinski, W., Leahy, J. J., Healy, M. G. and Lawlor, P. G. (2012), Effect of sawdust addition on composting of separated raw and anaerobically digested pig manure, *Journal of Environmental Management*, 111, pp. 70–7. DOI:10.1016/j.jenvman.2012.06.035.

Vaneeckhaute, C., Meers, E., Michels, E., Buysse, J. and Tack, F. M. G. (2013), Ecological and economic benefits of the application of bio-based mineral fertilizers in modern agriculture, *Biomass and Bioenergy*, 49, pp. 239–48. DOI:10.1016/j.biombioe.2012.12.036.

Vanek, S. J., Thies, J., Wang, B., Hanley, K. and Lehmann, J. (2016), Pore-size and water activity effects on survival of Rhizobium tropici in biochar inoculant carriers, *Journal of Microbial & Biochemical Technology*, 8, pp. 296–306.

Vasconcellos, R. L. F., Bonfim, J. A., Baretta, D. and Cardoso, E. J. B. N. (2016), Arbuscular mycorrhizal fungi and glomalin-related soil protein as potential indicators of soil quality in a recuperation gradient of the atlantic forest in Brazil, *Land Degradation & Development*, 27(2), pp. 325–34. DOI:10.1002/ldr.2228.

Velázquez, E., Carro, L., Flores-Félix, J. D., Martínez-Hidalgo, P., Menéndez, E., Ramírez- Bahena, M.-H., Mulas, R., González-Andrés, F., Martínez-Molina, E. and Peix, A. (2017), The legume nodule microbiome: A source of plant growth- promoting bacteria, in: Kumar, V., Kumar, M., Sharma, S., and Prasad, R. (Eds), *Probiotics and Plant Health*. Singapore: Springer Singapore, pp. 41–70.

Vessey, J. K. (2003), Plant growth promoting rhizobacteria as biofertilizers, *Plant and Soil*, 255(2), pp. 571–86. DOI:10.1023/A:1026037216893.

Vicario, J. C., Primo, E. D., Dardanelli, M. S. and Giordano, W. (2016), Promotion of peanut growth by co- inoculation with selected strains of Bradyrhizobium and Azospirillum, *Journal of Plant Growth Regulation*, 35(2), pp. 413–19. DOI:10.1007/s00344-015-9547-0.

Walpola, B. C. and Yoon, M.-H. (2012), Prospectus of phosphate solubilizing microorganisms and phosphorus availability in agricultural soils: A review, *African Journal of Microbiology Research*, 6(37), pp. 6600–5. DOI:10.5897/AJMR12.889

Wani, S. P. and Lee, K. K. (2002), Population dynamics of nitrogen fixing bacteria associated with pearl millet (*P. americanum* L.), in: *Biotechnology of Nitrogen Fixation in the Tropics*. University of Pertanian, Malaysia, pp. 21–30.

Wani, F. S., Latief Ahmad, T. A. and Mushtaq, A. (2015), Role of microorganisms in nutrient mobilization and soil health-A review, *Journal of Pure and Applied Microbiology*, 9(2), pp. 1401–10.

Wilkie, A. C., Riedesel, K. J. and Owens, J. M. (2000), Stillage characterization and anaerobic treatment of ethanol stillage from conventional and cellulosic feedstocks, *Biomass and Bioenergy*, 19(2), pp. 63–102. DOI:10.1016/S0961-9534(00)00017-9.

Yang, X., Wang, X., Wang, K., Su, L., Li, H., Li, R. and Shen, Q. (2015a), The nematicidal effect of Camellia seed cake on root-knot nematode Meloidogyne javanica of banana, *PLoS ONE*, 10 (4), pp. e0119700 (1–18). DOI:10.1371/journal.pone.0119700.

Yang, L., Zhao, F., Chang, Q., Li, T. and Li, F. (2015b), Effects of vermicomposts on tomato yield and quality and soil fertility in greenhouse under different soil water regimes, *Agricultural Water Management*, 160, pp. 98–105. DOI:10.1016/j.agwat.2015.07.002.

Yooyongwech, S., Samphumphuang, T., Tisarum, R., Theerawitaya, C. and Cha-um, S. (2016), Arbuscular mycorrhizal fungi (AMF) improved water deficit tolerance in two different sweet potato genotypes involves osmotic adjustments via soluble sugar and free proline, *Scientia Horticulturae*, 198, pp. 107–17. DOI:10.1016/j.scienta.2015.11.002.

Yu, T., Wang, L., Yin, Y., Wang, Y. and Zheng, X. (2008), Effect of chitin on the antagonistic activity of *Cryptococcus laurentii* against *Penicillium expansum* in pear fruit, *International Journal of Food Microbiology*, 122(1–2), pp. 44–8. DOI:10.1016/j.ijfoodmicro.2007.11.059

Yu, G., Ran, W. and Shen, Q. (2016), Compost process and organic fertilizers application in China, in: Larramendy, M. L. and Soloneski, S. (Eds), *Organic Fertilizers – From Basic Concepts to Applied Outcomes*. Rijeka: InTech, pp. 1–24.

Yusuf, R., Kristiansen, P. and Warwick, N. (2016), Effect of two seaweed products on radish (*Raphanus sativus*) growth under greenhouse conditions, *AGROLAND: The Agricultural Sciences Journal*, 2(1), pp. 1–7.

Zeng, D., Luo, X. and Tu, R. (2012), Application of bioactive coatings based on chitosan for soybean seed protection, *International Journal of Carbohydrate Chemistry*, 2012, pp. 104565 (1–5).

Zhang, H., Ding, W., He, X., Yu, H., Fan, J. and Liu, D. (2014), Influence of 20–year organic and inorganic fertilization on organic carbon accumulation and microbial community structure of aggregates in an intensively cultivated sandy loam soil, *PLoS ONE*, 9(3), pp. e92733. DOI:10.1371/journal.pone.0092733.

Zhao, R., Guo, W., Bi, N., Guo, J., Wang, L., Zhao, J. and Zhang, J. (2015), Arbuscular mycorrhizal fungi affect the growth, nutrient uptake and water status of maize (*Zea mays* L.) grown in two types of coal mine spoils under drought stress, *Applied Soil Ecology*, 88, pp. 41–9. DOI:10.1016/j.apsoil.2014.11.016.

Zhen, Z., Liu, H., Wang, N., Guo, L., Meng, J., Ding, N., Wu, G. and Jiang, G. (2014), Effects of manure compost application on soil microbial community diversity and soil microenvironments in a temperate cropland in China, *PLoS ONE*, 9(10), pp. e108555. DOI:10.1371/journal.pone.0108555.

Chapter 9

Improving water management in organic crop cultivation

Michael J. Goss, University of Guelph, Canada; Adrian Unc, Memorial University of Newfoundland, Canada; and Wilfried Ehlers, Georg-August University, Germany

1 Introduction
2 Key aspects of organic farming affecting availability and use of water
3 Developments in water management in organic agriculture
4 Conclusion
5 Where to look for further information
6 References

1 Introduction

1.1 Water on earth and in the atmosphere

The global water cycle gives a simple overview of how water associated with the soil, geological substrata, plants, animals and the atmosphere link to each other. As plants largely, but not uniquely, exploit water in the soil we need to understand how they are able to access this resource and know how large a volume is available. However, soil and plants also lose water to the atmosphere and hence the balance between these three parts of the cycle is critical for identifying whether and how the losses can be replenished, either in the short or long term. Water may seep through soil and into underlying rock strata, which can represent an important mechanism by which nutrients may be lost by leaching from the rooting zone of a crop. In some circumstances, this can be beneficial because it prevents the build-up of salts and the development of salinization. There are times when water needs to be removed from soils through constructed drainage schemes, for example, when there are barriers to natural drainage that cause waterlogging. This can prevent roots from exploring a sufficiently large volume of soil to anchor plants and capture enough mineral nutrients. If soils are not sufficiently permeable, incident rainfall may be so intense that water cannot infiltrate but flows from the field over the soil surface, which can lead to erosion and loss of mineral and organic matter fractions of the soil. We therefore have to consider both the static and dynamic aspects of water availability in the soil.

http://dx.doi.org/10.19103/AS.2017.0029.09

1.1.1 Water, weathering and soil formation

The first step in the process of soil formation is the weathering of parent rocks. Water contributes to this process through both physical and chemical actions. The process of physical weathering splits both rocks and minerals but does not change their chemical composition. Water plays a critical role in the splitting of rock by getting into cracks, where it expands on freezing. In arid regions, differential expansion of minerals under the heat of the sun can also split rocks. The fragments formed are transported down from higher elevations into valleys by surface water. Here the fine material can be deposited as the flow of water slows, covering the valley bottoms and effectively levelling the topographical features. After small rocks, large fragments and gravels are deposited, sand particles, which range in size from 20[1] to 2000 µm, settle out. Finer particles, ranging in size from 2 to 20 µm, silt, can also precipitate out. Silt particles tend to have a relatively large surface area-to-volume ratio and hence are easily subject to weathering.

Chemical weathering involves the breakdown of minerals by hydration, hydrolysis and dissolution. The direct effects of water are greatly enhanced by the presence of protons or hydronium ions (H_3O^+) that are derived from organic and inorganic acids. Thus, eventually, even the sparingly soluble silicates are finally broken down. The rate of mineral degradation increases with temperature. Consequently, many tropical soils are strongly developed at greater depths than those in cooler climates but at the same time they may only provide an inadequate supply of essential plant nutrients. In contrast, weathering and soil genesis in colder regions only penetrate to shallow depths of the mineral stratum.

During soil genesis, soil-specific minerals are formed, but at later stages of development these may themselves degrade. Disintegration of solid particles liberates solutes, such as cations, silicic acid, and iron and aluminium compounds. Water transports these solutes together with colloidal solids deeper into the soil or even into the rock strata below. Disintegration, displacement, precipitation and leaching are essential parts of the soil-forming process that depend on water. In temperate-humid regions, for example with limestone parent material, the first step towards soil formation is the decalcification of the rock fragments. Calcium ions leach out and the pH decreases, resulting in the decay of 'primary' minerals and the formation of 'secondary' silicate minerals that belong to the 'clay' fraction. These fine particles are on average smaller than 2 µm. During clay formation the soil colour turns brown, with the colour being dependent on the amount of organic matter and the level of oxidation of iron and manganese oxides. With time, the clay may start to migrate within the soil from near the surface to a lower layer, a soil-forming process called 'lessivage'. Another process, 'podzolization', involves the breakdown of clay by chemical weathering, and the release of sesquioxides of iron and aluminium, which can leach along with soluble humic acids to deeper layers.

In these ways, characteristic soil horizons are formed that are diagnostic of distinct soil types. In arid regions, where soil water is more likely to be lost by evaporation, water flows up towards the soil surface during the more prevalent rain-free periods. As a result, any solutes transported from the subsoil are deposited in the surface layers.

Soil genesis is accompanied by the formation of soil structure, which also is essentially dependent on soil water. Water causes some clay minerals to swell and shrink; furthermore, the soil matrix can become further divided by planes of weakness or fissures. Ice lenses,

1. In many countries, the lower limit for the size of sand is taken as 50 µm.

formed by freezing temperatures, can separate the soil matrix into aggregates of characteristic size and form. Aided by the action of plant roots and mycorrhizal fungi, aggregates made up of sand, silt and clay-sized particles can be formed, which are cemented together by inorganic oxides, amorphous aluminosilicates and organic matter (Goss and Kay, 2005). Both roots and fungal hyphae release organic mucilage that under dry conditions can enhance the stability of soil aggregates (Reid and Goss, 1981, 1982).

1.1.2 Water storage capacity and the availability of water in soil

The space between the solid particles of the soil, both within and between soil aggregates is made up of pores of different effective sizes. It is within this pore space that the storage and movement of water takes place. Just as a sponge, a clod of dry soil will absorb water. Initially, the rate that the water is soaked up may be relatively rapid but it declines with time and will eventually stop. At that point the forces in the soil sample causing the water to enter the pore space within the clod are neutralized. These forces originate on the surfaces of clay particles as 'short range' London–van der Waals forces, the negative electrostatic forces associated with clay minerals and from the adsorbed positively charged counter ions. In addition to these forces, there are capillary forces that are the result of the adhesion of water molecules to the surface of solid soil particles and the cohesive forces between water molecules. In most soils, water wets the surface of particles, which causes the surface of the water within a pore to form a concave meniscus at its boundary with the air. In some dry soils, the nature of the organic matter covering soil particles prevents the surfaces from wetting and the water may form a convex surface, effectively preventing it from entering the pores.

Because of the complexity of soil pores in terms of their structure, particularly the pore necks created where, for example, three touching particles form part of the pore, some air is entrapped as a soil profile wets up during a rainfall event. Prolonged rain will eventually result in the soil becoming saturated with water. When the rain stops, water filling the coarser pores will start to drain from the soil profile and this will continue for some time. The volume of water retained by the soil 'against gravity' after two days of drainage without further rain is commonly referred to as the field capacity. In reality, the soil commonly continues to drain slowly for longer than two days, so the water content at field capacity is really just a convenient concept for calculating water storage capacity. The other important value is the limit of water extraction by plants, the permanent wilting point. At that point, the adhesive forces are too strong for plants to absorb any of the residual water remaining in the soil. The difference between these two water contents, field capacity at the wet end of the scale and permanent wilting point at the dry end, is identified as the soil water content potentially 'available' to plants. Available water content varies with proportions of sand, silt and clay that constitute the soil. These proportions define the textural classes to which soils belong.

The total porosity of the soil is simply the proportion of the volume of voids in the total volume of soil. The sizes, shapes and volumes of individual pores are defined by the spatial distribution of the solid component of the soils and their propensity to form compact or loose aggregates. Small mineral particles, with greater surface electrostatic potentials per unit mass, associate more closely and this causes a greater proportion of the total soil porosity to occur as smaller pores. Thus, the average equivalent diameter of individual pores varies among soils. Smaller pores will retain water more strongly than larger pores; this phenomenon is known as capillarity. The energy by which a capillary can retain water can be simply expressed in terms of hydraulic head as (Or and Wraith, 2000):

$$h(m) = 14.84/r(\mu m) \tag{1}$$

where h is the height that water rises in a capillary with a radius r. Thus, for two soils with equal total porosity, the one with a larger proportion of pores in the small diameter range will retain more water. On the other hand, coarse-textured soils, such as sandy soils, which have most of their pores in the larger size domain (e.g. Unc and Goss, 2006) hold water less strongly than the smaller, capillary, pores. Thus, in general, sands and clay soils have the least available water, with silt and silty-clay loams commonly having the greatest amount of available water (Fig. 1).

1.1.3 Drivers of water movement

In the previous section we described how water moves into dry soil and now we need to consider how the forces at the surface of soil solids combine with others to cause water to move through the soil. We have also discussed the movement of water from upland areas down into the valley. In many parts of the world, where there is a lot of water cascading from one level to another, a part of the energy in the moving water is captured to generate hydroelectric power. The potential of water to do work is greater in the uplands than when it has fallen to the valley and part of that energy released can be used for turning the blades of turbines. In an electrical circuit, the rate at which the charge moves (the unit is coulomb per second) is the current I (unit is ampere) that is related to the potential difference V (unit is volts) between the battery terminals. We can write an equation to describe the movement of the charge:

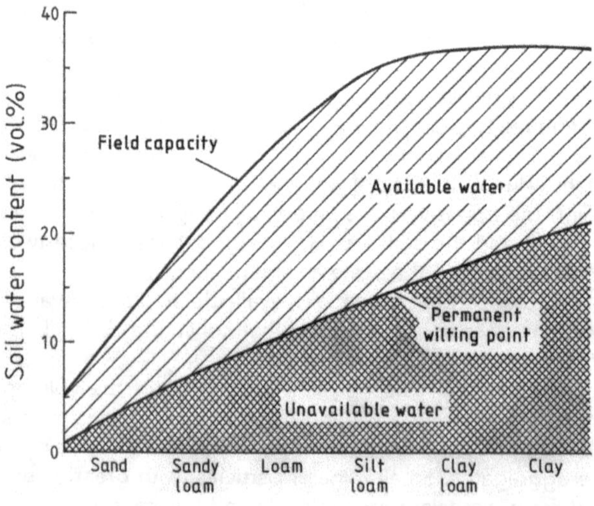

Figure 1 Changes in water content at field capacity and at the permanent wilting point as soil texture becomes finer from sand to clay. The maximum value of available water content occurs at the boundary between silt loam and clay loam texture.

$$I = V/R \quad (2)$$

where R is the resistance (unit is ohm = volt/ampere) in the wiring.

We can write a similar equation for water moving through the soil:

$$q = K(\Delta\phi/\Delta z) \quad (3)$$

where q is the volume of water moving through an area per unit of time (cm^3 H_2O per cm^2 cross-sectional area per unit of time), K is the reciprocal of the hydraulic resistance to flow, the hydraulic conductivity. The difference in water potential between two points in the soil is given by $\Delta\phi$, and this is calculated over a distance of Δz. If we make our calculations on the basis of the weight of water, this results in the unit for ϕ being centimetre, a unit of length. This can greatly simplify the calculations required to determine the flow rate of water, as the unit for K will be cm per unit of time.

As described above, ϕ is the total potential of the water. It is the summation of the matric potential, gravitational potential, the osmotic potential and the pressure potential. The matric potential results from the adsorption of water on the surfaces of mineral and organic materials constituting the solid matrix of the soil, as described above. As the soil dries, matric potential becomes the major component of the total potential. The gravitational potential results from the height of water above a datum position. Osmotic potential results from the presence of solutes dissolved in the water. For osmotic potential to play an important role the presence of a semipermeable membrane to separate two bodies of water is required. In the soil, knowledge of osmotic potential will therefore be important for water to be absorbed by plant roots, as the water has to cross the outer semipermeable membrane enclosing the cytoplasm of cells of the epidermis or the endodermis of a root (for more detail see Section 1.2). An air–water interface also acts as a semipermeable membrane, so under some circumstances the osmotic potential of the soil solution can be important for the loss of water by evaporation. However, even in soils subject to regular applications of soluble fertilizer, the concentration of ions in soil solutions is unlikely to influence the transport of water except in localized regions around fertilizer pellets applied when soils are relatively dry. Pressure potential develops if a water body is being compressed. In soils, the pressure potential is important for movement of water below the surface of the water table.

Equation 3 is known as the Darcy equation and establishes that a flow of water only occurs when there is a difference in its potential ($\Delta\phi$) between locations and the movement is from the point with the higher potential to that with the lower potential. The potential gradient ($\Delta\phi/\Delta z$) is the difference in potential divided by the distance (Δz) between locations, which could be, for example, the distance between two soil layers or soil horizons. If the topsoil is at equilibrium, no water movement takes place. If we measure the gravitational potential, Z, from the soil surface, the value of Z will decrease from Z = 0 cm to Z = –10 cm at a point 10 cm below the surface (Fig. 2). If we assume that ϕ depends only on Z and the matric potential (ψ), if there is no flow, there is no difference in total potential: $\Delta\phi = 0$ cm. Then ψ at the soil surface (ψ_0) must differ from the value at 10 cm below the surface (ψ_{-10}) by −10 cm, that is, $\psi_{-10} = \psi_0 + 10$ cm (Fig. 2a). If rainfall then occurs, the potential of water at the soil surface will increase. The rainwater will be attracted to the particle surfaces but that also reduces the strength of the attraction forces, which become less negative and hence the matric potential approaches 0. In consequence, the potential gradient will increase from the soil surface to the point 10 cm below. Water will then start to flow downwards from the soil surface. Depending on how dry the soil was at the start of the

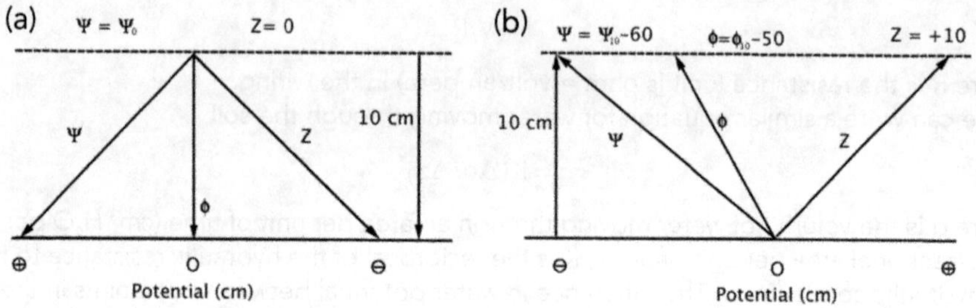

Figure 2 Components of the soil water potential under moist conditions. (a) Equilibrium conditions after rain. (b) Evaporation taking place at the soil surface.

rainstorm, the duration and intensity of the rainfall will determine the depth to which the soil wets. Some time after the end of the storm, the soil above a depth of 10 cm may reach a new equilibrium where no flow is taking place although the matric potential near the surface will again be 10 cm lower than that at the 10-cm depth. If water starts to be lost to the atmosphere by evaporation, the attractive forces at the surface of soil particles start to increase and the matric potential starts to decrease. In consequence, water will move upwards as the total potential gradient increases (becomes more negative) from the 10-cm depth (Fig. 2b).

We can now look at the field capacity and permanent wilting point in terms of matric potential. In soils with a deep unsaturated profile, and thus a thick layer above the watertable, the water held in pores at a matric potential of −330 cm is often taken as the amount at field capacity. This means that the largest water-filled pores are approximately 9 μm in diameter. At the permanent wilting point, the diameter of the largest water-filled pores is about 0.2 μm, so the matric potential is about −15000 cm. In shallower soils, the water potential at field capacity may be closer to −100 cm, but the potential at the wilting point may be less affected. Although in reality these matric potentials for the maximum and minimum ends of the plant available water in a field soil are only approximate, they are convenient parameters to use for comparing soils in the laboratory.

The growth and productivity of crops can be affected by both the excess and lack of available water.

1.1.4 Subsoil drainage schemes

When climatic conditions are such that rainfall dominates the hydrologic cycle, the soil profile can become saturated and water drains below the depth that plant roots are able to exploit the resource. This allows the water to contribute to groundwater resources. Some soils, particularly those dominated by clay content, drain so slowly that the profile can become waterlogged, and this greatly impairs crop growth and production. Field drainage schemes, which consist of perforated pipe drains that collect the water from the top metre of soil, transport this water off the field to natural watercourses or constructed drainage ditches. The required spacing of the perforated drains depends on the hydraulic conductivity of the subsoil, and in deep clay soils this can require close spacing, which can be prohibitively expensive. If the clay can be moulded, it is possible to create temporary round channels in the soil, 'mole drains', by use of a mole plough. In poorly permeable soils, effective drainage schemes with closely spaced drains can be created by back filling the trench formed when inserting the pipe drains with gravel and then drawing the mole plough through the trenches and perpendicular to them.

Mole drains can function for 3–4 years and allow a spacing of about 2 m. The evidence from fall-sown field crops suggests that the aim of the drainage scheme should be to prevent the top 50 cm of the soil from remaining waterlogged (Belford, 1981) for extended periods. Critical durations for the waterlogging taking place at depths shallower than 50 cm depend on the prevailing temperatures, which control the rate at which the oxygen levels in the soil decline (Cannell et al., 1985). For spring-sown crops, such as pea (*Pisum sativum* L.), even two days of waterlogging can result in significant reduction in shoot growth, pod formation and seed development (Cannell et al., 1979), in part because of the warmer temperatures that prevail after the crop is sown in early spring.

1.1.5 Irrigation systems

To augment the water available through rainfall and snow, it may be necessary to irrigate crops, choosing from a wide variety of schemes.

In *flood irrigation*, water flows into the soil from a supply channel at a sufficient rate to adequately cover the soil. This approach relies on the formation of a flat, even surface with a very small slope to ensure the whole field receives water. The main drawback with such schemes is the large loss of water by evaporation from supply channels and from the flooded soil.

Border strip irrigation involves forming dykes parallel to the slope that enclose strips into which the crop is seeded in rows aligned across the slope. The rows of the crop help to spread the water across the strip.

Furrow irrigation requires the formation of furrows or dykes between individual or groups of crop rows that are sown on ridges or raised beds. The water infiltrates laterally into the ridges or raised beds, which have a greater hydraulic conductivity than the flooded soil in the furrows.

In *basin irrigation* an area is surrounded by a dyke that allows water to pond but prevents it from running off the field. Ponding helps to ensure relatively uniform infiltration into soils that have poor permeability.

Sprinkler irrigation requires more technical equipment and the water is supplied under pressure so that fine droplets can be formed to simulate natural rainfall. Lines of sprinkler jets can move automatically up and down a field to apply a uniform distribution of water. Alternatively, the line of jets can be arranged to form the arm of a centre pivot, uniformly irrigating circular or semicircular areas over which the jets pass.

For crops of greater market value, *drip irrigation* can be used with water being supplied to individual plants through a drip nozzle. By limiting the area of wet soil at the surface, the system reduces the volume of water lost by evaporation before it can be used by the crop and can focus the irrigation on the soil volume containing the plant roots.

Another method for reducing unnecessary losses is to supply the water via subsurface pipes, *sub-irrigation*. Some combined systems have been developed that allow the water intercepted by a drainage scheme to be collected in a holding pond and then to be used to sub-irrigate the crop when the available water declines (Tan et al., 1999; Bonaiti and Borin, 2010).

All of these techniques for irrigation need a clear understanding of the local hydrological cycle as it pertains to the specific crop being produced. That in turn requires a knowledge of the processes involved in water use by each crop.

1.2 The soil–plant–atmosphere continuum

As identified in Section 1.1, there are three interactive components important to the hydrologic cycle: the soil, the plant and the atmosphere. The previous sections have provided an overview of the soil component and now we consider the plant component.

The water potential within plant cells is dependent on the osmotic potential, which is related to the presence of mineral ions and organic compounds, such as sucrose in the vacuole. These solutes are osmotically active and lower the water potential within the plant cell. The reference level for the osmotic potential is pure water with no solutes. Osmotic activity of solutes assumes the presence of cell membranes: the plasmalemma, which separates the protoplasm from the external medium, and the tonoplast between the protoplasm and the vacuole. These cell membranes seem to behave as if impermeable to solutes but permeable to water; however, they are capable of the facilitated transport of ions and other solutes that allow the build-up of a higher solute concentration inside than outside the cell. This difference in concentration facilitates the movement of water across the membrane through special structural units, aquaporins, within it and in consequence lowers the solute concentration inside the cell.

In a fully turgid plant cell, the protoplasm (the cellular content) is replete with water. A turgid cell, together with those in close proximity, gives the tissue some rigidity, which in turn determines the form of herbaceous plants. When the cell cannot absorb more water it is unable to induce water movement towards itself. Under those conditions the total water potential (ϕ) of the cell is zero. The osmotic potential, Ω, that has drawn water into the cell is balanced by the reaction to deformation exerted by the cell walls and surrounding turgid tissue, resisting further expansion that is required to accommodate more uncompressible water. The result is that the cell contents become pressurized. This positive pressure, the turgor pressure, increases the energy level of the water. Expressing turgor pressure in terms of the unit volume of water establishes another component of water potential, the pressure potential, P. It has as a reference level the value for water at atmospheric pressure. In a fully turgid cell at equilibrium, the pressure potential is at a maximum:

$$\phi = \Omega + P \tag{4}$$

When plant cells are subject to extreme drying, some water remains under tension within the microfibrillar structure of the cell walls. This means that we should write Equation 4 as:

$$\phi = \Omega + P + \psi \tag{5}$$

where ψ is the potential resulting from the water attraction to a surface (matric potential), this time the surfaces are within the cell wall. However, when a cell is fully turgid, ψ will be zero. Hence:

$$P = -\Omega \text{ when } \phi = 0 \tag{6}$$

The uptake of water by roots and its movement within the plant do not rely on the expenditure of metabolic energy. The water simply flows from sites of higher potential to sites of lower potential. Water flow within the plant, from the epidermis of the root through the cortex to the xylem in the central stele, and from the stele to various organs of the plant, and finally from the leaves to the atmosphere, is caused by differences in total water potential, just as in the case of water flow through the soil.

Evaporation is important in the transfer of water from the soil and from the plant to the atmosphere. Hence, before continuing with the overview of water moving through the plant, we need to consider the process of evaporation a little further. For evaporation from the soil, water changes its state from a liquid to vapour at or near the soil surface. This change requires the input of energy, most of which comes from the radiant energy of the sun. The water molecules diffuse into the air and away from the soil surface, depending on differences in the vapour pressure. The key values here are the saturated vapour pressure of the air (e_s), which is the maximum value the air can accommodate and which depends on air temperature, and the actual vapour pressure (e). The difference between e_s and e (e_s – e) is known as the saturation deficit. The larger the saturation deficit the more water vapour can be held in the air mass. In addition to diffusion, the transfer of water vapour away from the soil will be much quicker if it takes place as a mass flow, also referred to as convection. Convective vapour movement can result from local heating of the air near the soil surface or can be caused by wind. All of these determinants of evaporation, energy supply, saturation deficit and convection, depend on the climatic factors that create evaporative demand or the potential evaporation. Evaporation from moist soil is influenced to a much greater degree by atmospheric conditions than by the characteristics of the soil surface.

Solar energy reaching the earth's surface, the global or total radiation R_T, arrives as short wave radiation, ranging between 300 and 3000 nm, and may be a little less than half that reaching the outer part of the atmosphere in more humid regions but can reach up to 70% of that value in arid zones with little cloud. The instantaneous value of R_T varies according to the time of day, cloudiness, atmospheric turbidity, season, latitude, slope aspect and altitude. The fraction of R_T reflected from the surface, the fraction r, (the albedo) depends on the colour of the surface and its roughness. Of the radiation that is absorbed, a fraction is re-radiated to the atmosphere in the form of long wave radiation (R_L). The net radiation, R_N, available to keep the air and soil warm and for use by plants in photoassimilation (photosynthesis) is given by:

$$R_N = R_T(1-r) - R_L \tag{7}$$

$$R_N = G + H + LE + A \tag{8}$$

where G is the energy flux used to heat the soil and H is the fraction used to heat the air.

Only a small part of the energy flux is used for photosynthesis, A, and the remainder of the net radiation goes into evaporating water, LE. The term LE, the latent heat flux, is made up of the latent heat of vaporization, L, with a value of 2.45 kJ g^{-1} at 20°C, and the evaporation rate, E g cm^{-2} day^{-1}. The heat of vaporization is required to change liquid water into a gas. The energy flux used for evaporation is only identified if some of it is taken from the evaporating body itself, which will cool down.

Potential evaporation can be calculated from the energy balance equation (Eq. 8) but when taking into account the ease of measurement it is better to combine it with the aerodynamic aspect. Penman (1948) developed such a combination method. The Penman equation for the daily evaporation from a water surface is given by:

$$LE_P = \frac{(\Delta/\gamma)(R_N - G) + LE_a}{\Delta/\gamma + 1} \tag{9}$$

for the potential evaporation rate, E_p (g cm^{-2} day^{-1}), we can write:

$$E_P = \{[(\Delta/\gamma)(R_N - G)/L] + E_a\}/(\Delta/\gamma + 1) \quad (10)$$

where Δ is the slope of the function of saturated vapour pressure versus temperature (mbar °C^{-1}); γ is the psychrometric constant (mbar °C^{-1}); E_a (mm day^{-1}) is a ventilation–humidity term, which takes into consideration the influence of wind speed and the vapour pressure deficit of the air on evaporation; and L, RN and G are as defined for Eq. 8.

As with the process of soil evaporation, three conditions have to be met for evaporation from a plant leaf. Within the leaf, water has to change from the liquid to the vapour phase, an energy-consuming process. Again, the energy comes from the radiant energy of the sun. Secondly, a drop in vapour pressure is necessary to start the diffusion of the water molecules in the vapour phase from the intercellular spaces of a leaf through stomatal pores and out from the confines of the plant. Finally, the water vapour must be removed from the leaf surface to the atmosphere. The latter occurs by diffusion across a thin boundary layer, but that transport is greatly enlarged by the mass flow driven by the wind.

Liquid water is supplied to the leaf through the vascular bundles, which are branched subdivisions of the conducting tissue of the stems and end in the mesophyll as a fine network. Vascular bundles contain the two conducting systems of a plant, the phloem that transports organic and inorganic solutes and the xylem that mainly conducts water and mineral ions. The main conducting units of the xylem in most crop plants are the vessels. As water comes to the end of a vascular bundle in a leaf it moves into the mesophyll cells. In some plants there is a really distinct layer of closely packed columnar cells with many chloroplasts, the palisade layer, and a layer of more loosely packed cells, the spongy mesophyll. In both cases, some water moves through the cells of the mesophyll to the epidermal cells. This outer layer of cells is typically covered with a waxy layer, the cuticle, that is not impervious to water and is responsible for some 3–5% of the total water lost from the leaf. However, the larger proportion of the water in the mesophyll evaporates from the cell walls into the intercellular spaces. The largest of these spaces are associated with structural pores in the epidermal layer, the stomates, and form the substomatal cavity. Stomates provide the main conduits for water vapour to escape from the leaf. Importantly, stomates are the essential openings for carbon dioxide (CO_2) to enter the leaves to be available for photosynthesis. Two specialized cells of the epidermis, the guard cells, form stomatal pores. Because of differential thickening of the cell wall, the guard cells subtend a large pore between them when fully turgid, but as the cells lose turgor pressure the pore size gets smaller. By closing their stomates, plants can greatly reduce their water loss as the supply from the soil declines. However, the closing of stomates also cuts off the supply of CO_2 for photosynthesis.

We can now track the pathway of water through the soil to the root surface, into the root and from there through the stem into the leaves and finally into the atmosphere, the transpiration stream. The component potentials in plant cells vary over a much greater range of values than do those in moist soil. Depending on how turgid the cells are, ϕ for plant cells is much lower (approximately –10000 cm H_2O) than ϕ for moist soil (ranges between −100 and −1000 cm H_2O). As a consequence of the potential drop between soil and plant, the water will tend to move automatically from the soil towards the root. Once at the root surface it can be absorbed at any point along the root, although the specific water uptake rate, UR, will be greater in the region of the root hair zone, where on the one hand cells are not suberized as a means of impeding radial flow, and on the other hand the surface area of the root hair cell is enhanced by the presence of the hair and by the fact that the root hair can penetrate some distance into the soil away from the drier interface with the root.

Water may be absorbed into root hairs of the epidermal cells or, if the epidermis has been lost, by entering into passage cells of the exodermis as the root ages. In either case, water crosses the semipermeable plasmalemma of the cells and enters the cytoplasm. Once it is within the cytoplasm it can pass from cell to cell by diffusion through the plasmodesmata, which are cytoplasmic connecting strands that link cell through adjoining cell walls. This is the symplastic pathway. Alternatively, water can move through the cell walls and intercellular spaces of the root but not cross a plasmalemma without immediately entering the cells. As a bulk flow, water can be conducted between cells, along neighbouring cell walls or through intercellular air spaces. This route is called the apoplastic pathway. Some water may move across the root via a mixture of the two routes but always crosses the plasmalemma or even the membrane surrounding the internal vacuole of a cell, the tonoplast. This is known as the transmembrane pathway.

Regardless of which of these three pathways carries the water across the root it will eventually arrive at the innermost layer of the cortex, the endodermis, where it has to enter the cell cytoplasm because the apoplastic pathway is impeded by suberization of the cell wall. The flow of water into the xylem of the root from the endodermis is via the transmembrane pathway, where water moves because of the osmotic potential. The vessels of the xylem have no protoplasm and the water enters the apoplastic pathway. In this part of the apoplastic pathway it is the cohesion between water molecules that maintains a constant column of water between the root and the leaves. The evaporation of water driven by the vapour pressure deficit lowers the water potential in the leaves that then drives the water movement through the plant.

It is appropriate at this point to identify how much water is involved in evapotranspiration (ET) from a field over the main growing season. A potential ET of 700 mm per season is fairly typical in Europe and North America. So, from a 1-ha field that ET is equivalent to 7×10^6 L or 7000 t of water. In herbaceous plants, about 80% of the fresh weight is due to water, which is used to provide rigidity to the structure and a relatively small amount, about 1%, is used in photosynthesis. The main role of transpiration and evaporation from the leaf is to keep the leaf temperature from exceeding that which causes the efficiency of the CO_2 fixation process to decline. Evaporation from the soil will cool the soil surface and atmospheric boundary layer above it but will have less benefit for processes in the leaf compared with transpiration.

1.3 The soil water balance

All the water that is lost by evaporation from the soil (E) and is transpired by the plants (T) originates from the precipitation (W) falling on a field. Some precipitation falls onto leaf surfaces and evaporates directly rather than entering the soil; this is referred to as interception (I). Of the water that enters the soil some may drain to below the rooting zone of the crop and be lost (D), in quantities governed by soil hydraulic parameters as described in Sections 1.1.2 and 1.1.3. Some (R) may simply run off over the surface of the soil and enter a watercourse. Depending on the soil water storage capacity, there may be changes in the amount of water (ΔS) in the soil at any time. All these components can be expressed in terms of the equivalent depth (mm) of water

$$W = E + T + I + D + R + \Delta S \tag{11}$$

Determining E and I separately from T is difficult when the crop covers the soil and it is convenient to combine the values into the ET of the plants. From the field water balance, if run-off does not occur, then:

$$ET = W - \Delta S - D \tag{12}$$

In Table 1 the components of the water balance, identified in Eq. 12, are shown for a clay soil under winter wheat. The values were obtained from measurements in the field comparing crops grown following removal of the harvest residues or their incorporation by ploughing. The main difference was the greater water loss to field drains when the residues were incorporated, possibly because of greater conductivity of the topsoil resulting from the associated tillage. The Penman equation used to calculate potential ET assumes loss from a short grass sward and underestimates losses from a relatively tall crop, such as wheat by approximately 15% (Ehlers and Goss, 2016). In the experiment described, the values were between 0 and 21% underestimated.

1.4 Efficiency of water use by crops

The fact that stomates are the point of control for water leaving the plant, and consequently for CO_2 entering, draws attention to the efficiency with which the plant makes use of these two resources. The ratio of the dry matter produced by a crop (DM) to the amount of water used (WU) provides a measure of that efficiency. If we simply consider the amount of water that moves through the plant in transpiration (T), the ratio DM/T is the transpiration efficiency (TE), which can be expressed in units of grams of plant dry matter per litre of water transpired or kg dry matter per m^3 H_2O. However, from a crop management perspective it is easier to determine the ratio of amount of shoot dry matter produced to the total water lost from the field in ET. That is the water use efficiency (WUE). We could also take into account the other ways that the water, which landed on the soil, has been redistributed and left the field without directly involving the plants of the crop.

Table 1 Components of the field water balance of a clay soil under winter wheat. The harvest residues were removed or ploughed under. Measurements of soil water content and soil water potential were made from the end of tiller formation in spring until harvest. The potential evapotranspiration was calculated from an automatic weather station on the site and changes in water storage determined by neutron scattering. Drainage was calculated from downward flows based on soil water potential gradients. The water used by the crop was equated to the evapotranspiration term in the water balance equation

	Harvest residues removed	Harvest residues incorporated
Precipitation (mm)	148	148
Change in soil storage (mm)	−107 ± 10.5	−105 ± 2.2
Drainage	30 ± 1.3	44 ± 2.4
Water used by crop from water balance (mm)	225 ± 11.8	209 ± 4.6
Potential evapotranspiration using the Penman equation (mm)	204	204

Starting with the water use efficiency defined above we can write:

$$WUE = \frac{DM}{ET} \tag{13}$$

Remembering that the transpiration efficiency (TE) is DM/T and therefore DM = TET, we can rewrite Eq. 13 as:

$$WUE = \frac{TET}{ET} = \frac{TET}{E+T} = \frac{TE}{1+E/T} \tag{14}$$

In this equation E/T is the ratio of evaporation from the soil to the transpiration through the plant. We can therefore add the water lost to run-off and drainage into the water balance equation and can describe the water use efficiency for plant production WUE_p:

$$WUE_P = \frac{TE}{1+(E+R+D)/T} \tag{15}$$

and finally, if we consider water that has been used by competing weeds T_{weed} we can write:

$$WUE_P = \frac{TE}{1+(E+R+D+T_{weed})/T} \tag{16}$$

Equation 16 allows different management options to be considered to ensure that the water available to a crop can be used effectively. Ensuring that the maximum amount of water possible can be stored in the soil and not lost in run-off is clearly an important objective as is reducing evaporation when water is scarce.

2 Key aspects of organic farming affecting availability and use of water

The understanding of how water is stored in soil and used by crops allows us to consider how management practices can affect these key aspects.

2.1 Soil management

Clearly, any changes made to the soil can have considerable impact on both the potential for water to enter the soil and to be held available to crops. As indicated in Section 1.1.2 a key aspect in the storage of water is the distribution of pores of different sizes within the soil. These pores can be located within and between the structural units of the soil.

2.1.1 Soil structure

Organic matter is essential for the formation and stabilization of soil aggregates because of its binding and cementing properties. Increasing the organic matter content of soil is generally associated with improved soil structural stability because the aggregates are more resistant to tension-free water, so that pores between and within them are more stable. Fungal hyphae and roots can enhance the creation of soil aggregates by enmeshing soil mineral particles (Miller and Jastrow, 2000). Arbuscular mycorrhizal fungi (AMF) form symbiotic relationships with about 90% of land plants, including most but not all crop plants. The dead hyphae of these fungi, along with decaying root material, can form the core of soil aggregates (Goss and Kay, 2005), but living hyphae produce glycoproteins, glomalin and glomalin-related soil proteins (Wright and Upadhyaya, 1996; Rillig, 2004b), which enhance stability through their cementing properties. 'Sticky' by-products, such as polysaccharides, of the activity of both bacteria and fungi as well as the release of root exudates cement aggregates of soil mineral and organic particles and thus lead to greater aggregate stability. Any management practices that enhance microbial activity and thus the production of polysaccharides tend to give soil greater structural resilience. Importantly, when plants dry the soil during transpiration, the stabilization of soil can increase. However, just adding organic matter to soil is not sufficient to increase stability (Fig. 3a), so there is also an organic matter quality aspect. The increase in aggregate stability reduces their disruption (Fig. 3b) and enhances the infiltration rate (Fig. 3c). After rainfall, the infiltration was found to decline with bigger changes occurring in soil with greater stability (Fig. 3c). A small amount of particle detachment from aggregates, possibly in combination with organic matter, had a large impact on the ability of soil to allow water to infiltrate (Fig. 3d). The greater reduction in infiltration likely resulted from blocked pores, whereas the disruption of the more stable aggregates still permitted good infiltration (Fig. 3d). However, Reinhart et al. (2015) found that the relationship between stability of aggregates and infiltration was not constant and for larger aggregates it could even be negative. The effects of the variation in aggregate stability resulting from the different quality organic material on infiltration are evident from Fig. 4; the results shown in Figs. 3 and 4 are from the same study by Ekwue (1992).

Many comparative assessments of organic farms with conventional operations suggest that the soil organic matter (SOM) content is generally greater under organic agricultural systems, although that does not always have to be the case (Huntley et al., 1997; Trewavas, 2004; Crittenden and de Goede, 2016; see also Watson, Chapter 3). The general benefits to water management of maintaining a large SOM in soil are also evident in terms of water availability. SOM has the net effect of reducing soil bulk density (Hudson, 1994) and attracts water molecules to its surfaces. Even though some of the water associated with organic matter may not be available to plants, Hudson (1994) established that for soils ranging from sands to silty clay loams, the water content at field capacity was increased more by an increase in SOM than it was at the permanent wilting point, thus increasing the available water content. An increase in SOM not only enhances water storage in soil micropores, but the greater development of stable aggregates will also lead to a larger proportion of inter-aggregate spaces, soil mesopores. These pores hold water at relatively large potentials between −2.5 and −300 cm (Regelink et al., 2015) and hence are able to drain water more efficiently (Boyle et al., 1989). The presence of free-draining inter-aggregate spaces and channels formed by soil fauna, such as earthworm, reduces the risk of ponding with all the negative effects on infiltration, soil aeration and related chemical and biological functions.

A decline in aggregate stability can lead to a number of changes in soil structure, including the formation of surface crusts and a dense plough layer which may then be more easily compacted further by heavy machinery. Surface crusts are commonly produced by the energy of raindrops impacting soil surface aggregates and by the slaking and dispersion

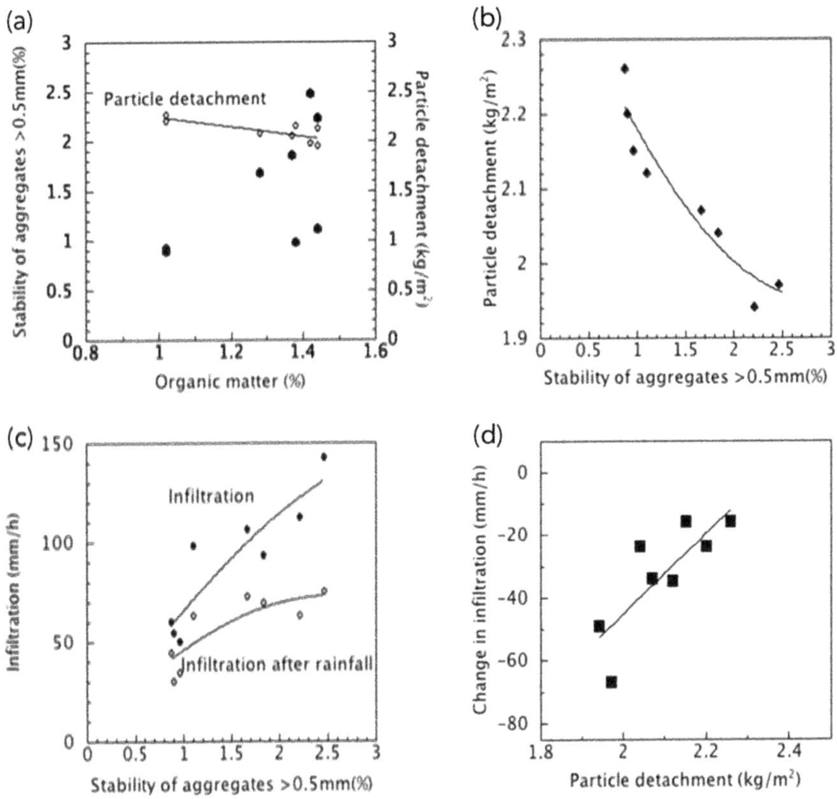

Figure 3 Importance of organic matter for soil structural stability and infiltration. (a) Relationship between soil organic matter and the stability of soil aggregates (closed circles). Although the stability tends to be greater at larger soil organic matter content, the values were not correlated. Nevertheless, the particle detachment (open circles) was inversely related to organic matter content (R2 = 0.89, $p < 0.001$). (b) Particle detachment is strongly related to aggregate stability. (c) Relationship between infiltration into soil and aggregate stability (closed circles) and after rainfall (open circles), the infiltration declined with bigger changes occurring in soil with greater stability. (d) The change in infiltration was inversely related to the amount of particle detachment. Based on Ekwue (1992).

(Folorunso et al., 1992; Carrizo et al., 2015) of soil particles in the presence of tension-free water on the soil surface. Importantly, surface crusts can greatly impair the infiltration of water from rainfall, irrigation or snowmelt, resulting in more loss in run-off from the soil. Surface sealing can have the same effect and can result from the surface application of liquid manure rather than solid manure (Unc and Goss, 2006). Increases in soil density in the plough layer horizon can result as the soil naturally consolidates with changes in soil water content, but it can also result from compaction due to wheeled traffic associated with sowing, pest and weed control, nutrient application and harvesting (Hamza and Anderson, 2005). The treading of soil by grazing animals, as well as when they are herded between paddocks, can result in soil compaction of grassland (Hamza and Anderson, 2005).

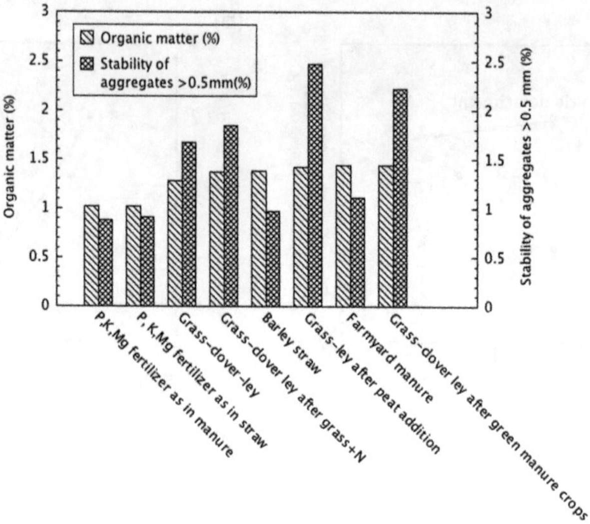

Figure 4 The variation in organic matter and associated stability of soil aggregates resulting from contrasting organic and inorganic amendments applied in agronomic treatments. Based on Ekwue (1992).

2.1.2 Tillage

Tillage systems have been designed to produce a suitable seedbed for crops as well as control weeds and incorporate organic amendments. What has been clear for more than a century (Sturtevant, 1887) is that the use of inversion tillage, for example, mouldboard ploughing, for weed control and seedbed preparation is detrimental to the soil. Much of the negative impact comes from the increased breakdown of SOM and the reduction in beneficial soil organisms, including AMF and earthworms.

In North and South America and the Mediterranean basin, no-till techniques have been developed and refined to plant crops without prior tillage and restricting any soil disturbance to the layer above the depth of seed placement. As no-till soils are cooler and wetter in the early spring season, this can lead to retarded and reduced mineralization and nitrification. Consequently, in Northern Europe emphasis has been more on shallow or reduced tillage practices (e.g. Crittenden et al., 2015). In both cases, the lack of inversion allows retention of harvest residues on the soil surface to protect it from raindrop impact, slaking and dispersion, thereby enhancing infiltration and reducing the risk of both wind and water erosion of the soil.

Although tillage may lead to an immediate increase in total soil porosity in the tilled layer, mainly in form of larger inter-aggregate pores (Lipiec et al., 2006), this is often counterbalanced by more compaction in the subsurface soil layer in comparison with a no-till management (Tebrügge and Düring, 1999). Moreover, tillage-induced soil porosity is not resilient as the plough layer compacts easily during the growth season especially when using heavy machinery. On the other hand, repeated mechanical disturbance and enhanced aeration due to repeated tillage increase the breakdown of SOM and thus breakdown of soil aggregates with the concurrent loss of the biological functions associated with the lost aggregates (Jiang et al., 2011). This can cause soils to consolidate and become compact more readily after tillage (Tebrügge and Düring, 1999), eventually causing them to have a smaller water storage capacity. Moreover, consistent tillage at a given depth can lead to formation of a compact layer, a plough pan, that delays or stops water infiltration into the deeper soil layers, reducing the water storage in the entire soil profile, often leading to water saturation and erosion of the top layer. Nevertheless, by omitting intensive tillage, efficient weed control, especially of perennials, is limited under conditions of organic farming in temperate-humid climates (Zikeli and Gruber, 2017). Future strategies involving the occasional no-till seeding of grain legumes (Massucati and Köpke, 2011) and the use of 'bioherbicides' (Giepen et al., 2014) applied with precision farming measures (Ammann, 2009) may help to overcome these limitations.

Earthworms are commonly classified according to their lifestyle. Epigeic earthworms live close to the surface, where they are important for incorporating surface litter into the soil. In contrast, anecic species live within the soil and characteristically form vertical burrows that are permanent and can be important in the rapid infiltration of water to depth (Ehlers, 1975). The anecic group will collect material from the soil surface, and pull it deep into their burrows before eating it, but they will explore the different layers for nutrients. Endogeic earthworms also tend to live below the surface but their burrows are mainly horizontal within a soil layer and are not permanent. Differences in earthworm numbers and in the predominance of different groups can exist within fields under the same management treatments, largely because of the impacts of tillage (Crittenden and de Goede, 2016). Differences in earthworm populations between management systems also tend to reflect the intensity of tillage (Ehlers, 1975; Barnes and Ellis, 1979; Berry and Karlen, 1993; Crittenden and de Goede, 2016), the length of leys (Scullion et al., 2002; Han et al., 2015) and extent of natural and improved grassland (Fraser et al., 1994). Scullion et al. (2002) found in an extended survey that the most consistent difference between organic and conventional farmland was the biomass of the earthworm population, which was more frequently greater under organic soil management. However, the smaller biomass under conventional management comprised fewer but more mature individuals. No consistent differences between these management systems were found in the deep burrowing anecic worms, which contrasts with the consistently smaller number in ploughed soil compared with no-till land (Ehlers, 1975; Barnes and Ellis, 1979).

One other important tillage practice is the use of equipment to minimize the slope and to guide water to areas where it can be used, thus managing water storage and preventing run-off. Repeated contour tillage enables a semblance of terracing to be developed. Forming laneways that can be put under permanent cover (grassed waterways) can prevent uncontrolled and potentially erosive water loss. On a smaller scale run-off zones can be guided into pits prepared for individual or groups of plants, called the Zaï farming approach.

2.1.3 Nutrient supply

A key aspect of the field water balance is the proportion of water passing to the atmosphere in transpiration relative to that evaporated from the soil surface. In the leaf, there is a balance between the water that passes out through the stomates and the inwards diffusion of CO_2. The factor that influences the relative water loss from the soil through evaporation and transpiration is the proportion of the soil that is shaded by leaves, strictly the area of green leaf covering unit area of soil, the leaf area index (LAI). As a plant canopy develops, the area of ground covered increases with each new leaf but quite quickly the newer leaves shade a part of the older leaves so the LAI increases more slowly. In field crops, once the ground is fully shaded by the leaves, the radiant energy will largely be used in transpiration of water by the plant rather than evaporation from the soil. The rate of development of LAI at the start of the main growing season is therefore important for the effective use of water in the soil. When the LAI has a value of 1, the transpiration rate of the plant is about 40% of its maximum potential.

Depending upon the crop, as LAI reaches a value between 3.5 and 4.5 its transpiration is about 90% of its maximum potential rate. Over this same range of LAI, the net radiation component of soil evaporation may decline by 3–5% of its maximum potential rate, with a further reduction associated with the aerodynamic term (Ritchie, 1972). Both the LAI and the net assimilation rate of the leaves depend on the nutritional status of the plant. Although nutrients can be supplied to conventionally farmed crops as very soluble mineral fertilizers, organic operations depend on materials that need to be mineralized or are only sparingly soluble. In consequence, this could be an important constraint to optimizing production if the start of the growing season is cool and wet. As organic matter releases its nutrients more slowly over the growth season this can limit the development rate of LAI. Improved yields have also been the result of extending the period for which leaves remain green, which may also require enhancing the nutrient supply towards the end of season or preventing early senescence from pest and disease attack.

The selection of crops and the means to protect them against competition, pests and diseases is the purpose of cropping systems.

2.1.4 Crop rotations

Crop rotation is the ordered and planned cycling of crops over time in an arable field; the crops tending to be grown in pure stand with the plant species involved being annuals, biennials or perennials (Ehlers and Goss, 2016). Ideally, the crops selected will make optimum use of the local environmental conditions, taking account of solar radiation, temperature, water supply and soil fertility. Consideration needs to be given to having species that contrast in rooting habit, such as depth and form (tap or fibrous root system). In this way, a crop can take up nutrients that leached below the rooting system of a predecessor. Taproots can generate continuous pores in the soil that can provide pathways for roots of subsequent crops (Cresswell and Kirkegaard, 1995; Han et al., 2016) to access water held in subsoil horizons by avoiding the penetration resistance of the bulk soil (Gaiser et al., 2012). The biopores produced may also allow water to infiltrate rapidly to depth (Cresswell and Kirkegaard, 1995; Kautz, 2014). Fibrous root systems are commonly considered to support soil aggregation and the stabilization of structure, thereby increasing water-holding capacity.

The crops selected may be sequences of plants, all providing saleable components, such as seed, edible leaf, stem fibres and storage roots, or include crops that are grown to enhance or protect the soil. The latter may not be grown 'in season', so that they will not survive over a cold winter, but their shoots will still protect the soil from raindrop impact and the nutrients taken up during their growth period will be recycled. Staple crops, such as wheat, barley, maize, oilseed and grain legumes, may form the main focus of the rotation, but mixtures, such as alfalfa-grass or grass-clover, may also be grown for fodder and support sustainability goals.

In North America and Europe, winter cover crops have been extensively investigated as a means of protecting the soil from the erosive forces of wind and water as well as to increase SOM, thereby enhancing water infiltration and storage. Some of the species used include those that are grown out of season, such as spring oat, but others, such as rye, have to be killed before the next crop can be sown. There is evidence that, with the appropriate selection and prolonged use of winter cover crops, there can be significant increases in SOM, particularly if associated with no-till (Blanco-Canqui et al., 2011, 2015). The greater SOM content can also be associated with improved stability of soil aggregates (Hermawan and Bomke, 1997). Folorunso et al. (1992), working with soils that tended to form a surface crust, observed considerable reductions in the surface strength in the presence of cover crops compared with their absence. Abdollahi and Munkholm (2014) found that the resistance to penetrometer was significantly ($p < 0.05$) smaller between 32 and 38 cm below the soil surface where the brassica cover crop fodder radish (*Raphanus sativus* L. var. *oleiformis*) had grown than where it was not. In their experiment, the compacted zone was associated with a plough pan, considered to have predated the reported experiment (Abdollahi and Munkholm, 2014). In a related experiment, total porosity and air-filled porosity for pores >30 μm diameter at 12 to 16-cm depth in the plough layer horizon were significantly reduced under reduced tillage systems (no-till and harrowing). Irrespective of the tillage treatment, Abdollahi et al. (2014) found that growing a winter cover crop created continuous macropores, which enhanced permeability to air and water and reduced the impedance to root growth. Chen and Weil (2010) investigated root growth of cover crops through compacted soil and found that the best penetration was observed for fodder radish followed by rapeseed, whereas rye was the least effective. In their experiments, Folorunso et al. (1992) found that either the steady-state rate of infiltration or the cumulative total of water infiltrated over 6 h was greater with cover crops than without. The residues from cover crop shoots can provide a mulch to reduce soil evaporation, aid infiltration as well as reducing run-off (Blanco-Canqui et al., 2012) and provide protection from erosive forces (Unger and Vigil, 1998). The plants can also help dry the soil surface layers in spring to aid early planting (Unger and Vigil, 1998).

Cover crops can also help in controlling weeds by competing with their seedlings for light, water and nutrients (Teasdale and Mohler, 2000; Kolodziejek and Patykowski, 2015) or by the release of allelopathic chemicals (Reberg-Horton, 2005; Blanco-Canqui et al., 2015). The physical presence of mulch comprised of cover crop material can adversely affect weed emergence, depending on the proportion of the ground being covered (comparable with LAI) and the solids component of the material (Teasdale and Mohler, 2000).

Pathogen and pest control are important factors in selecting crops and the order in which they are to be grown. Selections can be made to prevent pests and diseases building up in the soil and locality by separating two susceptible crops with one that is not susceptible.

3 Developments in water management in organic agriculture

Many aspects of agricultural production require change if there is to be sufficient food to meet the demands of another almost two billion people by 2050. The land area required for production is declining because of urbanization and land degradation. Salinization, compaction and desertification of soils are increasing and we recognize the damage to the earth that is associated with the clearing of forest. Furthermore, the land that could be brought under production is often marginal for arable agriculture. The conclusion is that the productivity per unit area of land has to increase. At the same time, farms have to grow more produce but without further impacting the wider environment or creating food that is itself of poor or even harmful quality. The build-up of toxic metals in soil and activities that can lead to increased levels of antibiotic resistance genes in soil microbes have to be prevented. Solutions have to be found to prevent invasive species taking over as new weeds and halting the threats to crops or stored produce from migrating insects. But this cannot be at the risk of harmful effects to wildlife or humans. Contamination of water resources with plant nutrients, pharmaceuticals and hormonally active compounds has to stop. More importantly, the concerns over water resources for agricultural production continue to mount as the competition with the supply of potable water for cities becomes more acute.

The required intensification of production must extend to those areas, where at present it is resource limited and levels must be maintained where the impacts of agriculture have contributed to global warming and environmental degradation. Plant breeding is vitally important so that improved water use efficiency can come from enhanced transpiration efficiency as well as from more efficient irrigation, crop protection and weed control. Additionally there is a need to enable plants to be more effective in growing in soils with increasing levels of salinity.

3.1 Managing soil limitations through enhanced soil biodiversity

One of the most exciting developments in the last twenty years has been the ability to identify the different groups of organisms that interact within the rooting zone of plants. The new generation of molecular techniques allows the components of local food webs, previously hidden in the soil, to be identified and quantified. The understanding of the roles that different groups have in the soil is beginning to make possible the identification of practices, which can be adopted in arable agriculture (Goss et al., 2017). Local soil and environmental conditions and plant species will likely influence the range, diversity and functioning of microorganisms across the soil–plant continuum (Tahtamouni et al., 2016). Soil management practices that favour stable microbial ecosystems may be critical for long-term sustainability of agricultural production (Lucero et al., 2014). Landmark papers have identified the critical activities of AMF in natural ecosystems that determine the community dynamics of higher plants (van der Heijden et al., 1998; Helgason et al., 1998; van der Heijden and Horton, 2009). Meanwhile, the contribution that AMF make to enable their host plants respond positively to threats from both biotic and abiotic stresses is becoming clearer, as is the range of abiotic stresses that plants can gain protection against through their symbiotic community. These contributions include resistance to drought (Augé et al., 2001; Augé, 2004), salinity (Liu et al., 2016), toxic metals (Ahmed et al., 2006) and metalloids (Alho et al., 2015), enhanced supplies of nutrients (Al-Karaki and Clark, 1998) and defence against soil-borne pathogens (Akhtar and Siddiqui, 2008). There is also evidence that weed competition can be affected (Qiao et al., 2016). But we also know that there are interactions in the soil between AMF and other organisms, particularly the so-called helper-bacteria, which enhance the potential for AMF to colonize new plants and improve their ability to support the host. A key development has been the recognition that, via the fungal component of a mycorrhiza present in the soil, many host plants can be linked together and communicate with each other. It appears possible that knowledge of one host being attacked is communicated to all linked hosts, which then start the process of initiating production of compounds that can counter the perceived threat.

The practices that provide most support for AMF can be divided into those that make the soil more amenable to their survival and those that provide more mycotrophic hosts (Goss et al., 2017). Reduced non-inversion tillage and no-till cultivation dominate the first category together with the retention of harvest residues. Critical in the second category is the development of rotations that sequence crops, which are members of the grass family (Poacea), with dicots, particularly legumes (Fabaceae), together with using cover crops to prevent soil being devoid of growing plants for extended periods. Non-mycorrhizal break crops in the rotation have to be followed by strongly mycorrhizal plants to maintain the efficiency of the whole system.

4 Conclusion

Improvement in water management in organic production systems requires the adoption of practices that would be considered appropriate for building soil quality with respect to nutrients: build the organic matter levels in the soil and protect it from the effects of erosion and excess wetness. The move away from inversion tillage, always using the mouldboard plough, to shallower soil disturbance, which keeps residues on the surface, balances the adoption of selected rotations that can include cover crops. The evidence points to these practices building SOM; supporting stable, functional, microbial communities and preventing and alleviating compaction and erosion. These approaches help to reduce run-off and support better infiltration and storage of rainwater and irrigation water. The choice of tillage system, crops and cropping sequences can also help in the integration of AMF into the farming system, which can help protect the crops not only from water shortage but also from many other biotic and abiotic stresses.

5 Where to look for further information

The following organizations are involved in research related to soil conditions and plant growth applicable to organic agricultural practices:

ETH Zurich University (https://www.ethz.ch/en/the-eth-zurich/sustainability/research-for-sustainable-development/natural-resources.html)
Rodale Institute (https://rodaleinstitute.org)
The Organic Agriculture Centre of Canada - Dalhousie University (https://www.dal.ca/faculty/agriculture/oacc/en-home.html)
USDA Agricultural Research Service National Laboratory for Agriculture and The Environment (https://www.ars.usda.gov/midwest-area/ames/nlae/)
USDA Natural Resources Conservation Service (https://www.nrcs.usda.gov/wps/portal/nrcs/main/national/water/)

Specific information on water and plant production is contained in Ehlers, W. and Goss, M. (2016). *Water Dynamics in Plant Production*. 2nd Edition. CAB International, Wallingford, UK. This book contains an extensive reference list on water in the soil–plant–atmosphere continuum and also gives an account of the processes of water requirements together with water uptake and use by crop plants.

The potential for supporting the role that mycorrhizal fungi play under modern agricultural practices is evaluated in the *Functional Diversity of Mycorrhiza and Sustainable Agriculture. Management to Overcome Biotic and Abiotic Stresses*, written by Goss, M. J., Carvalho, M. and Brito, I. (2017). Academic Press, Elsevier, San Diego, CA, USA. Further developments will be available through the Institute of Mediterranean Agriculture and Environmental Sciences, University of Évora, Portugal (http://www.icaam.uevora.pt.).

6 References

Abdollahi, L. and Munkholm, L. J. (2014). Tillage system and cover crop effects on soil quality: I. Chemical, mechanical, and biological properties. *Soil Science Society of America Journal* 78, 262–70.

Abdollahi, L., Munkholm, L. J. and Garbout, A. (2014). Tillage system and cover crop effects on soil quality: II. Pore characteristics. *Soil Science Society of America Journal* 78, 271–9.

Ahmed, F. R. S., Killham, K. and Alexander, I. (2006). Influences of arbuscular mycorrhizal fungus *Glomus mosseae* on growth and nutrition of lentil irrigated with arsenic contaminated water. *Plant and Soil* 283, 33–41.

Akhtar, M. S. and Siddiqui, Z. A. (2008). *Glomus intraradices, Pseudomonas alcaligenes*, and *Bacillus pumilus*: effective agents for the control of root-rot disease complex of chickpea (*Cicer arietinum* L.). *Journal of General Plant Pathology* 74, 53–60.

Al-Karaki, G. N. and Clark, R. B. (1998). Growth mineral acquisition and water use by mycorrhizal wheat grown under water stress. *Journal of Plant Nutrition* 21, 263–76.

Alho, L., Carvalho, M., Brito, I. and Goss, M. J. (2015). The effect of arbuscular mycorrhiza fungal propagules on the growth of subterranean clover (*Trifolium subterraneum* L.) under Mn toxicity in *ex situ* experiments. *Soil Use and Management* 31, 337–44.

Ammann, K. (2009). Why farming with high tech methods should integrate elements of organic agriculture. *New Biotechnology* 25, 378–88.

Augé, R. M. (2004). Arbuscular mycorrhizae and soil/plant water relations. *Canadian Journal of Soil Science* 84, 373–81.

Augé, R. M., Stodola, A. J. W., Tims, J. E. and Saxton, A. M. (2001). Moisture retention properties of a mycorrhizal soil. *Plant and Soil* 230, 87–97.

Barnes, B. T. and Ellis, F. B. (1979). Effects of different methods of cultivation and direct drilling, and disposal of straw residues, on populations of earthworms. *Journal of Soil Science* 32, 323–33.

Belford, R. K. (1981). Response of winter wheat to prolonged waterlogging under outdoor condition. *Journal of Agricultural Science, Cambridge* 97, 557–68.

Berry, E. C. and Karlen, D. L. (1993). Comparison of alternative farming systems. II. Earthworm population density and species diversity. *American Journal of Alternative Agriculture* 8, 21–6.

Blanco-Canqui, H., Claassen, M. M. and Presley, D. R. (2012). Summer cover crops fix nitrogen, increase crop yield, and improve soil–crop relationships. *Agronomy Journal* 104, 137–47.

Blanco-Canqui, H., Mikha, M. M., Presley, D. R. and Claassen. M. M. (2011). Addition of cover crops enhances no-till potential for improving soil physical properties. *Soil Science Society of America Journal* 75, 1471–82.

Blanco-Canqui, H., Shaver, T. M., Lindquist, J. L., Shapiro, C. A., Elmore, R. W., Francis, C. A. and Hergert, G. W. (2015). Cover crops and ecosystem services: Insights from studies in temperate soils. *Agronomy Journal* 107, 2449–74.

Bonaiti, G and Borin, M. (2010). Efficiency of controlled drainage and subirrigation in reducing nitrogen losses from agricultural fields. *Agricultural Water Management* 98, 343–52.

Boyle, M., Frankenberger, W. T. and Stolzy, L. H. (1989). The influence of organic matter on soil aggregation and water infiltration. *Journal of Production Agriculture* 2, 290–9.

Cannell, R. Q., Belford, R. K., Blackwell, P. S., Govi, G. and Thomson, R. J. (1985). Effects of waterlogging on soil aeration and on root and shoot growth and yield of winter oats (*Avena sativa* L.). *Plant and Soil* 85, 361–73.

Cannell, R. Q., Gales, K., Snaydon, R. W. and Suhail, B. A. (1979). Effects of short-term waterlogging on the growth and yield of peas (*Pisum sativum*). *Annals of Applied Biology* 93, 327–35.

Carrizo, M. E., Alesso, C. A., Cosentino, D. and Imhoff, S. (2015). Aggregation agents and structural stability in soils with different texture and organic carbon contents. *Scientia Agricola* 72, 75–82.

Chen, G. and Weil, R. R. (2010). Penetration of cover crop roots through compacted soils. *Plant and Soil* 331, 31–43.

Cresswell, H. P. and Kirkegaard, J. A. (1995). Subsoil amelioration by plant roots: The process and the evidence. *Australian Journal of Soil Research* 33, 221–39.

Crittenden, S. J. and de Goede, R. G. M. (2016). Integrating soil physical and biological properties in contrasting tillage systems in organic and conventional farming. *European Journal of Soil Biology* 77, 26–33.

Crittenden, S., Poot, N., Heinen, M., van Balen, D. and Pulleman, M. (2015). Soil physical quality in contrasting tillage systems in organic and conventional farming. *Soil & Tillage Research* 154, 136–44.

Ehlers, W. (1975). Observations on earthworm channels and infiltration in tilled and untilled loess soil. *Soil Science* 119, 242–9.

Ehlers, W. and Goss, M. (2016). *Water Dynamics in Plant Production*. 2nd Edition. CAB International, Wallingford, UK.

Ekwue, E. I. (1992). Effect of organic and fertiliser treatments on soil physical properties and erodibility. *Soil & Tillage Research* 22, 199–209.

Folorunso, O. A., Rolston, D. E., Prichard, T. and Louie, D. T. (1992). Soil surface strength and infiltration rate as affected by winter cover crops. *Soil Technology* 5, 189–97.

Fraser, P. M., Haynes, R. J. and Williams, P. H. (1994). Effects of pasture improvement and intensive cultivation on microbial biomass, enzyme activities, and composition and size of earthworm populations. *Biology and Fertility of Soils* 17, 185–90.

Gaiser, T., Perkons, U., Küpper, P. M., Uteau Puschmann, D., Peth, S., Kautz, T., Pfeifer, J., Ewert, F., Horn, R. and Köpke, U. (2012). Evidence of improved water uptake from subsoil by spring wheat following lucerne in a temperate humid climate. *Field Crops Research* 126, 56–62.

Giepen, M., Neto, F. S. and Köpke, U. (2014). Controlling weeds with natural phytotoxic substances (NPS) in direct seeded soybean. In: Rahmann, G. and Aksoy, U. (Eds), *Building Organic Bridges*, Volume 2, Thuenen Report, no. 20. Johann Heinrich von Thünen-Institut, Braunschweig, Germany, pp. 469–72.

Goss, M. J. and Kay, B. D. (2005). Chapter 9: Soil aggregation. In: Wright, S. F. and Zobel, R. W. (Eds), *Roots and Soil Management: Interactions between Root and the Soil*. Agronomy Monograph No 48. ASA, CSSA, and SSSA, Madison, WI, USA, pp. 163–80.

Goss, M. J., Carvalho, M. and Brito, I. (2017). *Functional Diversity of Mycorrhiza and Sustainable Agriculture. Management to Overcome Biotic and Abiotic Stresses*. Academic Press, Elsevier, San Diego, CA, USA.

Hamza, M. A. and Anderson, W. K. (2005). Soil compaction in cropping systems: A review of the nature, causes and possible solutions. *Soil & Tillage Research* 82, 121–45.

Han, E., Kautz, T., Perkons, U., Lüsebrink, M. and Köpke, U. (2015). Quantification of soil biopore density after perennial fodder cropping. *Plant and Soil* 394, 73–85.

Han, E., Kautz, T. and Köpke, U. (2016). Precrop root system determines root diameter of subsequent crop. *Biology and Fertility of Soils* 52, 113–18.

Helgason, T., Daniell, T. J., Husband, R., Fitter, A. H. and Young, J. P. W. (1998). Ploughing up the wood-wide web? *Nature* 394, 431.

Hermawan, B. and Bomke, A. A. (1997). Effects of winter cover crops and successive spring tillage on soil aggregation. *Soil & Tillage Research* 44, 109–20.

Hudson, B. D. (1994). Soil organic matter and available water capacity. *Journal of Soil and Water Conservation* 49, 189–94.

Huntley, E. E., Collins M. E. and Swisher, M. E. (1997). Effects of farm management on soil quality. In: Gallaher, R. N. and McSorley, R. (Eds), *Proceedings of the 20th Annual Southern Conservation Tillage Conference for Sustainable agriculture*. Institute of Food and Agricultural Sciences, University of Florida, Florida USA, pp. 184–8. Available at http://ufdc.ufl.edu/UF00053902/00001/197j (Accessed 5 May 2017).

Jiang, X., Wright, A. L., Wang, J. and Li, Z. (2011). Long-term tillage effects on the distribution patterns of microbial biomass and activities within soil aggregates. *Catena* 87, 276–80.

Kautz, T. (2014). Research on subsoil biopores and their functions in organically managed soils: A review. *Renewable Agriculture and Food Systems* 30, 318–27.

Kolodziejek, J. and Patykowski, J. (2015). Effect of Environmental Factors on Germination and Emergence of Invasive *Rumex confertus* in Central Europe. *The Scientific World Journal*, 2015, 10. Available at http://dx.doi.org/10.1155/2015/170176.

Lipiec, J., Kuś, J., Słowińska-Jurkiewicz, A. and Nosalewicz, A. (2006). Soil porosity and water infiltration as influenced by tillage methods. *Soil & Tillage Research* 89, 210–20.

Liu, S., Guo, X., Feng, Gu., Maimaitiaili, B., Fan, J. and He, X. (2016). Indigenous arbuscular mycorrhizal fungi can alleviate salt stress and promote growth of cotton and maize in saline fields. *Plant and Soil* 398, 195–206.

Lucero, M., DeBolt, S., Unc, A., Ruiz-Font, A., Reyes, L., McCulley, R., Alderman, S. and Dinkins, R. (2014). Advancing food security through long term research networks that explore agroecosystems across disciplines and scales. In: Oelbermann, M. (Ed.), *Sustainable Agroecosystems in Climate Change Mitigation*. Wageningen Academic Publishers, Wageningen, The Netherlands, pp. 183–202.

Massucati, L. F. P. and Köpke, U. (2011). Faba bean (*Vicia faba* L.): response to occasional direct seeding into straw residues under temperate climate conditions. In: Neuhoff, D., Halsberg, N., Rasmussen, I. A., Hermansen, J., Ssekyewa, C. and Mok Sohn, S. (Eds), *Proceedings of the 3rd Scientific Conference of ISOFAR-Organic is Life-knowledge for Tomorrow*, Namyangju, South Korea, 28 September to 1 October 2011, IOL, Volume 1, pp. 198–201.

Miller, R. M. and Jastrow, J. D. (2000). Mycorrhizal fungi influence soil structure. In: Kapulnik, Y. and Douds, D. D. (Eds), *Arbuscular Mycorrhizas: Physiology and Function*. Springer Netherlands, Dordrecht, pp. 3–18.

Or, D. and Wraith, J. M. (2000). Soil water content and water potential relationships. In: Sumner, M. E. (Ed.), *Handbook of Soil Science*. CRC Press, Boca Raton, FL, USA, pp. A53–A81.

Penman, H. L. (1948). Natural evaporation from open water, bare soil and grass. *Proceedings of the Royal Society of London, Series A* 193, 120–46.

Qiao, X., Bei, S., Li, H., Christie, P., Zhang, F. and Zhang, J. (2016). Arbuscular mycorrhizal fungi contribute to overyielding by enhancing crop biomass while suppressing weed biomass in intercropping systems. *Plant and Soil* 406, 173–85

Reberg-Horton, S. C., Burton, J. D. Danehower, D. A. Ma, G. Monks, D. W. Murphy, J. P. Ranells, N. N., Williamson, J. D. and Creamer, N. G. (2005). Changes over time in the allelochemical content of ten cultivars of rye (*Secale cereale* L.). *Journal of Chemical Ecology* 31, 179–93.

Regelink, I. C., Stoof, C. R., Rousseva, S., Weng, L., Lair, G. J., Kram, P., Nikolaidis, N. P., Kercheva, M., Banwart, S. and Comans, R. N. J. (2015). Linkages between aggregate formation, porosity and soil chemical properties. *Geoderma* 247–48, 24–37.

Reid, J. B. and Goss, M. J. (1981). Effect of living roots of different plant species on the aggregate stability of two arable soils. *Journal of Soil Science* 32, 521–41.
Reid, J. B. and Goss, M. J. (1982). Interactions between soil drying due to plant water use and decreases in aggregate stability caused by maize roots. *Journal of Soil Science* 33, 47–53.
Reinhart, K. O., Nichols, K. A., Petersen, M. and Vermeire, L. T. (2015). Soil aggregate stability was an uncertain predictor of ecosystem functioning in a temperate and semiarid grassland. *Ecosphere* 6(11), 238.
Rillig, M. C. (2004). Arbuscular mycorrhizae, glomalin, and soil aggregation. *Canadian Journal of Soil Science* 84, 355–63.
Ritchie, J. T. (1972) Model for predicting evaporation from a row crop with incomplete cover. *Water Resources Research* 8, 1204–13.
Scullion, J., Neale, S. and Philipps, L. (2002). Comparisons of earthworm populations and cast properties in conventional and organic arable rotations. *Soil Use and Management* 18, 293–300.
Sturtevant, E. L. 1887. *Weeds in Corn*. Fifth Annual Report, New York Agricultural Experimental Station, Geneva, pp. 50–1.
Tahtamouni, M. E., Khresat, S. e., Lucero, M., Sigala, J. and Unc, A. (2016). Diversity of endophytes across the soil-plant continuum for *Atriplex* spp. in arid environments. *Journal of Arid Land* 8, 241–53.
Tan, C. S., Drury, C. F., Ng, H. Y. F. and Gaynor, J. D. (1999). Effect of controlled drainage and subirrigation on subsurface tile drainage and nitrate loss and crop yield at the farm scale. *Canadian Water Resources Journal* 24, 177–86.
Teasdale, J. R. and Mohler, C. L. (2000). The quantitative relationship between weed emergence and the physical properties of mulches. *Weed Science*. 48, 385–92.
Tebrügge, F. and Düring, R. A. (1999). Reducing tillage intensity – a review of results from a long-term study in Germany. *Soil & Tillage Research* 53, 15–28.
Trewavas, A. (2004). A critical assessment of organic farming-and-food assertions with particular respect to the UK and the potential environmental benefits of no-till agriculture. *Crop Protection* 23, 757–81.
Unc, A. and Goss, M. J. (2006). Impact of land application of organic wastes on soil hydraulic properties and on water partitioning. *Journal of Environmental Engineering & Science* 5, 243–51.
Unger, P. W. and Vigil, M. F. (1998). Cover crop effects on soil water relationships. *Journal of Soil and Water Conservation* 53, 200–7.
van der Heijden, M. G. A. and Horton, T. R. (2009). Socialism in soil? The importance of mycorrhizal fungal networks for facilitation in natural ecosystems. *Journal of Ecology* 97, 1139–50.
van der Heijden, M. G. A., Klironomos, J. N., Ursic, M., Moutoglis, P., Streitwolf-Engel, R., Thomas Boller, T., Andres Wiemken, A. and Sanders, I. R. (1998). Mycorrhizal fungal diversity determines plant biodiversity, ecosystem variability and productivity. *Nature* 396, 69–72.
Wright, S. F. and Upadhyaya, A. (1996). Extraction of an abundant and unusual protein from soil and comparison with hyphal protein of arbuscular mycorrhizal fungi. *Soil Science* 161, 575–86.
Zikeli, S. and Gruber, S. (2017). Reduced tillage and no-till in organic farming systems, Germany – status quo, potentials and challenges. *Agriculture* 7, 35.

Part 2

Crop pests and diseases

Chapter 10

Disease and pest management in organic farming: a case for applied agroecology

M. R. Finckh, S. M. Junge, J. H. Schmidt and O. D. Weedon, Universität Kassel, Germany

1 Introduction
2 General principles of plant protection in organic agriculture
3 Case study: soil regeneration and effects on potato health
4 Integrating diversity through evolutionary breeding
5 Requirements for agroecology-based ecological plant protection
6 Conclusion and future trends
7 Acknowledgements
8 Where to look for further information
9 References

1 Introduction

Pest and disease management in conventional systems usually focuses on issues of direct control, while cultural practices and prevention are considered as general agronomical practices. The usually very brief sections about plant protection in organic farming books typically stress the importance of resistances, point to plant protection products allowed in organic farming and describe measures that can enhance natural disease and insect control. The rest of plant protection is presumed to be taken care of by rotations and these are allocated in the agronomy sections. In Part 1 of this book almost all chapters deal with methods that contribute to improved plant health. Nevertheless, organic growers are frequently confronted with insect pests and plant pathogens.

Organic disease management has been dealt with in several recent reviews and a textbook (Finckh et al. 2015b; van Bruggen et al. 2016; van Bruggen and Finckh 2016; Döring et al. 2012b). Many disease problems are difficult to solve or may even prevent organic production of certain crops in certain areas. Examples are Black Sigatoka (caused by *Mycosphaerella fijiensis*) on bananas (Pocasangre et al. 2015), many downy mildew diseases on vegetables (Koike 2015) or potato late blight (caused by *Phytophthora infestans*) (Finckh et al. 2015a). For natural insect pest control, several reviews and books are available (Altieri 1994; Letourneau et al. 2011; Letourneau 1997; Tooker and Frank 2012).

http://dx.doi.org/10.19103/AS.2017.0029.11

Organic pesticides are discussed in detail in a separate chapter (see also Chellemi et al. 2015; Koch and Groot 2015; Tamm et al. 2015). Unfortunately, despite being mostly biodegradable, there is a lack of selectivity in many organic pesticides such as the pyrethrins, and the two most commonly used 'organic' fungicides based on sulphur and copper. Sulphur is non-selective and affects beneficial as well as harmful insects and mites. Other non-selective compounds like vinegar and salts are usually repellent and rarely kill the target organisms. In the case of copper, there are well-known environmental problems when used regularly in high amounts that make its use more than questionable in organic farming (Flemming and Trevors 1989). Like sulphur, copper may also harm the beneficial 'microflora' in the 'phyllosphere' and thus interferes with natural biocontrol (Jaros-Su et al. 1999; Stirling et al. 1999; Ruano-Rossil et al. 2004).

Much more important than organic pesticides, albeit hard to quantify for their effects on pests and pathogens, are system imminent processes that enhance what has been termed 'natural biocontrol mechanisms' (Alabouvette et al. 2006) and that can only be found when studying the system as a whole. In this chapter, we briefly summarize the general principles of plant protection in organic agriculture (Section 2) and will then present two agroecological approaches to system health and sustainability that we have been exploring in an attempt to approach the topic of plant protection in an as holistic way as possible. Section 3 describes how a potato cropping system, aimed at soil regeneration and long-term system effects, may at the same time benefit potato health itself. Enhancement of system health through breeding for diversity in wheat is described in Section 4. We attempt to outline a complex system view where we identify important research questions that need to be addressed to support the agroecological approach to plant and system health in Section 5 and conclude with some general remarks. For the rest of this chapter, whenever we deal with both insect pests and plant pathogens, the term 'pest' has been used. The distinction of insect pests or pathogens is made where necessary.

2 General principles of plant protection in organic agriculture

The basic principle for all pest management approaches is to identify the most vulnerable moments in a pest's life cycle in order to reduce the risk of pest multiplication. Normally, the aim is regulation rather than eradication. For the system to work, it has to be based on a longer-term implementation plan, especially with respect to soil fertility and rotation planning, that also addresses weed management issues. In addition, vertebrate and invertebrate pests, such as voles, birds and slugs, among others need to be considered when managing agroecosystems. In general, the same pests play a role in organic and conventional agriculture. However, certain factors affect the importance of pests in organic agricultural systems. For example,

- changed frequencies at which host crops are cultivated due to extended rotations strongly affect specialized soil- and residue-borne pests
- differences in plant nutrition affect the susceptibility of plants to pests and
- higher biodiversity affects pests above and below ground.

Because few direct control measures are available, the whole system has to be based on strong preventive measures, especially with respect to seed health.

The main pest management strategies (Table 1) rarely function alone; the combination of the approaches allows the system to achieve functionality and resilience. None of these strategies is new and they are generally seen as the backbone of any solid agricultural operation. However, consequences of their neglect are much larger in organic than in conventional agriculture as no potent pesticides are available to correct mistakes made during a season.

Seed health is obviously the starting point for plant protection, and without genetic resistances, no agricultural system could work. Unfortunately, not all pathogens of importance to organic farming are considered in current seed certification procedures

Table 1 Main strategies and approaches for insect pest and disease management in organic farming systems

Strategy	Function/examples and/or agroecosystems effects
Seed health	Reduction of inoculum load Avoid introduction of pests
Resistance	Reduction of regional inoculum load
Rotations	Break life cycles of specialized pathogens Increase microbial activity and diversity Improve nutritional balance of plants
Soil quality and plant nutrition	Improved soil structure for better water-holding capacity, drainage, soil life Improved nutritional balance enhances resistance Certain organic amendments may induce resistance
Avoidance	Planting time to avoid overlap with the same crop and coincidence with phases of pest increase Physical barriers to insects (netting) and other animals Quarantines
Plant diversity	Distance, barrier, induced and associated resistance effects Regional reduction of inoculum load Effects on plant nutritional balance Effects on microclimate Effects on microbial diversity
Manipulation of the environment	Climate control in greenhouses to reduce pest conduciveness Microclimate manipulation in the field through pruning schemes, distance between plants, direction of rows in field, soil drainage, drip irrigation, timing of irrigation, etc. Good soil structure management for optimum water dynamics and root health
Biocontrol	Introduction of organisms without establishment Enhancement of natural biocontrol through agroecological approaches
Soil disinfestation	Especially applicable in greenhouses Depending on the method, may improve soil health
Direct control	Organic pesticides as a last measure, usually with no long-term effects on agroecosystem and should be as target/pest specific as possible

because chemical seed treatments are taken for granted. Thus, there are no strict limits for *Fusarium* levels in cereals. Much effort has been invested in the determination of innovative biological and physical seed treatment methods (Koch and Groot 2015).

As many diseases and often also insect pests favour high nitrogen inputs (Huber and Thompson 2007), there is a general tendency towards lower disease pressure in organic field crops due to obligate pathogens such as rusts and powdery mildews, and some other pathogens. However, in organic vegetable production, nutrient inputs are generally high and disease pressure may increase proportionally. The effects of the various nutrients on plant diseases have been treated systematically (Datnoff et al. 2007).

For soil-borne and less mobile pests, rotation design and soil health management combined with plant resistances and sometimes the application of biocontrol and soil disinfestation are most important. In well-managed organic farming systems, the general tendency is towards a reduction of pests and pathogens (van Bruggen et al. 2016; Chellemi et al. 2015; van Bruggen and Semenov 2015; Leoni et al. 2015; Litterick et al. 2004). This has been attributed to soil suppressiveness, which in turn is linked to organic amendments, a highly active and diverse microflora and good soil structure in general (Litterick et al. 2004; Kinkel et al. 2011; van Bruggen and Semenov 2015; Mazzola and Manici 2012). Although rotations are most effective against pests with host specialization, they also work against less specialized pests because different crops provide soils with different exudates, enhancing microbial diversity and soil biological activity and fertility.

There is widespread concern over the damage caused by many current agricultural practices to soil structure and the ecosystem services provided, as well as general soil and organic matter losses (Bellamy et al. 2008; Montgomery 2007). On the one hand, this directly affects soil quality and functions (Allison 1973; Dexter 2004) that are, among others, related to soil suppressiveness to diseases (Weller et al. 2002; Kinkel et al. 2011; Mazzola 2004; Litterick et al. 2004). On the other hand, organic matter may also influence plant health through resistance induction (Finckh and Tamm 2015; Vallad and Goodman 2004). Thus, loss of soil organic matter may directly increase plant susceptibility to diseases.

One approach to overcome this problem is reduced tillage and conservation agriculture (CA). The FAO (http://www.fao.org/ag/ca/index.html) describes CA as '… an approach to managing agro-ecosystems for improved and sustained productivity, increased profits and food security, while preserving and enhancing the resource base and the environment'. CA is characterized by three linked principles, namely:

1 Continuous minimum mechanical soil disturbance
2 Permanent organic soil cover and
3 Diversification of crop species grown in sequences or associations.

For organic operations, reduced or no-till poses challenges with respect to weed control and plant nutrition (Sans et al. 2011; Berner et al. 2008; Peigné et al. 2016). Weeds in reduced tillage can be managed relatively well by judicious use of cover crops, grass-clover leys and appropriate crop rotations in the long run. Problems with adequate and timely mineralization of nutrients in undisturbed soils are more serious in temperate climates. Such soils may have problems warming up in spring, and inadequate soil structure and insufficient amount of soil organic matter are generally problematic, not only in conventional, but also in organic farming systems (Soane et al. 2012). A systems approach to soil improvement and pest management is described in Section 3.

In contrast to soil-borne pests, avoiding the coincidence of crop and highly mobile air-borne pests is often not easily feasible due to climatic restrictions and because concerted regional management is rarely done (Finckh and Tamm 2015). For example, organic oilseed rape production is usually not possible in areas with a high density of conventional oilseed rape production. In such areas, populations of pollen beetles (*Meligethes aeneus*) are high due to high levels of insecticide resistance, including pyrethrins (Slater et al. 2011), making their organic management impossible.

Regional management can be very effective especially when combined with the judicious use of resistances. For example, growers may agree not to plant before a certain date, to maintain certain distances between fields or to make use of certain resistances at certain times, but not others. Also, the use of different resistances in spring-grown and winter-grown crops of the same species or regionally in breeding programmes can be very effective (Finckh and Wolfe 2015; Browning et al. 1969; Knott 1972). Cover crops or green manure crops planted directly after a main crop do not only enhance soil fertility, but often also contribute to field hygiene and thus to avoid coincidence of host and pest. Regional management is practised to a certain extent in the Netherlands in order to reduce disease pressure by *P. infestans*. Inoculum from conventional and organic operations has been shown to start local epidemics in the Netherlands (Zwankhuizen et al. 1998). The same strict rules apply to organic and conventional potato operations in the Netherlands. Trash piles have to be destroyed and/or covered by April 15 and volunteer crops and disease foci in crops where disease exceeds about 5% diseased leaf area also have to be destroyed or covered in order to keep inoculum loads down (Anon. no year). Successful application of biocontrol agents may also depend on a regional approach. For example, it should be possible to drastically reduce cherry fruit flies (*Rhagoletis cerasi*) by targeting the larvae when emerging from the cherries before they pupate by the application of entomopathogenic nematodes (Herz et al. 2006). However, this only makes sense if producers of the whole region participate as the flies are highly mobile.

Diversity for resistance (gene deployment) within the same field by using either lines or cultivars with different resistances in mixtures or by simply mixing differentially resistant plant species is very successful in reducing airborne pathogens and often also insect pests (Finckh and Wolfe 2015, 2006; Letourneau et al. 2011; Vidal et al. 2017). The ecological interactions leading to positive mixture effects beyond disease control are being increasingly recognized and their exploitation has recently been brought into a more formal (agro)ecological framework (Barot et al. 2017). For example, how intra- and interspecific diversity can be achieved through evolutionary breeding within the agroecosystem. The approach, some results and some of the many important open questions are presented in Section 4.

The environmental conditions are manipulated intensively in greenhouse and plastic tunnel operations to reduce pests, for example, by supplying extra heating to avoid dew formation and through measures that influence humidity and light quality (van Bruggen 2015). In arable crops, soil drainage, irrigation, planting direction, reduced crop density and/or pruning are important measures (Finckh and van Bruggen 2015). Hedgerow management can influence the local field microclimate. Coppicing hedges in the years when potatoes or other crops very susceptible to high humidity are grown in an adjacent field can reduce the site conduciveness to late blight, for example (Finckh et al. 2015a). As hedges may be the overwintering places for certain insect pests, it may be necessary to manage the habitat in between hedges and production fields, for example, by using barrier or trap crops (Finckh and van Bruggen 2015).

Biocontrol is very successful in glasshouse operations. In the field, introduction of biocontrol organisms may depend on regional approaches as described above for the cherry fruit fly. Biocontrol is available for field applications against certain insect pests [Neem (*Azadirachta indica*), Bt (*Bacillus thuringiensis*), Pyrethrum, etc.] as well as against a few pathogens (based on *Bacillus, Pseudomonads, Trichoderma, Coniothyrium, etc.*). It is very important to enhance the naturally occurring biocontrol through proper habitat management, rotations, soil fertility management and general enhancement of soil quality. Overall, the application of microbe-based biocontrol appears to be more effective in soils with lower microbial activity due to lower competition (Hiddink et al. 2005). For example, sales of biocontrol agents such as *Trichoderma*- or *Beauveria-bassiana*-based products are especially lucrative in the low-organic matter soils of large-scale operations in Matto Grosso in Brazil (M. Finckh personal observation).

Soil disinfestation can be achieved with physical measures (heat, water), as well as with biological/biochemical measures such as biofumigation or biological (or anaerobic) soil disinfestation and has been reviewed recently (van Bruggen and Finckh 2016; Gamliel and van Bruggen 2016). However, many of these measures, including many organic pesticides, are expensive and therefore only applied as a last measure. Direct control of soil and air-borne pathogens in organic systems was recently described in detail (Chellemi et al. 2015; Tamm et al. 2015).

3 Case study: soil regeneration and effects on potato health

Potato crops notoriously reduce soil humus contents and are prone to soil erosion due to repeated massive soil movements during their cultivation. Typical organic potato production systems in Germany commonly rely on ploughing and regular energy-intensive mechanical weed management practices, which often result in more intense soil disturbance than in conventional systems. The consequence is usually a serious loss of soil organic matter and humus content during potato production that needs to be remedied afterwards. Besides these general ecological concerns, potatoes worldwide regularly suffer from late blight often in dimensions that are not manageable under organic conditions (Finckh et al. 2015a). Copper-based fungicides are widely applied in organic production systems but cannot be considered truly sustainable (see above) despite being listed in the IFOAM-Organics International norms (IFOAM 2014). Colorado potato beetles (*Leptinotarsa decemlineata*) are an important insect pest under temperate conditions, often requiring the application of Neem or Bt-based biological insecticides. From a holistic point of view, organic potato production cannot be considered sustainable with respect to maintaining and increasing soil quality, and, overall, there is much room for improvement in order to create a system more in line with the principles of health, ecology, fairness and care as stipulated by IFOAM-Organics International (IFOAM no year).

Roughly over the last 15 years, (mainly) organic growers have been working on the development of an alternative potato-growing system based on the use of mulch (Kayser

2002; Brunner 2017; Steinert 2016). Some growers have claimed large effects in reducing or delaying late blight and potato beetles through mulching. In 2013, we started to study the system with surprising effects for the whole agricultural system, including pest development. The work we describe is a sequence of adaptations for system optimization conducted mostly in replicated field trials, where possible, for two consecutive years. As late blight did not occur in all years and the project was only partially financed, data for all years are not available. Some of the main results with respect to soil quality, disease and insect pest suppression are presented.

From 2013 to 2016, two two-year trials and a preliminary trial on the effects of reduced tillage and/or mulch were conducted (Table 2). OSCAR I and II (Table 2) were conducted within a trial setup to follow the conversion to minimum tillage based on non-inversion tillage with a heavy cultivator (10–15 cm soil depth), compared with a mouldboard ploughed system (25–30 cm deep) (OSCAR 2016). The cover crops grown before potatoes were sown in year three of the experiments, and potatoes were planted in year four. Non-inversion tillage was combined with a dead mulch as recommended by farmers practising minimum tillage in potatoes (Steinert 2016) (Fig. 1). Since it is a comparison of two systems only, there was neither ploughed system with mulch nor minimum tillage without mulch as this would have doubled the size of the experiments, which was not possible. Thus, no field that had not been ploughed for four years was available to compare non-inversion tillage with and without mulch. For the ploughed system, the OSCAR III and IV trials were set up comparing different mulches in 2015 and 2016 (Table 2).

In contrast to the mulch application which requires technical innovation and potentially investments for farmers, the incorporation of the standing or previously mulched cover crop only requires an efficient rototilling equipment with sharp knives, running at higher than typical speed. This system allows for potato production in soils that are tilled only

Table 2 Experiments conducted with potatoes on the effects of reduced tillage and/or mulch

Year, name*	Factors compared	Comments
2013–14, OSCAR I	Tillage and mulch: plough versus reduced tillage plus mulch (rye/winter pea, 124 t DM ha^{-1}) Cover crops: weedy fallow, winter vetch, oilseed radish/black oat mix	Minimum tillage plots without ploughing since 2010, relatively cool and wet spring
2014–15, OSCAR II	Tillage and mulch: plough versus reduced tillage plus mulch (tricale/vetch, 265 t DM ha^{-1}) Cover crops: weedy fallow, winter vetch, oilseed radish/black oat mix	Minimum tillage plots without ploughing since 2011, vole infestation killed much of the cover crops, extremely dry spring, no *P. infestans* in 2015
2015 and 2016, OSCAR III/IV	Three mulch types: none, grass clover, triticale/vetch in a ploughed system	Extremely dry spring, no *P. infestans* 2015, 2016 dry spring, moderate epidemic *P. infestans*
2016, RePCro	Minimum tillage system incorporating a precrop of triticale/vetch by fast 4-cm-deep rototilling in top soil without mulch versus with 516 t DM ha^{-1} of triticale mulch, no hilling of potatoes	Minimum tillage only in that year 2016 dry spring, moderate epidemic *P. infestans*

*OSCAR I, II and OSCAR III were financed by the EU-project OSCAR (OSCAR 2016).

Figure 1 Top, left: application of freshly cut mulch material 8–10 cm deep with a manure spreader on the potato plots in mid-May. Top, right: the potatoes spread their roots into the mulch in order to access water and nutrients. Bottom: during a dry period in 2015, canopy closure is achieved in the mulched potatoes (B). In the unmulched potatoes (A), the plants remain stagnant in their development and are unable to compete with weeds, contributing to a reduced yield. Photos: S. Junge.

4–5 cm deep and practitioners reported that it results in an increase of soil organic matter through the season and improved soil structure. A preliminary trial was set up in 2016 [(RePCro, regenerative potato cropping system), Table 2, Fig. 2] in a small field that was available with a suitable cover crop of triticale/vetch. The cover crop was mulched and rototilled to 4 cm soil depth, and eight rows of potatoes (two varieties) were planted into this material without mulch. Another eight rows were planted with a 15-cm mulch cover of triticale with a C/N ratio of 52:1. Data were collected in four pseudo-replicates of four rows by 10 m to obtain an estimate of the variance.

Late blight epidemics occurred in 2014 and 2016. In these trials, cumulative disease severity measured as Area Under the Disease Progress Curve (AUDPC) was significantly

Figure 2 Regenerative potato cropping (RePCro) from left to right: after the breakdown of the cover crops with the rotary hoe to a depth of 5 cm, the potatoes were planted in the layer of rotting plant material. After the application of mulch, the potatoes emerge through the mulch layer. The soil covered with mulch builds up biologically bound aggregates until harvest. The succeeding crop benefits from this increased soil fertility: Rhizobacteria settle directly on the roots, protect against soil-borne pathogens and aid nutrient accumulation. Photos: S. Junge.

reduced by about one-third in the mulched plots (Table 3). Disease reductions were clearly visible (Fig. 3). Epidemics always started later and proceeded more slowly. The same phenomenon was observed in parallel trials conducted in south Brazil (OSCAR 2016).

In addition to effects on *P. infestans*, in June 2014, detailed assessments were conducted on potato beetle (*L. decemlineata*) adults, egg clusters and larvae in mulched and bare plots. The number of beetles per hectare did not differ among treatments with 1025 and 1300 in unmulched versus mulched plots, respectively (Junge et al. unpublished data). This suggests that there was no effect on the host plant finding process of the adult beetles that migrate from the previous year's areas. In contrast, however, only about half the number of egg clusters were found in mulched plots (2400 ha^{-1}) compared with unmulched plots (4750, $P < 0.05$) and only 12475 larvae per hectare were found with mulch compared to 65400 without mulch ($P < 0.05$). These effects were repeatedly found in mulch experiments from 2015, 2016 and 2017 (e.g. Fig. 4). Phelan (1997) reported that plants grown in organic soils displayed a better nutrient balance, which reduces the fertility of the European corn borer in comparison to soils with conventional chemical inputs. Alyokhin et al. (2005) mentioned that organic soil amendments can reduce populations of the Colorado potato beetle. A better nutrient availability through the mulch layer (Riley et al. 2003) could mean that an optimal nutrient balance in the plant is established, making the plants less susceptible to the potato beetle. In 2016, damaged egg clusters were only found in the mulched plots (Junge et al. 2017). This could be an indication of increased predation. It could also be that mulch, as a structural element, serves as a barrier for the movement of the larvae and beetles, as reported by Szendrei et al. (2009).

In the OSCAR I and II trials, we also studied the weed seed bank and nematode dynamics. Where cover crops were grown before the potato crops, the weed seed banks in the top 15 cm of the soil in mulched and minimally tilled plots were not significantly different from the ploughed plots. However, if no cover crops had been used, the seed banks increased significantly, despite the clear suppressive effects of the mulch on annual weeds (Schmidt 2017). In contrast to annual weeds, perennial weeds, especially couch

Table 3 Late blight (area under the disease progress curve, AUDPC) and marketable yields in the trials described in Table 2

Trial/year	Tillage	Cover crop	Mulch	AUDPC (Rel)	SED[1]	Yield (t ha^{-1}) (Rel)	SED[1]
OSCAR I 2014	Plough	Weedy fallow	None	1376a (1.00)	87.9/91.0	30.4ab (1.00)	2.19/2.19
		Vetch		1342a (0.98)		34.8a (1.14)	
		Black oat/radish		1249a (0.91)		33.9a (1.12)	
	Reduced	Weedy fallow	Pea/rye	899b (0.65)		26.1b (0.86)	
		Vetch		891b (0.65)		32.7a (1.08)	
		Black oat/radish		834b (0.61)		29.7ab (0.98)	
OSCAR II 2015	Plough	Weedy fallow	None	–[2]		24.6b (1.00)	1.97/1.99
		Vetch[3]		–		25.7b (1.04)	
		Black oat/radish[3]		–		25.2b (1.02)	
	Reduced	Weedy fallow	Vetch/triticale	–		30.2a (1.23)	
		Vetch[3]		–		30.7a (1.25)	
		Black oat/radish[3]		–		28.7ab (1.17)	
OSCAR III 2015	Plough	Weedy fallow	None	–		26.1a (1.00)	5.25
		Weedy fallow	Vetch/triticale	–		25.8a (0.99)	
		Weedy fallow	Grass/clover	–		23.6a (0.90)	
OSCAR IV 2016	Plough	Weedy fallow	None	2111a (1.00)	123.3	24.5b (1.00)	1.86
		Weedy fallow	Vetch/triticale	1361b (0.64)		31.8a (1.30)	
		Weedy fallow	Grass/clover	1532b (0.73)		30.9a (1.26)	
RePCro 2016	Reduced	Vetch/triticale	None	1207a (1.00)	105.7	18.7a (1.00)	2.68
	Reduced	Vetch/triticale	Triticale	748b (0.62)		20.6a (1.10)	

In parentheses the values relative to the control without mulch are given. Numbers within trial followed by the same letters were statistically not significantly different at $P < 0.05$ (Tukey test).

[1] Standard error of the difference between means (within/between tillage factor levels).

[2] No late blight occurred in 2015.

[3] In the winter 2014/15, a vole infestation severely damaged all cover crops. This resulted in low biomass remaining, especially for the preferred leguminous vetch.

Figure 3 Left: aerial photograph of OSCAR III on 19 July 2016 (ploughed system with two types of mulch and unmulched control, see Table 2 for details of experiment and Table 3 for disease and yield data). Replications are labelled with roman numerals. White frames mark the untouched parts of the plots. Bare areas were sampled destructively. (A) clover grass mulch, (B) unmulched, (C) vetch triticale mulch. Middle and right: 26 June 2016: Rain splash on unmulched plants (middle) in contrast to no splash on mulched plants (right). Photos: S. Junge.

Figure 4 OSCAR IV, 11 July 2016: Colorado potato beetle–infested plants in the unmulched variants on the left, in comparison to the less infested mulched plants on the right. Photo: S. Junge.

grass (*Elymus repens*) and docks (*Rumex* spp.), may profit from mulching depending on the weather. Thus, in the wet year of 2014, such weeds doubled in the mulch system, while in 2015, they were halved in OSCAR I and II, respectively (Junge et al. 2016). Such a system will only be successful in fields where perennial weeds have been well managed over time with appropriate crop rotations and management, reducing overall perennial weed pressure. In organic systems ploughing as well as black fallow are made use of in case of heavy infestations. Nevertheless, besides avoiding the build-up of problem perennial weeds, the aim has to be to avoid such measures that are overall destructive to soil life. Figure 5 shows the cause/effect relationships in the suppression of weeds by mulch. Plant parasitic nematodes tended to increase in the minimum tillage plots over time,

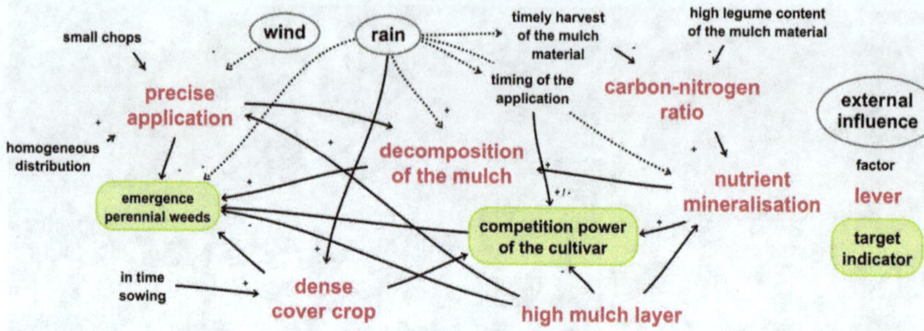

Figure 5 Effects and relationships of the controllable (solid line) and unmanageable (dotted line) factors of the mulch system. The + and – indicate whether these influences increase or decrease the pointed factors, levers or target indicators. The situation will change with every season, including the importance of each influence. Experience and knowledge can prevent mistakes in this complex system and help to ensure agronomic success.

but their increase could be counteracted by the use of an oilseed radish/black oat cover crop (Schmidt et al. 2017b). The beneficial effect of brassicas with respect to nematode infestation was also shown in similarly set up experiments across Europe (Schmidt et al. 2017a). Both studies confirmed oilseed radish (*Raphanus sativus* var. *oleiformis*) and black oats (*Avena strigosa*) as non-host crops of root-lesion nematodes (*Pratylenchus* spp.), and their use as cover crops therefore provides important control options in organic, as well as conventional, production systems. In addition, the populations of free-living nematodes were determined to be significantly higher, with 3100 individuals per 100 ml of soil under minimum tillage with mulch compared to 1700 in the ploughed system (Schmidt et al. 2017b). These nematodes are beneficial for the soil food web and serve as indicators for the farming system sustainability (Freckman 1988; Neher 2010).

The effects of reduced tillage and/or mulch on yields depended on year and weather conditions. Despite the considerable suppression of late blight in 2014, on average, yields with reduced tillage and mulch were 30 t ha^{-1} compared with 33 t ha^{-1} in the ploughed system. However, there was a tillage/mulch and cover crop interaction, and when vetch was used as a cover crop, there were no significant yield differences between tillage systems (Table 3). In the cool and wet spring of 2014, mineralization processes were slow under the mulch, resulting in delayed development of the potatoes, especially where no cover crops had been used. In contrast, in 2015, the spring was extremely dry. Under mulch, more extensive root systems better enabled the potatoes to cope with drought stress. Thus, despite great losses to the cover crops due to a vole infestation, mulching combined with reduced tillage led to much better water availability and overall significantly increased yields (more than 20%) under mulch in OSCAR II. The mulch shades the soil, reducing evaporation, thereby increasing soil water availability and thus nutrient uptake. Triticale/vetch mulch also increased yields by 10% in the ploughed OSCAR III experiment in 2015, but these yield increases were much more dramatic (29%) in OSCAR IV in 2016, where not only a dry spring but also late blight played a role (Table 3). Likewise, potato yields were 10% higher in the triticale mulch than in unmulched plots in the preliminary RePCro trial of 2016, thus underlining the results observed in OSCAR II in 2015 and OSCAR IV in 2016 (Table 3).

Besides the direct effects on pests and yield, the mulch system, especially when combined with reduced tillage, has a considerable impact on the soil structure and subsequent crops. There were visibly more active earthworms in the system during the potato season. The beneficial effects on soil life in the succeeding crop were clearly visible in 2017. Wheat grew extremely well after the RePCro potatoes and soil remained attached to the roots of the wheat when digging them up (Fig. 2). Probable reasons could be changes in the root architecture and competition for space by other microorganisms that may have been enhanced by changes in root exudates that, in turn, benefit a specific rhizobacterial community (Gosling et al. 2006). Such phenomena have also been ascribed to the effects of large numbers of rhizobacteria that with their slime produce a conglomerate of soil and bacteria in the rhizosphere, that is, truly living soil (Pieterse et al. 2016). The microbial biomass carbon (C_{mic}) was greater under reduced tillage with mulch (337 μg C g^{-1} soil) than in the plough system without mulch (270 μg C g^{-1} soil) in OSCAR I and II after potatoes (Schmidt unpublished data). Within the microbial community, arbuscular mycorrhizal fungi (AMF) profit most from reduced tillage due to maintenance of the mycorrhizal networks. Besides their beneficial effects on nutrient uptake by plants, AMF have been shown to reduce damage caused by various pathogens (Gosling et al. 2006).

Overall, the work on mulching and reduced tillage in potatoes under organic conditions has been very promising for pest control, enhancing populations of beneficial nematodes and other microbes, and remedying the negative soil impacts caused by excessive and disturbing potato production. The system impacts go beyond pests and in Section 5 we make an attempt to describe the complexity of the whole system with respect to agroecological interactions.

4 Integrating diversity through evolutionary breeding

The benefits of biodiversity for plant protection have recently been reviewed (Finckh and Wolfe 2015; Vidal et al. 2017; Ostergaard et al. 2009; Tooker and Frank 2012). Crop diversification can be achieved by mixing different species or cultivars or by breeding for (resistance) diversity within a crop (Finckh and Wolfe 2015; Finckh 2008; Wolfe and Döring 2016; Döring et al. 2011; Murphy et al. 2005). An approach to the development of high-yielding, but also highly diverse and adaptable heterogeneous, plant populations through composite crosses (CCs), that is, the crossing in all possible combinations of several to many parental lines (Jain and Qualset 1975), was developed in California about 100 years ago (Harlan and Martini 1929). This was later termed 'evolutionary plant breeding' (Suneson 1956). Genetically diverse populations created through this approach may be subjected to mass selection under different environmental conditions, thus mimicking the development of landraces, while integrating the advantages of modern high-yielding and high-quality varieties. For example, phaseolus beans that had evolved within a composite cross were more amenable to mixed cropping than beans selected early on as pure lines (Allard 1961). The multiplication and conservation of genetically diverse populations *in situ* and in differing environments has been referred to as 'dynamic management' (Allard 1990; Le Boulc'h et al. 1994; Paillard et al. 2000; Porcher et al. 2004). Research papers on a number of wheat CCs have described adaptive changes within populations in response to local selection pressures while, simultaneously, maintaining or increasing genetic variation, resulting in the conservation of genetic diversity (Enjalbert et al. 1999; Goldringer et al.

2006; Rhoné et al. 2008). Such materials have also been termed 'modern landraces' (Murphy et al. 2005).

In order to start the development of modern wheat landraces, a set of three winter wheat CCs were started in 2001 in the United Kingdom (Döring et al. 2015; Phillips and Wolfe 2005) based on a half-diallel intercross of 20 varieties (Jones et al. 2010; Döring et al. 2015) (Fig. 6). Since then the F_5, seed of the UK CCs has been grown and maintained at our research farm in Germany under both conventional and organic conditions in two parallel populations within plots of at least 100 m^2 (Table 4). The maintenance of parallel populations within the same management scheme allows for the separation of accidental changes from management effects (e.g. organic versus conventional) (Goldringer et al. 2006). In 2016/17, the populations were in the F_{16} phase. In addition to the populations, in most years, three or more pure wheat varieties that were of importance to organic farming, and/or represented specific susceptibilities to certain diseases, were grown alongside the CCs in order to assess disease pressure and the agronomic performance of the CCs compared to current varieties.

Throughout the years, the CC yields were similar to those of the reference varieties under organic farming conditions with some notable exceptions that could partly be related to extreme climatic events (Table 5). An extreme climatic event occurred in 2012

	Crossings for YQ Population				Crossings for Q Population								Crossings for Y Population							
		Quality											Yield							
		Cadenza	Hereward	Maris Widgeon	Mercia	Monopol	Pastiche	Renan	Renesansa	Soisson	Spark	Thatcher	Buchan	Claire	Deben	High Tiller Line	Norman	Option	Tanker	Wembley
Quality	Bezostaya	(YQ	(YQ	(YQ	(YQ	(YQ	(YQ	(YQ	(YQ	(YQ	(YQ	(YQ	(YQ	(YQ	(YQ	(YQ	(YQ	(YQ	(YQ	(YQ)
	Cadenza		Q	Q	Q	Q	Q	Q	Q	Q	Q	Q	YQ	YQ	YQ	YQ	YQ	YQ	YQ	YQ
	Hereward			Q	Q	Q	Q	Q	Q	Q	Q	Q	YQ	YQ	YQ	YQ	YQ	YQ	YQ	YQ
	Maris Widgeon				Q	Q	Q	Q	Q	Q	Q	Q	YQ	YQ	YQ	YQ	YQ	YQ	YQ	YQ
	Mercia					Q	Q	Q	Q	Q	Q	Q	YQ	YQ	YQ	YQ	YQ	YQ	YQ	YQ
	Monopol						Q	Q	Q	Q	Q	Q	YQ	YQ	YQ	YQ	YQ	YQ	YQ	YQ
	Pastiche							Q	Q	Q	Q	Q	YQ	YQ	YQ	YQ	YQ	YQ	YQ	YQ
	Renan								Q	Q	Q	Q	YQ	YQ	YQ	YQ	YQ	YQ	YQ	YQ
	Renesansa									Q	Q	Q	YQ	YQ	YQ	YQ	YQ	YQ	YQ	YQ
	Soisson										Q	Q	YQ	YQ	YQ	YQ	YQ	YQ	YQ	YQ
	Spark											Q	YQ	YQ	YQ	YQ	YQ	YQ	YQ	YQ
	Thatcher												YQ	YQ	YQ	YQ	YQ	YQ	YQ	YQ
Yield	Buchan													Y	Y	Y	Y	Y	Y	Y
	Claire														Y	Y	Y	Y	Y	Y
	Deben															Y	Y	Y	Y	Y
	High Tiller Line																Y	Y	Y	Y
	Norman																	Y	Y	Y
	Option																		Y	Y
	Tanker																			Y

Figure 6 Crossing scheme underlying the wheat composite cross (CC) populations used in the study. Cultivars were released between 1934 and 2000 and are characterized by either high baking quality (Q) or high yielding ability (Y). The half-diallel includes Q*Q, Y*Y and Y*Q crosses. Bezostaya is listed twice as it was both of high quality and high yielding at the time of release.

Table 4 Description of the development of the composite cross wheat populations growing in Neu-Eichenberg (C = Conventional, O = Organic, YQ = Yield × Quality, Q = Quality and Y = Yield)

Generation	Year	Composite cross wheat population development in Witzenhausen											
F_5[1]	2005/06	CYQ		OYQ		CQ		OQ		CY		OY	
F_6[2]	2006/07	CYQI	CYQII	OYQI	OYQII	CQI	CQII	OQI	OQII	CYI	CYII	OYI	OYII
F_7	2007/08	CYQI	CYQII	OYQI	OYQII	CQI	CQII	OQI	OQII	CYI	CYII	OYI	OYII
F_8	2008/09	CYQI	CYQII	OYQI	OYQII	CQI	CQII	OQI	OQII	CYI	CYII	OYI	OYII
....													
F_{16}	2016/17	CYQI	CYQII	OYQI	OYQII	CQI	CQII	OQI	OQII	CYI	CYII	OYI	OYII

[1] YQ, Q and Y populations were split into organic (O) and conventional (C).
[2] CYQ, OYQ, CQ, OQ, CY and OY were split into two parallel populations (I and II).

Table 5 Outstanding environmental events during the maintenance of the wheat CCs at the University of Kassel, 2005–16

Year	Type	Description
2009	Brown rust	Epidemic clearly visible on susceptible cultivar 'Bussard'
2012	Frost	Extreme bare frost for two weeks in February after a warm winter followed by six weeks of warm and dry weather
Since 2013	Yellow rust	More or less severe epidemics most likely due to the newly introduced 'warrior' races (Hovmøller et al. 2016)
2014/15	Drought	Very dry spring

when after a frost-free winter until 31 January, temperatures dropped by about 25°C within one night followed by two weeks of bare frost. This left the soil frozen to a depth of 50 cm. Immediately afterwards, unusually high temperatures followed (mean temperatures in March were 7.5°C, i.e. 3.3°C above the 14-year trend since 2000) combined with a six-week dry spell. This resulted in severe winter kill that seriously affected most of the population parental cultivars that had to be ploughed under. In that year, all 12 CCs, listed in Table 4, were grown for comparison in a replicated field trial and yielded on average 4.0 t ha^{-1}. This was significantly lower than three reference cultivars that were more winter hardy than the parents (mean yield 4.9 t ha^{-1}), but also significantly higher than the physical mixture of the parents (2.9 t ha^{-1}). Thus, despite the parental genetics not giving much winter hardiness, the CCs were able to produce a moderate yield. In the following year, when no winter kill occurred, yields of the CCs were equal to the three reference cultivars at 6.1 t ha^{-1} and again significantly higher than the yield of the physical mixture (5.3 t ha^{-1}) (Brumlop et al. 2017). The experimental year of 2015/16 was characterized by a very dry spring, resulting in moderate to low yields across experiments. In this experimental year, the organically managed CCs outyielded four reference cultivars (Achat, Butaro, Capo and Wiwa) grown for comparison. The mean yield of the reference cultivars was 3.9 t ha^{-1} in comparison to the mean yield of the CCs of 4.2 t ha^{-1}. The best-performing CC was the OYQI (see Table 4) with a yield of 4.6 t ha^{-1}, followed by the pure line cultivar Achat (4.5 t ha^{-1}). The remaining CCs all achieved yields between 4.4 and 3.9 t ha^{-1}, with the pure line cultivars Wiwa and Butaro yielding lowest with 3.7 and 3.4 t ha^{-1}, respectively (data not published).

CCs shown were not only highly resilient in terms of yield under challenging biotic and abiotic stresses but also displayed superior yield stability. A number of stability measures were calculated for the organically and conventionally managed CCs over five years (2008, 2010, 2012, 2014 and 2015). The two reference cultivars Achat and Capo were constant during these five years, grown as comparisons to the organically managed CCs. The stability measures included the coefficient of variation (CV) (Francis and Kannenberg 1978), Finlay–Wilkinson regression slope (b) (Finlay and Wilkinson 1963), coefficient of determination (R^2) (Pinthus 1973), mean square error of the residuals (MSE) (Eberhart and Russel 1966), environmental variance (S^2) (Roemer 1917) and Wricke's Ecovalence (W^2) (Wricke 1962). The organic YQ population tended towards the highest stability for three parameters (R^2, MSE and W^2). This population achieved a mean yield of 4.9 t ha^{-1} over the five years, with only the reference cultivar Achat achieving a higher mean yield of 5.1 t ha^{-1}. The conventionally managed YQ population displayed the best stability values for mean yield (4.8 t ha^{-1}) and for S^2, indicating a low variance in yield achieved over the years and describing the potential of this population to buffer GxE interactions. This buffering capacity supports the hypothesis that increasing genetic diversity should increase the capacity and resilience of populations or systems to cope with changing and challenging abiotic and biotic stresses (Wolfe and Döring 2016; Döring et al. 2011; Dwivedi et al. 2017; Raggi et al. 2017). The YQ population is a cross of all 20 parents and therefore, the most genetically diverse in comparison to the Q and Y populations (see Fig. 6). Of great interest are the separate evolutions that the populations seem to take, depending on the management system in which they are grown. The Y populations, for example, differed in terms of their regression coefficient (b) depending on whether they were grown under organic or conventional conditions. Under conventional conditions, the Y population (b = 1.24) was better able to take advantage of the mineral N inputs, thereby illustrating a better proficiency and adaptability to conventional inputs and readily available N. The regression coefficient of the organically managed Y population was closer to 1, reflecting the poorer ability of the population, in comparison to the conventional Y population, to take advantage of available N in an organic management system where mineralization is slower and nutrients not so readily available. A similar pattern was found by Bertholdsson et al. (2016), whereby the Y CCs displayed a tendency towards lower early vigour in the F_{11} based on root weight in a hydroponic experiment. The organic Y CCs, however, had adapted to the organic system through a significantly larger root system in comparison to their conventional counterparts, highlighting the potential of CCs to adapt to particular environments and management systems. The reference cultivars Achat and Capo were found to be less stable for the majority of stability parameters in comparison to the organic CCs, although the cultivar Achat tended towards the highest mean yield over the five years (Weedon et al. 2016b). The separate evolutions of CCPs are of great relevance to CC research and on-farm implementation as they indicate the potential of adaptation to differing management and farming systems, which include not only differing nutrient regimes, but also minimal tillage and mixed cropping systems. Undoubtedly, genetically diverse populations provide a novel and efficient way to create heterogeneous crops suited to unique and changing environments. Nevertheless, further breeding and selection is needed in this area.

In addition to severe meteorological events, several rust epidemics occurred (Table 5). During the severe brown rust (caused by *Puccinia tritici*) epidemic in 2009, mean severity in the organic CCs was 4% on June 29, while severity on the most susceptible cultivar *Bussard* was 14% (Fig. 7). By July 9, the mean severity on the organic CCs was 59%, while

Figure 7 Brown rust on 7 July 2009 on cultivar Bussard (left) and one of the YQ CCs (right). Photos: M. R. Finckh.

on the cultivars, with the exception of *Tamaro* that was very resistant, the severities ranged between 56% and 93% (Table 6). The variation in cumulative disease was much lower for the CCs than for the references, but also for the parents grown in an adjacent field (Fig. 8, top). During the 2015 yellow rust (caused by *Puccinia striiformis*) epidemic, observations were very similar with the range of severities very predictable for the CCs, but not for the references and parents (Fig. 8 bottom).

In addition to the air-borne diseases, foot rot pathogens such as *Pseudocercosporella herpotrichoides* and *Fusarium* sp. occurred, usually at very low to moderate severity levels. As expected, there were no significant disease-reducing effects through the CCs against these foot rot pathogens because there are no race-specific resistances known to these pathogens.

Besides the generally low to moderate problems with pathogens, the CCs also appear to be very competitive against weeds. Data are currently being collected, but during the spring of 2017, which was a year favouring poppy (*Papaver rhoeas*) and corn flower (*Cyanus segetum*) germination, the YQ CC growing in a farmer's field, sown on the same day by the same farmer with the same equipment, clearly demonstrated its ability to suppress these weeds (Fig. 9).

The CCs have been very resilient with respect to abiotic and biotic environmental stresses. A number of CCs were also tested in several other European countries since 2013 in an EU-Core Organic consortium (see www.cobra-div.eu, last access 15 July 2017) with generally very good results in terms of yield and resilience, and these CCs remained stable even under extreme environmental changes from year to year (Weedon et al. 2016a).

Table 6 Brown rust severity (% diseased leaf area) in organic CCs and five reference cultivars on two dates in 2009 (unpublished data)

Entry	29 June	9 July
Composite crosses		
F_8-OYQI	4.3	66.0
F_8-OYQII	2.8	66.7
F_8-OQI	7.1	72.2
F_8-OQII	3.1	45.0
F_8-OYI	1.5	38.1
F_8-OYII	4.3	67.2
Mean CC	3.9	59.2
Commercial pure line cultivars		
Achat	9.7	93.3
Batis	4.2	55.9
Bussard	14.1	78.5
Capo	2.7	Senescent
Tamaro	1.2	8.9

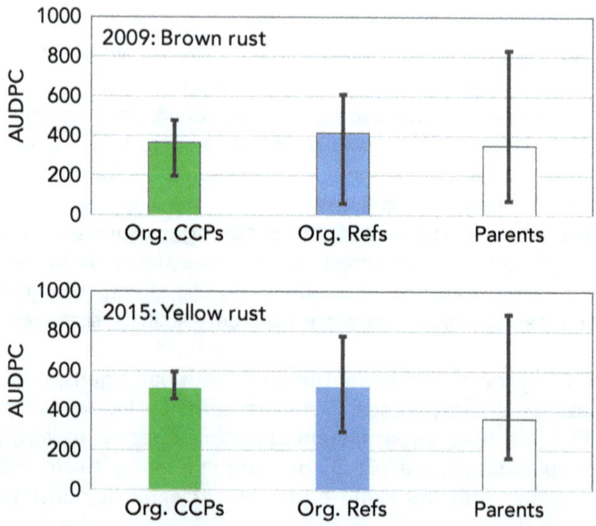

Figure 8 Mean cumulative disease (AUDPC = area under the disease progress curve) in 2009 (top) and 2015 (bottom) for the organic CC populations, the reference varieties grown in the respective years and the parental cultivars. The bars represent the range of values observed for the respective groups.

Figure 9 Farmers' field in 2017 planted with a pure line cultivar (left) and the YQ CCP F_{16} (right) showing clear weed suppression by the CC. Photo: S. Heinrich.

5 Requirements for agroecology-based ecological plant protection

The 'Organic 3.0' discussion paper (Arbenz et al. 2015) clearly indicates that organic farming needs 'continuous improvement towards best practice' and that overall, a much more agroecology-centred focus is needed in order to make advances in organic farming. Such an approach is multidisciplinary with plant health as a focus, coupled together with enhancing long-term ecological services and systems effects and addressing socio-economic questions (Wolfe and Döring 2016).

Changing the perspective towards how management affects pathogens and pests changes the focus from an agronomic to an agro-ecological approach that has been suggested for more than 30 years (Altieri 1987; Altieri and Liebman 1986; Risch et al. 1983; Altieri and Letourneau 1982; Vilich and Sikora 1996). The agroecology approach has also been named 'Biological Systems Management' in the context of plant protection (Vilich and Sikora 1996), focusing on the long-term effects on soils and spatio-temporal effects on pests and beneficials. Alabouvette et al. (2006) pointed out that methods for enhancing biocontrol within the system are needed and that '*...this strategy is much more difficult to follow than a systematic application of pesticides or BCAs* (biological control agents, added by authors). *It will not provide ready-to-use tools applicable in any situation. On the contrary, it involves ecological studies to determine the most suitable practices in a given geographical region. It also implies that the cropping system at the farm level and* *probably at the regional level must be taken into consideration, rather than with individual crops separately*'.

During the last century, a decline in soil organic matter worldwide and ever-increasing erosion problems have become evident (Bellamy et al. 2008; Lal 2005; Capriel 2013). Turbé et al. (2010) pointed out that 'while in general chemical treatments and tillage aimed at improving soil fertility trade off with soil carbon storage and decontamination services, in contrast mulching, composting and crop rotations all contribute to improve soil structure, water transfer and carbon storage'. The main aim of organic farming is to conserve and improve soil quality (IFOAM 2014). This aim can only be achieved by taking a systems approach based on agroecological interactions that need to be managed and guided for maximum plant health and productivity, a prerequisite for the plant's ability to return carbon to the soil. Recently, therefore, a movement for 'regenerative agriculture' has emerged, calling for methods to restore what has been lost (e.g. Soloviev and Landua 2016), a focus that is also imminent to agroecology research.

Agroecology research means looking at systems holistically, while studying all important factors intimately to gain a better understanding of the interactions. Generally, longer-term trials are needed to determine system changes and resilience. The potato-mulch system described above is used here to point out the main factors that lead to system changes and their potential effects on system resilience with respect to plant health, soil quality and productivity. Figure 5 illustrates the complex interactions in the system as affected by mulch properties, cover crops and weather. Target indicators in Fig. 5 are weeds. The interactions with respect to diseases and insect pests would have to be added. As discussed above, mulch applications affect the microclimate and potentially the dispersal of pests. In addition, effects on beneficials and plant nutritional status may play a role. For example, N-acquisition and dynamics in the mulched crops change dramatically compared to unmulched crops (Forge et al. 2003), and there are open questions about the nitrate versus ammonia contents and their effects on pests (Smiley 1978). If sustainability in the sense of *meeting the needs of the present without compromising the ability of future generations to meet their own needs; conserving an ecological balance by avoiding depletion or destruction of natural resources* (Arbenz et al. 2015) is aimed at, systems analysis will be about process optimization rather than maximization.

Optimization will need to include questions about the best plant genotypes and the best level of diversity. This includes breeding for diversity and plant mixtures, for weed suppression, adaptation to nutrition within the organic cycle and probably some other important properties. First results of the work on CCs above show that populations apparently do adapt to the different management systems. Thus, early root and shoot lengths were significantly increased after five years of organic management compared with the conventional management (Bertholdsson et al. 2016). In contrast, long-term breeding trends have led to decreases in competitiveness in cereals (Bertholdsson 2003) and since the middle of the twentieth century, dependence on mycorrhiza in wheat has declined (Hetrick et al. 1993). CCs consist of genotypes differing considerably in their ability to produce progeny, which affects the effective population size, a factor that needs to be taken into account when maintaining populations (Brumlop 2017; Goldringer et al. 2006). The large inter-genotypic variation may be important in contributing to yield stability because a very strong connection apparently exists between increasing diversity in yielding ability of single mixture components in genotype mixtures (Kiær et al. 2009).

Two important factors need special attention when developing systems for enhanced agroecological benefits and interactions based on soil regeneration and biodiversity if such systems are to be made use of in future agriculture. On the one hand, there is a need to adapt and develop agricultural technology for managing such systems. Reduced tillage and mulch-based agriculture requires specially adapted planting equipment, while mixed cropping requires special mechanical weed control, harvesting and separation technology. Precision-type equipment often based on robotics for recognition of problem weeds or slugs, for example, is likely to make important contributions in the future.

On the other hand, policy and legislation need to take into account agroecological approaches. The greatest limiting factor in the development of and a wider scale implementation of agroecology and sustainable food systems is the lack of significant research funding for multidisciplinary collaboration that prioritizes whole-systems research (Carlisle et al. 2017; Dwivedi et al. 2017). Weiner (2017) discusses the limiting framework of 'political and economic structures' inhibiting the application of agroecological approaches and argues that the goal of sustainable food production can only be achieved once agriculture is no longer treated as an economic commodity, but sustainability is prioritized as an actual goal. Breeding and selling of diversified germplasm is impeded by legal hurdles (Brumlop et al. 2013), while in the current subsidy system in Europe, it is often quite difficult to have mixed cropping supported. The marketing of heterogeneous populations is currently possible for wheat, oats, maize and barley under the Council Directive 2014/150/EU until 31 December 2018 (Commission 2014). The use of heterogeneous populations has been hampered by the misconception that these populations cannot fill the requirements for cultivar registration, particularly with regard to the DUS (distinctness, uniformity, stability) criteria of the International Union for the Protection of New Varieties of Plants (UPOV 2002). These are more easily applicable to genetically uniform lines and F_1 hybrid cultivars (Dwivedi et al. 2017; Raggi et al. 2017). Methods for characterization of heterogeneous populations are needed in order to sufficiently describe these populations. This, in turn, will ensure ease of heterogeneous population registration and marketing. So far, mainly organic breeders and researchers have supported and pursued heterogeneous crop populations as a viable alternative to conventionally bred cultivars (Döring et al. 2015; Raggi et al. 2017).

The organic seed sector faces three main issues in terms of seed production. First, only a low number of cultivars that are suitable for organic production are available. Secondly, little agronomic information is available on a number of cultivars under organic management and lastly, there is an insufficient volume of organically certified seed currently being produced (Döring et al. 2012a). As the organic seed sector is still facing these issues, it is more than likely that the supply of sufficient seed of heterogeneous crop populations will also be hampered initially, until supply can meet market demand. Other hindrances to the widespread use of CCs include the reluctant acceptance of CCs by millers, in the case of wheat CCs, or other processors down the value chain. A number of projects and breeders are working on these particular aspects in order to involve the entire value chain and to pave the way for easier market acceptance. The use of a specific label on CC products, such as a 'Diversity' label may make acceptance easier. This labelling idea has been pursued within organic agriculture through labels such as 'Bioverita' (www.bioverita.ch, last access July 2017), for example, in order to promote organically bred cultivars. Farmers that are able to process the CCs themselves in the form of farm mills and bakeries have an advantage in this sense, and can use the CCs as a novel marketing concept. The testing of winter wheat CCs by bakers is of great interest. Already there has been sensory testing of organic cultivars and heritage grains such as emmer, in order to find unique and desired tastes (Kucek et al. 2016). Some quality and sensory testing has been done with winter wheat CCs with promising results (Wolfe, personal communication). CCs may likely play an important role in the future, as the merits of diversifying not only food systems, but also diet become known (Dwivedi et al. 2017). Rahmanian et al. (2014) refer to reports from Italian and Iranian farmers whose bread, produced from wheat CCs, could be eaten by consumers with gluten intolerances.

The financing of CCs is another question that needs to be addressed. Breeders that have created these CCs need to be compensated for their work through the cost of the seed and the breeder's fee that follows every year should a farmer decide to re-sow saved seed. This becomes a contentious issue when one considers the ability of the CCs to adapt and change to differing environments. At what point has a CC changed enough that it no longer belongs to the breeder, but to the farmer having shaped its evolution? Possible solutions could include a time limit on fees for farm-saved seed to be paid to breeders or a premium on CC products paid by the end consumer, but used to support the work of CC breeders.

In our future experiments, we want to combine the advantages of the presented systems. Evolutionary populations adapted to minimum tillage systems could be selected for weed competitive ability, while at the same time develop a co-evolutionary relationship with related rhizobacterial communities (Pérez-Jaramillo et al. 2016), which improve nutrient uptake and contribute to living soils (Dessaux et al. 2016). In this way, the main disadvantages of reduced tillage systems, such as restricted nutrient uptake and greater weed occurrence (Carr et al. 2012), could be counteracted. The cultivation of a living soil through regenerative agriculture can enhance soil organic carbon content and soil macroaggregation (Zhang et al. 2017), increase water capacity (Vogeler et al. 2006) and suppress pathogens (Peters et al. 2003). Less soil disturbance through management practices such as minimum tillage, reduced pesticide use and application of mineral fertilizers will lead to a more resilient and regenerative agriculture, but alternative options and management systems need to be further explored and tested.

6 Conclusion and future trends

The most effective soil protection and building is achieved by 'farming systems where soil protection objectives are fully integrated into the production philosophy and the farm plan' (SoCo 2009). The example described above on the development of a soil regenerative potato cultivation system RePCro is a good example of agroecology as science, movement and practice as described by Wezel and co-workers (Wezel et al. 2009). It is from the practice of single farmers that we started our scientific studies. At the same time there is a current movement by very enthusiastic farmers in Germany on the use of mulch techniques engaging in teaching each other and spreading the information with modern media (see: http://www.bodenfruchtbarkeit.net/, (in German) last access, 17 July 2017). Science is also making more and more use of modern communication technology to spread the ideas (Crossland et al. 2015; OSCAR 2016) (see www.covercrops.eu for direct access to the toolbox).

The benefits of biodiversity for plant protection, yield stability and soil conservation in general have been widely documented for intra-specific, as well as for interspecific diversity (Finckh and Wolfe 2015; Vidal et al. 2017; Ostergaard et al. 2009; Tooker and

Frank 2012). The current ecological view is that there are positive correlations between diversity and stability, on the one hand, and between diversity and productivity on the other hand (Loreau 2010; Tilmann et al. 1996; Barot et al. 2017; Newton et al. 2010). Where landraces do not play an important role any more, dynamically managed populations can provide the necessary material for farmer participatory approaches to breeding and selection for low-input agriculture producing on the long run locally adapted materials (i.e. modern landraces) (Dawson et al. 2008). This would contribute to an overall diversification of the agricultural system and thus to dynamic development and conservation of genetic resources. It is clear that there are many hurdles yet to be overcome before the widespread acceptance of CCs is cemented into common markets. What is clear is that CCs provide an interesting and promising alternative option to conventionally bred cultivars, and that they may help to address climate change, decentralize the breeding process, increase yield stability and diversity within the agricultural landscape and provide novel and diverse food products. Despite this knowledge, current agricultural production usually relies on single genetically homogeneous crops at a given time. Reasons for this are that there (i) is a lack of main crop cultivars specifically adapted to more diversified growing systems; (ii) is a lack of appropriate technologies for dealing with diversity and (iii) are legal hurdles and policies that need to be changed in order to support agricultural sustainability and its further development in practice.

The two main factors allowing for plant production in agriculture are the availability of fertile soils and the availability of genetic resources that are adapted to the given agro-ecosystem. Both are intrinsically connected to plant health management. The relationship between soil quality and plant health indicates that an agroecological approach to plant disease management has to include a strong focus on the improvement of soil quality, especially soil organic matter content. This, incidentally, is one of the main pillars of organic agriculture as defined by IFOAM/Organics International (IFOAM 2014). Crop cultivars have to be capable of coping with the environmental variation occurring from year to year and besides intra-varietal diversity, mixed cropping *per se* is expected to play a key role in improving crop performance and overall tolerance to stress (Barot et al. 2017; Newton et al. 2010). Future agricultural systems must not only allow for plant adaptation to every environment, whether this be climatic or management-based, but the regeneration of soils must be the baseline for these systems. Regenerating soils is no mean feat and agriculture will need to focus on 'High Biomass Cropping Systems', in order to improve and recreate soil organic matter that is 'labile, biologically active, chemically and biologically diverse and stable' (Weiner 2017).

7 Acknowledgements

Part of the work on RePCro (2014–15) was financed by the European Union FP7 Project no. 289277: OSCAR (Optimizing Subsidiary Crop Applications in Rotations). Some of the work on CCs has been financed by the German Ministry programme BÖLN (Project no: 2810OE082, and 2812OE21 (EU core Organic II COBRA) and by the BMBF grant INSUSFAR (Grant no: 031A350C). S.J. is being supported by a University of Kassel doctoral stipend.

8 Where to look for further information

So far, the only comprehensive treatise of plant disease management in organic agriculture has been published in 2015 (Finckh et al. 2015b). An important reason for this is because disease and pest management in organic agriculture is a question of systems research that often is considered agronomical in nature. Any research for organic farming has to be conducted within long-term organic systems as the interactions of soil conditions and plant reactions will otherwise give spurious results. This has been recognized by funding organizations focusing on organic farming who by now require such setups. Long-term organic research institutions exist in many countries in Europe, the United States and other regions, and many universities support organic research with at least one faculty addressing specifically these issues. Education for organic farming and research at the BSc and MSc level is offered in many European countries and more and more around the world.

9 References

Alabouvette C., Olivain C. and Steinberg C. (2006). Biological control of plant diseases: The European situation. *European Journal of Plant Pathology* 114:329–41.

Allard R. W. (1961). Relationship between genetic diversity and consistency of performance in different environments. *Crop Science* 1:127–33.

Allard R. W. (1990). The genetics of host- pathogen coevolution: Implications for genetic resource conservation. *Journal of Heredity* 81:1–6.

Allison F. E. (1973). *Soil Organic Matter and Its Role in Crop Production*. Elsevier, New York, NY.

Altieri M. A. (1987). *Agroecology. The Scientific Basis of Alternative Agriculture*. Westview Press and IT Publications, Boulder, CO and London, UK.

Altieri M. A. (1994). *Biodiversity and Pest Management in Agroecosystems*. Haworth Press, New York, NY.

Altieri M. A. and Letourneau D. K. (1982). Vegetation management and biological control in agroecosystems. *Crop Protection* 1:405–30.

Altieri M. A. and Liebman M. (1986). Insect, weed and plant disease management in multiple cropping systems. In: Francis C. A. (Ed.), *Multiple Cropping Systems*. Macmillan, New York, NY, pp. 183–218.

Alyokhin A., Porter G., Groden E. and Drummond F. (2005). Colorado potato beetle response to soil amendments: A case in support of the mineral balance hypothesis? *Agriculture, Ecosystems & Environment* 109(3):234–44.

Anon. (no year). *Teeltvoorschrift Phytophthora Infestans*, Nederlandse Voedsel-en Warenautoriteit Ministerie van Economische Zaken, Utrecht, The Netherlands.

Arbenz M., Gould D. and Stopes C. (2015). Organic 3.0 for truly sustainable farming & consumption. Launched at the ISOFAR International Organic EXPO 2015, Goesan County, South Korea. IFOAM – Organics International and Sustainable Organic Agriculture Action Network (SOAAN), Bonn, Germany, 28pp.

Barot S., Allard V., Cantarel A., Enjalbert J., Gauffreteau A., Goldringer I., Lata J.-C., Le Roux X., Niboyet A. and Porcher E. (2017). Designing mixtures of varieties for multifunctional agriculture with the help of ecology. A review. *Agronomy for Sustainable Development* 37(2):13. doi:10.1007/s13593-017-0418-x.

Bellamy P. H., Loveland P. L., Bradley R. I., Lark R. M. and Kirk G. J. D. (2008). Carbon losses from all soils across England and Wales 1978–2003. *Nature* 437:245–8.

Berner A., Hildermann I., Flibbach A., Pfiffner L., Niggli U. and Mader P. (2008). Crop yield and soil fertility response to reduced tillage under organic management. *Soil & Tillage Research* 101:89–96.

Bertholdsson N. O. (2003). Variation in allelopathic activity over 100 years of barley selection and breeding. *European Weed Research Society Weed Research* 44:78–86.

Bertholdsson N. O., Weedon O., Brumlop S. and Finckh M. R. (2016). Evolutionary changes of weed competitive traits in winter wheat composite cross populations in organic and conventional farming systems. *European Journal of Agronomy* 79:23–30.

Browning J. A., Simons M. D., Frey K. J. and Murphy H. C. (1969). Regional deployment for conservation of oat crown rust resistance genes. *Special Report, Iowa Agriculture and Home Economics Experiment Station* 64:49–56.

Brumlop S. (2017). Agrobiodiversity through evolutionary plant breeding – System specific adaptation of agronomic and morphological traits and effective populations size in composite cross populations of winter wheat. Dissertation, University of Kassel, Kassel, Germany.

Brumlop S., Reichenbecher W., Tappeser B. and Finckh M. R. (2013). What is the SMARTest way to breed plants and increase agrobiodiversity? *Euphytica* 194:53–66.

Brumlop S., Pfeiffer T. and Finckh M. R. (2017). Evolutionary effects on morphology and agronomic performance of three winter wheat composite cross populations maintained for six years under organic and conventional conditions. *Organic Farming* 3(1):34–50. doi:10.12924/of2017.03010034.

Brunner F. (2017). Kartoffel mulchen. Available at: https://humus-macht-leben.com/bodenfrucht-barkeit/kartoffel-mulchen/. Accessed 2 October 2017.

Capriel P. (2013) Trends in organic carbon and nitrogen contents in agricultural soils in Bavaria (South Germany) between 1986 and 2007. *European Journal of Soil Science* 64:445–54.

Carlisle L., Miles A., Delonge M. S. and Carlisle L. (2017). Triggering a positive research and policy feedback cycle to support a transition to agroecology and sustainable food systems. *Agroecology and Sustainable Food Systems* 41(7):855–79. doi:10.1080/21683565.2017.1331179.

Carr P. M., Mäder P., Creamer N. G. and Beeby J. S. (2012). Editorial: Overview and comparison of conservation tillage practices and organic farming in Europe and North America. *Renewable Agriculture and Food Systems* 27 (Special Issue 01):2–6.

Chellemi D. O., van Bruggen A. H. C. and Finckh M. R. (2015). Direct control of soilborne diseases. In: Finckh M. R., van Bruggen A. H. C. and Tamm L. (Eds), *Plant Diseases and their Management in Organic Agriculture*. APS Press, St. Paul, MN, pp. 217–26.

Commission E (2014). Commission implementing decision of 18 March 2014 on the organisation of a temporary experiment providing for certain derogations for the marketing of populations of the plant species wheat, barley, oats and maize pursuant to Council Directive 66/402/EEC. *Official Journal of European Union L* 82:29–36.

Crossland M., Fradgeley N., Creissen H., Howlett S., Baresel P., Finckh M. and Girling R. (2015). An online toolbox for cover crops and living mulches. *Aspects of Applied Biology* 129:1–5.

Datnoff L. E., Elmer W. H. and Huber D. M. (2007). *Mineral Nutrition and Plant Disease*. APS Press, St. Paul, MN.

Dawson J. D., Murphy K. M. and Jones S. S. (2008). Decentralized selection and participatory approaches in plant breeding for low-input systems. *Euphytica* 160:143–54.

Dessaux Y., Grandclément C. and Faure D. (2016). Engineering the Rhizosphere. *Trends in Plant Science* 21 (3):266–78. doi:10.1016/j.tplants.2016.01.002.

Dexter A. R. (2004) Soil physical quality: Part I. Theory, effects of soil texture, density, and organic matter, and effects on root growth. *Geoderma*, 120:201–14.

Döring T. F., Knapp S., Kovacs G., Wolfe M. S. and Murphy K. (2011). Evolutionary plant breeding in cereals – into a new era. *Sustainability* 3:1944–71.

Döring T. F., Bocci R., Hitchings R., Howlett S., Lammerts van Bueren E. T., Pautasso M., Raaijmakers M., Rey F., Stubsgaard A., Weinhappel M., Wilbois K. P., Winkler L. R. and Wolfe M. S. (2012a).

The organic seed regulations framework in Europe-current status and recommendations for future development. *Organic Agriculture* 2(3–4):173–83. doi:10.1007/s13165-012-0034-7.

Döring T. F., Pautasso M., Wolfe M. S. and Finckh M. R. (2012b). Pest and disease management in organic farming: Implications and inspirations for plant breeding. In: Lammerts van Bueren E. T. and Myers L. D. (Eds), *Organic Crop Breeding*. Wiley-Blackwell, Chichester, UK, pp. 39–60.

Döring T. F., Annicchiarico P., Clarke S., Haigh Z., Jones H. E., Pearce H., Snape J., Zhang J. and Wolfe M. S. (2015). Comparative analysis of performance and stability among composite cross populations, variety mixtures and pure lines of winter wheat in organic and conventional cropping systems. *Field Crops Research* 183:235–45.

Dwivedi S. L., Bueren E. T. L. V., Ceccarelli S., Grando S., Upadhyaya H. D. and Ortiz R. (2017). Diversifying food systems in the pursuit of sustainable food production and healthy diets. *Trends in Plant Science* 22:842–56. doi:10.1016/j.tplants.2017.06.011.

Eberhart S. A. and Russel W. A. (1966). Stability parameters for comparing yield. *Crop Science* 6:36–40.

Enjalbert J., Goldringer I., Paillard S. and Brabant P. (1999). Molecular markers to study genetic drift and selection in wheat populations. *Journal of Experimental Botany* 50:283–90.

Finckh M. R. (2008). Integration of breeding and technology into diversification strategies for disease control in modern agriculture. *European Journal of Plant Pathology* 120:399–409.

Finckh M. R. and Tamm L. (2015). Organic management and airborne diseases. In: Finckh M. R., van Bruggen A. H. C. and Tamm L. (Eds), *Plant Diseases and their Management in Organic Agriculture*. APS Press, St. Paul, MN, pp. 53–66.

Finckh M. R. and van Bruggen A. H. C. (2015). Organic production of annual crops. In: Finckh M. R., van Bruggen A. H. C. and Tamm L. (Eds), *Plant Diseases and their Management in Organic Agriculture*. APS Press, St. Paul, MN, pp. 25–32.

Finckh M. R. and Wolfe M. S. (2006). Diversification strategies. In: Cooke B. M., Gareth Jones D. and Kaye B. (Eds), *The Epidemiology of Plant Disease*. Springer, Dordrecht, The Netherlands, pp. 269–308.

Finckh M. R. and Wolfe M. S. (2015). Biodiversity enhancement. In: Finckh M. R., van Bruggen A. H. C. and Tamm L. (Eds), *Plant Diseases and their Management in Organic Agriculture*. APS Press, St. Paul, MN, pp. 153–74.

Finckh M. R., Tamm L. and Bruns C. (2015a). Organic potato disease management. In: Finckh M. R., van Bruggen A. H. C. and Tamm L. (Eds), *Plant Diseases and their Management in Organic Agriculture*. APS Press, St. Paul, MN, pp. 239–57.

Finckh M. R., van Bruggen A. H. C. and Tamm L. (2015b). *Plant Diseases and their Management in Organic Agriculture*. APS Press, St. Paul, MN, 414pp.

Flemming C. A. and Trevors J. T. (1989). Copper toxicity and chemistry in the environment: A review. *Water, Air, and Soil Pollution* 44(1):143–58. doi:10.1007/bf00228784.

Finlay K. W. and Wilkinson G. N. (1963). The analysis of adaptation in a plant-breeding programme. *Australian Journal of Agricultural Research* 14:742–54.

Forge T. A., Hogue E., Neilsen G. and Neilsen D. (2003). Effects of organic mulches on soil microfauna in the root zone of apple: Implications for nutrient fluxes and functional diversity of the soil food web. *Applied Soil Ecology* 22(1):39–54. doi:10.1016/S0929-1393(02)00111-7.

Francis T. R. and Kannenberg L. W. (1978). Yield stability studies in short-season maize. I. A descriptive method for grouping genotypes. *Canadian Journal of Plant Science* 58:1029–34.

Freckman D. W. (1988). Bacterivorous nematodes and organic-matter decomposition. *Agriculture, Ecosystems & Environment* 24(1):195–217. doi:10.1016/0167-8809(88)90066-7.

Gamliel A. and van Bruggen A. H. C. (2016). Maintaining soil health for crop production in organic greenhouses. *Scientia Horticulturae* 208:120–30. doi:http://dx.doi.org/10.1016/j.scienta.2015.12.030.

Goldringer I., Prouin C., Rousset M., Galic N. and Bonnin I. (2006). Rapid differentiation of experimental populations of wheat for heading-time in response to local climatic conditions. *Annals of Botany* 98:805–17.

Gosling P., Hodge A., Goodlass G. and Bending G. D. (2006). Arbuscular mycorrhizal fungi and organic farming. *Agriculture, Ecosystems & Environment* 113(1–4):17–35. doi:10.1016/j.agee.2005.09.009.

Harlan H. V. and Martini M. L. (1929). A composite hybrid mixture. *Journal of American Society of Agronomy* 21:487–90.

Herz A., Köppler K., Vogt H., Elias E., Katz P. and Peters A. (2006). Biological control of the cherry fruit fly, *Rhagoletis cerasi* L. (Diptera, Tephriti-dae) by use of entomopathogenic nematodes: First experiences towards practical implementation. In: Boos M. (Ed.), *Ecofruit – 12th International Conference on Cultivation Technique and Phytopathological Problems in Organic Fruit-Growing: Proceedings to the Conference from 31st January to 2nd February 2006 at Weinsberg/Germany*. Fördergemeinschaft Ökologischer Obstbau e.V. (FÖKO)Traubenplatz 5, D-74189 Weinsberg, Germany, pp. 67–72.

Hetrick B. A. D., Wilson G. W. T. and Cox T. S. (1993), Mycorrhizal dependence of modern wheat cultivars and ancestors: A synthesis. *Canadian Journal of Botany* 71(3):512–18. doi:10.1139/b93-056.

Hiddink G. A., van Bruggen A. H. C., Termorshuizen A. J., Raaijmakers J. M. and Semenov A. M. (2005), Effect of organic management of soils on suppressiveness to *Gaeumannomyces graminis* var. *tritici* and its antagonist, *Pseudomonas fluorescens*. *European Journal of Plant Pathology* 113:417–35.

Hovmøller M. S., Walter S., Bayles R. A., Hubbard A., Flath K., Sommerfeldt N., Leconte M., Czembor P., Rodriguez-Algaba J., Thach T., Hansen J. G., Lassen P., Justesen A. F., Ali S. and de Vallavieille-Pope C. (2016). Replacement of the European wheat yellow rust population by new races from the centre of diversity in The near-Himalayan region. *Plant Pathology* 65:402–11.

Huber D. M. and Thompson I. A. (2007). Nitrogen and plant disease. In: Datnoff L. E., Elmer W. H. and Huber D. M. (Eds), *Mineral Nutrition and Plant Disease*. APS Press, St. Paul, MN, pp. 31–44.

IFOAM (no year). *Principles of Organic Agriculture*. Available at: www.ifoam.bio/en/organic-landmarks/principles-organic-agriculture. Accessed 2 October 2017.

IFOAM (2014). *The IFOAM Norms for Organic Production and Processing*. www.ifoam.bio/en/ifoam-norms. Accessed 2 October 2017.

Jain S. K. and Qualset C. O. (1975). New developments in the evaluation and theory of bulk populations. In: *Third International Barley Genetics Symposium*. Karl Thiemig, München, Germany, pp. 739–49.

Jaros-Su J., Groden E. and Zhang J. (1999). Effects of selected fungicides and the timing of fungicide application on *Beauveria bassiana*-induced mortality of the Colorado potato beetle (Coleoptera: Chrysomelidae). *Biological Control* 15:259–69.

Jones H., Clarke S., Haigh Z., Pearce H. and Wolfe M. (2010). The effect of the year of wheat variety release on productivity and stability of performance on two organic and two non-organic farms. *The Journal of Agricultural Science* 148(3):303–17. doi:10.1017/S0021859610000146.

Junge S., Reinhard-Kolempa M., Schmidt J. H. and Finckh M. R. (2016). Mulch und reduzierte Bodenbearbeitung im Öko-Anbau. *Kartoffelbau* 67(7):30–2.

Junge S., Saucke H. and Finckh M. R. (2017). Reduktion der Kartoffelkäferpopulation durch Minimalbodenbearbeitung und Transfermulch (Reduction of Colorado potato beetle population through reduced tillage combined with dead mulch materials). In: *Deutsche Pflanzenbautagung*, 26–29 September 2017, Witzenhausen, Germany.

Kayser B. (2002). *Mulch und Untersaaten im Gemüseanbau ohne Bodenbearbeitung – Ein kleines Handbuch für Selbstversorger und Erwerbsgärtner* (Ebook). 2 edn. Auflage self-published, Minden, Germany.

Kiær L. P., Skovgaard I. M. and Østergård H. (2009). Grain yield increase in cereal variety mixtures: A meta-analysis of field trials. *Field Crops Research* 114 (3):361–73. doi:http://dx.doi.org/10.1016/j.fcr.2009.09.006

Kinkel L. L., Bakker M. G. and Schlatter D. L. (2011). A coevolutionary framework for managing disease-suppressive soils. *Annual Review of Phytopathology* 49:47–67.

Knott D. R. (1972). Using race-specific resistance to manage the evolution of plant pathogens. *Journal of Environmental Quality* 1:227–31.

Koch E. and Groot S. P. C. (2015). Health management for seeds and other organic propagation material. In: Finckh M. R., van Bruggen A. H. C. and Tamm L. (Eds), *Plant Diseases and their Management in Organic Agriculture*. APS Press, St. Paul, MN, pp. 189–203.

Koike S. T. (2015). Organic vegetable disease management. In: Finckh M. R., van Bruggen A. H. C. and Tamm L. (Eds), *Plant Diseases and their Management in Organic Agriculture*. APS Press, St. Paul, MN, pp. 311–17.

Kucek L. K., Dyck E., Russell J., Clark L., Hamelman J., Burns-leader S., Senders S., Jones J., Benscher D., Davis M., Roth G., Zwinger S., Sorrells M. E. and Dawson J. (2016). Evaluation of wheat and emmer varieties for artisanal baking, pasta making, and sensory quality. *Journal of Cereal Science* 74(2017):19–27. doi:10.1016/j.jcs.2016.12.010.

Lal R. (2005). Soil erosion and carbon dynamics. *Soil and Tillage Research* 81:137–42.

Le Boulc'h V., David J. L., Brabant P. and de Vallavieille-Pope C. (1994). Dynamic conservation of variability: Responses of wheat populations to different selective forces including powdery mildew. *Genetics, Selection and Evolution* 26(suppl. 1):221–40.

Leoni C., Rossing W. and van Bruggen A. H. C. (2015). Crop rotation. In: Finckh M. R., van Bruggen A. H. C. and Tamm L. (Eds), *Plant Diseases and their Management in Organic Agriculture*. APS Press, St. Paul, MN, pp. 127–40.

Letourneau D. K. (1997). Plant-arthropod interactions in agroecosystems. In: Jackson L. E. (Ed.), *Ecology in Agriculture*. Academic Press, London, UK and New York, NY, pp. 239–90.

Letourneau D. K., Armbrecht I., Rivera B. S., Lerma J. M., Carmona E. J., Daza M. C., Escobar S., Galindo V., Gutierrrez C., López S. D., Mejíja J. L., Rangel A. M. A., Rangel J. H., Rivera L., Saavedra C. A., Torres A. M. and Trujillo A. R. (2011). Does plant diversity benefit agroecosystems? A synthetic review. *Ecological Applications* 21:9–21.

Litterick A. M., Harrier L., Wallace P., Watson C. A. and Wood M. (2004). The role of uncomposted materials, composts, manures, and compost extracts in reducing pest and disease incidence and severity in sustainable temperate agricultural and horticultural crop production: A review. *Critical Reviews in Plant Sciences* 23(6):453–79.

Loreau M. (2010). Linking biodiversity and ecosystems. *Philosophical Transactions of the Royal Society B* 365:49–60.

Mazzola M. (2004). Assessment and management of soil microbial community structure for disease suppression. *Annual Review of Phytopathology* 42:35–59.

Mazzola M. and Manici L. M. (2012). Apple replant disease: Role of microbial ecology in cause and control. *Annual Review of Phytopathology* 50:45–65.

Montgomery D. R. (2007). Soil erosion and agricultural sustainability. *Proceedings of the National Academy of Sciences* 104(33):13268–72. doi:10.1073/pnas.0611508104.

Murphy K., Lammer D., Lyon S., Brady C. and Jones S. S. (2005). Breeding for organic and low-input farming systems: An evolutionary-participatory breeding method for inbred cereal grains. *Renewable Agriculture and Food Systems* 20:48–55.

Neher D. A. (2010). Ecology of plant and free-living nematodes in natural and agricultural soil. *Annual Review of Phytopathology* 48:371–94. doi:10.1146/annurev-phyto-073009-114439.

Newton A. C., Gravouil C. and Fountaine J. M. (2010). Managing the ecology of foliar pathogens: Ecological tolerance in crops. *Annals of Applied Biology* 157:343–59.

OSCAR (2016). *Optimising Subsidiary Crop Applications in Rotations, Final Report*, pp. 39. http://cordis europa eu/docs/results/289/289277/final1-final-report-complete.pdf.

Ostergaard H., Finckh M. R., Fontaine L., Goldringer I., Hoad S. P., Kristensen K., Lammerts van Bueren E. T., Mascher F., Munk L. and Wolfe M. S. (2009). Time for a shift in crop production: Embracing complexity through diversity at all levels. *Journal of the Science of Food and Agriculture* 89:1439–45.

Paillard S., Goldringer I., Enjalbert J., Doussinault G., de Vallavieille-Pope C. and Brabant P. (2000). Evolution of resistance against powdery mildew in winter wheat populations conducted under dynamic management. I – Is specific resistance selected?. *TAG* 101:449–56.

Peigné J., Casagrande M., Payet V., David C., Sans F. X., Blanco-Moreno J. M., Cooper J., Gascoyne K., Antichi D, Bàrberi P., Bigongiali F., Surböck A., Kranzler A., Beeckman A., Willekens K., Luik A., Matt D., Grosse M., Heß J., Clerc M., Dierauer H. and Mäder P. (2016). How organic farmers practice conservation agriculture in Europe. *Renewable Agriculture and Food Systems* 31(01):72–85. doi:10.1017/S1742170514000477.

Pérez-Jaramillo J. E., Mendes R. and Raaijmakers J. M. (2016). Impact of plant domestication on rhizosphere microbiome assembly and functions. *Plant Molecular Biology* 90:635–44.

Peters R. D., Sturz A. V., Carter M. R. and Sanderson J. B. (2003). Developing disease-suppressive soils through crop rotation and tillage management practices. *Soil and Tillage Research* 72(2):181–92.

Phelan P. L. (1997). Soil-management history and the role of plant mineral balance as a determinant of maize susceptibility to the European corn borer. *Biological Agriculture & Horticulture* 15(1–4):25–34. doi:10.1080/01448765.1997.9755179.

Phillips S. L. and Wolfe M. S. (2005). Evolutionary plant breeding for low input systems. *Journal of Agricultural Science* 143:245–54.

Pieterse C. M. J., de Jonge R. and Berendsen R. L. (2016). The soil-borne supremacy. *Trends in Plant Science* 21(3):171–3. doi:10.1016/j.tplants.2016.01.018.

Pinthus M. (1973). Estimates of genotypic value: A proposed method. *Euphytica* 22:345–51.

Pocasangre L., Vicente L. P. and Ferris H. (2015). Organic banana disease management. In: Finckh M. R., van Bruggen A. H. C. and Tamm L. (Eds), *Plant Diseases and their Management in Organic Agriculture*. APS Press, St. Paul, MN, pp. 351–66.

Porcher E., Gouyon P.-H. and Lavigne C. (2004). Dynamic management of genetic resources: Maintenance of outcrossing in experimental metapopulations of a predominantly inbreeding species. *Conservation Genetics* 5(2):259–69. doi:10.1023/b: coge.0000030009.31028.99.

Raggi L., Ciancaleoni S., Torricelli R., Terzi V., Ceccarelli S. and Negri V. (2017). Evolutionary breeding for sustainable agriculture: Selection and multi-environmental evaluation of barley populations and lines. *Field Crops Research* 204:76–88. doi:10.1016/j.fcr.2017.01.011.

Rahmanian M., Salimi M., Razavi K., Haghparast R. and Ceccarelli S. (2014). Evolutionary populations: Living gene banks in farmers' fields. *Farming Matters*, 24–9.

Rhoné B., Remoué C., Galic N., Goldringer I. and Bonnin I. (2008). Insight into the genetic bases of climatic adaptation in experimentally evolving wheat populations. *Molecular Ecology* 17:930–43.

Riley H., Løes A-K, Hansen S. and Dragland S. (2003). Yield responses and nutrient utilization with the use of chopped grass and clover material as surface mulches in an organic vegetable growing system. *Biological Agriculture & Horticulture* 21(1):63–90. doi:10.1080/01448765.2003.97552 50.

Risch S. J., Andow D. and Altieri M. A. (1983). Agroecosystem diversity and pest control: Data, tentative conclusions and new research directions. *Environmental Entomology* 12:625–9.

Roemer T. (1917). Sind die ertragsreichen Sorten ertragssicherer? *Mitt DLG* 32:87–9.

Ruano-Rossil J. M., Radcliffe E. B. and Ragsdale D. W. (2004). Disruption of entomopathogenic fungi of green peach aphid, *Myzus persicae* (Sulzer), by fungicides used to control potato late blight. In: Simon J. C., Dedryver C. A., Rispe C. and Hullé M. (Eds), *Aphids in a New Millennium*. INRA, Versailles, France, pp. 365–70.

Sans F. X., Berner A., Armengot L. and Maeder P. (2011). Tillage effects on weed communities in an organic winter wheat-sunflower-spelt cropping sequence. *Weed Research* 51(4):413–21.

Schmidt J. H. (2017). Conservation agriculture in organic farming: Impacts on weeds and plant-parasitic nematodes during the transition. Dissertation, University of Kassel, Kassel, Germany.

Schmidt J. H., Bergkvist G., Campiglia E., Radicetti E., Wittwer R., Finckh M. R. and Hallmann J. (2017a). Effect of tillage, subsidiary crops, and fertilization on plant-parasitic nematodes in a range of agro-environmental conditions within Europe. *Annals of Applied Biology* 171:477–89.

Schmidt J. H., Finckh M. R. and Hallmann J. (2017b). Oilseed radish/black oat subsidiary crops can help regulate plant-parasitic nematodes under non-inversion tillage in an organic wheat-potato rotation. *Nematology* 19:1135–46.

Slater R., Ellis S., Genay J. P., Heimbach U., Huart G., Sarazin M., Longhurst C., Müller A., Nauen R., Rison J. L. and Robin F. (2011). Pyrethroid resistance monitoring in European populations of pollen beetle (*Meligethes* spp.): A coordinated approach through the Insecticide Resistance Action Committee (IRAC). *Pest Management Science* 67:633–8.

Smiley R. W. (1978). Antagonists of *Gaeumannomyces graminis* from the rhizoplane of wheat in soils fertilized with ammonium- or nitrate-nitrogen. *Soil Biology and Biochemistry* 10(3):169–74. doi:10.1016/0038-0717(78)90092-5.

Soane B. D., Ball B. C., Arvidsson J., Basch G., Moreno F. and Roger-Estrade J. (2012). No-till in northern, western and south-western Europe: A review of problems and opportunities for crop production and the environment. *Soil and Tillage Research* 118:66–87.

SoCo (2009). *Final Report on the Project 'Sustainable Agriculture and Soil Conservation (SoCo)'*. European Commission, 171pp. Available at: http://publications.jrc.ec.europa.eu/repository/handle/JRC51775

Soloviev, E. R. and Landua G. (2016). Levels of regenerative agriculture. Version 1.0 – September 2016. Terra Genesis International. Available at: http://www.terra-genesis.com/wp-content/uploads/2017/03/Levels-of-Regenerative-Agriculture-1.pdf. Last access 7 October 2017.

Steinert K. (2016). Pfluglose Biolandwirtschaft und Mischfruchtanbau im Saarland – Säen und in Ruhe lassen. *Landwirtschaft ohne Pflug* (11/2016):4–14.

Stirling A. M., Stirling G. R., Pegg K. G. and Hayward A. C. (1999). Effect of copper fungicide on *Colletotrichum gloeosporioides* and other microorganisms on avocado leaves and fruit. *Australian Journal of Agricultural Research* 50:1459–68.

Suneson C. A. (1956). An evolutionary plant breeding method. *Agronomy Journal* 48:188–91.

Szendrei Z., Kramer M. and Weber D. C. (2009). Habitat manipulation in potato affects Colorado potato beetle dispersal. *Journal of Applied Entomology* 133 (9–10):711–19. doi:10.1111/j.1439-0418.2009.01429.x.

Tamm L., Speiser B. and Holb I. (2015). Direct control of airborne diseases. In: Finckh M. R., van Bruggen A. H. C. and Tamm L. (Eds), *Plant Diseases and their Management in Organic Agriculture*. APS Press, St. Paul, MN, pp. 205–16.

Tilmann D., Wedin D. and Knops J. (1996). Productivity and sustainability influenced by biodiversity in grassland ecosystems. *Nature* 379:718–20.

Tooker J. F. and Frank S. D. (2012). Genotypically diverse cultivar mixtures for insect pest management and increased crop yields. *Journal of Applied Ecology* 49(5):974–85. doi:10.1111/j.1365-2664.2012.02173.x.

Turbé A., De Toni A., Benito P., Lavelle P., Lavelle P., Ruiz N., Van der Putten W., Labouze E. and Mudgal S. (2010) *Soil Biodiversity: Functions, Threats and Tools for Policy Makers*. Report for European Commission (DG Environment). Bio Intelligence Service, IRD, and NIOO, 250pp. Available at: http://ec.europa.eu/environment/archives/soil/pdf/biodiversity_report.pdf. Last access 2 October 2017.

UPOV (2002). General introduction to the examination of distinctness, uniformity and stability and the development of harmonized descriptions of new varieties of plants, pp. 1–26.

Vallad G. E. and Goodman R. M. (2004). Systemic acquired resistance and induced systemic resistance in conventional agriculture. *Crop Science* 44:1920–34.

van Bruggen A. H. C. (2015). Organic greenhouse production. In: Finckh M. R., van Bruggen A. H. C. and Tamm L. (Eds), *Plant Diseases and their Management in Organic Agriculture*. APS Press, St. Paul, MN, pp. 43–50.

van Bruggen A. H. C. and Semenov A. M. (2015). Soil health and soilborne diseases in organic agriculture. In: Finckh M. R., van Bruggen A. H. C. and Tamm L. (Eds), *Plant Diseases and their Management in Organic Agriculture*. APS Press, St. Paul, MN, pp. 67–89.

van Bruggen A. H. C. and Finckh M. R. (2016). Plant diseases and management approaches in organic farming systems. *Annual Review of Phytopathology* 54:25–54.

van Bruggen A. H. C., Gamliel A. and Finckh M. R. (2016). Plant disease management in organic farming systems. *Pest Management Science* 72:30–44.

Vidal T., Boixel A-L, Durand B., de Vallavieille-Pope C., Huber L. and Saint-Jean S. (2017). Reduction of fungal disease spread in cultivar mixtures: Impact of canopy architecture on rain-splash dispersal and on crop microclimate. *Agricultural and Forest Meteorology* 246:154–61. doi:https://doi.org/10.1016/j.agrformet.2017.06.014.

Vilich V. and Sikora R. A. (1996). Diversity in soilborne microbial communities a tool for biological system management of root health. In: Boland G. J. and Kuykendall D. L. (Eds), *Plant-Microbe Interactions and Biological Control*. Marcel Dekker, Inc., New York, NY, pp. 1–14.

Vogeler I., Horn R., Wetzel H. and Krümmelbein J. (2006), Tillage effects on soil strength and solute transport. *Soil and Tillage Research* 88 (1–2):193–204. doi:10.1016/j.still.2005.05.009

Weedon O., Brumlop S., Heinrich S., Boening A., Elsner M., Finckh M. R., Lammerts van Bueren E., Nuijten E., Messmer M., Baresel J. P., Goldringer I., Miko P., Megyeri M., Borgen A., Rasmussen S., Pearce B. and Wolfe M. S. (2016a). Agronomic performance of two generations (F12 and F13) of thirteen winter wheat composite cross wheat populations with differing cultivation histories in 2014/15. In: Kölliker R. and Boller B. (Eds), *Plant Breeding: The Art of Bringing Science to Life. Abstracts of the 20th EUCARPIA General Congress. 29 August – 1 September 2016*. Agroscope, Zürich, Switzerland, p. 192.

Weedon O., Brumlop S., Heinrich S. and Finckh M. R. (2016b). Yield stability analysis for three winter wheat composite cross populations under organic and conventional management over five years. In: Kölliker R. and Boller B. (Eds), *Plant Breeding: The Art of Bringing Science to Life. Abstracts of the 20th EUCARPIA General Congress. 29 August – 1 September 2016*. Agroscope, Zürich, Switzerland, p. 193.

Weiner J. (2017). Applying plant ecological knowledge to increase agricultural sustainability. *Journal of Ecology* 105:865–70. doi:10.1111/1365-2745.12792.

Weller D. M., Raaijmakers J. M., McSpadden Gardener B. B. and Tomashow L. S. (2002). Microbial populations responsible for specific soil suppressiveness to plant pathogens. *Annual Review of Phytopathology* 40:309–48.

Wezel A., Bellon S., Doré T., Francis C., Vallod D. and David C. (2009). Agroecology as a science, a movement and a practice. A review. *Agronomy for Sustainable Development* 29(4):503–15.

Wolfe M. and Döring T. (2016). The increasing need for an agroecological approach to wheat breeding. In: Bonjean A., Angus W. and Van Ginkel M. (Eds), *World Wheat Book*. Lavoisier, Cachan, France, pp. 845–72.

Wricke G. (1962). On a method of understanding the biological diversity in field research. *Z Pflanzenzücht* 47:92–6.

Zhang X., Xin X., Zhu A., Zhang J. and Yang W. (2017). Effects of tillage and residue managements on organic C accumulation and soil aggregation in a sandy loam soil of the North China Plain. *CATENA* 156:176–83. doi:10.1016/j.catena.2017.04.012.

Zwankhuizen M. J., Govers F. and Zadoks J. C. (1998). Development of potato late blight epidemics: disease foci, disease gradients, and infection sources. *Phytopathology* 88:754–63.

Chapter 11

Direct plant protection in organic farming

Bernhard Speiser, Hans-Jakob Schärer and Lucius Tamm, Research Institute of Organic Agriculture (FiBL), Switzerland

1 Introduction
2 Current practices in organic plant protection
3 Currently authorised materials
4 Dynamics of authorisation of materials in the EU
5 Efforts to replace copper fungicides
6 Future trends and conclusion
7 Where to look for further information
8 References

1 Introduction

1.1 Importance of direct and indirect plant protection in organic systems

In conventional agriculture, plant protection is strongly focused on the application of pesticides. By contrast, organic farming has a much broader view of plant protection, which also includes the maintenance of plant health through adapted management practices. The best-known examples of such indirect plant protection measures include the choice of adapted species and varieties, crop rotation, cultivation techniques, thermal methods and the protection and/or release of natural enemies. Such measures are highly specific for individual crops, pests/pathogens, regions and cropping systems. Readers with an interest in such practices may consult the book of Finckh et al. (2015), where an overview of such practices given by Van Bruggen (2015) and the roles of crop rotation (Leoni et al., 2015), water management (Koike and Cahn, 2015), biodiversity enhancement (Finckh and Wolfe, 2015) and crop resistance (de Milliano et al., 2015) in controlling plant diseases are discussed in detail. Detailed case studies for organic potato, small grains, rice, temperate legumes, vegetables, apple, grape, banana and coffee illustrate the potential and limitations of cultural practices versus direct plant protection in different crops and site conditions.

The principles of organic farming clearly state that priority must be given to indirect plant protection. Nevertheless, there are crops for which indirect measures alone do not provide sufficient protection. These crops would suffer heavy losses without additional

http://dx.doi.org/10.19103/AS.2017.0029.12

direct plant protection measures. This is particularly the case in perennial crops such as fruit trees and grapevine (Tamm et al., 2015).

This chapter focuses largely on direct plant protection and more precisely on the substances which may be used for this purpose. Nevertheless, we underline the importance of indirect plant protection measures for organic farming systems.

1.2 Permitted/prohibited substances: two complementary points of view

Organic direct plant protection could be described by listing those substances which may be used, or by those substances which are prohibited for use. The two approaches are complementary and reflect the fact that pesticides have both a positive and a negative reputation, depending on the stakeholders questioned.

The negative reputation of pesticides is associated with their unintentional and undesirable occurrence in the environment and in foods, and with their negative side effects on non-target organisms and human health. Some pesticides have turned out to be very stable in the environment and some accumulate in the food chain. Today, traces of pesticides can be detected in almost every environmental compartment (e.g. watercourses, soils) and many foods. Some pesticides pose a risk to non-target organisms (e.g. neonicotinoids on bees) and have been associated with negative impacts on human health and/or the environment (Pimentel et al., 1992; Malik, 2014; Bourguet and Guillemaud, 2016). From a consumer's point of view, it is important to know that organic farming avoids substances which are responsible for such undesirable effects. In a recently published survey among Swiss consumers, the non-use of synthetic pesticides was the second most important motivation to buy organic food (Stolz et al., 2017). In this context, it makes sense to describe organic plant protection by the substances that are not used in it, for example, 'organic farming does not use synthetic pesticides' (for an explanation of the term 'synthetic pesticides', see Section 4.3).

However, there are also positive aspects of pesticide use. From a farmer's point of view, pesticides are sometimes necessary to save crop health and are essential for economically viable production (Tamm et al., 2015). For an organic farmer, it is not enough to know which substances are prohibited, but he needs to know precisely which substances he is allowed to use. The greater part of this chapter is dedicated to describing the allowed substances (see Sections 3.2–3.7).

1.3 Regulation of pesticides as part of organic farming standards

In the context of organic standards, the range of allowed pesticides is only one aspect among many other rules for organic production. From the standard setter's point of view, the range of allowed pesticides can be regulated in the same way as the range of other kinds of 'inputs' such as fertilisers, disinfectants, feed materials and food or feed additives. Although the substances authorised for each use are different, the evaluation criteria and the legal processes for defining which substances are allowed are similar.

To avoid any ambiguity, most standards for organic production contain positive lists of allowed substances and specify that all substances not mentioned in the positive list shall be prohibited. In the European Union (EU), the substances allowed for plant protection are listed in Annex II of Regulation 889/2008. This Annex is regularly updated, because new substances are regularly developed to adapt to changes in pesticide legislation and in the authorised use of existing plant protection products. For this reason, we have chosen not to present a complete list of all authorised substances in this chapter, as this might be outdated before the book goes to print. Instead, we describe the major categories of substances which are eligible for authorisation and explain the evaluation criteria for new substances. The principles of organic production are universal, and the range of authorised substances is similar all over the world. By contrast, the legal structure and precise wording differ substantially between different laws and standards. For simplicity, this chapter is based on the European organic legislation.

2 Current practices in organic plant protection

Numerous farmer guides and handbooks provide advice on the organic management of individual crops, including detailed information on direct and indirect plant protection measures. Regarding disease control in several crops of major importance, the book of Finckh et al. (2015) contains dedicated chapters. The following paragraphs provide a very brief summary over current control practices for weeds, diseases and pests, so that the range of allowed substances can be found in the appropriate context. Growth regulators are not discussed in this chapter, because they are of little importance in organic farming. The only substances authorised for this purpose are ethylene (used for flowering induction in pineapple, for fruit ripening and for sprouting inhibition of tubers) and plant oils such as caraway oil and mint oil (used for sprouting inhibition of potatoes).

2.1 Control of weeds

Weed control on organic farms relies entirely on management practices. In arable crops, the main method is an adapted crop rotation. In grassland, mowing and/or grazing is the main method for weed control, but these methods are only effective with appropriate timing and intensity. In between the rows of perennial crops, these methods are also important, but the choice of species for undersowing also contributes to weed control. In annual crops, the choice of machinery and the timing and intensity of soil cultivation play an important role. In some high-value crops (mainly vegetables), hand weeding is used, but labour costs limit the extent of its use. Robots for weeding are not yet widely used, while interest in precision farming techniques is rapidly increasing and novel machinery becomes available. The use of smart machinery is fully compliant with the current regulations for organic production.

In contrast, no herbicides are authorised in EU organic production. For substances such as fatty acids and plant oils, their use as insecticides is authorised, while their use as herbicides is explicitly prohibited.

2.2 Control of plant diseases

The use of resistant or tolerant varieties is important to reduce the impact of plant diseases. However, this approach is often limited by the fact that resistant cultivars have different properties from other varieties. For example, scab-resistant apple varieties need far less fungicide sprays than typical varieties. Despite this positive trait, the introduction of resistant or tolerant apple varieties into the market takes many years, because consumers tend to know their favourite apples by the cultivar and avoid unknown new cultivars. In Switzerland this problem could be overcome with new marketing concepts which emphasise taste instead of cultivar denomination (Silvestri et al., 2008).

Among the substances of plant or animal origin, fennel oil, laminarin, chitosan hydrochloride, willow bark and horsetail extract may be used for the control of plant diseases. Under current European legislation, they are classified as fungicides, stimulators of the plants' natural defence system or basic substances (see Section 4.4). Copper compounds, sulphur, lime sulphur, potassium hydrogen carbonate, calcium hydroxide and a range of microbial biocontrol agents (e.g. *Aureobasidium pullulans*, *Coniothyrium minitans*, *Gliocladium catenulatum*) may be used to control plant diseases. Copper fungicides are discussed in detail in Section 5.

2.3 Control of pests

Invertebrate biocontrol agents are widely used, particularly in greenhouses. Eilenberg et al. (2001) distinguished 'inoculation biocontrol' and 'inundation biocontrol'. In the former, specimens for initial release are purchased from a commercial source, but will form reproducing populations inside the greenhouse after release. In the latter, the biocontrol agent does not form reproducing populations, but has to be released repeatedly. In practice, many biocontrol agents reproduce to some extent, but populations have to be augmented with additional releases from time to time. To improve their reproduction, many growers use 'banker plant systems' which provide shelter and complementary food to the biocontrol agent (Frank, 2010). Insect and mite biocontrol agents are mostly used in greenhouses. The main exception is *Trichogramma brassicae*, which is widely used in maize against the European corn borer. A few species can also be used in buildings for food storage such as silos and mills. Nematodes are used in covered crops and in the open field, but their use is restricted to horticultural crops for financial reasons. There are also several important microbial biocontrol agents, for example, *Bacillus thuringiensis*, *Beauveria* spp., *Metarhizium anisopliae* and *Cydia pomonella granulosis* virus. Many vertebrates (amphibians, reptiles, mammals and birds) include pests in their diet, and might thereby contribute to limiting outbreaks of pest populations. However, it is difficult to utilise vertebrates for targeted pest control. Perhaps the best-known example is the use of runner ducks (*Anas platyrhynchos*) for slug control (Speiser et al., 2001). This practice is usually not economic, and it is limited to smaller areas, due to the ducks' requirements for human attendance (fencing, water and shelter).

Methods such as mating disruption and mass trapping are among the most preferred ones. Pheromones are generally authorised for organic production, along with several other attractants. For the control of two particularly harmful pests, *Bactrocera oleae* and *Ceratitis capitata*, traps containing deltamethrin or lambda-cyhalothrin are currently authorised (these two substances are strictly limited to the use in traps and to the control of these two pests).

For direct control of insect pests, a number of substances of plant or mineral origin are authorised, for example, azadirachtin, pyrethrins, *Quassia* extract, rapeseed oil, paraffin oil, spinosad, aluminium silicate and soft soap (referred to as 'fatty acids'). Storage pests may be controlled with carbon dioxide or kieselgur. Pest slugs or snails may be controlled with ferric phosphate. Vertebrate pests such as deer and hares may be repelled with sheep fat or quartz sand.

3 Currently authorised materials

3.1 Legal background (European Union legislation)

Plant protection on organic farms is subject to the legislation on organic production as well as the legislation on plant protection. Only substances and practices which comply with both the legislations may be used. This is described for the EU in the following section.

In the context of plant protection legislation, the EU Regulation No 540/2011 (Annex I) contains a list of all substances which may be used as constituents of plant protection products (so-called active substances). In any given plant protection situation, however, only those plant protection products which are registered in the specific country for the specific use on a given crop may be applied. Pesticide registration also specifies details of use such as the dosage, the number and timing of applications and the pre-harvest interval. Pesticide legislation applies to all pesticides, regardless of whether they are used in conventional or organic agriculture, and it also covers plant extracts (e.g. azadirachtin), pheromones and microbial biocontrol agents (e.g. *B. thuringiensis*). Since a few years, the EU pesticide legislation recognises 'basic substances' as a separate category; for more details see Section 4.4.

In the context of organic legislation, Regulation No 889/2008 (Annex II) contains a list of all active substances which may be used in EU organic production. Only a small proportion of all the pesticides authorised for general agriculture are permitted for use in organic farming. The following sections give an overview of the authorised substances/organisms.

3.2 Invertebrate biocontrol agents

Predatory and parasitic insects, predatory mites and entomopathogenic nematodes are commercially available and are widely used for plant protection. Such organisms are collectively referred to as 'invertebrate biocontrol agents', and sometimes also as 'beneficials'. Examples of predatory insects include *Aphidoletes aphidimyza* (Diptera, Cecidomyiidae), *Adalia bipunctata* (Coleoptera, Coccinellidae) and *Chrysoperla carnea* (Neuroptera, Chrysopidae). Examples of parasitic insects include *T. brassicae* (Hymenoptera, Trichogrammatidae). Examples of predatory mites include *Amblyseius swirskii* (Gamasida, Phytoseiidae). Examples of entomopathogenic nematodes include *Steinernema carpocapsae*.

The use of invertebrate biocontrol agents is one of the most preferred methods of plant protection in organic farming (Speiser et al., 2006). Under EU legislation, they are not considered to be plant protection products and are therefore not listed in Annex II of Regulation 889/2008. Nevertheless, they may be used in EU organic farming. National regulations concerning the import and release of invertebrate biocontrol agents are not currently harmonised across Europe (Hunt et al., 2011).

3.3 Microbial biocontrol agents

Microbial biocontrol agents include bacteria (e.g. *B. thuringiensis*, *B. subtilis*), fungi (e.g. *Beauveria brongniartii*, *M. anisopliae*, *A. pullulans*) and viruses (e.g. *C. pomonella granulosis* virus). The longest known and most used microbial biocontrol agent is *B. thuringiensis*. Some strains can be used to control Lepidoptera, while others may be used to control Diptera (mainly aquatic disease vectors) or Coleoptera such as the potato beetle (*Leptinotarsa decemlineata*) (Bravo et al., 2011). Genes from *B. thuringiensis* are often used to develop insect-resistant transgenic crops (soy, maize, rape and cotton), but transgenic crops may not be used in organic production.

Microorganisms are generally authorised for organic production, provided they are not genetically modified organisms (GMOs). In the EU, microbial preparations are regarded as plant protection products, and only registered products may be used.

3.4 Substances of plant origin

Substances of plant origin have traditionally been used for plant protection in organic farming. All plant oils are authorised (e.g. rapeseed or sesame oil as insecticide, fennel oil as fungicide and caraway or mint oil as sprouting inhibitors; for a discussion on this, see Section 4.4). The following substances are also authorised: azadirachtin, laminarin, pyrethrins and *Quassia* extract. Since 2016, substances of plant or animal origin which are approved as 'basic substances' (see Section 4.4) and which are food may also be used. This group comprises lecithins, sucrose, fructose, vinegar and horsetail extract. Nicotine extract and rotenone were authorised earlier, but are not any more due to toxicological concerns.

3.5 Substances of animal origin

Substances of animal origin have also traditionally been used for plant protection in organic farming. All pheromones are authorised (used for mating disruption or mass trapping). Beeswax is used as a pruning agent, sheep fat as a repellent for game animals, hydrolysed proteins as attractants for pest insects and whey and chitosan hydrochloride to stimulate plants' natural defences.

3.6 Substances of microbial origin

At present, spinosad is the only substance of microbial origin which may be used for plant protection in organic farming. It is an insecticide with well-known effectivity, and it is also widely used in conventional production. When spinosad was approved for EU organic farming, the experts clarified that microbial products are not automatically approved as a group, but need to be evaluated and authorised individually (Forster et al., 2008). Derivatives of microbial products (e.g. strobilurins) are not authorised for organic farming.

3.7 Other substances

The group of 'other substances' is a heterogeneous assemblage of substances, many of which have traditionally been used in organic farming. Some substances are directly obtained from nature as minerals (aluminium silicate, kieselgur and quartz sand). Paraffin oil is a natural constituent of petroleum, but undergoes substantial purification. Other substances occur in nature, but are obtained with chemical processes for practical reasons (often to ensure the necessary purity). This group includes carbon dioxide, ethylene, copper compounds, ferric phosphate, sulphur, lime sulphur and potassium hydrogen carbonate. Calcium hydroxide and soft soap are obtained with 'simple chemical processes', but have traditionally been used.

3.8 Availability of authorised substances by country

As explained earlier, only those plant protection products which are registered in a specific country for a specific use on a specific crop may be applied. If this is not the case, the product may not be used. There are marked differences between EU member states in which products are registered, leading to very heterogeneous availability of products. Because only few substances are authorised for organic farming, the non-availability of one substance can often not be compensated, because no other substance with a comparable effect is authorised. In some countries, the non-availability of substances at national level has been an important bottleneck for production. The situation in 2004 was described by Speiser and Schmid (2004). Although this report is outdated in terms of individual substances, countries and uses, it illustrates the general pattern and the mechanisms behind it. The following paragraphs briefly discuss the main reasons why plant protection products are not always registered for all uses which are technically possible.

Distributors of plant protection products are private companies which consider registration as an investment to open new business opportunities. Before engaging in the registration procedure, they will estimate registration costs and potential payback from product sales. If estimated registration costs and financial risks outweigh the potential gains, they will not attempt registration. Such economic mechanisms have greatly limited the availability of plant protection products for organic farmers in many European countries (Ehlers, 2011).

Registration costs include costs for dossier preparation and registration fees. For complex mixtures of substances such as plant extracts, it may be very expensive to determine which substance(s) contribute to the pesticidal effect and to carry out all toxicological studies for each component of potential toxicological concern. For pheromones acting by mating disruption, the main challenge lies in establishing efficacy trials, because they act on large surfaces. An EU-funded project has elaborated proposals on how pesticide registration could be improved in the case of baculoviruses (Hauschild, 2011), bacteria and fungi (Strauch et al., 2011), plant extracts (Tamm et al., 2011a) and pheromones (Speiser et al., 2011).

In the past decades, registration of plant extracts and microorganisms in the EU has sometimes taken very long (Ehlers, 2011). This is another obstacle to registration, because it increases the time span between investment (costs for product development and registration) and payback (product sales).

The investments for registration must be proportionate to the market volume. The market volume depends on the surface of the crop in a given country, the importance of the pest or disease which is controlled, the efficacy of the pesticide and the presence or absence of competitor products in the country. Organic fields usually make up only a small percentage of the total area of a crop grown in a given country. Thus organic farming is only a small target market. In recent years, however, the demand for biocontrol agents and pheromones has greatly risen due to the fact that non-organic farmers are under pressure to reduce the levels of pesticide residues. This growing demand has stimulated a more rapid development of such products, which benefits organic farmers.

3.9 Grey areas in legislation

Plant health is the result of complex interactions between plants and their environment. Substances which are used for direct control of pests and diseases and which are therefore legally classified as pesticides have been discussed earlier. In addition, however, a number of disinfectants, fertilisers, trace elements, microorganisms, 'plant strengtheners' or other products also have the potential to influence plant health indirectly (Tamm et al., 2011b). In many cases, the mode of action is unknown, and reports on efficacy are controversial. European countries have taken different regulatory approaches to the use of such products (Speiser and Schmid, 2004). Thus, there is a grey zone where several legislations (national and EU) interact. The final decision whether such a product is allowed for organic farming depends on the individual constellation, and may vary between different countries. This is illustrated with a few examples in the following sections.

Herbal decoctions have traditionally been used for the maintenance of plant health. As soon as a plant protection claim is made (i.e. control of a specific pest or disease is mentioned on the product label), such a product has to be registered as a pesticide, which is a long and costly process (see the previous section). If no plant protection claim is made such products may be marketed without pesticide registration.

All trace elements which are authorised for conventional production may be used in organic farming. The EU fertilisers' legislation recognises substances such as 'copper salt', copper hydroxide and copper oxychloride as trace element fertilisers. The foliar application of such fertilisers is an obvious overlap with copper fungicides. Again, the key is whether or not a plant protection claim is made.

Disinfectants also have a role in maintaining plant health. This applies in particular to the disinfection of greenhouse equipment and equipment for growing mushrooms. Also, the disinfection of cutting tools is important to prevent the spread of bacterial diseases (such as fire blight). At the time of writing, the use of disinfectants in EU organic plant production is not yet regulated. According to a proposal by the Expert Group for Technical Advice for Organic Production (EGTOP) (see Section 4.2), similar substances should be authorised as in organic animal husbandry (EGTOP, 2016a).

4 Dynamics of authorisation of materials in the EU

The list of authorised pesticides is constantly evolving, as illustrated in the following section. Examples for the authorisation history of individual substances (including reasons for listing or de-listing) can be found in Speiser et al. (2014) or Tamm et al. (2015).

4.1 Developments in the last 25 years

When the European 'Organic Regulation' was first published in 1991 (EC, 1991), it contained 19 entries of individual substances or groups of substances which were authorised for plant protection. In 2016, the organic regulation contained 26 entries of authorised substances. A quantitative comparison of the two lists is not possible, because some items were split up, while others were pooled. Instead, the development is qualitatively described in the following sections.

Among the items which were cancelled, the majority had to be de-listed for reasons not related to organic production or organic principles (mainly alignment with pesticide legislation). Only few substances were de-listed because they were considered not to be suitable for organic farming any more.

4.1.1 Traditional substances still in use

Pyrethrins, extract of *Quassia amara*, sulphur, soft soap, pheromones and paraffin oil were continuously authorised from 1991 to 2016. Diatomaceous earth was authorised in 1991, then de-listed and finally reintroduced in 2016. Sodium bicarbonate was authorised in 1991 and later de-listed. In 2016, EGTOP has recommended its reintroduction, but by the time of writing, this has not yet taken place. However, the similar substance potassium bicarbonate was authorised. Among the microbial biocontrol agents, only *B. thuringiensis* and granulosis viruses were mentioned in 1991, while all microorganisms were later authorised. Microbial biocontrol agents were always considered as acceptable for organic farming, and the difference merely reflects the poor availability of such products in 1991. Copper fungicides were also authorised continuously, but not the same substances were allowed. In 1991, Bordeaux mixture (a mixture of copper sulphate and slaked lime) and Burgundy mixture (a mixture of copper sulphate and sodium carbonate) were authorised. In 2016, Bordeaux mixture is still authorised together with copper hydroxide, oxychloride, oxide and sulphate, while Burgundy mixture is not authorised any more. Stone meal was generically authorised in 1991. In 2016, this is not the case anymore. Instead, kaolin and quartz sand are listed as two separate items in Annex II. In 1991, plant and animal oils were authorised. In 2016, the entry has been changed to 'plant oils'. As there are no pesticides based on animal oils, these are not mentioned any more.

4.1.2 Historically used substances no longer in use

Propolis and sodium silicate were mentioned in 1991, but not any more in 2016. These substances were de-listed, because they are not considered to be pesticides under EU legislation. Extract of *Ryania speciosa* was mentioned in 1991, but not any more in 2016, because it is no longer registered as a pesticide in general agriculture as a result of the re-evaluation of pesticides under pesticide legislation. Extract of *Derris elliptica* (also known as rotenone) was also mentioned in 1991, but not any more in 2016. This substance was authorised as a pesticide in general agriculture, but the registration was not prolonged due to human health concerns. In all these cases, the substances were de-listed from the organic regulation due to factors outside the control of the organic sector.

Among the molluscicides, metaldehyde was authorised in 1991. In 2016, this substance is not authorised for organic production any more, but ferric phosphate is authorised. Metaldehyde is a synthetic substance, but it was traditionally authorised, because slugs can cause severe damage and there were no alternatives available for their control. Ferric phosphate is an example of a substance which occurs in nature but is chemically manufactured for practical reasons. When ferric phosphate became available, it was judged to be more acceptable than metaldehyde.

4.1.3 New substances approved since 1991

Several new substances are novel pesticides which were not developed in 1991. This group includes azadirachtin, spinosad, laminarin and sheep fat, as well as some of the basic substances authorised today.

Other substances were not authorised in 1991, but had to be authorised later because of urgent need. The first example is ethylene. Ethylene is mainly used for degreening bananas after transport from Central or South America to the final market in Europe. Its use in banana was considered as essential, and therefore authorised. Later, the use in kiwis, kakis, citrus (only as part of a strategy for the prevention of fruit fly damage), pineapple, potatoes and onions was also authorised. In 2016, all restrictions of the use of ethylene were deleted. This was done for legal considerations, and not because further uses were considered as essential. The second example is the pyrethroids deltamethrin and lambda-cyhalothrin. These two substances were authorised for the control of the olive fruit fly (*B. oleae*) and the Mediterranean fruit fly (*C. capitata*). Because these two substances are not well in line with the principles of organic plant protection (see Section 4.3), their use was limited to these two pests, and to the use inside traps (no spray applications). Also, the organic sector discusses whether they could be further restricted or phased out again (EGTOP, 2011).

Finally there is a group of substances which were not explicitly authorised in 1991, because they were not considered as pesticides at that time. These may seem to be 'new' substances, while in reality they have been traditionally used. Examples are horsetail extract (used for strengthening plants), lime sulphur (used against apple scab) and beeswax (used as pruning agent). For calcium hydroxide (used as trunk paint), hydrolysed proteins (used as attractants in traps) and carbon dioxide (used for conservation of food from storage pests), there is little evidence about their status in 1991, but we assume that they also belong to this group.

4.2 Authorisation process

This chapter describes the process for authorisation of new substances in EU organic farming. The process for de-listing of an existing substance or for changing the conditions for its use follows the same pattern. This section is specific for the situation in the EU at the time of writing, while the processes under other legal or private standards may be very different. In the EU, the authorisation process is governed by Article 16 of Regulation 834/2007. Three main steps can be identified: request; discussion and decision; and implementation.

4.2.1 Request

At EU level, the process officially starts with a request from an EU member state. However, national administrations usually make such a request only if the organic sector clearly expresses a need for that substance (for a discussion on necessity, see the following section). Thus, the request is usually preceded by lobbying activities. How this is done depends on the organisation of the organic sector and its connections with the administration, and may vary from country to country and from case to case. The organic sector and the manufacturers may support the administration by providing data on the substance or by drafting texts for the request.

4.2.2 Discussion and decision

Discussion of the requests requires a high degree of specialist knowledge in several disciplines such as organic practices, crop protection, chemistry and environmental sciences. The European Commission has therefore decided to seek technical advice from independent experts. In 2009, the European Commission created the EGTOP. Since then, requests for the authorisation of new substances were usually subjected to a technical evaluation by EGTOP.

EGTOP is consulted for different subjects related to plant production, animal husbandry and food processing. Until now, it has produced three reports related to plant protection products. All EGTOP reports are public. They are published on the European Commission's website[1]. When the technical evaluation by EGTOP is available, the European Commission discusses the request with the member state delegates and then takes a decision.

4.2.3 Implementation

Whenever a change of the current policy has been decided, an amendment of Annex II of Regulation 889/2008 is necessary. This is done with a separate Commission Implementing Regulation. This process takes several months. It is not specific for organic farming, but follows the general pattern for changes in European legislation. It is therefore not described in detail here.

4.2.4 Duration of the process

The entire process from the preparation of a dossier to the authorisation of a new substance may take several years. Input manufacturers as well as organic farmers are often frustrated by this duration, and have repeatedly proposed to establish a 'fast-track' procedure. However, there is a risk that with such a procedure, substances which are controversial within the organic sector or which might compromise the reputation of organic farming could be authorised inadvertently. Therefore, no fast-track procedure was established until now. However, the general authorisation of basic substances of plant or animal origin which are food was decided as a compromise.

4.3 Authorisation criteria

This section describes the major requirements which a new substance has to fulfil. The quintessence of these requirements is very similar for all organic farming standards, but the precise wording differs from one standard to another.

4.3.1 Defining 'natural' and 'synthetic' substances

It is intuitively clear that only 'natural' substances are eligible for use in organic farming, but what does this mean precisely? Substances produced by plants (e.g. pyrethrins), animals (e.g. beeswax) or microorganisms (e.g. spinosad) are eligible for organic production. Naturally occurring mineral substances such as quartz sand or kaolin (a clay mineral) are also eligible. By contrast, synthetic molecules which do not occur in nature (so-called 'synthetic pesticides') are not eligible.

In certain cases, the naturally occurring materials cannot be used for plant protection, because they are not available in sufficient quantities or in appropriate quality. In such cases, the substances may be obtained by chemical synthesis (so-called 'nature-identical' materials). This exception applies, for example, to pheromones, sulphur and bicarbonate. Nature-identical substances may be used for plant protection, but not as fertilisers. There are sufficient naturally occurring materials which may be used as fertilisers; therefore there is no need to authorise nature-identical materials for fertilisation.

1. https://ec.europa.eu/agriculture/organic/eu-policy/expert-advice/documents/final-reports_en.

It could be argued that all materials are ultimately derived from nature, by more or less complex processes. At the moment, the EU organic legislation does not explicitly specify which processing steps are allowed and which are not. However, the general understanding is that 'chemical processes' (i.e. processes which change the chemical structure) are prohibited, while 'physical processes' such as drying, milling and extraction with water are allowed.

4.3.2 Major authorisation criteria

The EU organic legislation defines the following authorisation criteria:

- Substances must be of plant, animal, microbial or mineral origin. This means that they must be natural in the sense discussed earlier.
- Organisms must not be GMOs, and substances must not be of GMO origin. In the context of plant protection, this concerns mainly microbial biocontrol agents, rapeseed oil and lecithin.
- Substances must not have inacceptable effects on the environment, and they must be harmless for human health. In the EU, effects on human health and the environment are assessed during pesticide registration, and the organic sector should not duplicate these efforts. In exceptional cases, however, the organic sector has a more strict view than general agriculture about which environmental effects are acceptable. For example, the EU organic legislation tolerates lower amounts of copper than pesticide registration (see Section 5).
- The substances and their use must be necessary. The use of a plant protection product can be considered necessary (i) if it controls pests or diseases at least partially, (ii) if there are no authorised alternative methods or products, or at least none which are practical and economic and (iii) if the pest or disease it controls has some relevance for organic farming.
- The substances and their use must comply with the traditions and principles of organic farming and with consumer expectancies. This point covers any additional aspects which might occur with certain substances or uses. Examples include aspects of animal welfare, food quality and/or authenticity, social effects, vegan nutrition, etc.

The aforementioned criteria are evaluated as a whole. For example, limited negative side effects may be tolerated, if a product is highly necessary (e.g. side effects of spinosad on certain non-target organisms). These criteria apply for the authorisation of new substances. For substances which have traditionally been used in organic farming, there is no formal requirement that they must also fulfil these criteria, but most of them do. Those few substances which do not fulfil the criteria are closely observed, and if possible replaced by new substances which comply better with the criteria. The best-known example is copper, which is discussed in detail in Section 5.

4.4 Substances exempt from individual authorisation in EU organic farming

When a new substance for plant protection is developed, it first has to be approved under pesticide legislation. Once this has been achieved, it may be used in conventional agriculture, but not in organic farming. Use in organic farming is only possible after it has also passed the authorisation process described earlier and is included in Annex II of Regulation 889/2008. However, the following four entries in Annex II refer to entire groups of substances (with some restrictions):

- microorganisms (not from GMO origin)
- pheromones (only in traps and dispensers)
- plant oils (all uses authorised, except herbicide)
- basic substances (only those which meet the definition of 'foodstuff' and have plant or animal origin)

Thus, any new microorganism which has been approved for use in general farming may be applied in organic production (as long as it is not a GMO). The same applies for any new pheromone or plant oil.

The entry of 'basic substances' requires some explanations. The category of 'basic substances' has been introduced in EU pesticides legislation a few years ago. It describes substances which may be used for plant protection but have a predominant use for another purpose. Registration of basic substances is easier, cheaper and faster than for ordinary pesticides, but does not offer any exclusivity of commercialisation to the applicant. Typically, requests for basic substances are made by non-profit organisations such as growers' associations and public-funded research institutes, while commercial companies prefer ordinary pesticide approval. Examples of substances approved as basic substances in the EU include calcium hydroxide, lecithin, sucrose and vinegar (note that rapeseed oil and garlic oil also have a predominant use outside plant protection, but are not categorised as basic substances, because this was not requested by the applicants). Basic substances are automatically approved for organic farming only (i) if they meet the definition of 'foodstuff' and (ii) if they have plant or animal origin. An example of such automatic approval is lecithin, which is a basic substance and is also food and of plant origin. By contrast, calcium hydroxide is a basic substance and food, but not of plant or animal origin. Calcium hydroxide is therefore not automatically authorised under the entry of basic substances. It may nevertheless be used in organic farming, because it is separately mentioned in Annex II.

4.5 Authorisation of commercial products

Pesticides are not used in the form of active substances, but as formulated products sold under a trade name. For older active substances without patent protection, there are numerous commercial products. As a practical guide for organic farmers, there are so-called 'input lists' in many countries, which list all authorised products, often together with the authorised uses. Input lists may be prepared by organic certifiers, authorities or specialised institutes.

As a minimum requirement, input lists must ensure that products comply with national pesticide legislation, that they contain only active substances mentioned in Annex II of Regulation 889/2008 and that these substances are not obtained from GMOs (e.g. rapeseed oil). Other input lists (e.g. the input lists for Switzerland) are based on criteria which go further and include not only the active substances but also all co-formulants. For example, insecticides based on pyrethrins often contain piperonyl butoxide as a 'synergist'. In many countries, such insecticides have traditionally been allowed, because synergists are not part of the evaluation criteria. By contrast, the Swiss input lists were always opposed to piperonyl butoxide, and manufacturers replaced it with other synergists such as sesame oil or rapeseed oil already in 1997–98. In 2014, the EGTOP has recommended to phase out the use of piperonyl butoxide in EU organic farming (EGTOP, 2014b).

5 Efforts to replace copper fungicides

Copper is one of the pesticides that are longer in use. Its effectiveness against grape downy mildew was discovered already in 1882. Today, it is used in various crops such as grapes, fruits, berries, vegetables, potatoes and ornamentals against a wide array of oomycete and ascomycete fungi as well as bacteria. Copper is of special importance for organic farming, because for a number of important diseases, copper is the only fungicide, or at least the most effective one. This is particularly true for downy mildew of grapevine (*Plasmopara viticola*), for apple scab (*Venturia inaequalis*) and for potato late blight (*Phytophthora infestans*). However, copper is still widely used in conventional agriculture, and the great majority of all copper fungicides are applied on conventionally managed surfaces (Hillenbrand et al., 2005).

The environmental fate and ecotoxicology of copper is complex. Copper naturally occurs in soils, water bodies and all living organisms. Small quantities of copper are essential for plants and animals, and copper deficiencies lead to severe symptoms. On the other hand, excessive amounts of copper are toxic. As copper is not metabolised in the environment, it gradually accumulates in the soil as a result of continuous applications. Although the precise dynamics of copper in the soil are not yet fully understood, there is general agreement that applications of copper fungicides should be limited as much as possible, to minimise any risks. Organic farming has long realised the potential negative impact of copper fungicides. Until now, however, copper is essential for several crops and its use in organic production could therefore not be prohibited.

The applied doses of copper have greatly varied over time. The greatest amounts of copper were applied between 1920 and 1955, where average doses of 10–15 kg/ha/yr and maximum doses of 80 kg/ha/yr of pure copper were applied in Eastern Switzerland. Before 1920 and after 1955, doses were more modest (average below 5 kg/ha/yr, maximum around 10–20 kg/ha/yr) (Räz et al., 1987). Many modern vineyards were already planted with grapevine around 1920–55, and have been subject to the high dosages of copper mentioned earlier. Even if very little copper is applied today, the copper levels in these soils are strongly influenced by historic copper applications (Strumpf et al., 2009, 2011).

National registrations of copper compounds set different maximum doses in different EU member states. In the context of the organic legislation, a uniform limit for copper use was introduced in 2002. It was originally set at 8 kg/ha/yr elementary copper, and then successively reduced to 6 kg/ha/yr in 2006. On many organic farms, today's copper use is considerably lower than this maximum limit. For example, Swiss organic winegrowers use approximately 3 kg/ha/yr on average. In other crops, doses vary between 0.1 kg/ha/yr (cucumbers) and 2.8 kg/ha/yr (potatoes) (Speiser et al., 2015).

5.1 Minimisation of copper use

Organic farmers throughout Europe have made great efforts to reduce copper use on their farms while retaining healthy crops. There is little scope to reduce the number of applications (which mainly depends on climatic conditions), but it has been possible to reduce the dosage per application significantly. There is no general agreement about the lowest effective dose. Many winegrowers successfully use doses around 200–300 g/ha per application, but some pioneers use doses of 100 g/ha or less. The following approaches have proven most successful (and are often applied in combination):

- optimisation of the timing of applications by use of forecasting models,
- optimisation of the coverage of foliage/fruits with a copper film through improved application technology (sprayers and/or nozzles),
- use of modern, rainfast copper formulations,
- tank mixes of copper fungicides with spreaders/stickers to reduce washing off, and
- adaptation of the spraying pattern (how often, how much) to the situation of the individual farm, taking into account all of the above points as well as local climatic conditions and disease pressure.

Last but not least, the susceptibility of the crop or cultivar plays a key role. However, it is not always possible for farmers to grow resistant/tolerant cultivars, because the market sometimes demands other varieties. Particularly in viticulture, the market strongly demands wines made from traditional varieties, and wines made from interspecific, resistant varieties are difficult to market. The survey mentioned earlier illustrates the potential of resistant varieties: while traditional grape varieties were treated with 2.9 kg/ha/yr copper, interspecific grape varieties are meanwhile treated with 0.5 kg/ha/yr only (Speiser et al., 2015).

5.2 Alternatives to copper fungicides

For many years, there has been an intensive search for substances which could replace copper fungicides. A number of substances have shown some effectiveness. These substances are at various stages of development or use.

- Potassium hydrogen carbonate is authorised for organic production and is commercially available. It can partially replace copper in fruit and vegetable crops.
- Lime sulphur is authorised for organic production and commercially available in most European countries. It is mainly used to control apple scab, but it also has insecticidal and acaricidal properties.
- Aluminium sulphate has the potential for partial replacement of copper in grapevine and apple, but is available in a few countries only. In Switzerland, it is registered as a plant protection product and authorised for organic production. In Germany, it is marketed with the claim of being a plant strengthener, and may therefore be used in organic farming.
- Phosphonates are commercially available in most countries. They could partially replace copper in a number of crops including grapevine, fruit, berries and vegetables. However, they are not authorised for organic production. For a number of years, there was a discussion on whether they should be authorised. However, EGTOP has recommended to not authorise them (EGTOP, 2014b). The main reason for rejection was that their use inevitably causes substantial residues in the food. Its synthetic manufacture and systemic mode of action were also seen as problematic.

- A number of new substances (plant extracts and microbial biocontrol agents) are currently under development (Dagostin et al., 2011). For reasons of intellectual property, the precise substances are usually published with some delay. For example, Prokop et al. (2016) reported effectiveness of a fatty acid against downy mildew of grapevine, apple scab, late blight on potato and tomato and peach leaf curl. For phenolics extracted from olive pomace, an antifungal effect was also reported (Winkelhausen et al., 2005). Among the microbial biocontrol agents, *Cladosporium cladosporioides* strain H39 is effective against apple scab (Köhl, 2015) and *Lysobacter capsici* strain AZ78 against downy mildew of grapevine (Segarra et al., 2016).

Until now, no single substance which could completely replace copper on its own has been found. However, several of these substances may partially replace copper on at least one crop. From what is known today, this is likely the case also for the new substances currently under development.

5.3 Other copper sources in agricultural production

Until now, the discussion on copper reduction has mainly focused on copper fungicides. However, copper fungicides are not the only sources which contaminate soils with copper. Other major sources are:

- Animal feeds are usually complemented with copper. As copper is typically added in excess to the animals' needs, it is excreted and ends up in the manure. There is considerable scope for reduction of copper in feed, particularly for young piglets. For many years, the private standards of Bio Suisse have greatly limited the copper ration in pig feed, without any signs of undersupply (EGTOP, 2014a).
- Many mineral fertilisers (e.g. rock phosphate), all composts and manures contain copper. The content of copper in manure is directly related to the dosage of copper in animal feeds. The case of copper as a trace element fertiliser is discussed in Section 3.9.
- Water naturally contains traces of copper, but irrigation equipment may greatly increase its copper content, depending on the alloys used.
- Finally, there are several non-agricultural sources of copper contamination of the environment, such as traffic and construction (Hillenbrand et al., 2005). In contrast to the aforementioned sources, organic farming has no influence on these sources.

Overall strategies to protect soils from copper pollution should not only consider the minimisation/replacement of copper fungicides, but also the minimisation of copper contamination via feed, fertilisers and irrigation water.

6 Future trends and conclusion

New active substances are continuously developed and authorised for use in general agriculture. In the case of substances of natural origin (see Section 4.3), the question whether they should also be used in organic farming arises. The EU organic legislation will therefore have to be updated regularly, as a reaction to progress in research and development. Examples are sodium hydrogen carbonate, willow bark extract ('Salix spp. cortex'), chitosan hydrochloride, COS-OGA (a complex of oligosaccharides and oligogalacturonans), garlic extract and diammonium phosphate (used as an attractant in traps). For these substances, EGTOP has recommended that they should be authorised for EU organic production (EGTOP, 2016b). However, no legal act has been published yet.

There are also substances which are not well in line with the principles of organic production, but had to be authorised due to the lack of effective alternatives. In these cases, further restrictions or a phase-out are likely in the longer term. Examples are copper fungicides (see Section 5) and the pyrethroids deltamethrin and lambda-cyhalothrin (see Section 4.1).

7 Where to look for further information

Information on approval of plant protection products for general agriculture is given by the competent authorities. The 'EU pesticides database' (http://ec.europa.eu/food/plant/pesticides/eu-pesticides-database) provides an overview on all active substances approved at EU level. Many countries, including all EU Member States, maintain registers of plant protection products registered for use in that country. These registers also inform on the conditions for use of the products, such as the crops on which they may be used, dosage, timing, number of applications, precautionary measures, pre-harvest intervals etc. The registers range from simple tables to sophisticated databases and most of them are only available in the national language(s).

Information on authorisation of substances for organic farming at EU level is provided by Regulation 889/2008, Annex II (http://eur-lex.europa.eu/legal-content/en/ALL/?uri=CELEX:32008R0889). The EU Commission maintains a 'consolidated version' which incorporates all changes which were made since it was published in 2008 (to be found on the same website). Lists of commercial products authorised for organic production can be found in so-called 'input lists', for example the input lists for Switzerland (www.betriebsmittelliste.ch), Germany (www.betriebsmittelliste.de), Austria (http://www.infoxgen.com/produktsuche-biologisch.html) and The Netherlands (www.dutchinputlist.fibl.org). A similar list exists also for the USA (https://www.omri.org/omri-lists). Information on the reasons why a given substance should or should not be authorised for EU organic farming can be found in the reports of EGTOP (https://ec.europa.eu/agriculture/organic/eu-policy/expert-advice/documents_en).

Information on new active substances under development is scarce, because the involved companies safeguard their intellectual property. Within these constraints, the 'Annual Biocontrol Industry Meeting' (ABIM) provides an excellent forum, where recent developments worldwide are reported (http://www.abim.ch). The 'European Copper Conference' organised annually by the Julius-Kühn institute provides an update on all questions relating to the reduction or replacement of copper fungicides (http://kupfer.julius-kuehn.de).

Practical guidance for organic farmers on how to control specific pests or diseases is provided by national agricultural extension services and by private consultants. For example, a large selection of 'technical guides' for organic farmers can be downloaded from https://shop.fibl.org.

8 References

Bourguet, D. and Guillemaud, T., 2016. The hidden and external costs of pesticide use. *Sustainable Agriculture Reviews* 19, 35–120.

Bravo, A., Likitvivatanavong, S., Gill, S. S. and Soberón, M., 2011. Bacillus thuringiensis: A story of a successful bioinsecticide. *Insect Biochemistry and Molecular Biology* 41, 423–31.

Dagostin, S., Schärer, H. J., Pertot, I. and Tamm, L., 2011. Are there alternatives to copper for controlling grapevine downy mildew in organic viticulture? *Crop Protection* 30, 776–88.

de Milliano, W. A. J., Lammerts van Bueren, E. and Myers, J. R., 2015. Resistance and resistance breeding for organic farming. In: Finckh, M. R., van Bruggen, A. H. C. and Tamm, L. (Eds.), *Plant Diseases and their Management in Organic Agriculture*. APS Press, St. Paul, Minnesota, pp. 175–87.

EC, 1991. Council regulation No 2092/91 of June 24 1991 on organic production of agricultural products and indications referring thereto on agricultural products and foodstuffs. *Official Journal of the European Communities* L198, 1–15.

EGTOP, 2011. *Final Report on Plant Protection Products*. European Commission, Directorate-General for Agriculture and Rural Development, B.4 Organics. https://ec.europa.eu/agriculture/organic/sites/orgfarming/files/docs/body/final_report_egtop_on_plant_protection_products_en.pdf.

EGTOP, 2014a. *Final Report Feed Mandate II*. European Commission, Directorate-General for Agriculture and Rural Development, B.4 Organics. https://ec.europa.eu/agriculture/organic/sites/orgfarming/files/docs/body/egtop-final-report-feed-ii_en.pdf.

EGTOP, 2014b. *Final Report On Plant Protection Products (II)*. European Commission, Directorate-General for Agriculture and Rural Development, B.4 Organics. http://ec.europa.eu/agriculture/organic/eu-policy/expert-advice/documents/final-reports/egtop-final-report-on-ppp-ii_en.pdf.

EGTOP, 2016a. *Final Report on Cleaning and Disinfection*. European Commission, Directorate-General for Agriculture and Rural Development, B.4 Organics. https://ec.europa.eu/agriculture/organic/sites/orgfarming/files/docs/body/final_report_egtop_on_cleaning_disinfection_en.pdf.

EGTOP, 2016b. *Final report on Plant Protection (III)*. European Commission, Directorate-General for Agriculture and Rural Development, B.4 Organics. https://ec.europa.eu/agriculture/organic/sites/orgfarming/files/egtop_plant_protection_iii-2016-10-26-definitive_version.pdf.

Ehlers, R. U., 2011. Cost-benefit, risk and trade-off analysis of regulation. In: Ehlers, R. U. (Ed.), *Regulation of Biological Control Agents, first edition*. Springer Science+Business Media B.V., Dordrecht, Heidelberg, London, New York, pp. 139–53.

Eilenberg, J., Hajek, A. and Lomer, C., 2001. Suggestions for unifying the terminology in biological control. *Biocontrol* 46, 387–400.

Finckh, M. R., van Bruggen, A. H. C. and Tamm, L., 2015. *Plant Diseases and their Management in Organic Agriculture*. APS Press, St. Paul, Minnesota.

Finckh, M. R. and Wolfe, M. S., 2015. Biodiversity enhancement. In: Finckh, M. R., van Bruggen, A. H. C. and Tamm, L. (Eds), *Plant Diseases and their Management in Organic Agriculture*. APS Press, St. Paul, Minnesota, pp. 153–74.

Forster, R., Micheloni, C., Regouin, E., Speiser, B. and Viñuela, E., 2008. *Report of the Ad-hoc Expert Group on Pesticides in Organic Food Production*. Meeting on 22–23 January 2008. European Commission, Directorate F.5. Organic Farming, Brussels.

Frank, S. D., 2010. Biological control of arthropod pests using banker plant systems: Past progress and future directions. *Biological Control* 52, 8–16.

Hauschild, R., 2011. Facilitations in the regulation of plant protection products containing baculoviruses. In: Ehlers, R. U. (Ed.), *Regulation of Biological Control Agents*. Springer Science+Business Media B.V., pp. 259–66.

Hillenbrand, T., Toussaint, D., Böhm, E., Fuchs, S., Scherer, U., Rudolphi, A. and Hoffmann, M., 2005. *Einträge von Kupfer, Zink und Blei in Gewässer und Böden - Analyse der Emissionspfade und möglicher Emissionsminderungsmaßnahmen*. Texte 19/05. Umweltbundesamt, Dessau.

Hunt, E. J., Loomans, A. J. M. and Kuhlmann, U., 2011. An international comparison of invertebrate bioligical control agent regulation: what can Europe learn? In: Ehlers, R. U. (Ed.), *Regulation of Biological Control Agents, first edition*. Springer, Dordrecht, Heidelberg, pp. 79–112.

Köhl, J., 2015. Toward an integrated use of biological control by Cladosporium cladosporioides H39 in apple scab (Venturia inaequalis) management. *Plant Disease* 99, 535–43.

Koike, S. T. and Cahn, M. D., 2015. Water management. In: Finckh, M. R., van Bruggen, A. H. C. and Tamm, L. (Eds), *Plant Diseases and their Management in Organic Agriculture*. APS Press, St. Paul, Minnesota, pp. 141–51.

Leoni, C., Rossing, W. and Van Bruggen, A. H. C., 2015. Crop rotation. In: Finckh, M. R., van Bruggen, A. H. C. and Tamm, L. (Eds), *Plant Diseases and their Management in Organic Agriculture*. APS Press, St. Paul, Minnesota, pp. 127–40.

Malik, P., 2014. Environmental pollution by pesticides: Sources and solution. *International Journal of Innovative Research & Development* 3, 242–6.

Pimentel, D., Acquay, A., Biltonen, M., Rice, P., Silva, M., Nelson, J., Lipner, V., Giordano, S., Horowitz, A. and D'Amore, M., 1992. Environmental and economic costs of pesticide use. *BioScience* 42, 750–60.

Prokop, A., Kuttig, S. and Baumjohann, P., 2016. The new fatty acid fungicide Neu 1143 F: A contribution to copper reduction in organic farming?. In: *European Copper Conference*, 17–18 November 2016. JKI, Berlin-Dahlem. http://kupfer.julius-kuehn.de/index.php?menuid=36&downloadid=94&reporeid=0.

Räz, B., Schüepp, H. and Siegfried, W., 1987. Hundert Jahre Plasmopara-Bekämpfung und Kupfereintrag in die Rebberge. *Schweizerische Zeitschrift für Obst- und Weinbau* 123, 272–7.

Segarra, G., Puopolo, G., Porcel-Rodriguez, E., Giovannini, O. and Pertot, I., 2016. Monitoring Lysobacter capsici AZ78 using strain specific qPCR reveals the importance of the formulation for its survival in vineyards. *FEMS Microbiology Letters* 363, 1–6.

Silvestri, G., Egger, S., Székely, T. and Weibel, F., 2008. Schorfresistente Sorten überzeugen Konsumentinnen und Konsumenten. *Schweizerische Zeitung für Obst- und Weinbau* 11/08, 10–13.

Speiser, B., Glen, D., Piggott, S., Ester, A., Davies, K., Castillejo, J. and Coupland, J., 2001. *Slug Damage and Control of Slugs in Horticultural Crops*. FiBL, Frick. https://core.ac.uk/download/pdf/10919097.pdf.

Speiser, B., Mieves, E. and Tamm, L., 2015. Kupfereinsatz von Schweizer Biobauern in verschiedenen Kulturen. *Agrarforschung Schweiz* 6, 160–5.

Speiser, B. and Schmid, O., 2004. Current evaluation procedures for plant protection products used in organic agriculture. In: *Proceedings of a Workshop Held 25–26 September 2003 in Frick, Switzerland*. FiBL, Frick, Switzerland.

Speiser, B., Tamm, L. and Mattock, S., 2011. Proposals for regulation of semiochemicals. In: Ehlers, R. U. (Ed.), *Regulation of Biological Control Agents, first edition*. Springer Science+Business Media B.V., Dordrecht, Heidelberg, London, New York, pp. 305–21.

Speiser, B., Tamm, L. and Weibel, F. P., 2014. Regulatory framework for plant protection in organic farming. In: Bellon, S., Penvern, S. (Eds), *Organic Farming, Prototype for Sustainable Agricultures*. Springer Netherlands, the Netherlands, pp. 65–82.

Speiser, B., Wyss, E. and Maurer, V., 2006. Biological control in organic production: First choice or last option? In: Eilenberg, J. and Hokkanen, H. (Eds), *An Ecological and Societal Approach to Biological Control*. Springer, Dordrecht, The Netherlands, pp. 27–46.

Stolz, H., Blattert, S., Rebholz, T., Stolze, M., 2017. Biobarometer Schweiz: Wovon die Kaufentscheidung für Biolebensmittel abhängt. *Agrarforschung Schweiz* 8, 62–9.

Strauch, O., Strasser, H., Hauschild, R. and Ehlers, R. U., 2011. Proposals for bacterial and fungal biocontrol agents. In: Ehlers, R. U. (Ed.), *Regulation of Biological Control Agents, first edition*. Springer Science+Business Media B.V., Dordrecht, Heidelberg, London, New York, pp. 267–88.

Strumpf, T., Steindl, A., Strassemeyer, J., Riepert, F., 2011. Erhebung von Kupfergesamtgehalten in ökologisch und konventionell bewirtschafteten Böden. Teil 1: Gesamtgehalte in Weinbergsböden deutscher Qualitätsanbaugebiete. *Journal für Kulturpflanzen* 63, 131–43.

Strumpf, T., Stendel, U. and Vetter, C., 2009. Gesamtgehalte von Kupfer in Böden des Kernobstanbaus, Weinbergen und Hopfenanlagen. *Journal für Kulturpflanzen* 61, 117–25.

Tamm, L., Speiser, B. and Holb, I., 2015. Direct control of airborne diseases. In: Finckh, M. R., van_Bruggen, A. H. C. and Tamm, L. (Eds), *Plant Diseases and their Management in Organic Agriculture*. APS Press, St. Paul, Minnesota, pp. 205–16.

Tamm, L., Speiser, B. and Mercier, T., 2011a. Proposals for regulation of botanicals. In: Ehlers, R. U. (Ed.), *Regulation of Biological Control Agents, first edition*. Springer Science+Business Media B.V., Dordrecht, Heidelberg, London, New York, pp. 289–304.

Tamm, L., Thürig, B., Fliessbach, A., Goltlieb, A. E., Karavani, S. and Cohen, Y., 2011b. Elicitors and soil management to induce resistance against fungal plant diseases. *NJAS - Wageningen Journal of Life Sciences* 58, 131–7.

Van Bruggen, A. H. C., 2015. Introduction to plant disease management in organic agriculture. In: Finckh, M. R., van Bruggen, A. H. C. and Tamm, L. (Eds), *Plant Diseases and their Management in Organic Agriculture*. APS Press, St. Paul, Minnesota, pp. 119–26.

Winkelhausen, E., Pospiech, R. and Laufenberg, G., 2005. Antifungal activity of phenolic compounds extracted from dried olive pomace. *Bulletin of the Chemists and Technologists of Macedonia* 24, 41–6.

Chapter 12

Integrated weed management in organic cropping systems

Paolo Bàrberi, Scuola Superiore Sant'Anna, Italy

1 Introduction
2 Key weed management challenges: intra-row weed control
3 Key weed management challenges: problematic weeds and weed management in reduced- and no-till organic systems
4 Integrated weed management and sustainable organic crop production
5 Future trends
6 Where to look for further information
7 References

1 Introduction

Many farmers recognise weed management as the most difficult technical challenge facing organic crop producers (Turner et al., 2007), and some perceive it as a major obstacle for conversion from conventional to organic production. In general, weed interference is not much of a problem in fruit crops, except during the nursery phase and the early production years (Abu-Irmaileh, 1993). In forage and grassland systems most weeds can be controlled through timely cuts for hay and managed grazing, although some creeping perennials and unpalatable species may become progressively dominant in the sward (Naylor, 2002). Instead, weeds are often a major problem in arable, vegetable herbs and spice crops wherever synthetic herbicides cannot be used. This is due to the limited competitiveness of some of these crops and/or to the strict weed-free requirements for high-quality produce, for example, herbal preparations (Pannacci et al., 2017).

In terms of weed control effectiveness, a direct comparison between synthetic herbicides commonly used in conventional agriculture and alternative (usually non-chemical) control methods used in organic agriculture would usually be in favour of the former (Melander et al., 2005). This evidence suggests that weed management in organic crop production cannot be successful if it is just based on an input substitution approach. This implies that organic farmers should embrace a paradigm shift and acknowledge the necessity to combine several types of interventions within a larger spatial and temporal context. integrated weed management (IWM) provides the theoretical foundations as well as technical solutions to meet this challenge.

http://dx.doi.org/10.19103/AS.2017.0029.13

Taking inspiration from the already established concept of integrated pest management (IPM), IWM was first termed by Shaw (1982) as a directed agroecosystem approach for the management and control of weed at threshold levels to prevent economic damage in the current and future years. IWM is based on the deployment of several tactics across the entire cropping system and not just in a given crop and a given growing season. The rationale of IWM is that long-lasting effective weed management can better be achieved by combining tactics of different nature and with different effects that would result in only partial weed control if applied alone. The combination between 'many little hammers' (Liebman and Gallandt, 1997) results in diversified weed management strategies that reduce the risk for weed species to adapt to agricultural disturbance and hence to become dominant. Organic farming, due to the absence of synthetic herbicides exerting a high selection pressure on weed communities, represents an ideal playground for the application of IWM.

The individual measures ('tactics') composing an IWM strategy have been classified in various ways. Here a distinction between preventive, cultural and direct methods is used (Bàrberi, 2002), which is based on the application time and the main effect expected. Preventive methods are applied before a crop is grown with the main goal of reducing the number of weeds emerging during the growing cycle of that crop and hence crop/weed competition. Examples of tactics belonging to this group are crop rotation, primary tillage, the false/stale seedbed technique, soil solarisation, soil steaming and the use of cover crops as green manures or surface (dead) mulches, in the latter case also exerting an allelopathic effect. In addition, basic preventive measures are the use of clean (weed free) crop seeds and farm hygiene to avoid weed seed imports, for example, the cleaning of tillage/cultivation tools before entering a new field and the use of clean organic manures and composts.

Cultural methods are applied during a crop growing cycle with the main goal of increasing the competitive ability of crop against weeds, for example, through a preferential supply of resources to the crop at the detriment of weeds. Examples are the use of competitive genotypes, increased crop seeding rates, optimised seeding/planting pattern and row spacing, localised fertilisation and irrigation, intercropping and the use of cover crops as living mulch.

Direct methods are techniques used during a crop growing cycle with the specific aim to kill the weeds that emerged in the crop. In organic farming, they include mechanical (e.g. various harrowing and hoeing systems), thermal (e.g. flame weeding), biological (e.g. use of fungal pathogens and animals like insects, molluscs, fishes, ducks, poultry, sheep and goats to selectively control weeds in various crops) and chemical methods ('natural herbicides', e.g. aqueous extracts or essential oils of Brassicaceae and aromatic plants). A summary of the main preventive, cultural and direct methods with relevant examples can be found in Table 1.

Organic farmers should plan their IWM system by including tactics from each of these three groups and combine them in a long-term cropping system perspective. The importance of a systems approach for effective and long-lasting weed management has long been recognised (Liebman and Davis, 2000). This implies being aware of carry-over effects of a given tactic across two or more subsequent growing cycles, which can be exploited to improve the effectiveness of weed management strategies. For example, Melander and Rasmussen (2000) grew spring barley in widely spaced rows leaving a 25-cm free band every 50 cm. Soil in this band was repeatedly hoed to stimulate weed emergence hence depleting the weed seedbank. In the following year, a poorly competitive crop (sugar beet or kale) was directly drilled in the cultivated bands, where weed emergence was reduced by 62–92% (with consequent reduction of competition from weeds), thanks to the residual effect of hoeing applied in the previous year.

Table 1 Main tactics that can be included in integrated weed management strategies for organic farming, with the indication of category, main effect(s) and relevant examples

Tactic	Category	Main effect	Example
Crop rotation	Preventive	Reduction of weed emergence	Alternation between winter and summer crops
Soil tillage/cultivation	Preventive (P) + direct (D)	Reduction of weed emergence + weed destruction	Ploughing (P), discing (P), false/stale seedbed technique (P), photocontrol (P), hoeing (D), cultivation (D)
Cover crops	Preventive (P) + cultural (C)	Reduction of weed emergence and/or competition	Green manures (P), living mulches (C)
Mulching	Preventive (P) + cultural (C)	Reduction of weed emergence and/or competition	Dead mulches (P), in-crop plastic mulches, for example, for transplants (C)
Flame weeding	Preventive (P) + direct (D)	Reduction of weed emergence + weed destruction	Last operation in stale seedbed technique (P), in-crop use of (shielded) LPG-propelled burners (D)
Soil solarisation	Preventive	Reduction of weed emergence	Use in the summer before crop sowing/transplanting
Soil steaming	Preventive	Reduction of weed emergence	Broadcast or banded (in-row) application
Intercropping	Cultural	Reduction of weed competition	Pea/barley, celery/leek, various relay cropping systems
Crop sowing density	Cultural	Reduction of weed competition	Increased sowing density in winter cereals
Crop spatial arrangement	Cultural	Reduced weed competition	Reduced between-row or in-row distances, transplanting
Fertilisation	Cultural	Reduced weed competition	Localised (in-row) application of fertilisers
Irrigation	Cultural	Reduced weed competition	Trickle/drip irrigation
Mechanical weeding	Direct	Weed destruction	Intra-row weeders (finger-, brush-, torsion weeders, robotic weeding)
Thermal weeding	Direct	Weed destruction	Selective hot water application
Chemical weeding	Direct	Weed destruction	Natural herbicides (e.g. brassica essential oils, clove oil)

IWM offers an ample choice of solutions to be tailored to the specific needs and problems of organic farmers. Nevertheless, some key challenges remain in organic farming with respect to weed management. Among them, three that are frequently cited are (i) intra-row weed control; (ii) the management of problematic weeds (perennials and parasitic) and (iii) weed management in reduced- and no-till organic systems. In recent years, organic research has been active to find solutions to these challenges. Relevant evidence of these challenges as well as of successful case studies addressed through dedicated research is reported and discussed in the following section.

2 Key weed management challenges: intra-row weed control

Direct in-crop weed control is usually less efficient with non-chemical methods than with synthetic herbicides, many of which have a wide activity spectrum targeting species with different morphological features. Preventive and cultural methods can help reduce weed abundance and weed competition in organic crops; yet in many situations it is still needed to apply direct methods to keep weed populations below an acceptable level. Use of natural herbicides extracted from plants known to contain allelopathic compounds (Dayan et al., 2012; Evans and Bellinder, 2009) is usually not a viable option due to their limited persistence, lack of selectivity, potential phytotoxicity on crop, high cost and limited diffusion. In addition, they encourage an input substitution approach which is not in line with the spirit of IWM and of overall organic crop management.

In the vast majority of cases, direct weed control in organic crops relies on the use of mechanical methods like harrows and hoes. In cereal and small grain crops and, in general, in crops that are grown in narrow rows, the spring (flex)-tine harrow is a commonly used tool (Bàrberi et al., 2000; Cirujeda et al., 2003). Although it treats the whole soil surface regardless of the crop presence, some selectivity can be reached by taking into account the relative size of crop and weeds at the time of pass, and tine adjustment (Rasmussen and Svenningsen, 1995; Kurstjens et al., 2000). However, this harrow does not guarantee adequate weed control in all cases, so growing a cereal crop in paired or wider rows has been proposed as an alternative crop management practice allowing the use of hoeing (Melander et al., 2003). In this situation, as well as in the more general case of row crops where hoes are the most widespread tools for direct weed control, a major problem is that weeds growing very close to the crop row cannot be controlled without risking significant damage to the crop. If left uncontrolled, intra-row weeds can become very competitive just because of their proximity to the crop. This is unacceptable in most organic crops, especially high-value ones like vegetables, forcing the farmer to using hand-hoeing or hand-weeding. Manual weeding raises production costs considerably in developed countries and takes the lion's share of family labour in countries where labour cost is lower (Gbehounou and Bàrberi, 2016). It has been shown that there is a linear relationship between the number of intra-row weeds and the hours spent on hand-weeding in organic vegetables, which can be as high as 152 and 177 h ha^{-1} in poorly competitive crops like carrot and onion, respectively (van der Weide et al., 2008). As such, direct non-chemical methods able to improve intra-row weed control are clearly needed in organic crop production.

In the last two decades, some novel machines specifically aimed to allow intra-row weed control have been proposed. Examples are finger weeders, brush weeders and torsion weeders. Finger weeders are made of two pairs of ground-driven truncated steel cones pointing towards the soil at either sides of the crop row. These cones are equipped with rubber or plastic spikes ('fingers') that pull out small weeds without damaging the crop, provided it is well rooted. This machine is often used in vegetable crops, both transplanted and seeded (van der Schans et al., 2006). Brush weeders are made of flexible fibreglass brushes, usually operated by the tractor power take-off, that rotate around a vertical or horizontal axis. Implements can be front or rear mounted: in the latter case, an operator can steer the brushes to ensure they work close to the crop row. Speed and rotation direction of vertical-axis brushes can be set as to accumulate soil on the crop row or to remove it (and weeds) away from the row (Cloutier et al., 2007). In the first case, the ridging effect buries the small weeds, but seeds present on soil left on the row may facilitate subsequent in-row weed emergence flushes. Torsion weeders (Fig. 1) are made of pairs of angled spring tines connected to a rigid frame and bent towards the row so to enable sol cultivation in parallel. Weed seedlings are uprooted, while well-anchored crop plants are skipped by the tines, thanks to the spring action (Bleeker et al., 2002). The larger the size of the spring tines, the more aggressive the treatment. Similar to spring (flex)-tine harrows, torsion weeder tines can be adjusted to crop, weed and soil conditions. Similar to brush weeders, the torsion weeder is mainly used on vegetable crops like carrot and onion. In transplanted onion, one weed harrowing and three passes with torsion weeder reduced

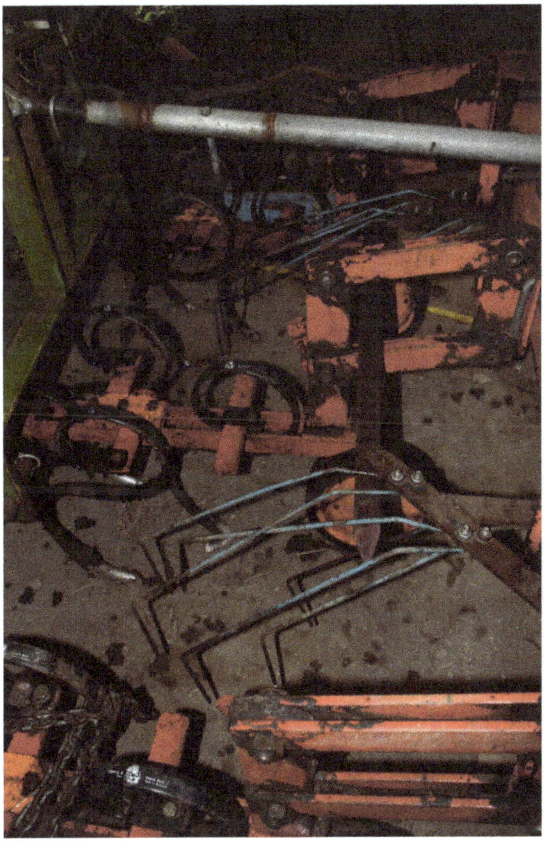

Figure 1 Cultivator equipped with torsion weeders.

intra-row weed numbers by 85% and hand-weeding time by 73% compared with one weed harrowing and standard inter-row cultivation (Ascard and Fogelberg, 2008). The best weed management strategy in transplanted onion increased the yield by 96% compared with the best strategy in seeded onion, demonstrating the importance of integrating cultural and direct methods to improve IWM.

High-tech opportunities to control intra-row weeds come from robotic weeding. Advanced machine automation and improved sensor technology to visually detect plants allow mechanical weeding with blades operated by an actuator only in the part of the rows where no crop plant is present (Fig. 2). Robotic weeders like this have a high accuracy and may also be operated unmanned (López-Granados, 2011; Nørremark et al., 2012). Clearly, their investment cost is high which obviously can be borne only by large-scale organic farms.

Simpler intra-row weeders like finger, brush or torsion weeders are relatively cheap and can be easily coupled with standard inter-row hoeing, hence making the overall direct weed control more cost-effective. However, the need to drive their operating tools precisely towards the crop row usually reduces their working speed and therefore their working capacity, unless laser-guided sowing/planting systems to obtain straight rows are used (van Zuydam et al., 1995). In addition, these devices need to be operated on a very smooth soil with adequate moisture conditions to prevent dust formation or soil clogging around the weeding tools. They should also be operated on small weeds (ideally at the cotyledon or first true leaves stage) and on crop plants grown enough to withstand contact with the tines/spikes without damage. As such, the concurrent application of preventive methods that reduce and/or delay weed emergence (e.g. the false/stale seedbed technique) and of cultural methods that increase the size gap between crop and weed plants at an early stage (e.g. transplanting) is needed to maximise weed control rates from intra-row weeders

Figure 2 Image showing an intra-row robotic weeder.

(van der Weide et al., 2008). This shows that a well-planned IWM strategy is the best way to improve weed management in organic crop production. Further improvement may be obtained by controlled traffic systems, that is, by precisely locating the position of crop rows and associated weed and crop management interventions across a whole crop sequence (Vermeulen et al., 2010). Interestingly, hand-held versions of these machines are available for use by small-scale farmers who may not be able to afford the standard versions.

More recently, intra-row weed control with thermal methods has been proposed as an alternative to mechanical methods. In particular, band steaming, that is, the application of hot water vapour to a narrow soil band where the crop is later sown or transplanted, has shown good promises (Melander and Kristensen, 2011). Steam injection into soil can effectively control a range of soilborne pests and diseases, inactivate weed seeds and kill emerging weed seedlings (Peruzzi et al., 2012). Steaming is applied in a band 5–6 cm wide and 6–7 cm deep, which dramatically reduces fuel consumption compared to broadcast soil steaming, where up to 5000 L ha^{-1} of diesel may be needed (Ascard et al., 2007). The effect of soil steaming against weeds is further improved by adding compounds like CaO or KOH, which drive an exothermic reaction (Fig. 3), thus boosting soil temperatures in the upper soil layers and keeping them for longer (Bàrberi et al., 2009). Band steaming can be considered a preventive weed management method addressing intra-row weeds, whereby inter-row weeds are usually controlled by hoeing. The advantage of intra-row thermal weed control methods relative to mechanical methods is that the soil is left undisturbed until crop sowing/transplanting, which prevents further weed emergence. If soil disturbance associated to crop establishment does not go beyond the steaming depth, little subsequent weed emergence can be expected. However, it should be noted

Figure 3 Machine for broadcast soil steaming with the addition of activating compounds. Note (right) the plastic mulch film laid down immediately after steaming.

that not all weed species are equally sensitive to steaming, with large-seeded ones being less sensitive (Vidotto et al., 2013). In addition, seedling emergence in some species (e.g. *Lamium purpureum*) may be stimulated by soil steaming (Bàrberi et al., 2009).

3 Key weed management challenges: problematic weeds and weed management in reduced- and no-till organic systems

3.1 Management of problematic weeds

In organic crop production, perennial weeds like couchgrass (*Elymus repens*), thistles (e.g. *Cirsium arvense*) and docks (*Rumex* spp.) are increasingly abundant, for example, in Central and Northern Europe (Rydberg and Milberg, 2000). The likely cause of this is the incorrect choice of tillage or cultivation implements. When soil seedbed preparation or stubble cultivation is done with rotary implements (e.g. hoes) powered by the tractor power take-off, soil is finely crumbled, and vegetatively propagating modified stems (e.g. rhizomes or stolons) and roots of perennial weeds are broken into small pieces and dispersed nearby.

These fragments contain buds that help perennial weeds colonise new areas in a quicker and more effective way (Bourdôt et al., 2006), although this risk is limited under dry soil conditions. Presence of a ley phase (e.g. as grass-clover sward) in crop rotations also facilitates biennial and perennial weeds due to reduced disturbance by primary tillage compared to annual crop rotations and to the advantage that these species have versus annual weeds which are incapable of regrowth after mowing.

Although the refinement of tillage and cultivation methods can help contain the spread of perennial weeds, by itself it is unlikely to reduce abundance where it is problematic. Once more, the solution lies in the integration between different tactics. The success of IWM in organic crop production starts from the knowledge of biological and ecological characteristics of target weeds. In the case of perennials, the key point is to know what is the weakest phase in their life cycle. In temperate environments, perennials survive the winter season via dormant buds present on their underground propagules. When soil temperature rises, these buds generate new sprouts that form the vegetative seasonal phase of the plant. These plants can set seeds and, towards the end of the summer, start to transfer the photosynthates produced in the growing season to their underground storage organs to get ready for the oncoming unfavourable season. Repeated cultivation can weaken the plants and reduce the amount of energy reserves that perennial weeds can store below ground (Graglia et al., 2006). In addition, growth of a competitive cover crop or sward that is repeatedly mown will help exhaust their reserves, which would mainly be used for immediate regrowth as well as to compete with the companion crop instead of being stored (Lukashyk et al., 2008).

In Denmark, tailored combination of a competitive cover crop with timely rotary cultivation and subsequent rhizome removal has reduced the abundance of *Elytrigia repens* by 84–97% (Melander et al., 2013). In Norway, novel equipment for stubble cultivation has been developed to improve perennial weed control in organic spring cereals in combination with undersown clovers (Brandsæter et al., 2012). In Switzerland, selective treatment by means of a commercial hot-water high-pressure cleaner with a rotating nozzle operating at >80°C has been demonstrated as an effective (>80% control) thermal direct weed control tactic against *Rumex obtusifolius* (Latsch et al., 2017).

Parasitic weeds are a serious problem especially in Mediterranean, subtropical and tropical environments (Parker and Riches, 1993). Since many of them are hosts specific to some crops, in areas and soils where they are present, a well-targeted and diversified crop rotation should help keep them at bay. This is generally easier to establish under organic or agroecological management than under conventional industrial management. Also in the case of parasitic weeds, knowledge of their biology and ecology is key to design successful IWM systems. For example, witchweed (*Striga* spp.), a troublesome parasitic weed of staple cereals in Africa and Asia, can be kept under control by increasing soil fertility, for example, through the use of cover crops and mulches or the introduction of legume crops in the rotation (Ayongwa et al., 2011). Furthermore, preventive management of *Striga* spp. can be enhanced by the use of resistant or tolerant maize or sorghum varieties (Badu-Apraku and Lum, 2007).

3.2 Weed management in reduced- and no-till organic systems

Reduced tillage and especially no-till systems have long been considered very hard to be used under organic crop production, mainly because of the probability of the weed populations to increase up to an uncontrollable level (Peigné et al., 2007). This is because non-inversion tillage concentrates weed seeds close to the soil surface, from where they can more easily germinate (Légère et al., 2011). However, a good and diversified crop rotation can limit this problem by preventing the majority of weed species to complete their life cycle and impeding the selection of weeds well adapted to a recurrent management regime, thanks to the alternation of disturbance types (Amossé et al., 2013; Bàrberi, 2002). Further improvement is expected by the inclusion of suppressive cover crops and mulches (Fig. 4 and 5) in diversified rotations based on reduced/no-till, thereby exploiting the full potential of conservation agriculture (www.fao.org/ag/ca/1a.html; Hobbs et al., 2008).

One main technical challenge associated with the use of cover crops in organic no-till systems is the termination of the cover crop mechanically to prepare the seedbed for the next cash crop, given that chemical termination with systemic herbicides (mainly glyphosate) is obviously not an option. Mowing implies the risk of regrowth of the cover crop, which then may become a weed in the next cash crop cycle. A better solution seems to be the use of a roller crimper (Fig. 6). This machine, first developed at the Rodale Institute in Pennsylvania (USA), consists of a wide heavy roll with V-shaped blades aimed to minimise bumping on soil and hence maximise vegetation squeezing and consequent devitalisation. Excellent results have been obtained by using the roller crimper on a rye (*Secale cereale* L.) catch crop followed by no-till soybean or vegetable crop (Canali et al., 2013; Smith et al., 2011), although control of perennial weeds may at times be challenging (Mirsky et al., 2012). When used on cover crops with higher regrowth potential, care should be paid to application timing. For example, in the case of hairy vetch (*Vicia villosa* Roth), the roller crimper should not be passed earlier than the full bloom stage: when passing from the vegetative to the reproductive stage, plant tissues become gradually lignified, which reduces regrowth. An additional interesting feature of the roller crimper is that it allows cover crop termination and cash crop sowing in just one pass, by mounting a no-till driller to the rear of the tractor.

Although reduced- and especially no-till systems usually result in higher weed density and in weed community shifts towards higher proportion of grasses, wind-disseminated annuals,

Figure 4 Image showing excellent weed suppression exerted by a cover crop mixture.

Figure 5 Residual weed suppression exerted by a *Trifolium squarrosum* dead mulch on organic aubergine.

Figure 6 Roller-crimper for mechanical termination of cover crops.

biennials and perennials (Armengot et al., 2015; Bàrberi and Lo Cascio, 2001), they do not necessarily result in reduced yield. A recent meta-analysis conducted on 26 published studies comparing inversion with non-inversion tillage in organic systems (Cooper et al., 2016) showed that, despite an overall 50% increased weed density, some non-inversion tillage methods were able to maintain crop yields as relative to deep or shallow inversion tillage.

Similar results were obtained by Bàrberi et al. (2014) who pooled data from 13 European long-, mid- and short-term experiments involving reduced tillage with or without cover crops. Reduced tillage increased weed abundance and diversity as compared to ploughing, but crop yields were not significantly different in 69% of the cases.

One advantage of diversifying crop rotation is that effective weed management in one crop can compensate for suboptimum management in another, thus avoiding build-up of abundant weed populations. Crop rotation also allows hybrid systems where, for example, no-till is used for some crops and reduced tillage for other crops (Carr et al., 2013). Tailored diversification in crop rotation, reduced tillage and cover crop/mulches results in effective IWM strategies which can facilitate the adoption of conservation agriculture methods in organic farming.

4 Integrated weed management and sustainable organic crop production

Although the conceptual foundations of IWM have long been established and several success stories based on the 'many little hammers' approach are in place, the implementation of strategies exploiting in full the potential of integrating different tactics in a coherent weed management strategy is still untapped. As a matter of fact, the majority of organic farmers focus on the application of direct – mainly mechanical – methods, whereas preventive and cultural methods are only partly applied (Riemens et al., 2010). This may be due to short-term economic pressures that largely dictate the choice of crops, their sequence and their management practices. As a consequence, comprehensive agroecological knowledge for the design of organic cropping systems is rarely implemented, especially in large-scale farms where an input substitution approach dominates (Darnhofer et al., 2010).

Recently, it has been hypothesised that 'true' IWM could be realised only by synergistically combining current and emerging technological innovations with information systems and decision tools for real-time use (Young et al., 2017). Improved automation and robotic technologies (see e.g. Åstrand and Baerveldt, 2002; Christensen et al., 2009; van der Weide et al., 2008) could be coupled with information and decision support systems in integrated platforms readily available to farmers and technicians. So far, high-tech applications for IWM have been addressed in large-scale farms in which the farmers can afford to buy equipment. Tailoring these solutions to the needs and requirements of organic systems and of small-scale family farmers – which are the vast majority (90%) of farms in the world (www.fao.org/family- farming/themes/small-family-farmers/en) – and making them affordable to a variety of users would be one of the great challenges of the future.

A thrust for the adoption of comprehensive IWM strategies may also come from the elucidation of crop–weed competitive interactions. It has been shown that the common assumption that crop yield should be proportionally reduced with increasing weed biomass does not always hold true, especially in organic systems (Ryan et al., 2010). Based upon this evidence, a novel theory, the Resource Pool Diversity Hypothesis (RPDH), has been formulated to explain how soil resource pool diversity can mediate competition between crops and weeds (Smith et al., 2009). The assumptions of the RPDH are that (i) the intensity of interspecific competition in plant communities depends on the degree to which niche differentiation and resource partitioning occur among species and that (ii) agricultural practices like crop rotation, fertilisation and weed management determine the soil resource pools from which crops and weeds may partition resources. According to the RPDH, crop yield loss due to competition from weeds per unit of weed density should decrease with increasing cropping system diversity. In addition, the degree of overlap in soil resource niche breadth between crops and weeds – which depends on species-specific functional traits for resource acquisition – will determine the reduction of crop–weed competition with an increase in resource pool diversity. Better resource use complementarity in organic systems might then explain why crop yield – within certain limits – is maintained at higher weed abundance. An increasing body of evidence seems to support the validation of the RPDH. For example, soil collected from organic farms showed a higher relative abundance of mutualistic than pathogenic organisms compared to conventional soil, which decreased crop–weed competition and increased facilitation (Johnson et al., 2017). Identification of management practices promoting positive soil biota effects could then minimise crop–weed competition and reduce the need for intensive weed management. This new scientific evidence, if duly disseminated to advisors and farmers, may foster the adoption of comprehensive IWM systems by giving the message that 'zero tolerance' against weeds is not needed, especially in organic farming.

5 Future trends

In the quest for sustainable weed management strategies, a new concept has been proposed, with clear similarities with IWM and potential important implications for organic farming: that of 'ecological weed management' (EWM). This term was explicitly used for the first time in a review paper about a decade ago (Bastiaans et al., 2008), although it was recalled in a previous book (Liebman et al., 2001). In that paper, EWM was defined as '... any adjustment or modification to the general management of the crop or cropping systems that contributes to the regulation of weed populations and reduces the negative impact of weeds on crop production...' and involves both the tactical and strategic levels of decision-making. The authors highlighted three mechanisms linked to EWM that can be activated through appropriate management practices, mainly falling in the preventive and cultural management categories of IWM, as defined earlier.

The first mechanism is reduced weed seedling emergence, which can be obtained for example, by (i) photocontrol, that is, soil cultivation performed at night or under a light-proof cart to prevent triggering germination of light-sensitive weed seeds (Riemens et al., 2007; Hartmann, 2016) or by (ii) the use of natural or artificial mulches as soil cover (Davis, 2010; Marles et al., 2010).

The second mechanism is improved crop competitiveness, for example, through adjustment of crop sowing or planting pattern (Anderson, 2009; O'Donovan et al., 2013), use of transplants (Bàrberi, 2002; Tillet et al., 2008), localised fertilisation and/or irrigation (Petersen, 2005), use of competitive and/or allelopathic crop genotypes (Andrew et al., 2015; Bertholdsson, 2010; Worthington et al., 2013) or polycultural systems, for example, intercropping, relay cropping or agroforestry (Picasso et al., 2008).

The third mechanism of EWM is the reduction in weed seedbank size, which can be obtained by limiting seed inputs and favouring seed losses, for example, through increased seed predation, decay and germination (Gallandt, 2006). Weed seed predation mainly occurs after weed seeds have been shed on soil via consumption by insects and rodents (Davis et al., 2013; Westerman et al., 2011) and can be enhanced by manipulation of agricultural habitats, for example, no-till, delayed stubble cultivation and field margins (Landis et al., 2005). Weed seed decay refers to creation of soil conditions conducive to increased seed mortality, for example, by fungal attack. These conditions are species-specific (Gómez et al., 2014) and, if better understood, may lead to the development of specific management solutions against target weeds. Increased weed seed germination can be obtained for example, by the application of the false and stale seedbed techniques (Cloutier et al., 2007) which also encourages fatal germination, that is, when weed seeds germinate in a soil layer too deep to allow seedling emergence (Fenner and Thompson, 2005).

Many of these are in fact preventive or cultural weed management practices that, if complemented by the addition of non-chemical direct tactics, would make EWM and IWM basically identical. In the context of organic farming, both approaches point out the importance of preventive and cultural methods in an overall strategy. To strengthen the message it intends to deliver, IWM for organic systems could be defined as 'a combination of methods aimed to achieve long-term weed suppression through the use of ecological interactions between crop, weeds, soil and/or other taxa, fostered by appropriate agroecosystem management with the least possible use of direct non-chemical weed control methods'. Bridging the gap between IWM and EWM may facilitate the design and adoption of weed management strategies for organic farming more aligned with the agroecological approach, that is, the true, original spirit of organic agriculture.

Finally, further advancement and fine-tuning of IWM can be achieved from reorientation towards transdisciplinary and participatory weed research, which are expected to facilitate knowledge sharing and engagement of experts from different disciplines, as well as of farmers and other stakeholders from the planning stage of research. Transdisciplinary weed research has been termed as 'an integrated process of inquiry and action that addresses complex weed problems in the context of broader efforts to improve economic, environmental and social aspects of ecosystem sustainability' (Jordan et al., 2016). The basis of this approach is that weed management can only be tackled in the larger perspective of agricultural sustainability and with the involvement of a multi-actor team. Participatory weed management can be favoured by the application of tailored methodologies aimed to highlight similarities and differences between farmers, advisors and scientists in problem perception and priority actions. For example, a 'mental model' approach has been applied to weed management in organic farming (Jabbour et al., 2013). It allowed to synthesise expert knowledge based on interviews with scientists, extension specialists and farmers, highlighting the emphasis they place on knowledge of EWM as a driver of their decisions. In that study, farmers were aware of the EWM concepts discussed by experts, but placed less emphasis on ecological complexity and more emphasis on the role of weeds as indicators of soil nutrient conditions and of knowledge gained from their own and other farmers' experience as opposite to that gained from scientific research. In addition, farmers emphasised the importance of cost of time and labour for weed management operations, whilst scientists mainly emphasised crop yield loss. Since it has been shown that the effectiveness of IWM in organic systems depends on farmers' attitude towards tactics, especially preventive ones (Riemens et al., 2010), systematic incorporation of their views in future research should result in the design of more effective, tailor-made IWM systems.

6 Where to look for further information

6.1 Textbooks

Booth, B. D., Murphy, S. D. and Swanton, C. J. (2003), *Weed Ecology in Natural and Agricultural Systems*, CABI.

Buhler, D. D. (Ed.) (1999), *Expanding the Context of Weed Management*, The Haworth Press.

Chauhan, B. S. and Mahajan, G. (Eds) (2014), *Recent Advances in Weed Management*, Springer.

Liebman, M., Mohler, C. L and Staver, C. P. (2001), *Ecological Management of Agricultural Weeds*, Cambridge University Press.

Radosevich, S. R., Holt, J. S. and Ghersa, C. M. (Eds) (2007), *Ecology of Weeds and Invasive Plants*, John Wiley & Sons.

Upadhyaya, M. K. and Blackshaw, R. E (Eds) (2007), *Non-chemical Weed Management: Principles, Concepts and Technology*, CABI.

van der Schans, D. and Bleeker, P. (Eds) (2006), *Practical Weed Control in Arable Farming and Outdoor Vegetable Cultivation Without Chemicals*, Wageningen UR, Applied Plant Research.

6.2 Websites

Extension.org – Preventative Practices in Organic Weed Management: http://articles.extension.org/pages/59489/preventative-practices-in-organic-weed- management

Extension.org – Twelve Steps Toward Ecological Weed Management in Organic Vegetables: http://articles.extension.org/pages/18539/twelve-steps-toward-ecological-weed-management-in-organic-vegetables

European Weed Research Society – Working Group Physical and Cultural Weed Control: www.ewrs.org/physical_and_cultural_weed_control.asp

PennState Extension – Creating a Weed Management Plan for Your Organic Farm: http://extension.psu.edu/business/start-farming/vegetables/factsheets/creating-a-weed- management-plan-for-your-organic-farm

Rodale's Organic Life – Weed Control, The Organic Way: www.rodalesorganiclife.com/garden/weed-control-organic-way

The Weed's Network: http://invasivespecies.org.au/traction.

6.3 Research centres

Aarhus University, Denmark: http://pure.au.dk/portal/en/bo.melander@agrsci.dk

Cornell University, USA – Weed Ecology and Management Laboratory: https://weedecology.css.cornell.edu

Iowa State University, USA – Department of Agronomy: www.agron.iastate.edu/personnel/userspage.aspx?id=646

United States Department of Agriculture (USDA): www.ars.usda.gov/people- locations/person/?person-id=42691

University of Maine, USA – Weed Ecology and Management: http://umaine.edu/weedecology.

7 References

Abu-Irmaileh, B. E. (1993), Weed management in fruit trees. *Emir. J. Agric. Sci.* 5, 13–37.

Amossé, C., Jeuffroy, M.-H., Celette, F., Goulevant, G. and David, C. (2013), Relay-intercropped forage legumes help to control weeds in organic grain production. *Eur. J. Agron.* 49, 158–67.

Anderson, R. L. (2009), Impact of preceding crop and cultural practices on rye growth in winter wheat. *Weed Tech.* 23, 564–8.

Andrew, I. K. S., Storkey, J. and Sparkes, D. L. (2015), A review of the potential for competitive cereal cultivars as a tool in integrated weed management. *Weed Res.* 55, 239–48.

Armengot, L., Berner, A., Blanco-Moreno, J. M., Mäder, P. and Sans, F. X. (2015), Long-term feasibility of reduced tillage in organic farming. *Agron. Sust. Dev.* 35, 339–46.

Ascard, J. and Fogelberg, F. (2008), Mechanical in-row weed control in transplanted and direct-sown bulb onions. *Biol. Agric. Hort.* 25, 235–51.

Ascard, J., Hatcher, P. E., Melander, B. and Upadhyaya, M. K. (2007), Thermal weed control. In: *Non-chemical Weed Management. Principles, Concepts and Technology* (M. K. Upadhyaya and R. E. Blackshaw (Eds)). CAB International, Delhi, pp. 155–75.

Åstrand, B. and Baerveldt, A. J. (2002). An agricultural mobile robot with vision-based perception for mechanical weed control. *Auton. Robots*, 13, 21–35.

Ayongwa, G. C., Stomph, T. J., Belder, P., Leffelaar, P. A. and Kuyper, T. W. (2011), Organic matter and seed survival of Striga hermonthica – mechanisms for seed depletion in the soil. *Crop Prot.* 30, 1594–600.

Badu-Apraku, B. and Lum, A. F. (2007), Agronomic performance of Striga resistant early- maturing maize varieties and inbred lines in the savannas of West and Central Africa. *Crop Sci.* 47, 737–50.

Bàrberi, P. (2002), Weed management in organic agriculture: Are we addressing the right issues? *Weed Res.* 42, 176–93.

Bàrberi, P. and Lo Cascio, B. (2001), Long-term tillage and crop rotation effects on weed seedbank size and composition. *Weed Res.* 41, 325–40.

Bàrberi, P., Aendekerk, R., Antichi, D., Armengot, L., Berner, A., Bigongiali, F., Blanco-Moreno, J. M., Carlesi, S., Celette, F., Chamorro, L., Crowley, O., Döring, T., Grosse, M., Haase, T., Heß, J., Huiting, H., José-María, L., Klaedtke, S., Kranzler, A., Luik, A., Peigné, J., Sukkel, W., Surböck, A., Talgre, L. and Sans, F. X. (2014), Reduced tillage and cover crops in organic arable systems preserve weed diversity without jeopardising crop yield. In: *Proceedings 18th IFOAM Organic World Congress*, Istanbul, 13–15 October (G. Rahmann and U. Aksoy(Eds)). IFOAM, Istanbul, pp. 765–7.

Bàrberi, P., Moonen, A. C., Peruzzi, A., Fontanelli, M. and Raffaelli, M. (2009), Weed suppression by soil steaming in combination with activating compounds. *Weed Res.* 49, 55–66.

Bàrberi, P., Silvestri, N., Peruzzi, A. and Raffaelli, M. (2000), Finger harrowing of durum wheat under different tillage systems. *Biol. Agric. Hort.* 17, 285–303.

Bastiaans, L., Paolini, R. and Baumann, D. T. (2008), Focus on ecological weed management: What is hindering adoption? *Weed Res.* 48, 481–91.

Bertholdsson, N.-O. (2010), Breeding spring wheat for improved allelopathic potential. *Weed Res.* 50, 49–57.

Bleeker, P., van der Weide, R. and Kurstjens, D. (2002), Experiences and experiments with new intra-row weeders. In: *Proceedings 5th EWRS Workshop on Physical and Cultural Weed Control*, Pisa, 11–13 March. Scuola Superiore Sant'Anna, Pisa, pp. 97–100.

Bourdôt, G. W., Hurrell, G. A., Saville, D. J. and Leathwick, D. M. (2006), Impacts of applied *Sclerotinia sclerotiorum* on the dynamics of a *Cirsium arvense* population. *Weed Res.* 46, 61–72.

Brandsæter, L. O., Thomsen, M. G., Wærnhus, K. and Fykse, H. (2012), Effects of repeated clover undersowing in spring cereals and stubble treatments in autumn on *Elymus repens, Sonchus arvensis* and *Cirsium arvense*. *Crop Prot.* 32, 104–10.

Canali, S., Campanelli, G., Ciaccia, C., Leteo, F., Testani, E. and Montemurro, F. (2013), Conservation tillage strategy based on the roller crimper technology for weed control in Mediterranean vegetable organic cropping systems. *Eur. J. Agron.* 50, 11–18.

Carr, P. M., Gramig, G. G. and Liebig, M. A. (2013), Impacts of organic zero tillage systems on crops, weeds, and soil quality. *Sustainability* 5, 3172–201.

Christensen, S., Søgaard, H. T., Kudsk, P., Nørremark, M., Lund, I., Nadimi, E. S. and Jørgensen, R. (2009), Site-specific weed control technologies. *Weed Res.* 49, 233–41.

Cirujeda, A., Melander, B. and Rasmussen, I. A. (2003), Relationship between speed, soil movement into the cereal row and intra-row weed control efficacy by weed harrowing. *Weed Res.* 43, 285–96.

Cloutier, D. C., van der Weide, R. Y., Peruzzi, A. and Leblanc, M. L. (2007), Mechanical weed management. In: *Non-chemical Weed Management. Principles, Concepts and Technology* (M. K. Upadhyaya and R. E. Blackshaw (Eds)). CAB International, Delhi, pp. 111–34.

Cooper, J., Baranski, M., de Lange, M. N., Bàrberi, P., Fließbach, A., Peigné, J, Berner, A., Brock, C., Casagrande, M., Crowley, O., David, C., de Vliegher, A., Döring, T. F., Dupont, A., Entz, M., Grosse, M., Haase, T., Halde, C., Hammerl, V., Huiting, H., Leithold, G., Messmer, M., Schloter, M., Sukkel, W., van der Heijden, M., Willekens, K., Wittwer, R. and Mäder, P. (2016), Shallow non- inversion tillage in organic farming maintains crop yields and increases soil C stocks: A meta- analysis. *Agron. Sust. Dev.* 36, 22.

Darnhofer, I., Lindenthal, T., Bartel-Kratochvil, R. and Zollitsch, W. (2010), Conventionalisation of organic farming practices: From structural criteria towards an assessment based on organic principles. A review. *Agron. Sust. Dev.* 30, 67–81.

Davis, A. D. (2010), Cover-crop roller-crimper contributes to weed management in no-till soybean. *Weed Sci.* 58, 300–9.

Davis, A. D., Taylor, E. C., Haramoto, E. C. and Renner, K. A. (2013), Annual postdispersal weed seed predation in contrasting field environments. *Weed Sci.* 61, 296–302.

Dayan, F. E., Owens, D. K. and Duke, S. O. (2012), Rationale for a natural products approach to herbicide discovery. *Pest Manag. Sci.* 68, 519–28.

Evans, G. J. and Bellinder, R. R. (2009), The potential use of vinegar and a clove oil herbicide for weed control in sweet corn, potato, and onion. *Weed Tech.* 23, 120–8.

Fenner, M. and Thompson, K. (2005), *The Ecology of Seeds*. Cambridge University Press, Cambridge.

Gallandt, E. R. (2006), How can we target the weed seedbank? *Weed Sci.* 54, 588–96.

Gbehounou, G. and Bàrberi, P. (2016), Weed management. In: *Mainstreaming Ecosystem Services and Biodiversity into Agricultural Production and Management in East Africa*. FAO Technical Guidance Document, pp. 29–45.

Gómez, R., Liebman, M. and Munkvold, G. (2014), Weed seed decay in conventional and diversified cropping systems. *Weed Res.* 54, 13–25.

Graglia, E., Melander, B. and Jensen, R. K. (2006). Mechanical and cultural strategies to control *Cirsium arvense* in organic arable cropping systems. *Weed Res.* 46, 304–12.

Hartmann, K. M. (2016), Photo-biological control of weed germination. *Med. Aromat. Plants* 5, 247.

Hobbs, P. R., Sayre, K. and Gupta, R. (2008), The role of conservation agriculture in sustainable agriculture. *Phil. Trans. Royal Soc. B: Biol. Sci.* 363, 543–55.

Jabbour, R., Zwickle, S., Gallandt, E. R., McPhee, K. E., Wilson, R. S. and Doohan, D. (2013), Mental models of organic weed management: Comparison of New England US farmer and expert models. *Renew. Agric. Food Syst.* 29, 319–33.

Johnson, S. P., Miller, Z. J., Lehnhoff, E. A., Miller, P. R. and Menalled, F. D. (2017), Cropping systems modify soil biota effects on wheat (*Triticum aestivum*) growth and competitive ability. *Weed Res.* 57, 6–15.

Jordan, N., Schut, M., Graham, S., Barney, J. N., Childs, D. Z., Christensen, S., Cousens, R. D., Davis, A. S., Eizenberg, H., Ervin, D. E., Fernandez-Quintanilla, C., Harrison, L. J., Harsch, M. A., Heijting, S., Liebman, M., Loddo, D., Mirsky, S. B., Riemens, M., Neve, P., Peltzer, D. A., Renton, M., Williams, M., Recasens, J. and Sønderskov, M. (2016), Transdisciplinary weed research: New leverage on challenging weed problems? *Weed Res.* 56, 345–58.

Kurstjens, D. A. G., Perdok, U. D. and Goense, D. (2000), Selective uprooting by weed harrowing on sandy soils. *Weed Res.* 40, 431–47.

Landis, D. A., Menalled, F. D., Costamagna, A. C., and Wilkinson, T. K. (2005), Manipulating plant resources to enhance beneficial arthropods in agricultural landscapes. *Weed Sci.* 53, 902–8.

Latsch, R., Anken, T., Herzog, C. and Sauter, J. (2017), Controlling *Rumex obtusifolius* by means of hot water. *Weed Res.* 57, 16–24.

Légère, A., Stevenson, F. and Benoit, D. (2011), The selective memory of weed seedbanks after 18 years of conservation tillage. *Weed Sci.* 59, 98–106.

Liebman, M. and Davis, A. S. (2000), Integration of soil, crop, and weed management in low- external-input farming systems. *Weed Res.* 40, 27–47.

Liebman, M. and Gallandt, E. R. (1997), Many little hammers: Ecological management of crop- weed interactions. In: *Ecology in Agriculture* (L. E. Jackson (Ed.)). Academic Press, Cambridge, pp. 291–343.

Liebman, M., Mohler, C. L. and Staver, C. P. (2001), *Ecological Management of Agricultural Weeds*. Cambridge University Press, Cambridge.

López-Granados, F. (2011), Weed detection for site-specific weed management: Mapping and real-time approaches. *Weed Res.* 51, 1–11.

Lukashyk, P., Berg, M. and Köpke, U. (2008), Strategies to control Canada thistle (Cirsium arvense) under organic farming conditions. *Renew. Agr. Food Syst.* 23, 13–18.

Marles, S. M., Warkentin, T. D. and Holm, F. A. (2010), Field pea seed residue: A potential alternative weed control agent. *Weed Sci.* 58, 433–41.

Melander, B. and Kristensen, J. K. (2011), Soil steaming effects on weed seedling emergence under the influence of soil type, soil moisture, soil structure and heat duration. *Ann. Appl. Biol.* 158, 194–203.

Melander, B. and Rasmussen, K. (2000), Reducing intrarow weed numbers in row crops by means of a biennial cultivation system. *Weed Res.* 40, 205–18.

Melander, B., Cirujeda, A. and Jørgensen, M. H. (2003), Effects of inter-row hoeing and fertilizer placement on weed growth and yield of winter wheat. *Weed Res.* 43, 428–38.

Melander, B., Rasmussen, I. A. and Bàrberi, P. (2005), Integrating physical and cultural methods of weed control – examples from European research. *Weed Sci.* 53, 369–81.

Melander, B., Nørremark, M. and Kristensen, E. F. (2013), Combining mechanical rhizome removal and cover crops for *Elytrigia repens* control in organic barley systems. *Weed Res.* 53, 461–9.

Mirsky, S. B., Ryan, M. R., Curran, W. S., Teasdale, J. R., Maul, J., Spargo, J. T., Moyer, J., Grantham, A. M., Weber, D., Way, T. R. and Camargo, G. G. (2012), Conservation tillage issues: Cover crop-based organic rotational no-till grain production in the mid-Atlantic region, USA. *Renew. Agr. Food Syst.* 27(Special Issue 01), 31–40.

Naylor, R. E. L. (2002), *Weed Management Handbook*. Blackwell Science, Berlin.

Nørremark, M., Griepentrog, H. W., Nielsen, J. and Søgaard, H. T. (2012), Evaluation of an autonomous GPS-based system for intra-row weed control by assessing the tilled area. *Precision Agric.* 13, 149–62.

O'Donovan, J. T., Harker, K. N., Turkington, T. K. and Clayton, G. W. (2013), Combining cultural practices with herbicides reduces wild oat (*Avena fatua*) seed in the soil seed bank and improves barley yield. *Weed Sci.* 61, 328–33.

Pannacci, E., Lattanzi, B. and Tei, F. (2017), Non-chemical weed management strategies in minor crops: A review. *Crop Prot.* 96, 44–58.

Parker, C. and Riches, C. R. (1993), *Parasitic Weeds of the World: Biology and Control*. CAB International, Wallingford.

Peigné, J., Ball, B. C., Roger-Estrade, J. and David, C. (2007), Is conservation tillage suitable for organic farming? A review. *Soil Use Manag.* 23, 129–44.

Peruzzi, A., Raffaelli, M., Frasconi, C., Fontanelli, M. and Bàrberi, P. (2012), Influence of an injection system on the effect of activated soil steaming on *Brassica juncea* and the natural weed seedbank. *Weed Res.* 52, 140–52.

Petersen. J. (2005), Competition between weeds and spring wheat for 15N-labelled nitrogen applied in pig slurry. *Weed Res.* 45, 103–13.

Picasso, V. D., Brummer, E. C., Liebman, M., Dixon, P. M. and Wilsey, B. J. (2008), Crop species diversity affects productivity and weed suppression in perennial polycultures under two management strategies. *Crop Sci.* 48, 331–42.

Rasmussen, J. and Svenningsen, T. (1995), Selective weed harrowing in cereals. *Biolog. Agric. Hort.* 12, 29–46.

Riemens, M. M., Groeneveld, R. M. W., Kropff, M. J. J., Lotz, L. A. P., Renes, R. J., Sukkel, W. and van der Weide, R. Y. (2010), Linking farmer weed management behavior with weed pressure: More than just technology. *Weed Sci.* 58, 490–6.

Riemens, M. M., van der Weide, R. Y., Bleeker, P. O. and Lotz, L. A. P. (2007), Effect of stale seedbed preparations and subsequent weed control in lettuce (cv. Iceboll) on weed densities. *Weed Res.* 47, 149–56.

Ryan, M. R., Mortensen, D. A., Bastiaans, L., Teasdale, J. R., Mirsky, S. B., Curran, W. S., Seidel, R., Wilson, D. O. and Hepperly, P. R. (2010), Elucidating the apparent maize tolerance to weed competition in long-term organically managed systems. *Weed Res.* 50, 25–36.

Rydberg, N. T. and Milberg, P. (2000), A survey of weeds in organic farming in Sweden. *Biol. Agric. Hort.* 18, 175–85.

Shaw, W. C. (1982). Integrated weed management systems technology for pest management. *Weed Sci.* 30, Suppl. 1, 2–12.

Smith, A. N., Reberg-Horton, S. C., Place, G. T., Meijer, A. D., Arellano, C. and Mueller, J. P. (2011), Rolled rye mulch for weed suppression in organic no-tillage soybeans. *Weed Sci.* 59, 224–31.

Smith, R. G., Mortensen, D. A. and Ryan, M. R. (2009), A new hypothesis for the functional role of diversity in mediating resource pools and weed-crop competition in agroecosystems. *Weed Res.* 50, 37–48.

Tillett, N. D., Hague, T., Grundy, A. C. and Dedousis, A. P. (2008), Mechanical within-row weed control for transplanted crops using computer vision. *Biosyst. Eng.* 99, 171–8.

Turner, R. J., Davies, G., Moore, H., Grundy, A. C. and Mead, A. (2007), Organic weed management: A review of the current UK farmer perspective. *Crop Prot.* 26, 377–82.

van der Schans, D., Bleeker, P., Molendijk, L., Plentinger, M., van der Weide, R., Lotz, B., Bauermeister, R., Total, R. and Baumann, D. T. (2006) Practical weed control in arable farming and outdoor vegetable cultivation without chemicals. PPO publication 532, Applied Plant Research Wageningen University, Lelystad, the Netherlands, 77p.

van der Weide, R. Y., Bleeker, P. O., Achten, V. T. J. M., Lotz L. A. P., Fogelberg, F. and Melander, B. (2008). Innovation in mechanical weed control in crop rows. *Weed Res.* 48, 215–24.

van Zuydam, R. V., Sonneveld, C. and Naber, H. (1995), Weed control in sugar beet by precision guided implements. *Crop Prot.* 14, 335–40.

Vermeulen, G. D., Tullberg, J. N. and Chamen, W. C. T. (2010), Controlled traffic farming. In: *Soil Engineering* (R. K. Gupta, I. P. Abrol, C. W. Finkl, M. B. Kirkham and M. C. Arbestain (Eds)). Springer, Berlin, pp. 101–-20.

Vidotto, F., de Palo, F., and Ferrero, A. (2013), Effect of short-duration high temperatures on weed seed germination. *Ann. Appl. Biol.* 163, 454–65.

Westerman, P. R., Luijendijk, C. D., Wevers, J. D. A. and van der Werf, W. (2011), Weed seed predation in a phenologically late crop. *Weed Res.* 51, 157–64.

Worthington, M. L., Reberg-Horton, S. C., Jordan, D. and Murphy, J. P. (2013), A comparison of methods for evaluating the suppressive ability of winter wheat cultivars against Italian ryegrass (Lolium perenne). *Weed Sci.* 61, 491–9.

Young, S. L., Pitla, S. K., van Evert, F. K., Schueller, J. K. and Pierce, F. J. (2017), Moving integrated weed management from low level to a truly integrated and highly specific weed management system using advanced technologies. *Weed Res.* 57, 1–5.

Part 3

Standards and monitoring

Chapter 13

Setting and reviewing standards for organic farming

Susanne Padel, The Organic Research Centre, UK

1 Introduction
2 Historic development of organic standards and control systems
3 Principles of organic production
4 Regulation of organic production
5 The future of organic standards development
6 Future trends and conclusion
7 Where to look for further information
8 References

1 Introduction

Organic farming has developed rapidly on a global scale, with both markets and areas of production continuously growing (Willer and Lernoud, 2017). One of the cornerstones of this development is the fact that, unlike many other initiatives aiming to improve farming towards greater sustainability, organic farming is governed by clear standards and inspection/certification procedures, which in many parts of the world are protected by legislation.

One could describe these organic standards and regulations on farming and food production as the basis of a contract, setting out the rules for an agreement between the producers, who undertake to follow the rules, and the consumers, who have a guarantee that these rules have been followed. This process between the producers and consumers is mediated by a certification body. Entry into the system by both producers and consumers is voluntary, even if the organic standard is covered by regulation. Organic standards are therefore part of a growing number of voluntary sustainability standards of food and agriculture, and they are described by Giovanucci et al. (2014) as the 'grandfather' of such standards.

The aim of '*sustaining the health of soils, ecosystems and people*' is part of the definition of organic agriculture (http://www.ifoam.bio/en/our-library/organic-basics), and there is a growing body of scientific literature documenting the positive sustainability impacts of organic farming (e.g. Mäder et al., 2002; Lampkin, 2007; Reganold and Wachter, 2016; Schader et al., 2012b). However, the current standards and certification procedures are based on pass/fail criteria for minimum requirements and do not actively encourage organic

http://dx.doi.org/10.19103/AS.2017.0029.14

operators to strive for further improvements of their systems. Several important sustainability outcomes that are covered by the principles of organic agriculture (e.g. biodiversity, soil fertility and fairness), are currently not directly addressed in most organic standards. This leads to criticism that the process-oriented organic farming standards and the associated certification processes do not fully safeguard the sustainability performance of individual organic farms, and the question arises as to how this should be addressed going forward.

To help address this, this chapter provides a brief history of organic standards development, and introduces the values and principles of organic farming that have guided this process. A section on how organic farming is regulated introduces the different types of rules that are currently used to achieve common objectives, such as prohibitions or obligations to use specified inputs and practices. The outlook includes a very brief summary of trends in sustainability assessment, before exploring what contribution research can make to further organic standards development.

2 Historic development of organic standards and control systems

The ideas on which organic agriculture is based can be traced back along different lines. During the first half of the twentieth century, several pioneers from a range of backgrounds (farmers, researchers and medical doctors) began to develop 'biological' cropping methods, based on the idea that healthy soils would be the key to producing healthy crops, and thus help to improve human health. The schools of thought that have influenced today's organic farming include biodynamic farming as taught by Rudolf Steiner, organic farming ideas of the living soil promoted by Lady Eve Balfour, (inspired by Albert Howard and Robert McCarrison), bio-organic farming according to Müller and Rusch, the biological farming methods of Lemaire–Boucher, and Bob Rodale's regenerative farming (Lampkin, 2002; Vogt, 2007). Each has contributed some aspects, often in different parts of the world, but mainly contributing to a common concept. There are also several other movements that have shaped organic farming, such as the growing awareness of the widespread use and impact of pesticides (e.g. *Silent Spring* by Rachel Carson), Schumacher's *Small Is Beautiful*, the 'back-to-the-land' and the environmental movements (see Lockeretz, 2007 for a more comprehensive guide to the history of organic farming).

The first organic standards were set by the private sector. The Demeter co-operative set the first very short private norms in 1928 in Germany, as part of a contract for farmers who wanted to use the Demeter label, first registered in 1931. The Müller–Rusch movement of organic-biological agriculture in Switzerland also drafted its first tentative norms in 1946, which later led to the first Swiss organic standards. In the United Kingdom, the first Soil Association standards were published in 1967. Their structure served as an example for other standards, while the first private French standards date back to 1972. During the 1970s, groups of farmers in different parts of the United States developed certification systems to assure consumers that products labelled as organic were produced according to their standards, for example the California Certified Organic Farmers (established in 1973).

Early standards were often in the form of recommendations, rather than certification requirements, and the farmers received feedback on their practices through peers or consultants. However, with the growth of the organic market and increased trade, the relationship between consumer and producer became less personal. There was a need for a more rigorous independent quality assurance system to protect both parties, as well as other members of the value chain (Schmid, 2007). The organic sector was therefore one of the first agricultural sectors to engage with, and develop, a system of third-party inspection/certification or voluntary sustainability standards in the food sector (Giovannucci et al., 2014). In the 1980s, a system of private standard setting and third-party inspection and certification evolved. Some standard owners and control bodies grew out of producer/consumer groups and several maintained a strong representation of consumers within the organisations, whilst others were more farmer-led. Most of these organisations were engaged in several other activities besides certification. Several more specialised organisations dedicated to certification also started in this period, such as Skal (Netherlands, which started in the early 1990s) and KRAV (Sweden, with the first standard agreed in 1985).

A next step for organic farming standards was their inclusion in national legislation. Among the first countries to develop national organic farming legislation were the United States: Oregon (1973), and in Europe: France (1980), Austria (1983) and Denmark (1987) (Lampkin et al., 1999b; Ellsworth, 2001). The European Union introduced its first Regulation on organic production in 1991, in response to a growing consumer demand for certified organic products (EC, 1991). The main role of the legislation was to provide a legal definition of organic farming to protect both consumers and organic farmers against false and misleading organic claims. Prior to its introduction, products claiming to be produced in a more natural way but without clear production standards were commonly found, for example, on the shelves of supermarkets in Germany (Padel and Zerger, 1994). This and other EU-wide common rules helped to ensure transparency at all stages of production and processing as well as fair competition. Organic standards were also taken up by the US Government with the establishment of an Organic Food Protection Act in 1990. This aimed to establish national standard governing, assure consumers that organically produced products meet a consistent standard, and thus facilitate US-wide trade in organic products. Defining in detail what organic farming meant was achieved several years later with the USDA Standards of the National Organic Programme (NOP) that took effect in 2002 (Schmid, 2007; Ellsworth, 2001).

Both the European Regulation and the NOP have influenced the development of the organic sector worldwide. Typically, regulations cover rules that govern production of raw materials, food processing, the control system, how reference to organic can be made in labelling, and cross-border recognition of organic status (e.g. what minimal requirements imported organic products have to fulfil (see also below)). The control rules include provisions for the supervision and accreditation of control bodies, whilst making reference to the ISO standard for product-based controls (ISO 65). The development of national and international regulations for organic farming, also made the business of organic certification of greater interest for commercial certification companies (Rundgren, 2002).

Like all other producers, organic producers also have to comply with the general food and feed legislations of their country, even if these requirements are not explicitly repeated in the organic rules. This is relevant for standard comparisons across different regions. For example, producers in Europe have to comply with the labour laws that protect workers' rights. Some aspects of worker protection are therefore not specifically mentioned in the European organic standards. However, some standard owners that orient themselves towards importing products into the EU include such rules, for example the Naturland Fair standards (http://www.naturland.de/en/naturland/what-we-do/naturland-fair/criteria-standards.html).

In the EU, the official recognition of organic standards in a European Regulation also paved the way for the financial support of organic farmers and the conversion process under the Regulation (EC) 2078/92 on agri-environmental schemes (EC, 1992). This support continues to this day, with some changes in most EU member states, as part of the Rural Development Programmes. Most support schemes – with some initial exceptions like in Sweden – require organic management to be controlled according to Regulation (EC) 2092/91 by an approved control body (Lampkin et al., 1999a). This organic certification is provided under government supervision, although in some cases (e.g. Denmark) the state gained the status of a professional service for regulatory compliance (Wai, 2007). In the most recent CAP reform of 2013 organic producers were further recognised by qualifying automatically for the 'Greening' part of the basic support payments.

Organic farming was among the first agricultural sectors to introduce voluntary standards and inspection and certification procedures, which later became part of national and international legislation. Organic standards are not static, and different areas are continuously revised in light of new developments on farms, because of research on organic farming systems, general trends in the agricultural industry, and importantly, changing consumer concerns.

3 Principles of organic production

Organic farming is clearly value based. However, the ideas and values of organic farming originate from different movements. The core values could therefore not easily be traced back to one clear set of principles, that regulators could consider when organic standards became part of national and international regulations. IFOAM (the International Federation of Organic Agriculture Movements) established a task force that consulted with its members on core values in the early 2000s, and presented four Principles of Organic Agriculture (Box 1) that were debated and democratically accepted by IFOAM's members in 2005 (Luttikholt, 2007; Padel et al., 2009). Comparing the value elements of the four principles with a range of key sources outlining organic values, showed that these four IFOAM Principles of Health, Ecology, Fairness and Care are a good representation of the value basis of organic agriculture as described in the literature (Padel et al., 2009).

Box 1 The four principles of organic agriculture

Principle of Health: Organic agriculture should sustain and enhance the health of soil, plant, animal and human as one and indivisible.

Principle of Ecology: Organic agriculture should be based on living ecological systems and cycles, work with them, emulate them and help sustain them.

Principle of Fairness: Organic agriculture should build on relationships that ensure fairness with regard to the common environment and life opportunities.

Principle of Care: Organic agriculture should be managed in a precautionary and responsible manner to protect the health and well-being of current and future generations and the environment.

Source IFOAM (2014).

Breaking down the four principles into the underlying value elements also facilitates comparison with other documents, such as standards and regulations. This shows that several outcomes considered important for overall farm sustainability (e.g. biodiversity, soil fertility, fairness) are not fully covered by the current legal standards (Padel et al., 2009). Nevertheless, there is good scientific evidence that organic farming delivers in many of these areas.

For example, Reganold and Wachter (2016) examined the performance of organic farming in light of four key sustainability metrics: productivity, environmental impact, economic viability and social well-being. They concluded that organic farming systems are (when compared with mainstream farming) more profitable despite lower yields. They are also more environmentally friendly, and deliver equally or more nutritious foods that contain less (or no) pesticide residues and contaminants. There is also evidence indicating that organic agricultural systems deliver greater ecosystem services and social benefits (Reganold and Wachter, 2016). However, the lack of explicit coverage of certain outcomes in the standards has led to criticism that a process-oriented assessment approach does not fully safeguard the sustainability performance of an individual farm, first expressed by Buck et al. (1991) in relation to organic vegetable production in California.

Articles that state principles and objectives of organic farming were introduced in the EU Regulation as part of the total revision (EC, 2007). Reports and articles refer to all four IFOAM principles, but are less far reaching in relation to some values such as fairness and biodiversity. The stating of objectives and principles explicitly in one place in the EU Regulations has contributed to a common understanding of the core concept of organic farming. However, there is uncertainty as to whether the statements are legally binding, and whether and how the principles and objectives should be considered in the interpretation of any more detailed rules. For example, the principles state that organic livestock production should be 'land-based', but the detailed rules require only a limited part of the feed to be produced on the farm itself, or in the same region and the term region is not clearly defined. This enables some organic livestock production to develop independently from crop production (Padel et al., 2013).

Standards and regulations should clearly state the principles and values of organic farming on which they are based. Values are *per se* in need of interpretation. It is therefore equally important to consider the decision-making structures and processes of any standard owner or regulator as part of the principles of organic farming. Standards should be based on consultation with all affected stakeholders from the farmer to the consumer, and the processes should facilitate a coherent interpretation of the objectives and principles in implementation of more detailed rules (Padel et al., 2009, 2007).

4 Regulation of organic production

The overall aim of any standard or regulation is to maintain and enhance the quality of a product and to provide guarantees to the consumer. Producers signing up to organic farming therefore make a commitment to follow these standards. Good standards also give guidance to producers on how organic principles can be applied to the different enterprises on the farm. This provides a guarantee to consumers that the rules have been followed, for example, that non-permitted fertilisers and pesticides have not been applied. Certification bodies effectively act as mediators between producers and consumers by verifying that rules have been adhered to.

It is important to be clear that any producer entering into organic farming does so voluntarily, and that the responsibility that the rules are followed lies, in the first instance, with the producer, who also agrees that his/her farm is scrutinised by an inspection through a control body. Organic farming is thus different from food legislation that all producers must adhere to, even when the organic standard is part of national/international law. Producers can withdraw from organic production, and one of the reasons quoted for withdrawals or re-conversions is the administrative burden arising from certification (Koesling et al., 2012; Kuhnert et al., 2013). Also the consumers can withdraw their custom at any time, and may do so if they no longer agree with the rules. Thus, in setting organic standards or regulations, the interests of the primary producers, processors and consumers need to be considered, and a compromise found. Gershunny (2016) in her personal memoirs about her role in developing the USDA standards, highlighted that the aim of making organic standards very 'pure' can have many unintended consequences, in particular damaging the farmers' willingness to convert to organic production because the rules become so complicated that they are impossible to follow.

Current organic standards can be divided into two main groups:

- Legally binding regulations that protect the term organic (e.g. EU Organic Regulations, USDA programme) with the overarching FAO *Codex Alimentarius* guidelines acting as a global guidance document. When selling in a country where the term 'organic' is legally protected it can only be used when the rules valid in that country have been followed and operators are inspected, otherwise an operator can be taken to court, although cases are rare. In the EU the terms 'biological' and 'ecological' are also protected by the Regulation.
- Private standards of symbol holders (e.g. Demeter, Soil Association and Bioland) for which the IFOAM underlying principles act as a global reference point. In this case the right to use the symbol or trademark depends on following these specific rules.

Organic regulations and standards apply equally to all areas of organic production and also to food processing and labelling. They mainly contain two different types of rules:

- Rules for primary producers and food processing operators that seek to be certified for producing according to the agreed standards, and
- Rules for labelling that govern how the organic status of products can be communicated to the buyer.

Regulations also contain rules for inspection or control bodies (both public and private), and rules for authorities and organisations that approve, accredit and supervise certification bodies both public and private (ITF, 2006; Padel, 2010).

The following sections focus on the rules and standards for organic production at the farm level, and explore how the organic standards translate the principles, aims and objectives into rules that can be audited through inspection and certification processes.

At present, four main types of rules are used, alone or in combination, which will be discussed with some examples below. In essence these are:

- Prohibition of specific inputs (e.g. synthetic nitrogen fertiliser and herbicides);
- Restrictions on the use of some external inputs that cannot be fully removed (e.g. copper as a plant protection agent);
- Preference for the use of inputs from organic origin (e.g. using only feed materials that originate from other organic farms);
- Obligations to use certain 'good practices' (e.g. practising a diversified crop rotation) (Padel et al., 2013).

Table 1 illustrates how these different instruments have been used in the EU Regulation of organic farming of 2007, and relates this to the objectives and principles as they are stated in early articles of the same Regulation. In the monitoring of sustainability of farming systems and of animal welfare, the measurement of desirable outcomes (e.g. greenhouse gas balance, animal welfare score) is used as a way to assess achievement. However, previously this approach of stating desirable outcomes has not been widely used in organic standards. These different types of rules are also a reflection of how organic farming is understood, as a way of farming that uses only 'organic(ally) approved non-synthetic materials' or as a farming system that uses the 'best agronomic practices'. Gershunny (2016) referred to this as the 'material or practice' conundrum.

4.1 Prohibitions and restrictions

Prohibitions and restrictions are important types of organic rules, common to all standards, and organic farming is often misconceived as being only about the non-use of synthetic chemicals in agriculture. The best known example is probably the ban on synthetic nitrogen fertiliser. Another example of a clearly stated prohibition in organic farming is the use of GMOs or of using battery cages in organic chicken farming in Europe. Prohibitions are either included in the form of stating the prohibition or restriction directly (e.g. this input shall not be used) or as stating a positive list of inputs that can be used (e.g. organic farmer can only use the following inputs for fertilisation). Both approaches can be found in the different organic regulations.

The EU Regulation mainly uses positive lists (in Annexes) and this can create *de facto* prohibitions that are in place, but not explicitly stated. For example, the ban on using herbicides is not stated as a clear prohibition, but is implicit in the obligation that only permitted substances can be used for crop protection that are specified in one of the annexes. The ban on herbicides exists because no herbicide is included in the substances in the annex. The benefits of organic farming for biodiversity might be easier if the ban on herbicides was more clearly stated. Nevertheless, it might be an advantage for developing/using natural phytotoxic ingredients for weed control under special site conditions (Giepen, 2017). Alongside clear prohibitions, there are also strong restrictions on several other inputs, for example, most pesticides and many soluble fertilisers, although there are some permitted exemptions. The revised EU Regulation of 2007 stated criteria for inclusion of substances (see Article 16) and established an expert group to provide advice on substances to be included (EC, 2007). For more details on the expert group including all reports on recommendations, see https://ec.europa.eu/agriculture/organic/eu-policy/expert-advice_en.

Prohibitions and restrictions on the use of inputs, specific practices, or the listings of permitted inputs are usually easy to audit and control, compared with the overall

Table 1 Monitoring instruments used in the EU Regulation of organic farming

Production rules*	Respect natures systems/cycles	Contribute to biodiversity	Make responsible use of natural resources			
			Energy	Water	Soil	Air and climate
Prohibitions (A: 4 (a) iii and (c))						
No mineral nitrogen fertilisers (A: 12.1 (e))	√	√	√	√	√	√
No herbicides, only authorised products can be used (A: 12 (h), B: Annex II)	√	√	√	√	√	√
No landless livestock production (B: 16)	√		√			√
No hydroponic production (B: 4)	√			√	√	
No use of GMOs (A: 9)	√					
Strict control of external inputs (A: 4 (b)), minimisation of the use of non-renewable resources (A: 5 (b)) and recycling of wastes and by-products (A: 5 (c))						
Only permitted fertilisers: low-soluble mineral fertiliser (A: 4 (b) iii) and soil conditioners when need proven (B: 3, Annex I)	√	√			√	
Only authorised plant protection products when established threat (A: 12.1 (h), B: Annex II)	√	√			√	√
Feed primarily from holding or same region (with exceptions) (A: 14.1 (d))	√		√			
Stocking density and use of livestock manure restricted to maximum of 170 kg N/ha and year (B: 3 &15.1)	√	√	√	√	√	√

Preference for inputs from organic origin (A: 4b with exceptions (A: 4d))						
Manage entire holding organically (with exceptions) (A: 11)	√	√	√	√	√	√
Only organic seed (with exceptions) (A: 12.1)	√					
Only organic feed (with 5% exceptional rule for monogastrics) (A: 14 (d) ii)	√					
Obligations to use good husbandry practices and prevention (A: 4 (a) iv and 5)						
Multi-annual crop rotation including legumes and other green manures (A: 12.1 (b))	√	√	√	√	√	
Tillage and cultivation practices that maintain organic matter, and protect soil (A: 12.1 (a))	√	√	√	√	√	
Maintain crop health through prevention (natural enemies, the choice of species and varieties, crop rotation) cultivation techniques and thermal processes (A: 12.1 (g))	√	√	√		√	
Numbers of livestock limited to minimise overgrazing, poaching, soil erosion or pollution (A: 14.1 (b) iv)	√	√		√	√	√

[*] Article numbers refer to (A) Council Regulation (EC) 834/2007 and (B) Commission Regulation (EC) 889/2008.
Source: Based on Padel et al. (2013) adapted.

sustainability and environmental impact of a farming system. Prohibitions on certain inputs are also easy to communicate to consumers, and indeed if consumers are asked what they understand about organic farming, a frequent answer will be 'no fertiliser, no pesticides' (e.g. Zanoli, 2004). However, just replacing one damaging input with another more benign one is often not sufficient to make an organic farming system function well. In many cases there is a need to redesign the system. For example, introducing crop rotations with fertility building crops rather than trying to just replace non-permitted fertiliser with organically accepted ones, such as livestock manure. This has been expressed in the efficiency–substitution–redesign framework (MacRae et al., 1990; Hill, 1985) where farmers might start with adopting more efficient use of inputs, then substitute certain inputs with other more benign ones, and finally adopt a system redesign/management approach to reduce the need for inputs in the first place.

Overall, the focus on input prohibitions and restrictions represents a means to an end, rather than the end in itself, although there are clear risks if ends and means get confused (Lampkin et al., 2017). How organic farming is understood as either using only permitted organic inputs or as using good husbandry practices to achieve the desired outcomes is part of this debate (Gershunny, 2016). Prohibiting certain inputs says nothing about what organic management involves instead, and why certain technologies and practices are preferred over others (Lampkin et al., 2017). For example, instead of relying on the use of synthetic nitrogen fertiliser, organic farms use biological nitrogen fixation through legumes. On a cropping farm this implies the need to include legumes in the crop rotations. This is why the standards also contain obligation rules (see below).

4.2 Use of inputs from organic farming systems

In addition to the restriction of external inputs, many organic standards also include rules specifying that certain agricultural inputs must be sourced from other organic farms. This applies in particular to seeds and feedstuff. The rule makes organic farming less dependent on inputs from the non-organic sector and therefore less vulnerable to contamination (e.g. with GMOs). Like prohibitions and restrictions, such rules are potentially easy to verify in the inspection and certification process. However, because of a lack of availability of organically produced inputs in regions with small organic sectors, derogations have been built into the rules, which allow farmers to revert to the use of non-organic inputs under certain circumstances. In the external evaluation of the EU Regulation, Padel et al. (2013) concluded that this system of exceptional rules, established to allow regional differences in climate, stage of sector development and specific husbandry practices to be taken into account, seems to be inadequate. Across member states these rules have been interpreted very differently creating potentially unfair competition. Also, the need to apply for and grant derogations also creates additional administrative demands for both the operators (e.g. farmers, growers and food processers) and the control bodies. Despite the aim to reduce the sector's reliance on non-organic inputs, there are several cases where timetables for step-wise reductions of the permitted percentages were not kept, and limited or no progress has been made.

Whether there is progress in achieving a lower reliance on non-organic inputs is difficult to monitor. Padel et al. (2013) found that there is insufficient data available on the scale of non-organic feed use in organic agriculture and the current availability of organic feed to carry out a quantitative analysis of the coverage and uptake of this rule. There is a general consensus among many organic producer organisations, processors and traders, that organic protein crop production (at EU level) is insufficient to meet the demand for the monogastric livestock. If transition to 100% organic diets cannot be realised using European organic supply chains, the standard demand is likely to lead to increased reliance on extra EU imports in the short and medium term. This would contradict the aim of a land-based livestock sector and the local/regional sourcing of feed. Alternative protein supplies are already being developed to help address this, including methods to produce methionine via enzymatic fermentation based on organic raw materials, or the use of insect larvae or algae as a protein source for feed. Whilst these new techniques seem to be promising, they are not ready for broad practical use yet (Padel et al., 2013). One of the topics of discussion, as part of the process of revision of organic rules in Europe since 2014, has been the question about how the use of regional feed can be defined. Any definition has to be acceptable to both the producers, who have to meet it, and to the control bodies that have to verify it. Any solution based on a specific distance that the feed can travel is unlikely to be suitable for the inspection and certification processes.

4.3 Commitment to good practice

As already stated, simply not using synthetic inputs and doing nothing else (organic farming by default) is likely to lead to failures in terms of productivity, and financial and environmental sustainability. As a result many areas are regulated by the expectation to use certain practices. This makes good sense if there is a clear connection between the 'good practice' and the principle or objective that is intended through implementation of the rule. This can be illustrated through the example of the rule requiring a multi-annual crop rotation. The principle aim that the rule wants to encourage is for a producer to grow legumes, for biological N fixation and to achieve greater diversity of cropping and create greater resilience. Whilst crop rotations are a perfectly good practice for mixed and arable farms, a blanket statement of the obligation to use a crop rotation does not work on farms that consist mainly of permanent crops or pastures, and for farms that use systems of multi- and intercropping (such as crop mixtures or agro-forestry). Such farms achieve diversity in space with several different species grown at the same time (including legumes), rather than over time as would be the case with the crop rotation. Indeed, ploughing up permanent grassland to introduce a rotation would have negative effects for biodiversity, and in reality control bodies are known not to enforce this rule where it is not appropriate. This example illustrates that care needs to be taken when stating good practice obligation, and that the conditions under which they have to apply are also clearly stated, that is, practising crop rotation only applies to arable and horticultural land but not to permanent cropping systems. Specialised organic operators, such as horticultural, vine, pig and poultry producers, can find good practice obligations particularly challenging. Several attempts have been made in Europe to develop specific rules for organic production in glass houses, but this has not yet been considered in the drafting of the EU Regulation.

4.4 Defining outcomes of organic production in the rules

Another option to regulate organic farming would be to state the outcomes that an organic operator has to achieve to qualify for organic status, in line with the principles of organic production. This could potentially apply to all areas where organic agriculture states aspirations, such as protection of soil health, biodiversity, a balanced system and so on (see also Section 3). However, including such outcomes in the standards and certification requires well-established and widely accepted metrics, indicators and thresholds, and a good understanding of how these can be verified as part of the certification system. The example of residue thresholds shown in Box 2 illustrates some of the issues faced in following this approach. However, this is not to say that monitoring for residues and contaminants cannot and should not be one of the instruments used in the inspection system.

Box 2 Proposal for setting residue and contaminant thresholds for organic products in the proposal for the revision of the EU Regulation

In the proposal for a revision of the EU Regulation, the European Commission proposed to introduce a de-certification threshold for non-allowed products and substances. The proposal stated that *the level above which the product cannot be sold as organic will be established, taking into account the baby food directive (0.010 mg/kg)*. This part of the proposal was one of the major areas of controversy in the three year negotiation and was not included as such in the compromise text.

Early on, IFOAM-EU expressed great concern about this part of the proposal. It was argued that, whilst organic production aims to be free of pollutants and substances not allowed in organic production such as GMOs, pesticides and fertilizers, and regulates this by using rules that prohibit using these substances, organic producers operate in a contaminated world and contamination can frequently occur by accident. By focusing on threshold levels as a key organic certification tool, rather than inspecting for input use, the standards would move away from the process-based approach that has prevailed so far. There is also a danger that organic food producers would have to pay for additional laboratory testing and be subject to the risk of contamination by conventional neighbours.

Source IFOAM-EU (2014).

5 The future of organic standards development

5.1 Aligning organic rules with sustainability assessments

At present, there is no internationally accepted consensus on how to make agriculture more sustainable and how to measure the sustainability of food production. The various sustainability assessment approaches take diverging perspectives on the notion of 'sustainability', concerning what should be assessed, where the system boundaries are drawn, and how it should be measured. There are various tools and initiatives emerging aimed at measuring the sustainability of farming and food, including in organic farming, but these are separate from organic certification and standards. Some of these efforts focus only on one environmental impact category (e.g. climate change) while others cover all dimensions of sustainability. Attempts can be broadly grouped into those driven by gaining a better scientific understanding and/or developing policy advice, initiatives that aim to better communicate sustainability in the market place (including as part of CSR reporting of food businesses) and initiatives aimed at improving practices on farms and in the food sector. Assessments include both broad assessments, aimed at a 'quick and dirty' assessment of a wide range of sustainability outcomes in a less in-depth way, and comprehensive approaches that look in detail at one or more specific sustainability outcome, such as greenhouse gas emissions or biodiversity. One of the main barriers for wider utilisation of the various sustainability assessment tools is cost. Generally a broad and shallow approach is cheaper to carry out as it requires less time and/or data to complete, whereas a comprehensive approach requires more time and/or data and therefore is generally more expensive, especially if including multiple objectives (Schader et al., 2012a). The FAO has published Guidelines for Sustainability Assessment of Food and Agriculture systems (SAFA) which specify the procedure, principles and minimum requirements for a sustainability impact assessment (FAO, 2012). These SAFA Guidelines provide for the first time one common language for the assessment of sustainability in agriculture.

Initiatives specifically seeking to evaluate sustainability of organic farming systems include studies of individual product assessments using life cycle assessment (LCA) (e.g. Tuomisto et al., 2012; Hietala et al., 2015) or assessments balancing different sustainability metrics (Reganold and Wachter, 2016). There are also tools aimed at communication with market partners (see e.g. http://www.fibl.org/en/themes/sustainability-assessment/smart-en.html) or communication platforms, such as Nature & More (http://www.natureandmore.com) aiming to communicate the commitment and effort that individual growers make towards the planet and its people. There are also some tools that are aimed at awareness raising and advice at the farm level, for example RISE (Häni et al., 2003) or the public goods tool of ORC (Gerrard et al., 2011) (for comparisons of tools see also Schader et al., 2012a; de Olde et al., 2016).

IFOAM, the international umbrella organisations of the organic movement, has also engaged in this field. The 'Sustainable Organic Agriculture Action Network (SOAAN)' of IFOAM was set up for developing a programme that positions organic agriculture and its related supply chains as a holistic, sustainable approach to the production of food and fibre for all of human society (http://www.ifoam.org/growing_organic/Best_Practice_Program/SOAAN_Structure.php). This has now led to the publication of two relevant documents, the Organic 3.0 guide (Arbenz et al., 2017) and a Best Practice Guideline for Agriculture & Value Chains (SOAAN, 2013).These documents are intended to broaden the scope of organic farming beyond defining it purely by its standards and certification procedures (see Box 3). However, the documents are less clear on what measures can be used to reach an outcome-based system of organic farming that aims for continuous improvement.

Box 3 Organic 3.0 – for truly sustainable farming and consumption

With Organic 3.0 IFOAM aims to respond to the many challenges and opportunities and develop farming systems that are ecologically sound, economically viable, socially just, culturally diverse and transparently accountable.

The overall goal of Organic 3.0 is to enable a widespread uptake of truly sustainable farming systems and markets, based on organic principles and imbued with a culture of innovation, of progressive improvement towards best practice, of transparent integrity, of inclusive collaboration, of holistic systems and of true value pricing.

Organic agriculture is a lighthouse for truly sustainable agriculture and agricultural production systems. Organic 3.0 expands the participation options, and positions organic as a modern, innovative farming system that holistically integrates the local and regional context including its ecology, economy, society, culture and accountability. Regeneration of resources, responsibility in production, sufficiency in consumption, and the ethical and spiritual development of human values, practices and habits are concepts that guide the building of a new organic culture that can drive societal development.

The core of Organic 3.0 is the living relationships between consumers and producers, which include the stories of products and production and the multiple benefits of organic agriculture. At its heart, Organic 3.0 is not prescriptive but descriptive: instead of enforcing a set of minimum rules to achieve a final static result, this model is outcome-based and continuously adaptable to local context. Organic 3.0 is still grounded upon clearly defined minimum requirements such as the ones maintained by many government regulations and private schemes around the world, and in the objectives of the IFOAM Standards Requirements. However, it also expands outward from these base requirements: it calls for a culture of continuous improvement through private- and stakeholder-driven initiatives towards best practices based on local priorities, as described in the Best Practices Guidelines of IFOAM – Organics International.

Source Arbenz et al. (2017).

Current organic standards address only a small part of wider sustainability aims, even though these wider aspects can be indirectly addressed if the rules are followed. For example, organic rules contain only a small number of direct provisions to enhance biodiversity, but increases in biodiversity on organic farms compared to non-organic farming systems is a well-proven outcome that has been confirmed in the scientific literature (e.g. Reganold and Wachter, 2016).

As yet there is no commonly agreed standard for organic/ecological agriculture that covers all aspects of sustainability, and one of the reasons is clearly that the conditions under which organic farming is practised around the world varies considerably. Organic standard setters have been developing new elements in the standards in response to concerns of organic stakeholders (e.g. social or ethical standards and environmental standards) and in some cases these are also used to differentiate products in the market place (see e.g. Zander and Hamm, 2010; Zander et al., 2010). Arbenz et al. (2017) called for continuous improvement of the system and for further development, beyond certified organic, and Rahmann et al. (2016) showed very clearly the important role that research must play in achieving these aims. In contrast, Seufert et al. (2017) argued for a change in the regulations by placing more emphasis on environmental best practice, in line with the original aims of the pioneers that had a holistic understanding of systems and how to achieve environmental improvements. Interestingly, Seufert et al. (2017) do not support the route of stating environmental outcomes and thresholds in organic standards and recommend stating accepted good environmental practices instead. There should be sound evidence that confirms 'good practices' to be effective in delivering the desired sustainability outcomes (i.e. reducing biodiversity loss, land degradation, climate change and water use) and do not lead to negative outcomes in other important areas of sustainability.

6 Future trends and conclusion

One of the cornerstones of organic farming is the fact that it is governed by clear standards of production. 'Organic' is today one of the most recognised food labels, and the basic meaning of it is understood by many consumers around the world (Seufert et al., 2017). The term 'organic' is legally protected in many parts of the world and is understood as a farming system that prohibits the use of artificial synthetic inputs that may endanger health and the environment. This has allowed markets to grow globally at a rate above other food markets (Willer and Lernoud, 2017). However, there is also a widely held view that organic is more than just certification and that there is a gap between certified organic (as defined by the current regulations, standards and certification procedures) and the ideas and principles of organic farming (see e.g. Lampkin et al., 2017; Seufert et al., 2017; Rahmann et al., 2016; Arbenz et al., 2017). The current certification system operates on a pass/fail system and does not encourage further development regarding the sustainability of the farms. Despite the administrative effort for both producers and control bodies to carry out organic inspection and certification on an annual basis, the control system contributes very little to building an evidence base to substantiate related claims.

Standard-setting is a balancing process that has to consider the views of various stakeholders, most notably farmers, food businesses and consumers, but also regulators and control institutions. The development of organic standards so far has focused mainly on areas that are easy to codify (Lockie et al., 2006; van der Grijp, 2006), such as inputs or practices that are prohibited or required. With growing market demand, there is a danger that globally organic standards will be defined by the common denominators. Seufert et al. (2017) argue that the standard-setting process has been mainly consumer-driven and highlight the danger in orienting standards towards one stakeholder group, which may lead to other stakeholders (e.g. producers) feeling under-represented. If organic standards prioritise the needs of consumers over those of producers, there is a danger that they become very difficult to comply with, and discourage those that might want to convert to organic. As a result, demand is likely to grow faster than supply, as is already the case in several parts of the developed world (Willer and Lernoud, 2017).

The challenge for the future is developing organic standards in a balanced way to meet the needs of all stakeholders, by encouraging organic production to become overall more sustainable and providing evidence of the wider benefits of organic farming. Focus must also be on maintaining the clarity of rules and keeping them practical for producers and verifiable by control bodies, so that more producers can benefit from the opportunities that organic farming offers.

6.1 Research to support the development of organic standards

There is a need to continue with the provision of evidence regarding the impact of organic principles and practices, including areas that are at present not clearly codified in organic standards. The SOAAN group of IFOAM has taken a first step in publishing best practice guidelines (SOAAN, 2013) but there is only a limited amount of scientific evidence on how these practices contribute to the sustainability outcomes that the organic principles aim to achieve. Related to this is the need to continue developing simple indicators that are robust but not too time consuming to implement, and which could therefore become part of organic certification in the future. One example that illustrates the case of bringing together the knowledge of researchers with that of control bodies is the AssureWel (http://www.assurewel.org/aboutassurewel) project that aims to develop a practical system of welfare outcome assessment for the major farm animal species, which can be used in farm assurance schemes. Similar efforts are needed to develop certification indicators for environmental practices adopted frequently on organic farms. Rahmann et al. (2016) challenged the research community as part of Organic 3.0 to contribute to the development of more integral system indicators for standards and certification that avoid overregulation, whilst encouraging participation from farmers of multiple scales and values. This highlights the need to consider the transaction costs that the organic farmers and consumers have to pay for the inspection and certification (e.g. costs for time of inspection and samples that are analysed) and to minimise the administrative burden of the control system. There is also a need to account more clearly for the externalities of organic farming compared with other farming systems.

It is also necessary to evaluate the impact of the current standards and thus inform their revisions, although the political process is not always strictly evidence based. The trust in the organic claim depends on simple and clearly structured standards and regulations that can be easily understood and followed by farmers, consumers, and others involved in organic supply chains. Whilst European Regulation (EC) 834/2007 contributed to the transparency of the legislative framework by clearly setting out the objectives and principles of organic production, the fact that many more detailed implementation rules are set out in a separate Regulation ((EC) 889/2008) still requires an operator to consult a high number of articles when interpreting particular issues (Sanders et al., 2013).

Finally, there is a need to better understand the barriers that prevent more farmers from converting to organic production, including the impact of organic farming on farm profitability. There is some evidence that organic farming is more profitable. Crowder and Reganold (2015) carried out a meta-analysis of the profitability of 55 crops in five continents and found that, if the premiums that producers receive for their organic status are considered, organic agriculture was significantly more profitable than conventional agriculture. However, in many organic communities profit is frequently associated with corporate agriculture, and farmers that want to stay profitable are accused of converting for the wrong reasons, although economic sustainability is clearly one of the pillars of sustainability frequently mentioned by farmers. Studies have also shown that the administrative burden of organic certification may be a barrier for entry (Kuhnert et al., 2013).

7 Where to look for further information

Primary information about organic standards and regulations comes in the first instance from the documents themselves, which can be accessed either through collections of legislation such as Eurolex or through the control bodies.

The FAO Codex Alimentarius Guidelines for the Production, Processing, Labelling and Marketing of Organically Produced Foods (AC/GL 32-1999) were last modified in 2013 and can be accessed at http://www.fao.org/fao-who-codexalimentarius/standards/thematic-compilations/en/.

The IFOAM Norms or Standards Requirements are also called the Common Objectives and Requirements of Organic Standards (COROS). These were developed as a joint venture of the IFOAM Organic Guarantee System and the GOMA Project (Global Organic Market Access) undertaken by FAO, IFOAM and UNCTAD. The main purpose of COROS is international equivalence assessments of organic standards and technical regulations, and provides the basis for assessing equivalence of standards for inclusion in the IFOAM Family of Standards. All documents can be accessed via http://www.ifoam.bio/en.

Systematic comparisons of organic standards are more limited: a comparison of the European Regulation with several private and other governmental standards as part of the Organic Revision Project can be found at http://www.organic-revision.org/, see also Padel et al. (2009). A new comparison of how organic ideas have been codified in eight different regulations around the globe was published by Seufert et al. (2017).

8 References

Arbenz, M., Gould, D. and Stopes, C. 2017. Organic 3.0 – the vision of the global organic movement and the need for scientific support. *Org. Agr.*, 7(3), 199–207.

Buck, D., Getz, C. and Guthman, J. 1991. From farm to table: The organic vegetable commodity chain of Northern California. *Sociol. Rural.*, 37, 3–20.

Crowder, D. W. and Reganold, J. P. 2015. Financial competitiveness of organic agriculture on a global scale. *Proc. Natl. Acad. Sci. USA*, 112, 7611–16.

De Olde, E. M., Oudshoorn, F. W., Sørensen, C. A. G., Bokkers, E. A. M. and De Boer, I. J. M. 2016. Assessing sustainability at farm-level: Lessons learned from a comparison of tools in practice. *Ecol. Indic.*, 66, 391–404.

EC. 1991. Council Regulation (EEC) No 2092/91 of 24 June 1991 on organic production of agricultural products and indications referring thereto on agricultural products and foodstuffs. *Official J. Eur. Commun.*, L198, 1–15.

EC. 1992. Council Regulation (EEC) No 2078/92 of the 30 June 1992 on agricultural production methods compatible with the requirements of the protection of the environment and the maintenance of the countryside. *Official J. Eur. Commun.*, L215, 85–90.

EC. 2007. Council Regulation (EC) No 834/2007 of 28 June 2007 on organic production and labelling of organic products and repealing Regulation (EEC) No 2092/91. *Official J. Eur. Union*, L189, 1–23.

Ellsworth, J. 2001. *The History of Organic Food Regulation* (2001 Third Year Paper). Harvard University, USA. Available from: http://nrs.harvard.edu/urn-3:HUL.InstRepos:8889458.

Gerrard, C. L., Smith, L., Padel, S., Pearce, B., Hitchings, R., Measures, M. and Cooper, N. 2011. OCIS Public Goods Tool Development. Organic Research Centre. Newbury.

Gershuny, G. 2016. *Organic Revolutionary: A Memoir of the Movement for Real Food, Planetary Healing and Human Liberation.* Joe's Brook Press, Barnet, VT.

Giepen, M. 2017. Occasional direct seeding of grain legumes in Organic Agriculture in Germany and Brazil: Fertilisation with P and S and weed control with natural herbicides. PhD Thesis, University of Bonn, Bonn.

Giovannucci, D., Von Hagen, O. and Wozniak, J. 2014. Corporate social responsibility and the role of voluntary sustainability standards. In: Schmitz-Hoffmann, C., Schmidt, M., Hansmann, B. and Palekhov, D. (Eds), *Voluntary Standard Systems*. Springer Verlag, Berlin and Heidelberg, pp. 359–84.

Häni, F., Braga, F., Stämpfli, A., Keller, T., Fischer, M. and Porsche, H. 2003. RISE, a tool for holistic sustainability assessment at the farm level. *International Food and Agribusiness Management Review*, 6(4), 78–90.

Hietala, S., Smith, L., Knudsen, M., Kurppa, S., Padel, S. and Hermansen, J. 2015. Carbon footprints of organic dairying in six European countries – real farm data analysis. *Org. Agr.*, 5, 91–100.

Hill, S. 1985. Redesigning the food system for sustainability. *Alternatives*, 12, 32–6.

IFOAM-EU 2014. IFOAM EU position paper on 'Residue threshold in organic products challenges and consequences of threshold levels as certification tool in organic production'. http://www.ifoam-eu.org/sites/default/files/ifoameu_reg_position_residues_20140226.pdf.

IFOAM 2014. The IFOAM norms for organic production and processing. version 2014. IFOAM. http://www.ifoam.bio/sites/default/files/ifoam_norms_version_july_2014.pdf.

ITF (Ed.) 2006. *Strategy on Solutions for Harmonizing International Regulations of Organic Certification*. UNTAC, FAO, IFOAM, Geneva, Rome and Bonn.

Koesling, M., Løes, A.-K., Flaten, O., Kristensen, N. H. and Hansen, M. W. 2012. Farmers' reasons for deregistering from organic farming. *Org. Agr.*, 2, 103–16.

Kuhnert, H., Behrens, G., Hamm, U., Müller, H., Sanders, J., Nieberg, H. and Strohm, R. 2013. Dauerhafte Ausweitung des ökologischen Landbaus in Deutschland: Analyse der Ausstiege von Betrieben und Entwicklung eines Konzepts zur nachhaltigen Vermeidung (Extension of organic farming in Germany: Determination of reasons for re-conversion of organic farms to conventional agriculture and development of a sustainable approach to prevent it). *Final Report BOELN*. Thünen-Institut, Institut für Betriebswirtschaft, Land und Markt, Universität Kassel, Fachgebiet Agrar- und Lebensmittelmarketing, D-Witzenhausen, Braunschweig, Hamburg and Witzenhausen.

Lampkin, N. (Ed.). 2002. *Org. Farm.* Old Pond Publishing Ltd, Ipswich, UK.

Lampkin, N. 2007. Organic farming's contribution to climate change and agricultural sustainability. In: *Welsh Organic Producers' Conference*. Builth Wells, Wales.

Lampkin, N., Foster, C. and Padel, S. 1999a. *The Policy and Regulatory Environment for Organic Farming in Europe*. University of Hohenheim, Hohenheim.

Lampkin, N., Foster, C. and Padel, S. 1999b. *The policy and regulatory environment for organic farming in Europe: Country reports*. University of Hohenheim, Hohenheim.

Lampkin, N. H., Smith, J. and Smith, L. 2017. Agroecology and organic farming as approaches to reducing the environmental impacts of agricultural chemicals. In: Hester, R. and Harrison, R. (Eds), *Agricultural Chemicals and the Environment: Issues and Potential Solutions*. The Royal Society of Chemistry, London. 94–113.

Lockeretz, W. (Ed.). 2007. *Organic Farming: An International History*, Wallingford: CAB International (CABI).

Lockie, S., Lyons, K., Lawrence, G. and Halpin, D. 2006. *Going Organic Mobilizing Networks for Environmentally Responsible Food Production*. CABI Publishing, Wallingford.

Luttikholt, L. W. M. 2007. Principles of organic agriculture as formulated by the International Federation of Organic Agriculture Movements. *NJAS Wagen. J. Life Sci.*, 54, 347–60.

MacRae, R. J., Hill, S. B., Mehuys, G. R. and Henning, J. 1990. Farm-scale agronomic and economic conversion from conventional to sustainable agriculture. *Adv. Agron.*, 43, 155–98.

Mäder, P., Fliessbach, A., Dubois, D., Gunst, L., Fried, P. and Niggli, U. 2002. Soil fertility and biodiversity in organic farming. *Science*, 296, 1694–7.

Padel, S. (Ed.). 2010. *The European Regulatory Framework and its Implementation in Influencing Organic Inspection and Certification Systems in the EU*. The Organic Research Centre - Elm Farm (ORC), Newbury, UK.

Padel, S. and Zerger, U. 1994. Economics of organic farming in Germany. In: Lampkin, N. H. and Padel, S. (Eds), *The Economics of Organic Farming*. CAB International, Wallingford.

Padel, S., Röcklinsberg, H. and Schmid, O. 2009. The implementation of organic principles and values in the European Regulation for organic food. *Food Policy*, 34, 245–51.

Padel, S., Röcklinsberg, H., Verhoog, H., Schmid, O., De Wit, J., Alrøe, H. F. and Kjeldsen, C. 2007. Balancing and integrating basic values in the development of organic regulations and standards: Proposal for a procedure using case studies of conflicting areas. In: *EEC2092/91 (Organic) Revision (SSPE-CT-2004-502397): Project report D 2.3.* University of Wales, Aberystwyth (UWA) and Danish Research Centre for organic Farming (DARCOF), Aberystwyth, Tjele.

Padel, S., Vieweger, A., Nocentini, L., Devot, A., Schmid, O. and Stolze, M. 2013. Adequacy of the production rules. In: Sanders, J. (Ed.), *Evaluation of the EU Legislation on Organic Farming–Study Report*. Thünen Institute of Farm Economics, Braunschweig.

Rahmann, G., Reza Ardakani, M., Bàrberi, P., Boehm, H., Canali, S., Chander, M., David, W., Dengel, L., Erisman, J. W., Galvis-Martinez, A. C., Hamm, U., Kahl, J., Köpke, U., Kühne, S., Lee, S. B., Løes, A.-K., Moos, J. H., Neuhof, D., Nuutila, J. T., Olowe, V., Oppermann, R., Rembiałkowska, E., Riddle, J., Rasmussen, I. A., Shade, J., Sohn, S. M., Tadesse, M., Tashi, S., Thatcher, A., Uddin, N., Von Fragstein Und Niemsdorff, P., Wibe, A., Wivstad, M., Wenliang, W. and Zanoli, R. 2016. Organic Agriculture 3.0 is innovation with research. *Org. Agr.*, 7, 169–97.

Reganold, J. P. and Wachter, J. M. 2016. Organic agriculture in the twenty-first century. *Nature Plants*, 2, 15221.

Rundgren, G. 2002. History of organic certification and regulation from ideology to standards. *Org. Stand.*, 14, 13–15.

Sanders, J., Padel, S., Nocentini, L., Stolze, M., Huber, B., Zander, K., Polakova, J. and Keenleyside, C. 2013. Towards an improved legislative framework for organic farming – Overall conclusions and recommendations. In Sanders, J. (Ed.), *Evaluation of the EU Legislation on Organic Farming–Study Report*. Thünen Institute of Farm Economics, Braunschweig.

Schader, C., Meier, M., Grenz, J. and Stolze, M. 2012a. The trade-off between scope and precision in sustainability assessments of food systems. In: *Proceedings of the Tenth European IFSA Symposium*, Aarhus, Denmark.

Schader, C., Stolze, M. and Gattinger, A. 2012b. Environmental performance of organic farming. In: Boye, J. I. and Arcand, Y. (Eds), *Green Technologies in Food Production and Processing*. Springer US, Boston, MA.

Schmid, O. 2007. Development of standard for organic farming. In: Lockeretz, W. (Ed.), *Organic Farming: An International History*. CABI, Wallingford.

Seufert, V., Ramankutty, N. and Mayerhofer, T. 2017. What is this thing called organic? – How organic farming is codified in regulations. *Food Policy*, 68, 10–20.

SOAAN 2013. *Best Practice Guideline for Agriculture & Value Chains. Public version 1.0.* SOAAN, Bonn.

Tuomisto, H. L., Hodge, I. D., Riordan, P. and Macdonald, D. W. 2012. Comparing energy balances, greenhouse gas balances and biodiversity impacts of contrasting farming systems with alternative land uses. *Agr. Sys.*, 108, 42–9.

Van Der Grijp, N. 2006. Private regulatory approaches and the challenge of pesticide use. In: *Ethics and the Politics of Food. Proceedings of EurSafe Congress 2006*. Oslo, Norway.

Vogt, G. 2007. The origins of organic farming. In: Lockeretz, W. (Ed.), *Organic Farming: An International History*. CAB International, Wallingford.

Wai, O. K. 2007. Cooperation between Conformity Assessment Bodies (CABs) in organic certification. In: *Background Papers of the International Task Force on Harmonization and Equivalence in Organic Agriculture*. UNCTAD/FAO/IFOAM, Geneva, Rome and Bonn.

Willer, H. and Lernoud, J. (Eds). 2017. *The World of Organic Agriculture, Statistics and Emerging Trends 2016*. IFOAM, FiBL, Bonn, Frick.

Zander, K. and Hamm, U. 2010. Consumer preferences for additional ethical attributes of organic food. *Food Qual. Prefer.*, 21, 495–503.

Zander, K., Hamm, U., Freyer, B., Gössinger, K., Hametter, M., Naspetti, S., Padel, S., Stolz, H., Stolze, M. and Zanoli, R. 2010. *Farmer Consumer Partnerships – How to Successfully Communicate the Values of Organic Food*. Self-published, Department of Agricultural and Food Marketing, University of Kassel, Witzenhausen.

Zanoli, R. (Ed.). 2004. *The European Consumer and Organic Food*. School of Management and Business, University of Wales, Aberystwyth.

Chapter 14

Measuring and improving the environmental performance of organic farming

A. Muller, Research Institute of Organic Agriculture (FiBL) and Institute for Environmental Decisions, Department of Environmental Systems Science, Federal Institute of Technology Zurich, Switzerland; M. Meier and C. Schader, Research Institute of Organic Agriculture (FiBL), Switzerland; A. Gattinger, Justus-Liebig University Giessen, Germany; and M. Steffens, Research Institute of Organic Agriculture (FiBL), Switzerland

1 Introduction
2 Notions of environmental performance
3 Efficiency, consistency and sufficiency
4 Methods for assessing environmental performance
5 Improving performance
6 Conclusions and future trends
7 Acknowledgements
8 Where to look for further information
9 References

1 Introduction

The literature provides many assessments comparing the environmental performance of organic and conventional farming in specific settings either on farm or on product level. The latter ones cover a wide range of crop and livestock commodities, such as for dairy farms, fruit production or soybean systems and many others for example (Müller-Lindenlauf et al. 2010; Longo et al. 2017; Pashaei Kamali et al. 2017). These individual assessments are often used to derive general conclusions on the environmental performance of organic versus conventional farming. Recently, several review articles have collected and synthesized this knowledge (Schader et al. 2012; Meier et al. 2015; Reganold and Wachter 2016; Seufert and Ramankutty 2017). As an example, we display the summary results from the most recent publication (Seufert and Ramankutty 2017) in Fig. 1.

These review studies consolidated the fact that organic farming tends to perform better than conventional farming if reference is made to environmental impacts per hectare, while it tends to perform often worse when addressing impacts per unit output. However, this

http://dx.doi.org/10.19103/AS.2017.0029.16

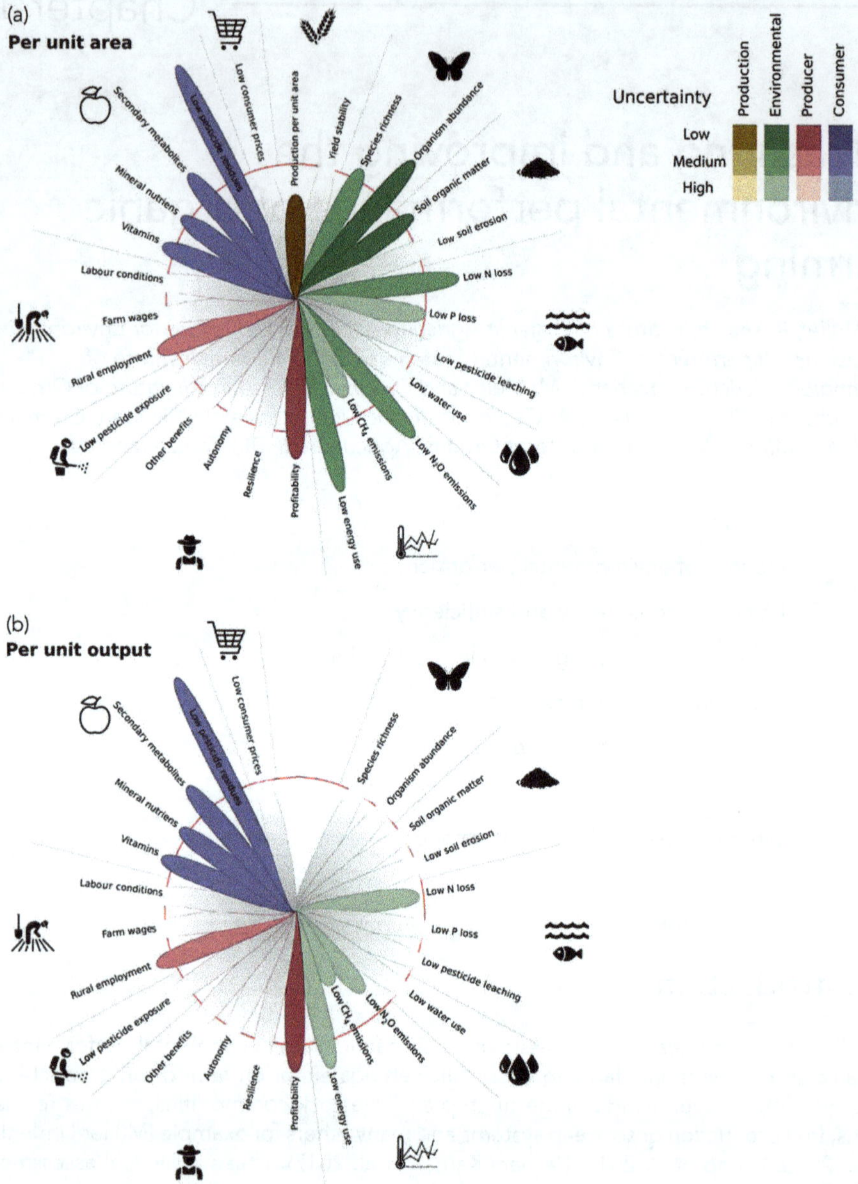

Figure 1 Performance of organic agriculture relative to conventional agriculture (red circle) per unit area (a) and per unit output (b). Source: Seufert and Ramankutty (2017).

performance and the differences between the farming systems also strongly depend on the environmental impact considered, for example greenhouse gases or some biodiversity indicator.

Impacts per product unit are usually assessed using life cycle assessment (LCA), which determines the environmental impact first per farmed area and then derives the environmental impact per product unit by dividing the impact per area by the yield output achieved on this area. A higher environmental impact per product in organic agriculture is then often explained by the well-established presence of the yield gap of, in general, about 20–25% between organic and conventional farming, with some differences between different crop classes, such as less than 10% for legumes or 30% for vegetables (Seufert et al. 2012; Ponisio et al. 2015). That this argument explains only some differences between organic and conventional products gained by LCA has been analysed in detail (Meier et al. 2015). In fact, the reasons why products from organic agriculture sometimes have a higher environmental impact are manifold and vary between truly lower efficiency and limitations within the assessment method itself. In addition, whether a product from organic agriculture has a higher environmental impact than the same product from conventional agriculture depends very much on the environmental impact category that is used for this comparison.

A truly lower efficiency can for example be observed for the global warming potential (GWP) of organic animal products. The GWP per head and per unit of output are often higher in organic farming, as animals take longer to attain the required slaughter weight or have lower yields for eggs and milk. This effect can exceed the positive effect of less concentrates being fed to the animals in many organic systems. However, despite the lower productivity in the organic animal production systems, ecotoxicity per product unit usually is considerably lower when analysing a wide range of comparative LCA studies (Meier et al. 2015, 2017). This can be explained by the ban on synthetic pesticides in organic agriculture. In this context, organic agriculture is often criticized for the use of copper compounds for plant protection. This is, however, an issue for a few crops only (mainly vineyards and other fruits). As mentioned above, organic agriculture performs better than conventional agriculture in relation to ecotoxicity, which considers all potentially harmful chemical compounds used and not only specific ones such as copper. Sometimes, it is also pointed out that less-effective plant protection substances in organic farming result in considerably more spraying activity with correspondingly higher energy use. On the other hand, organic farming tends to perform better regarding animal health aspects, the increased longevity and reduced disease pressure, and restricted use of antibiotics contributes to a better environmental performance regarding those indicators (Reganold and Wachter 2016). In the case of ruminants, greenhouse gas emissions could be higher, as the organic animals tend to consume less concentrate and more grassland-based feed than conventional animals with correspondingly higher methane emissions from enteric fermentation. Based on assessment results showing higher environmental impacts per product unit, strategies for improving the environmental performance of organic agriculture then generally focus on reducing the yield gap and thus emissions per unit output, or on reducing the emissions per hectare without affecting yields (Niggli 2015; Niggli et al. 2017).

Given all this knowledge and the availability of recent review articles on the environmental performance of organic agriculture, in this chapter, we want to locate this discussion on environmental performance of organic farming in a wider context. First, we want to emphasize the importance to discuss how 'performance' is defined – in particular, the variables and system boundaries it refers to – and how it is assessed. Eco-efficiency defined as environmental impacts per unit output addresses a specific aspect of the environmental performance of agriculture, while it does not allow for a final conclusion on the environmental sustainability of a farming system. In particular, focusing on one-dimensional indicators tends to neglect key aspects of production systems with an inherent complex multidimensional output bundle such as organic farming which is generally considered as multifunctional. Furthermore, there are several limitations within LCA when using this method to assess the eco-efficiency of agricultural products. These limitations need to be overcome in order to use LCA to conclusively compare the eco-efficiency of agricultural products produced in different production systems.

Second, we want to emphasize that the environmental performance of organic farming cannot adequately be assessed with reference to agricultural production alone. It needs a food systems perspective to fully capture the environmental performance of agricultural production systems. Consider, for example, food wastage – high yields are less crucial for good total environmental performance in case less of the output is wasted. With wastage reduction, the whole production system can be smaller, and in total, greenhouse gas emissions, for example, may be lower though production is organic with higher emission factors per unit output. These considerations complement the notion of 'efficiency' (addressing resource use per unit output), with the notions of 'consistency' addressing optimal resource use within a systemic context and 'sufficiency' (addressing the total aggregate level of resource use); see also Schader et al. (2015) and Muller et al. (2017).

Third, for an encompassing assessment of the environmental performance of organic farming, a single focus on either crop or livestock production is not sound. In organic farming, crop and livestock production are often closely linked, if only by using manure as fertilizers. There, for example, the question arises, whether emissions from manure management may be allocated to the livestock system, where the manure is sourced from, or to the crop production system where it is used as a fertilizer. Furthermore, legume leys in crop rotations and grasslands are important feed sources for animals. Thus, livestock plays a crucial role in thorough assessment of organic crop production, paying due attention to manure as fertilizers and to crop rotations. Clearly, bio-vegan production without livestock altogether is a possibility, but hardly the reality (Schmidt 2004; Hagemann and Potthast 2015). Therefore, we do not exclude livestock from the analysis presented in this chapter, albeit the volume this chapter contributes to focuses on crop production.

These are central issues to be addressed when discussing the environmental performance of farming systems and of organic farming in particular. It would be important to acknowledge them even more explicitly when addressing the environmental performance of organic farming and to attempt to also shape this discussion by suggesting adequate definitions and tools for environmental performance and its assessment that are able to capture the specific characteristics of all farming systems without bias. Thus, these aspects should be taken up in the ongoing discussion in the future of organic farming and on organic research agendas, as for example framed in the work on Organic 3.0 and the discussions in the context of the Technology Innovation Platform of IFOAM–Organics International (TIPI) (IFOAM 2015; Niggli et al. 2017).

The chapter is organized as follows: Section 2 addresses notions of 'environmental performance' and Section 3 presents the three complementary key concepts of 'efficiency', 'consistency' and 'sufficiency'. Section 4 deals with methods to assess environmental performance, Section 5 addresses strategies for improving the environmental performance and Section 6 concludes the chapter.

2 Notions of environmental performance

Environmental performance of agricultural production is often assessed by indicators that relate environmental impacts to output. Some indicators are less directly addressing the environmental impacts but rather build on proxies for impacts or bundles of impacts, for example when assessing input use, such as nitrogen, in relation to output, which expresses nitrogen use efficiency and relates to a number of environmental impacts such as to biodiversity loss, water eutrophication and GWP. Generally, these indicators provide measures for various types of 'efficiency' – often termed 'environmental' – or 'eco-efficiency' – and allow us to assess how much resources are used or how many environmental impacts are caused by a certain production system to produce one unit of output; in other words, how much output can be produced by unit of input and resource use and thus, per unit of environmental impact caused by this input and resource use. Such measures provide important information on the production system in question and allow for comparison between different production systems relating these indicators. However, several challenges arise: Even though by definition LCA has to cover all environmental indicators that are relevant within the context of a specific production system, LCAs of agricultural products still do not routinely provide an assessment of land-use-related impacts on either biodiversity or soil quality. However, in the context of agricultural production, impacts on biodiversity and soil quality are essential, as agriculture is a main driver for biodiversity loss and degradation of soils. Therefore, assessments of the environmental impacts of agricultural production using LCA are up to today always incomplete.

Eco-efficiency is definitely an important measure with regard to environmental sustainability. However, in the context of agricultural products, the sole focus on eco-efficiency does not allow for a conclusion whether the agricultural production system behind a product is environmental friendly or not. As the focus on eco-efficiency of an agricultural product favours products from intensive production systems with high yields, the conclusion would obviously be to intensify agricultural production in order to reduce its environmental impact. However, we know that environmental problems from agriculture are mainly caused by intensive agricultural practices. In fact, when relating environmental impacts of an intensive agricultural production system to the agricultural area instead of the product, high impacts per area become evident that are not visible in the product-related perspective. These high local impacts on the regional production level may be too high for the local ecosystems, thus jeopardizing their resilience and also affecting planetary boundaries. To date, efficiency indicators have difficulties in dealing with aspects of environmental impacts that do not relate to relative measures. Examples are the total water availability and water scarcity in a watershed, the critical nitrogen load in an ecosystem, or the total amount of greenhouse gas emissions that can be emitted to still meet the two-degree centigrade goal with a certain high probability. Efficiency measures are not sensitive to such restrictions on the total aggregated amounts of environmental impacts, resources and sink capacities. In order to overcome this problem, LCA of agricultural products needs to integrate local and global ecosystem boundaries, which determine the maximum acceptable intensity level per production area. Only products with an eco-efficiency within these boundaries can be considered to be produced within an environmental friendly production system. The importance of such choice of viewpoint is evident when assessing the environmental performance of organic farming, where it makes a big difference, whether impacts are assessed in relation to units of product (e.g. tons) or in relation to units of area (e.g. hectares) (Meier et al. 2015; Seufert and

Ramankutty 2017). In addition, the output may not only be measured in mass quantities only. Several alternatives also exist. Output can, for example, be measured in quantity of human-digestible protein, energy or nutrients.

Finally, trade-offs and synergies between different indicators and quantities, as well as their dynamics in response to changes of the wider context are difficult to grasp with such efficiency measures based on the ratio of two quantities. Some methods address these issues to a certain extent, though. Consequential life cycle assessment (CLCA; Earles and Halog 2011), for example, attempts to deal with some of those aspects, for example with those that are related to changes in co-product supply or demand in case a change in the main product is investigated (e.g. change in beef supply in case of additional milk production) – we will address this in more detail in Section 4 on assessment methods and proxy indicators. Other such trade-offs and synergetic aspects are not dealt with systematically, though, even in CLCA. One example is the rebound effect, which may occur in several forms. If, for example, output of a certain cropping system is higher due to some technological progress or management improvements, revenues on the areas under this cropping system may increase. This may trigger incentives to crop more areas than before, as there is a chance to increase revenues by selling more of this more efficient and thus cheaper production. Consequently, in a region of ongoing deforestation, such improvement may even lead to increased deforestation rather than decreased deforestation as one may expect on the first sight (Barretto et al. 2013). A more indirect instance of the rebound effect relates to disposable income. For illustration, let us assume that some production process becomes more efficient and thus cheaper, say, for example some staple crop production under new management with reduced input use and correspondingly lower environmental impacts. The reduced input costs may translate into lower consumer prices, while, being a staple crop, consumption may not increase that much. Thus, more of the income is available for other consumption, which in turn will lead to other environmental impacts. In sum, environmental impacts may even increase, although relative environmental impacts of the staple crop decreased (Valin et al. 2013; Font Vivanco and van der Voet 2014).

3 Efficiency, consistency and sufficiency

Given these challenges of efficiency approaches, we suggest to explicitly accept and highlight these limitations and gaps and to develop complementary approaches to optimally deal with them and to broaden the understanding of environmental performance. 'Consistency' is such a concept on the same level of abstractness as 'efficiency'. It focuses on optimal resource use within a given systemic context, thus covering aspects such as closed nutrient cycles or trade-offs and synergies between different resource allocations and different indicators. 'Sufficiency' is another such concept focusing on the total consumption, production and impact levels without reference to output or input and thus relative assessments. Thus, 'efficiency' can be framed as focusing on 'optimal relative resource use', while 'consistency' focuses on 'optimal allocation of resources within a systemic context'. This captures how some resources can be used for certain activities, while others cannot, and where resource use is also optimized in such a way as to minimize or altogether avoid losses on the level of the whole system. 'Sufficiency', finally, focuses on the 'total aggregate resource use level', without reference to outputs or different utilizations. We frame these concepts by means of the notion 'resource use', as this most generally captures input use and emissions and the environmental impacts resulting thereof, via the sink capacity of the environment as a resource to deal with those.

A comprehensive picture of the environmental performance of organic farming and other farming systems can thus be compiled by complementing efficiency indicators with indicators for the consistency of the production activities and also their sufficiency. An example is grassland-based ruminant, by-product and waste-based monogastric production. Grassland-based ruminant production is not very 'efficient', if assessed via GWP per kg output. However, it is 'consistent' in as much as it utilizes grasslands and other plant-based resources for food production that otherwise could not have been used for such. And it is linked to sufficiency, as this kind of production would result in overall lower animal sourced food quantities supply (Schader et al. 2015). This then establishes the link from production to the wider food systems perspective, as such a reduction in output needs to be in line with corresponding changes in consumption. Otherwise, considerable leakage of environmental impacts may occur, in case the missing supply is just replaced by imports with potentially even worse environmental impacts than the original production. Thus, a comprehensive assessment of the environmental performance of organic farming along the lines of efficiency, consistency and sufficiency necessitates a combined assessment of the production and consumption aspects – the holistic food systems perspective. Thus, working on improving the environmental performance of organic farming and agriculture in general cannot focus on the plot and farm levels alone, but it needs to embark on a societal discussion also taking the consumers and their consumption patterns and lifestyles, as well as our societal organization and means for influencing changes on a societal level into account (Muller and Huppenbauer 2016).

4 Methods for assessing environmental performance

In this section, we shortly discuss several methods to assess environmental performance of farming systems. They loosely relate to efficiency, consistency and sufficiency in as much as they are adequate to best address aspects on these respective concepts. The methods discussed are LCA, productivity analysis and food systems modelling. We close this section by some discussion on simplifications via proxy indicators and aggregation of indicators.

4.1 Life cycle assessment

LCA is often the method of choice for assessing environmental performance with a focus on eco-efficiency and we have already addressed it in previous sections. Nemecek et al. (2011) may serve as an illustrative example for comparisons of organic and other agricultural production systems, and Meier et al. (2015, 2017) add to the methodological discussion related to this, as discussed above. Based on the material and energy flows (i.e. elementary flows) involved in a product system compiled in the so-called life cycle inventories, the LCA assesses the encompassing environmental impact and resource use of a specific product. Attributional LCA (ALCA) attributes elementary flows and potential environmental impacts to the functional unit of a specific product system, thus assuming system boundaries that cover the production process and its upstream supply chain. In case products result from a production process, impacts are allocated among the main and the co-products by a normative rule. This is often done based on their relative economic value or relative weight. Examples are rapeseed meal as a by-product of rape oil production, straw as a by-product of wheat grains or beef meat as a by-product of dairy production. ALCA thus delivers detailed efficiency measures within a static setting. ALCA can be used to compare the state of system A with the state of system B under the chosen allocation rule. One important use of LCA is hotspot analysis, as it allows identifying the hotspots for environmental impacts along the value chain, that is to identify where changes in production practices may result in biggest improvements.

The criticism of the static setting is at the core of the CLCA approach that attempts to improve on the drawbacks of the static setting (Thomassen et al. 2008) as CLCA includes activities in the product system of interest as far as they are expected to change as a consequence of a change in demand for the functional unit. Thus, CLCA studies the environmental consequences of possible changes between alternative product systems. It is thus a system modelling approach in which activities in a product system are linked so that other activities are included in the product system to the extent that they are expected to change as a consequence of a change in demand for the functional unit of the product of interest. If, for example, wheat is produced organically instead of conventionally, the ALCA would provide a comparison of the performance of the two product systems. CLCA attempts to account for two dynamic aspects. First, it aims at capturing the marginal producer, thus assessing the environmental impact of additional units of wheat that would be produced for the market in case demand increases by that amount. Thereby, it attempts to capture some economic aspects of the current business as usual situation. Second, it captures interlinkages between main and co-products, both on the output and input sides. For wheat, these effects are less relevant, as the co-product – straw – is economically not very important. However, it is relevant for other products, such as in particular for animal products. In a CLCA of milk, the co-product beef would enter the equation such that the additional milk production leads to a certain amount of beef meat, which thus would not need to be produced elsewhere – and environmental impacts related to this or other production can then be subtracted from the impacts of the production of interest, as this latter would – if implemented – result in such reduction (thereby substitutability of these products clearly needs to be scrutinized as quality differences may arise with respect to pure meat breeds). Thus, CLCA attempts to capture all product and co-product mass streams. Another example would be oil crops. For rapeseed oil, for example, this means that a CLCA of a new cultivar with higher oil content and thus lower rapeseed meal shares would need to account for the reduction in feed supply from this meal, as the missing amount of feed would then have to been sourced elsewhere – all these are the 'consequences' assessed in CLCA that are missing in ALCA. Therefore, CLCA can be used as decision support when comparing the environmental impact of alternative agricultural production systems.

The LCA approach is also the basis to calculate single indicator footprints (e.g. carbon, water or land use footprints), which are increasingly used for product labelling as means of communication to consumers. These are usually based on ALCA, and sometimes backed with unreliable numbers, as they are applied in a business context with clear marketing rather than research aims. These indicators and labels indicate for specific products by how much they are better than some standard alternative, or how high their absolute emissions per unit of product are – similar to other product information such as calorie, protein or fat contents.

4.2 Productivity analysis

Another means to assess environmental performance of agricultural production systems with a focus on efficiency is 'productivity analysis' (PA), a statistical approach frequently used in economics. Different to the model-based LCA that allows assessing a single product, PA is data driven. This means that it does not rely on data from inventories in combination with detailed description of the relevant processes, but rather on representative samples of the input and output characteristics of certain production situations of interest. Thus, PA sees agricultural production situations as production processes, where multiple inputs transformed into multiple outputs, which also include unintended or undesirable outputs such as pollution (Picazo-Tadeo and Prior 2009; Picazo-Tadeo et al. 2012; Kumbhakar and Malikov 2017). Based on a sample of such production units, for example farms, the multidimensional relation between these inputs and outputs is then statistically estimated, thus delivering information on how changes in inputs may affect outputs. Thereby, it treats the production process itself as a black box without the need to know how exactly the inputs are transformed into the outputs. This estimate in particular provides information on how many outputs may in principle be produced with a given input in an optimal situation – estimated as the 'production possibility frontier' based on the available data – and how a single production unit may perform in relation to this optimal production situation. Based on this approach, several estimates for eco-efficiencies can be derived. It can be assessed, by how much a certain input or pollutant could be reduced without reducing output and which relation between the input or pollutant and the output may be the best achievable one. By comparing the production and the performance of single farms in relation to those for a sample of organic and a sample of comparable conventional farms, differences in the environmental performance between these farming systems can be identified and analysed. On the other hand, PA has a strong focus on the production side and does not cover upstream impacts.

PA does not rely on many assumptions on the functional form of the production frontier as it does not explicitly assume an underlying production model. It, however, requires data on inputs and outputs from a representative sample of production units for the production system and related context of interest. By this, it is sensitive to local characteristics of the situation investigated. Furthermore, PA is aimed at investigating the performance of whole production units with, in principle, multiple outputs, thus naturally being able to capture trade-offs and synergies between different inputs and outputs, while LCA allows for product-specific assessments. If allocation of inputs or impacts to single products is, however, possible within the production units, then PA also allows for assessment of single products. As PA estimates a frontier over a sample of production units, it is also able to capture aspects of 'consistency'. It addresses trade-offs and synergies of input use in a concrete wider production systems context.

Regrettably, this strand of the literature is rather disregarded in the LCA context and by the recent reviews on the environmental performance of organic farming (Crowder and Reganold 2015; Reganold and Wachter 2016; Seufert and Ramankutty 2017). Results from PA do not contradict the LCA findings and are able to complement those with additional information, in particular also on economic performance and economic indicators of environmental aspects, such as shadow costs of pollutants and their reductions.

4.3 Multi-criteria assessment approaches

Like LCAs, multi-criteria assessments have been used to assess the environmental performance of farming systems and specific technologies, innovations and policies (Dodgson et al. 2001). Contrary to the purely quantitative LCAs, multi-criteria assessments can include semi-quantitative and qualitative aspects of environmental impacts. This implicitly requires a weighting procedure between the different aspects that determine the environmental performance of a farming system. At the same time, these tools are suitable to also cover social and economic aspects, which often cannot be assessed in a purely quantitative way. There is a multitude of different multi-criteria assessment methods which can be used for the assessment of organic farming (Schader et al. 2014).

One of the multi-criteria assessment tools most frequently used for comparing the performance of organic and conventional farming systems is the SMART-Farm Tool (Sustainability Monitoring and Assessment Routine) (Schader et al. 2016). The SMART-Farm Tool measures the degree of goal achievement against 58 sustainability goals as specified in the Guidelines for Sustainability Assessment of Agriculture and Food Systems (SAFA) (FAO 2014). In the environmental dimension, these are Greenhouse Gases, Air Quality, Water Withdrawal, Water Quality, Soil Quality, Land Degradation, Ecosystem Diversity, Species Diversity, Genetic Diversity, Material Use, Energy Use, Waste Reduction & Disposal, Animal Health and Freedom from Stress of Animals. It is important to note that the SMART-Farm Tool does not express the performance related to a functional unit but to a universal goal. For instance, the goal for the theme of Water Quality is specified as 'The release of water pollutants is prevented and water quality is restored' (FAO 2014). This means that SMART-Farm Tool rather takes the perspective of consistency than the perspective of efficiency and can be used complementary to LCA and other approaches (Schader et al. 2016). For most environmental themes, organic farms outperform conventional farms in a direct comparison; however, differences between farm types and regions tend to be much more substantial than the differences between organic and conventional farms. Furthermore, the degree of differences between organic and conventional farms depends very much on the specific farming system and can range from substantial to only marginal differences (Ssebunya et al. 2016).

4.4 Food systems modelling

While PA allows assessing certain aspects beyond efficiency, other methods are needed to fully account for consistency and sufficiency in agricultural production and food systems. Mass- and nutrient flow food system modelling is the method of choice here, as this allows to fully capture all food systems interlinkages on a physical and an agronomic level. It enables assessing the option space of scenarios comparing different production systems such as organic and conventional farming. The environmental performance on a per unit product or area basis – with the potential limitations as indicated above – is then complemented with the aggregated view, addressing the total environmental impacts (sufficiency), as well as the system consistency view, addressing the utilization of the available resources. The main addition of the food system approach is that it complements the efficiency measures with hitherto missing. A prime example is the assessment of organic agriculture as undertaken in Muller et al. (2017). Here, the option space for organic farming is assessed in the context of food wastage reduction and grassland- and wastage-based livestock production with correspondingly reduced supply and consumption of animal sourced food (Fig. 2). It illustrates how the environmental performance of the whole food system compares for different scenarios, and that organic agriculture can play an important role albeit its higher land use and emissions per unit output. The key to arrive at sustainable solutions is to not merely focus on the agricultural production and its efficiency but to account for the consumption side and aggregate resource use aspects such as regional scarcities as well. Another example of such an assessment of option spaces for sustainable food systems is Schader et al. (2015) that focuses on the role that grassland-based ruminant production and monogastrics fed on by-products only may play in sustainable food systems. These scenarios reduce concentrate feed and other food-competing feed and similarly find that in combination with changes on the consumption side (i.e. sufficiency aspects), these less intensive production systems can well play an important role. They thus account, for example, for the role of grassland-based ruminant production in contexts where grassland is widely available while cropland is a scarce resource. A third example is Erb et al. (2016), focusing on food systems with minimal deforestation impacts (Fig. 3).

% Wastage reduction	% reduction in food-competing feed	Climate change impact on yields: zero, % organic 0	20	40	60	80	100	medium, % organic 0	20	40	60	80	100	high, % organic 0	20	40	60	80	100
0	0	25	21	15	10	4	-3	23	19	14	8	3	-3	21	16	12	7	1	-4
	50	20	16	12	7	2	-4	18	14	10	6	1	-4	17	13	9	4	0	-5
	100	15	11	7	3	-1	-5	13	10	7	3	-1	-5	12	9	5	2	-2	-6
25	0	23	19	14	8	2	-4	21	17	12	7	1	-4	19	15	10	5	0	-5
	50	18	14	10	6	1	-5	17	13	9	5	0	-5	15	11	7	3	-1	-5
	100	13	10	6	2	-2	-6	12	9	5	2	-2	-6	11	8	4	1	-3	-6
50	0	21	17	12	7	1	-5	19	15	10	5	0	-6	17	13	9	4	-1	-6
	50	16	12	8	4	0	-6	15	11	7	3	-1	-6	14	10	6	2	-2	-6
	100	11	8	5	1	-3	-7	10	7	4	1	-3	-7	10	7	3	0	-3	-7

Figure 2 Illustration of the food systems approach to environmental performance showing the option space regarding nitrogen surplus for a conversion to organic agriculture in combination with reduced food wastage and with a livestock system with reduced food-competing feed. Source: Muller et al. (2017). N-surplus (positive values) or deficit (negative values) in kg N/ha. Scenarios differ in: organic shares (0–100%), climate change impacts (low, medium, high), food-competing feed reductions (0, 50, 100% reduced from the levels in the reference scenario) and wastage reduction (0, 25, 50% compared to the reference scenario). Colour code for comparison to the reference scenario value (i.e. 0% organic agriculture, no changes in livestock feed and food waste, dotted grey): >10 kg/ha: red (unsustainably high), between 10 and 5 kg/ha: blue (optimum, reduction from current average surplus by 60–80%), between 5 and −2 kg/ha: yellow (critical, rather low), <−2 kg/ha: orange 337 (deficit).

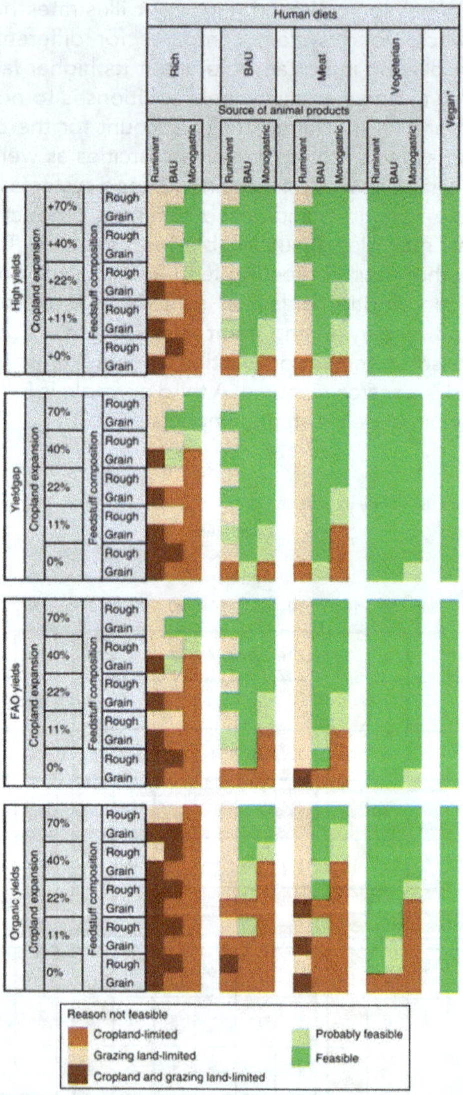

Figure 3 Illustration of the food systems approach to environmental performance showing the option space regarding land use for reduced deforestation under several dietary scenarios ('BAU' stands for 'business as usual'). Source: Erb et al. (2016). The option space (green cells) results from the combination of four yield variants, five cropland expansion variants, two variants of the feed basis for livestock at the supply side (lines) and five human diet variants combined with three variants of the origin of livestock products at the demand side (columns). Each cell in the option space represents a scenario. Scenarios affected by cropland limitations represent cases where the global demand for cropland products exceeds global supply by more than 5% and scenarios affected by grazing limitations when global grazing intensity exceeds ecological 348 thresholds. 'Probably feasible' are scenarios for which demand and supply differ by <5%. *For the VEGAN diet, the source of livestock products is not relevant.

Such mass- and nutrient flow models explicitly disregard economic aspects. They aim at assessing the biophysical and agronomic viability of different scenarios and the related environmental impacts. Such an approach allows to assess the environmental performance of farming systems by indicating in which range environmental impacts may lay for various scenario assumptions. Which political and economic conditions and mechanisms then may support or rather hinder the realization of one or the other scenario is not part of the assessment. These models thus do not provide realistic forecasts on how a future food system may look like, but they rather line out which scenarios may have big potential for improvements. Furthermore, they show which scenarios may exhibit a large number of synergies between different goals, and which may rather be dominated by trade-offs and challenges. There are a number of well-established economic food systems models which allow to assess various scenarios and related environmental impacts (von Lampe et al. 2014). By design, those, however, result in a restricted representation of the option space – albeit a likely more realistic one with regard to common behavioural drivers – as the economic part provides strict decision rules regarding which scenario may be realized. As important as those tools are for assessing food systems projections, the mass- and nutrient flow models are more important to inform the discussion on environmental performance of different farming systems.

4.5 Proxy indicators and aggregations

LCA, PA and food systems modelling are methods of considerable complexity, if applied scrupulously. We face a general need for coarse-tuning rather than fine-tuning on the policy level to achieve true improvements regarding the challenges to global environmental boundaries. Thus, the assessment of the environmental performance of agricultural and food systems should also provide simple robust aggregate proxy indicators that may help in guiding society towards such bigger goals. Total nitrogen input, total fossil fuel use and total land use are three simple key indicators that can fulfil such a function. Somewhat more complex, the amount of food-competing feed or human appropriation of net primary production also can serve for this (Robinson et al. 2016). For coarse-tuning, these indicators are best assessed in aggregate levels, that is relating to sufficiency, and not as efficiency indicators, that is relative to some output or input measure.

5 Improving performance

Big potential for improvements lies with the farmers themselves, where always some innovators and inventive individuals will develop improvements. Those will focus on improving not only the business case, but also the environment, given the still high importance of values and intrinsic motivation among organic farmers. A second approach to foster improvements is to acknowledge the wide spread of performance with respect to many indicators within organic production systems. Thus, identifying and analysing best practices allows providing blueprints for improved management. Experts can help to identify such practices, but it is crucial that the farmers and extension officers themselves also work more systematically on identifying and spreading innovations and best practices.

While some diffusion of innovations and best practices will be governed by market dynamics, for other aspects, some external support for assuring widespread adoption of improvements is needed and such can take various forms. For example, the LCA inventories need to be improved to better reflect certain key aspects, such as detailed nitrogen modelling with one encompassing model addressing all related nutrient flows in a common and interlinked framework (Meier et al. 2015, 2017), and also to better reflect the particularities of organic farming as opposed to conventional farming. This latter issue is relevant for heavy metal contamination, for example, where values from conventional manure and so forth are used to also model organic production systems. This would improve knowledge on efficiency measures, and in a policy context aiming at increasing resource use efficiency and reducing environmental impacts, such is definitely needed. Work in improved plant protection is of the same quality, as farmers alone cannot provide such, and applied research needs to contribute substantially. This also fits well to the increasing importance of internalizing external costs, which itself tends to trigger improvements towards increasing efficiency, viz. lower environmental impacts and thus – after internalization – costs per unit product. Hotspots for improvements are yields, plant protection, recycling phosphorus and emissions from composting and other organic fertilizer management processes, including biogas slurry and so forth. Another support may focus on providing well-designed institutional platforms for identifying and spreading innovations and best practices among the farmers themselves – one key concept for this are 'innovation platforms' which, if well designed, can be very effective (Sanyang et al. 2016).

Thereby, it is important to not overly focus on efficiency. Organic farming should not continue to become an ever more intensive production system with somewhat lower impacts than conventional production. It should remain or come back to a largely more extensive production system that thus allows staying within local, regional and global ecosystem carrying capacity boundaries by explicitly acknowledging the relevance of consistency and sufficiency considerations, besides efficiency. Thus, better databases for the food system modelling and the role of organic farming in food systems need to be developed, to better assess the option space for organic farming in environmentally sustainable food systems. Even if such data are available, particular efforts are then needed to take up the relevant aspects of consistency and sufficiency strategies on the consumer side. Thus, to improve the environmental performance of organic farming understood in the broad sense presented in this chapter, it is unavoidable to work towards reductions in food wastage and animal product consumption and thus to work towards development and implementation of effective institutions to achieve such.

6 Conclusions and future trends

Before concluding, we want to emphasize that environmental aspects are only one part of a sustainability assessment. This means, in particular, that good environmental performance does not necessarily mean good sustainability performance – as this latter covers a range of additional indicators addressing social, economic and governance issues. We emphasize this, as in the general discussion this is often neglected, and sustainability assessment often has a strong focus on the aspects. Thus, working on the environmental performance of organic farming is important and promising, but we should not forget to improve the performance regarding social issues (e.g. labour rights, wages, etc.) and other aspects (e.g. animal welfare) as well.

As conclusions, we highlight the importance to go beyond the production level when assessing the environmental performance of farming systems such as organic farming. Otherwise, it is not possible to address consistency and sufficiency aspects of environmental performance that emerge in combination with consumption only. Acknowledging this provides the full option space for organic agriculture in environmentally sustainable food systems, in particular providing the means to go beyond the challenges of higher impacts per unit output in organic farming. Adopting a food systems perspective provides the space for less intensive production with lower yields but a number of other benefits, without comprising food security or aggregate environmental performance.

To achieve this, the organic sector should work much more towards playing a key role in defining how performance of agro-ecosystems and food systems should be assessed. As soon as organic farming accepts to be judged on the basis of yields and environmental performance per unit output, it willingly refrains from pointing out a number of its advantages, as those indicators, as important as they are, draw an incomplete picture only. Organic farming thrives as a multifunctional production system, where focusing on one of them only always leads to an incomplete underestimation of performance. Thus, it ideally fits to be assessed within a whole food systems context, which is a most providing reference frame for an encompassing assessment of environmental performance. Only in this way, a balance between the complementary aspects of efficiency, consistency and sufficiency can be reached.

When working on the food systems level, the scope is naturally broad. Thus, the focus of assessment rather shifts from the performance of a production system being fully organic as opposed to one that is conventional towards the assessment of which role organic farming may play within a whole food system and which extent of organic production may go along with which aggregate outputs and environmental impacts. Thus, organic agriculture is not to be understood as the better and only acceptable alternative, but rather as a blueprint and landmark for environmentally sustainable production systems, that may also help to improve environmental performance of agriculture as a whole.

Being a landmark for environmentally sustainable agriculture, organic farming also needs to adopt a more open position towards new technologies. This is a challenging discussion, and certification, labelling and marketing aspects are crucial with this respect as well, but regarding an underlying attitude, organic farming research should scrutinize and option that may emerge, to take an explicit and well-thought through decision in favour or against it.

Finally, the biggest challenge to achieve improved environmental performance of organic farming within a food systems context is to get the consumers on board and to truly work and develop towards increased sufficiency in consumption. Coarse-tuning rather than fine-tuning policies are important and a reference to the liberal society to avoid overly paternalistic approaches. The key is thus to work towards improvement both on the field and societal levels, as otherwise, the high aims of a two-degree goal or getting the nitrogen surplus down to an acceptable level are in many countries way out of reach.

7 Acknowledgements

Funding from the Mercator Foundation Switzerland is gratefully acknowledged.

8 Where to look for further information

Key review papers on the subject are Schader et al. (2012), Meier et al. (2015), Reganold and Wachter (2016) and Seufert and Ramankutty (2017).

Key journals that frequently publish papers on the subject are the *Journal of Cleaner Production* (https://www.journals.elsevier.com/journal-of-cleaner-production/) and the *International Journal of Life Cycle Assessment* (http://www.springer.com/environment/journal/11367) and relevant papers are also presented at the LCAFood conferences. The proceedings of this can be found here (http://lcafood2014.org/proceedings/LCA_Food_2014_Proceedings.pdf; https://www6.inra.fr/lcafoodconferencearchives/Past-conferences/LCA-Food-2016).

9 References

Barretto, A. G. O. P., G. Berndes, G. Sparovek and S. Wirsenius. (2013). 'Agricultural intensification in Brazil and its effects on land-use patterns: An analysis of the 1975–2006 period'. *Global Change Biology*, 19(6): 1804–15.

Crowder, D. W. and J. P. Reganold. (2015). 'Financial competitiveness of organic agriculture on a global scale'. *Proceedings of the National Academy of Sciences*, 112(24): 7611–16.

Dodgson, J., M. Spackman, A. Pearman and L. Phillips. (2001). DTLR multi-criteria analysis manual. London: National Economic Research Associates (NERA). Available online at: http://iatools.jrc.ec.europa.eu/public/IQTool/MCA/DTLR_MCA_manual.pdf (Accessed on 16 July 2016).

Earles, J. M. and A. Halog. (2011). 'Consequential life cycle assessment: A review'. *The International Journal of Life Cycle Assessment*, 16(5): 445–53.

Erb, K.-H., C. Lauk, T. Kastner, A. Mayer, M. C. Theurl and H. Haberl. (2016). 'Exploring the option space for feeding the world without deforestation'. *Nature Communications*, 7: 11382.

FAO. (2014). Sustainability Assessment of Food and Agriculture systems (SAFA). Rome: Food and Agriculture Organization of the United Nations (FAO). Available online at: http://www.fao.org/nr/sustainability/sustainability-assessments-safa/en/ (Accessed on 31 July 2014).

Font Vivanco, D. and E. van der Voet. (2014). 'The rebound effect through industrial ecology's eyes: A review of LCA-based studies'. *The International Journal of Life Cycle Assessment*, 19(12): 1933–47.

Hagemann, N. and T. Potthast (2015). Necessary new approaches towards sustainable agriculture – innovations for organic agriculture. In: Dumitras, D. E., Jitea, I. M. and Aerts, S. (Eds), *Know Your Food: Food Ethics and Innovation*. Wageningen, the Netherlands: Wageningen Academic Publishers, pp. 107–13.

IFOAM. (2015). Organic 3.0 – for truly sustainable farming & consumption. Bonn, Germany: International Federation of Organic Agriculture Movements IFOAM-Organic International.

Kumbhakar, S. C. and E. Malikov. (2017). 'Good modeling of bad outputs: Editors' introduction'. *Empirical Economics*, 54(1): 1–6.

Longo, S., M. Mistretta, F. Guarino and M. Cellura. (2017). 'Life cycle assessment of organic and conventional apple supply chains in the North of Italy'. *Journal of Cleaner Production*, 140: 654–63.

Meier, M. S., F. Stoessel, N. Jungbluth, R. Juraske, C. Schader and M. Stolze. (2015). 'Environmental impacts of organic and conventional agricultural products – Are the differences captured by life cycle assessment?' *Journal of Environmental Management*, 149: 193–208.

Meier, M. S., S. Hörtenhuber, C. Schader and M. Stolze. (2017). Organic farming and sustainability: Life cycle assessments of organic foods. Fact Sheet, Frick, Switzerland: Research Institute of Organic Agriculture FiBL.

Müller-Lindenlauf, M., C. Deittert and U. Köpke. (2010). 'Assessment of environmental effects, animal welfare and milk quality among organic dairy farms'. *Livestock Science*, 128(1–3): 140–8.
Muller, A. and M. Huppenbauer. (2016). 'Sufficiency, liberal societies and environmental policy in the face of planetary boundaries'. *GAIA – Ecological Perspectives for Science and Society*, 25(2): 105–9.
Muller, A., C. Schader, N. El-Hage Scialabba, J. Brügemann, A. Isensee, K.-H. Erb, P. Smith, K. Klocke, F. Leiber, M. Stolze and U. Niggli. (2017). 'Strategies for feeding the world more sustainably with agriculture'. *Nature Communications*, 8: 1290. doi:10.1038/s41467-017-01410-w.
Nemecek, T., D. Dubois, O. Huguenin-Elie and G. Gaillard. (2011). 'Life cycle assessment of Swiss farming systems: I. Integrated and organic farming'. *Agricultural Systems*, 104(3): 217–32.
Niggli, U. (2015). 'Incorporating agroecology into organic research – an ongoing challenge'. *Sustainable Agriculture Research*, 4(3): 149–57. doi:10.5539/sar.v4n3p149.
Niggli, U., C. Andres, H. Willer and B. P. Baker. (2017). 'Building a global platform for organic farming research, innovation and technology transfer'. *Organic Agriculture*, 7(3): 209–24.
Pashaei Kamali, F., M. P. M. Meuwissen, I. J. M. de Boer, C. E. van Middelaar, A. Moreira and A. G. J. M. Oude Lansink. (2017). 'Evaluation of the environmental, economic, and social performance of soybean farming systems in southern Brazil'. *Journal of Cleaner Production*, 142: 385–94.
Picazo-Tadeo, A. J. and D. Prior. (2009). 'Environmental externalities and efficiency measurement'. *Journal of Environmental Management*, 90(11): 3332–9.
Picazo-Tadeo, A. J., M. Beltrán-Esteve and J. A. Gómez-Limón. (2012). 'Assessing eco-efficiency with directional distance functions'. *European Journal of Operational Research*, 220(3): 798–809.
Ponisio, L. C., L. K. M'Gonigle, K. C. Mace, J. Palomino, P. de Valpine and C. Kremen. (2015). 'Diversification practices reduce organic to conventional yield gap'. *Proceedings of the Royal Society of London B: Biological Sciences*, 282(1799): 20141396.
Reganold, J. P. and J. M. Wachter. (2016). 'Organic agriculture in the twenty-first century'. *Nature Plants*, 2: 1–8.
Robinson, N. P., C. M. Cox and J. Koo. (2016). Harnessing net primary productivity data for monitoring sustainable development of agriculture. Washington DC, USA: International Food Policy Research Institute IFPRI.
Sanyang, S., S. J.-B. Taonda, J. Kuiseu, N. T. Coulibaly and L. Konaté. (2016). 'A paradigm shift in agricultural research for development: The role of innovation platforms'. *International Journal of Sustainability*, 14(2): 187–213.
Schader, C., M. Stolze and A. Gattinger. (2012). Environmental performance of organic farming. In: Boye, J. I. and Arcand, Y. (Eds), *Green Technologies in Food Production and Processing*. New York, NY: Springer, pp. 183–210.
Schader, C., J. Grenz, M. S. Meier and M. Stolze. (2014). 'Scope and precision of sustainability assessment approaches to food systems'. *Ecology and Society*, 19(3): 42.
Schader, C., A. Muller, N. El-Hage Scialabba, J. Hecht, A. Isensee, K.-H. Erb, P. Smith, H. Makkar, P. Klocke, F. Leiber, P. Schwegler, M. Stolze and U. Niggli. (2015). 'Impacts of feeding less food-competing feedstuffs to livestock on global food system sustainability'. *Journal of the Royal Society Interface*, 12: 20150891.
Schader, C., L. Baumgart, J. Landert, A. Muller, B. Ssebunya, J. Blockeel, R. Weisshaidinger, R. Petrasek, D. Mészáros and S. Padel. (2016). 'Using the Sustainability Monitoring and Assessment Routine (SMART) for the systematic analysis of trade-offs and synergies between sustainability dimensions and themes at farm level'. *Sustainability*, 8(3): 274.
Schmidt, H. (2004). Viehloser Ackerbau im ökologischen Landbau – Evaluierung des derzeitigen Erkenntnisstandes anhand von Betriebsbeispielen und Expertenbefragungen. Kassel, Germany: University of Kassel. http://orgprints.org/5020.
Seufert, V. and N. Ramankutty. (2017). 'Many shades of gray—The context-dependent performance of organic agriculture'. *Science Advances* 3(3): e1602638.
Seufert, V., N. Ramankutty and J. A. Foley. (2012). 'Comparing the yields of organic and conventional agriculture'. *Nature*, 485(7397): 229–32.

Ssebunya, B., L. Baumgart, J. Landert and C. Schader. (2016). Using SMART to compare social, economic, environmental and governance impacts of certification at smallholder farm level. Montpellier, France: Le Corum, p. 232.

Thomassen, M. A., R. Dalgaard, R. Heijungs and I. de Boer. (2008). 'Attributional and consequential LCA of milk production'. *International Journal of Life Cycle Assessment*, 13: 339–49.

Valin, H., P. Havlík, A. Mosnier, M. Herrero, E. Schmid and M. Obersteiner. (2013). 'Agricultural and greenhouse gas emissions: Trade-offs or synergies between mitigation and food?' *Environmental Research Letters*, 8(3): 035019.

von Lampe, M., D. Willenbockel, H. Ahammad, E. Blanc, Y. Cai, K. Calvin, S. Fujimori, T. Hasegawa, P. Havlik, E. Heyhoe, E. Kyle, H. Lotze-Campen, D. Mason d'Croz, G. C. Nelson, R. D. Sands, C. Schmitz, A. Tabeau, H. Valin, D. van der Mensbrugghe and H. van Meijl. (2014). 'Why do global long-term scenarios for agriculture differ? An overview of the AgMIP Global Economic Model Intercomparison'. *Agricultural Economics*, 45: 3–20.

Chapter 15

Eco-functional intensification of organic farming

Thomas F. Döring, University of Bonn, Germany

1 Introduction
2 Material and energy flows
3 Ecological regulation
4 Technical developments
5 Challenges
6 Conclusion
7 Where to look for further information
8 References

1 Introduction

Agricultural production is facing several serious challenges at the global level. Food security needs to be achieved for a growing world population (Gerland et al., 2014) that is fast changing its eating habits, with increasing demand for animal protein (Lassaletta et al., 2016). Simultaneously, food production needs to reduce its increasing ecological footprint (Crist et al., 2017), especially the negative impact of intensive high-input agriculture on biodiversity (Wilting et al., 2017). Resources on which agriculture depends are limited, so that resource use of agricultural production systems needs to decrease. Soil degradation and desertification (Bloem and Dunbar, 2016) as well as climate change pose additional complex factors limiting and disrupting the potential of agricultural systems to deliver food security and to decrease the likelihood of undernourishment (Dawson et al., 2016).

Organic agricultural systems have been shown to have positive effects on resource use via lower inputs, and positive impacts on biodiversity (Bengtsson et al., 2005; Tuck et al., 2014). However, several studies have also highlighted a general yield gap between organic and conventional agriculture (de Ponti et al., 2012; Seufert et al., 2012; Ponisio et al., 2015). It has been shown that this yield gap depends on crop species and there is some evidence that diversification of cropping systems can reduce the gap (Ponisio et al., 2015). Beyond choice of crop species and diversification, research has dedicated much to increase organic productivity and to reduce resource use, for example by optimizing reduced tillage operations in organic farming (Cooper et al., 2016). In general, however, organic farming faces the same dilemma of agricultural production as mentioned above,

http://dx.doi.org/10.19103/AS.2017.0029.17

as organic farming aims to increase productivity while maintaining a positive impact on biodiversity and using minimal resources. An urgent question of global importance is therefore how organic agricultural systems can be improved with regard to the various functions of producing food, saving biodiversity and reducing resource use – in short, how organic farming can increase its multifunctional performance.

Several different approaches have been proposed over the past few years in response to these challenges (Bommarco et al., 2013; Tittonell, 2014). Some have been bundled together under the label 'eco-functional intensification' (EFI; Halberg et al., 2015), a term introduced by Niggli et al. (2008). According to these authors, EFI means 'activating more knowledge and achieving a higher degree of organization per land unit', and emphasizes 'the deliberate minimal use of external chemically based inputs' (Halberg et al., 2015). The document introducing the concept mentions six specific characteristics of EFI. The first four of them refer to ecological processes:

1 'It intensifies the beneficial effects of ecosystem functions, including biodiversity, soil fertility and homeostasis.
2 It uses the self-regulating mechanisms of organisms and of biological or organizational systems in a highly intensive way.
3 It closes materials cycles in order to minimize losses (e.g. compost and manure).
4 It searches for the best match between environmental variation and the genetic variability of plants and crops' (Niggli et al., 2008, p. 34).

The last two characteristics refer to livestock welfare and human labour, with indirect though potentially large significance for the design of farming systems:

5 'It also means increased livestock welfare, with a positive impact on the health and productivity of animals.
6 It uses and provides more farm labour per land unit, principally such of high quality and professional satisfaction' (Niggli et al., 2008, p. 34).

Primarily, EFI is seen as a potential strategy to increase production through intensifying ecological processes rather than through increasing input of material and energy. Halberg et al. (2015, p. 128) summarize EFI as 'intensification of land use and agriculture by means of improved knowledge and application of biological principles and agro-ecological methods and increased cooperation and synergy between different components of agro-eco systems and food systems'.

What complicates the discussion about EFI is the fact that different labels and terms have been applied to the same measures as listed above under the heading of EFI. Also, there is a degree of fuzziness in what EFI means, or could mean. Despite this fuzziness, or because of it, EFI has been taken up by policy makers and has also gained traction among scientists of various disciplines. EFI has become a buzzword, though one among many others in the context of agriculture. Although the terminological boundaries of EFI are relatively unclear, and its meaning will overlap with that of other buzzwords, there is probably no need to develop a new or stricter definition of EFI.

Yet, as this chapter attempts a critical appraisal of the potential role of EFI in organic farming, the concept of EFI requires a closer look. As a starting point, it is useful to break the term down into its components – ecology, function and intensification. In the context of EFI, the specification of the process as 'ecological' can be seen as a contrast to chemical, input-related approaches, instead focusing on interactions between organisms and the (agricultural) environment. A classic example is pest regulation by antagonists, for example through natural predators and parasitoids. As a term, however, 'ecology' is not entirely unambiguous, as it can also mean a purely descriptive science, which might equally well refer to the ecological interactions in an input-driven system.

Similar considerations apply to the term 'functional'. According to Jax (2005), several different meanings can be attributed to the term in the context of ecology, including processes and interactions, at organism and system levels, roles and services. As 'functional' has no sharp boundaries and often means little more than 'useful for something', it provides relatively little inspiration. In the context of EFI, the aspect of functionality refers to the roles and services of different objects of agro-ecosystems, such as pest regulation, pollination, biomass production and nutrient cycling. However, the emphasis on functions – understood as roles or services – is not entirely uncontroversial. Building on the points presented by Peterson et al. (2010), it can be argued that focusing on functions (as services) reflects an anthropocentric attitude that may favour the commodification of nature and ecological functions, disregarding intrinsic values of nature that are not easily translated into economic gains.

As for intensification, this term directly refers to another buzzword, sustainable intensification, which has been discussed controversially within the organic sector. Like conventional intensification, sustainable intensification has been seen as too much depending on increased material inputs (Halberg et al., 2015). If intensification approximately means 'more of the same in the same space or time' the concept of intensification seems relatively straightforward and can also be applied to ecological approaches. Consider the example of predation, referring to point (2) above. In the context of EFI, intensification could mean higher predatory activity, more predators or higher predation rate. Similarly, taking the example of resource use (see point (3) above), intensification could refer to higher resource use efficiency through reduced losses.

Although the ideas of EFI are – at least in principle – applicable to many types of farming systems, the main area of potential application is organic farming (Halberg et al., 2015). There is, however, a large diversity of organic farming systems, depending on regions and landscapes, people and markets, historical developments and politics. This chapter follows the IFOAM definition of organic farming and mainly concentrates on certified systems within temperate climates. The principles of organic farming (IFOAM, 2005) serve as a tacit background for assessing the appropriateness of EFI measures. The chapter starts with a discussion of opportunities for and measures of EFI in organic farming (Sections 2, 3 and 4) and summarizes and discusses general challenges of EFI (Section 5). Exploring the opportunities for EFI follows the general question: What can be intensified in an ecological way, within organic farming systems? Three approaches are discussed, namely (1) intensifying the exploitation of underused resources in the realm of material and energy flows; (2) intensifying ecological regulation processes; and (3) intensifying the use of novel digitally based technologies.

2 Material and energy flows

One way of specifying potential cases for EFI is the identification of underused resources. Here, the general aim is to intensify resource use in an ecological way. An opportunity to locate resources that are presently underused is to identify spatial or temporal gaps in the growth of crops in the field. For example, in crop rotations temporal gaps can be identified between main crops which can then be filled with fast growing cover crops such as mustard, or through dual cropping systems or undersowing. Cover crops and living mulches can reduce weed pressure (Hiltbrunner et al., 2006) and protect the soil against erosion; they fulfil this protective function better than weeds within the crop (Arnhold et al., 2014). As far as organic farming is concerned, cover cropping and undersowing are already widely adopted.

Spatial gaps within the field occur especially in row crops such as potatoes, maize or soybean. The latter two species have relatively high temperature demands which can lead to slow growth of the young plants during spring, thereby exposing the soil to risk of erosion. This risk can be exacerbated in organic farming because of lower plant cover compared to conventional systems (Arnhold et al., 2014). Leaving this unused space to weeds which then need to be mechanically removed is not an efficient use of resources. Open soil will also increase unproductive evaporation. Instead, these row crops can be undersown with species suppressing weeds and providing additional benefits, for example floral resources to pollinators. While the undersown species will exert competitive pressure on the main crop, total resource use is higher than in the monocultures.

In this context, weeds can in general be seen as indicators of underused resources. They do not only indicate ecological conditions of the location where they are found, but wherever they grow, resource use by crops is incomplete. While the idea of replacing weeds by (additional) crops to increase and intensify overall resource use is compelling, it needs to be borne in mind that organic agriculture also strives to protect weed diversity (van Elsen, 2000).

Opportunities for EFI may not only be found in uncropped space or time, but may also refer to specific biotic and abiotic resources. An abiotic resource which may often be underused is light. For example, the spatio-temporal distribution of light use in temperate arable systems is characterized by a (short) temporal mismatch of near-maximal light supply during summer and, simultaneously, low light demand of cereal crops during the last ripening stage. Exploiting this mismatch, the idea is to grow another plant species that is partly complementary in its temporal distribution of light demand in order to maximize total light use. In particular, this can be achieved in agroforestry systems, where the tree canopy makes use of the incoming radiation during final cereal ripening. However, the time period during which grain yield and quality of cereals are insensitive to lowered light levels does not begin before the late grain filling stage (up to growth stage 87), and is relatively short. Between growth stage 89 and harvest there are typically only 7–14 days. Also, shading can reduce grain yield substantially during grain filling (Beed et al., 2007). On the other hand, when only little light is shaded out (8%), no significant effect on grain yield was observed in a study on different wheat cultivars, even when shading was present throughout the entire growing season (Li et al., 2010). The insensitivity of grain yields to low levels of shading plus the temporal but almost complete insensitivity to shading before harvest provides the opportunity to add another, complementary plant species. Indeed, agroforestry systems can be designed such that shading of the cereal is minimized, for example by a north-south orientation of the tree rows, or by choosing a species with late bud burst such as walnut (Dufour et al., 2013). In an agroforestry system of walnut and cereals, plant growth models revealed a higher total light use in the agroforestry system than the respective monocultures (Dupraz and Talbot, 2012).

Agroforestry is one example for diversification, a strategy that has been mentioned as a key component of EFI (Jensen et al., 2015), because it offers multiple benefits for agricultural system development (Isbell et al., 2017; Råberg et al., 2017). Apart from agroforestry, other measures of diversification include intercropping of arable crops such as cereals and grain legumes (Yu et al., 2016), but also the mixture of different cultivars of the same crop species (Barot et al., 2017) or the use of composite cross populations (Döring et al., 2015). There is an extensive literature on these approaches and recent meta-analyses reveal that the yield gains that can be obtained by legume–cereal intercropping (Yu et al., 2016) or cultivar mixtures (Kiær et al., 2009; Reiss and Drinkwater, 2018) can be substantial but are also quite variable. There are several mechanisms that have been identified to be responsible for higher yields in diversified systems than in the average of the respective monocultures (Loreau and Hector, 2001; Bedoussac et al., 2015). One of these mechanisms is complementary resource use, that is overall resource use is larger in the intercrop, thereby making use of otherwise unused or underused resources. However, as already observed in the nineteenth century, this more intensive resource use also leads to higher resource depletion rates (Wollny, 1885). In any case, because of the high variability of mixture effects reported so far, there is an urgent need to optimize mixture composition and to determine success factors (Barot et al., 2017).

Not all strategies to tap underused resources are necessarily based on diversification. One interesting approach is the biologically intensified use of the subsoil. Roots and biological activity in agricultural soils are mainly concentrated in the topsoil where tillage takes place. Making use of nutrients and water in the subsoil (Kautz et al., 2013) mainly relies on two biological mechanisms. One is the cultivation of deep-rooted crops. Species such as chicory (*Cichorium intybus*), lucerne (*Medicago sativa*) or lupin (*Lupinus angustifolius*) are characterized by deep rooting systems that will take up nutrients from below the zones that are usually reached by species with shallower rooting systems. As plant residues of deep-rooted species are incorporated into the topsoil, the cultivation of deep-rooted species leads to a slow pumping up of nutrients, especially P and K, from the subsoil to the topsoil. In addition, the thicker roots of deep-rooted species are responsible for the generation of biopores. It appears that these root-derived biopores are relatively stable after the decay of the roots. Biopores are preferred routes for the growth of the new roots of subsequent crops. As a consequence, uptake of water has been shown to be improved in a cereal crop following deep-rooted lucerne (Gaiser et al., 2012).

A second mechanism of generating biopores is the activity of deep-burrowing earthworms such as *Lumbricus terrestris* (Athmann et al., 2017). In these earthworm biopores, concentration of major plant nutrients is strongly amplified in relation to the surrounding bulk soil. Therefore, and possibly for biomechanical reasons, earthworm-generated biopores are also preferred places of root growth. While it is currently unclear whether higher biopore density also leads to improved yields, it is plausible that especially under dry or nutrient-poor conditions, advantages below ground will translate into higher or more stable yields, as well.

Ultimately, however, it is necessary to recognize that any mechanism tapping into the underused resources of the subsoil, particularly nutrients, will also lead to a depletion of this source. In the long run, this strategy is only sustainable if there is a mechanism to return nutrients to the soil (whether subsoil or topsoil) at the same rate as they are extracted. As in all organic systems, cycles need to be closed as far as possible. There are several measures to prevent or reduce leakage of nutrients from organic systems, for example the cropping of species with strong fast growing root system, the use of cover crops or rotation design. However, one study showed that long-term trends of nutrient contents in soils under organic management were negative (Bachinger et al., 2015).

3 Ecological regulation

One of the key elements of EFI is the intensification of ecological regulation mechanisms, in particular improving the regulation of pests and diseases (see point (2) cited in the Introduction). In terms of ecological pest control, several approaches have been intensively investigated over the past decades (Gurr et al., 2017). Many of these approaches aim to support the populations of natural predators by providing additional food resources (nectar, pollen or alternative prey) as well as habitat structures. Examples include strips of flowering crop and non-crop plant species directly next to the target field (Kienegger et al., 2003), widening of field margins (Marshall et al., 2006), hedges or grassy beetle banks between fields (Collins et al., 2002).

There is ample evidence that in the majority of cases these non-crop areas support arthropod predators and other antagonists (Griffiths et al., 2008). For instance, populations of lycosid spiders were shown to be supported by field margins (Marshall et al., 2006). Parasitoid wasps can benefit from flower strips (Lavandero et al., 2005). Many carabid species found in non-cropped areas such as beetle banks can be seen as relevant predators of aphids (Collins et al., 2003). There is also evidence that ground-dwelling predators (such as carabid beetles) as well as aerial aphid antagonists (such as ladybirds, hoverfly larvae or parasitoids) have a regulating effect on cereal aphid populations (Collins et al., 2002; Schmidt et al., 2003).

However, what has also been demonstrated is the large number of factors affecting the size and even the direction of these generally beneficial effects. For example, Marshall et al. (2006) showed that for many target taxa the effects of 6 m wide field margins were dependent on landscape type. In addition, the age of field margin strips was shown to be relevant for predator/pest ratio in a study on experimental farms in Germany, where 6-year strips were associated with higher predator/pest ratios than 1-year strips (Denys and Tscharntke, 2002). Further, floral composition of flower strips is highly relevant for providing appropriate and usable food resources to parasitoids. This is partly because of differences in the chemical composition of the nectar (Wäckers, 2001), but also due to differences in olfactory attractiveness and nectar accessibility (Wäckers, 2004; Vattala et al., 2006). Variable results were obtained with regard to the width of non-crop strips. It may be plausible that populations of antagonists increase with the area of suitable habitat provided, but in a study comparing 2 m and 6 m wide strips no evidence was found for the wider field margins increasing antagonist populations in the adjacent crop field (Holland et al., 2008). This was thought to be a consequence of the relatively high mobility of the affected beneficial insects.

What has also become clear is that the effects obtained by the mentioned measures of ecological pest control are subject to a large degree of variability. For example, although one study showed that, on average, parasitoids can have a reducing effect on cereal aphid populations within the growing season if parasitism rate exceeds about a third of the host population, there was also a tremendous variation, that is, even for parasitism rates above one half of the host population there were several cases where aphid population growth was still positive (Schmidt et al., 2003). Also, not all pests are regulated equally well after establishing structures for conservation biological control. In broccoli (*Brassica oleracea* var. *italica*), mixed results were obtained for the effect of flower strips on insect pests. In a two-year study, aphid densities on broccoli were reduced in the vicinity of flower strips, but in both study years larvae of two lepidopteran pests diamondback moth (*Plutella xylostella*) and the cabbage white (*Pieris rapae*) tended to be found in higher densities in broccoli plots with flower strips than in control plots (Kienegger et al., 2003).

In addition, there are also trade-offs when trying to optimize the management of field structures established for biological control. On the one hand, older field margin strips were found to support higher predator/pest ratios, but on the other hand, there was a significant increase of the cover of the perennial grass weed *Elymus repens* in the older strips (Denys and Tscharntke, 2002), from where it may invade into the field if not managed appropriately.

Summarizing, there is a large number of factors affecting the success of ecological pest control as well as a large variation observed even in those studies that control for these factors. It is therefore not surprising that researchers still see a demand for proving that resource provisioning for antagonists is indeed 'strong enough to improve herbivore regulation within crop fields' (Pollier et al., 2018). Moreover, even if this was firmly established, there is not sufficient data to confirm that ecological pest control translates through to yield improvement (Griffiths et al., 2008). Finally, many biocontrol studies refer to model systems that have little relevance in organic crops. Ecological control of cereal aphids is well-studied, but aphids only rarely affect organic cereals, partly because of the lower nitrogen levels found in organic arable farming. Conversely, the Colorado potato beetle (*Leptinotarsa decemlineata*) causes serious crop damage in organic potatoes, but in Europe where this insect was introduced, ecological control through natural enemies is not a promising option (Döring and Lynch, 2018).

However, there are also many successful cases of implementing ecological pest regulation, especially where structures and plant species grown for supporting beneficial insects simultaneously serve multiple purposes. An example is the use of broad strips sown with a mixture of flowering species on the headlands of arable land on the experimental organic farm Wiesengut of the University of Bonn. The plants in the flower mixture produce nectar and pollen, thereby supporting a large diversity of beneficial insects such as hoverflies, whose larvae prey on the aphids in the neighbouring crop. On the headlands, yield expectations of the faba bean crop are low mainly because of soil compaction. Therefore, this headland can more easily be sacrificed for the flower strips. In addition, soil compaction on the headlands is counteracted by deep rooting species in the flower mixture. Finally, leguminous species in the mix increase soil fertility through nitrogen fixation, so that the flower strip serves as a good precrop for a subsequent main crop.

4 Technical developments

EFI emphasizes the importance of knowledge input in agricultural management. Opportunities for intensifying this input through digital technologies in agriculture are currently being generated at an unprecedented pace. In particular, rapid advances are made in technologies dealing with (1) large-scale data acquisition, (2) data interpretation and (3) interference with objects in the field. The developments in digital and related technologies, discussed below for the context of EFI, are likely to accelerate over the coming decades. The organic farming sector, whether or not it strives for EFI, will need to investigate carefully the upcoming opportunities provided by these technologies.

The first area of digital developments refers to data acquisition in agriculture and comprises advanced sensor development, positioning systems and autonomous mobile sensor carriers. Within the last few years, data acquisition in the field has been revolutionized in terms of higher spectral, spatial and temporal resolution (Fiorani et al., 2012; Fiorani and Schurr, 2013; Behmann et al., 2016). Of particular relevance is the development of non-invasive sensing, for example through multispectral cameras, hyperspectral imaging but also through the simultaneous use of cheap conventional off-the-shelf RGB cameras. Active fluorescence sensors for use from a distance are being investigated for implementation in agriculture (Raesch et al., 2014); in addition, passive approaches using imaging spectrometers capture the sun-induced fluorescence signal of crops (Pinto et al., 2016). All these techniques allow non-destructive and non-invasive, repeated measurement of crop properties over time, and with resolutions varying from below the millimetre scale up to several square kilometres. Monitoring of crop growth and fine scale mapping through space and time has also become easier through developments in precise GPS positioning. Further, unmanned aerial vehicles (UAVs) (Burkart et al., 2014) and satellite missions (Drusch et al., 2017) have increased the range of spatial scales that can be covered in crop monitoring. In addition, non-invasive sensing can also be based on smartphone applications, which greatly enhances the practicability for farmers, as in the case of identifying sugar beet diseases in the field (Hallau et al., 2018).

The increasing data streams from agricultural fields can only be analysed meaningfully with developments in a second area of digital technologies, namely data processing through machine learning, crop modelling and similar approaches. These can be used for supporting crop management decisions, for characterization and selection of crops in plant breeding or for better understanding of processes and interactions underlying the growth of crops or disease dynamics (Pacilly et al., 2018). Current research analyses the multispectral data streams with advanced automatic classification algorithms for identifying plant diseases (Mahlein, 2016; Wahabzada et al., 2016), weed species (Pérez-Ortiz et al., 2016; McCool et al., 2017; Tang et al., 2017) or measuring plant nutritional status or stress effects (Singh et al., 2016). It is expected that applications based on these technologies of data analysis will substantially expand their flexibility, the range of recognizable objects and the general practicability in the future. In addition, crop modelling may help not only with understanding processes of crop growth, but also with real-time autonomous decision-making on the ground (Kuflik et al., 2009; Small et al., 2015). In particular, process-based crop models simulate growth and development of crops, weeds and diseases in relation to meteorological variables. Models may quantify various limitations to plant growth by water, nutrients or other factors; they are now widely being used for example in farm decision support and climate change impact assessment (Ewert et al., 2015).

The third area in which digital technologies are likely to make relevant contributions to EFI refers to the translation of extracted patterns and recognized objects into action in the field. Currently, actuators, mounted on lightweight autonomous robots for use in agriculture are being developed (Bechar and Vigneault, 2016, 2017). Different actuators can fulfil various functions ranging from weed removal to harvesting. Because field robots are often designed as lightweight vehicles, application in agriculture offers the opportunity to decrease soil compaction through agricultural machinery. This has recently been identified as an opportunity for yield increase in organic agriculture (Röös et al., 2018).

More generally, the combined advances in these areas offer tremendous potential for application in organic farming and EFI (Hamm et al., 2017; Maurel and Huyghe, 2017), particularly in automatic plant disease detection, automatic weed identification, the determination of plant nutrient status and organic plant breeding. Digitally based technologies, comprising sensing and monitoring, machine learning and crop modelling, as well as robotics, are likely to make direct (organic) plant protection possible in a much more targeted and selective way than is currently the case. As in conventional agriculture, organic fields are typically treated with constant application levels, both in plant protection and in (organic) fertilization. Thereby, local microclimate, topography, spatial heterogeneity of soil properties and clumped distribution of weeds, pests and diseases are usually largely ignored, despite attempts previously made in so-called precision farming. Integrating crop growth models with locally obtained sensor information for the detection of disease symptoms or weeds is a promising pathway to make application of plant protection agents more targeted and locally selective. For conventional agriculture, this idea was already developed in the early 1990s (Thompson et al., 1991) and is expanding fast as several previous technical challenges have been overcome. For organic agriculture research in these areas is only just starting (Knoll et al., 2016; Hamm et al., 2017).

Digital technologies constitute a potential way to massively reduce the use of biocides, also within organic agriculture, mainly because only the affected part of the field is treated. Currently however, it is unknown what long-term impacts the emerging technologies will have in the ecological, economic and societal domains of agricultural and food systems. Empirical evidence for biodiversity effects of digital technologies in agriculture is presently completely lacking. With regard to weeding, where technological advances are closest to larger-scale implementation in practice, robots are expected to enable more targeted weed management. Yet it is neither likely nor desirable that robots will achieve complete control of weeds on a given farm. Current technological barriers to complete weed control by robots include (1) automatic recognition of weeds that deviates from 100% accuracy (Sujaritha et al., 2017) and (2) incomplete control (no complete kill, regrowth after treatment). In addition, there may be economic constraints to achieving complete weed control. This will depend on the way in which costs increase with robotic treatment intensity. While incomplete control may to some degree be unavoidable, it may also be a programmed and deliberate feature of robotic weeding, with the aim of protecting biodiversity while simultaneously minimizing yield losses and sustaining plant health.

One of the explicit aims of EFI is to optimize the match between environmental variation and plant genetic variability. Digital technologies offer several pathways for making progress towards this aim. First, environmental variation may be recorded with higher resolution and accuracy through digital technologies. Second, genetic variation may better be characterized by the new technologies approaches. Specifically, organic and other plant breeders targeting the organic sector can be supported by advanced sensor ensembles which allow fast and high-resolution phenotyping of crops in the field. Ultimately, this may help to increase the diversity of genotypes that can be phenotyped by expanding this approach for example to landraces (Murphy et al., 2005; Chable et al., 2009) and to minor crop species (Ebert, 2014), or by incorporating specific management conditions such as high density for improved weed control (Weiner et al., 2010). Of particular interest to organic farmers and breeders are root properties, partly because of nutrient limitations inherent in many organic farming systems. New techniques are currently being developed for root phenotyping (Weigand and Kemna, 2017; Cseresnyés et al., 2018; Liu et al., 2018), although these approaches are not yet fully capable of detecting the complexities of root growth in real fields, they may make root assessment for (organic) breeders easier and more informative in the near future. Similarly, the deployment of different genotypes within a field, depending on soil properties, may become a possibility once advanced autonomous sowing machinery have been developed (Khandait et al., 2018).

5 Challenges

Several existing and upcoming opportunities for EFI in organic farming have been elaborated in the preceding sections. Unfortunately, however, these opportunities are not entirely unbounded. This section discusses five areas representing challenges for EFI, namely upscaling and transferability, indeterministic system behaviour, general limits to optimization, socioeconomic framework and concept fuzziness.

5.1 Upscaling and transferability

EFI in organic farming (and elsewhere) will only have a substantial impact if experimental insights gained in research, and observations obtained on farms can, at least partly, be transferred from one farming environment to another, so that EFI solutions can spread among practitioners. However, complex interactions (e.g. genotype × environment × management interactions) and the emergence of novel phenomena at higher scales tend to impede the general transferability of results in agricultural systems. Even for relatively simple questions such as cultivar choice within a given region, transferability of results from experimental stations to farms may be low, thereby complicating decision-making (Klepatzki et al., 2014).

Crop diversification as a key strategy of EFI is even more complex, adding to this transferability problem. A recent review of legume–cereal intercropping found, on average, large potential yield gains through intercropping, as measured by the land equivalent ratio (LER) (Yu et al., 2016). However, neither average nor potential are relevant in a given farming environment, that is, in a particular year on a specific, unchangeable site. Instead, this environment is hit by the full weight of the individual data point. The large variance and uncertainty observed across intercropping experiments means that more comprehensive and more systematic research is necessary in order to identify factors that determine the success of this diversification approach in a given environment (Döring, 2016). Thus, advancing optimization and ensuring transferability for intercropping and other measures of EFI requires substantial investment in infrastructure for practice-oriented research and knowledge exchange.

5.2 Indeterministic system behaviour

This second point is related to the previous one, but is more fundamental. The emphasis on the importance of knowledge input in EFI appears to suggest that many, if not most, problems might be solved if only the right knowledge was there, that more research and more knowledge will lead us to the golden grail of ecological farming. Many phenomena in ecology, however, are simply, and in principle, unknowable. From the mid-1970s on, it was increasingly recognized that even simple sets of deterministic rules can generate unpredictable system behaviour. For example, the Australian ecologist Robert M. May conclusively demonstrated that a simple deterministic recursive equation capturing the properties of logistic population growth can generate chaotic behaviour (May, 1975), essentially leading to complete unpredictability of population development over time. It is unlikely that identifying more and more factors, and quantifying their effects, for example in ecological pest control studies (Plećaš et al., 2014), will overcome this general phenomenon of low predictability in agro-ecosystems.

Even within organic agriculture research, however, there seems to be strong pressure to follow the paradigms of predictability and controllability. While it is clear that progress can be made towards better understanding and improving predictions, there will always be fundamental limitations to the predictability and controllability of agro-ecosystems (Döring, 2011). One potential solution is the diversification of different EFI approaches, thereby spreading the risk of failure rather than attempting to optimize single solutions.

5.3 General limits to optimization

EFI is a knowledge-intensive way of optimizing agricultural systems. In the words of its originators, 'EFI offers a huge opportunity to produce more food without compromising the quality of the environment' (Niggli et al., 2008, p. 35). However, optimization can only be taken to a certain level. In fact, trade-offs between different aims are often inevitable, and because of these trade-offs, there are general limitations to the optimization of agricultural production (Sadras and Denison, 2016). For example, improving ecological control of pests through supporting antagonists as part of EFI requires additional food resources to be provided for the insects, for example through flower strips or beetle banks. Although such measures will improve multifunctionality as they also generate benefits elsewhere, some of the target functions will exhibit reduced performance, for example because of land lost to cash crops, because of additional costs for seeds or because the new semi-natural habitat also provides shelter for other species of crop pest (Gurr et al., 2017).

While there are certainly opportunities for improvement of multifunctionality through EFI (Storkey et al., 2015), only very few measures will exhibit no trade-offs. Pushing the boundaries of optimization may become harder and harder. The main challenge, then, may be to bring expectations of what can be achieved through EFI down to realistic levels, without losing enthusiasm for continuously improving EFI. In the face of dramatic crises of climate disruption, biodiversity loss and resource depletion, however, looking for optimization and synergies will not carry us the whole way ahead. Beyond the application of refined ecological techniques, substantial effort is required for re-defining and re-negotiating the very aims of agricultural and food systems.

5.4 Socioeconomic framework

The economic pressures and fast social developments of today, the structures of the agricultural and food system, and the rules of global agricultural trading deeply impact on the decisions of many (organic) farmers. EFI can therefore not be seen in isolation from the larger socioeconomic framework. Consequently, EFI was conceptualized by its originators as part of a larger strategy that includes the empowerment of rural economies (Niggli et al., 2008). However, making EFI really work may require more than that. Taxes on nitrogen inputs or pesticides, decoupling direct farm payments from farm size, a ban on food speculation, have all been discussed as potential regulations that may result in better prospects for ecological agriculture. Going further, to make the necessary steps towards true sustainability, societies may have to adopt a more radical approach by renouncing the economic growth paradigm (Jackson, 2016).

But also if the fundamental rules of the current socioeconomic framework are left unchallenged, important questions remain. For example, as elaborated above, one area with particular potential for EFI is the development of digital technologies; currently however, it is unclear whether, or to which degree, these technologies will be open to all types and kinds of farmers. While the organic sector strives for fairness and equal opportunities, the pace and direction of technological development in agriculture is not in the hands of the organic farming community. It will be important to ensure participation in technical progress and to avoid a digital divide at the sector level.

5.5 Concept fuzziness

EFI started off in a policy document from and for the organic farming sector. As stated in the introduction, its boundaries are somewhat blurry, but that may not be a disadvantage. One of the problems, however, is that any innovative farm that can serve as an example for EFI will also be an example for other labels. In fact, ecological pest regulation through conservation biological control has been claimed as a key element of both *Eco-functional Intensification* and *Sustainable Intensification* (Gurr et al., 2017). Even more overlap is evident with the concept of *Ecological Intensification* (Bommarco et al., 2013; Tittonell, 2014). Similarly, it could be argued that EFI is equivalent to *best practice organic farming*. As a label, therefore, EFI appears to be relatively unspecific. If this is so, it remains to be seen if EFI can generate enough inspiration beyond the tedious groundwork of continuously trying to improve multifunctional system performance, little by little.

6 Conclusion

EFI is a concept emphasizing the importance of knowledge input for optimizing ecological interactions in farming systems. It provides powerful opportunities for improving organic farming systems. Identifying underused resources, intensifying ecological regulation and applying novel digital technologies for more targeted crop management are just a few areas that provide prospects for advancing organic agriculture. It is evident that making use of these opportunities requires substantial investment in knowledge generation through systematic research, both on and off farm. Further, during the last few years, gradual changes in research priorities and policies have occurred which have brought aims of stakeholders from within and outside the organic sector closer together. In its task of progressing EFI (or similar concepts labelled differently), the organic sector is therefore faced with a multitude of new opportunities for collaboration and exchange.

At the same time however, it should be kept in mind that EFI does not generate a trade-off free zone. Mostly, optimization will be tedious and slow. Controlling ecological processes is impeded by the (desirable) complexity of agricultural systems. Implementation of EFI measures on farms depends on many factors, and decisions and practices will be messier than the ideas and concepts thought up by ecologists and agronomists. Finally, promoting EFI, or similar approaches under different labels, will require changes at the socioeconomic level. The aims of agricultural production need to be re-negotiated: towards greater multifunctionality!

7 Where to look for further information

Examples of and inspiration for eco-functional intensification can be found in many sources, but first-hand experience on organic farms is probably unbeatable. Farmers with ecological innovations might be identified through regional organic farmers' networks, for example the certification bodies. For practically minded readers, diverse information has been compiled on the Agricology website (https://www.agricology.co.uk).

8 References

Arnhold, S., Lindner, S., Lee, B., Martin, E., Kettering, J., Nguyen, T. T., Koellner, T., Ok, Y. S. and Huwe, B. 2014. Conventional and organic farming: Soil erosion and conservation potential for row crop cultivation. *Geoderma* 219–220, 89–105.

Athmann, M., Kautz, T., Banfield, C., Bauke, S., Hoang, D. T., Lüsebrink, M., Pausch, J., Amelung, W., Kuzyakov, Y. and Köpke, U. 2017. Six months of *L. terrestris* L. activity in root-formed biopores increases nutrient availability, microbial biomass and enzyme activity. *Applied Soil Ecology* 120, 135–42.

Bachinger, J., Bloch, R., Reckling, M. and Stein-Bachinger, K. 2015. 20 Jahre Modellbetrieb für Ökolandbau in Müncheberg. Teil 1: Entwicklung von Humus und Nährstoffgehalten. In: Häring, A. M., Hörning, B., Hoffmann-Bahnsen, R., Luley, H., Luthardt, V., Pape, J. and Trei, G. (Eds), *13. Wissenschaftstagung Ökologischer Landbau*, Eberswalde.

Barot, S., Allard, V., Cantarel, A., Enjalbert, J., Gauffreteau, A., Goldringer, I., Lata, J.-C., Le Roux, X., Niboyet, A. and Porcher, E. 2017. Designing mixtures of varieties for multifunctional agriculture with the help of ecology. A review. *Agronomy for Sustainable Development* 37, 13.

Bechar, A. and Vigneault, C. 2016. Agricultural robots for field operations: Concepts and components. *Biosystems Engineering* 149, 94–111.

Bechar, A. and Vigneault, C. 2017. Agricultural robots for field operations. Part 2: Operations and systems. *Biosystems Engineering* 153, 110–28.

Bedoussac, L., Journet, E.-P., Hauggaard-Nielsen, H., Naudin, C., Corre-Hellou, G., Jensen, E. S., Prieur, L. and Justes, E. 2015. Ecological principles underlying the increase of productivity achieved by cereal-grain legume intercrops in organic farming. A review. *Agronomy for Sustainable Development* 35, 911–35.

Beed, F., Paveley, N. and Sylvester-Bradley, R. 2007. Predictability of wheat growth and yield in light-limited conditions. *The Journal of Agricultural Science* 145, 63–79.

Behmann, J., Mahlein, A.-K., Paulus, S., Dupuis, J., Kuhlmann, H., Oerke, E.-C. and Plümer, L. 2016. Generation and application of hyperspectral 3D plant models: Methods and challenges. *Machine Vision and Applications* 27, 611–24.

Bengtsson, J., Ahnström, J. and Weibull, A.-C. 2005. The effects of organic agriculture on biodiversity and abundance: A meta-analysis. *Journal of Applied Ecology* 42, 261–9.

Bloem, J. and Dunbar, M. 2016. Land degradation and desertification. In: Orgiazzi, A., Bardgett, R. D. and Barrios, E. (Eds), *Global Soil Biodiversity Atlas*. European Union, Luxembourg, pp. 130–1.

Bommarco, R., Kleijn, D. and Potts, S. G. 2013. Ecological intensification: Harnessing ecosystem services for food security. *Trends in Ecology & Evolution* 28, 230–8.

Burkart, A., Cogliati, S., Schickling, A. and Rascher, U. 2014. A novel UAV-based ultra-light weight spectrometer for field spectroscopy. *IEEE Sensors Journal* 14, 62–7.

Chable, V., Goldringer, I., Dawson, J., Bocci, R., Lammerts van Bueren, E. T., Serpolay, E., González, J. M., Valero, T., Levillain, T., Van der Burg, W. J., Pimbert, M., Pino, S. and Kik, C. (Eds). 2009. *Farm Seed Opportunities: A Project to Promote Landrace Use and Renew Biodiversity*. Bioversity International, Rome.

Collins, K., Boatman, N., Wilcox, A., Holland, J. and Chaney, K. 2002. Influence of beetle banks on cereal aphid predation in winter wheat. *Agriculture, Ecosystems & Environment* 93, 337–50.

Collins, K., Boatman, N., Wilcox, A. and Holland, J. 2003. A 5-year comparison of overwintering polyphagous predator densities within a beetle bank and two conventional hedgebanks. *Annals of Applied Biology* 143, 63–71.

Cooper, J., Baranski, M., Stewart, G., Nobel-de Lange, M., Bàrberi, P., Fließbach, A., Peigné, J., Berner, A., Brock, C. and Casagrande, M. 2016. Shallow non-inversion tillage in organic farming maintains crop yields and increases soil C stocks: A meta-analysis. *Agronomy for Sustainable Development* 36, 1–20.

Crist, E., Mora, C. and Engelman, R. 2017. The interaction of human population, food production, and biodiversity protection. *Science* 356, 260–4.

Cseresnyés, I., Rajkai, K., Takács, T. and Vozáry, E. 2018. Electrical impedance phase angle as an indicator of plant root stress. *Biosystems Engineering* 169, 226–32.

Dawson, T. P., Perryman, A. H. and Osborne, T. M. 2016. Modelling impacts of climate change on global food security. *Climatic Change* 134, 429–40.

Denys, C. and Tscharntke, T. 2002. Plant-insect communities and predator-prey ratios in field margin strips, adjacent crop fields, and fallows. *Oecologia* 130, 315–24.

de Ponti, T., Rijk, B. and Van Ittersum, M. K. 2012. The crop yield gap between organic and conventional agriculture. *Agricultural Systems* 108, 1–9.

Döring, T. F. 2011. Potential and limitations of plant virus epidemiology: Lessons from the Potato virus Y pathosystem. *Potato Research* 54, 341–54.

Döring, T. F. 2016. Re-thinking research on intercropping systems. *Mitteilungen der Gesellschaft für Pflanzenbauwissenschaften* 28, 32–3.

Döring, T. F. and Lynch, D. H. 2018. Organic potato cultivation. In: Wale, S. (Ed.), *Achieving Sustainable Cultivation of Potatoes. Vol. 2 Production, storage and crop protection.* Burleigh Dodds Science Publishing, Cambridge, UK.

Döring, T. F., Annicchiarico, P., Clarke, S., Haigh, Z., Jones, H. E., Pearce, H., Snape, J., Zhan, J. and Wolfe, M. S. 2015. Comparative analysis of performance and stability among composite cross populations, variety mixtures and pure lines of winter wheat in organic and conventional cropping systems. *Field Crops Research* 183, 235–45.

Drusch, M., Moreno, J., Del Bello, U., Franco, R., Goulas, Y., Huth, A., Kraft, S., Middleton, E. M., Miglietta, F. and Mohammed, G. 2017. The FLuorescence EXplorer Mission Concept – ESA's Earth Explorer 8. *IEEE Transactions on Geoscience and Remote Sensing* 55, 1273–84.

Dufour, L., Metay, A., Talbot, G. and Dupraz, C. 2013. Assessing light competition for cereal production in temperate agroforestry systems using experimentation and crop modelling. *Journal of Agronomy and Crop Science* 199, 217–27.

Dupraz, C. and Talbot, G. 2012. Evidences and explanations for the unexpected high productivity of improved temperate agroforestry systems. In: Mosquera-Losada, M. R., Pantera, A., Rosati, A., Amaral, J., Smith, J., Rigueiro-Rodríguez, A., Watté, J. and Dupraz, C. (Eds), *1st European Agroforestry Conference*, 9 October 2012. Priorities for European Agroforestry, Brussels, Belgium.

Ebert, A. W. 2014. Potential of underutilized traditional vegetables and legume crops to contribute to food and nutritional security, income and more sustainable production systems. *Sustainability* 6, 319–35.

Ewert, F., Rötter, R. P., Bindi, M., Webber, H., Trnka, M., Kersebaum, K. C., Olesen, J. E., van Ittersum, M. K., Janssen, S. and Rivington, M. 2015. Crop modelling for integrated assessment of risk to food production from climate change. *Environmental Modelling & Software* 72, 287–303.

Fiorani, F. and Schurr, U. 2013. Future scenarios for plant phenotyping. *Annual Review of Plant Biology* 64, 267–91.

Fiorani, F., Rascher, U., Jahnke, S. and Schurr, U. 2012. Imaging plants dynamics in heterogenic environments. *Current Opinion in Biotechnology* 23, 227–35.

Gaiser, T., Perkons, U., Küpper, P. M., Puschmann, D. U., Peth, S., Kautz, T., Pfeifer, J., Ewert, F., Horn, R. and Köpke, U. 2012. Evidence of improved water uptake from subsoil by spring wheat following lucerne in a temperate humid climate. *Field Crops Research* 126, 56–62.

Gerland, P., Raftery, A. E., Ševčíková, H., Li, N., Gu, D., Spoorenberg, T., Alkema, L., Fosdick, B. K., Chunn, J. and Lalic, N. 2014. World population stabilization unlikely this century. *Science* 346, 234–7.

Griffiths, G. J., Holland, J. M., Bailey, A. and Thomas, M. B. 2008. Efficacy and economics of shelter habitats for conservation biological control. *Biological Control* 45, 200–9.

Gurr, G. M., Wratten, S. D., Landis, D. A. and You, M. 2017. Habitat management to suppress pest populations: Progress and prospects. *Annual Review of Entomology* 62, 91–109.

Halberg, N., Panneerselvam, P. and Treyer, S. 2015. Eco-functional intensification and food security: Synergy or compromise? *Sustainable Agriculture Research* 4, 126.

Hallau, L., Neumann, M., Klatt, B., Kleinhenz, B., Klein, T., Kuhn, C., Röhrig, M., Bauckhage, C., Kersting, K. and Mahlein, A. K. 2018. Automated identification of sugar beet diseases using smartphones. *Plant Pathology* 67, 399–410.

Hamm, U., Häring, A. M., Hülsbergen, K.-J., Isermeyer, F., Lange, S., Niggli, U., Rahmann, G. and Horn, S. 2017. Research strategy of the German Agricultural Research Alliance (DAFA) for the development of the organic farming and food sector in Germany. *Organic Agriculture* 7, 225–42.

Hiltbrunner, J., Liedgens, M., Bloch, L., Stamp, P. and Streit, B. 2006. Legume cover crops as living mulches for winter wheat: Components of biomass and the control of weeds. *European Journal of Agronomy* 26(1), 21–9.

Holland, J., Oaten, H., Southway, S. and Moreby, S. 2008. The effectiveness of field margin enhancement for cereal aphid control by different natural enemy guilds. *Biological Control* 47, 71–6.

IFOAM. 2005. The Principles of Organic Agriculture. http://www.ifoam.org/about_ifoam/principles/.

Isbell, F., Adler, P. R., Eisenhauer, N., Fornara, D., Kimmel, K., Kremen, C., Letourneau, D. K., Liebman, M., Polley, H. W. and Quijas, S. 2017. Benefits of increasing plant diversity in sustainable agroecosystems. *Journal of Ecology* 105, 871–9.

Jackson, T. 2016. *Prosperity Without Growth: Foundations for the Economy of Tomorrow*. Routledge, London.

Jax, K. 2005. Function and 'functioning' in ecology: What does it mean? *Oikos* 111, 641–8.

Jensen, E. S., Bedoussac, L., Carlsson, G., Journet, E.-P., Justes, E. and Hauggaard-Nielsen, H. 2015. Enhancing yields in organic crop production by eco-functional intensification. *Sustainable Agriculture Research* 4, 42.

Kautz, T., Amelung, W., Ewert, F., Gaiser, T., Horn, R., Jahn, R., Javaux, M., Kemna, A., Kuzyakov, Y., Munch, J.-C., Pätzold, S., Peth, S., Scherer, H. W., Schloter, M., Schneider, H., Vanderborght, J., Vetterlein, D., Walter, A., Wiesenberg, G. L. and Köpke, U. 2013. Nutrient acquisition from arable subsoils in temperate climates: A review. *Soil Biology and Biochemistry* 57, 1003–27.

Khandait, P., Sariyam, S., Agarkar, D., Dhongde, V., Doble, S., Makhe, R. and Dhakate, R. 2018. Seed sowing robot. *International Journal of Engineering Science* 8(3), 16564.

Kiær, L. P., Skovgaard, I. M. and Østergård, H. 2009. Grain yield increase in cereal variety mixtures: A meta-analysis of field trials. *Field Crops Research* 114, 361–73.

Kienegger, M., Kromp, B. and Kahrer, A. 2003. The effect of strips of flowers on pests and beneficial arthropods in adjacent broccoli plots. *IOBC WPRS Bulletin* 26, 61–70.

Klepatzki, J., Döring, T. F., Macholdt, J. and Ellmer, F. 2014. Comparing the reliability of maize variety data from on-farm trials and experimental stations. *Journal für Kulturpflanzen* 66, 389–95.

Knoll, F. J., Holtorf, T. and Hussmann, S. 2016. Investigation of different sensor systems to classify plant and weed in organic farming applications. *SAI Computing Conference (SAI)*, 2016. IEEE, pp. 343–8.

Kuflik, T., Prodorutti, D., Frizzi, A., Gafni, Y., Simon, S. and Pertot, I. 2009. Optimization of copper treatments in organic viticulture by using a web-based decision support system. *Computers and Electronics in Agriculture* 68, 36–43.

Lassaletta, L., Billen, G., Garnier, J., Bouwman, L., Velazquez, E., Mueller, N. D. and Gerber, J. S. 2016. Nitrogen use in the global food system: Past trends and future trajectories of agronomic performance, pollution, trade, and dietary demand. *Environmental Research Letters* 11, 095007.

Lavandero, B., Wratten, S., Shishehbor, P. and Worner, S. 2005. Enhancing the effectiveness of the parasitoid *Diadegma semiclausum* (Helen): Movement after use of nectar in the field. *Biological Control* 34, 152–8.

Li, H., Jiang, D., Wollenweber, B., Dai, T. and Cao, W. 2010. Effects of shading on morphology, physiology and grain yield of winter wheat. *European Journal of Agronomy* 33, 267–75.

Liu, X., Dong, X., Xue, Q., Leskovar, D. I., Jifon, J., Butnor, J. R. and Marek, T. 2018. Ground penetrating radar (GPR) detects fine roots of agricultural crops in the field. *Plant Soil* 423, 517–31.

Loreau, M. and Hector, A. 2001. Partitioning selection and complementarity in biodiversity experiments. *Nature* 412, 72–6.

Mahlein, A.-K. 2016. Plant disease detection by imaging sensors–parallels and specific demands for precision agriculture and plant phenotyping. *Plant Disease* 100, 241–51.

Marshall, E., West, T. and Kleijn, D. 2006. Impacts of an agri-environment field margin prescription on the flora and fauna of arable farmland in different landscapes. *Agriculture, Ecosystems & Environment* 113, 36–44.

Maurel, V. B. and Huyghe, C. 2017. Putting agricultural equipment and digital technologies at the cutting edge of agroecology. *OCL* 24, D307.

May, R. M. 1975. Biological populations obeying difference equations: Stable points, stable cycles, and chaos. *Journal of Theoretical Biology* 51, 511–24.

McCool, C., Perez, T. and Upcroft, B. 2017. Mixtures of lightweight deep convolutional neural networks: Applied to agricultural robotics. *IEEE Robotics and Automation Letters* 2, 1344–51.

Murphy, K., Lammer, D., Lyon, S., Carter, B. and Jones, S. S. 2005. Breeding for organic and low-input farming systems: An evolutionary-participatory breeding method for inbred cereal grains. *Renewable Agriculture and Food Systems* 20, 48–55.

Niggli, U., Slabe, A., Schmid, O., Halberg, N. and Schlüter, M. 2008. Vision for an organic food and farming research agenda 2025. Organic knowledge for the future. Technology Platform Organics. IFOAM Regional Group European Union (IFOAM EU Group), Brussels and International Society of Organic Agriculture Research (ISOFAR), Bonn, Germany.

Pacilly, F. C., Hofstede, G. J., van Bueren, E. T. L., Kessel, G. J. and Groot, J. C. 2018. Simulating crop-disease interactions in agricultural landscapes to analyse the effectiveness of host resistance in disease control: The case of potato late blight. *Ecological Modelling* 378, 1–12.

Pérez-Ortiz, M., Peña, J. M., Gutiérrez, P. A., Torres-Sánchez, J., Hervás-Martínez, C. and López-Granados, F. 2016. Selecting patterns and features for between-and within-crop-row weed mapping using UAV-imagery. *Expert Systems with Applications* 47, 85–94.

Peterson, M. J., Hall, D. M., Feldpausch-Parker, A. M. and Peterson, T. R. 2010. Obscuring ecosystem function with application of the ecosystem services concept. *Conservation Biology* 24, 113–19.

Pinto, F., Damm, A., Schickling, A., Panigada, C., Cogliati, S., Müller-Linow, M., Balvora, A. and Rascher, U. 2016. Sun-induced chlorophyll fluorescence from high-resolution imaging spectroscopy data to quantify spatio-temporal patterns of photosynthetic function in crop canopies. *Plant, Cell & Environment* 39, 1500–12.

Plećaš, M., Gagić, V., Janković, M., Petrović-Obradović, O., Kavallieratos, N. G., Tomanović, Ž., Thies, C., Tscharntke, T. and Ćetković, A. 2014. Landscape composition and configuration influence cereal aphid–parasitoid–hyperparasitoid interactions and biological control differentially across years. *Agriculture, Ecosystems & Environment* 183, 1–10.

Pollier, A., Guillomo, L., Tricault, Y., Plantegenest, M. and Bischoff, A. 2018. Effects of spontaneous field margin vegetation on the regulation of herbivores in two winter crops. *Basic and Applied Ecology* 27, 71–82.

Ponisio, L. C., M'Gonigle, L. K., Mace, K. C., Palomino, J., de Valpine, P. and Kremen, C. 2015. Diversification practices reduce organic to conventional yield gap. *Proceedings of the Royal Society of London B: Biological Sciences* 282, 20141396.

Råberg, T. M., Carlsson, G. and Jensen, E. S. 2017. Productivity in an arable and stockless organic cropping system may be enhanced by strategic recycling of biomass. *Renewable Agriculture and Food Systems* (online). doi:10.1017/S1742170517000242.

Raesch, A. R., Muller, O., Pieruschka, R. and Rascher, U. 2014. Field observations with laser-induced fluorescence transient (LIFT) method in barley and sugar beet. *Agriculture* 4, 159–69.
Reiss, E. R. and Drinkwater, L. E. 2018. Cultivar mixtures: A meta-analysis of the effect of intraspecific diversity on crop yield. *Ecological Applications* 28, 62–77.
Röös, E., Mie, A., Wivstad, M., Salomon, E., Johansson, B., Gunnarsson, S., Wallenbeck, A., Hoffmann, R., Nilsson, U. and Sundberg, C. 2018. Risks and opportunities of increasing yields in organic farming. A review. *Agronomy for Sustainable Development* 38, 14.
Sadras, V. O. and Denison, R. F., 2016. Neither crop genetics nor crop management can be optimised. *Field Crops Research* 189, 75–83.
Schmidt, M. H., Lauer, A., Purtauf, T., Thies, C., Schaefer, M. and Tscharntke, T. 2003. Relative importance of predators and parasitoids for cereal aphid control. *Proceedings of the Royal Society of London B: Biological Sciences* 270, 1905–9.
Seufert, V., Ramankutty, N. and Foley, J. A. 2012. Comparing the yields of organic and conventional agriculture. *Nature* 485, 229–32.
Singh, A., Ganapathysubramanian, B., Singh, A. K. and Sarkar, S. 2016. Machine learning for high-throughput stress phenotyping in plants. *Trends in Plant Science* 21, 110–24.
Small, I. M., Joseph, L. and Fry, W. E. 2015. Development and implementation of the BlightPro decision support system for potato and tomato late blight management. *Computers and Electronics in Agriculture* 115, 57–65.
Storkey, J., Döring, T. F., Baddeley, J., Collins, R., Roderick, S., Jones, H. and Watson, C. A. 2015. Engineering a plant community to deliver multiple ecosystem services. *Ecological Applications* 25, 1034–43.
Sujaritha, M., Annadurai, S., Satheeshkumar, J., Sharan, S. K. and Mahesh, L. 2017. Weed detecting robot in sugarcane fields using fuzzy real time classifier. *Computers and Electronics in Agriculture* 134, 160–71.
Tang, J., Wang, D., Zhang, Z., He, L., Xin, J. and Xu, Y. 2017. Weed identification based on K-means feature learning combined with convolutional neural network. *Computers and Electronics in Agriculture* 135, 63–70.
Thompson, J., Stafford, J. and Miller, P. 1991. Potential for automatic weed detection and selective herbicide application. *Crop Protection* 10, 254–9.
Tittonell, P. 2014. Ecological intensification of agriculture – sustainable by nature. *Current Opinion in Environmental Sustainability* 8, 53–61.
Tuck, S. L., Winqvist, C., Mota, F., Ahnström, J., Turnbull, L. A. and Bengtsson, J. 2014. Land-use intensity and the effects of organic farming on biodiversity: A hierarchical meta-analysis. *Journal of Applied Ecology* 51, 746–55.
van Elsen, T. 2000. Species diversity as a task for organic agriculture in Europe. *Agriculture, Ecosystems & Environment* 77, 101–9.
Vattala, H., Wratten, S., Phillips, C. and Wäckers, F. 2006. The influence of flower morphology and nectar quality on the longevity of a parasitoid biological control agent. *Biological Control* 39, 179–85.
Wäckers, F. 2001. A comparison of nectar and honeydew sugars with respect to their utilization by the hymenopteran parasitoid Cotesia glomerata. *Journal of Insect Physiology* 47, 1077–84.
Wäckers, F. 2004. Assessing the suitability of flowering herbs as parasitoid food sources: Flower attractiveness and nectar accessibility. *Biological Control* 29, 307–14.
Wahabzada, M., Mahlein, A.-K., Bauckhage, C., Steiner, U., Oerke, E.-C. and Kersting, K. 2016. Plant phenotyping using probabilistic topic models: Uncovering the hyperspectral language of plants. *Scientific Reports* 6, 22482.
Weigand, M. and Kemna, A. 2017. Multi-frequency electrical impedance tomography as a non-invasive tool to characterize and monitor crop root systems. *Biogeosciences* 14, 921–39.
Weiner, J., Andersen, S. B., Wille, W. K. M., Griepentrog, H. W. and Olsen, J. M. 2010. Evolutionary Agroecology: The potential for cooperative, high density, weed-suppressing cereals. *Evolutionary Applications* 3, 473–9.

Wilting, H. C., Schipper, A. M., Bakkenes, M., Meijer, J. R. and Huijbregts, M. A. 2017. Quantifying biodiversity losses due to human consumption: A global-scale footprint analysis. *Environmental Science & Technology* 51, 3298–306.
Wollny, E. 1885. Saat und Pflege der landwirthschaftlichen Kulturpflanzen – Handbuch für die Praxis. Verlag Paul Parey, Berlin.
Yu, Y., Stomph, T.-J., Makowski, D., Zhang, L. and van der Werf, W. 2016. A meta-analysis of relative crop yields in cereal/legume mixtures suggests options for management. *Field Crops Research* 198, 269–79.

Chapter 16

Biodiversity as a prerequisite of sustainable organic farming

Lukas Pfiffner and Laura Armengot, Research Institute of Organic Agriculture (FiBL), Switzerland

1 Introduction
2 Biodiversity and land-use intensity
3 Impact of organic farming on biodiversity
4 Biodiversity at different spatial scales
5 Impact of organic farming on selected functional groups
6 Future trends and conclusion
7 References

1 Introduction

For hundreds of years, agriculture has substantially contributed to the diversity of species and habitats, and has formed many of today's landscapes. However, over the last century, the intensification of agriculture with high inputs of synthetic pesticides and fertilisers combined with monocrop specialisation has been detrimental to the diversity of genetic resources of crop varieties and livestock breeds, to the wild flora and fauna species and to the diversity of ecosystems (Tscharntke et al. 2005). Among terrestrial vertebrates, 322 species have become extinct since the 1500s. Populations of the remaining species show 25% average decline in abundance (Dirzo et al. 2014). The patterns of invertebrates are equally dire: 67% of monitored invertebrates show a 45% mean abundance decline. A global assessment of land-use change and related pressures has already reduced local biodiversity intactness (Newbold et al. 2016). The average proportion of biodiversity remaining in local ecosystems is beyond its recently proposed planetary boundary across 58% of the world's surface, where 71% of human beings live. The large net losses of biodiversity in many regions that have already occurred potentially lead to harm ecosystem functions and services (Hooper et al. 2012; Newbold et al. 2016).

Organic farming is dependent upon stabilising agroecosystems, maintaining ecological services, developing agroecological measures to their optimum and linking agricultural activities with the conservation of biodiversity (Stockdale et al. 2001). Wild species perform a variety of ecological services within farming systems, for example natural pollination, pest control and maintenance of soil fertility. Thus, higher levels of biodiversity can strengthen farming systems and their resilience under changing environmental conditions. Enhancing

http://dx.doi.org/10.19103/AS.2017.0029.18

functional biodiversity is a key ecological strategy to bring sustainability to production on organic farms. Organic systems also substantially reduce external inputs and do not use mineral N fertilisers, pesticides and genetically modified organisms (Reganold and Wachter 2016). By respecting the natural capacity of plants, animals and the landscape, organic agriculture aims to optimise quality in all aspects of agriculture and the environment, and diversification practices can help to reduce the organic to conventional yield gap (Ponisio et al. 2015).

Organic agriculture is thus committed to the conservation and enhancement of biodiversity within agricultural systems, both from a philosophical perspective and from the practical viewpoint of maintaining productivity. The importance of biodiversity as part of a well-balanced organic system is part of the IFOAM International Basic Standards for Organic Agriculture Production and Processing for Organic Agriculture (Anon. 2017) that have been developed worldwide for organic farming; guidelines for biodiversity conservation have been updated to face current challenges (Bosshard et al. 2009).

Biodiversity is generally assessed at three distinct levels:

- Genetic diversity: the variation between individuals and between populations within a species;
- Species diversity: the different types of plants, animals and other life forms within a region; and
- Ecosystem diversity: the variety of habitats found within an area (e.g. grassland, marsh, forest and cropland).

These levels have been used in this chapter to show the links between biodiversity and organic agriculture, with focus on species and habitat diversity. In general, the degree of biodiversity in agroecosystems depends on four main characteristics of the agroecosystems:

1 The intensity of management/farming (e.g. pesticides, fertilisers, soil tillage).
2 Land use (crop types: perennial, annual crops) and the permanence of various crops within the agroecosystem.
3 The diversity of vegetation within and around the agroecosystem (semi-natural and non-crop habitats).
4 Landscape infrastructure (cleared, simple, complex).

We give an overview in this chapter of the impact of organic farming on biodiversity. The available data of these comparison studies originate mostly from Western and Northern Europe and North America, while studies from most other regions, especially Africa, Asia and Australia, are largely lacking. Finally, we discuss some examples of the application of functional agro-biodiversity to boost some key ecosystem services on organic farms.

2 Biodiversity and land-use intensity

The conservation of biodiversity in agroecosystems is closely related to the type and intensity of agricultural practices, that is, to the land-use intensity at both local and landscape scales.

High-intensity land use is one of the major causes of biodiversity loss. Land-use intensity derives from the different agricultural practices applied in a certain field or area, that is, at local scale: through, for example fertilisation, pesticides, soil tillage, crop rotations; and at landscape scale: through, for example the area of semi-natural habitats and field margins. Reliable methods for assessing the land-use intensity of cropping systems are essential because they should allow the comparison of farms or regions and the implementation of adequate measures for species conservation strategies (Bockstaller et al. 2008).

In general, floral and faunal diversity is influenced at local scale by both anthropogenic factors (e.g. type and intensity of farming practices) and non-anthropogenic factors (e.g. site conditions such as soil conditions, altitude or microclimate); and also at landscape scale, that is, the quality of the landscape infrastructure (e.g. area and diversity of semi-natural habitats). Farmers have many options to enhance wildlife on their farms using agri-environmental schemes and biodiversity-related in-field options, for example low-input farm practices (Stoeckli et al. 2017). Biotic factors such as the supply of food resources and floral, structural diversity in crop and non-crop habitats influence the diversity and abundance of these organisms. Figure 1 shows exemplary relevant factors altering invertebrate populations at farm level in a simplified scheme.

In this section, we discuss the wide range of land-use intensity and farm practices in the context of organic vs. conventional farms.

2.1 Land-use intensity in conventional and organic farming

Conventional farming is usually more input intensive than organic farming. The most distinct differences in farm practices between organic and conventional farming are found in plant protection (less harmful products to non-target organisms and non-use of herbicides), fertilisation management (lower nitrogen input and mainly organic fertilisers) and crop rotations. Moreover, organic farms were found to be more likely mixed farms and include grass in the rotational practices (Norton et al. 2009). However, it has been found that land-use intensity can vary widely within both farming systems; that is, the organic and conventional fields could not always be well differentiated by their land-use intensity (Armengot et al. 2011). Therefore, it might be an oversimplification to always equate

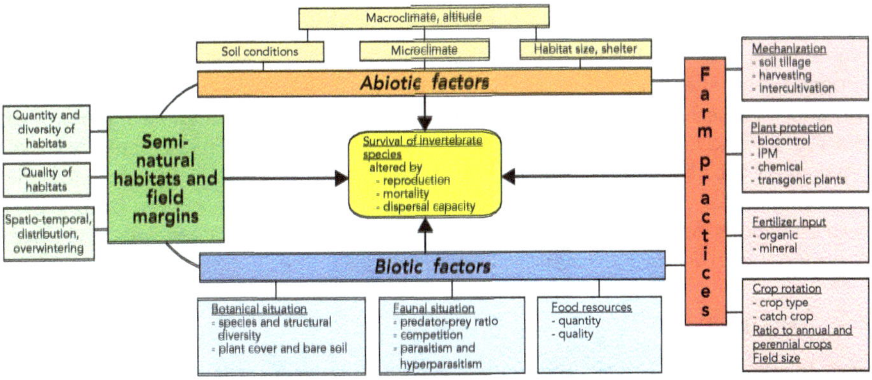

Figure 1 Most relevant factors affecting invertebrate populations at farm level (simplified scheme).

organic farming with low-intensity management or vice versa. For instance, under both farming systems, the specialisation of farms and continuous cultivation are associated with management intensification in crop habitats. Moreover, farming practices and land use are reported to be more heterogeneous and more diversified among organic farmers (Clough et al. 2007a; Armengot et al. 2011), which create a huge gradient in land-use intensity.

2.2 Agricultural practices and biodiversity

Different indexes for assessing land-use intensity and its effect on biodiversity, using mainly agricultural inputs related to land productivity such as fertilisation, pesticide applications and tillage intensity, have been developed (Armengot et al. 2011; Flohre et al. 2011a). Here, we relate specific farming practices and their potential effect on biodiversity, focusing on the main differences between organic and conventional practices. However, the processes and factors relating agriculture and biodiversity are numerous and often with many interactions at different scales. As a result, it is difficult to ascribe a particular biodiversity response to an individual agricultural practice. Our main focus here is on arable crops. However, similar to arable crops, meadow and pasture management has significantly changed in most of Central Europe transitioning from extensive to intensive farming, which has affected biodiversity (Dumont and Tallowin 2011). For instance, increase in fertiliser inputs or in mowing frequencies has had negative effects on plant diversity (Herben and Huber-Sannwald 2002).

2.2.1 Crop rotations

Crop rotations are not exclusive to organic farmers, but simpler rotations and a higher percentage of monocultures are usually found among conventional farmers (Armengot et al. 2011). For instance, the monitoring of 20 fields for five years after the conversion to organic farming in northern Spain revealed that the number of cultivated crops increased from 1, wheat monoculture, to 17, including ancient and local varieties of cereal and legumes (Chamorro et al. 2014). Crop diversity has been positively associated with the diversity of weeds and arthropods (Stevenson et al. 1997; Billeter et al. 2008). A diverse crop rotation, which introduces different crop growth periods and management practices, might favour a higher species diversity in a temporal scale, or might increase the species pool. And at the same time, mainly in the case of weed species, it might prevent the abundance of very few species. The effects of crop rotation on biodiversity depend upon the choice of crops grown and their sequence within the rotation. Crop rotations might have higher influence on biodiversity and the associated functional traits than other farming practices, for example the intensity of the tillage (Stevenson et al. 1997; Armengot et al. 2016). Despite the benefits of crop rotation for biodiversity, including grass clover meadows and leys, factors such as climate conditions, for example water availability in Mediterranean areas, might hamper the establishment of diverse crop rotations in such regions.

2.2.2 Fertilisation

Further to the well-known environmental problems of high rates of mineral fertilisation applied in conventional agriculture, such as eutrophication of surface water, fertilisation has been associated with a reduction of biodiversity. For instance, high input of nitrogen fertilisers has detrimental effects on species richness of weeds and birds (Billeter et al.

2008; Storkey et al. 2012; Rotchés-Ribalta et al. 2016). The negative effect of fertilisation on weed species is usually indirect. Fertiliser stimulates crop growth, which reduces light availability for weeds and in consequence, the weed growth. Although most of the detrimental effects of fertilisation have been documented when using high inputs of mineral fertilisers, a negative correlation between weed species richness and total external nitrogen fertilisers applied has been observed under organic farming, where only organic fertilisers are allowed (Rotchés-Ribalta et al. 2016; Armengot et al. 2011). This correlation can even be stronger when slurry was applied, since it contains higher and more available forms of nitrogen than other organic solid fertiliser sources (Rotchés-Ribalta et al. 2016).

However, there are some studies in Mediterranean areas showing that fertilisation rates do not have any effect on the richness and biomass of weed species (Garcia-Martin et al. 2007; Guerrero et al. 2010; José-María et al. 2013), contrary to the general tendency observed in Central Europe (Billeter et al. 2008; Storkey et al. 2012). Then, limiting fertiliser applications may have no benefit for weed conservation in Mediterranean areas, which contrast with the conservation programmes launched in temperate regions.

2.2.3 Plant protection

In 2015, the IUCN report 'Worldwide Integrated Assessment of the Impacts of Systemic Pesticides on Biodiversity and Ecosystems' (van Lexmond et al. 2015), authored by 29 independent scientists, assessed over a thousand peer-reviewed studies and concluded that systemic pesticides have serious negative impacts on pollinators and other terrestrial and aquatic invertebrates, amphibians and birds, and on ecosystem functioning and services (Chagnon et al. 2015). The European Academies' Environmental Science Council came to broadly similar conclusions in their comprehensive review (EASAC 2015). However, the risk varies with the toxicity of the particular pesticide to different species and with the level of exposure according to farm practice, and the phenology, behaviour and habitat use of the non-target organisms. Pesticides can have direct sublethal or lethal effects on non-target organisms. Prominent indirect effects of pesticides act on food webs and species composition through removal of prey or important plant food resources. Impact of pesticides on individuals, populations up to community level have been reviewed by Köhler and Triebskorn (2013).

Geiger et al. (2010a) analysed 13 components of intensification and found that insecticides and fungicides have had the most consistent negative effects. Biopesticides, living organisms or products derived from them, have the ability to suppress or reduce pests (Deravel et al. 2014). They are classified according to their origin, that is, microbial, plant or animal, and can be used in both conventional and organic farming (Deravel et al. 2014). Although having the reputation of being less effective than the chemical counterparts, interest in them is growing, especially to counteract the negative effects of the chemical pesticides on biodiversity and health. However, several biopesticides can also be detrimental to different faunal taxa (Ivie et al. 2002; Bahlai et al. 2010).

Weeds are still often recognised as a major constraint for crop production (Milberg and Hallgren 2004). Weed harrowing with flex-tine harrows is one of the most widely used direct methods for controlling weeds in organically managed fields, especially in cereal fields, as an alternative to herbicides. They till the soil surface, uprooting and/or covering weed seedlings with soil. Weed harrowing in organic cereal fields has been proved to be an efficient alternative to herbicides (Armengot et al. 2013). Moreover, harrowing allows the maintenance of high levels of weed diversity. However, weed harrows can cause mechanical damage to the crop, and then factors such as the phenological stage of weeds and crops, soil type and moisture or weather conditions must be carefully assessed before harrowing (Barberi 2002). Moreover, mechanical weed control can cause damage to non-target organisms, such as soil nesting birds. In contrast to mechanical methods, herbicide applications, widely used among conventional farmers, diminish plant species richness and change species composition of non-crop plants (Hyvönen 2007). Moreover, plants outside crop fields can be affected by herbicide drift and runoff from sprayed fields towards adjacent non-target areas (Jobin et al. 1997; Bahlai et al. 2010). The effect of herbicide misplacement and its ecological consequences (e.g. Boutin et al. 2014), as well as the increasing number of cases of herbicide-resistant weeds, which diminish the efficiency of herbicides controlling weeds (Heap 2017), have been widely reported.

2.2.4 Tillage operations

Tillage by mouldboard ploughs, disc plough or spading machine, that is, tools which invert soil layers at a depth of usually 20–35 cm, is traditionally considered a key preventive weed control method for arable crops, especially in organic farming, where herbicides cannot be applied. Conservation tillage techniques, that is, reduced tillage (non-inversion tillage) or no till, are primarily applied in conventional cropping systems and are less accepted and implemented by organic farmers because of the potential of increased weed infestation (Casagrande et al. 2016).

Recent studies on the effect of reduced tillage practices under organic farming in long-term trials have shown higher weed abundance and overall higher weed species richness compared with inversion tillage (Armengot et al. 2016). It has also been reported that the functional traits of weed communities vary according to the type of tillage. For instance, weed communities under reduced tillage were potentially less competitive because they were shorter and had less affinity to nutrients. On the other hand, weed communities under conventional tillage had potentially less seed production and a lower abundance of perennial species (Armengot et al. 2016).

Tillage operations also affect soil fauna such as earthworm populations, which play an important role in many soil functions (Kuntz et al. 2013). Crittenden et al. (2014) found a decrease in earthworm abundance after mouldboard ploughing in both conventional and organic farming. Reducing tillage intensity in conventional farming affected earthworms differently than in organic farming; that is, mean abundance did not differ between reduced tillage and mouldboard farming in conventional farming, but it did under organic farming.

3 Impact of organic farming on biodiversity

The impact of organic farming on biodiversity has been analysed through multiple meta-analyses (Bengtsson et al. 2005; Tuck et al. 2014; Lichtenberg et al. 2017) and review articles (Hole et al. 2005; Gomiero et al. 2011; Rahmann et al. 2011). On-farm biodiversity components depend on farming practices at farm and landscape scales, but this strongly interacts with fine- and coarse-scale variables. Taxa groups respond differently to agricultural practice (Fuller et al. 2005) and to land use at different spatial scales and often at multiple spatial scales (Gabriel et al. 2010; Lichtenberg et al. 2017). The variation found in many studies can be due to (1) different responses of taxa to changes in farm practice (Fuller et al. 2005); (2) the impact of the landscape composition and regional context (Holzschuh et al. 2007; Rundlöf et al. 2008); (3) the variation of conversion time to organic farming (Jonason et al. 2011) and (4) the extent of organic farming in a given landscape (Henckel et al. 2015). Therefore, aspects such as similar or equal site conditions (soil type), similar surroundings (local and regional) and a sufficient long conversion period (depending on colonisation speed of taxa) have to be considered when assessing the impact of farm management on biodiversity. However, it is crucial to separate the impact of farming systems from non-anthropogenic site variables and the surrounding landscape infrastructure. Different metrics such as abundance, local richness and evenness, and regional richness and evenness have been applied to assess community responses to farm practices.

A recent hierarchical meta-analysis of 94 studies comparing organic and conventional farming found that on average, organic farming increased species richness of plants, arthropods, microbes and birds by about 30% (Tuck et al. 2014). They concluded this based on species richness of plants, arthropods, microbes and birds, and this result is robust over the last 30 years of published studies and does not diminish at larger scale. The average effect size and response of organic farming depend on functional guild, taxonomic group and crop type. Plants benefit the most, while arthropods, birds and soil biota [earthworms, microorganisms, arbuscular mycorrhizal fungus (AMF)] show a positive response with higher varying levels within the taxa. Lichtenberg et al. (2017) conducted a global meta-study based on 60 studies and found that organic farming increased arthropod abundance (+45% change), local richness (+19%) and regional richness (+11%), but arthropod communities on organic farms had marginally lower local evenness (−6%) and regional evenness (−8%). The benefits of organic farming are much more pronounced in annual than in perennial crops. Significant differences of organic farming among crop types were found, in cereals (most of studies) and in mixed farming large effects were detected, and moderate positive effect sizes in vegetable crops and orchards and the smallest in grassland habitats (Tuck et al. 2014).

3.1 Flora

3.1.1 Weed flora in arable fields

Weed species, the most predominant flora in agroecosystems, are those that thrive in crop fields without being intentionally sown. The development and spread of agriculture has been associated with the evolution of an adapted and diverse flora (and fauna) (Potts 1997). Consequently, arable weed populations depend on human activity, since they are related to agricultural management and the continuity of farms. Although arable weed species display some biological adaptations to overcome the pressure of agricultural practices, intensive land use has been detrimental to weed communities, which has resulted in reduced arable weed abundance and diversity (Robinson and Sutherland 2002; Storkey et al. 2012; Chamorro et al. 2016).

Compared to conventional farming, organic farming has been proved to increase weed cover, species richness and diversity of both above-ground vegetation and seed bank (Romero et al. 2008; José-Maria and Sans 2011; Armengot et al. 2012; Rader et al. 2014). Moreover, higher relative abundance of rare, nature-value and threatened weed species has been reported in organic fields (van Elsen 2000; Aavik and Liira 2010; Chamorro et al. 2016). When compared with other taxonomic groups, weeds benefited the most from organic farming (Tuck et al. 2014). However, relatively low values of weed richness and biomass have been found in organic fields (José-Maria et al. 2013), and no differences have been found in weed diversity between both farming systems (Weibull and Östman

2003; Winfree et al. 2008), which could be related to the existence of a gradient within organic and conventional management practices (Armengot et al. 2011). Intensification of non-chemical weed management in organic farms has been observed to impact negatively on weed species diversity (van Elsen 2000).

From the nature conservation and service provision point of view, the overall significant increase of abundance and diversity in arable weeds under current organic management might still be insufficient to recover weed populations to the same level as the values that were observed before the widespread agricultural intensification (Chamorro et al. 2016; Richner et al. 2015). And this is especially important for the perennials and non-nitrophilous species, as well as for the characteristic and rare species (Hyvonen 2007). Despite the implementation of less-intensive management practices under organic farming, many arable plants, especially rare weed species, are no longer present due to previous intense management which has exhausted regional and/or local species seed bank pools. Consequently, the reintroduction of weed species has also been explored to increase biodiversity in agroecosystems (Epperlein et al. 2014). In a study evaluating rare species among organic fields, Rotchés-Ribalta et al. (2015) found that their occurrence is more determined by stochastic factors related to the local species pool depending on the history of fields and farms rather than on specific current farming practices. However, the presence of organic farms in the landscape is associated with higher local weed diversity in conventional farmed fields, which may reach similar diversity level to organic fields in the margins and benefit from plant species diversity of organic arable fields (Henckel et al. 2015).

Arable weed community composition and their functional traits also differ between organic and conventional management (Armengot et al. 2012; Rader et al. 2014). For instance, in general, higher abundance of grass species has been reported in conventional fields (Romero et al. 2008; José-Maria et al. 2011) mainly due to the extended use of auxin herbicides in cereal fields to control broadleaved weeds and the presence of tolerant and herbicide-resistant grasses. The higher relative incidence of broadleaved species in organic as opposed to conventional fields suggests that these species are less able to tolerate the intensive weed control measures and denser crop swards of herbicide-treated, heavily fertilised conventional arable fields (Rydberg and Milberg 2000; Hyvönen et al. 2003). Furthermore, legumes are more abundant in organic fields (Romero et al. 2008; José-María et al. 2011; Armengot et al. 2012). This could be related not only to their higher sensitivity to herbicides but also to their recuperation after cessation of mineral fertilisation (van Elsen 2000).

The relative importance of insect-pollinated weeds is also higher in organic fields, mainly related to lower relative cover of grasses, which are mainly wind-pollinated species (Gabriel and Tscharntke 2007; Romero et al. 2008; José-María et al. 2011), but also to the lower availability of pollinators due to the smaller amount of floral diversity and food resources in conventional fields (Holzschuh et al. 2007; Morandin and Winston 2005). The elevated proportion of high insect-pollinated weeds extends the food potential of arthropods and is thus a source of complexity for the agroecosystem (Marshall et al. 2003). Moreover, Chamorro et al. (2016) found that organic management enhanced weed richness for species of interest to birds, pollinators and other invertebrates by a factor of 1.7, 1.6 and 1.8, respectively, and in the weed abundance by a factor of 1.6, 2.4 and 3.6, respectively, over conventional crops.

3.1.2 Flora in grassland

Pallett et al. (2016) monitored biomass production and plant species diversity in grassland of southern England before and after a change from conventional to organic farming for

18 years. Immediately after the conversion, they found a decrease in biomass production, up to 50%, and an increase in plant species richness. In the subsequent years, yields remain stable, reaching 31–66% of the levels recorded before the conversion, whereas species richness increased by 300%. Changes in the productivity were explained by changes in plant composition; that is, high-yielding nitrophilous plants were outcompeted by lower yielding grasses and broadleaved weeds, which showed a trade-off between increasing species richness and decreasing productivity.

However, an 18-year study in Germany, representing the range of gradient in land-use intensity of Central Europe, showed that vascular plant diversity and also yield and fodder quality did not differ between organic and conventional grasslands (Klaus et al. 2013), although arthropod diversity was higher under organic farming. Considering plant diversity, Batáry et al. (2012) found that it was higher in conventional meadows, but the cover of broadleaved weeds was higher in organically managed meadows. Power et al. (2012) also found that organic grasslands contained more insect-pollinated forbs than conventional ones. In vineyards and in grassland strips within them, Nascimbene et al. (2012) reported that organic farming promoted plant species richness and the effect was more pronounced for perennial than for annual species.

3.2 Arthropods

3.2.1 Bees

The conservation of bees is highly dependent on the availability of sufficient food and nesting resources. Moreover, the habitat heterogeneity and intensity of farm practices greatly influence the bee populations. Honeybees, solitary wild bees and bumblebees benefit from the higher coverage and diversity of wild flora in and around organic fields (Gabriel and Tscharntke 2007; Rundlöf et al. 2008). With increasing organic farming areas, populations of wild bees, honeybees and bumblebees may also markedly benefit in the surrounding farmland and semi-natural areas (Holzschuh et al. 2007). Holzschuh et al. (2008) found that an increase in organic cropping in the surrounding landscape from 5% to 20% enhanced bee species richness in fallow strips by 50%, density of solitary bees by 60% and bumble bee density by 150%. Bee species richness and bumble bee density responded strongest to organic cropping in landscape sectors with 500 m radius and solitary bee density to organic cropping in landscape sectors with 250 m radius. Thus, organic agriculture can contribute to a better natural pollination of wild and crop plants in the surrounding environment of organic farms (Gabriel and Tscharntke 2007). However, landscape effects on bees are mainly resource-mediated and altered by farming intensity. Comparing different crops, bees were found to be significantly more abundant and species rich in organic compared with conventional fields as shown in winter wheat (Holzschuh et al. 2007; Clough et al. 2007a) and canola (Morandin and Winston 2005), and were found to have, on-farm scale, higher species richness on organic farms (Forrest et al. 2015). The reproduction rates (more nest cells as a proxy) of a solitary bee (*Osmia lignaria*) can be increased on organic farms in a simple landscape (Williams and Kremen 2007).

The general limited use of pesticides in organic farming may contribute to higher abundances and diversity of wild bees. Many bee species forage also on many flowering crops (e.g. rape, berries, orchards) in the agricultural landscape and in consequence may be exposed to different pesticides. There is growing evidence that the widespread use of systemic insecticides such as neonicotinoids and pyrethroids in Europe has strong effects on the survival of bees (Rundlöf et al. 2015; EASAC 2015). Gill et al. (2012) showed that combined pesticide exposure severely affects individual- and colony-level traits in bees. They detected that chronic exposure of bumblebees to two pesticides (neonicotinoid and pyrethroid) at field-level concentrations impairs natural foraging behaviour and increases worker mortality, leading to significant reductions in brood development and colony success. Furthermore, they found that worker foraging performance, particularly pollen collecting efficiency, was significantly reduced with observed knock-on effects for forager recruitment, worker losses and overall worker productivity. Goulson et al. (2015) found that the decline of bee populations is driven by combined stress from parasites, pesticides and lack of flowers. In addition to the immediate lethal effects of pesticides (e.g. neonicotinoids), a variety of sublethal impacts on the health, the reproduction and the behaviour of honeybees, bumblebees and stingless bees have been detected (Whitehorn et al. 2012; Laycock et al. 2012; EASAC 2015).

3.2.2 Beneficial, polyphagous beetles

Most studies reported a generally higher abundance and some evidence for greater species richness of polyphagous predators such as carabids on organic arable fields (reviewed in Hole et al. 2005; Bengtsson et al. 2005). A few studies found that the landscape infrastructures are more important than the farming system for the species richness of carabids (Weibull and Östman 2003; Purtauf et al. 2005). In contrary, others found little evidence that the observed farming system differences for spiders or carabid beetles can be explained by non-crop habitat differences (hedgerows or quality of non-crop habitats) (Feber et al. 2015). Evidence suggested that the distribution and higher abundance and species richness of carabids in organic arable fields can be explained by the greater plant richness, plant cover and more favourable microclimatic conditions and by omitting pesticides (Kromp et al. 1999; Diehl et al. 2012). Many carabid species especially phyto-zoophagous species such as *Amara* spp. and *Harpalus* spp. benefit from a greater weed flora because they forage on grass and weed plant seeds. Diekötter et al. (2016) found a granivorous carabid species (*Harpalus affinis*) to be three and a half fold more abundant in organic fields and showed the potential of this species to control arable weeds at local and landscape scales. Kromp et al. (1989) found a positive relation between weed density and the abundance of *A. consularis* and *H. rufipes* in organically farmed potato fields. This may explain the higher abundance of not only these phyto-zoophagous species on organic fields, but also xerophilous species. Partly endangered field species and large-size carabids (*Carabus* spp.) may profit from organic farming and a few may benefit from conventional farming as *Loricera pilicornis* (Pfiffner and Luka 2003; Döring and Kromp 2003). Similar effects on species were reported from Purtauf et al. (2005), who found that seven or more species had a preference for either conventional or organic fields. Species preferring conventional fields were mainly hygrophilous, likely due to the dense stands of cereals. However, Feber et al. (2015) found inconsistent farming effects on carabids. The lack of agreement between some studies on the effects of organic farming may also indicate the large range in farm practices applied on-farm (Norton et al. 2009; see Section 2.1). Varying soil types is assumed to be a further confounding factor (Irmler 2003).

Rove beetles (staphylinids), a further important group of polyphagous predators, showed no clear correlation with farming systems. Some studies even found lower abundance and species richness on organic fields (Krooss and Schaefer 1998) or similar species richness (Clough et al. 2007b). In contrast to the carabids, staphylinids may respond much less to the landscape infrastructure and farming intensity, but at a local scale, differences of species richness at edge vs. centre were found (Clough et al. 2007b). In respect to functional groups, there were some differences between farming systems: the abundance and species richness of the functional group of detritivorous staphylinids were higher in the organic fields, but not those of predators or fungivores. This indicates that resource quality and quantity in the fields appear to play a major role for this functional group. Detritivorous staphylinids may profit from the pure organic fertilisation applied in organic fields. Surprisingly, in the study of Clough et al. (2007b), they found no significant effect of the landscape context on the overall staphylinid fauna.

3.2.3 Spiders

Spiders as polyphagous predators play an important role in most cropping systems (Nyffeler and Birkhofer 2017). Some studies reported higher abundance and species richness of spiders under organic farming (Feber et al. 1998; Pfiffner and Luka 2003). The positive impact of organic farming on spider is related to a richer arable flora with greater associated prey items, providing greater structural complexity. Interestingly, highly relevant predators such as wolf spiders (lycosids such as *Pardosa agrestis*, *P. palustris* and *Trochosa ruricola*) are often found to be more abundant in organically arable fields (Glück and Ingrisch 1990; Pfiffner and Luka 2003; Schmidt et al. 2005). Schmidt et al. (2005) found linyphiids (*Oedothorax* spp.) and lycosids (*Pardosa* spp.) to be 88% and 103%, respectively, more abundant in organic arable fields. However, effects of farming system on species richness show a non-consistent pattern. Other more recent studies found that the heterogeneity in the landscape is a key factor influencing the spider fauna (Clough et al. 2005; Schmidt et al. 2005). Therefore, they concluded that the landscape infrastructure as the amount of non-crop habitats as semi-natural habitats and field margins (location within fields) may alter the richness, diversity and abundance of spiders more than the farming system. At field edges, on average three and a half times higher densities were found compared to the field centre, showing the location as significant factor for abundance and species richness (Clough et al. 2005). Feber et al. (2015) found that the farming system effects in the cropped area were probably largely attributable to differences in crop management; reduced inputs of pesticides (herbicides and insecticides) and fertilisers are possible influences, and they found some evidence for a positive effect of non-crop plant species richness in organic fields on hunting spider activity density. The inconsistencies of these results may be partly explained by the grade of interspersion of organic and conventional fields in the landscape. Many spider species are known to rapidly recolonise habitats over greater distances through their aerial mode of dispersal from perennial area to arable fields.

3.2.4 Butterflies

Several studies reported higher abundance and species richness of butterflies on organic arable farms (Feber et al. 1997; Rundlöf and Smith 2006; Rundlöf et al. 2008; Jonason et al. 2011), while Puig-Montserrat et al. (2017) found no differences in Spanish vineyards. The study of Rundlöf and Smith (2006) showed that both organic farming and landscape heterogeneity significantly increased butterfly species richness and abundance. This can be partly explained by variation in local availability of nectar and host-plant resources. They also found a significant interaction between farming practice and landscape heterogeneity, because organic farming only significantly increased butterfly species richness and abundance in homogeneous rather than in heterogeneous landscapes. However, local butterfly species richness can be also positively affected by a large proportion of organic farming in the surrounding landscape, independent of the local farming practice (Rundlöf et al. 2008). Local and landscape farming practices interacted such that the farming practice within fields had a larger effect on butterfly abundance if surrounded by conventionally rather than organically managed fields. But the area of organic farming in the surrounding landscape can have additive (species richness) or an interactive effect (abundance) (Rundlöf et al. 2008). They concluded that the gain in butterfly abundance from organic farming would be largest if farms in regions where biodiversity is threatened by agricultural intensification were converted to organic management, because these are the areas where biodiversity currently is low and where the effect of organic farming on biodiversity is high.

Jonason et al. (2011) examined the effects of time since transition to organic farming on butterfly species richness and abundance. Surveys were conducted in cereal fields and adjacent field margins on 60 farms, 20 conventional and 40 organic, in two regions in Sweden. They found butterfly species richness 20% higher on organic farms and butterfly abundance about 60% higher, compared with conventional farms. The time since organic transition affected butterfly abundance gradually over the 25-year period, resulting in a 100% increase. In contrast, no time since transition effect on butterfly or plant species richness was found, indicating that the main effect took place immediately after the transition to organic farming. The effect of organic farming on diversity was rapid for plant and butterfly species richness, whereas butterfly abundance increased gradually with time since transition. As flower-visiting insects, butterflies gain from the greater abundance and diversity of food plants in crop (e.g. grassland rich in plant species) and non-crop habitats and the lack of systemic pesticides on organic farms (Feber et al. 1997; Gibson et al. 2007; Schader et al. 2008).

3.3 Soil biota

It has been suggested that organic farming has large effects on soil conditions (Sandhu et al. 2010), for example content of organic matter (Gattinger et al. 2012), and its effects on soil organisms are ambiguous and often underestimated. Below-ground biodiversity is an important part of the agroecosystem and interacts in a diverse way with above-ground biodiversity (Poveda et al. 2006). Studies on the interacting effects of local and landscape intensification on the below-ground detritivore community, including bacteria, fungi, earthworms and other soil fauna taxa, are rare. At a long-term experimental site (14 years), Henneron et al. (2015) found a significant increase in abundance and biomass of all soil organisms in organic farming and conservation agriculture compared to conventional fields. Whereas in intensively used grassland, no consistent impact of organic farming was found (Younie and Armstrong 1995). In general, organic farming appears to be more efficient in conserving above-ground than below-ground diversity (Flohre et al. 2011a) and can counteract detrimental effects of agricultural intensification on farmland soil biota.

3.3.1 Earthworms

Populations of earthworms are strongly influenced by different cultural practices, such as soil tillage, use of pesticides, fertilisation and crop rotations (crop residues, leys) (see review: Blouin et al. 2013). Evidence from most comparative studies under arable regime and similar soil conditions shows a general lower density and biomass of earthworms in conventionally farmed soils (Reganold et al. 1993; Pfiffner and Mäder 1997; Scullion et al. 2002; Pfiffner and Luka 2007; Henneron et al. 2015). Organic farming often enhances anecic earthworm species (e.g. *Lumbricus terrestris, Nicodrilus longus, Nicodrilus nocturnus*) abundance and biomass (Pfiffner and Mäder 1997; Pfiffner and Luka 2007; Henneron et al. 2015), and these agroecologically relevant species help to improve soil physical properties through their burrowing activities, such as water infiltration, drainage and aggregate stability (Siegrist et al. 1998). Under low-input arable farming (integrated crop management using no insecticides and fungicides and growth regulators), smaller earthworm populations showing lower biomass, abundance and species richness were found compared to organic fields (Pfiffner and Luka 2007). However, Pelosi et al. (2009) found a direct seeding cropping system with living mulch to increase abundance of anecic and epigeic species compared to organic farming; but over all three investigation years, no consistent pattern was found. Excessive tillage can have a serious negative impact on earthworms (Ernst and Emmerling 2009). Crittenden et al. (2014) found a decrease in earthworm abundance after mouldboard ploughing in both conventional and organic farming. Reducing tillage intensity in conventional farming affected earthworms differently than in organic farming. Mean abundance did not differ between reduced tillage and mouldboard farming in conventional farming, but it did in organic farming. Kuntz et al. (2013) showed that reduced tillage can be an effective approach to benefit earthworms in organic arable cropping on heavy clay soils. Agricultural practices such as soil tillage, fertiliser management and plant protection are assumed to be the main factors influencing earthworm populations. Moreover Flohre et al. (2011b) showed that the landscape context plays a significant role in shaping effects of organic vs. conventional farming on soil biota. Earthworm species richness in simple landscapes was enhanced by organic farming (with enhanced predation), whereas in complex landscapes, conventional farming (with reduced predation) fostered earthworm species richness.

3.3.2 Microbial organisms

Microbial communities play an important role in agroecosystems due to their involvement in many different soil processes and functions. In soils, microbial communities are the key driver for nutrient transformation and release, as well as being directly and indirectly involved in many other ecosystem services such as erosion control by formation of stable soil aggregates and soil structuring or pest and disease regulation.

The soil meta-analysis of Lori et al. (2017) showed that overall organic farming enhances total microbial abundance and activity in agricultural soils on a global scale. They found that organic systems had 32–84% greater microbial biomass carbon, microbial biomass nitrogen, total phospholipid fatty acids and dehydrogenase, urease and protease activities than conventional systems. Exclusively, the metabolic quotient as an indicator for stresses on microbial organisms remained unaffected by the farming systems. The categorical subgroup analysis revealed that crop rotation, the inclusion of legumes in the crop rotation and organic inputs are important farming practices affecting soil microbial community size and activity. Furthermore, they showed that differences in microbial size and activity between organic and conventional farming systems vary as a function of land use (arable, orchards and grassland), plant life cycle (annual and perennial) and climatic zone.

The biomass of soil bacteria and fungi can be negatively affected by conventional compared to organic farming (Glover et al. 2000; Fließbach et al. 2007; Kong et al. 2011; Henneron et al. 2015) with often pronounced effects on community composition (Birkhofer et al. 2012; Hartmann et al. 2014). Fließbach and Mäder (2000) reported that microbial biomass C and N as well as their ratio to the C and N pools were higher in a long-term comparison trial of organic and two conventional treatments (purely mineral and mixed fertilised including manure). Increasing rotational diversity by organic farming may greatly change soil microbial community structure and activity, with positive effects on aggregate formation and soli organic matter accrual (Tiemann et al. 2015). Hartmann et al. (2014) found that organic farming increased richness in soil microbiota compared to conventional purely minerally fertilised fields in a long-term trial. The application and quality of the organic fertilisers were shown as two of the major determinants of microbial diversity. Furthermore Klingen et al. (2002) showed on a functional level that soil pests can be strongly reduced in organic soils than in conventional soils due to a richer fungal fauna. Otherwise, the study of Wickings et al. (2016) in the USA showed that soil microbial traits did not differ between organic farming and conventional farming. Lynch (2015) argued that the soil microbial community composition may be resilient to changes in farming system, and variation in soil type and structure may be more important for microbial soil organisms in general than the farming system itself.

Positive effects on arbuscular mycorrhizal fungal (AMF) communities are reported in several studies with higher species diversity, an increased AMF inoculum in soils and greater crop colonisation (reviewed in Gosling et al. 2006; Oehl et al. 2003). The organic management enhances the diversity of AMF communities which are richer and more diverse and more similar to low-input grassland than conventional fields (Verbruggen et al. 2010).

3.4 Vertebrates

3.4.1 Birds and mammals

Overall, organic farming increases the richness and abundance of birds in arable field agroecosystems (Batáry et al. 2010; Chamberlain et al. 2010; Smith et al. 2010; Fischer et al. 2011a; Beecher et al. 2002) and grassland (Batáry et al. 2010). This could be explained mainly by the reduced food availability and pesticide use in conventional farming (Geiger et al. 2010a,b). Hallmann et al. (2014) also reported declines in insectivorous birds due to high concentrations of neonicotinoids.

However, the effect and interaction with the landscape characteristics is assumed to be much stronger than for other taxa. For instance, the positive effects of organic farming on farmland birds proved to be limited to simplified landscapes (Geiger et al. 2010a). The effectiveness of organic farming and agri-environmental schemes depends on the landscape composition. More farmland birds were observed in areas with more stubble, pasture and green manure crops. Species richness was higher in areas with more pasture. In Finland, Piha et al. (2007) found that landscape structure and agricultural land use were the main drivers for bird assemblage, and organic farming was only favourable to skylark and lapwing, but not to overall density, richness or biomass of birds. Agricultural grasslands strongly determined the bird assemblage. In Sweden, Smith et al. (2010) also found an interaction between farm management and landscape. They reported a positive relationship on the bird richness and organic farming only in homogeneous landscapes, mainly because of the passerine invertebrate feeder birds. However, species richness of non-passerines was positively related to organic farming independent of the landscape complexity. But bird abundance was positively related to landscape heterogeneity but not to farm management. In a Mediterranean context, Morales et al. (2015) reported similar results, as they found that a greater proportion of variance in winter bird community composition was better explained by landscape-related factors compared with field management factors (71% vs 29%, respectively). Species richness of winter birds was favoured by diverse landscapes, containing patches of natural vegetation. However, for the case of skylark, total abundance increased as landscape became more homogeneous and dominated by cereals and arable land. In a study in the Netherlands comparing field-breeding birds on organic and conventional arable farms, densities of most bird species did not differ (Kragten and de Snoo 2008), with the exception of skylark and lapwing. For skylarks, difference in crop types grown in organic and conventional farms was the only factor explaining for the differences in their densities. Rahmann et al. (2006) reported that the number of territories for skylarks increased considerably after the conversion from conventional to organic farming in Germany. On the contrary, the number of yellowhammer territories fluctuated largely in relation to the availability of field margin strips, both on conventional and organic lands.

For small mammals, Fisher et al. (2011b) reported that complex landscapes significantly increased abundance and marginally species richness and diversity, but only in conventional fields, whereas organic farming increased small mammals in simple landscapes. In the case of the hare, the monitoring of Lang and Godt (2009) shows that hares have benefited from the conversion to organic farming. After ten years of conversion, the density of hares increased from 8 to 55 km^{-2}, and main reasons were concluded: a better food supply through high abundance and cover of year-around forages and legume leys and many nesting places around the crops.

3.5 Benefits for rare species

Organic farming is often not sufficient to conserve rare species that require more specific habitat conditions to survive in the agricultural landscape. To preserve rare and endangered species, adapted species protection programmes are frequently necessary. The usual schemes within agri-environmental programmes are not sufficient. But organic farming in combination with non-crop habitats and semi-natural areas could substantially contribute to improving species richness of beneficial arthropods (Pfiffner and Luka 2003). Skylarks, a typical species that have been suppressed through intensification of farming, as well as the now rare lapwings, partridges and whinchats, achieve higher population densities on organically managed farms (NABU 2004; Neumann et al. 2007). Moreover, organic farming has also been proved to increase rare and threatened weed species in arable land (van Elsen 2000; Chamorro et al. 2016).The recent global review of Lichtenberg et al. (2017) shows that organic farming increases abundance and richness of both rare and common arthropods at the local and regional scales. At the local scale, they found an increase of arthropod richness by promoting rare taxa (27% increase) more strongly than common taxa (14% increase), but with weaker effects on evenness. This was mainly due to the promotion of rare species by organic farming. Birkhofer et al. (2014) found that increasing proportions of organically managed land in Central or Northern Europe promote the number of species (carabids and butterflies) that consume plant material, and some of these species are of conservation concern.

3.6 Habitat diversity on organic farms

Next to farming intensity, the proportion of non-crop areas as semi-natural habitats and field margins on the farms is a central factor for preserving biodiversity. Hedges, species and structurally rich meadows, and uncropped field margins, serve as habitat and as a temporary refuge and overwintering site for many invertebrates (Pfiffner and Luka 2000). Comparisons in England (Gibson et al. 2007), Denmark (Aude et al. 2003) and Switzerland (Schader et al. 2008) show that the amount of semi-natural areas and non-crop habitats on organic farms is higher than that on conventional farms. Organic farms had smaller field sizes and higher and wider hedgerows with less gaps and are subject to less frequent management and characterised by a higher field and farm complexity (Norton et al. 2009). Even in such heterogeneous landscape types, the organic farms produced greater field and farm complexity than non-organic farms. Gibson et al. (2007) found in complex landscapes larger areas of semi-natural habitats on organic farms, but no differences in plant abundance or richness in these elements. A comprehensive analysis of all Swiss farms found that on average, organic farms have 22% more semi-natural areas and valuable low-input crop habitats compared to conventional farms with 13%. Organic farms therefore implemented two-thirds more measures than their conventional counterparts. The largest differences were found in extensive and less intensively used meadows as well as hedges, and standard fruit trees in lowlands and hilly regions (Schader et al. 2008).

4 Biodiversity at different spatial scales

Disentangling relationships between biodiversity components at local and regional scales can reveal patterns of community assembly and mechanisms that shape community structure (Gering et al. 2003). The interactive effects of landscape complexity and on-farm management may differ across functional groups that have varying capacity to move across landscapes (Tscharntke et al. 2005; Chaplin-Kramer et al. 2011; Feber et al. 2015). The effect of field or farm management may have greater influence on taxa with low mobility such as earthworms and plants, whereas the landscape-scale management have more influence on the high mobile taxa with different resource requirements, for example syrphids, butterflies, bumblebees and birds. Thus, a dominance of highly mobile species can be found in homogeneous landscapes (Jonason et al. 2011). The larger species pool in complex landscapes and the proximity to source habitats facilitate the (re)colonisation of species.

Several studies suggested that the effects of organic farming were likely to be greatest in simple, intensively managed landscapes (Bengtsson et al. 2005), varying between functional groups; for example pollinators respond stronger to organic farming in simple landscapes than predators (Tuck et al. 2014). The difference in biodiversity between organic and conventional farming generally increased with increasing proportion of arable land in a landscape (Tuck et al. 2014). For instance, the effect of organic farming on predators is greater in intensively managed landscapes, whereas the effect of organic farming on pollinators does not increase much with land-use intensity as shown in the meta-study of Tuck et al. (2014). On the contrary, a global meta-study of Lichtenberg et al. (2017) outlines that effect sizes of arthropods in response to organic farming do not differ between the local and regional scales. Landscape complexity did not compromise the influence of organic farming on arthropod abundance or evenness. Greater richness gains in complex than simple landscapes were driven predominantly by rare taxa. In complex landscapes, organic farming had stronger positive effects on local richness of rare (organic +44%) than of common (organic +21%) arthropod taxa. Organic farming within complex landscapes also increased local abundance and regional richness of rare taxa (+78% and +17%, respectively) to a greater extent than common taxa (+33% and +4%, respectively).

The relative importance of farming intensity and landscape complexity affecting biodiversity of plants also vary depending on the location within the field, which can be attributed to differences in the farm practices. For instance, José-Maria et al. (2010) reported that management is the main factor in explaining differences in plant species richness among field centres, whereas changes at boundaries are mainly due to landscape characteristics (simple vs complex landscapes), and at edges both factors are relevant. The same pattern was found for the weed composition, and the effects of the management and the landscape complexity varied in relation to the different positions within the field. The effect of management increased from the boundary to the centre of the field (boundary: 1.9%; edge: 2.7%; centre: 6%), whereas the percentage explained by the surrounding landscape decreased (14.5%, 13.6%, 10.2%). The comparison of arable weed communities in Mediterranean and temperate agroecosystems revealed very similar patterns in additive diversity components (alfa-, beta- and gamma-diversity) as affected by farming practices, despite substantial differences in species composition across these contrasting regions (Armengot et al. 2012). Diversity components at both farm systems and regional scales (for all weeds and for the different functional groups) showed that farm/region contributed most to overall species richness, indicating substantial differences in weed species composition among farms. The same pattern has been observed for arthropods and birds (Flohre et al. 2011a) and in different regions (Rundlöf et al. 2008). Heterogeneity in weed community composition among organic farms (beta-diversity) has been found to be much higher than that for conventionally managed farms and accounts for the majority of species diversity in agricultural landscapes (Gabriel et al. 2006). The largest contribution of organic farming to overall weed species richness may be related to the less-intensive agricultural practices and higher heterogeneity in management among farms compared with conventional farming (Clough et al. 2007a; Armengot et al. 2011). Flohre et al. (2011a) found that the effects of management intensity at different spatial scales were taxon specific. High farming intensity significantly decreased alpha and beta diversity of weeds at all scales, while beta diversities of carabids were only decreased at the region scale, and bird beta diversities were decreased at the farm and region scales.

The amount of organically managed agricultural land has a beneficial effect and may function as a habitat source for species altering local–regional species pool interactions (Henckel et al. 2015; Lichtenberg et al. 2017). In central western France, Henckel et al. (2015) found that organic farming leads to higher local weed diversity and that the presence of organic farming in an intensely farmed landscape was also associated with higher weed diversity for conventionally farmed fields. This means a significant benefit for the whole landscape and poses a significant public good of organic farming. Conservation management needs to target multiple spatial scales considering the multiple benefits of low-input farming (e.g. organic farming). This results in synergies for a sustainable agriculture protecting environmental resources (e.g. less eutrophication, nutrient loss, agro-chemical input). A more detailed understanding of agricultural intensification effects on diversity patterns of various taxa and at multiple spatial scales is crucial to develop more efficient agri-environmental schemes in agroecosystems.

5 Impact of organic farming on selected functional groups

The impact of organic and conventional farming system on wildlife is different in relation to taxonomic group and crop type but also varies on functional groups. Lichtenberg et al. (2017) showed among functional groups in a global meta-study that organic farming had the greatest effect on pollinators and predators without augmenting herbivore (pest) populations. They found that organic farming increased the abundance (+90%), local richness (+55%) and regional richness (+32%) of pollinator communities, but did not impact pollinator evenness. For predator communities, organic farming increased abundance (+38%) and local richness (+14%), but did not affect regional richness, and lowered local (−9%) and regional (−14%) evenness. For herbivore or detritivore (arthropods), no impact on abundance, local or regional richness, or local or regional evenness was found. Most functional groups such as pollinators, predators, herbivores and producers (plants) are more diverse in organic farming, with the exception of soil microbial decomposers (Tuck et al. 2014). However, Hartmann et al. (2014) found distinct microbial diversity under long-term organic and conventional farming. Organic farming increased richness and decreased evenness of the soil microbiota compared to conventionally farmed fields receiving only mineral fertilisers. The diversity of functional traits in a community shows more evidence in understanding ecosystem services than species richness alone. Functional biodiversity is an important basis for many processes in our ecosystems to function. Agricultural habitats with high functional diversity are shown to better adapt to environmental changes. Critical ecological processes are influenced by the higher biodiversity and larger population densities of various species seen on organic farms. Organic farming can enable improvements for functions and services such as natural pollination, pest control, decomposition and reduction of erosion (see Table 1).

5.1 Biological pest control

Biological pest control on organic farms relies on maintaining healthy populations of pest predators and parasitoids. The higher diversity of flora and fauna also encourages beneficial organisms that 'naturally' reduce pests (Zehnder et al. 2007). Organic farming can lead to a significantly more balanced number of beneficial insects, which reduce pests and yield losses in potato crops as shown by Crowder et al. (2010). Higher predator species evenness on organic farms has been shown to translate into increased natural pest control, with the potential to reduce yield gaps compared with conventional agriculture (Crowder et al. 2010). A more diverse flora and fauna found in organic soils (Mäder et al. 2002; Birkhofer et al. 2008; Hartmann et al. 2014) may lead to a stronger reduction in soil pests than in conventional soils due to richer fungal fauna (Klingen et al. 2002). Poveda et al. (2006) found below-ground effects of organic and conventional farming on above-ground plant–herbivore and plant–pathogen interactions. Cereal aphid abundance and fungus infection of *Septoria* were reduced by defaunation but only in organic soils. They suggested that soil organisms in organic farming systems are more important for aphid performance and the infection rate by *Septoria* than in conventional systems. Krauss et al. (2011) found a decreased functional diversity and biological pest control in conventional compared to organic cereal fields. The abundance of cereal aphids was five times lower in organic fields, while predator abundances were three times higher and predator-prey

Table 1 Examples of studies comparing the impact of organic (ORG) versus non-organic (NON) farming on certain ecosystem services and important agroecological processes

Agroecological process	ORG > NON	ORG = NON	NON > ORG
Natural pollination			
i) Annual crops	Andersson et al. (2012)	Brittain et al. (2010)	
	Holzschuh et al. (2008)[a]		
	Morandin and Winston (2005)[a]		
ii) Wild plants	Gabriel and Tscharntke (2007)		
Natural control of organisms			
i) Animals			
Above-ground	Birkhofer et al. (2016)	Winqvist et al. (2011)	Rusch et al. (2015)
Below-ground	Klingen et al. (2002)		
ii) Pathogens	Van Bruggen et al. (2016)	Van Bruggen et al. (2016)	Van Bruggen et al. (2016)
iii) Weeds	Navntoft et al. (2009)	Jonason et al. (2013)	
	Diekötter et al. (2016)		
Decomposition			
i) Organic matter	Dominguez et al. (2014)	Diekötter et al. (2010)	
ii) Dung in pastures	Hutton and Giller (2003)		
Reduction in soil erosion	Erhard and Hartl (2010)		Arnhold et al. (2014)
	Siegrist et al. (1998)		
Carbon sequestration and soil organic matter	Gattinger et al. (2012)		

[a] Abundance and diversity of pollinators.

ratios twenty times higher in organic fields, indicating a significantly higher potential for biological pest control in organic fields. Insecticide treatment in conventional fields had only a short-term effect on aphid densities, while later in the season, aphid abundances were even higher and predator abundances lower in treated compared with untreated conventional fields. The data of this study indicate that insecticide treatment kept aphid predators at low abundances throughout the season, thereby significantly reducing top-down control of aphid populations. A Swedish study showed that organic farming can positively affect the biological control of aphids and yields in spring barley independent of landscape complexity (Birkhofer et al. 2016). Lower aphid densities and higher aphid predation by spiders were determined in organic fields compared to conventional fields.

Macfadyen et al. (2009) analysed 20 food webs containing 374 plant, 370 herbivore and 193 parasitoid species on organic and conventional farms in England. They showed that organic farms have significantly more species at three trophic levels (plant, herbivore and parasitoid) and significantly different network structures. Herbivores on organic farms were attacked by more parasitoid species on organic farms than on conventional farms.

However, differences in network structure did not translate into differences in robustness to simulated species loss, and they found no difference in percentage parasitism across a variety of host species.

A study in California comparing conventional with organic tomato fields showed a higher natural enemy abundance and greater species richness in the organic tomato field (Letourneau and Goldstein 2001). There was no significant difference for any type of damage to tomato foliage or fruit, showing that the organic system achieves the same levels of pest control without having to apply synthetic chemical pesticides. In a Europe-wide study in eight West and East European countries, Geiger et al. (2010a) found important negative effects of agricultural intensification on wild plant, carabid and bird species diversity and on biological pest control – estimated from the number of aphids taken by predators. Of the 13 measured farm practices of intensification, they found that the use of insecticides and fungicides had consistent negative effects on biodiversity. Insecticides also reduced the biological control potential. Organic farming and other agri-environment schemes aiming to mitigate the negative effects of intensive farming on biodiversity increased the diversity of wild plant and carabid species, but – contrary to their expectations – not the diversity of breeding birds. Geiger et al. (2010a) concluded that despite decades of European policy to ban harmful pesticides, the negative effects of pesticides on wild plant and animal species persist, at the same time reducing the opportunities for biological pest control.

Mixed effects of organic farming and landscape complexity on biological control were found in another Europe-wide study on 153 farms in five European regions (Winqvist et al. 2011). Predation on aphids was highest in organic fields in complex landscapes and declined with increasing landscape homogeneity. The biological control potential in conventional fields was not affected by landscape complexity, and in homogenous landscapes, it was higher in conventional fields than in organic fields. These mixed results stress the importance of taking both local management and regional landscape complexity into account when developing efficient agri-environment schemes. Both local farming practices and landscape-scale intensification (quality and structure of surrounding landscape) are assumed to affect communities of natural enemies (Roschewitz et al. 2005; Feber et al. 2015).

5.2 Pollination

Wild pollinators such as solitary bees, bumblebees and hoverflies are responsible for the majority of pollination services. A British study showed that the UK honeybee population only provides at most one-third of pollination services, with the remainder being supplied by wild pollinators (Breeze et al. 2011). Another study showed that flower-visiting wild bees and hoverflies enhance fruit set of crops even where honeybees are frequent (Garibaldi et al. 2011). The pollination by managed honeybees supplements, rather than substitutes for pollination by wild insects, was also demonstrated in a global study comparing the pollination services provided by honeybees and other wildflower visitors in 41 crop systems worldwide (Garibaldi et al. 2013). Insect pollination, needed for 75% of all the food crops, is estimated to be worth about 10% of the economic value of the world's entire food production (Gallai et al. 2009). Pollinators appear to be strongly declining globally in both diversity and abundance (Potts et al. 2010). Among functional groups, the meta-analysis of Tuck et al. (2014) detected largest effect size of organic farming for pollinators compared to herbivores, predators, producers and decomposers. There are some studies indicating the potential of organic farms for pollination services in different cropping systems. Insect-pollinated plants of the arable flora benefit disproportionately from organic farming, which appeared to be related to higher pollinator densities in organic fields (Gabriel and Tscharntke 2007). However, Kehinde and Samways (2012) found in the Cape Floristic Region of South Africa that species richness of monkey beetles, but not the species richness of bees, benefited from organic compared to conventional vineyard management. Monkey beetles are among the most important pollinator guilds as they help to sustain the endemic vegetation of this region. The absence of pollination often limits seed production, and therefore it may greatly simplify the plant community structure. Hence, this disruption of plant–pollinator interactions due to agricultural intensification may cause important shifts in plant community structure. Crop and non-crop habitats are strongly connected via bee foraging at the landscape scale. Holzschuh et al. (2008) detected an important crop to non-crop relationship; bees that depend on nest sites in fallow strips benefited from the more abundant flower resources provided by broadleaved weeds in surrounding organic crop fields. Thus, the incorporation of organic field crop into conventionally managed agricultural landscapes can provide food resources needed to sustain greater pollinator species richness in non-crop habitats and may help to improve pollination. The relative importance of habitat type and farming system varied among functional groups (pollinators, predators, parasitoids), suggesting that their perception of crop–non-crop boundaries or the availability of their food resources differed (Holzschuh et al. 2010). Interestingly, habitat connectivity appeared to be enhanced by both higher edge densities and organic field management. Andersson et al. (2012) found that pollination success (yield quantity and quality) and the proportion of fully pollinated berries were higher on organic compared to conventional farms, and this difference was already evident 2–4 years after conversion to organic farming. Changes in land-use intensity and habitat quality can result in changed trophic interactions, but the directions of changes are often not generally predictable (Holzschuh et al. 2010). Organic farming can substantially help to sustain pollination services by generalist bee species in agricultural landscapes, and further measures are required to conserve more specialised bees by high-value semi-natural habitats and tailored agri-environmental schemes.

6 Future trends and conclusion

The benefits of organic farming for biodiversity are greatly context dependent and vary among taxa groups. However, organic agriculture aims at sustaining healthy soils, ecosystem services through intra-crop biodiversity and diversity in non-crop habitats and site-specific farming intensity. Given the environmental risks of intensive agriculture, farming practices should be modified to decrease the risks. Organic farming can be a biodiversity-friendly alternative with less input, but novel concepts and individual farming practices need to be improved by ecological intensification to meet higher productivity. Further growth of the organic farming sector could contribute in mitigating the further loss of biodiversity in agricultural landscapes.

At the same time, diversified small farms are increasingly being replaced by large, simplified and intensive production systems (Tscharntke et al. 2005; Bennett et al. 2012). This is problematic because intensified farming impacts the long-term sustainability of agroecosystems, threatening long-term global food security (Ray et al. 2012). Mitigating ecological harm from agricultural intensification through farm management should also support the long-term sustainability of agricultural systems and ecological resilience (Oliver et al. 2015; Lichtenberg et al. 2017). Biodiversity can help maintain stability of ecosystem processes through time and processes such as response diversity and functional redundancy (Cardinale et al. 2006). The biodiversity benefits of organic farming could provide insurance against the loss of ecosystem services within crop fields by supporting multiple species that perform similar functional roles (Elmqvist et al. 2003).

Resilient systems must also simultaneously exhibit multiple functions (multifunctionality). Functional redundancy alone insufficiently supports multifunctionality under changing environmental conditions (Isbell et al. 2011). An increase of species richness may contribute to multifunctionality including the promotion of rare taxa. Rare taxa are increasingly being found to be critical for multifunctionality and even single ecosystem processes (Soliveres et al. 2016). Organic farming can take a leading role and shows many ways with inspiring approaches to link food production whilst supporting biodiversity and fundamental ecosystem services.

The role of biodiversity for multiple ecosystem services is not yet fully understood. The loss of biodiversity could have dramatic consequences for the agroecosystems. There is still a substantial lack of knowledge about different animal taxa (mammals, amphibians and arthropod herbivores and detritivores) in the context of different farming systems and about species occurrence of high nature conservation relevance. A multidimensional biodiversity research approach, including from the field to the landscape perspective, and considering the functional aspects, is then necessary to mitigate the contemporary mass extinction of flora and fauna.

Further research testing and developing farming systems aiming to conserve biodiversity and reducing degradation of functional biodiversity on-farm is also needed. They should be based on extensive experiences of whole farm approach of organic agriculture. One of the current challenges is how to make a transition to a broad-scale adoption of more sustainable farming practices while avoiding big yield losses. A key issue is the reduction of agrochemical inputs in agricultural landscapes, in order to avoid the decline of ecosystem services and at the same time maximising the potential of crop yields. A sound strategy towards ecological intensification of the systems is necessary, in which knowledge on multiple management inputs and multiple ecosystem services should help to both maximise biodiversity conservation and ecosystem services and crop yields.

7 References

Aavik, T. and Liira, J. 2010. Quantifying the effect of organic farming, field boundary type and landscape structure on the vegetation of field boundaries. *Agriculture, Ecosystem and Environment*, 135, 178–86.

Andersson, G. K., Rundlöf, M. and Smith, H. G. 2012. Organic farming improves pollination success in strawberries. *PLoS ONE*, 7(2), e31599.

Anon. 2017. The IFOAM norms for organic production and procession. Version 2014. https://www.ifoam.bio/en/ifoam-norms.

Armengot, L., Jose-Maria, L., Blanco-Moreno, J. M., Bassa, M., Chamorro, L. and Sans, F. X. 2011. A novel index of land use intensity for organic and conventional farming of Mediterranean cereal fields. *Agronomy for Sustainable Development*, 31, 699–707.

Armengot, L., Sans, F. X., Fischer, C., Flohre, A., Jose-Maria, L., Tscharntke, T. and Thies, C. 2012. The β-diversity of arable weed communities on organic and conventional cereal farms in two contrasting regions. *Applied Vegetation Science*, 15, 571–9.

Armengot, L., Jose-Maria, L., Chamorro, L. and Sans, F. 2013. Weed harrowing in organically grown cereal crops avoids yield losses without reducing weed diversity. *Agronomy for Sustainable Development*, 33, 405–11.

Armengot, L., Blanco-Moreno, J. M., Bàrberi, P., Bocci, G., Carlesi, S., Aendekerk, R., Berner, A., Celette, F., Grosse, M., Huiting, H., Kranzler, A., Luik, A., Mäder, P., Peigné, J., Stoll, E., Delfosse, P., Sukkel, W., Surböck, A., Westaway, S. and Sans, F. X. 2016. Tillage as a driver of change in weed communities: A functional perspective. *Agriculture, Ecosystems & Environment*, 222, 276–85.

Arnhold, S., Lindner, S., Lee, B., Martin, E., Kettering, J., Nguyen, T. T., Koellner, T., Ok, Y. S. and Huwe, B. 2014. Conventional and organic farming: Soil erosion and conservation potential for row crop cultivation. *Geoderma*, 219, 89–105.

Aude, E., Tybirk, K. and Bruus Pedersen, M. 2003. Vegetation diversity of conventional and organic hedgerows in Denmark. *Agriculture, Ecosystems, & Environment*, 99, 135–47.

Bahlai, C. A., Xue, Y., McCreary, C. M., Schaafsma, A. W. and Hallett, R. H. 2010. Choosing organic pesticides over synthetic pesticides may not effectively mitigate environmental risk in soybeans. *PLoS ONE*, 5, 1–7.

Barberi, P. 2002. Weed management in organic agriculture: Are we addressing the right issues? *Weed Research*, 42, 177–93.

Batáry, P., Matthiesen, T. and Tscharntke, T. 2010. Landscape-moderated importance of hedges in conserving farmland bird diversity of organic vs. conventional croplands and grasslands. *Biological Conservation*, 143, 2020–7.

Batáry, P., Holzschuh, A., Orci, K. M., Samu, F. and Tscharntke, T. 2012. Responses of plant, insect and spider biodiversity to local and landscape scale management intensity in cereal crops and grasslands. *Agriculture, Ecosystems & Environment*, 146(1), 130–6.

Beecher, N. A., Johnson, R. J., Brandle, J. R., Case, R. M. and Young, L. J. 2002. Agroecology of birds in organic and nonorganic farmland. *Conservation Biology*, 16(6), 1620–31.

Bengtsson, J., Ahnström, J. and Weibull, A. C., 2005. The effects of organic agriculture on biodiversity and abundance: A meta-analysis. *Journal of Applied Ecology*, 42, 261–9.

Bennett, A. J., Bending, G. D., Chandler, D., Hilton, S. and Mills, P. 2012. Meeting the demand for crop production: The challenge of yield decline in crops grown in short rotations. *Biological Reviews*, 87, 52–71.

Billeter, R., Liira, J., Bailey, D., Bugter, R., Arens, P., Augenstein, I., Aviron, S., Baudry, J., Bukacek, R., Burel, F., Cerny, M., De Blust, G., De Cock, R., Diekötter, T., Dietz, H., Dirksen, J., Dormann, C., Durka, W., Frenzel, M., Hamersky, R., Hendrickx, F., Herzog, F., Klotz, S., Koolstra, B., Lausch, A., Le Coeur, D., Maelfait, J. P., Opdam, P., Roubalova, M., Schermann, A., Schermann, N., Schmidt, T., Schweiger, O., Smulders, M. J. M., Speelmans, M., Simova, P., Verboom, J., van Wingerden, W. K R. E., Zobel, M. and Edwards, P. J. 2008. Indicators for biodiversity in agricultural landscapes: a pan-European study. *Journal of Applied Ecology*, 45(1), 141–50. doi: 10.1111/j.1365-2664.2007.01393.x.

Birkhofer, K., Bezemer, M. T., Bloem, J., Bonkowski, M., Christensen, S., Dubois, D., Ekelund, F., Fliessbach, A., Gunst, L., Hedlund, K., Mäder, P., Mikola, J., Robin, C., Setälä, H., Tatin-Froux, F., van der Putten, W. H. and Scheu, S. 2008. Long-term organic farming fosters below and aboveground biota: Implications for soil quality, biological control and productivity. *Soil Biology & Biochemistry*, 40(9), 2297–308.

Birkhofer, K., Bezemer, T. M., Hedlund, K. and Setälä, H. 2012. Community composition of soil organisms under different wheat-farming systems. In: Cheeke, T. E., Coleman, D. C. and Wall, D. H. (Eds), *Microbial Ecology in Sustainable Agroecosystems*, pp. 89–111. Boca Raton, FL: CRC Press.

Birkhofer, K., Ekroos, J., Corlett, E. B. and Smith, H. G. 2014. Winners and losers of organic cereal farming in animal communities across Central and Northern Europe. *Biological Conservation*, 175, 25–33.

Birkhofer, K., Arvidsson, F., Ehlers, D., Mader, V. L., Bengtsson, J. and Smith, H. G. 2016. Organic farming affects the biological control of hemipteran pests and yields in spring barley independent of landscape complexity. *Landscape Ecology*, 31(3), 567–79.

Blouin, M., Hodson, M. E., Delgado, E., Baker, G., Brussaard, L., Butt, K. R., Dai, J., Dendooven, L., Pérès, G., Tondoh, J., Cluzeau, D. and Bru, J.-J. 2013. A review of earthworm impact on soil function and ecosystem services. *European Journal of Soil Science*, 64, 161–82.

Bockstaller, C., Guichard L., Makowski D., Aveline A., Girardin P. and Plantureux S. 2008. Agri-environmental indicators to assess cropping and farming systems. A review. *Agronomy for Sustainable Development*, 28, 139–49.

Bosshard, A., Reinhard, B. R. and Taylor S. (Eds). 2009. *Guide to Biodiversity and Landscape Quality in Organic Agriculture*. Bonn, Germany: IFOAM.

Boutin, C., Strandberg, B., Carpenter, D., Mathiassen, S. K. and Thomas, P. J. 2014. Herbicide impact on non-target plant reproduction: What are the toxicological and ecological implications? *Environmental Pollution*, 185, 295–306.

Breeze, T. D., Bailey, A. P., Balcombe, K. G. and Potts, S. G. 2011. Pollination services in the UK: How important are honeybees? *Agriculture, Ecosystems & Environment*, 142, 137–43.

Brittain, C., Bommarco, R., Vighi, M., Settele, J. and Potts, S. G. 2010. Organic farming in isolated landscapes does not benefit flower-visiting insects and pollination. *Biological Conservation*, 143(8), 1860–7.

Cardinale, B. J., Srivastava, D. S., Duffy, J. E., Wright, J. P., Downing, A. L., Sankaran, M. and Jouseau, C. 2006. Effects of biodiversity on the functioning of trophic groups and ecosystems. *Nature*, 443, 989–92.

Casagrande, M., Peigné, J., Payet, V., Mäder, P., Sans, F. X., Blanco-Moreno, J. M., Antichi, D., Bàrberi, P., Beeckman, A., Bigongiali, F., Cooper, J., Dierauer, H., Gascoyne, K., Grosse, M., Heß, J., Kranzler, A., Luik, A., Peetsmann, E., Surböck, A., Willekens, K. and David, C. 2016. Organic farmers' motivations and challenges for adopting conservation agriculture in Europe. *Organic Agriculture*, 4, 281–95.

Chagnon, M., Kreutzweiser, D., Mitchell, E. A., Morrissey, C. A., Noome, D. A. and Van der Sluijs, J. P. 2015. Risks of large-scale use of systemic insecticides to ecosystem functioning and services. *Environmental Science and Pollution Research*, 22(1), 119–34.

Chamberlain, D. E., Joys, A., Johnson, P. J., Norton, L., Feber, R. E. and Fuller, R. J. 2010. Does organic farming benefit farmland birds in winter? *Biology Letters*, 6, 82–4.

Chamorro, L., Armengot, L., José-María, L. and Sans, F. X. 2014. Organic farming enhances the recovery of ancient crops and segetal weeds in Catalonia (NE of Spain). In: Rahmann, G. and Aksoy, U. (Eds), *Proceedings of the 4th ISOFAR Scientific Conference*. 'Building Organic Bridges', at the Organic World Congress 2014, 13–15 October, Istanbul, Turkey (eprint ID 23844).

Chamorro, L., Masalles, R. M. and Sans, F. X. 2016. Arable weed decline in Northeast Spain: Does organic farming recover functional biodiversity? *Agriculture, Ecosystems & Environment*, 223, 1–9.

Chaplin-Kramer, R., O'Rourke, M. E., Blitzer, E. J. and Kremen, C. 2011. A meta-analysis of crop pest and natural enemy response to landscape complexity. *Ecology Letters*, 14, 922–32.

Clough, Y., Kruess, A., Kleijn, D. and Tscharntke, T. 2005. Spider diversity in cereal fields: Comparing factors at local, landscape and regional scales. *Journal of Biogeography*, 32, 2007–14.

Clough, Y., Holzschuh, A., Gabriel, D., Purtauf, T., Kleijn, D., Kruess, A., Deweter, I.-S. and Tscharntke, T. 2007a. Alpha and beta diversity of arthropods and plants in organically and conventionally managed wheat fields. *Journal of Applied Ecology*, 44, 804–12.

Clough, Y., Kruess, A. and Tscharntke, T. 2007b. Organic versus conventional arable farming systems: Functional grouping helps understand staphylinid response. *Agriculture, Ecosystems & Environment*, 118, 285–90.

Crittenden, S. J., Eswaramurthy, T., de Goede, R. G. M., Brussaard, L. and Pulleman, M. M. 2014. Effect of tillage on earthworms over short- and medium-term in conventional and organic farming. *Applied Soil Ecology*, 83, 140–8.

Crowder, D. W., Northfield, T. D., Strand. M. and Snyder, W. E. 2010. Organic agriculture promotes evenness and natural pest control. *Nature*, 46, 109–12.

Deravel, J., Krier, F. and Jacques, P. 2014. Biopesticides, a complementary and alternative approach to the use of agrochemicals. *Biotechnologie Agronomie Societé et Environnement*, 18, 220–32.

Diehl, E., Wolters, V. and Birkhofer, K. 2012. Arable weeds in organically managed wheat fields foster carabid beetles by resource-and structure-mediated effects. *Arthropod-Plant Interactions*, 6, 75–82.

Diekötter, T., Wamser, S., Wolters, V. and Birkhofer, K. 2010. Landscape and management effects on structure and function of soil arthropod communities in winter wheat. *Agriculture, Ecosystems & Environment*, 137(1), 108–12.

Diekötter, T., Wamser, S., Dörner, T., Wolters, V. and Birkhofer, K. 2016. Organic farming affects the potential of a granivorous carabid beetle to control arable weeds at local and landscape scales. *Agricultural and Forest Entomology*, 18, 167–73.

Dirzo, R., Young, H. S., Galetti, M., Ceballos, G., Isaac, N. J. and Collen, B. 2014. Defaunation in the Anthropocene. *Science*, 345(6195), 401–6.

Domínguez, A., Bedano, J. C., Becker, A. R. and Arolfo, R. V. 2014. Organic farming fosters agroecosystem functioning in Argentinian temperate soils: Evidence from litter decomposition and soil fauna. *Applied Soil Ecology*, 83, 170–6.

Döring, T. F. and Kromp, B. 2003. Which carabid species benefit from organic agriculture? – a review of comparative studies in winter cereals from Germany and Switzerland. *Agriculture, Ecosystems & Environment*, 98, 153–61.

Dumont, B. and Tallowin, J. R. B. 2011. Chapter 14: Interactions between grassland management and species diversity. In: Lemaire, G., Hodgson, J. and Chabbi, A. (Eds), *Grassland Productivity and Ecosystem Services*, pp. 129–37. Wallingford, UK: CAB International.

EASAC. 2015. Ecosystem services, agriculture and neonicotinoids. Policy report 26, European Academies, Science Advisory Board (EASAC), 70p. Available at www.easac.eu.

Elmqvist, T., Folke, C., Nyström, M., Peterson, G., Bengtsson, J., Walker, B. and Norberg, J. 2003. Response diversity, ecosystem change, and resilience. *Frontiers in Ecology and the Environment*, 1, 488–94.

Epperlein, L. R. F., Prestele J. W., Albrecht, H. and Kollmann, J. 2014. Reintroduction of a rare arable weed: Competition effects on weed fitness and crop yield. *Agriculture, Ecosystems & Environment*, 188, 57–62.

Erhart, E. and Hartl, W. 2010. Soil protection through organic farming: A review. In: Lichtfouse, E. (Ed.), *Organic Farming, Pest Control and Remediation of Soil Pollutants*, pp. 203–26. Dordrecht, the Netherlands: Springer.

Ernst, G. and Emmerling, C. 2009. Impact of five different tillage systems on soil organic carbon content and the density, biomass, and community composition of earthworms after a ten year period. *European Journal of Soil Biology*, 45, 247–51.

Feber, R. E., Firbank, L. G., Johnson, P. J. and Macdonald, D. W. 1997. The effects of organic farming on pest and non-pest butterfly abundance. *Agriculture, Ecosystems & Environment*, 64(2), 133–9.

Feber, R. E., Bell, J. R., Johnson, P. J., Firbank, L. G. and Macdonald, D. W. 1998. The effects of organic farming on surface-active spider (Araneae) assemblages in wheat in Southern England, UK. *The Journal of Arachnology*, 26, 190–202.

Feber, R. E., Johnson, P. J., Bell, J. R., Chamberlain, D. E., Firbank, L. G., Fuller, R. J., Manley, W., Mathews, F., Norton, L. R., Townsend, M. and Macdonald, D. W. 2015. Organic farming: Biodiversity impacts can depend on dispersal characteristics and landscape context. *PLoS ONE*, 10(8), e0135921.

Fischer, C., Flohre, A., Clement, L. W., Batáry, P., Weisser, W. W., Tscharntke, T. and Thies, C. 2011a. Mixed effects of landscape structure and farming practice on bird diversity. *Agriculture, Ecosystems & Environment*, 141, 119–25.

Fischer, C., Thies, C. and Tscharntke, T. 2011b. Small mammals in agricultural landscapes: Opposing responses to farming practices and landscape complexity. *Biological Conservation*, 144, 1130–6.

Fließbach, A. and Mäder, P. 2000. Microbial biomass and size-density fractions differ between soils of organic and conventional agricultural systems. *Soil Biology and Biochemistry*, 32, 757–68.

Fließbach, A., Oberholzer, H. R., Gunst, L. and Mäder, P. 2007. Soil organic matter and biological soil quality indicators after 21 years of organic and conventional farming. *Agriculture, Ecosystems & Environment*, 118, 273–84.

Flohre, A., Fischer, C., Aavik, T., Bengtsson, J., Berendse, F., Bommarco, R., Ceryngier, P., Clement, L. W., Dennis, C. and Eggers, S. 2011a. Agricultural intensification and biodiversity partitioning in European landscapes comparing plants, carabids, and birds. *Ecological Applications*, 21, 1772–81.

Flohre, A., Rudnick, M., Traser, G., Tscharntke, T. and Eggers, T. 2011b. Does soil biota benefit from organic farming in complex vs. simple landscapes? *Agriculture, Ecosystems & Environment*, 141, 210–14.

Forrest, J. R. K., Thorp, R. W., Kremen, C. and Williams, N. M. 2015. Contrasting patterns in species and functional-trait diversity of bees in an agricultural landscape. *Journal of Applied Ecology*, 52, 706–15.

Fuller, R. J., Norton, L. R., Feber, R. E., Johnson, P. J., Chamberlain, D. E., Joys, A. C., Mathews, F., Stuart, R. C., Townsend, M. C., Manley, W. J., Wolfe, M. S., Macdonald, D. W. and Firbank, L. G. 2005. Benefits of organic farming to biodiversity vary among taxa. *Biology Letters*, 1, 431–4.

Gabriel, D. and Tscharntke, T. 2007. Insect pollinated plants benefit from organic farming. *Agriculture, Ecosystems & Environment*, 118, 43–8.

Gabriel, D., Roschewitz, I., Tscharntke, T. and Thies, C. 2006. Beta diversity at different spatial scales: Plant communities in organic and conventional agriculture. *Ecological Applications*, 16, 2011–21.

Gabriel, D., Sait, S. M., Hodgson, J. A., Schmutz, U., Kunin, W. E. and Benton, T. G. 2010. Scale matters: The impact of organic farming on biodiversity at different spatial scales. *Ecology Letters*, 13(7), 858–69.

Gallai, N., Salles, J. M., Settele, J. and Vaissière, B. E. 2009. Economic valuation of the vulnerability of world agriculture confronted with pollinator decline. *Ecological Economics*, 68(3), 810–21.

Garcia-Martin, A., Lopez-Bellido, R. J. and Coleto, J. M. 2007. Fertilisation and weed control effects on yield and weeds in durum wheat grown under rain-fed conditions in a Mediterranean climate. *Weed Research*, 47, 140–8.

Garibaldi, L. A., Steffan-Dewenter, I., Kremen, C., Morales, J. M., Bommarco, R., Cunningham, S. A. and Klein, A. M. 2011. Stability of pollination services decreases with isolation from natural areas despite honey bee visits. *Ecology Letters*, 14, 1062–72.

Garibaldi, L. A., Steffan-Dewenter, I., Winfree, R., Aizen, M. A., Bommarco, R., Cunningham, S. A. and Klein, A. M. 2013. Wild pollinators enhance fruit set of crops regardless of honey bee abundance. *Science*, 339, 1608–11.

Gattinger, A., Muller, A., Haeni, M., Skinner, C., Fliessbach, A., Buchmann, N., Mäder, P., Stolze, M., Smith, P., Scialabba, N. E.-H. and Niggli, U. 2012. Enhanced top soil carbon stocks under organic farming. *Proceedings of the National Academy of Sciences*, 109, 18226–31.

Geiger, F., Bengtsson, J., Berendse F., Weisser, W. W., Emmerson, M., Morales, M. B., Ceryngier, P., Liira, J., Tscharntke, T., Winqvist, C., Eggers, S., Bommarco, R., Pärt, T., Bretagnolle, V., Plantegenest, M., Clement, L. W., Dennis, C., Palmer, C., Oñate, J. J., Guerrero, I., Hawro, V., Aavik, T., Thies, C., Flohre, A., Hänke, S., Fischer, C., Goedhart, P. W. and Inchausti, P. 2010a. Persistent negative effects of pesticides on biodiversity and biological control potential on European farmland. *Basic and Applied Ecology*, 11, 97–105.

Geiger, F., de Snoo, C. R., Berendse, F., Guerrero, I., Morales, M. B., Oñate, J. J., Eggers, S., Pärt, T., Bommarco, R., Bengtsson, J., Clement, L. W., Weisser, W. W., Olszewski, A., Ceryngier, P., Hawro, V., Inchausti, P., Fischer, C., Flohre, A., Thies, C. and Tscharntke, T. 2010b. Landscape composition influences farm management effects on farmland birds in winter: A pan-European approach. *Agriculture, Ecosystems & Environment*, 139, 571–7.

Gering, J. C., Crist, T. O. and Veech, J. A. 2003. Additive partitioning of species diversity across multiple spatial scales: Implications for regional conservation of biodiversity. *Conservation Biology*, 17, 488–99.

Gibson, R. H., Pearce, S., Morris, R. J., Symondson, W. O. C. and Memmott, J. 2007. Plant diversity and land use under organic and conventional agriculture: A whole-farm approach. *Journal of Applied Ecology*, 44, 792–803.

Gill, R. J., Ramos-Rodriguez, O. and Raine, N. E. 2012. Combined pesticide exposure severely affects individual- and colony-level traits in bees. *Nature*, 491, 105–9.

Glover, J. D., Reganold, J. P. and Andrews, P. K. 2000. Systematic method for rating soil quality of conventional, organic, and integrated apple orchards in Washington State. *Agriculture, Ecosystem & Environment*, 80, 29–45.

Glück, E. and Ingrisch, S. 1990. The effect of bio-dynamic and conventional agriculture management on Erigoninae and Lycosidae spiders. *Journal of Applied Entomology*, 110, 136–48.

Gomiero, T., Pimentel, D. and Paoletti, M. G. 2011. Environmental impact of different agricultural management practices: Conventional vs. organic agriculture. *Critical Reviews in Plant Sciences*, 30(1–2), 95–124.

Gosling, P., Hodge, A., Goodlass, G. and Bending, G. D. 2006. Arbuscular mycorrhizal fungi and organic farming. *Agriculture, Ecosystem & Environment*, 113, 17–35.

Goulson, D., Nicholls, E., Botías, C. and Rotheray, E. L. 2015. Bee declines driven by combined stress from parasites, pesticides, and lack of flowers. *Science*, 347, 1255957.

Guerrero, I., Martinez, P., Morales, M. B. and Onate, J. J. 2010. Influence of agricultural factors on weed, carabid and bird richness in a Mediterranean cereal cropping system. *Agriculture Ecosystems & Environment*, 138, 103–8.

Hallmann, C. A., Foppen, R. P. B., van Turnhout, C. A. M., de Kroon, H. and Jongejans, E. 2014. Declines in insectivorous birds are associated with high neonicotinoid concentrations. *Nature*, 511(7509), 341–3. doi:10.1038/nature13531

Hartmann, M., Frey, B., Mayer, J., Mäder, P. and Widmer, F. 2014. Distinct soil microbial diversity under long-term organic and conventional farming. *International Society for Microbial Ecology Journal*, 9(5): 1177–94.

Heap, I. 2017. The international survey of herbicide resistant weeds. Available at www.weedscience.org (Accessed on 1 July 2017).

Henckel, L., Börger, L., Meiss, H., Gaba, S. and Bretagnolle, V. 2015. Organic fields sustain weed metacommunity dynamics in farmland landscapes. *Proceedings of the Royal Society B Biological Science*, 282, 1808.

Henneron, L., Bernard, L., Hedde, M., Pelosi, C., Villenave, C., Chenu, C., Bertrand, M., Girardin, C. and Blanchart, E. 2015. Fourteen years of evidence for positive effects of conservation agriculture and organic farming on soil life. *Agronomy for Sustainable Development*, 35, 169–81.

Herben, T. and Huber-Sannwald, E. 2002. Effects of management on species richness of grasslands: Sward scale processes lead to large-scale patterns. In: Durand, J. L., Emile, J. C., Huyghe, C. and Lemaire, G. (Eds), *Proceedings of the 19th General Meeting of the European Grassland Federation on Multifunction Grasslands, Quality Forages, Animal Products and Landscapes*. Reading, UK: British Grassland Society, pp. 635–43.

Hole, D. G., Perkins, A. J., Wilson, J. D., Alexander, I. H., Grice, P. V. and Evans, A. D. 2005. Does organic farming benefit biodiversity? *Biological Conservation*, 122, 113–30.

Holzschuh, A., Stefan-Dewenter, I., Kleijn, D. and Tscharntke, T. 2007. Diversity of flower-visiting bees in cereal fields: Effects of farming system, landscape composition and regional context. *Journal of Applied Ecology*, 44, 41–9.

Holzschuh, A., Stefan-Dewenter, I. and Tscharntke, T. 2008. Agricultural landscapes with organic crops support higher pollinator diversity. *Oikos*, 117, 354–61.

Holzschuh, A., Steffan-Dewenter, I. and Tscharntke, T. 2010. How do landscape composition and configuration, organic farming and fallow strips affect the diversity of bees, wasps and their parasitoids? *Journal of Animal Ecology*, 79, 491–500.

Hooper, D. U., Adair, E. C., Cardinale, B. J., Byrnes, J. E., Hungate, B. A., Matulich, K. L. and O'Connor, M. I. (2012). A global synthesis reveals biodiversity loss as a major driver of ecosystem change. *Nature*, 486(7401), 105–8.

Hutton, S. A. and Giller, P. S. 2003. The effects of the intensification of agriculture on northern temperate dung beetle communities. *Journal of Applied Ecology*, 40, 994–1007.

Hyvönen, T. 2007. Can conversion to organic farming restore the species composition of arable weed communities? *Biological Conservation*, 137, 382–90.

Hyvönen,T., Ketoja, E., Salonen, J., Jalli, H. and Tiainen, J. 2003. Weed species diversity and community composition in organic and conventional cropping of spring cereals. *Agriculture, Ecosystems & Environment*, 97, 131–49.

Irmler, U. 2003. The spatial and temporal pattern of carabid beetles on arable fields in northern Germany (Schleswig-Holstein) and their value as ecological indicators. *Agriculture, Ecosystems & Environment*, 98, 141–51.

Isbell, F., Calcagno, V., Hector, A., Connolly, J., Harpole, W. S., Reich, P. B., Scherer-Lorenzen, M., Schmid, B., Tilman, D., Van Ruijven, J., Weigelt, A., Wilsey, B. J., Zavaleta, E. S. and Loreau, M. 2011. High plant diversity is needed to maintain ecosystem services. *Nature*, 477, 199–202.

Ivie, M. A., Pollock, D. A., Gustafson, D. L., Ivie, L. L., Rasolomandimby, J. and Swearingen, W. D. 2002. Field-based evaluation of biopesticide impacts on native biodiversity: Malagasy coleoptera and anti-locust entomopathogenic fungi. *Journal of Economic Entomology*, 95, 651–60.

Jobin, B., Boutin, C. and Des Granges, J. L. 1997. Effects of agricultural practices on the flora of hedgerows and woodland edges in southern Quebec. *Canadian Journal of Plant Science*, 77(2), 293–99.

Jonason, D., Andersson, G. K., Öckinger, E., Rundlöf, M., Smith, H. G. and Bengtsson, J. 2011. Assessing the effect of the time since transition to organic farming on plants and butterflies. *Journal of Applied Ecology*, 48(3), 543–50.

Jonason, D., Smith, H. G. and Birkhofer, K. 2013. Landscape simplification promotes weed seed predation by carabid beetles (Coleoptera:Carabidae). *Landscape Ecology*, 28, 487–94.

José-Maria, L. and Sans, F. X. 2011. Weed seedbanks in arable fields: Effects of management practices and surrounding landscape. *Weed Research*, 51, 631–40.

Jose-Maria, L., Armengot, L., Blanco-Moreno, J. M., Bassa, M. and Sans, F. X. 2010. Effects of agricultural intensification on plant diversity in Mediterranean dryland cereal fields. *Journal of Applied Ecology*, 47, 832–40.

José-Maria, L., Blanco-Moreno, J. M., Armengot, L. and Sans, F. X. 2011. How does agricultural intensification modulate changes in plant community composition? *Agriculture, Ecosystems & Environment*, 145, 77–84.

José-María, L., Armengot, L., Chamorro, L. and Sans, F. X. 2013. The conservation of arable weeds at crop edges of barley fields in northeast Spain. *Annals of Applied Biology*, 163, 47–55.

Kehinde, T. and Samways, M. J. 2012. Endemic pollinator response to organic vs. conventional farming and landscape context in the Cape Floristic Region biodiversity hotspot. *Agriculture, Ecosystems & Environment*, 146, 162–7.

Klaus, V. H., Kleinebecker, T., Prati, D., Gossner, M. M., Alt, F., Boch, S., Gockel, S., Hemp, A., Lange, M. and Müller, J. 2013. Does organic grassland farming benefit plant and arthropod diversity at the expense of yield and soil fertility? *Agriculture, Ecosystems & Environment*, 177, 1–9.

Klingen, I., Eilenberg, J. and Meadow, R. 2002. Effects of farming system, field margins and bait insect on the occurrence of insect pathogenic fungi in soils. *Agriculture, Ecosystems & Environment*, 91, 191–8.

Köhler, H. R. and Triebskorn, R. 2013. Wildlife ecotoxicology of pesticides: Can we track effects to the population level and beyond? *Science*, 341(6147), 759–65.

Kong, A. Y., Scow, K. M., Córdova-Kreylos, A. L., Holmes, W. E. and Six, J. 2011. Microbial community composition and carbon cycling within soil microenvironments of conventional, low-input, and organic cropping systems. *Soil Biology and Biochemistry*, 43(1), 20–30.

Kragten, S. and de Snoo, R. G. 2008. Field-breeding birds on organic and conventional arable farms in the Netherlands. *Agriculture, Ecosystems & Environment*, 126, 270–4.

Krauss, J., Gallenberger, I. and Steffan-Dewenter, I. 2011. Decreased functional diversity and biological pest control in conventional compared to organic crop fields. *PLoS ONE*, 6, e19502.

Kromp, B. 1989. Carabid beetle communities (Carabidae, Coleoptera) in biologically and conventionally farmed agroecosystems. *Agriculture, Ecosystems & Environment*, 27, 241–51.

Kromp, B. 1999. Carabid beetles in sustainable agriculture: A review on pest control efficacy, cultivation impacts and enhancement. *Agriculture Ecosystems & Environment*, 74, 187–228.

Krooss, S. and Schaefer, M. 1998. The effect of different farming systems on epigeic arthropods: A five-year study on the rove beetle fauna (Coleoptera: Staphylinidae) of winter wheat. *Agriculture, Ecosystems & Environment*, 69, 121–33.

Kuntz, M., Berner, A.,Gattinger, A., Scholberg, J. M., Mäder, P. and Pfiffner, L. 2013. Influence of reduced tillage on earthworm and microbial communities under organic arable farming. *Pedobiologia*, 56, 251–60.

Lang, J. and Godt, J. 2009. Profitiert der Feldhase vom ökologischen Landbau? 10. Wissenschaftstagung Ökologischer Landbau, Zürich. *Band*, 1, 434–7.

Laycock, I., Lenthall, K. M., Barratt, A. T. and Cresswell, J. E. 2012. Effects of imidacloprid, a neonicotinoid pesticide, on reproduction in worker bumble bees (Bombus terrestris). *Ecotoxicology*, 21, 1937–45.

Letourneau, K. and Goldstein, B. 2001. Pest damage and arthropod community structure in organic vs. conventional tomato production in California. *Journal of Applied Ecology*, 38, 557–70.

Lichtenberg, E. M., Kennedy, C. M., Kremen, C., Batáry, P., Berendse, F., Bommarco, R., Bosque-Pérez, N. A., Carvalheiro, L. G., Snyder, W. E., Williams, N. M., Winfree, R., Klatt, B. K., Åström, S., Benjamin, F., Brittain, C., Chaplin-Kramer, R., Clough, Y., Danforth, B., Diekötter, T., Eigenbrode, S. D., Ekroos, J., Elle, E., Freitas, B. M., Fukuda, Y., Gaines-Day, H. R., Grab, H., Gratton, C., Holzschuh, A., Isaacs, R., Isaia, M., Jha, S., Jonason, D., Jones, V. P., Klein, A.-M., Krauss, J., Letourneau, D. K., Macfadyen, S., Mallinger, R. E., Martin, E. A., Martinez, E., Memmott, J., Morandin, L., Neame, L., Otieno, M., Park, M. G., Pfiffner, L., Pocock, M. J. O., Ponce, C., Potts, S. G., Poveda, K., Ramos, M., Rosenheim, J. A., Rundlöf, M., Sardiñas, H., Saunders, M. E., Schon, N. L., Sciligo, A. R., Sidhu, C. S., Steffan-Dewenter, I., Tscharntke, T., Veselý, M., Weisser, W. W., Wilson, J. K. and Crowder, D. W. 2017. A global synthesis of the effects of diversified farming systems on arthropod diversity within fields and across agricultural landscapes. *Global Change Biology*, 23, 4946–57.

Lori, M., Symnaczik, S., Mäder, P., De Deyn, G. and Gattinger, A. 2017. Organic farming enhances soil microbial abundance and activity – A meta-analysis and meta-regression. *PLoS ONE*, 12(7), e0180442.

Lynch, D. H. 2015. Nutrient cycling and soil health in organic cropping systems— importance of management strategies and soil resilience. *Sustainable Agriculture Research*, 4, 80–8.

Macfadyen, S., Gibson, R., Polaszek, A., Morris, R., Craze, P., Planqué, R., Symondson, W. and Memmott, J. 2009. Do differences in food web structure between organic and conventional farms affect the ecosystem service of pest control? *Ecology Letters*, 12, 229–38.

Mäder, P., Fließbach, A., Dubois, D., Gunst, L., Fried, P. and Niggli, U. 2002. Soil fertility and biodiversity in organic farming. *Science*, 296, 1694–97.

Marshall, E. J. P., Brown, V. K., Boatman, N. D., Lutman, P. J. W., Squire, G. R. and Ward, L. K. 2003. The role of weeds in supporting biological diversity within crop fields. *Weed Research*, 43, 77–89.

Milberg, P. and Hallgren, E. 2004. Yield loss due to weeds in cereals and its large-scale variability in Sweden. *Field Crop Research*, 86, 199–209.

Morales, M. B., Oñate, J. J., Guerrero, I. and Meléndez, L. 2015. Influence of landscape and field-level agricultural management on a Mediterranean farmland winter bird community. *Ardeola*, 62(1), 49–65.

Morandin, L. A. and Winston, M. L. 2005. Wild bee abundance and seed production in conventional, organic, and genetically modified canola. *Ecological Applications*, 15, 871–81.

NABU. 2004. *Vögel der Agrarlandschaft – Bestand, Gefährdung, Schutz*. Berlin: Naturschutzbund Deutschland e.V., p. 44.

Nascimbene, J., Marini, L. and Paoletti, M. 2012. Organic farming benefits local plant diversity in vineyard farms located in intensive agricultural landscapes. *Environmental Management*, 49, 1054–60.

Navntoft, S., Wratten, S. D., Kristensen, K. and Esbjerg, P. 2009. Weed seed predation in organic and conventional fields. *Biological Control*, 49, 11–16.

Neumann, H., Loges, R. and Taube, F. 2007. Fördert der ökologische Landbau die Vielfalt und Häufigkeit von Brutvögeln auf Ackerflächen? *Berichte über Landwirtschaft*, 85, 272–99.

Newbold, T., Hudson, L. N., Arnell, A. P., Contu, S., De Palma, A., Ferrier, S., Hill, S. L. L., Hoskins, A. J., Lysenko, I. and Phillips, H. R. P. 2016. Has land use pushed terrestrial biodiversity beyond the planetary boundary? A global assessment. *Science*, 353, 288–91.

Norton, L., Johnson, P., Joys, A., Stuart, R., Chamberlain, D., Feber, R., Firbank, L., Manley, W., Wolfe, M., Hart, B., Mathews, F., Macdonald, D. and Fuller, R. J. 2009. Consequences of organic and non-organic farming practices for field, farm and landscape complexity. *Agriculture, Ecosystems & Environment*, 129, 221–7.

Nyffeler, M. and Birkhofer, K. 2017. An estimated 400–800 million tons of prey are annually killed by the global spider community. *The Science of Nature*, 104(3–4), 30.

Oehl, F., Sieverding, E., Ineichen, K., Mäder, P., Boller, T. and Wiemken, A. 2003. Impact of intensity on the species diversity of Arbuscular Mycorrhizal fungi in agroecosystems of Central Europe. *Applied and Environmental Microbiology*, 69, 2816–24.

Oliver, T. H., Isaac, N. J. B., August, T. A., Woodcock, B. A., Roy, D. B. and Bullock, J. M. 2015. Declining resilience of ecosystem functions under biodiversity loss. *Nature Communications*, 6, 10122.

Pallett, D. W., Pescott, O. L. and Schafer, S. M. 2016. Changes in plant species richness and productivity in response to decreased nitrogen inputs in grassland in southern England. *Ecological Indicators*, 68, 73–81.

Pelosi, C., Bertrand, M. and Roger-Estrade, J. 2009. Earthworm community in conventional, organic and direct seeding with living mulch cropping systems. *Agronomy for Sustainable Development*, 29, 287–95.

Pfiffner, L. and Luka, H. 2000. Overwintering of arthropods in soils of arable fields and adjacent semi-natural habitats. *Agriculture, Ecosystems & Environment*, 78, 215–22.

Pfiffner, L. and Luka, H. 2003. Effects of low-input farming systems on carabids and epigeal spiders – a paired farm approach. *Basic and Applied Ecology*, 4, 117–27.

Pfiffner, L. and Luka, H. 2007. Earthworm populations in two low-input cereal farming systems. *Applied Soil Ecology*, 37, 184–91.

Pfiffner, L. and Mäder, P. 1997. Effects of biodynamic, organic and conventional production systems on earthworm populations. *Entomological Research in Organic Agriculture*, 15, 3–10.

Piha, M., Tiainen, J., Holopainen, J. and Vepsäläinen, V. 2007. Effects of land-use and landscape characteristics on avian diversity and abundance in a boreal agricultural landscape with organic and conventional farms. *Biological Conservation*, 140, 50–61.

Ponisio, L. C., M'Gonigle, L. K., Mace, K. C., Palomino, J., Valpine, P. de and Kremen, C. 2015. Diversification practices reduce organic to conventional yield gap. *Proceedings of the Royal Society of London. Series B, Biological Sciences*, 282, 20141396.

Potts, G. R. 1997. Cereal farming, pesticides and grey partridges. In: Pain, D. J. and Pienkowski, M. W. (Eds), *Farming and Birds in Europe*, pp. 150–77. London, UK: Academic Press.

Potts, S. G., Biesmeijer, J. C., Kremen, C., Neumann, P., Schweiger, O. and Kunin, W. E. 2010. Global pollinator declines: Trends, impacts and drivers. *Trends in Ecology & Evolution*, 25, 345–53.

Poveda, K., Steffan-Dewenter, I., Scheu, S. and Tscharntke, T. 2006. Belowground effects of organic and conventional farming on aboveground plant–herbivore and plant–pathogen interactions. *Agriculture, Ecosystems & Environment*, 113, 162–7.

Power, E. F., Kelly, D. L. and Stout, J. C. 2012. Organic farming and landscape structure: Effects on insect-pollinated plant diversity in intensively managed grasslands. *PLoS ONE*, 7, e38073.

Puig-Montserrat, X., Stefanescu, C., Torre, I., Palet, J., Fàbregas, E., Dantart, J., Arrizabalaga, A. and Flaquer, C. 2017. Effects of organic and conventional crop management on vineyard biodiversity. *Agriculture, Ecosystems & Environment*, 243, 19–26.

Purtauf, T., Roschewitz, I., Dauber, J., Thies, C., Tscharntke, T. and Wolters, V. 2005. Landscape context of organic and conventional farms: Influences on carabid beetle diversity. *Agriculture Ecosystems & Environment*, 108, 165–74.

Rader, R., Birkhofer, K., Schmucki, R., Smith, H. G., Stjernman, M. and Lindborg, R. 2014. Organic farming and heterogeneous landscapes positively affect different measures of plant diversity. *Journal of Applied Ecology*, 51, 1544–53.

Rahmann, G. 2011. Biodiversity and organic farming: What do we know? *Agriculture and Forestry Research*, 3, 189–208.

Rahmann, G., Paulsen, H., Hotker, H. Jeromin, K., Schrader, S., Haneklaus, S. and Schnug, E. 2006. Contribution of organic farming to conserving and improving biodiversity in Germany – the example avi-fauna. *Aspects of Applied Biology*, 79, 187.

Ray, D. K., Ramankutty, N., Mueller, N. D., West, P. C. and Foley, J. A. 2012. Recent patterns of crop yield growth and stagnation. *Nature Communications*, 3, 1293.

Reganold, J. P. and Wachter, J. M. 2016. Organic agriculture in the twenty-first century. *Nature Plants*, 2, 15221.

Reganold, J. P., Palmer, A. S., Lockhart, J. C. and Macgregor, A. N. 1993. Soil quality and financial performance of biodynamic and conventional farms in New Zealand. *Science*, 260, 344–9.

Richner, N., Holderegger, R., Linder, H. P. and Walter, T. 2015. Reviewing change in the arable flora of Europe: A meta-analysis. *Weed Research*, 55, 1–13.

Robinson, R. A. and Sutherland W. J. 2002. Post-war changes in arable farming and biodiversity in Great Britain. *Journal of Applied Ecology*, 39, 157–76.

Romero, A., Chamorro, L. and Sans, F. X. 2008. Weed diversity in crop edges and inner fields of organic and conventional dryland winter cereal crops in NE Spain. *Agriculture, Ecosystems & Environment*, 124, 97–104.

Roschewitz, I., Gabriel, D., Tscharntke, T. and Thies, C. 2005. The effects of landscape complexity on arable weed species diversity in organic and conventional farming. *Journal of Applied Ecology*, 42, 873–82.

Rotchés-Ribalta, R., Blanco-Moreno, J. M., Armengot, L., José-María, L. and Sans, F. X. 2015. Which conditions determine the presence of rare weeds in arable fields? *Agriculture, Ecosystems & Environment*, 203, 55–61.

Rotchés-Ribalta, R., Blanco-Moreno, J. M., Armengot, L. and Sans, F. X. 2016. Responses of rare and common segetal species to wheat competition and fertiliser type and dose. *Weed Research*, 56, 114–23.

Rundlöf, M. and Smith, H. G. 2006. The effect of organic farming on butterfly diversity depends on landscape context. *Journal of Applied Ecology*, 43, 1121–7.

Rundlöf, M., Nilsson, H. and Smith, H. G. 2008. Interacting effects of farming practice and landscape context on bumble bees. *Biological Conservation*, 141, 417–26.

Rundlöf, M., Andersson, G. K. S., Bommarco, R., Fries, I., Hederström, V., Herbertsson, L., Jonsson, O., Klatt, B. K., Pedersen, T. R., Yourstone, J. and Smith, H. G. 2015. Seed coating with a neonicotinoid insecticide negatively affects wild bees. *Nature*, 521(7550), 77–80.

Rusch, A., Delbac, L., Muneret, L. and Thiéry, D. 2015. Organic farming and host density affect parasitism rates of tortricid moths in vineyards. *Agriculture, Ecosystems & Environment*, 214, 46–53.

Rydberg, N. T. and Milberg, P. 2000. A survey of weeds in organic farming in Sweden. *Biological Agriculture and Horticulture*, 18, 175–85.

Sandhu, H. S., Wratten, S. D. and Cullen, R. 2010. Organic agriculture and ecosystem services. *Environmental Science & Policy*, 13(1), 1–7.

Schader, C., Pfiffner, L., Schlatter, C. and Stolze, M. 2008. Umsetzung von Ökomassnahmen auf Bio- und ÖLN-Betrieben. *Agrarforschung*, 15, 506–11.

Schmidt, M. H., Roschewitz, I., Thies, C. and Tscharntke, T. 2005. Differential effects of landscape and management on diversity and density of ground-dwelling farmland spiders. *Journal of Applied Ecology*, 42, 281–7.

Scullion, J., Neal, S. and Philipps, L. 2002. Comparison of earthworm populations and cast properties in conventional and organic arable rotations. *Soil Use and Management*, 18, 293–300.

Siegrist, S., Schaub, D., Pfiffner, L. and Mäder, P. 1998. Does organic agriculture reduce soil erodibility? The results of a long-term field study on loess in Switzerland. *Agriculture, Ecosystems & Environment*, 69(3), 253–64.

Smith, H. G., Danhardt, J., Lindstrom, A. and Rundlof, M. 2010. Consequences of organic farming and landscape heterogeneity for species richness and abundance of farmland birds. *Oecologia*, 162, 1071–9.

Soliveres, S., Manning, P., Prati, D., Gossner, M. M., Alt, F., Arndt, H., Baumgartner, V., Binkenstein, J., Birkhofer, K., Blaser, S. and Blüthgen, N. 2016. Locally rare species influence grassland ecosystem multifunctionality. *Philosophical Transactions of the Royal Society B*, 371(1694), 20150269.

Stevenson, F. C., Legere, A., Simard, R. R., Angers, D. A., Pageau, D. and Lafond, J. 1997. Weed species diversity in spring barley varies with crop rotation and tillage, but not with nutrient source. *Weed Science*, 45, 798–806.

Stockdale, E. A., Lampkin, N. H., Hovi, M., Keatinge, R., Lennartsson, E. K. M., Macdonald, D. W., Padel, S., Tattersall, F. H., Wolfe, M. S. and Watson, C. A. 2001. Agronomic and environmental implications of organic farming systems. *Advances in Agronomy*, 70, 261–327.

Stoeckli, S., Birrer, S., Zellweger-Fischer, J., Balmer, O., Jenny, M. and Pfiffner, L. 2017. Quantifying the extent to which farmers can influence biodiversity on their farms. *Agriculture, Ecosystems & Environment*, 237, 224–33.

Storkey, J., Meyer, S., Still, K. S. and Leuschner, C. 2012. The impact of agricultural intensification and land-use change on the European arable flora. *Proceedings of the Royal Society B-Biological Sciences*, 279, 1421–9.

Tiemann, L. K., Grandy, A. S., Atkinson, E. E., Marin-Spiotta, E. and McDaniel, M. D. 2015. Crop rotational diversity enhances belowground communities and functions in an agroecosystem. *Ecology Letter*, 18, 761–71.

Tscharntke, T., Klein, A. M., Kruess, A., Steffan-Dewenter, I. and Thies, C. 2005. Landscape perspectives on agricultural intensification and biodiversity – ecosystem service management. *Ecology Letters*, 8, 857–74.

Tuck, S. L., C. Winqvist, F. Mota, J. Ahnström, L. A. Turnbull and J. Bengtsson. 2014. Land-use intensity and the effects of organic farming on biodiversity: A hierarchical meta-analysis. *Journal of Applied Ecology*, 51, 746–55.

van Bruggen, A. H., Gamliel, A. and Finckh, M. R. 2016. Plant disease management in organic farming systems. *Pest Management Science*, 72, 30–44.

van Elsen, T. 2000. Species diversity as a task for organic agriculture in Europe. *Agriculture, Ecosystems & Environment*, 77, 101–9.

van Lexmond, M. B., Bonmatin, J. M., Goulson, D. and Noome, D. A. 2015. Worldwide integrated assessment on systemic pesticides. *Environmental Science and Pollution Research*, 22, 1–4.

Verbruggen, E., Roling, W. F., Gamper, H. A., Kowalchuk, G. A., Verhoef, H. A. and van der Heijden, M. G. 2010. Positive effects of organic farming on below-ground mutualists: Large-scale comparison of mycorrhizal fungal communities in agricultural soils. *New Phytology*, 186, 968–79.

Weibull, A.-C. and Östman, Ö. 2003. Species composition in agroecosystems: The effect of landscape, habitat, and farm management. *Basic and Applied Ecology*, 4, 349–61.

Whitehorn, P. R., O'Connor, S., Wäckers, F. L. and Goulson, D. 2012. Neonicotinoid pesticide reduces bumble bee colony growth and queen production. *Science*, 336, 351–2.

Wickings, K., Grandy, A. S. and Kravchenko, A. N. 2016. Going with the flow: Landscape position drives differences in microbial biomass and activity in conventional, low input, and organic agricultural systems in the Midwestern U. S. *Agriculture, Ecosystems & Environment*, 218, 1–10.

Williams, N. M. and Kremen, C. 2007. Resource distributions among habitats determine solitary bee offspring production in a mosaic landscape. *Ecological Applications*, 17, 910–21.

Winfree, R., N. M. Williams, H. Gaines, J. S. Ascher and C. Kremen. 2008. Wild bee pollinators provide the majority of crop visitation across land-use gradients in New Jersey and Pennsylvania, USA. *Journal of Applied Ecology*, 45, 793–802.

Winqvist, C., Bengtsson, J., Aavik, T., Berendse, F., Clement, L. W., Eggers, S., Fischer, C., Flohre, A., Geiger, F., Liira, J. and Pärt, T. 2011. Mixed effects of organic farming and landscape complexity on farmland biodiversity and biological control potential across Europe. *Journal of Applied Ecology*, 48, 570–79.

Younie, D. and Armstrong, G. 1995. Botanical and invertebrate diversity in organic and intensively fertilised grassland. In: Isart, J. and Llerana, J. J. (Eds), *Land Use and Biodiversity: The Role of Organic Farming*, pp.35–44. Barcelona, Spain: Multitext.

Zehnder, G., Gurr, G. M., Wade, M. R., Wratten, S. D. and Wyss, E. 2007. Arthropod pest management in organic crops. *Annual Review of Entomology*, 52, 57–82.

Chapter 17

The impact of organic agriculture on diet and health

Kirsten Brandt, Newcastle University, UK

1 Introduction
2 Associations between organic consumption and diet choices
3 Organic foods and health
4 Combined impact on health of organic foods in diet
5 Future trends
6 Conclusion
7 References

1 Introduction

The impact of organic agriculture on diet and health comprises a network of several interdependent factors, some of which are addressed in other chapters in this book. The present chapter focuses on the health of consumers who actively prefer organic foods and drinks, as compared with consumers who are just as happy to consume conventional products.

This impact primarily occurs through two different factors (see Fig. 1): The first factor is consumer behaviour, the associations of choosing organic products with a range of behaviours such as high intake of vegetables, low intake of fizzy drinks and spending more time on exercise. The second factor is product composition, the effect of organic production methods on the composition of a particular food, such as pesticide residues or phytochemicals, which may in some cases affect the health of those who consume this food.

Each set of factors and relationships has been studied separately, while a few studies have attempted to cover additional relationships as well. Due to the many different foods that make up a diet and the general difficulties of studying the effects of food composition on human health, a lot more research is required before the relative importance of every part is fully understood and quantified.

http://dx.doi.org/10.19103/AS.2017.0029.19

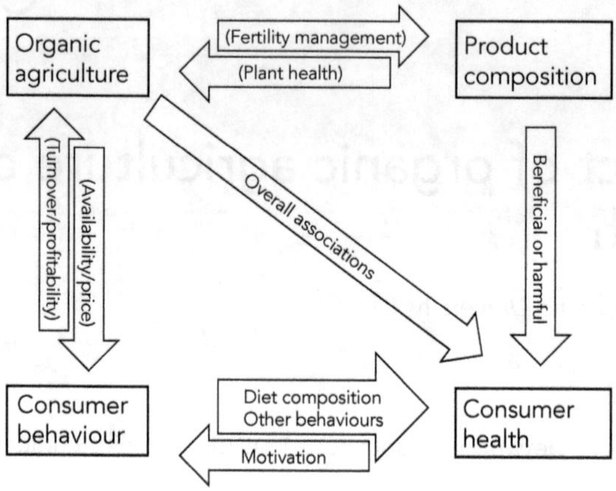

Figure 1 Interrelations between the production system organic agriculture and two key factors for consumer health: product composition and consumer behaviour. The present chapter focuses on the relations directly impacting consumer health, while other relations (in brackets) are only outlined briefly as background information.

All published studies of the 'overall association' (between organic food and human health) show that consumers of organic food score better for a range of health indicators than otherwise comparable consumers of conventional food, such as reduced occurrence of type 2 diabetes, high blood pressure and high cholesterol (Baudry et al., 2015), and less pre-eclampsia in pregnant women (Torjusen et al., 2014). Studies on children of mothers who prefer organic food have found fewer cases of undescended testes in newborn boys (Christensen et al., 2013; Brantsaeter et al., 2016) and reduced risk of childhood allergies (Alfven et al., 2006; Kummeling et al., 2008; Fagerstedt et al., 2016). However, these types of observational associations do not by themselves prove that the health is improved by the organic food, since consumption of organic food is only one of several indicators of a generally healthy lifestyle, and strongly confounded with other indicators. For example, 'participation in cultural activities' is such a confounded indicator, which also shows similarly consistent positive health associations (Nenonen et al., 2014; Bygren et al., 2009).

Due to this, it is important to study the individual factors and relations, preferably under conditions where such confounders can be controlled, to determine to what extent organic agriculture by itself may have a beneficial impact on diet and health, along with its more cultural role as a recognisable alternative to an unhealthy lifestyle. There are data indicating that consumption of organic vegetables and fruit promotes health more than that of the corresponding conventional produce (Brandt et al., 2011). However, it is also important to recognise the mutually supportive role of these two types of effects. The cultural recognition and support of the general benefits of organic agriculture increase the availability of organic products to consumers who are motivated to improve their diets, possibly facilitating this process of change (Pino et al., 2012). In this way, increased consumption of organic food may improve health among its consumers by more than the sum of the health benefit of each food item on its own.

2 Associations between organic consumption and diet choices

2.1 Diet composition

Studies of consumers of organic foods and drinks consistently show that inclusion of organic products in the diet is correlated with choice of a more healthy diet composition [as defined by, for example, WHO (2017)], that is, in terms of types of foods and drinks. Consumers of organic food are more likely to be vegetarians (Torjusen et al., 2012; Baudry et al., 2015; Petersen et al., 2013; Baudry et al., 2016), and non-vegetarian consumers of organic food have increased intake of vegetables and fruits and reduced intake of processed foods (Baudry et al., 2016; Denver and Christensen, 2015; Bradbury et al., 2014; Petersen et al., 2013; Torjusen et al., 2012) when compared with otherwise similar consumers who do not choose organic foods. This is not only the case for consumers who cook the food themselves; in the case of institutional kitchens such as in schools (He and Mikkelsen, 2014) and worksite canteens (Mikkelsen et al., 2006), the proportion of organic foods is correlated with how well the menu matches official dietary guidelines.

This effect on diet composition has two important consequences for the consumer, affecting their health and economy. The impact on health is not surprising. It is well known that increased consumption of fruits, vegetables and wholegrain foods (in some cases pooled as 'fibre') has substantial health benefits, as well as reductions in fat, salt and sugar, particularly from processed meat and fizzy drinks, while also reducing the risk of non-communicable diseases such as diabetes and cancer (Shay et al., 2012; Kushi et al., 2012). This basically means that a typical or average organic diet is better for health than a similarly average conventional diet, simply because it has a healthier composition in terms of the types of foods and drinks consumed. However, this benefit is not directly an impact of organic farming, since of course a conventional diet can easily have an equally healthy composition using conventional vegetables and so forth. The key question is: Why do most consumers of conventional products not adjust their diets, so they become equally healthy as the organic diets?

This leads to the other impact, on the consumer's economy. It is well known that organically produced foods and drinks are more expensive than conventional ones, everything being equal. Examples of websites with comparisons of consumer prices for a 'standard basket' of organic and conventional foods show average differences of 89% in the United Kingdom (Forster, 2016), 68% in the United States (Pillsbury, 2011) and 59% in Denmark (Swart et al., 2016). What is probably more surprising is that organic diets are often not more expensive than diets for similar families consuming a typical conventional diet (Denver and Christensen, 2015). This is simply caused by the differences in diet composition: vegetables are cheaper than meat, and raw ingredients are cheaper than processed products. So a diet where a lot of the food is purchased as raw vegetables, rice, pasta and so forth and then cooked is cheaper than a diet with a higher content of ready meals, meat and processed foods (Denver and Christensen, 2015). This calculation attempted to take into account neither the time spent cooking, nor the cost of having and using a kitchen (higher rent, increased use of electricity, etc.).

For caterers purchasing their ingredients wholesale, these differences are so large that they can offset much or all of the increased labour cost. The result is that the caterers feel that they are financially pressured to prepare foods with a higher proportion of the cheaper (but healthier) foods (Mikkelsen et al., 2006). In contrast, the consumers who prepare their meals individually or for their family do not generally mention financial pressure as a major reason for choosing a healthier diet composition [although it may still play a role (Denver and Christensen, 2015)]. For consumers it seems to be more a question of identity and health consciousness, that it 'feels wrong' to seek out organic raw materials and then use them to create an unhealthy diet (Pino et al., 2012; Grunert, 2005).

2.2 Other health determinants

Similarly, with diet, also, as in other contexts, consumers of organic food tend to have healthier habits and other circumstances than those who do not make this choice. They are less likely to smoke and more likely to exercise and have had longer education and jobs providing higher income (Baudry et al., 2016; Denver and Christensen, 2015; Bradbury et al., 2014; Torjusen et al., 2010). All of these factors are strongly correlated with most measures of human health, such as cardiovascular disease, cancer and diabetes (Shay et al., 2012; Kavanagh et al., 2010; Kushi et al., 2012), even when not considering organic food.

Obviously, this association is not caused by organic farming; organic food does not contain a 'magic' component that will make you stop smoking or get paid more for your work. Which means that in this case, the cause-and-effect relation must be in the other direction: The same factors, which help some people stop smoking or get promoted to a better paid job, must somehow also be involved in decisions regarding choice of organic foods.

Due to this, some studies have attempted to isolate the part of the association between organic food consumption and consumer health that is specific for the organic food itself. By recording all the other health indicators, such as smoking, social status, body mass index, exercise, fibre intake and so forth, a mathematical adjustment that calculates the risk if all these factors were equal can be made. In a study of cancer risk (Bradbury et al., 2014), the actual number of cancer cases in each group showed that the consumers reporting regular consumption of organic food had a 1% lower rate of cancer than those who reported never consuming organic foods; presumably this was not statistically significant (no statistics are provided). However, after adjustments for 9 such other health factors, the relative risk associated with only the food production system became similarly slightly higher, by 3%. This value was clearly stated as too small to reflect an actual difference, partly because the use of adjustments by itself increases the uncertainty. Still, this example illustrates how difficult or perhaps impossible it is to determine the actual impact of the organic produce on the health of the consumer, even when accurate data are available from thousands of consumers.

2.3 Belief in health benefits of organic food

So, what is it that makes a consumer choose organic food? Consumers of organic food (who are better educated and more well informed than the average consumer) usually report several reasons for why organic farming is a good idea in general; these include benefits for the environment, for local producers, animal welfare and so forth (Pieniak et al., 2010; Aertsens et al., 2011). However, the main reason reported for their choice is health benefits: the impression that an organic diet will benefit your health more than a conventional diet (Aertsens et al., 2009; Hsu et al., 2016; Denver and Christensen, 2015).

So, superficially it would appear that well-resourced health-conscious consumers select foods that they consider healthy along with all the other positive health behaviours, one of these being a 'health-halo' around organic foods (Prada et al. 2017).

As mentioned earlier, the good health of consumers of organic products may be caused by factors other than organic agriculture, since it is not currently feasible to test the contribution from each of the confounding factors separately. Often this is interpreted as: if a statement presents organic agriculture as causing food produced from it to have a health benefit, then this should be considered misleading advertising (Mulet, 2014). In the European Union, it is illegal to claim health benefits for products with no approved scientific evidence (European Parliament, 2006). Producers can be prosecuted for using claims breaching this rule, for example, about antioxidant effects (Hardy and Bacon, 2014; EFSA, 2011), 'de-tox' diets, etc.

This opinion is based on the assumption that consumers of organic food view the potential health benefits in the same way as consumers usually view general health claims on foods. If this was the case, the consumers of organic food would feel cheated if they discovered that the product did not deliver the health effect that they had paid for through the organic premium (Rittenhofer and Povlsen, 2015). However, studies of how consumers view health claims for either 'functional food' (with a specific approved health claim) compared with organic food (with an unspecific health 'halo') show that in fact many consumers do not perceive these claims in the same way (Goetzke et al., 2014). Several studies have shown that in particular consumers who actively seek organic products ('regular users') are less impressed by official health claims than consumers who only occasionally purchase organic products or not at all (Aschemann-Witzel et al., 2013; Vecchio et al., 2016). Interestingly, in a study asking consumers to choose between 'natural content' and 'health', where the claims regarding 'natural content' included phrases such as *free from artificial ingredients, additives and residues*, while those regarding 'health' focused on phrases such as *contains a lot of vitamins and minerals; rich in protein and fibre*, consumers of organic products found the 'natural content' much more important than the 'health' claims (Hasselbach and Roosen, 2015). So, the *avoidance* of unwanted artificial substances seems to be more important for these consumers than the possibility of benefits *beyond* what the basic food provides. In other words, consumers seem to choose organic products to avoid something they perceive as harmful, rather than to obtain particular benefits. This means that for consumers of products from organic agriculture, while the integrity (trustworthiness) of the certification system is extremely important (Rittenhofer and Povlsen, 2015), they are not particularly interested in scientific proof of benefits. This can be illustrated by two other independent indicators: One is that consumers who regularly choose organic products are almost twice as likely to consider themselves to be affected by food allergy/sensitivity than those who never consume organic products (Baudry et al., 2015). This indicates that these consumers may choose organic products in order to avoid conventional ones, since the reduced allergy risk for children growing up in households consuming organic food (Fagerstedt et al., 2016) shows that the increased occurrence of food allergy among adult consumers cannot be caused by the organic food itself.

Another indicator is that producers of functional foods are noticing that consumers are becoming increasingly interested in foods that are marketed based on absence of artificial constituents, using terms such as 'natural', 'free from' and 'organic', rather than the more conventional 'scientifically proven', 'fortified with nutrients' or equivalent concepts, so the 'natural' food market segment is growing faster than that of the 'functional' products (Sloan, 2016).

3 Organic foods and health

3.1 Health implications of differences in diet components

As mentioned earlier, diets that are rich in vegetables and fruits and low in processed meats and so forth are associated with better health than diets with lower intakes of these foods (Shay et al., 2012; Kushi et al., 2012). A meta-analysis of 16 studies, covering a total of 833 234 participants, who had provided information about their diet, recorded 56 423 deaths during periods of 5–26 years (Wang et al., 2014) and showed that for every daily portion of vegetables or fruit consumed, up to a total of 5 portions per day, the risk of dying within the study period was reduced by 5% (the hazard ratio was 0.95), mostly due to fewer cases of cardiovascular disease. Another example of a study using similar data has found that the lack of daily consumption of vegetables and fruits was responsible for 13.7% of the cases of heart attacks (acute myocardial infarction) (Yusuf et al., 2004). None of these studies took into account the fact of whether the produce was organically produced. However, on the one hand, this means that the approximately 25% higher intake of vegetables among consumers of organic products (Fig. 2a) will by itself improve health of those consumers substantially, compared with those who consume conventional diets with fewer vegetables and fruits. On the other hand, it also means that if there is any difference in the value for human health due to consumption of organic produce compared with that of conventional produce, then this will affect the consumers of organic food more than it affects those who prefer conventional diets.

A similar argument can be made regarding red meat and processed meat products, which are well known to be associated with bad health (higher risk of disease and death) (Shay et al., 2012; Kushi et al., 2012). Each additional 50 g of processed meat consumed per day can increase the risk of premature death by 25% (Abete et al., 2014), which can be compared with the difference between consumers of organic and conventional products shown in Fig. 2b.

In this context, it is important to keep in mind that all studies of this type of association are controlled for other factors, so for example the effect of increased consumption of vegetables and fruits is adjusted to control for the difference in meat consumption and vice versa (Abete et al., 2014; Wang et al., 2014). This can be difficult to achieve in cases where the participants tend to occur in 'clusters', each with a similar dietary pattern, rather than a random selection of different foods. For example, if all those who consume more vegetables also do not consume so much processed meat, it is not possible to determine how much of a health benefit is caused by each of the factors. Since this is exactly what is observed among consumers of organic or conventional foods (Baudry et al., 2016), the data for each food are probably not totally independent. However, it is still appropriate to expect that a group of consumers who have a high intake of vegetables and fruit and also a low intake of processed meat will have even better health than those who only meet one of these measures of a healthy diet.

3.2 Differences in composition of organic and conventional foods

3.2.1 Plant foods

The composition of plant foods from organic and conventional agriculture has been reviewed several times, most recently by Desjardins (2016). While there are disagreements

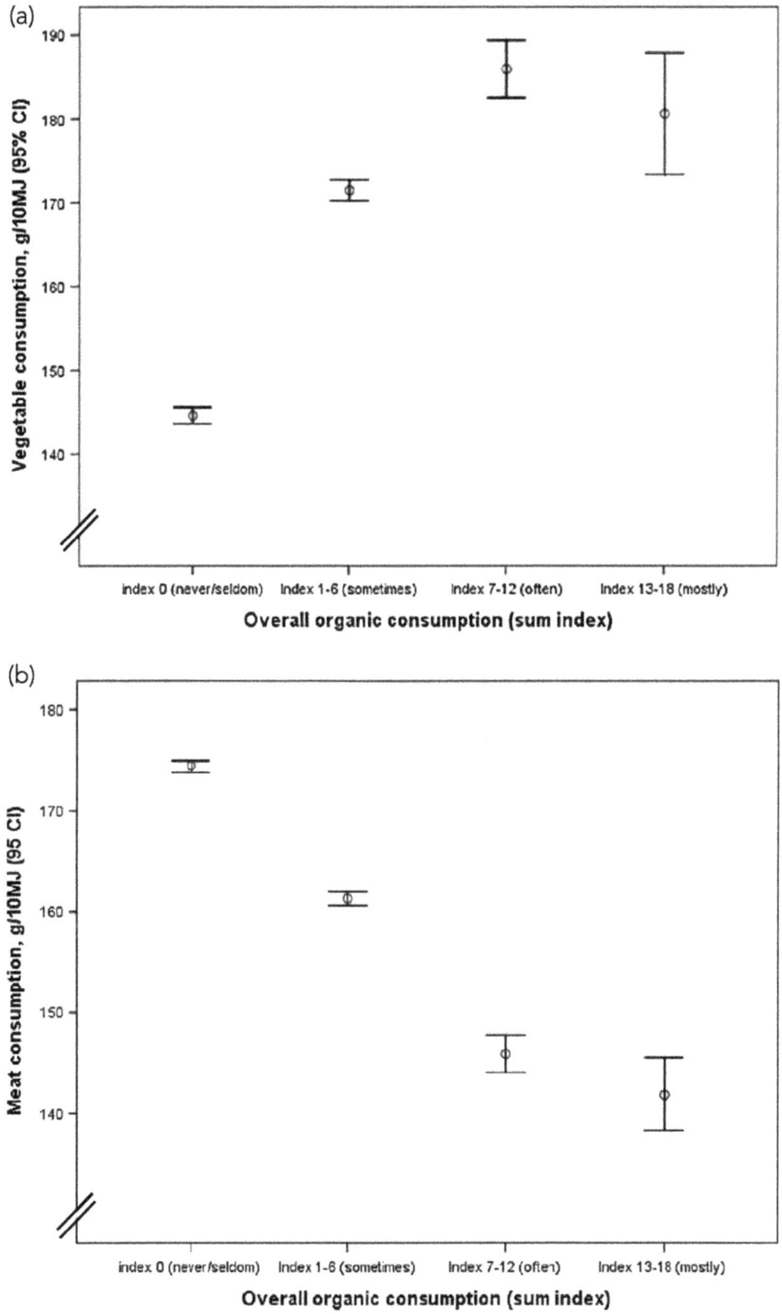

Figure 2 (a, b) Amount of vegetables and meat in the diet according to consumption of organic food. 10 MJ is 2390 kcal, so the values are similar to intake per day. From Torjusen et al. (2012).

regarding the optimal methodology, which particularly affects how significant the differences are, the various authors tend to identify the same general differences: Organically produced plant foods tend to have higher contents of vitamin C and of most phytochemicals, including some types of phenolic compounds. Vegetables and fruits also have increased dry matter content, which is often associated with higher levels of minerals such as magnesium, iron, etc. The conventionally produced plant foods have increased levels of pesticide residues and cadmium, higher content of storage protein in seeds, and higher levels of nitrate in vegetables.

These differences are all easy to explain, since they are direct consequences of the differences between the production processes:

In organic agriculture, the plants grow relatively slowly, using nutrients that are released gradually during the growing season, provided from a previous crop or slowly decomposing manure. Due to the slower growth, the plants accumulate more dry matter, minerals and phytochemicals (Brandt et al., 2011), in particular compounds involved in defence against pests and diseases (de Jong, 1995), because this is the optimal adaptation to a situation where growth rate is limited by nutrient availability, most often nitrogen. It is important to keep in mind that this adaptation is probably also optimal for the farmer. If an organic farmer provides too intensive fertilisation, for example, using more than its share of manure on the most valuable crop in the rotation, the increased growth rate will lead to a reduction in the content of natural defence compounds (de Jong, 1995), thus making this very crop become more susceptible to damage by pests or diseases.

In conventional agriculture, plant nutrients are provided as soluble minerals that the plants take up soon after application, enabling the plants to grow at maximum rate for much of the growing season. Conventionally farmed plants often take up more nitrate than required for growth so it is either used to make additional storage protein or accumulates in tissues as nitrate. Cadmium uptake from soil into plants is inhibited by organic matter (humus) in the soil, which can explain its low levels in organically produced plants, and it is found in conventional fertilisers as an unintended ingredient, which separately explains its higher levels in conventionally produced plants (Roberts, 2014). The more frequent occurrence and higher levels of residues of synthetic pesticides in conventional plant foods is an obvious consequence of the conventional farmers using these products, which are not permitted in organic agriculture. It must, however, be emphasised that even in conventional plant foods, only around half of the tested samples contain measurable levels of pesticide residues, and only a small percentage of tested samples contain more than the allowed level of residues of each pesticide (Baranski et al., 2014).

3.2.2 Animal foods

For animal foods, recent meta-analyses have summarised studies on meat and dairy products. For meat, the main differences that have been studied relate to fatty acid composition, where meat from organic agriculture tends to contain higher proportions of polyunsaturated fatty acids, in particular n-3 (omega-3) lipids, than corresponding products from conventional agriculture (Srednicka-Tober et al., 2016a). For milk and dairy products, the same differences were found, and additionally the organic milk contained more iron, less iodine and selenium, and more conjugated linoleic acid than conventional milk (Srednicka-Tober et al., 2016b). For both meat and milk, the observed differences could be fully explained by differences in the feeding regimes used in organic and conventional agriculture, particularly the requirements in organic farming for outdoor grazing and other on-farm production of animal feed. Fresh forage such as grass provides an additional source of polyunsaturated fatty acids, and together with the additional exercise supports accumulation of higher iron stores in the animals. Similarly, preferential use of local feed sources reduces the supply of selenium and iodine outside regions where the soil contains high levels of these minerals (Srednicka-Tober et al., 2016a,b).

A particular issue is bacteria and other pathogens with resistance to antibiotics. Due to the greater restrictions on the use of antibiotics and other drugs in organic farming, the pathogens in conventional animals are more likely to contain one or more of the many genes that confer resistance to antibiotics (Osterberg et al., 2016).

3.2.3 Processed foods

Processed foods are both cured meats such as sausages or cured ham, and any of a wide variety of foods that are prepared in one place, stored and then consumed somewhere else. The regulations defining organic food production ban or severely limit most additives such as preservatives, emulsifiers, colourants, sweeteners, phosphates, nitrite and so forth, which are routinely used in most conventional processed foods and drinks. So, while the actual contents in food of these substances are largely unknown, it is certain that the vast majority are found in conventional products, and only very small amounts in products sold as organic.

3.3 Health implications of compositional differences

3.3.1 Beneficial components and doses

One of the most frequently mentioned components of foods presumed to have health benefits are antioxidants. All living tissues contain antioxidants, which neutralise excess free radicals and thus prevent these free radicals from destroying other molecules in the tissues. If the tissue dies, and therefore no longer is able to regenerate the antioxidants, the free radicals may ravage unchecked, causing formation of rancid off-tastes that can ruin food quality during storage. Since a publication by Harman (1956) defined the 'free radical theory of ageing', it has been widely believed that a similar destructive process taking place inside living human cells was essentially responsible for ageing, and that consumption of foods with high levels of antioxidants was necessary and sufficient to keep this process in check and thus allow us to live longer than otherwise (Harman, 1956). This theory superficially matched the available evidence, for example, the consistent association of higher intake of vegetables and fruits (Wang et al., 2014) with lower mortality could be due to these foods providing most of the antioxidants in the diet. However, in all cases where the antioxidant theory of ageing has been rigorously tested in placebo-controlled supplementation studies, it has failed to show the expected simple correlation between intake of dietary antioxidants and longevity in animals and humans. Some examples are feeding short-lived animals (field voles) diets high in vitamin C or vitamin E (Selman et al., 2013) (see Fig. 3) or giving supplements high in the vitamin A-precursor beta-carotene to humans at elevated risk of lung cancer (Albanes et al., 1996), in both cases resulting in significant increases in mortality when compared with the 'normal'-antioxidant control diets.

Also, other approaches to test this theory have failed to support it, even to the extent of identifying contexts where increased generation of free radicals directly benefits health (Ristow and Schmeisser, 2011).

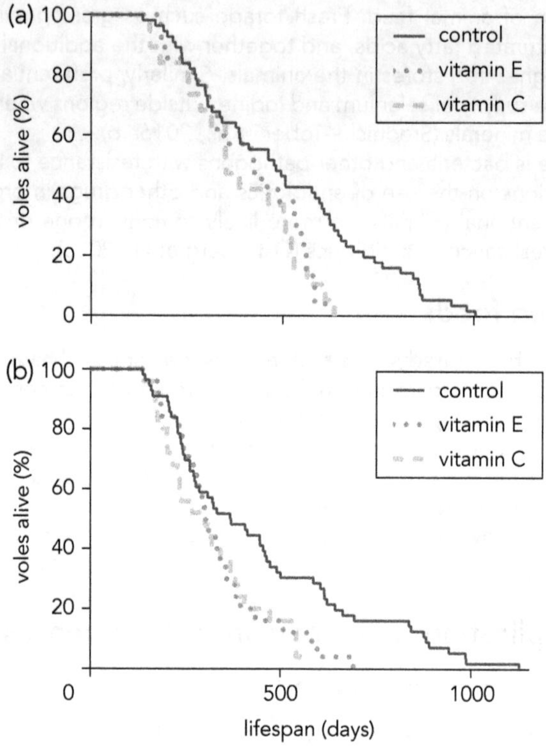

Figure 3 Kaplan–Meier survival curves of voles maintained in the cold (a) 7 ± 2°C or warm (b) 22 ± 2°C and given access to either a control diet, a vitamin E-supplemented diet or a vitamin C-supplemented diet from two months of age. Solid lines denote control, stippled lines denotes vitamin E and broken lines denotes vitamin C groups. From Selman et al. (2013).

It is clear that every food component has an optimum dose, so negative effects of excessively high doses of something does not rule out that other doses may be beneficial. Regarding antioxidant vitamins, the deleterious effects of very low intake are well known and understood; however, as these examples show, the optimum intake is probably not higher than what average consumers obtain in average diets (in developed countries), and symptoms of deficiency are very rare. Increased doses are only beneficial if the optimum dose is much higher than the average intake, and this does not seem to be the case for antioxidant vitamins or other compounds acting only by neutralising free radicals (Gruber et al., 2008).

In contrast, most phytochemicals (compounds produced only by plants) are not vitamins, meaning that none of these compounds are specifically required for a healthy diet. In addition to nutrients such as starch, fat, protein and fibre, plants produce many thousands of different chemical compounds with a bewildering variety of biological effects. Most of these compounds are defined as 'secondary metabolites', which means that they are not essential for the cellular metabolism, not even in the plant that makes them. Instead they function for the plant as agents that provide colour and fragrance, and protection against ultraviolet light or evaporation (waxes) or against diseases and pests. For the humans or other animals who consume plants, the secondary metabolites can be indigestible bulk (fibres and waxes), can be 'anti-nutrients' that reduce the nutritional value of the food, for example, by making proteins more difficult to digest, or can be toxicants, which are phytochemicals that each inhibit one or more of a wide range of cellular functions and, in high doses, can be toxic for the consumer. Even the most biologically active toxicants can be beneficial as medicines (for those who suffer from a corresponding disease). However, for obvious reasons, plants used for food tend to have lower levels of toxicants, which have only minor or no observable effects on human health (Brandt, 2016). Examples of food toxicants are glucosinolates, which are responsible for the pungent taste of mustard, broccoli and radish, or caffeine, which causes tea or coffee to increase alertness in those who consume it.

Due to the pervasiveness of the 'free radical theory of ageing', most phytochemicals have only been tested for antioxidant effects. It is now clear that the benefits of consuming plant foods, which originally focused the attention on phytochemicals (Hertog et al., 1993), could not be replicated using only the antioxidant properties of the phytochemicals (Hollman et al., 2011). However, recent studies have started to focus on the above-mentioned inhibitory effects as the possible key property of this large and diverse group of compounds (Brandt, 2016). If the doses normally found in vegetables, fruits and wholegrain foods cause a partial inhibition of a range of different cellular functions, thus interfering with the processes that turn these cells into cancer cells or promote inflammation associated with the metabolic syndrome (Ford et al., 2013), then this could explain many of the beneficial effects of consuming foods containing phytochemicals. Interestingly, these inhibitory or 'anti-nutrient' effects often result in lower weight gain, similar to what is known as 'caloric restriction' (Brandt and Molgaard, 2001), which is already known to improve health and increase lifespan in animals and humans. It is quite possible that some phytochemicals have greater beneficial effect than others; however, since until now very few scientists have studied these compounds from this point of view, we do not really have data to say which ones are important and which ones less so. However, from a consumer's health point of view, the phytochemicals are probably the most important components of the vegetables and fruits (Brandt, 2016). Due to this, it is reasonable to expect that conditions that increase the contents of all phytochemicals, such as organic agriculture with its lower nitrogen input (see Fig. 4), probably also increase the health benefits of these foods by a similar percentage, which is around 16% (Brandt et al., 2011).

In animal foods, the (presumed) beneficial compounds, which are affected by the agricultural system, are the fatty acids (polyunsaturated fatty acids, in particular omega-3) and the trace minerals iodine and selenium. Regarding unsaturated fatty acids, while the contents differ substantially between the organic and conventional production systems, the total content even in organic products is quite low, typically 5% of the total fat (Srednicka-Tober et al., 2016a,b). A small amount of rapeseed oil in the diet, or an even smaller amount of flaxseed, can provide these same nutrients with much less calories. Similarly, the most concentrated sources of the minerals iodine and selenium are seaweed or Brazil nuts, respectively. So, while milk is a good source of both minerals and conventional milk provides more minerals than organic milk, the overall intake of these minerals still depends more on the composition of the diet in terms of the types of foods included, than whether these foods are produced in organic or conventional farming systems.

A particular class of compounds, conjugated linoleic acids (often abbreviated to CLAs), are found in meat and milk from grass-fed ruminants and are almost absent from any other

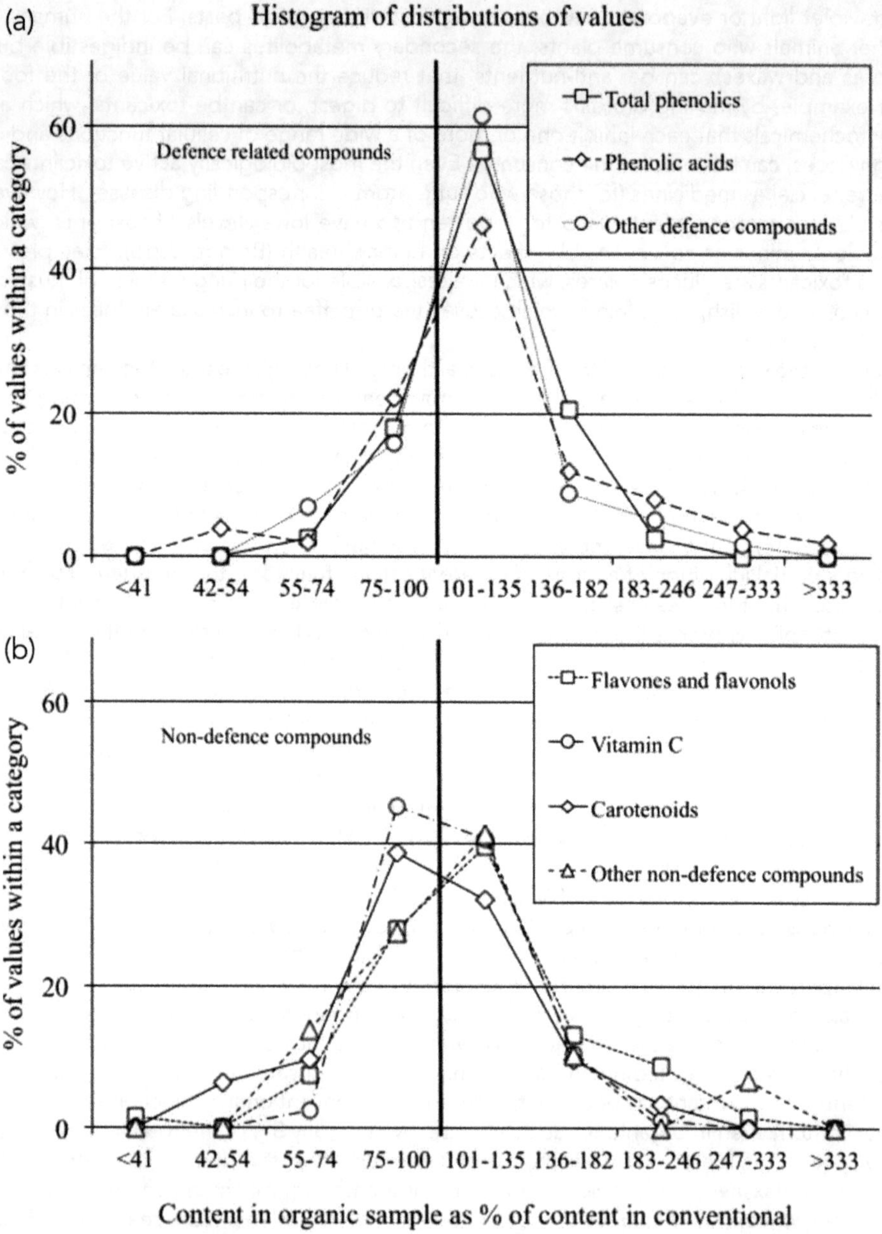

Figure 4 Graphical representation of the distribution of ratios of content in organic and conventional fruits and vegetables, for different categories of compounds. The vertical line indicates 100% (where the concentrations are equal). From Brandt et al. (2011).

food including similar foods from animals fed mostly concentrates without fresh forage (as it is common in conventional farming) (Srednicka-Tober et al., 2016a,b). Early studies of these compounds found that people with a high intake of CLA consistently had better health outcomes than people who did not consume this compound, and due to this it was believed that CLA caused those health benefits. Only when the different types of fats were compared in fully randomised controlled studies was it discovered that CLAs have the same detrimental effects on health as any other trans fats (Brouwer et al., 2013) (the most famous trans fats are those that were widely used in margarine to optimise the texture). The research about CLA was originally based on the observation that consumers with a high intake of these compounds have better health than others, for example, low risk of cancer (Aro et al., 2000). However, these studies did not take into account that people who choose organic (or other natural) food when feasible, are over-represented among consumers of butter and meat from grass-fed cows. Since this diet choice (organic diet) is associated with better health, the correlation between CLA intake and good health could be caused by something that is correlated with both (probably 'a generally healthy lifestyle' – or may be 'a diet with food from organic agriculture'?). In this way, the 'story' of CLAs is a good example of how difficult it is to find out which components in the diet are important for our health and which ones are just a matter of taste. It also shows how important it is to find this out! While the naturally occurring doses of CLA in food is too low to cause much damage or benefit in any case, the previously assumed benefits of CLA have resulted in a thriving industry producing synthetic food supplements that can provide much higher intake than the doses occurring naturally in foods. So rather than an 'academic' issue driven by curiosity to understand the mechanisms of nutrition and health, it is now about protecting consumers against unsubstantiated claims from a supplement industry that has seized a new market opportunity, with a risk of not only failing to deliver the promised health benefits, but possibly making matters worse (Brouwer et al., 2013).

In recent years, the view on nitrate in plant foods has changed radically. While it was previously considered a dangerous contaminant to be avoided where feasible (Bruning-Fann and Kaneene, 1993), it is now generally considered a beneficial compound, for example, for athletic performance (Affourtit et al., 2015). Previous concerns about possible cancer risks have not been confirmed, for example, high intake of leafy green vegetables (the main dietary source of plant-derived nitrate) is associated with reduced risk of cancer, and it is therefore no longer considered a major health risk (Xie et al., 2016). However, while conventional agriculture tends to cause slightly higher concentrations of nitrate in relevant crops (Baranski et al., 2014), the differences from one crop to another is very much larger (10s or 100s of times), so the choice of diet (in particular intake of green leafy vegetables) is probably much more important for intake of nitrate than the choice of production system. No studies have published data that would allow a comparison of the intake of plant-derived nitrate between consumers who prefer organic or conventional foods.

3.3.2 Harmful components and doses

Dietary pesticide residues occur primarily on and in plant foods, such as vegetables, fruits and wholegrain foods, either due to intentional use of pesticides on this plant or as contaminants from other sources. Pesticides are applied to the plant, for example, by spraying, and for each pesticide there are specific rules for the time after application until it is considered safe to harvest or consume the edible part of a treated plant. In conventional farming, the principle is to keep the level of the pesticides and their degradation products (residues) below a defined level considered safe for consumption. If the rules are adhered to and the weather corresponds to the estimates in the models used to calculate these values, then the pesticides will not cause acute (noticeable) toxicity. There are no cases reported in the literature where consumption of foods containing allowed levels of pesticide residues has resulted in illness clearly caused by these compounds. The allowed levels of residues are typically less than 25μg per kg body weight per day (Curl et al., 2003, Jensen et al., 2015), lower than what could cause symptoms if someone was allergic to these compounds. In organic farming, the use of pesticides is restricted to a few natural substances (e.g. copper, soap and some plant extracts), with few or no known adverse effects, even in high concentrations.

In contrast to the situation for consumers of appropriately produced conventional foods, there are numerous reports of acute or chronic illness caused by synthetic pesticides in other contexts: Exposure to workers applying the pesticides, to people living near farms or other locations where pesticides are used, or exposure caused by accidents and negligence of various types (Mostafalou and Abdollahi, 2013). For some diseases such as Parkinson's and other brain disorders, exposure to pesticides from occupational or accidental sources may significantly affect the number and severity of cases (Dardiotis et al., 2013; Shelton et al., 2014). Obviously, the existence of organic agriculture reduces the numbers of such casualties, and the reduced health care costs substantially contributes to the social benefits of organic farming (Reganold and Wachter, 2016). Together with the reduced damage to non-target animals and plants, in particular in natural ecosystems, these benefits to the general society help to justify the higher costs of food production in organic systems compared with conventional ones.

Nitrite and nitrate in processed meat are allowed in organic products, but only in lower amounts than in corresponding conventional products. Nitrite is added to prevent growth and toxin production by the bacterium *Clostridium botulinum*, which can produce a toxin that causes the disease botulism. Growth and toxin production of *C. botulinum* is inhibited by acid (pH < 4.6), oxygen (air), salt or temperatures <12°C, so it only occurs in foods that have been kept for some time at high temperature after it was prepared, and which contains neither acid nor sufficiently high concentrations of salt (WHO, 2016). In the United States, botulism caused by toxin in food is fatal in approximately 5% of cases; it is quite rare (<40 cases per year, approx. two deaths) and primarily caused by homemade preserved foods (Centers for Disease Control and Prevention, 2016). It has been known for many years that nitrite inhibits growth of the *C. botulinum* bacteria, and this is one of the reasons for adding it to processed meats, while another reason being that nitrite reacts with myoglobin in muscle material to form a pink colour, which is considered important for consumer acceptance (Govari and Pexara, 2015). On the other hand, while nitrite by itself probably is relatively harmless and may even benefit human health in appropriate concentrations (Govari and Pexara, 2015), nitrite in processed food may react with amino acids to form nitrosamines (Herrmann et al., 2015), which are likely causes of the increased cancer risk observed among people with relatively high intake of processed meat products (Zhu et al., 2014).

However, nitrite is not essential for food safety – any one of the above-mentioned methods (exposure to low pH, low temperature, air or high salt concentration) is sufficient to prevent accumulation of toxin if applied appropriately (WHO, 2016). On the other hand, some consumers show a strong preference for the characteristic colours and tastes associated with nitrite-cured products such as bacon, and many organic producers are very innovative to identify natural ingredients such as celery extract that can provide substantial levels of nitrite in the product. So, choosing organic processed meat does not guarantee the food to be completely free of nitrite (de Gonzalez et al., 2012). If it looks like conventional bacon and tastes like conventional bacon, it probably contains just as much nitrite and nitrosamines as the conventional bacon.

Synthetic colourants, sweeteners, emulsifiers, preservatives and other additives in complex foods are generally not permitted in organic products. A few but widely publicised studies have indicated increased risk of hyperactive behaviour in children exposed to a mixture of synthetic food dyes often used in drinks and sweets (Arnold et al., 2012). This has caused some restrictions in the use of colourants, but most of the indicated compounds are still used in conventional foods and drinks at least in some countries (Oplatowska-Stachowiak and Elliott, 2016). For synthetic sweeteners, the opinions are even more divided: Some experts consider artificial sweeteners an important positive option, which benefits people who want to reduce their calorie intake, while others argue that since intake of artificial sweeteners is associated with the opposite effect (increased obesity), they may do more damage than good, even though the mechanisms are not yet understood (Borges et al., 2017).

An interesting recent study indicates that the hyperactivity problem may be caused by the use of sugar-sweetened drinks as well, rather than specific colours or artificial sweeteners in these drinks (Yu et al., 2016), in which case organic 'colas' and similar products made from all natural ingredients may be just as bad for the children as the more synthetic ones. However, it is important to keep in mind that consumers of organic products tend to purchase less of 'fizzy drinks', sweets and similar products (Torjusen et al., 2012), even when organic 'equivalents' are available.

For emulsifiers, a recent line of research indicates that a range of serious health issues may be caused by an indirect effect on the types of microbes, mostly bacteria, which naturally live in our guts (Cani and Everard, 2015). Studies with mice indicate that both obesity (Lecomte et al., 2016) and colon cancer (Viennois et al., 2017) can be promoted by adding these very common food additives to the animals' diets, causing substantial concern about their safety, although this research has not yet led to changes in safety regulations.

3.4 Health implications of social aspects of organic diets

The positive health behaviour patterns described in Section 2.1 match exactly the types of psycho-social effects that are the aims of almost every single government initiative for promotion of healthy lifestyles and that are considered essential for achieving long-term cost-effective improvements to public health (de Ridder et al., 2017). For example, the cost of weight loss interventions that aim for long-term change in the lifestyle of an individual, without affecting the wider context of the person's environment and self-perception, have been calculated as approx. $1000 per kilo lost after 12 months (Sun et al., 2017). The data presented above indicate that in countries where public promotion of organic farming (usually justified by the environmental benefits) has made organic foods widely available, this may indeed have supported the achievement of those public health aims, in contrast to several other (often much more costly) food system policy initiatives (Hyseni et al., 2017). It could be argued that the health-conscious people consuming organic diets might have been choosing other equally healthy diets if no organic foods had been available. However, to the extent that the concept of an 'organic lifestyle' supports the consumers 'to achieve a sense of self-actualisation' (Pino et al., 2012), this would increase the proportion who stick to their improved health behaviours in the long term (de Ridder et al., 2017) and reap the benefits in terms of better health. This topic has hardly been investigated; however, Siegrist et al. (2015) analysed changes in consumer perceptions of environmental benefits and health benefits of diets with reduced meat and found that increases in one tended to enhance the other, providing at least circumstantial support for this interpretation.

4 Combined impact on health of organic foods in diet

Combining the effect of an organic lifestyle on food choice (Fig. 2) with differences in the composition of organic and conventional foods (Fig. 4) and indications of which components are beneficial or harmful, the following trends emerge:

1. Consumers with a preference for organic foods tend to have a healthier lifestyle, not only regarding food, but also in terms of non-food activities such as smoking and physical activity, compared with consumers who prefer conventionally produced foods.
2. Organic vegetables are both consumed in higher amounts and with more concentrated phytochemicals, so any benefits of phytochemicals for human health will be multiplied twice for a typical consumer of organic food.
3. Organic meat is consumed in lower amounts, in particular processed products. It is less clear whether there are differences in nutritionally positive or negative constituents. However, any harmful effects are reduced by their lower intake.
4. Organic diets tend to contain much less of processed foods in general, and those processed foods that are consumed contain much less of food additives linked with the 'emerging' risks such as emulsifiers, sweeteners and food dyes.
5. Consumers who choose organic food tend to be more interested in avoiding hidden risks than in enhancing (possibly equally vacuous) claimed health benefits of 'functional foods'. This ties in very well with the general ethos of organic agriculture, while opposing other trends promoting new technologies, thus forming the basis for the 'alternative' production system financing the cost of keeping organic agriculture separated from the mainstream conventional supply chains.

Each of these trends on their own can explain why consumers of organic food tend to be healthier than others in all respects that have been measured. Only the increased intake of phytochemicals [as indicated in (2)] is partially directly caused by the organic agriculture, in the sense that if the organic farmers change their methods, for example, in order to increase yield, part of this health benefit may be compromised. All the other factors on this list are mainly controlled by the consumers themselves, rather than by the producers of organic food (Table 1).

In conclusion, the benefits (more phytochemicals, less additives) of individual organically produced foods are probably important as motivation for consumers to choose foods that are more expensive than the corresponding conventional foods, even though for the overall diet the difference in cost is smaller, or often even represent a similar cost, due to the higher proportion of relatively cheap plant foods. However, the data indicate that the primary health benefits of organic agriculture for consumers is to provide a community, a

Table 1 Qualitative overview of what we know (more or less) about the impact of organic agriculture on diet and health

Category	Use by consumers of organic products	Substance	Organic versus corresponding conventional product	Overall difference between diets (high vs low)	Overall impact of organic (good vs bad)
Vegetables and fruits	↑	Vitamin C	↑	↑↑	=
		Other vitamins (E, A, B9, K etc.)	Similar	↑	=
		Most minerals, dry matter	↑	↑↑	=
		Cadmium	↓	=	=
		Phytochemicals	↑↑	↑↑	↑↑
		Protein	↓	=	=
		Pesticide residues	↓↓	↓↓	=
		Nitrate	↓	=	=
Meat that is raw when purchased	↓	Polyunsaturated fatty acids	↑	=	=
		Antibiotics resistant bacteria	↓	↓↓	?
Dairy products	=	Polyunsaturated fatty acids	↑	=	=
		Conjugated linoleic acids (CLA)	↑	↑	=
Processed meat	↓↓	Nitrosamines etc. from nitrite	↓	↓↓	↑↑
Other processed foods	↓↓	Added emulsifiers	↓	↓↓	?
		Synthetic colorants	↓↓	↓↓	?
		Synthetic sweeteners, flavourings etc.	↓↓	↓↓	?
Lifestyle choices	↓	Smoking	↓		↑↑
	↑	Physical exercise	↑		↑
Other aspects		Accidental pesticide exposure ('by-stander')	↓↓		↑↑

Single arrow shows general direction of increase or decrease, and double arrows indicate consistent or large differences. An '=' sign indicates 'difference too small to matter' (in the author's opinion) and a ? 'not sufficient data'.

self-perception framework and a supply chain to support a holistic lifestyle, which values the joy and skill of cooking with raw ingredients and taking responsibility for ones' own health (and that of one's family).

5 Future trends

5.1 Effects of food composition

Some of the most interesting lines of future research are the potential unintended health risks posed by food additives, such as artificial sweeteners, colourants and emulsifiers. These substances have all been thoroughly tested for toxicity in animal studies, and there is no indication that they cause disease as such on their own. However, the emerging evidence that these substances affect behaviour, food preferences and the gut microbiome in unexpected ways, as described in Section 3.2.2, indicates that the initial suspicions, which caused the ban on food additives in organic foods, were more prescient than it has been known until now. All the identified problems are issues that would not have been identified in a standard toxicological testing procedure. Much more research is required in order to fully understand the consequences of using these food additives. The eventual outcome is likely to be restrictions for conventional food producers, resulting in healthier conventional products (which, somewhat ironically, will reduce the relative benefit for consumers of organic products).

The resistance to antibiotics in bacteria is on the one hand a recognised problem, which already receives substantial attention and research funds. This has led to some additional restrictions on conventional agriculture, although no improvements in the problem have yet been achieved (Hao et al., 2016). On the other hand, there is still a large challenge to measure how much it affects the risks of disease and death among consumers and to find out whether the use of organic agriculture methods can be a solution to the problem or only a minor delay in its development.

Phytochemicals have only been briefly described in this chapter; more details can be found elsewhere (Brandt et al., 2011; Brandt, 2016). Phytochemicals are not essential nutrients, and each of them probably affects the human body in a slightly different way, which also depends on the dose. So, it is an immense future research task to unravel the contribution from each phytochemical to human health. It is relatively easier to study the importance of each phytochemical for the health of the plant that makes it, because each plant species tends to make only a limited number of primary defence compounds (toxicants). When agricultural methods (both supply of plant nutrients, especially nitrogen, and choice of cultivars) affect the accumulation of phytochemicals, then it has direct and rapid effects on the susceptibility to diseases and pests. So the observation that certain agronomic factors increase or decrease plant health is an excellent place to start, for discovering the next important phytochemical in that plant, and ideally also connects this phytochemical with human health. Consuming plant food with a high content of phytochemicals first of all helps you to lose weight, so this can help to explain why consumers of organic food tend to be slimmer than others (Torjusen et al., 2012) even though they are less likely to be on a weight loss diet (Baudry et al., 2015).

5.2 Psycho-social effects of belonging to the organic community

The philosophy behind organic farming emphasises that 'the health of individuals and communities cannot be separated from the health of ecosystems – healthy soils produce healthy crops that foster the health of animals and people' (IFOAM, 2005). For many years this type of approach was viewed by many scientists as a sort of 'magical thinking', fundamentally contrasting with the core concepts of 'conventional science' (Hansson, 1991). Conversely, the 'conventional science' has been criticised for its focus on mechanistic explanations and standardised medical interventions for every ailment, that were seen as leading to patronising 'one-size-fits-all' programmes of education and fortification (see e.g. Hever (2011)), justifying the need for an organic movement to focus on connectedness rather than reductionism. However, in recent years movements towards a common ground has occurred from both sides (Hoffmann, 2003; Huber et al., 2006), to recognise that health is not only the absence of disease, and food quality is not only the sum of beneficial components minus the harmful ones. Still, there is a long way to go. Note for example, that the potentials for positive and mutually supportive interactions as described in Section 3.4 could only be construed by combining literature from different fields, in ways that their authors probably never anticipated, and may not even condone.

6 Conclusion

Therefore, there is an urgent need for new research activities, to test and assess multifactorial concepts on nutritious food and human health, which transcend such outdated traditions and dogmas. In particular it is important that research comparing organic and conventional methods and products and consumer choices and consumer health is designed by consortia of scientists involving people with appropriate expertise and credibility regarding both organic and conventional concepts, so the outcomes will be recognised as credible by both communities. This must focus on determining what actually works, rather than attempt to support what hypothetically should work (a fundamental principle of science), and it must support individuals and communities to take and achieve responsibility for their own lives (a fundamental principle of the organic movement), rather than leave this to 'rampant capitalism' or 'the nanny state' (pick your preference). These issues are common for all types of complex research that requires interdisciplinary approaches (Hall et al. 2017). So, in practice, in research about the impact of organic agriculture on diet and health, the researchers in such teams must agree on their methods in advance and specifically agree in advance to refrain from announcing conclusions based on the absence of evidence if it turns out that the conditions for testing the hypotheses cannot be fulfilled. For example, natural science researchers may want to investigate to what extent consumers who choose organic food reduce their exposure to dietary pesticide residues, but on their own they would only be able to assess health as potential toxic effects of the measured doses, and might conclude that an overall low maximal dose means that no effect is possible. Such a research group should from the start (while still designing the research) collaborate with relevant groups in social science and psychology to additionally assess and discuss how the knowledge about those pesticide residues may affect the consumers socially and psychologically, e.g. in terms of cognitive dissonance with a role as parent, since this may be much more relevant to the reasons why a consumer would want to know about pesticide residues in the first place. At present, some researchers in both natural and social science may consider the other type of research misleading or unscientific, because it is so different from their own perception of what science is (Grundmann 2017). However at least there is increasing awareness about this problem and many active initiatives to try to reduce it through improved education of the researchers of the future (Hall 2017).

Organic farming thus provides a good example (not necessarily unique) of a concept that has well-documented evidence of effectiveness in terms of fulfilling consumers' expectations for a healthy diet option. The present analysis (Table 1) confirms that even though from a natural science perspective, the benefits for consumer health are different from what most consumers (and researchers) believe, the overall outcome in terms of health is still more likely to be positive than negative. It is therefore relevant for national and international research institutions to include organic food and farming as an important component of most future research efforts combining social science with natural science to improve public health.

7 References

Abete, I., Romaguera, D., Vieira, A. R., de Munain, A. L. and Norat, T. 2014. Association between total, processed, red and white meat consumption and all-cause, CVD and IHD mortality: A meta-analysis of cohort studies. *British Journal of Nutrition*, 112, 762–75.

Aertsens, J., Verbeke, W., Mondelaers, K. and van Huylenbroeck G., 2009. Personal determinants of organic food consumption: A review. *British Food Journal*, 111, 1140–67.

Aertsens, J., Mondelaers, K., Verbeke, W., Buysse, J. and van Huylenbroeck, G. 2011. The influence of subjective and objective knowledge on attitude, motivations and consumption of organic food. *British Food Journal*, 113, 1353–78.

Affourtit, C., Bailey, S. J., Jones, A. M., Smallwood, M. J. and Winyard, P. G. 2015. On the mechanism by which dietary nitrate improves human skeletal muscle function. *Frontiers in Physiology*, 6, 211.

Albanes, D., Heinonen, O. P., Taylor, P. R., Virtamo, J., Edwards, B. K., Rautalahti, M., Hartman, A. M., Palmgren, J., Freedman, L. S., Haapakoski, J., Barrett, M. J., Pietinen, P., Malila, N., Tala, E., Liippo, K., Salomaa, E. R., Tangrea, J. A., Teppo, L., Askin, F. B., Taskinen, E., Erozan, Y., Greenwald, P. and Huttunen, J. K. 1996. alpha-tocopherol and beta-carotene supplements and lung cancer incidence in the Alpha-Tocopherol, Beta-Carotene Cancer Prevention Study: Effects of base-line characteristics and study compliance. *Journal of the National Cancer Institute*, 88, 1560–70.

Alfven, T., Braun-Fahrlander, C., Brunekreef, B., von Mutius, E., Riedler, J., Scheynius, A., van Hage, M., Wickman, M., Benz, M. R., Budde, J., Michels, K. B., Schram, D., Ublagger, E., Waser, M., Pershagen, G. and Group, P. S. 2006. Allergic diseases and atopic sensitization in children related to farming and anthroposophic lifestyle - the PARSIFAL study. *Allergy*, 61, 414–21.

Arnold, L. E., Lofthouse, N. and Hurt, E. 2012. Artificial food colors and attention-deficit/hyperactivity symptoms: Conclusions to dye for. *Neurotherapeutics*, 9, 599–609.

Aro, A., Mannisto, S., Salminen, I., Ovaskainen, M. L., Kataja, V. and Uusitupa, M. 2000. Inverse association between dietary and serum conjugated linoleic acid and risk of breast cancer in postmenopausal women. *Nutrition and Cancer-an International Journal*, 38, 151–7.

Aschemann-Witzel, J., Maroscheck, N. and Hamm, U. 2013. Are organic consumers preferring or avoiding foods with nutrition and health claims? *Food Quality and Preference*, 30, 68–76.

Baranski, M., Srednicka-Tober, D., Volakakis, N., Seal, C., Sanderson, R., Stewart, G. B., Benbrook, C., Biavati, B., Markellou, E., Giotis, C., Gromadzka-Ostrowska, J., Rembialkowska, E., Skwarlo-Sonta, K., Tahvonen, R., Janovska, D., Niggli, U., Nicot, P. and Leifert, C. 2014. Higher antioxidant and lower cadmium concentrations and lower incidence of pesticide residues in organically grown crops: A systematic literature review and meta-analyses. *British Journal of Nutrition*, 112, 794–811.

Baudry, J., Mejean, C., Peneau, S., Galan, P., Hercberg, S., Lairon, D. and Kesse-Guyot, E. 2015. Health and dietary traits of organic food consumers: Results from the NutriNet-Sante study. *British Journal of Nutrition*, 114, 2064–73.

Baudry, J., Touvier, M., Alles, B., Peneau, S., Mejean, C., Galan, P., Hercberg, S., Lairon, D. and Kesse-Guyot, E. 2016. Typology of eaters based on conventional and organic food consumption: Results from the NutriNet-Sante cohort study. *British Journal of Nutrition*, 116, 700–9.

Borges, M. C., Louzada, M. L., de Sa, T. H., Laverty, A. A., Parra, D. C., Garzillo, J. M. F., Monteiro, C. A. and Millett, C. 2017. Artificially sweetened beverages and the response to the global obesity crisis. *Plos Medicine*, 14, e1002195.

Bradbury, K. E., Balkwill, A., Spencer, E. A., Roddam, A. W., Reeves, G. K., Green, J., Key, T. J., Beral, V., Pirie, K. and Million Women Study, C. 2014. Organic food consumption and the incidence of cancer in a large prospective study of women in the United Kingdom. *British Journal of Cancer*, 110, 2321–6.

Brandt, K. and Molgaard, J. P. 2001. Organic agriculture: Does it enhance or reduce the nutritional value of plant foods? *Journal of the Science of Food and Agriculture*, 81, 924–31.

Brandt, K., Leifert, C., Sanderson, R. and Seal, C. 2011. Agroecosystem management and nutritional quality of plant foods: The case of organic fruits and vegetables. *Critical Reviews in Plant Sciences*, 30, 177–97.

Brandt, K. 2016. Chapter 50: Vegetables and fruit in the prevention of chronic age-related diseases. In: Malavolta, M. and Mocchegiani, E. (Eds), *Molecular Basis of Nutrition and Aging*. San Diego: Academic Press.

Brantsaeter, A. L., Torjusen, H., Meltzer, H. M., Papadopoulou, E., Hoppin, J. A., Alexander, J., Lieblein, G., Roos, G., Holten, J. M., Swartz, J. and Haugen, M. 2016. Organic food consumption during pregnancy and hypospadias and cryptorchidism at birth: The Norwegian Mother and Child Cohort Study (MoBa). *Environmental Health Perspectives*, 124, 357–64.

Brouwer, I. A., Wanders, A. J. and Katan, M. B. 2013. Trans fatty acids and cardiovascular health: Research completed? *European Journal of Clinical Nutrition*, 67, 541–7.

Bruning-Fann, C. S. and Kaneene, J. B. 1993. The effects of nitrate, nitrite and N-nitroso compounds on human health – a review. *Veterinary and Human Toxicology*, 35, 521–38.

Bygren, L. O., Johansson, S.-E., Konlaan, B. B., Grjibovski, A. M., Wilkinson, A. V. and SjöStröM, M. 2009. Attending cultural events and cancer mortality: A Swedish cohort study. *Arts & Health*, 1, 64–73.

Cani, P. D. and Everard, A. 2015. Keeping gut lining at bay: Impact of emulsifiers. *Trends in Endocrinology and Metabolism*, 26, 273–4.

Centers for Disease Control and Prevention, C. D. C. 2016. *National Botulism Surveillance* [Online]. Available: https://www.cdc.gov/botulism/surveillance.html [Accessed 22 March 2017].

Christensen, J. S., Asklund, C., Skakkebaek, N. E., Jorgensen, N., Andersen, H. R., Jorgensen, T. M., Olsen, L. H., Hoyer, A. P., Moesgaard, J., Thorup, J. and Jensen, T. K. 2013. Association between organic dietary choice during pregnancy and hypospadias in offspring: A study of mothers of 306 boys operated on for hypospadias. *Journal of Urology*, 189, 1077–82.

Curl, C. L., Fenske, R. A. and Elgethun, K. 2003. Organophosphorus pesticide exposure of urban and suburban preschool children with organic and conventional diets. *Environmental Health Perspectives*, 111, 377–82.

Dardiotis, E., Xiromerisiou, G., Hadjichristodoulou, C., Tsatsakis, A. M., Wilks, M. F. and Hadjigeorgiou, G. M. 2013. The interplay between environmental and genetic factors in Parkinson's disease susceptibility: The evidence for pesticides. *Toxicology*, 307, 17–23.

de Gonzalez, M. T. N., Osburn, W. N., Hardin, M. D., Longnecker, M., Garg, H. K., Bryan, N. S. and Keeton, J. T. 2012. Survey of residual nitrite and nitrate in conventional and organic/natural/uncured/indirectly cured meats available at retail in the United States. *Journal of Agricultural and Food Chemistry*, 60, 3981–90.

de Jong, T. J. 1995. Why fast-growing plants do not bother about defence. *Oikos*, 74, 545–8.

Denver, S. and Christensen, T. 2015. Organic food and health concerns: A dietary approach using observed data. *Njas-Wageningen Journal of Life Sciences*, 74–5, 9–15.

de Ridder, D., Kroese, F., Evers, C., Adriaanse, M. and Gillebaart, M. 2017. Healthy diet: Health impact, prevalence, correlates, and interventions. *Psychology & Health*, 32(8), 907–41. doi:10.1080/08870446.2017.1316849.

Desjardins, Y. 2016. Are organically grown fruits and vegetables nutritionally better than conventional ones? Revisiting the question with new eyes. *Acta Hort. (ISHS)*, 1137:187–200. https://doi.org/10.17660/ActaHortic.2016.1137.27.

EFSA 2011. Guidance on the scientific requirements for health claims related to antioxidants, oxidative damage and cardiovascular health. *EFSA Journal*, 9(12), 2474.

European Parliament, C. O. T. E. U. (Ed.). 2006. Regulation (EC) No 1924/2006 of the European Parliament and of the Council of 20 December 2006 on nutrition and health claims made on foods. http://eur-lex.europa.eu/legal-content/EN/ALL/?uri=CELEX:02006R1924-20121129.

Fagerstedt, S., Hesla, H. M., Ekhager, E., Rosenlund, H., Mie, A., Benson, L., Scheynius, A. and Alm, J. 2016. Anthroposophic lifestyle is associated with a lower incidence of food allergen sensitization in early childhood. *Journal of Allergy and Clinical Immunology*, 137, 1253–6.e3.

Ford, N. A., Lashinger, L. M., Allott, E. H. and Hursting, S. D. 2013. Mechanistic targets and phytochemical strategies for breaking the obesity-cancer link. *Frontiers in oncology*, 3, 209.

Forster, S. 2016. *Buying Organic Food Costs UK Consumers almost £900 Extra Per Year* [Online]. Available: https://www.voucherbox.co.uk/press/2016/01/14/buying-organic-food-costs-uk-consumers-almost-900-extra-per-year/ [Accessed 17 March 2016].

Goetzke, B., Nitzko, S. and Spiller, A. 2014. Consumption of organic and functional food. A matter of well-being and health? *Appetite*, 77, 94–103.

Govari, M. and Pexara, A. 2015. Nitrates and Nitrites in meat products. *Journal of the Hellenic Veterinary Medical Society*, 66, 127–40.

Gruber, J., Schaffer, S. and Halliwell, B. 2008. The mitochondrial free radical theory of ageing - Where do we stand? *Frontiers in Bioscience*, 13, 6554–79.

Grundmann, R. 2017. The problem of expertise in knowledge societies. *Minerva*, 55(1), 25–48. doi:10.1007/s11024-016-9308-7.

Grunert, K. G. 2005. Food quality and safety: Consumer perception and demand. *European Review of Agricultural Economics*, 32, 369–91.

Hall, T. E., Engebretson, J., O'Rourke, M., Piso, Z., Whyte, K. and Valles, S. 2017. The need for social ethics in interdisciplinary environmental science graduate programs: Results from a nation-wide survey in the United States. *Science and Engineering Ethics*, 23(2), 565–88. doi:10.1007/s11948-016-9775-0.

Hansson, S. O. 1991. Is anthroposophy science? *Conceptus*, 25, 37–49.

Hao, H. H., Sander, P., Iqbal, Z., Wang, Y. L., Cheng, G. Y. and Yuan, Z. H. 2016. The risk of some veterinary antimicrobial agents on public health associated with antimicrobial resistance and their molecular basis. *Frontiers in Microbiology*, 7, 1626.

Hardy, S. and Bacon, L. L. P. 2014. *Ad Watchdog Upholds Complaint Against Naked Juice Antioxidant Claims* [Online]. Available: http://www.lexology.com/library/detail.aspx?g=f67eef4f-e74d-46b8-b6f5-59ffc73f6bb6 [Accessed17 March 2016].

Harman, D. 1956. Aging: A theory based on free radical and radiation chemistry. *Journal of Gerontology*, 11, 298–300.

Hasselbach, J. L. and Roosen, J. 2015. Motivations behind preferences for local or organic food. *Journal of International Consumer Marketing*, 27, 295–306.

He, C. and Mikkelsen, B. E. 2014. The association between organic school food policy and school food environment: Results from an observational study in Danish schools. *Perspectives in Public Health*, 134, 110–16.

Herrmann, S. S., Duedahl-Olesen, L. and Granby, K. 2015. Occurrence of volatile and non-volatile N-nitrosamines in processed meat products and the role of heat treatment. *Food Control*, 48, 163–9.

Hertog, M. G. L., Feskens, E. J. M., Hollman, P. C. H., Katan, M. B. and Kromhout, D. 1993. Dietary antioxidant flavonoids and risk of coronary heart-disease – The Zutphen Elderly Study. *Lancet*, 342, 1007–11.

Hever, J. 2011. *Reducing Reductionism: A Big Picture View of Nutrition* [Online]. Available: http://www.onegreenplanet.org/vegan-health/reducing-reductionism-a-big-picture-view-of-nutrition/[Accessed15 July 2017].

Hoffmann, I. 2003. Transcending reductionism in nutrition research. *American Journal of Clinical Nutrition*, 78, 514S–6S.

Hollman, P. C. H., Cassidy, A., Comte, B., Heinonen, M., Richelle, M., Richling, E., Serafini, M., Scalbert, A., Sies, H. and Vidry, S. 2011. The biological relevance of direct antioxidant effects of polyphenols for cardiovascular health in humans is not established. *Journal of Nutrition*, 141, 989S–1009S.

Hsu, S. Y., Chang, C. C. and Lin, T. T. 2016. An analysis of purchase intentions toward organic food on health consciousness and food safety with/under structural equation modeling. *British Food Journal*, 118, 200–16.

Huber, M., Knottnerus, J. A., Green, L., van der Horst, H., Jadad, A. R., Kromhout, D., Leonard, B., Lorig, K., Loureiro, M. I., van der Meer, J. W. M., Schnabel, P., Smith, R., van Weel, C. and Smid, H. 2011. How should we define health? *British Medical Journal*, 343, d4163.

Hyseni, L., Atkinson, M., Bromley, H., Orton, L., Lloyd-Williams, F., McGill, R. and Capewell, S. 2017. The effects of policy actions to improve population dietary patterns and prevent diet-related non-communicable diseases: Scoping review. *European Journal of Clinical Nutrition*, 71(6), 694–711. doi:10.1038/ejcn.2016.234.

IFOAM. 2005. *Principles of Organic Agriculture* [Online]. Available: https://www.ifoam.bio/en/organic-landmarks/principles-organic-agriculture [Accessed16 July 2017].

Jensen, B. H., Petersen, A., Nielsen, E., Christensen, T., Poulsen, M. E. and Andersen, J. H. 2015. Cumulative dietary exposure of the population of Denmark to pesticides. *Food and Chemical Toxicology*, 83, 300–7.

Kavanagh, A., Bentley, R. J., Turrell, G., Shaw, J., Dunstan, D. and Subramanian, S. V. 2010. Socioeconomic position, gender, health behaviours and biomarkers of cardiovascular disease and diabetes. *Social Science and Medicine*, 71, 1150–60.

Kummeling, I., Thijs, C., Huber, M., Van de Vijver, L. P. L., Snijders, B. E. P., Penders, J., Stelma, F., van Ree, R., van den Brandt, P. A. and Dagnelie, P. C. 2008. Consumption of organic foods and risk of atopic disease during the first 2 years of life in the Netherlands. *British Journal of Nutrition*, 99, 598–605.

Kushi, L. H., Doyle, C., McCullough, M., Rock, C. L., Demark-Wahnefried, W., Bandera, E. V., Gapstur, S., Patel, A. V., Andrews, K., Gansler, T. and Amer Canc, S. 2012. American Cancer Society Guidelines on Nutrition and Physical Activity for Cancer Prevention Reducing the Risk of Cancer With Healthy Food Choices and Physical Activity. *Ca-a Cancer Journal for Clinicians*, 62, 30–67.

Lecomte, M., Couedelo, L., Meugnier, E., Plaisancie, P., Letisse, M., Benoit, B., Gabert, L., Penhoat, A., Durand, A., Pineau, G., Joffre, F., Geloen, A., Vaysse, C., Laugerette, F. and Michalski, M. C. 2016. Dietary emulsifiers from milk and soybean differently impact adiposity and inflammation in association with modulation of colonic goblet cells in high-fat fed mice. *Molecular Nutrition and Food Research*, 60, 609–20.

Mariner, W. K. 2016. Beyond lifestyle: Governing the social determinants of health. *American Journal of Law & Medicine*, 42, 284–309.

Mikkelsen, B. E., Bruselius-Jensen, M., Andersen, J. S. and Lassen, A. 2006. Are green caterers more likely to serve healthy meals than non-green caterers? Results from a quantitative study in Danish worksite catering. *Public Health Nutrition*, 9, 846–50.

Mostafalou, S. and Abdollahi, M. 2013. Pesticides and human chronic diseases: Evidences, mechanisms, and perspectives. *Toxicology and Applied Pharmacology*, 268, 157–77.

Mulet, J. M. 2014. Should we recommend organic crop foods on the basis of health benefits? Letter to the editor regarding the article by Baranski et al. *British Journal of Nutrition*, 112, 1745–7.

Nenonen, T., Kaikkonen, R., Murto, J. and Luoma, M. L. 2014. Cultural services and activities: The association with self-rated health and quality of life. *Arts & Health*, 6, 235–53.

Oplatowska-Stachowiak, M. and Elliott, C. T. 2016. Food colors: Existing and emerging food safety concerns. *Critical Reviews in Food Science and Nutrition*, 57, 524–48.

Osterberg, J., Wingstrand, A., Jensen, A. N., Kerouanton, A., Cibin, V., Barco, L., Denis, M., Aabo, S. and Bengtsson, B. 2016. Antibiotic resistance in Escherichia coli from pigs in organic and conventional farming in four European countries. *PLoS ONE*, 11.

Petersen, S. B., Rasmussen, M. A., Strom, M., Halldorsson, T. I. and Olsen, S. F. 2013. Sociodemographic characteristics and food habits of organic consumers – a study from the Danish National Birth Cohort. *Public Health Nutrition*, 16, 1810–19.

Pieniak, Z., Aertsens, J. and Verbeke, W. 2010. Subjective and objective knowledge as determinants of organic vegetables consumption. *Food Quality and Preference*, 21, 581–8.

Pillsbury, M. W. 2011. *Price Differences, Organic Versus Non-Organic; Store Versus Farmers' Market* [Online]. Maine Organic Farmers and Gardeners Association. Available: http://www.mofga.org/Publications/MaineOrganicFarmerGardener/Fall2011/PriceDifferences/tabid/1966/Default.aspx [Accessed 17 March 2016].

Pino, G., Peluso, A. M. and Guido, G. 2012. Determinants of regular and occasional consumers' intentions to buy organic food. *Journal of Consumer Affairs*, 46, 157–69.

Prada, M., Garrido, M. V. and Rodrigues, D. 2017. Lost in processing? Perceived healthfulness, taste and caloric content of whole and processed organic food. *Appetite*, 114, 175–86.

Reganold, J. P. and Wachter, J. M. 2016. Organic agriculture in the twenty-first century. *Nature Plants*, 2, 15221.

Ristow, M. and Schmeisser, S. 2011. Extending life span by increasing oxidative stress. *Free Radical Biology and Medicine*, 51, 327–36.

Rittenhofer, I. and Povlsen, K. K. 2015. Organics, trust, and credibility: A management and media research perspective. *Ecology and Society*, 20(1), 6.

Roberts, T. L. 2014. Cadmium and phosphorous fertilizers: The issues and the science. *Procedia Engineering*, 83, 52–9.

Selman, C., McLaren, J. S., Collins, A. R., Duthie, G. G. and Speakman, J. R. 2013. Deleterious consequences of antioxidant supplementation on lifespan in a wild-derived mammal. *Biology Letters*, 9, 20130432.

Shay, C. M., Ning, H. Y., Allen, N. B., Carnethon, M. R., Chiuve, S. E., Greenlund, K. J., Daviglus, M. L. and Lloyd-Jones, D. M. 2012. Status of cardiovascular health in US adults prevalence estimates from the National Health and Nutrition Examination Surveys (NHANES) 2003–2008. *Circulation*, 125, 45–U377.

Shelton, J. F., Geraghty, E. M., Tancredi, D. J., Delwiche, L. D., Schmidt, R. J., Ritz, B., Hansen, R. L. and Hertz-Picciotto, I. 2014. Neurodevelopmental disorders and prenatal residential proximity to agricultural pesticides: The CHARGE Study. *Environmental Health Perspectives*, 122, 1103–9.

Siegrist, M., Visschers, V. H. M. and Hartmann, C. (2015). Factors influencing changes in sustainability perception of various food behaviors: Results of a longitudinal study. *Food Quality and Preference*, 46, 33–9. doi:10.1016/j.foodqual.2015.07.006.

Sloan, A. E. 2016. Top 10 functional food trends. *Food Technology*, 70, 24–45.

Srednicka-Tober, D., Baranski, M., Seal, C., Sanderson, R., Benbrook, C., Steinshamn, H., Gromadzka-Ostrowska, J., Rembialkowska, E., Skwarlo-Sonta, K., Eyre, M., Cozzi, G., Larsen, M. K., Jordon, T., Niggli, U., Sakowski, T., Calder, P. C., Burdge, G. C., Sotiraki, S., Stefanakis, A., Yolcu, H., Stergiadis, S., Chatzidimitriou, E., Butler, G., Stewart, G. and Leifert, C. 2016a. Composition differences between organic and conventional meat: A systematic literature review and meta-analysis. *British Journal of Nutrition*, 115, 994–1011.

Srednicka-Tober, D., Baranski, M., Seal, C. J., Sanderson, R., Benbrook, C., Steinshamn, H., Gromadzka-Ostrowska, J., Rembialkowska, E., Skwarlo-Sonta, K., Eyre, M., Cozzi, G., Larsen, M. K., Jordon, T., Niggli, U., Sakowski, T., Calder, P. C., Burdge, G. C., Sotiraki, S., Stefanakis, A., Stergiadis, S., Yolcu, H., Chatzidimitriou, E., Butler, G., Stewart, G. and Leifert, C. 2016b.

Higher PUFA and n-3 PUFA, conjugated linoleic acid, alpha-tocopherol and iron, but lower iodine and selenium concentrations in organic milk: A systematic literature review and meta- and redundancy analyses. *British Journal of Nutrition*, 115, 1043–60.

Sun, Y., You, W., Almeida, F., Estabrooks, P. and Davy, B. 2017. The effectiveness and cost of lifestyle interventions including nutrition education for diabetes prevention: A systematic review and meta-analysis. *Journal of the Academy of Nutrition and Dietetics*, 117(3), 404–21. doi:10.1016/j.jand.2016.11.016.

Swart, M. U., Kjeldsen, N. P. and Poulsen, L. A. 2016. Stort pristjek: Så meget dyrere er det at købe økologisk. *Søndagsavisen* [Online]. Available: http://www.sondagsavisen.dk/test/testogindkob/2016-06-03-stort-pristjek-sa-meget-dyrere-er-okologi/.

Torjusen, H., Brantsaeter, A. L., Haugen, M., Alexander, J., Bakketeig, L. S., Lieblein, G., Stigum, H., Naes, T., Swartz, J., Holmboe-Ottesen, G., Roos, G. and Meltzer, H. M. 2014. Reduced risk of pre-eclampsia with organic vegetable consumption: Results from the prospective Norwegian Mother and Child Cohort Study. *BMJ Open*, 4(9), e006143.

Torjusen, H., Brantsaeter, A. L., Haugen, M., Lieblein, G., Stigum, H., Roos, G., Holmboe-Ottesen, G. and Meltzer, H. M. 2010. Characteristics associated with organic food consumption during pregnancy; data from a large cohort of pregnant women in Norway. *BMC Public Health*, 10, 775.

Torjusen, H., Lieblein, G., Naes, T., Haugen, M., Meltzer, H. M. and Brantsaeter, A. L. 2012. Food patterns and dietary quality associated with organic food consumption during pregnancy; data from a large cohort of pregnant women in Norway. *BMC Public Health*, 12, 612.

Vecchio, R., Van Loo, E. J. and Annunziata, A. 2016. Consumers' willingness to pay for conventional, organic and functional yogurt: Evidence from experimental auctions. *International Journal of Consumer Studies*, 40, 368–78.

Viennois, E., Merlin, D., Gewirtz, A. T. and Chassaing, B. 2017. Dietary emulsifier-induced low-grade inflammation promotes colon carcinogenesis. *Cancer Research*, 77, 27–40.

Wang, X., Ouyang, Y., Liu, J., Zhu, M., Zhao, G., Bao, W. and Hu, F. B. 2014. Fruit and vegetable consumption and mortality from all causes, cardiovascular disease, and cancer: Systematic review and dose-response meta-analysis of prospective cohort studies. *BMJ-British Medical Journal*, 349, g4490.

World Health Organisation (WHO). 2016. *Botulism Fact sheet* [Online]. Available: http://www.who.int/mediacentre/factsheets/fs270/en/[Accessed22 March 2017].

World Health Organisation (WHO). 2017. *Nutrition, 5 keys to a healthy diet.* [Online]. Available: http://www.who.int/nutrition/topics/5keys_healthydiet/en/[Accessed15 July 2017].

Xie, L., Mo, M., Jia, H. X., Liang, F., Yuan, J. and Zhu, J. 2016. Association between dietary nitrate and nitrite intake and site-specific cancer risk: Evidence from observational studies. *Oncotarget*, 7, 56915–32.

Yu, C. J., Du, J. C., Chiou, H. C., Feng, C. C., Chung, M. Y., Yang, W. N., Chen, Y. S., Chien, L. C., Hwang, B. T. and Chen, M. L. 2016. Sugar-sweetened beverage consumption is adversely associated with childhood attention deficit/hyperactivity disorder. *International Journal of Environmental Research and Public Health*, 13, 678.

Yusuf, S., Hawken, S., Ounpuu, S., Dans, T., Avezum, A., Lanas, F., McQueen, M., Budaj, A., Pais, P., Varigos, J., Liu, L. S. and Investigators, I. S. 2004. Effect of potentially modifiable risk factors associated with myocardial infarction in 52 countries (the INTERHEART study): Case-control study. *Lancet*, 364, 937–52.

Zhu, Y., Wang, P. P., Zhao, J., Green, R., Sun, Z., Roebothan, B., Squires, J., Buehler, S., Dicks, E., Zhao, J., Cotterchio, M., Campbell, P. T., Jain, M., Parfrey, P. S. and Mclaughlin, J. R. 2014. Dietary N-nitroso compounds and risk of colorectal cancer: A case-control study in Newfoundland and Labrador and Ontario, Canada. *British Journal of Nutrition*, 111, 1109–17.

Part 4

Organic crop cultivation in the developing world

Chapter 18

Supporting smallholders in organic crop cultivation: the case of East Africa

R. Onwonga, University of Nairobi, Kenya; K. P. Sibuga, Sokoine University of Agriculture, Tanzania; H. Nduku, University of Nairobi, Kenya; L. Sigsgaard, University of Copenhagen, Denmark; A. Saria and L. Shechambo, Sokoine University of Agriculture, Tanzania; M. Montoro, University of Copenhagen, Denmark; C. Chepkoech, Q. Genga and R. Wahome, University of Nairobi, Kenya; N. Halberg, Aarhus University, Denmark; and H. Høgh-Jensen, Technical University of Denmark, Denmark

1 Introduction
2 The field study approach
3 Challenges and innovations in pest, weed and soil fertility management
4 Conclusion
5 Future trends
6 Where to look for further information
7 References

1 Introduction

Several African countries have large numbers of certified organic farmers; for example 190 000 in Uganda, 148 000 in Tanzania and 203 000 in Ethiopia (Willer and Lernoud, 2017) that would potentially benefit from access to high-value markets either domestically or via export (Halberg and Muller, 2013). However, there is a lack of research and development of appropriate agroecological methods for organic crop production (Ssekyewa et al., 2013), and stakeholders have identified financing of agricultural value chain research as a significant barrier (UNCTAD, 2016).

The value chains of several vegetables and other crops in Eastern Africa are growing in volume and value and are potentially making an impact on the livelihood of the involved producers, processers, transporters, retailers and sales-sites. This is partly due to trends like urbanization, globalization of many food value chains and increasing global and local tourism. This is in addition to the increasing purchasing power of the global middle class and increased interest in healthy and safe foods produced using environmentally friendly methods such as in organic agriculture.

http://dx.doi.org/10.19103/AS.2017.0029.20
Published by Burleigh Dodds Science Publishing Limited, 2019.

Organic agriculture can improve smallholder farmers' livelihood through a combination of increased income (secure market access and price premiums) and reduced debts from purchasing chemical inputs (Panneerselvam et al., 2013). Organic agriculture is one form of the so-called sustainable intensification (Halberg et al., 2015; Tittonell, 2014) and its success as a production form depends on knowledge-based agroecological farming practices in combination with well-organized value chains.

There is a strong need to explore the potential of applying a value chain approach to boost organic smallholder cropping systems, viability and productivity. High-value crops are tended through suitable agroecological methods that lead to the product quality that the market requires. If this is possible then there is hope for smallholders to release themselves out of the negative spiral of producing bulk low-value products to a standard market on more and more degraded land.

The need for agroecological intensification methods that promote sustainable health of soils with regard to physical, chemical and biological properties supporting sustained productivity of crops and subsequently leading to healthy crop produce that fetches high market premiums (Bouagnimbeck, 2009; Ssekyewa et al., 2013) is of essence. The methods include among others timely and efficient application of organic-based fertilizers, weed and pest control measures, mulching, crop rotation, intercropping and so forth, with the aim of enhancing and synchronizing the supply of nutrients to the crops and prevent pests and diseases (Palm et al., 2001).

In order to develop, test and document such practices for the improvement of organic farming practices in a market perspective, a group of soil and crop scientists, and economists from East Africa (Kenya, Tanzania, Uganda) and Denmark universities together with the respective national organic organizations (e.g. KOAN for Kenya, TOAM for Tanzania: and NUGAMO for Uganda) developed the 'Productivity and Growth in Organic Value Chains' project (ProGrOV; Høgh Jensen, 2011).

This chapter will report the outcomes of selected ProGrOV case studies in East Africa involving farmers, students and scientists together with advisors from organic organizations in developing and testing agroecological practices for the intensification of kale, maize, sweet pepper and tomatoes for organic markets. We selected certified organic chains because of their traceability and inherent focus on agroecological intensification methods.

2 The field study approach

2.1 Study site description

On-station and on-farm field experiments were carried out in Kenya and Tanzania for a period of 2 years covering four cropping seasons.

The on-station experiments were conducted at Kabete field station of the University of Nairobi, located about 15 km North of Nairobi, 1940 m a.s.l. on latitude 1°15′ S and longitude 36°41′ E, agro-climatical zone III (Jaetzold et al., 2006). The site has a bimodal rainfall distribution (mid-March to May, long rains; October to December, short rains). Average annual precipitation is 1000 mm with a minimum and maximum mean temperature of 13.7°C and 24.3°C, respectively. The soils are predominantly deep red nitisols containing 60–80% clay particles (FAO-UNESCO, 1990).

The on-farm experiments were conducted at Kiserian, in Kajiado County, Kenya and Ubiri and Forest villages in Lushoto district of northern Tanzania. Kajiado County is located between 01.42254° south and 036.66036° east with an elevation of 2070 m a.s.l. and categorized under agroecological zone IV (Jaetzold et al., 2006). The climate is typically semi-arid with annual temperature varying from 20°C and 28°C and a total annual rainfall between 450 and 1200 mm (Braun and Van de Weg, 1977). Kiserian soils are classified as phaeozems (FAO-UNESCO, 1990).

Ubiri (4°50′29″ south, 38°21′59″ east) and forest villages (40 47′55″ south, 380°17′24″ east) are located at 900–1200 m a.s.l. in north-eastern Tanzania with a minimum temperature range of 15–18°C in August and a maximum temperature range of 19–23°C in January/February. Rainfall is bimodal (800–1600 mm per annum) received during the short rain period (October to December), and long rain period (March to May).

2.2 Experimental design and treatments

The experimental layout was a randomized complete block design with a split plot arrangement replicated four times. Organically based soil nutrient management practices include use of organic inputs such as farmyard manure (FYM) and Minjingu rock phosphate (MRP) (Place et al., 2003; Isah et al., 2014; Onwonga et al., 2015; Simpson et al., 2011). In Kenya, the main plots were cropping systems: monocropping, intercropping and crop rotation and the split plots were FYM and MPR. Each plot measured 4.8 × 3.75 m. The test crops were: tomato (*Lycopersicon esculentum* L., variety Rio Grande) and maize (*Zea mays* L., variety H513) intercropped or in rotation with chickpea. Land was prepared manually using hand hoes in late February and September of 2012 and 2013. After FYM characterization, 10 tons ha^{-1} FYM was applied. Since Minjingu phosphate rock contains 28% P_2O_5, 490 kg ha^{-1} of it was applied to supply the recommended 60 kg P ha^{-1}. Spacing of 30 × 75 cm for maize, 45 × 75 cm for tomatoes and 10 × 30 cm for chickpea pure stands were adopted. Chickpea was intercropped in rows at 10 cm for both test crops. Weeding was done at 3 weeks after transplanting and fruiting. The crops were planted in March and October during the long- and short rain seasons of the experimental years.

In Tanzania there were seven treatments: pine tree (*Pinus patula* Schiede ex Schltdl. & Cham) leaves mulch 10 or 15 cm deep, wild sunflower (*Tithonia diversifolia*) grey leaves 10 or 15 cm deep, solarization, weeded or unweeded as the control. All types of mulch were applied 1 week after transplanting. The test crops were determinate tomato (*Lycopersicon esculentum* L, variety CAL-J) and sweet pepper (*Capsicum annuum*, variety California wonder).

All subplots, except the solarized subplots, were cultivated manually by hand hoe one day prior to transplanting. Solarized plots were cultivated 4 weeks earlier, watered to field capacity and then covered with a 0.1 mm thick transparent polyethylene sheet. Four-week-old tomato seedlings were transplanted, one seedling per hole in rows 75 cm apart and an intra-row spacing of 50 cm, which is equivalent to 26 667 plants ha^{-1}. Six-week-old sweet pepper seedlings were transplanted, one seedling per hole in rows 60 cm apart and an intra-row spacing of 40 cm to provide 41 667 plants ha^{-1}. The crops were planted in March and October during the long and short rain seasons of the experimental years.

Further details on site description and experimental design can be found in Genga (2014), Ndukhu (2014) and Chepkoech (2015) for Kenya and Saria (2015) for Tanzania.

3 Challenges and innovations in pest, weed and soil fertility management

3.1 Use of mulch to boost vegetable yields and control weeds and pests

Traditional weed control practices like hand hoeing or uprooting by hand requires significant labour inputs and limits the area a farmer can cultivate (Chianu and Akintola, 2003). In the early days of organic farming this was a main solution, in combination with certain plant extracts to control insect pest populations.

The use of biodegradable and/or organic mulches is becoming an established practice for weed control in organic farms, though usually considered to be economical only for high-value or perennial crops or on small acreages (Runham and Town, 1995). In our study, labour used to gather the mulching materials and their application to cover the plots were the criteria used for input cost estimation on the mulching treatment.

The studies showed that the cost for weed control using mulching practice could be estimated to be 18–23% of total production cost. Usoroh (1983) estimated that weeding alone costs 30–45% of the total labour required for fruit and vegetable production under African conditions.

Mulch may be a multipurpose strategy for organic farmers in East Africa to control weeds, for insect pest management and for supplying nutrients to the crops (Jackson and Harrison, 2008; Sinkeviciene et al., 2009). It can enhance the population of predators by providing them with shelter and protecting them from desiccation and temperature extremes, and thus increases predation rates (Johnson et al., 2004; Prasifka et al., 2006). Several studies have reported higher numbers of beneficial insects under mulched plots compared to bare soil (Jackson and Harrison, 2008; Johnson et al., 2004; Riechert and Bishop, 1990; Schmidt et al., 2004).

Improved weed control from mulching leads to increased yields of the tomato and sweet pepper cultivated at two different locations (Fig. 1). Pine mulch at 15 cm thickness was identified and ranked by farmers as the best alternative due to its long-lasting character, compact ability and the fact that the 15 cm level of pine mulch lasted throughout the production period.

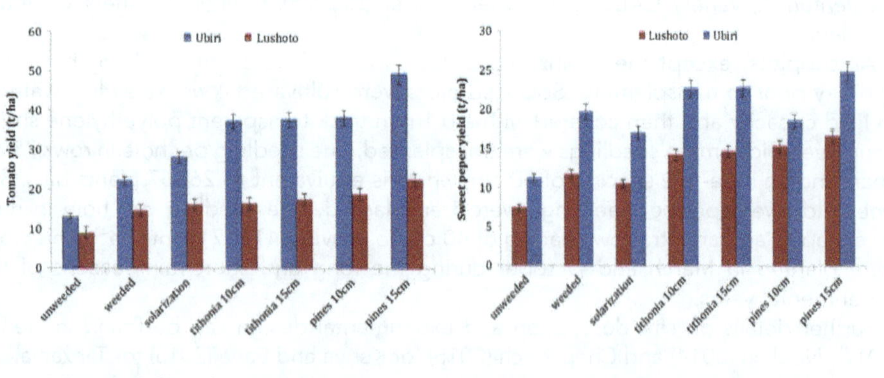

Figure 1 Tomato (left) and sweet pepper (right) yields (t ha^{-1}) in Ubiri and Lushoto in Tanzania.

Published by Burleigh Dodds Science Publishing Limited, 2019.

In tomato, unweeded plot (control) and hoe weeding were compared with solarization and two types of mulch material (Tithonia and Pines) each in 10 and 15 cm layers, respectively. Mulching resulted in up to 92% control of weeds compared to hoe weeding which controlled weeds by 81% in the tomatoes. Similarly, the case of sweet pepper crops, pine and wild lemon grass mulch controlled weeds by 83% (Fig. 2).

The results demonstrated that mulching also led to the protection of the produce against soil-borne pests and influenced the presence of beneficial arthropods such as spiders, ladybird beetles, bees and beneficial ants. Mulching attracted up to 24% more beneficial insects compared to bare soil plots. The beneficial insects observed were both natural enemies (ladybirds and spiders) and pollinators (bees and butterflies, Fig. 3).

The 15 cm pine mulch (*Pinus patula*) had the lowest aphid and thrips infestation. Somewhat higher numbers were observed with wild sunflower mulch (*Tithonia diversifolia*), but still pest infestation was lower than in hand-weeded and unweeded control. Cutworm densities in mulched plots were lower than those in unweeded controls.

The best performing mulch (15 cm pine) was subsequently tested in farmers' fields in the same area against 15 cm wild lemon grass, hand-weeded and unweeded control in both tomato and green pepper to assess the effects of treatments on ground-living predators. Mulches were found to increase the presence of beneficial arthropods such as spiders, ladybird beetles, bees and ants.

Further, both types of mulches had higher densities of the most common ground-active predator group, the wolf spiders (Lycosidae), and this was true both in the normal season and the off-season crops.

Though ground-living predators such as wolf spiders, which were more numerous in the mulched plots, do not climb up in the crop, they contributed to the observed reduction of aphids in the mulched plots through their own behaviour. Aphids regularly let themselves fall to the ground as a way of escaping plant-living natural enemies such as ladybirds, other predators or parasitoids and once on the ground the aphids can be captured and eaten by ground-living predators.

3.2 Integration of legumes, use of farmyard manure and locally available rock phosphate

The soil health/fertility challenge was addressed through testing various cropping systems involving legumes (chickpea and lupine) and the application of organic fertilizers/inputs

Figure 2 Tomato with pine mulch 10 cm (left) and 15 cm (right) at farm trials in Tanzania.

Published by Burleigh Dodds Science Publishing Limited, 2019.

Figure 3 Top left: Treatments, pitfall traps and baiting cards. Bottom left: Examples of the beneficial organisms recorded in tomato plots in 2014 [Coccinellidae, Carabidae, Carabidae and Lycosidae (wolf spider)]. Wolf spiders were the dominant epigeic predators.

(FYM and MRP). Vegetable crops (kales and tomatoes) and maize were used as test crops. The test crops were either grown as sole crops, in rotation and/or as intercrops (Fig. 4).

Other studies have shown that integration of legumes in cropping systems with application of organic fertilizers improves nutrient availability especially N and P and boosts crop productivity. For example, legume–cereal intercropping is practised widely in many countries, where it has been revealed that certain P-efficient grain legumes can mobilize sparingly available P, through rhizosphere processes (Giller, 2001), thereby enhancing its availability for the P-inefficient cereals (Dissanayaka et al., 2015; Li et al., 2003, 2010).

Among these species, the ability of white lupine (*Lupinus albus*) to enhance P availability for neighbouring cereals in P-deficient soils has been investigated in detail (Cu et al., 2005; Dissanayaka et al., 2015; Kamh et al., 1999; Lelei Joyce and Onwonga Richard, 2014; Li et al., 2010). White lupine, with its ability to form cluster roots to enhance nutrient acquisition from nutrient-poor soils and exude vast amounts of phosphate-mobilizing substances (Lambers et al., 2013), has been identified as an ideal plant species to thrive in soils where P remains in sparingly soluble forms.

Leguminous plants may, due to their ability to fix atmospheric nitrogen, improve soil fertility in both crop rotation and intercropping systems (Li et al., 2004). Intercropping maize and tomatoes with chickpea has been reported to increase light interception in the intercrops, reduce water evaporation and improve conservation of the soil moisture when compared with maize and tomato monocrops (Ghanbari et al., 2010).

Figure 4 Test crops at Kabete field station experimental site in Kenya.

Application of FYM and MRP with integration of legumes like white lupine and chickpea in kale cropping systems could provide a feasible low-cost alternative for rebuilding soil fertility (Opala et al., 2013; Lelei Joyce and Onwonga Richard 2014).

3.2.1 Effect of chickpea integration and organic inputs on tomato performance

Tomato fruit yields (fresh weight in t ha^{-1}) at two sites over two cropping seasons (season III and IV) were compared with unfertilized control with the application of FYM and MRP. The higher per ha yield of tomato fruits under sole and legume/tomato rotation than chickpea/tomato intercropping (Table 1) is mainly due to the lack of competition for resources common with intercropping. Similarly, the higher yields in the intercropping system compared with the monocrop is indicative of the fact that intercropping tomatoes with grain legumes and application of MRP improve plant-available soil phosphorus (Fig. 5) and subsequently tomato fruit yields. Grain legumes like chickpea and white lupine contribute to mobilization and availability of sparingly soluble phosphate. Promoting the use of these legumes, as agents of solubilizing rock phosphate, will therefore enhance P and N availability for crop uptake and subsequently enhance yield (Giller, 2001).

Table 1 Effects of cropping systems and organic inputs on tomato fruit yield at Kabete and Kiserian, Kenya

			Tomato fruit yield (t ha^{-1})	
Season	Cropping system	Organic inputs	Kabete	Kiseria
III	Intercrop	CONT	4.0^{c}	3.8^{c}
		MRP	5.6^{fg}	5.2^{fg}
	Monocrop	CONT	4.2^{d}	3.6^{b}
		FYM	7.0^{k}	7.0^{lm}
		MRP	5.8^{hi}	5.0^{e}
IV	Crop rotation	CONT	4.4^{d}	4.0^{d}
		FYM	8.0^{o}	7.6^{o}
		MRP	5.8^{hi}	5.6^{hi}
	Intercrop	CONT	3.8^{b}	4.0^{d}
		FYM	7.2^{lm}	6.8^{kl}
		MRP	5.2^{e}	5.6^{i}
	Monocrop	CONT	4.2^{d}	3.6^{b}
		FYM	7.2^{kl}	7.0^{lm}
		MRP	6.0^{j}	5.0^{e}
		Mean	4.6	4.4
		$LSD_{0.05}$	0.2	0.2

Abbreviations: CONT = Control, MRP = Minjingu Rock Phosphate, FYM = Farmyard Manure.
Means with the same letters within a column are not significantly different ($P \leqslant 0.05$).

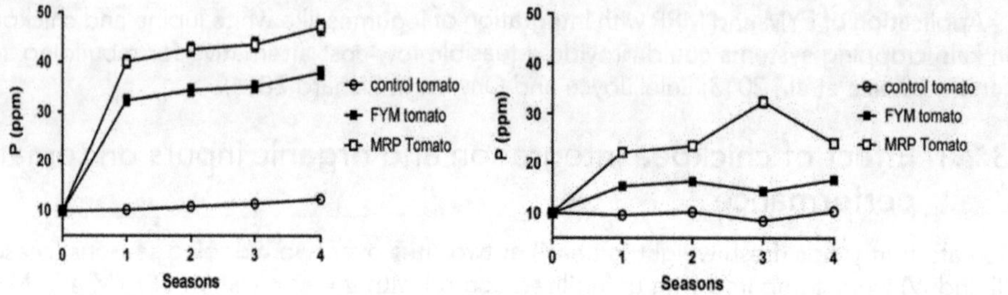

Figure 5 Plant-available P under chickpea–tomato rotation (left) and tomato monocrop (right) at Kabete field station in Kenya.

3.2.2 Effect of legume integration and organic inputs on maize performance

Maize grain yields (t ha^{-1}) under chickpea–maize rotation were significantly improved by adding FYM and/or MRP over a four-season study (Table 2). However, results for only

Table 2 Effects of cropping systems and organic inputs on maize yield for selected cropping seasons (III and IV) at Kabete and Kiserian experimental sites in Kenya

			Yield (t ha^{-1})	
Season	Cropping system	Organic inputs	Kabete	Kiserian
III	Intercrop	CONT	1.80^{b}	1.28^{b}
		FYM	5.30^{h}	4.77^{hi}
		MRP	4.31^{de}	3.78^{de}
	Monocrop	CONT	1.89^{b}	1.37^{b}
		FYM	5.45^{ij}	4.92^{hi}
		MRP	4.53^{def}	3.63^{d}
IV	Crop rotation	CONT	2.45^{c}	1.91^{c}
		FYM	5.67^{j}	5.09^{i}
		MRP	5.00^{gh}	4.44^{fg}
	Intercrop	CONT	1.82^{b}	1.29^{b}
		FYM	5.34^{ij}	4.82^{hi}
		MRP	4.35^{de}	3.83^{d}
	Monocrop	CONT	1.92^{b}	1.38^{b}
		FYM	5.49^{ij}	4.92^{hi}
		MRP	4.56^{e}	4.02^{e}
		Mean	3.30	2.83

Abbreviations: CONT = Control, MRP = Minjingu rock phosphate, FYM = Farmyard Manure.
Intercrop = Chickpea/Maize, Monocrop = Sole maize, Crop Rotation = Chickpea – Maize.
Means with the same letters within the column are not significantly different ($P < 0.05$).

seasons III and IV are shown.

Lathwell and Douglas (1990) reported that planting crops such as maize immediately following the incorporation of a legume green manure such as chickpea improved yields and if properly managed legume green manures have the potential to meet much of the N requirement of the succeeding non-legume crop. This corresponds well with other scholars who have studied the contribution that atmospherically derived nitrogen fixed by a leguminous crop may have on a succeeding crop (Adu-Gyamfi et al., 2007; Laberge et al., 2011; Sonbedjee et al., 2006). For example, Prasad (1996) found quantities in the range of 30–90 kg N ha^{-1} to be transferred to the succeeding crop and Sogbedji et al. (2006) reported that use of Mucuna (*Mucuna pruriens* L.) and pigeonpea (*Cajanus cajan* L.) after maize in the first year reduced N and P fertilizer need in the subsequent year.

The findings are in line with those of Nalatwadmath et al. (2003) who reported significant increases of grain yields when FYM and P fertilizers were used compared with control treatments. This implies the possibility of replacing chemical fertilizers with organic fertilizers on sites where organic fertilizers are available and of high quality.

3.2.3 Effect of legume integration and organic inputs on kale performance

Kale yields in the first month of long rainy season were highest in kale monocrop system (pure kale), followed by lupin/kale and chickpea/kale intercropping system with application of FYM and MRP, respectively (Fig. 6). The same trend was observed in the second (with chickpea/kale having higher yields) and third months. In the first month of short rainy season (SRS), pure kale (FYM), followed by lupin/kale and chickpea/kale (FYM and MRP) and chickpea/kale (MRP and FYM), had the highest yields.

In the second month of the SRS, the highest kale yields were recorded under lupin–kale, kale monocrop and lupin/kale (FYM and MRP) and were significantly not different from each other (Fig. 6).

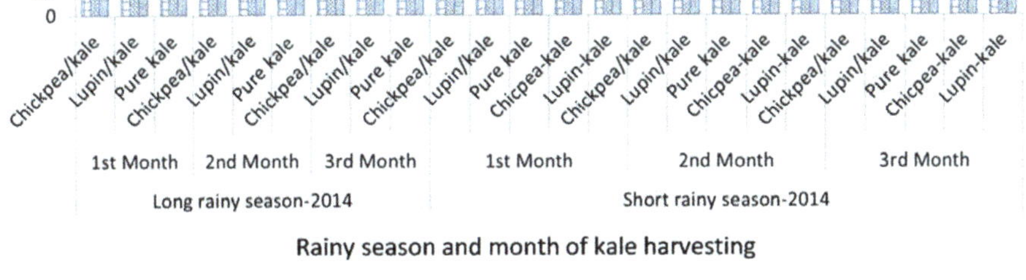

Figure 6 Kale yields as affected by cropping systems and organic inputs in the long and short rainy seasons of 2014 at Kabete field station, Kenya.

In the third month of kale harvesting, a slight decline in yield was realized; however, higher yields were recorded in lupine–kale rotation, chickpea–kale rotation and kale monocrop with application of FYM and MRP. The lowest yield in all cases was recorded in the control treatment, with lupin/kale intercrop (control) having the lowest yield (Fig. 6).

Synchronization of legume residue nutrient release with kale nutrient demand is imperative for enhanced soil fertility and kale productivity. The study further demonstrated that for the kale to benefit from the nutrients released by legume residues, dry chickpea and lupine legume residues requires it to be applied simultaneously with kale planting (Fig. 7).

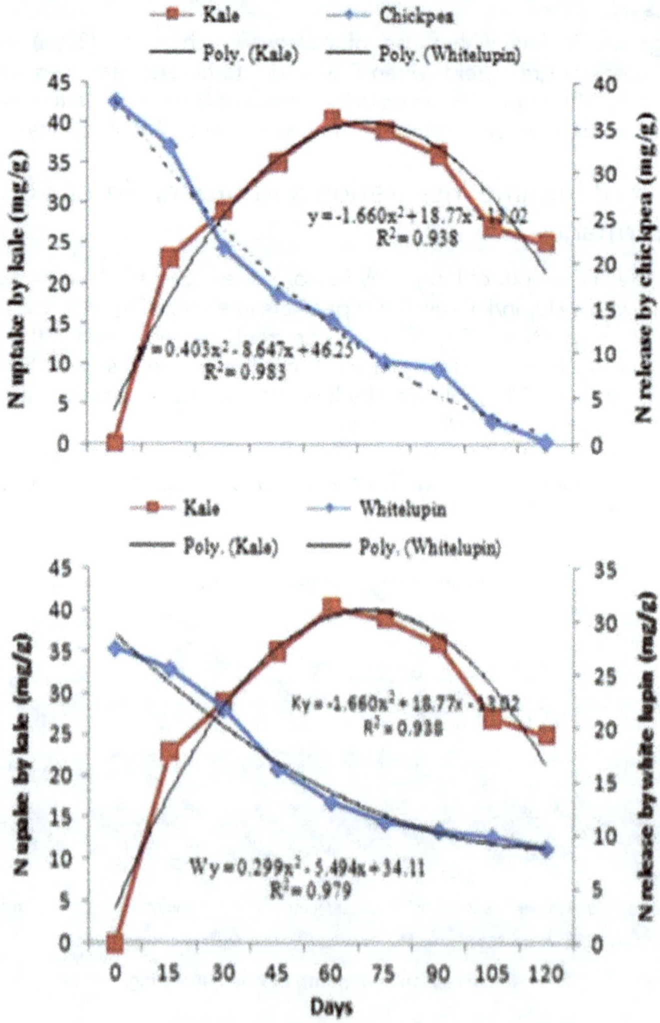

Figure 7 Legume nitrogen release in relation to kale nitrogen uptake at Kabete, Kenya.

4 Conclusion

Higher yield in crops treated with dead organic mulches positively affected natural enemies of pests (beneficial arthropods), reduced pest pressure and improved weed control, thereby contributing to the observed higher yields in both tomato and sweet pepper, potentially leading to increased net income to organic farmers' share and incentivize farmers to produce tomatoes and sweet pepper during the normal season as well as during the off-season.

Our case studies on mulch, in Tanzania, have therefore demonstrated that locally available plant material can be used successfully as dead mulch and that both the plant material and the thickness of the mulch layer will affect the outcome of mulching on weeds, insect pests and their natural enemies. As mulching is already in use and is a multipurpose strategy, it has the potential to become even more widely adopted and not only benefit the organic producers but also provide conventional growers with an alternative that can help them reduce pesticide use. Adopting the use of these mulching materials would provide an opportunity to produce vegetables without relying on either high-labour input or pesticide application and with higher or similar yields compared to the current farmer practice. A wider adoption of such practices could increase the quantity of organically produced vegetables and thus contribute to increasing its market.

In Kenya, maize, kale and tomato crops were grown as intercrops and/or in rotation with legumes and supplemented with locally available rock phosphate and FYM. The legumes contributed nitrogen to the accompanying crop. Thus, the total yield of the intercrops was higher than or similar to those of monocrops with addition of manure. However, for intercropping to be an interesting practice to farmers, the legume grown should be (as it is the case for lupin and chickpea) either locally consumed or suitable for a market accessible to the farmers.

Based on the control treatments (i.e. with no organic inputs), the experiments thus demonstrated that it is possible to double or even triple yields from the farmers' fields and give the same effect as using recommended amounts of FYM, which is limited on most farms.

These agroecological methods (legume integration with use of organic inputs) are not new. They are well demonstrated and in many studies and locations under varying conditions around the globe.

However, to use them effectively requires knowledge. Overall, the current case studies demonstrate that smallholder farmers may improve crop yields and quality substantially by using more knowledge on crops and their management and by intensifying use of locally available resources.

Adapting agroecological practices to target specific markets and quality requirements will allow for a significant intensification of vegetable production among smallholder farmers in East Africa. This is only likely to happen if the actors in value chains improve the governance in order to secure market access and supply in order to incentivize farmers to continue taking up and further develop such knowledge-intensive practices.

5 Future trends

The ProGrOV project has demonstrated the potential for improving and intensifying organic crop production under smallholder farming systems using agroecological practices. The research will further provide the farmers and value chain actors with more research-based evidence of the ecological advantages of agricultural intensification based on improved knowledge and use of locally available resources rather than intensification based on purchasing expensive external inputs.

The research capacity and interest in this area established among a group of young agronomists and economists in East Africa universities and the links established with the industry and the National Organic Movements in Kenya and Tanzania further provide a potential for professionalization of the agroecological and organic farming movement. This can improve the credibility and science-based policy development within the organizations, possibly leading to increased support for agroecologically based livelihood improvement of smallholder farmers and the further market-oriented development of the organic sector in East Africa.

The need for strengthening the research capacity and interest in agroecological practices can therefore be met in cross-disciplinary consortia between universities and stakeholder organizations by engaging students and scientists in participatory research in farmers' fields. The innovations also need to take into account the market demands for product quality.

ProGrOV developed a concept for linking research and development across value chains of selected products. However, more capacity in analysing and providing innovative solutions for improvements in value chain governance is still underdeveloped. Actors in the value chains need new methods for securing long-term engagement and building trust between them.

6 Where to look for further information

Further information on the application of agroecological practices for intensification and enhanced crop production can be accessed from the various postgraduate student works as published in respective thesis, published papers in refereed articles and presented in conferences, policy brief workshops and brief papers on the ProGrOV website (http://icrofs.dk/en/research/international-research/progrov/), as well as in Organic E-prints.

7 References

Adu-Gyamfi, J. J., Myaka, F. A., Sakala, W. D., Odgaard, R., Vesterager, J. M. and Høgh-Jensen, H. (2007). Biological nitrogen fixation and nitrogen and phosphorus budgets in farmer-managed intercrops of maize–pigeonpea in semi-arid southern and eastern Africa. *Plant and Soil* 295, 127–36.

Bouagnimbeck, H. (2009). Organic farming in Africa. In: H. Willer und L. Kilcher (Eds), *The World of Organic Agriculture. Statistics and Emerging Trends*. IFOAM, Bonn and FiBL, Frick, ITC, Geneva.

Braun, H. M. H. and Van de Weg, R. F. (1977). Proposal for rating of land qualities (2nd approximation). Report no. IC 7. Kenya Soil Survey, Nairobi.

Chepkoech, C. (2015). Effect of legume integration and application of organic fertilizers on soil nutrient status and kale yield in Kabete, Kenya. MSc Thesis, University of Nairobi, Nairobi, Kenya.

Chianu, J. N. and Akintola, J. O. (2003). Labour needs for weeding: Implications for the adoption of improved fallow management systems in the derived savannas of south-western Nigeria. *Nigerian Journal of Weed Science* 16, 1–7.

Published by Burleigh Dodds Science Publishing Limited, 2019.

Cu, S. T., Hutson, J. and Schuller, K. A. (2005). Mixed culture of wheat (*Triticum aestivum* L.) with white lupin (*Lupinus albus* L.) improves the growth and phosphorus nutrition of the wheat. *Plant and Soil* 272(1), 143–51.

Dissanayaka, D. M. S. B., Maruyama, H., Masuda, G. and Wasaki, J. (2015). Interspecific facilitation of P acquisition in intercropping of maize with white lupin in two contrasting soils as influenced by different rates and forms of P supply. *Plant and Soil* 390(1–2), 223–36.

FAO-UNESCO (1990). Soil map of the world. Revised Legend, World soil resources report no. 60. FAO, Rome.

Genga, Q. (2014). Effect of organic-based fertility management strategies on soil nutrient status and marketable quality of kale in Kabete, Kenya. MSc Thesis, University of Nairobi, Nairobi, Kenya.

Ghanbari, A., Dahmardeh, M., Siahsar, B. A. and Ramroudi, M. (2010). Effect of maize (*Zea mays* L.) – cowpea (*Vigna unguiculata* L.) intercropping on light distribution, soil temperature and soil moisture in arid environment. *Journal of Food, Agriculture and Environment* 8(1), 102–8.

Giller, K. E. (2001). *Nitrogen Fixation in Tropical Cropping Systems*. CAB International, Wallingford, UK.

Halberg, N. and Muller A. (Eds) (2013). *Organic Agriculture for Sustainable Livelihoods*. Routledge, Oxon, UK.

Halberg, N., Panneerselvam, P. and Tryer, S. (2015). Eco-functional intensification and food security: Synergy or compromise? *Sustainable Agriculture Research* 4(3), 126–39.

Høgh Jensen, H. (2011). Innovation research in value chains. *Icrofs News* (2), 10–13. Available at: http://icrofs.dk/fileadmin/icrofs/ProGrOV/I_news_2011_2_page_11_13.pdf

Isah, A. S., Amans, E. B., Odion, E. C. and Yusuf, A. A. 2014. Growth rate and yield of two tomato varieties (*Lycopersicon esculentum* Mill) under green manure and NPK fertilizer rate Samaru northern Guinea savanna. *International Journal of Agronomy* 2014, Article ID 932759, 8p. http://dx.doi.org/10.1155/2014/932759

Jackson, D. M. and Harrison, H. F. (2008). Effect of a killed-cover crop mulching system on sweet potato production, soil pests, and insect predators in South Carolina. *Journal of Economic Entomology* 101(6), 1871–80.

Jaetzold, R., Schmidt, H., Hrnetz, B. and Shisanya, C. (2006). *Farm management handbook Vol II*, Part C, East Kenya. Subpart C1, Eastern Province. Ministry of Agriculture, Nairobi, Kenya.

Jemo, M., Abaidoo, R. C., Nolte, C., Tchienkoua, M., Sanginga, N. and Horst, W. J. (2006). Phosphorus benefits from grain-legume crops to subsequent maize grown on acid soils of southern Cameroon. *Plant and Soil* 284(1), 385–97.

Johnson, J. M., Hough-Goldstein, J. A. and Vangessel, M. J. (2004). Effects of straw mulch on pest insects, predators, and weeds in watermelons and potatoes. *Environmental Entomology* 33(6), 1632–43.

Kamh, M., Horst, W. J., Amer, F., Mostafa, H. and Maier, P. (1999). Phosphate by cover crops. *Plant and Soil* 211, 19–27.

Laberge, G., Haussmann, B., Ambus, P. and Høgh-Jensen, H. (2011). Cowpea rhizodeposition and belowground transfer of nitrogen to millet in intercrop and rotation in a sandy soil of the Sudano-Sahelian eco-zone of Niger. *Plant and Soil* 340, 360–82.

Lambers, H., Clements, J. C. and Nelson, M. N. (2013). How a phosphorus-acquisition strategy based on carboxylate exudation powers the success and agronomic potential of lupines (Lupinus, Fabaceae). *American Journal of Botany* 100(2), 263–88.

Lelei Joyce, J. and Onwonga Richard, N. (2014). White lupin (*Lupinus albus* L. cv. *Amiga*) increases solubility of Minjingu phosphate rock, phosphorus balances and maize yields in Njoro Kenya. *Sustainable Agriculture Research* 3(3), 37–49.

Li, L., Zhang, F. S., Li, X. L., Christie, P., Sun, J. H., Yang, S. C. and Tang, C. (2003). Interspecific facilitation of nutrient uptake by intercropped maize and faba bean. *Nutrient Cycling in Agroecosystems* 68, 61–71.

Li, F. M., Song, Q. H., Jjemba, P. K. and Shi, Y. C. (2004). Dynamics of soil microbial biomass C and soil fertility in cropland mulched with plastic film in a semiarid agro-ecosystem. *Soil Biology and Biochemistry* 36(11), 1893–902.

Li, H., Shen, J., Zhang, F., Marschner, P., Cawthray, G. and Rengel, Z. (2010). Phosphorus uptake and rhizosphere properties of intercropped and monocropped maize, faba bean, and white lupin in acidic soil. *Biology and Fertility of Soils* 46(2), 79–91.

Nalatwadmath, S. K., Rao, M. S. R. M., Patil, S. L., Jayaram, N. S., Bhola, S. N. and Prasad, A. (2003). Long term effects of integrated nutrient management on crop yields and soil fertility status in vertisols of Bellary. *Indian Journal of Agricultural Research* 37(1), 64–7.

Ndukhu, H. (2014). Evaluating the influence of organic amendments and legume integration on soil chemical properties and crop performance in Central Kenya. MSc Thesis, University of Nairobi, Nairobi, Kenya.

Onwonga, R. N., Namoi, N. L. and Lelei, J. J. (2015). Influence of organic based technologies on soil nutrient status in semi-arid yatta sub-County, Kenya. *Journal of Agricultural Science* 7(8), 56.

Opala, P. A., Okalebo, J. R. and Othieno, C. (2013). Comparison of effects of phosphorus sources on soil acidity, available phosphorus and maize yields at two sites in western Kenya. *Archives of Agronomy and Soil Science* 59(3), 327–39.

Palm, C., Giller, K., Mafongoya, P. and Swift, M. (2001). Management of organic matter in the tropics: Translating theory into practice. *Nutrient Cycling in Agroecosystems*, 61(2), 63–75.

Panneerselvam, P., Halberg, N. and Lockie, S. (2013). Consequences of organic agriculture for smallholder farmers' livelihood and food security. In: N. Halberg and A. Müller (Eds), *Organic Agriculture for Sustainable Livelihoods* (Chapter 2, pp. 21–44). Routledge, London and New York.

Place, F., Christopher, B., Barrett, H., Freeman, A. and Ramisch, J. J. and Vanlauwe, B. (2003). Prospects for integrated soil fertility management using organic and inorganic inputs: Evidence from smallholder African agricultural systems. *Food Policy* 28, 365–78.

Prasad, R. (1996). Cropping system & agricultural sustainability. *Indian Farming* 46, 39–44.

Prasifka, J. R., Schmidt, N. P., Kohler, K. A., O'Neal, M. E., Hellmich, R. L. and Singer, J. W. (2006). Effects of living mulches on predator abundance and sentinel prey in a corn–soybean–forage rotation. *Environmental Entomology* 35(5), 1423–31.

Riechert, S. E. and Bishop, L. (1990). Prey control by an assemblage of generalist predators: Spiders in garden test systems. *Ecology* 71(4), 1441–50.

Runham, S. R. and Town, S. J. (1995). An economic assessment of mulches in field scale vegetable crops. *Proceedings of the Brighton Crop Protection Conference – Weeds*, Brighton, pp. 925–30.

Saria, J. (2015). Incidence of emerging and innovative climate change adaptation practices for smallholder farmers in Nachingwea District, Southern Tanzania. *International Journal of Environmental Monitoring and Analysis*, 3(5), 303–11.

Schmidt, M. H., Thewes, U., Thies, C. and Tscharntke, T. (2004). Aphid suppression by natural enemies in mulched cereals. *Entomologia Experimentalis et Applicata* 113(2), 87–93.

Simpson, R., Oberson, A., Culvenor, R., Ryan, M. H., Veneklaas, E. J., Lambers, H., Lynch, J. P., Ryan, P. R., Delhaize, E., Smith, S. E., Harvey, P. R. and Richardson, A. E. (2011). Strategies and agronomic interventions to improve the phosphorus-use efficiency of farming systems. *Plant and Soil*, 349, 89–120.

Sinkeviciene, A., Jodaugiene, D. and Pupaliene, R. (2009). The influence of organic mulches on soil properties and crop yield. *Agronomy Research*, 7, 485–91.

Ssekyewa, C., George, F. and Muller, A. (2013). Research needs for development of organic agriculture in Sub-Saharan Africa. In: N. Halberg and A. Muller (Eds), *Organic Agriculture for Sustainable Livelihoods*. Routledge, London, UK.

Tittonell, P. (2014). Ecological intensification of agriculture – sustainable by nature. *Current Opinion in Environmental Sustainability*, 8, 53–61.

UNCTAD (2016). United Nations Conference on Trade and Development: Annual Report 2016. http://unctad.org/en/PublicationsLibrary/dom2017_en.pdf

Usoroh, N. (1983). Weed problems and control strategies in fruit and vegetable production in Nigeria. *Acta Horticulture*, 123, 299–304.

Willer, H. and Lernoud, J. (2017). *The World of Organic Agriculture: Statistics and Emerging Trends 2017*. Research Institute of Organic Agriculture (FiBL) and IFOAM – Organics International, Bonn, Germany.

Chapter 19

Improving organic agriculture in the developing world: Africa

Victor Olowe, Institute of Food Security, Environmental Resources and Agricultural Research (IFSERAR), Federal University of Agriculture, Abeokuta (FUNAAB), Nigeria

1 Introduction
2 Status of organic agriculture in Africa
3 EOA-I in Africa
4 Steps to improving organic agriculture in Africa
5 Conclusion
6 Where to look for further information
7 References

1 Introduction

Africa is the second largest continent in size after Asia and is rich in natural resources. Agriculture still remains the backbone of the continent's economy since it accounts for about 58% of the economically-active population, 34% of gross domestic product and one-third of economic growth (FAO, 2014). The continent accounts for 25% of the world's arable land and yet produces just 10% of global agricultural output (Jayaram et al., 2010). In spite of this, and with 11 of its countries (such as Burundi, Nigeria, Rwanda and others) featuring among the 25 most arable countries in the world, the majority of these countries still remain food insecure (Fig. 1). Of the 850 million people suffering from hunger in the world, 239 million (28.1%) are in Africa. Sub-Saharan Africa (SSA) remains a food-insecure region of the continent partly because about 75% of the land is degraded with less than 4% under irrigation. In spite of these problems, agriculture in SSA still contributes between 30 and 50% of national income in the region and has significant potential to generate greater income if properly practised (Toenniessen et al., 2008). McIntire (2014) has listed the challenges (and potential solutions) facing agriculture as follows:

1 raising agricultural productivity to accelerate the delayed shift of labour and national product into industry and services;

http://dx.doi.org/10.19103/AS.2017.0029.21

2 slowing high population growth, which leads to greater land pressure and progressively blocks the traditional avenue of rural growth via land expansion;
3 seizing the opportunity provided by economic growth in the service, industrial and natural resource sectors, to expand demand and induce technical and land use changes, including both farm consolidation and farm fragmentation;
4 addressing food insecurity and malnutrition and their relation to the fiscal and management capacities of African states;
5 mitigating climate change, the effects of which are projected to be especially adverse for Africa's agricultural potential; and
6 developing public policies – investments and incentives – to stimulate agriculture's potential.

Against the backdrop of these challenges, the African heads of state took a decision (EX. CL/Dec. 621 (XVII) in 2011) to mainstream organic agriculture into the agricultural systems of all member states by the year 2020. The full details of the decision are presented in Annexe 1. Following this decision, the concept of ecological organic agriculture (EOA) was developed as a holistic system. At present, the global market of organic products stands at US$80 billion (Willer and Lernoud, 2015). Despite the thriving market, certified organic agriculture is still very insignificant in terms of its share of global farmland (<1%) relative to total global farmland. Similarly, overall funding for organic research is less than 1% of the total fund allocated to agricultural research even though organic farmland in Europe is over 4% of total farmland in the continent (Rahmann and Aksoy, 2014). A recent survey identified that lack of funding was a particular constraint facing organic agriculture in Africa (UNCTAD, 2016).

Figure 1 Countries in Africa with greatest proportion of arable land.

Annexe 1 African Union Council's Decision on Organic Farming

The African Union Council took a decision during the 18th Ordinary Session, 24–28 January 2011, EX.CL/Dec.621 (XVIII) as follows:

The Executive Council,

1 TAKES NOTE of the Report of the Conference of Ministers of Agriculture held in Lilongwe, Malawi, on 28 and 29 October 2010 on organic farming, and ENDORSES the Resolution contained therein;
2 EXPRESSES concern over the current practice of exploitation of the organic farmers in Africa;
3 REQUESTS the Commission and its New Partnership for Africa's Development Planning and Coordinating Agency (NPCA) to:
    - Initiate and provide guidance for an African Union (AU)-led coalition of international partners on the establishment of an African organic farming platform based on available best practices; and
    - Provide guidance in support of the development of sustainable organic farming systems and improve seed quality.
4 CALLS UPON development partners to provide the necessary technical and financial support for the implementation of this decision;
5 REQUESTS the Commission to report regularly on the implementation of this decision.

2 Status of organic agriculture in Africa

Global organic farmland measures around 47.3 million hectares, distributed as follows: Africa (3%), the United States and Canada (7%), Latin America (15%), Asia (8%), Europe (27%) and Oceania (40%) (Willer and Lernoud, 2015). Organic agricultural land in Africa spreads over 1.3 million ha of land, with most (455 000 ha) concentrated in the East African Community (Uganda, Burundi, Kenya, Rwanda and Tanzania). The entire organic agricultural land in the community totals 35% of the organic farmland in the continent. Uganda has the greatest area of organic farmland in Africa (231 157 ha) and correspondingly the highest number of organic producers (191 000), second only to India (650 000). Tanzania and Ethiopia also boast 149 000 and 136 000 organic producers, respectively (Willer and Lernoud, 2015).

Of the 455 000 ha under organic production in the continent, about 213 850 ha (47%) is dedicated to permanent tree crop cover (olives, tea, coffee, cacao, nuts, and tropical and subtropical fruits), whilst 86 450 ha (19%) is used for arable land crops including oil seeds, fibre plants, medicinal plants, cereals and pulses. Around 25 480 (5.6%) is for pastoral land.

African organic agriculture is both certified and non-certified (Parrott et al., 2006). Certification is carried out by the participatory guarantee system (PGS), designed for local or regional markets. Local markets where certified organic products are thriving include Egypt, South Africa, Uganda, Kenya, Tanzania and lately Nigeria. There is a strong demand for certified organic products such as organic pineapples from Africa by markets in North America and Europe, which account for 97% of global revenue in organic farming (Willer et al., 2008). In order to explore this vast potential market for exports, international donors were implored to support the eight participating African countries (Ethiopia, Kenya, Tanzania, Uganda, Nigeria, Mali, Senegal and Benin) in the ongoing African Union

Annexe 2 Farmers' declaration during the third African Organic Conference in Lagos, Nigeria

Declaration of the farmers' forum for organic and ecological agriculture

At the third African Organic Conference, 5–7 October 2015, Lagos, Nigeria

We are delegates mandated by organizations of small-scale farmers, including peasants, local communities, family farmers, pastoralists, fish breeders, animal breeders and bee keepers. We represent part of the millions of farming families producing over 70% of the food consumed by humanity. As investors in agriculture we are very proud to do it in the most sustainable way including economic, social and environmental dimensions of development.

During the farmers' forum we discussed, discovered and agreed upon the following points:

a Our peasant communities benefit from an incredible diversity of production adapted to our distinct agricultural and food production systems.
b We are innovative when it comes to food and fibre production in hostile conditions.
c We are resilient despite climate crisis we are able to produce food and preserve seed for next season.
d Organic farmers are efficient in the use of space and water: in a small piece of land we can maintain fertility, produce various crops at once and ensure food security and food sovereignty.
e Diversity of our seed and products can ensure healthy and nutritious food for people including specific diets for special needs.
f Organic farmers, especially women, are professionals in the production, sustainability, conservation and multiplication of seed.
g We develop and organize dynamic and self-managed networks firmly anchored to our rural and urban bases.
h We organize regional and intercontinental links to exchange our knowledge, agricultural practices and rich biodiversity in an independent manner.

Despite our inherent rich knowledge and potential, we note with regret the following points:

a That there is a lack of support for the work we have been doing for centuries and that continuously improves our livelihoods.
b That there is a lack of support for the promotion and expansion of our knowledge and practices.
c That the initiatives of peasant communities are not being upheld, despite their leading to real food security and food sovereignty.
d That peasant seeds are being criminalized by current seed laws promoting the interests of certain seed companies.
e That so-called improved seeds are being promoted despite the knowledge that they are often poorly adapted to local conditions and lead to seed dependency.
f That agro-industrial systems with intensive use of agrochemicals result in negative impacts on soils and biodiversity, in poor human nutrition and increased public health costs.
g That farmers are indebted to the agro-industrial system in which they are tied financially.

Due to our awareness of all the richness, knowledge and practices and the sustainability of organic and ecological farmers' practices, we, the participants of the 3rd AOC, recommend and call on the public institutions of Africa, and the whole world to:

- Recognize organic and ecological agriculture as the only solution for sustainable livelihood and to give it proper support.
- Support self-organized organic farmers' structure including domestic marketing.
- Define and promote policy that supports local and ecological food enterprises.
- Define and promote policy that supports local and ecological food enterprises.
- Improve support for organic and ecological farmers both for agricultural inputs and for access to resources.
- Define and promote policy and educational programmes to ensure that youth undertakes sustainable agricultural and environmental knowledge.
- Define better policies enabling conservation of organic seed and biodiversity.
- Make it clear that cohabitation between conventional and ecological organic agriculture is not a policy option, because the former makes the latter practically impossible.

Commission-led EOA Initiative (EOA-I) during the third African Organic Conference held in Lagos, Nigeria, on 5–9 October 2015. Participants were drawn from 28 countries in the world. The declaration by farmers at the forum is presented in Annexe 2.

3 EOA-I in Africa

EOA has been defined as a holistic system that sustains the health of ecosystems and relies on functional cycles adapted to local conditions, rather than the use of synthetic inputs that have adverse effects on total health (human, animal, plant and environmental). It also operates on the four principles of organic agriculture as prescribed by the International Federation of Organic Agriculture Movements (IFOAM-ORGANICS), namely Principles of Care, Ecology, Health and Fairness. The overall goal of the initiative is to mainstream EOA into national agricultural production systems by 2025 in order to improve environmentally sound agricultural productivity, food security, access to markets and sustainable development in Africa (EOA-I Strategic Plan).

The key objectives of EOA-I are as follows:

- i Agricultural productivity;
- ii Food security, conserved and protected biodiversity;
- iii Practices and technologies that contribute to sustainable agricultural productivity, complementary and synergetic effects, environmental and human resilience etc.;
- iv Functioning markets for organic produce and products to optimize value of investments; and
- v Drawing on tradition, innovation and science to benefit the shared environment and promote fair relationships and a good quality of life for all involved.

The initiative operates in these key priority areas:

1. Research, training and extension
2. Information and communication
3. Value chain and market development
4. Networking and partnerships
5. Policy and programme development
6. Institutional capacity development

The initiative is being funded by the Swiss Agency for Development and Cooperation's Global Programme for Food Security, the Swedish Society for Nature Conservation (SSNC-SIDA) and the African Union Commission. The modest achievements of the initiative within the continent since its inception are highlighted in Box 1.

4 Steps to improving organic agriculture in Africa

The steps outlined in Sections 4.1–4.10 are recommended for the improvement of organic agriculture in Africa.

Box 1: Achievements of EOA-I (2011–2015)

- Key operational and functional structures firmly put in place:
  - i The EOA-I Continental Steering Committee (functional)
  - ii The Regional EOA-I Steering Committee (established)
  - iii The National EOA-I Structures (functional)
- Curriculum templates at three levels are developed and are being mainstreamed into:
  - i Certificate level
  - ii Diploma level
  - iii Degree (undergraduate and graduate levels)
- EOA-I Strategic Plan (2015–2025) in place.
- Alignment to the AU-Department of Rural Economy and Agriculture agenda.
- Alignment with the Comprehensive African Agriculture Development Plan Results framework.
- Alignment with the Malabo declaration and Agenda 2063.
- Alignment to the UN-Sustainable Development Goals Nos. 1, 2, 3, 12, 13, 14, 15 and 17.
- Organization of regional training workshops on organic standards and certification systems, organic production, marketing and extension support.
- Documentation of indigenous knowledge of organic agriculture practitioners in Southwest Nigeria in a booklet format.
- Information and communication materials on EOA practices developed.
- Regular training of trainers (extension agents, farmers and researchers).
- Establishment of more farmers' groups and organic markets.
- Appointment of trained organic desk officers in Agricultural Development Programs of six states in the south-western region of Nigeria
- The Organic Unit has been upgraded to a Division within the Department of Input Support Services of the Federal Ministry of Agriculture and Rural Development (FMA&RD), Nigeria.
- Many farmers' groups have been certified using the PGS. Twelve enterprises were certified in Nigeria between 2014 and 2015.

Source: Based on EOA-I Strategic Plan for 2015–2025.

4.1 Advocating the merits of organic agriculture

Although organic agriculture has been recognized as the fastest growing global food sector for over a decade (with a growth rate of 20% per annum), most African countries have not yet entered into the organic market. The principal reason for this is uncertainty regarding the term 'organic', with regard to the scope of organic farming, and regulations guiding its practice, as well as confusion with concepts such as 'natural', 'wholesome' and 'ecological', combined with the misuse of organic labels on products (Bello, 2008). There is an urgent need to educate the public on the basic principles of organic agriculture, and the health advantages of organic foods over conventionally produced foods.

4.2 Developing resilient varieties of major staple crops

For organic agriculture to be significantly improved in Africa, there is a need to develop resilient cultivars of the continent's staple food crops such as cassava, millet, sorghum, yam, rice, maize, sweet potatoes and beans. It has been shown that foreign technologies are not always directly transferable in order to improve productivity of staple crops (Moyo et al., 2015). The need to develop local resilient cultivars of staple crops using local technologies, taking into consideration the African environment and the vagaries of tropical weather conditions, is obvious. However, recent studies have reported success in the development of high-yielding crop varieties in Africa such as a 40–60% increase in yield of drought-tolerant cassava in Malawi and Rwanda (FAO, 2013), and a 5.5% increase in the continent's maize production between 2000 and 2010 through the cultivation of high-yielding varieties (Cardoni and Angelucci, 2013). Cultivating beans that can tolerate warm temperatures in East Africa resulted in a 53% and 60% increase in yield in Rwanda and Uganda, respectively (Pabra, 2015). NERICA rice varieties (a combination of African and Asian rice varieties) which are drought-tolerant, high-yielding, early maturing and resistant to major pests and diseases have resulted in an approximately 30% yield increase in rice in SSA (AfricaRice, 2015). It is suggested that other crop varieties should be developed for specific soils and locations or regions for increased productivity and their adoption promoted through effective extension services. Crops that could be improved include sesame, soybean and sunflower since they are in high demand in the international market and are highly suitable for cultivation as sole crops or mixed with other crops in the humid tropics (Olowe and Adeyemo, 2009; Olowe and Adebimpe, 2009; Olowe and Adeniregun, 2010, 2011; Adekunle et al., 2014).

4.3 Enhancing irrigation agriculture and wetlands

Rain-fed agriculture is not ideally suited to African countries due to rainfall being infrequent or erratic. Hence the need for irrigation agriculture is obvious. Based on world averages, Africa has less than one-quarter of the world's irrigated land, with only around 4% (95 m ha) of SSA's crop areas irrigated (Svendsen et al., 2009). Poor management of water is the main constraint - heavy downpours occur in the rainy season, but this is not stored in dams or reservoirs for use during the dry season (Moyo et al., 2015). Whilst Africa has the least strategies to mitigate the effects of climate change, the advantages of utilizing irrigated agriculture systems as opposed to rain-fed have been reported to be over 90% when averaged over 31 countries (Fugile and Rada, 2013). Furthermore, reported returns on irrigation projects in Africa vary between 6.6 and 28.0% on average, which is quite promising (You et al., 2011). It is therefore suggested that most African countries allocate more resources to irrigation agriculture to redress over-reliance on rain-fed farming and also to make conscious efforts to tap into the evolving 'climate-smart agriculture' aimed at addressing food security challenges in the world. According to Bergkamp (2000), the total land area under wetlands in SSA is about 228 million ha. Unfortunately, less than 10% of the wetland (50 million ha) is being put to use in agriculture in West Africa (Thenkabail et al., 1995). However, the inclusion of upland rice between a lowland rice and cowpea-vegetable sequence (Adigbo et al., 2007) and ratooned rice crop in a rice-rice-vegetable cropping sequence has been demonstrated as a feasible technology capable of increasing agricultural productivity of the wetland (Adigbo et al., 2012). This should be properly explored for organic production of short-duration vegetables in order to boost the productivity of the wetlands.

4.4 Increasing production of organic soil amendments and bio-pesticides

The inherently poor fertility status of most African soils, coupled with widespread soil nutrient mining accounts for the low productivity of the soils in the continent. Today, on average the rate of fertilizer application is 8 kg/ha with about 16 kg/ha in southern Africa and 3 kg/ha in Central Africa (Morris et al., 2007). Inorganic fertilizers easily damage the soil if poorly managed, so management strategies for replacing plant nutrients via biological processes have to be used and improved (e.g. rock phosphate via composting processes). Appropriate use of economically viable, ecologically sound, environmentally benign and socially acceptable soil enhancements such as crop residue management, green manure, composting, farmyard manure, agro-forestry technologies, alley farming, planted fallow, cover crops and cereal legumes intercropping or rotation have been reported to improve soil structure, enhance nutrient availability to crops and build up soil fertility (Brady and Weil, 2002; Edmeades, 2003; Herencia et al., 2007). In Africa, it has been estimated that farmers lose up to eight million tons of soil nutrients (equivalent to US$4 billion annually) due to soil erosion, volatilization and so on (Sanchez and Swaminathan, 2005). In order to boost the use of organic fertilizers by farmers in Africa, efforts must be geared towards processing adequate amounts of organic fertilizers on-farm, and producing manures and organic fertilizers at the community level, in order to make them readily available for farmers to purchase at affordable prices. The commencement of the Growth Enhancement Support Scheme (GESS) in Nigeria in 2011 has stimulated the establishment of over 31 organic fertilizer plants across Nigeria. These factories are ready to sell different types of organic fertilizers to farmers to boost their production outputs (Tiri et al., 2014) (see Box 2).

If this scheme is to have relevance to organic agriculture, other organic inputs must be included in the package. Bio-pesticides from plants that are readily available such as neem (*Azadirachta indica* L.) seeds, wild sunflower (*Tithonia diversifolia*) and weed species (*Chromolaena odorata* L.) also possess some anti-fungicidal effects that can potentially replace the synthetic pesticides used in conventional agriculture. Such developed products include organic neem insecticide such as 'Ozoneem' Trishul (www.ozonebiotech.com/ozoneemoil.pdf), which is a broad-spectrum bio-insecticide and effective on a wide range of crops including vegetables, ornamentals and fruits. Improved control of some foliar diseases of sesame have been achieved using plant extracts (Enikuomehin et al., 2016). In 2016, the US Environmental Protection Agency declared 60% of all herbicides, 90% of all fungicides and 30% of all insecticides carcinogenic (CCMOF, 2016). Most of these products are being sold freely on-shelf in Africa. If organic agriculture is to improve in the continent, efforts must be geared towards developing organic herbicides and bio-pesticides.

4.5 Increasing government investment in agriculture

According to the FAO (2015), investment in agriculture reduced from 4.5% of total expenditure in 2001 to 2.5% in 2012 despite the pledge made by African heads of state in the Maputo Declaration to spend at least 10% of annual budget on agriculture in 2003 and their subsequent reaffirmation to promote agricultural development at the 2014 Malabo Declaration on Agricultural Growth and Transformation (http://pages.au.int/sites/default/files/Malabo%20Declaration%202014_11%2026-.pdf). Therefore, in order to improve

Box 2: Overview of GESS Programme in Nigeria

The Federal Government of Nigeria, under the auspices of the Agricultural Transformation Agenda, initiated the GESS in 2011 to enhance agricultural productivity through timely, efficient and effective delivery of yield-increasing farm inputs, mainly fertilizers and seeds to registered farmers through the electronic wallet system that enables them to receive subsidized electronic vouchers on their mobile phones.

i Target of 5 million farmers per year for 2011, 2012, 2013 and 2014.
ii Projected to generate 5–10 times returns in increased production.
iii In 2012, 1.5 million smallholder farmers got their subsidized seeds and fertilizers.
iv 10 million farmers have been registered and now possess identity cards.
v In 2013, 3.4 million farmers got subsidized inputs.
vi Allocation to GESS programme declined from N15.9 billion (US$52 million) in 2012 to N11.07 billion (US$36 million) in 2013 and N6.3 billion (US$21 million) in 2014.
vii Between 2012 and 2014, the government of Nigeria allocated N18.6 billion (US$61 million) to inorganic fertilizers, N455.6 million (US$1.5 million) to organic fertilizers and N14.2 billion (US$47 million) to seeds/seedlings and cuttings. Allocation to organic fertilizers was low because inclusion of the input only started recently.

Recommendations for improvement of GESS

i Need for an Act on GESS to guarantee sustainability.
ii Need to improve on delivery time by improving on settlements with agro-dealers.
iii Overhaul redemption process in order to fast-track it.
iv Increase the number of redemption centres and locate them strategically to ease access by stakeholders.
v Carry along farmers and agro-dealers in key decision making.
vi Involve Departments of Agriculture at the local government areas to periodically register and update data on stakeholders.
vii Conduct an independent study on smallholder farmers' accessibility to agricultural credit facilities.
viii Increase government budget to agriculture as prescribed by Maputo Declaration (10% of budget to agriculture) since the highest allocation (1.9%) was last made in 2011.

Source: The Need to Cultivate Our Farmers' Future. An Assessment of Nigeria's GESS Programme. Actionaid. www.actionaid.org/nigeria.

organic agriculture in the continent, African states must invest more into the sector in order to address poor domestic production and provide all necessary infrastructures required (such as tractors, irrigation facilities, processing facilities and packaging materials) to boost the productivity of stakeholders along the value chains (AfDB, 2014). For example, grain reserves established at strategic locations can help to stabilize prices in case of natural disasters during the production season (Toenniessen et al., 2008). Governments can also assist in subsidizing the cost of major inputs such as seeds, fertilizers and bio-pesticides for farmers to encourage organic farming. According to the FAO (2016), smallholders and family farmers produce up to 80% of the food supply in SSA. This category of farmers should be empowered such that they can develop and be more competitive with large-scale farmers.

4.6 Subsidization and certification

Since organic agriculture in Africa is currently both certified and uncertified (Parrott et al., 2006), efforts must be geared towards organizing the farmers into cooperative societies to assist them in getting certified, either by the third-party certification body or even the PGS. This needs to take into account the fact that certification is a very expensive exercise, especially third-party certification. There is a long list of certification bodies operating in Africa by IFOAM-Organics International (www.ifoam.bio/en/list-certification-bodies-operating-africa). Some indigenous certification bodies in Africa are no longer functional such as TANCERT – Tanzania Certification Body, and UGOCERT – Uganda Organic Certification LTD (www.psfuganda.org). Organic practitioners should be encouraged to get certified under the PGS to meet the local demand for organic produce and later gain third-party certification through government assistance. In April 2007, the East African Community comprising of Burundi, Kenya, Rwanda, Tanzania and Uganda adopted 'The East African Organic Products Standards' which guide organic sector production under the label 'Kilimohai' which in Swahili means 'living agriculture' (https://www.oecd.org/aidfortrade/47719232.pdf). This type of initiative should be extended to other regions in the continent if organic agriculture is to improve. It is believed that this will foster trans-regional trade links within the continent.

4.7 Evolving policies for the value chain

At present, organic agriculture is not being given prime position in the scheme of activities in the agriculture sector by the policy-makers of most SSA countries. The underlying reasons for this scenario could be attributed to indifference, lack of proper understanding as a production system capable of alleviating poverty and hunger in society, scepticism and probably hostility. However, there are some success stories – for example, organic coffee has been recognized by the Coffee Development Authority (UCDA) in Uganda (ugandacoffee.go.ug). In Nigeria, an organic division was recently created in the Federal Ministry of Agriculture & Rural Development (FMA&RD) under the Department of Input Support Services to address matters relating to organic agriculture. It is believed that a lack of national standards or regulations could be responsible for the delayed response of governments in recognizing organic agriculture. Appropriate policies are to be put in place if smallholder farmers in Africa are to break into international markets where higher standards are required. When properly integrated into agricultural value chains at the global level, African organic farmers will be challenged to add value to their export products, particularly cash crops. However, farmers must have access to financial assistance in order to meet the required standards (AfDB, 2014). Close cooperation with international organic trade companies should be officially encouraged and successful strategies communicated to the broader public.

4.8 Reforming landownership laws and policies

Landownership rights in Africa have always been a contentious issue, closely related to poverty and land degradation. About 202 million ha of arable land is uncultivated in Africa with about 33 million small farms having an average of less than 2 ha (Moyo et al., 2015), attributable to poor land governance and ownership based on customary laws (World Bank, 2015). Land is traditionally allocated through inheritance or other regulatory and

---

Box 3: Land reform in Nigeria: Progress, problems & prospects (2009)

A Presidential Technical Committee was created in 2009 to undertake a reform of the land tenure situation in Nigeria arising from various problems emanating from the Land Use Act of 1978, rectify the lapses in the Act and provide registrable titles to all landowners, including farmers.

Key recommendations:

- Setting up of Land Reform Commission to handle the process nationwide.
- Unification of land laws, operating regulations and procedures for land transactions.
- Securing property rights and land titles for all socio-economic classes, especially farmers.
- Land reform programmes to be properly funded by the three tiers of government.
- Land reform bill proposed to the National Assembly for consideration.

Source:siteresources.worldbank.org/EXTARD/Resources/336681-1236436879081/5893311-127120.pdf.

---

distributive mechanisms with women having lower status than men (Boon, 2007). However, this has changed in the recent past as land has largely become state property and is being reallocated to prospective farmers along lines of wealth, race and gender, resulting in conflicts between user groups (UNEP, 2002). In order to proffer lasting solutions to the frequent conflicts, an all inclusive land reform system has been suggested by Moyo et al. (2015). The new system will support property rights, ensure security of land tenure and enable land to be used as collateral in accessing loans from banks. It will also allow redistribution of large corporate farmlands to resource-constrained farmers' groups such that it can harness the synergies between smallholder and large-scale farming. If this type of reform is put in place, it will drive the adoption of organic agriculture by more farmers across the continent when supported by efficient extension services and participatory approaches, preferably at community level. Key findings by a committee on land reform in Nigeria are highlighted in Box 3.

4.9 Use of ICT in all aspects of farming

The use of information communication technology (ICT) in agriculture is rapidly gaining importance globally. ICT can be an efficient extension service tool when properly managed, allowing farmers in rural areas to access information on crops in demand, current prices, weather forecasts, locations to buy essential inputs, and so on (Kariuki, 2011). ICT also has the potential to provide stakeholders with information on land preparation and appropriate sowing dates, crop management, and information on market prices, thereby improving their bargaining power (Moyo et al., 2015). One such example is the Ethiopia Commodity Exchange (ECX) which provides information on products' grades and prices and organizes the buyers and sellers. Through the ECX body, farmers have reported increases in income of 10–30% (Adewunmi, 2012). Similarly, the Nigerian government initiated a scheme in 2011 called GESS, which utilizes an electronic wallet system to allow farmers to receive subsidized electronic vouchers for fertilizers and seeds on their mobile phones (Tiri et al.,

2014). More recently in Nigeria, due to the high percentage of small- to medium-scale farmers in rural areas, coupled with limited internet services in such remote locations, a mobile platform was developed that delivers agricultural extension services to organic agriculture practitioners. Details of the device are available on www.orgfarmob.org.

4.10 The Network for Organic Agricultural Research in Africa

Research is usually regarded as the backbone of any successful programme. The Network for Organic Agricultural Research in Africa (NOARA) was first established during the Organic World Congress in June 2008 at Modena, Italy, and later formally launched in Kampala, Uganda, in May 2009 during the first African Organic Conference. The Network was further strengthened during the second African Organic Conference in Lusaka, Zambia, in 2012 with the formation of sub-regional committees (Eastern Africa, Southern Africa, Western Africa (Anglophone), Central Africa and Northern Africa). Six baseline studies have been carried out across the continent to assess the level of involvement of researchers in EOA-I, and key thematic areas for research have been identified and a research agenda developed for the continent (Juma et al., 2013). It is therefore suggested that NOARA be reinvigorated to establish the research agenda, discuss it at various relevant meetings, harvest and harmonize ideas, and disseminate to the sub-committees for implementation in the sub-regions. NOARA is expected to actively drive the adoption of organic agriculture under the EOA-I by collaborating with FiBL-led research programmes in West and East Africa and the International Centre of Insect Physiology and Ecology Forum for Agricultural Research in Africa activities. The dearth of manpower in key areas of organic agriculture in Anglophone West Africa, as identified by Aiyelaagbe et al. (2016) should also be addressed.

5 Conclusion

Through the exploits of EOA-I, organic agriculture is becoming a rapidly growing sector of agriculture in Africa. However, the sector lacks adequate funding to enable it to fast-track its development in the key areas of certification, demand-driven research, and equipment purchase. Concerted efforts should therefore be geared towards aligning the initiative with other relevant and recognized programmes, such as the Comprehensive Africa Agricultural Development Programme (CAADP), in order to bridge the funding gap and thereby improve the development of organic agriculture in Africa. This chapter has identified steps to improving organic agriculture in Africa, and if properly followed, organic agriculture will enhance the productivity, resilience and profitability of smallholder farming in Africa.

6 Where to look for further information

Recent developments in organic agriculture in the world (including Africa) can be found in the latest FiBL survey on certified organic agriculture worldwide (Willer and Lernoud, 2017). The document covers every aspect of organic agriculture including latest developments in organic agriculture in Africa (pages 162–166). Full details can be accessed at http://www.organic-world.net/yearbook-2017/pdf.html

A recent policy brief on 'Financing Organic Agriculture in Africa: Mapping the Issues' was prepared by the United Nations Conference on Trade and Development (UNCTAD). It revealed that the financing required to develop organic agriculture is not being met sufficiently and discussed the various barriers, trends and opportunities regarding financing, as well as market developments and price premiums, and the perceived evolution of access to finance. For full details please visit www.unctad.org/en/PublicationsLibrary/webdicted2016d6_en.pdf

The third African Organic Conference (3AOC) was successfully hosted by Nigeria in Lagos on October 5–7, 2015. A total of 52 papers were accepted for oral presentation and four as posters. In all, 231 participants from 28 countries within and outside Africa attended the conference. The Scientific Track of the Conference was jointly sponsored by the International Society of Organic Agriculture Research (ISOFAR), Swiss Aids and the African Organic Network (AfroNet), and the full proceedings are available on www.isofar.org.

The Food and Agriculture Organization of the United Nations (the FAO) compiled the proceedings of the Plenary Presentations of the Conference to 3AOC on the theme 'Achieving Social and Economic Development in Africa through Ecological and Organic Agricultural Alternatives'. The 18 presentations that covered all aspects of organic agriculture were divided into sub-themes: Progress, Research, Markets, and Challenge and opportunities. For more details, please visit www.prodinra.inra.fr/ft?id=65E154F2-DC63-4C10-BA3A-961EA294244.

Another very informative publication by the FAO on 'Innovative Markets for Sustainable Agriculture: How innovations in market institutions encourage sustainable agriculture in developing countries' was published in 2016. It contains 17 chapters and case studies from five African countries (Namibia, Nigeria, Uganda, Benin and Tanzania) that feature prominently in the publication. Full details are available at www.fao.org/3/a-i5907e.pdf.

The Ecological Organic Agriculture-Initiative (EOA-I) 2015–2025 Strategic Plan contains the full details of the initiative and the expected outcomes from the various highlighted activities. The strategic plan is a joint product from all relevant stakeholders across the continent of Africa. Please visit www.kilimohai.org/fileadmin/02_documents/EOA/THE_EOA_STARTEGIC_PLAN_2010-2025.

7 References

Adekunle, Y. A., Olowe V. I., Olasantan, F. O., Okeleye, K. A., Adetiloye, P. O. and Odedina, J. N. (2014). Mixture productivity of cassava-based cropping system and food security under humid tropical condition. *Food Energy Sec.* 3:43–60.

Adewunmi, F. (2012). Four ways how technology can boost African Farming. Available at http://www.howwermadeitnafrica.com/four-ways-technology-can-boost-african-farming/22908/.

Adigbo, S. O., Okeleye, K. A. and Adigbo, V. B. (2007). Performance of upland rice fitted into lowland rice-vegetable/cowpea sequence in rainfed inland valley. *Agronomy J.* 99:377–83.

Adigbo, S. O., Olojede, M. O., Harris, P. J. C. and Ajayi, O. (2012). Ratooned lowland neriac rice varieties as an option for triple cropping in inland valleys of derived Savannah in Nigeria. *Exp. Agr.* 48:551–62.

Aiyelaagbe, I. O. O., Harris, P. J. C and Olowe, V. I. O. (2016). Skills gaps in organic agriculture and SWOT analysis in higher educational institutions (HEIs) in Anglophone west Africa. *Org. Agr.* 6:109–18.

AfDB (African Development Bank. (2014). Global value chains and Africa's integration into the global economy. Chapter 2 in Annual report 2013. AfDB: Tunis, Tunisia.

AfricaRice. (2015). Available at http://africarice.org/warda/newsrel-average-yield-Mar-13.asp.

Atungwu, J. J., Agbonahor, M. U., Aiyelaagbe, I. O. O. and Olowe, V. I. O. (2016). Community-based farming scheme in Nigeria: enhancing sustainable agriculture. In A. Loconto, A. S. Poist and P. Santtacolma (Eds), *Innovative Markets for Sustainable Agriculture. How Innovations in Market Institutions Encourage Sustainable Agriculture in Developing Countries*. FAO and INRA: Rome, 57–78.

Bello, W. B. (2008). Problems and prospect of organic farming in developing countries. *Ethiop. J. Environ. Stud. Manag.* 1:36–43.

Bergkamp G., Pirot, J. Y. and Hostettler, S. (2000). *Integrated Wetlands and Water Resources Management*. IUCN: Gland, Switzerland.

Boon, E. K. (2007). Regional sustainable development review: Africa – food security in Africa: challenges and prospects. Available at -http://www.eolss.net/sample-chapter/c16/E1-48-02-00pdf.

Brady, N. C. and Weil, R. R. (2002). *The Nature and Properties of Soil*. Macmillian: New York, pp. 616–60.

Cardoni, P. and Angelucci, F. (2013). *Analysis of Incentives and Disincentives for Maize in Nigeria*. Technical notes series. Monitoring African food and agricultural policies (MAFAP), FAO: Rome.

Churn Creek Meadow Organic Farm (CCMOF) (2016). Reasons to buy organic. Available at www.ccmof.com/ccmof_website_008.htm. Accessed on 10 October 2016.

Edmeades, D. C. (2003). The long-term effects of manures and fertilizers on soil productivity and quality: a review. *Nutr. Cycl. Agroecosys.* 66: 165–80.

Food and Agriculture Organisation (FAO). (2015). Government expenditure on agriculture: Experimental series. Available at http://www.fao.org/economic/ess/esseconomic/expenditure/fr/.

Food and Agriculture Organisation (FAO). (2016). Factsheet_smallholders and family farmers. http://www.fao.org/fileadmin/templates/nr/sustainability_pathways/docs/Factsheet_SMALLHOLDERS.pdf Assessed on 08 October 2016.

Fuglie, K. O. and Rada, N. E. (2013). *Resources, Policies and Agricultural Productivity in Sub-Saharan Africa. ERR-145*. US Department of agriculture, economic research service: Washinghton, DC.

Herencia, J. F., Ruiz-Porras, J. C., Melero, S., Garcia-Galavis, P. A., Morillo, E. and Maqueda, C. (2007). Comparison between organic and mineral fertilization for soil fertility levels, crop macronutrient concentrations and yield. *Agron. J* 99:973–83.

Jayaram, K., Riese, J. and Sanghvi, S. (2010). Agriculture: Abundant opportunities. *McKinsey Quarterly*. Summer 2010.

Juma, M., Kadzere I. and Ssekyewa, C. (2013). Deepening organic agriculture research in Africa. Available at www.ifoam.bio/sites/default/files/mwatima_juma_deepening_organic_agriculture_research_in_africa_3_july.pdf.

Kariuki, J. G. (2011). The Future of agriculture in Africa. Boston University. The Federick S. Pardee Centre for the study of the longer range future. The Pardee Paper No. 15. Africa 2060. p. 46.

McIntire, J. C. (2014). Transforming african agriculture. *Glob. J. Emerg. Markets Econ.* 6:145–79.

Morris, M., Kelly, V. A. Kopicki, R. J. and Byerlee, D. (2007). *Fertilizer Use in African Agriculture. Lessons Learned and Good Practice Guidelines*. The World bank: Washington, DC, p. 162.

Moyo, J. M., Bah, E. M. and Verdier-Chouchane, A. (2015). The African competitiveness report: Transforming Africa's agriculture to improve competitiveness. Chapter 2.1, pp. 37–52.

Olowe, V. I. O. and Adeyemo, A. Y. (2009). Enhanced crop productivity and compatibility through intercropping of sesame and sunflower varieties. *Ann. Appl. Biol. J.* 155:285–91.

Olowe, V. I. O. and Adebimpe, O. A. (2009). Intercropping sunflower with edible soybeans enhances total crop productivity. *Biol. Agr. Hortic.* 26:365–77. https://doi.org/10.1080/01448765.2009.9755095.

Olowe, V. I. O. and Adeniregun, O. O. (2010). Seed yield, yield attributes and oil content of newly released sesame (*Sesamum indicum* L.) varieties. *Arch. Agron. Soil Sci.* 56:201–10.

Olowe, V. I. O. and Adeniregun, O. O. (2011). Appropriate threshing procedure guarantees enhanced grain yield of dehiscent sesame (*Sesamum indicum* L.). *Agricultura Tropica et Subtropica* 44:56–8.

Pabra (Pan-African Bean Research Alliance). (2015). Available at http://www.pabra-africa.org/impact.

Parrott, N., Ssekyewa, C, Maunike, C. and Ntabi, S. M. (2006). Organic farming in Africa. Archived at http://www.orgprints.org/5161.

Rahmann, G. and Aksoy, U. (2014). Forward of proceedings of the 4th ISOFAR scientific conference at the organic world congress held on 13–15 October 2014 in Istanbul, Turkey.

Sanchez, P. A. and Swaminathan, M. S. (2005). Hunger in Africa: the link between unhealthy people and unhealthy soil. *Lancet* 265: 442–4.

Svendsen, M., Ewing, M. and Msangi, S. (2009). *Measuring Irrigation Performance in Africa*. IFPRI Discussion Paper 00894. International Policy Research Institute: Washington, DC.

Thenkabail, S., Prasad, S. and Nolte, C. (1995). Mapping and characterizing inland valley agroecosystems of West and central Africa: A methodology integrating remote sensing, global positioning system, and ground truth data in a geographic information systems framework, RCMD Monograph, No. 16. IITA: Ibadan, Nigeria.

Tiri, C. D., Ojoko, E. A. and Aruwajo, A. (2014). Growth enhancement support scheme (GESS) and the challenges of food security in Nigeria: A Review. *APRN J. Agr. Biol. Sci.* 9: 226–32.

Toenniessen, G., Adesina, A. and DeVries, J. (2008). Building an alliance for a green revolution in Africa. *Ann. N. Y. Acad. Sci.* 1136: 233–42.

UNEP (2002). *Africa Environmental Outlook: Past, Present and Future Perspectives*. UNEP: Nairobi, p. 422.

United Nations Conference on Trade and Development (UNCTAD). (2016). Financing organic agriculture in Africa: Mapping the issues. UNCTAD/WEB/DITC/TED/2016/6, p. 8.

Willer, H., Yussefi-Menzler, M. and Sorensen, N. (Eds). (2008). *The World of Organic Agriculture: Statistics and Emerging Trends 2008*. FiBL-IFOAM Report. Research institute of organic agriculture (FiBL), Frick and IFOAM-Organics International.

Willer, H. and Julia, L. (Eds) (2015). *The World of Organic Agriculture. Statistics and Emerging Trends 2015*. FiBL-IFOAM Report. Research institute of organic agriculture (FiBL), Frick and IFOAM-Organics International.

World Bank. (2015). *World Development Indicators (online)*. World bank: Washington, DC.

You, L., Ringler, C., Wood-Sichra, U., Roberston, U., Wood, S., Zhu, T., Nelson, G., Guo, Z. and Sun, Y. (2011). What is the irrigation potential for Africa? A combined Biophysical and Socio-economic approach. *Food Policy* 36: 770–82.

Chapter 20

Improving organic agriculture in the developing world: Asia

Sang Mok Sohn, Dankook University, South Korea

1 Introduction
2 Key crops grown in Asia
3 Organic production in Asia
4 Rice production systems
5 Types of organic rice farming
6 Organic vegetable production
7 Improving organic agriculture in Asia
8 Conclusion and future trends
9 References

1 Introduction

Asia is the largest continent by land mass but represents only 9% of the world's organic agricultural land. This amounted to a total of 4.9 million ha in 2017, but represented only 0.3% of the total agricultural area of the continent (all of the following data were derived from Willer and Lernoud, 2018). The largest country by area is China (2.28 million ha). Timor-Leste has the highest proportion of organic agricultural land at almost 7.4%.

In 2016, about 49% of all organic farmland in Asia was used for arable crops (2.4 million ha), 1% for grassland (almost 32000 ha) and 16% (776000 ha) for permanent crops cultivation. As information is lacking on around 34% of the total area and detailed information on horticultural land use is not available, all categories must be considered with caution.

2 Key crops grown in Asia

2.1 Cereals

Asia's key arable crops are cereals, traditionally rice (36% of the total organic cereal area), wheat (26%) and grain maize (18%). About 75% of the organic rice areas are located in China (276000 ha). Almost all the organic wheat in Asia is grown in two countries, China

http://dx.doi.org/10.19103/AS.2017.0029.22

(almost 147000 ha) and Kazakhstan (120000 ha). Oilseeds (predominantly soybean) are grown on at least 582000 ha (mainly in China and India) and represent about 1% of the total oilseed area in Asia.

2.2 Permanent crops, shrimp, fish and textiles

Permanent crops are dominated by coconut, which is grown on about 199000 ha (mainly in the Philippines), coffee (112000 ha, mainly in Indonesia and Timor-Leste), tea (at least 93000 ha, almost all in China) and temperate region fruit (about 100000 ha, Willer and Lernoud, 2018). Organic wild collection plays a major role in India and China. Although the majority of production is in fresh produce and field crops, the production of processed products is increasing rapidly. Aquaculture (shrimp and fish production) shows a strong growth in China, Indonesia, Vietnam, Thailand, Malaysia and Myanmar. Organic textiles are another important growth area.

3 Organic production in Asia

About 1.1 million organic producers are active in Asia. In India, there are an estimated 835000 producers, and in the Philippines, almost 166000. At 5.9 billion Euros, China represents the fourth largest market for organic products worldwide (the total for Asia being 7.3 billion Euros). Sector growth in China is increasingly driven by imports, especially from Oceania and North America. However, domestic markets have been developed in the Southern and Eastern parts of the region (Japan, South Korea, Taiwan and Singapore). Megacities such as Kuala Lumpur, Manila, Bangkok, Beijing, Shanghai, Jakarta, Delhi and Bangalore also demonstrate an increase in the internal consumption of organic products.

3.1 Governmental support and certification

Chang (2018) has compiled an overview of governmental support for organic agriculture in several Asian countries. In China, governments and national and local groups strongly support the evolution of organic agriculture. India has introduced a common logo for organic foods. Community-supported agriculture and participatory guarantee systems (PGS) are attracting more interest across different Asian countries. PGS certification provides small and subsistence farmers with better access to certified organic markets. An increase in active involvement of more than 120 local municipal mayors in the development of organic agriculture has been reported from the Philippines. Through making organic farming central to their policies, some of these municipalities have become recognised internationally for their development of organic agriculture as a tool for poverty alleviation and sustainable city development.

Organic agriculture was listed among the top five 'urgent agendas' of the revamped Ministry of Agriculture and Cooperative in Thailand (Ong, 2016). In 2016, there were 23 organic certification bodies active in China (Hu, 2016) and 68 organic certification bodies in the Republic of Korea. According to Willer and Lernoud (2017), 15 countries (Armenia, China, Georgia, India, Indonesia, Iran, Israel, Japan, South Korea, Lebanon, Malaysia, Saudi Arabia, Taiwan, Thailand and United Arab Emirates) have fully implemented national regulations on organic agriculture. Four countries (Bangladesh, Jordan, Kyrgyzstan and

Pakistan) are reported to be in the process of drafting regulations. Nine countries (Bahrain, Bhutan, Brunei, Hong Kong, Kuwait, Laos, Oman, Qatar and Vietnam) have national standards but no national legislation. In several countries, work on national standards and regulations is still in progress which will result in further growth of the Asian organic sector.

3.2 Characteristics of Asian agriculture

In contrast to continents such as North America, East Asia has less agricultural area per capita. Farms are small and have traditionally developed along a labour-intensive and less capital-intensive path. However, East Asia has experienced a history of intensified agricultural evolution, entailing highly developed horticulture and vegetable production typically characterised by low mechanisation until the 1990s. These farms depend on high levels of nutrient input and generally have only a small amount of grassland.

The Asian continent has a great diversity of environments and landscapes, site conditions and cultural backgrounds. However, it has only a small number of cropping systems. The first group is based on rice, which is the most important staple food in Asia. The second group focuses on high value crop production, that is vegetables which are typical of East Asian land use systems. These systems are represented by the leading International Rice Research Institute (IRRI, Philippines) and the World Vegetable Center (AVRDC, Taiwan).

Changing to organic farming methods does not automatically guarantee improved biodiversity. Organic farming practices still require active participation from the farmer and creating arable farmland which will generate biodiversity depends upon an understanding of conservation ethics. However, more diverse production systems have potential market benefits where they reflect the diversity of Asian dishes based on rice and vegetables. This is particularly the case with Korean dishes which have a broad spectrum of (often fermented) vegetables, such as Kimchi. As low-calorie, protein-rich, low-carb food such as seaweed, algae, seafood and fish are significant contributors to low levels of obesity, South Korea might serve as a blueprint for Asian organic health food rich in phytochemicals and a low content of undesirable ingredients and residues. Thus, the eating culture could be an important driver of planned diversity in the fields and landscapes of Asia.

4 Rice production systems

More than 20% of the global population and 70% of Asians derive their carbohydrate and around 15% of their protein intake from rice. Rice is the major staple food in Asia and rice-based cropping patterns are prevalent in much of East Asia which is dominated by wetland (lowland) culture. The general method of wetland rice cultivation in Asia has remained largely unchanged from the traditional method of rice farming established around 2000–1500 BC. The resilience of rice, which has been cultivated as a monocrop for centuries, is illustrated by the Ifugao rice terraces in Banaue, Philippines, which are thought to be around 2000 years old.

Between 1998 and 2004, the standards for organic rice cultivation in Asia were discussed at Dankook University, Cheonan, Republic of Korea, by stakeholders including academics, scientists and leaders of organic farmer groups in Asia, as an initiative of the Asian Research Network of Organic Agriculture (ARNOA) and the Rural Development Administration (RDA, Republic of Korea). These were finalised in 2004 as ARNOA Organic Rice Standards.

4.1 Crop rotation and diversity

It is possible that the extended cultivation of a more or less resilient rice monocrop has hindered the evolution of a clear rotation design in East Asia. However, in Southern regions where more than one harvest per year is possible, a second cash crop can be grown. For instance in the southern parts of the Republic of Korea (below 38°30'N), strawberries, watermelons, potatoes, onions, leaf and root vegetables, tomatoes, barley and wheat are grown. The revenue based on these second cash crops is often higher when compared with rice and therefore reduces the likelihood of farmers growing green manure as a second crop.

A detailed description of the evolution of rice production systems in Asia, including diversification strategies, has been published by Becker and Angulo (2018). Growing a greater diversity of crops would allow farmers to reap the environmental and energy advantages of longer, more complex crop rotations. Having regard to nutrient management and the control of pests and diseases as well as revenues, it seems worth investing more research in combining rice and vegetable production. This seems particularly advisable in countries like South Korea, where the soils are overloaded with nutrients (especially nitrogen (N)) due to historic high N inputs partly caused by imported feed-stuff from the United States. The high N demand of vegetables could interact in a complementary manner with modified rice production.

The diversified wetland rice cultivation systems which have been developed during the last five decades could meet these aims. Some of them, which fit with classical organic farming approaches, are described in Section 5, including those which imitate the pioneer mixed farming system approach, although these are not based on forage crops and ruminants as they are in Europe. As weeds are one of the major constraints on crop production in organic farming systems, the diversified production systems of organic rice farmers in Korea, Japan, China, Vietnam and the Philippines are assessed and valued primarily on their capacity to control weed competition in paddy fields. The competitiveness of planted rice seedlings is generally higher when compared with pre-germinated seeds. Moreover, seedlings and hills allow for better mechanical weed control compared to pre-germinated seeds which are often broadcast-distributed or sown in narrow rows. Paddy field-adapted legumes and vegetables, *Azolla*-rice, snail-rice, fish-rice, duck-rice (weeding) systems and their potential combinations will also be described.

4.2 Green manure

The rice yield levels currently achieved by Asian farmers depend on large N inputs. *Sesbania* spp. and *Aeschynomene afraspera* have been found to be promising green manure crops which fix high amounts of N symbiotically and grow quickly in flooded soil conditions. The aquatic fern *Azolla* spp., which lives symbiotically with *Anabaena azollae* cyanobacteriam, is among the most effective systems showing the highest rates of biological nitrogen fixation, as well as rapid growth. All these green manure crops may successfully be used as biofertilisers (Ventura et al., 1987) and have been shown to generate residual cumulative effects on soil fertility and rice yields over time (Ventura and Watanabe, 1993).

Azolla has the potential to fulfil the N needs of an eight tons per hectare rice harvest. A fertiliser N equivalence of more than 100 kg N/ha (applied as urea) has been determined (Gutbrod, 1987). Depending on climatic conditions, it may either be grown as a green manure pre-crop or as an intercrop, acting as a support in the growing rice stand, or as a combination of both. Besides delivering N, under tropical conditions, the dense *Azolla* mat enhances rice growth and yield as a function of lowering the water temperature, as well as by competing against weeds. Wide double row stands enable the use of rotary harrows and the incorporation of the *Azolla* mat during mechanical weeding.

The standard book by Lumpkin and Plucknett (1982), although giving a detailed overview only on the *Azolla* knowledge of the 1970s, remains a rich source and basis for developing efficient organic *Azolla*-rice production systems. The constraints and risks of *Azolla*-based rice production systems are outlined by Gutbrod (1987) together with the most significant obstacles to be overcome. These are:

1 The high P demand of *Azolla* can be delivered only with super phosphate, a measure which is not currently permitted under the rules of certified organic agriculture.
2 The high labour costs required for the propagation and conservation of the *Azolla* inoculum as well as its incorporation in the field.
3 Its low tolerance of high temperatures and intensive irradiation (this can be minimised by the intercropping strategy).
4 Damage by insects, fungi and frost.
5 Adequate water management.

Because of the generally easy access to commercial organic manures and composts, organic farmers are not currently seeking to develop this labour-intensive cultivation method.

5 Types of organic rice farming

5.1 Rice bran farming

Rice bran residues are produced when rice grains are polished. The bran is rich in protein and vitamins and is a potentially high value food. As South Korean diets are based on polished rice, high amounts of rice bran are produced. With a content of 1.5–2.5% N and 2% phosphorus (P), the bran has a market value as feed for domestic animals and as an organic fertiliser. Up to 2 tons/ha (which is the equivalent of 30–50 kg N/ha and 30–40 kg P/ha) are applied 1 day after planting seedlings in irrigated fields. This results in a vigorous growth of algae which competes efficiently with weeds. After about 30 days, the algae layer sinks to the ground and releases nutrients to the rice plants.

As weeds may develop before the algae layer falls, additional manual or mechanical weeding is often necessary. For this reason, the method is not widely used by farmers except where there is a low water table, and in combination with snail farming (Section 5.3).

5.2 Rice–duck farming

Rice–duck farming originated in Japan and has been practised in several Asian countries since the 1990s. It is an effective form of ecological engineering where there is a well-developed market for duck meat, as in China (Xi and Qin, 2009).

The rice–duck farming system has been spreading throughout Korea since 1992, but further research is required for its successful adaptation and management. The optimal population of duck per rice paddy field unit, the release time of the ducklings, duck management after release, a strategy for managing the required breed, and the processing and marketing of the meat all need further investigation (Sohn, 2000a,b). This system is labour intensive and increased income from the meat production has to be balanced both against the construction of duck sheds and fences to protect the birds from predators, and the additional feeding which includes a final fattening period. The popularity of duck farming in South Korea fell after the H5N1 avian flu outbreak of 2008 and the majority of the country's organic farms currently prefer rice–snail farming.

5.3 Rice–snail farming

The Golden Apple Snail (GAS), *Pomacea canaliculata* (Lamarck) which originated in South America, is listed as one of the hundred worst invasive Aryan species (ISSG, www.issg.org/database). Due to its high rate of reproduction, adaptability, feeding activity and lack of predators within Asia, it has become a serious pest in the Philippines, Cambodia, Thailand and Vietnam where it is not properly managed (Joshi et al., 2005). However, organic rice farmers in Korea, Japan and the Philippines are utilising GAS as a biological weeding agent which has proved far more effective than ducks or fish in paddy rice fields (Yusa, et al., 2003).

5.4 Rice–fish farming

Rice–fish farming has a long history in the rice growing areas of South-East Asia. Fish are trapped in the rice fields, grow along with the crop and remain after harvesting. This traditional practice of captive rice–fish culture has evolved into an aquaculture farming system in which herbivorous fishes are used to regulate weeds.

5.5 Integrated organic rice systems

The integration of fish and the N fixing aquatic fern *Azolla* has shown potential for increasing the production of organic rice. Cagauan et al. (2000) have shown that this combination can result in nutrient enhancement, feed supplementation and biological weed and pest control, as well as increasing the overall productivity of a farm. Nutrient cycling in an integrated rice–fish–*Azolla*–duck farming system is more efficient and productive compared with rice–duck or rice–fish farming systems alone (Cagauan et al., 1996).

Duck sheds constructed over a fishpond refuge situated contiguous to the rice field allows spilled feed to fall directly into the fishpond. The duck manure serves as a fertiliser for plankton production while the spilled feed can be directly consumed by the fish. *Azolla* can be an inexpensive food source for the Nile tilapia (*Oreochromis niloticus*) grown in rice fields. Increased fish production has been demonstrated in integrated beneficial feed systems. In other production systems, *Azolla* served as a beneficial *in situ* fresh food for macro-phytophagous fish (Liu and Liu, 1995).

6 Organic vegetable production

Most organic vegetable production in the tropics is currently intended for the high priced export market. (Kenya is a well-developed example.) China is the world's largest producer of fruit and vegetables, accounting for about a third of global production (FAO, 2007) and is developing its organic industry. According to Zhou (2003), vegetables are the most profitable part of its organic food market. The domestic market in China takes 20% of the country's total organic vegetable production. Prices paid to farmers for organic vegetables are between 10% and 100% higher than those for conventional vegetables (UN, 2003; FAO, 2007).

6.1 Local and domestic markets

As vegetables are an important component and major source of micronutrients and phytochemicals in Asian diets, especially for poorer people, local and domestic markets may become important for the further evolution of organic farming systems in several regions of Asia. In terms of production, the value of vegetables was reported to be nearly twice as high as that of rice in Taiwan and Thailand (Danuwat, 2006). These high value crops can therefore provide an important source of income, especially for smaller farms. The labour-intensive nature of vegetable production also benefits local economies by providing additional employment (Zhang et al., 2007).

6.2 Pesticides

Pesticide poisoning is a serious problem worldwide. This is especially the case in Asia, where pesticide residues in fruits and vegetables, which are often eaten fresh and unprocessed, may have a large negative physiological impact on consumers. Concern over this has raised the profile of organic fruit and vegetables when compared with organic staple foods such as cereals. Consequently, it is reasonable to assume a future growing market for vegetables with lower pesticide residues and an organic farming blueprint will help farmers to meet the increasingly stringent vegetable production standards introduced by the food industry.

6.3 Production in tropical regions

Organic vegetable production in the tropics faces unique challenges when compared with organic production in temperate areas. Production constraints are imposed by climatic stresses such as heat, seasons of drought, tropical cyclones and temporary floods. A higher incidence of pests and diseases exists and often remains present throughout the year. Further obstacles include a lack of seeds and adapted cultivars, limited markets for organic inputs and the predominance of smallholder production with limited access to knowledge and technical and financial resources (Juroszek et al., 2008a).

Organic vegetable production currently consists of only a small number of crops for which there is an assured market, such as tomato, vegetable soybean, sweet pepper, cabbage, onion, garlic, beans, broccoli, cauliflower, kohlrabi and cucumber. In contrast with Africa, there has to date been little use of indigenous vegetables, which in some cases were shown to be well adapted to tropical conditions and offered the potential for domestic marketing opportunities (Juroszek et al., 2008a).

6.4 Nutrient management in organic vegetable production

In order to achieve high crop yields, horticultural soils must provide sufficient nutrients for vegetables during their relatively short vegetation period. Most vegetables need high inputs of nutrients and plant protection interventions to reach satisfactory levels of yield and quality.

If soil fertility is improved by careful management of soil structure and nutrient availability as well as by water availability, nutrient amounts by area applied with compost and manure may often be higher when compared with arable crops. Thus, organic farming is not necessarily a low-input system, particularly in vegetable growing. An increasing number of organic farmers in Asia are sourcing their organic inputs off-farm, a choice facilitated by the growing input market for horticultural production. This contrasts with the European practice, where the approach of pioneer organic farmers was to organise a mixed farm (with ruminants) which closely approximates a closed system.

Organic farmers are rarely sufficiently informed about the nutrient content or quantities of purchased organic amendments (especially in regard to N and P), due to incomplete data, and poorly informed or non-existent extension services. As a result, organic farmers in Asia often apply too high an amount of organic fertiliser. Many organic vegetable farms may thus be considered as high-input oriented, sometimes showing excessively positive nutrient balances. In some areas of China and Korea, organic farmers apply around 20–40 t/ha/year and compost maximum amounts of 80–100 t/ha/year have been reported (Sohn, 2016). In Indonesia, up to 190 t/ha/year of organic fertiliser is usually applied on organic vegetable farms (Moeskopsa et al., 2010).

Using excessive amounts of organic fertiliser results in a high risk of leaching from the over-accumulation of N in the soils (Gutser et al., 2005). Seo et al. (1997) showed that N applied above the optimum level is not taken up by crops and is leached into the subsoil. Sohn (2016) indicated overuse of organic fertilisers resulting in negative environmental impacts such as salt accumulation (especially under greenhouse conditions), phosphate accumulation, as well as nitrate leaching in organic farming systems in Korea.

6.5 Organic manuring and conventional fertilisers

More than 95% of organic farming is practised under greenhouse conditions in South Korea. Sohn and Kücke (2008) investigated two greenhouse soils in order to study the effects of organic manuring and the conventional use of mineral fertilisers. By using passive capillary samplers (PCAPS), their research team found that a single application of organic compost in conventional farming systems (COF) increased the nitrate concentration only slightly when compared with conventional mineral fertilisation (CMF), while regular compost application (Organic, OF) caused a dramatic increase of nitrate concentrations in the leachate ranging from 87 to 329 ppm NO_3^--N. In plots where no organic fertilisers were applied (control, CMF), a concentration of less than one ppm organic N in all seepage water samples was shown. In comparison to this, the concentrations reached values up to 11 ppm organic N in plots with one single compost application (COF), and concentrations up to 23 ppm organic N in OF treatment. In the OF field, the amount of total N (TN) in the leachate was maintained in a range of 40–100 kg/ha during the sampling dates, which was summed up at the end of the experiment as 300 kg N/ha. The comparison of treatments resulted in OF > COF > CMF > control. The amount of leached N increased with an increase in both types of fertilisation. However, the effect in organic fertilisation was more apparent.

Based on results of the case study by Sohn and Kücke (2008), it may be concluded that the Korean Organic Farming system may raise some ecological risks in terms of groundwater pollution by nitrate but also by organic N and P. With regard to this study, it can be assumed that leaching of N and P may frequently occur in organically farmed soils under greenhouse conditions. This is not consistent with the aim of organic farming which is to protect the environment (Köpke, 1989). To avoid groundwater pollution through the leaching of N and P, internationally recognised organic farming regulations such as Codex guidelines and IFOAM Basic Standards, which suggest crop rotation including the cultivation of green manure, legumes and deep rooting crops, must be practised in Asian farming systems (Sohn, 1999a,b, 2000a,b).

Although controlled amounts of organic residues, sewage sludge and animal manures enhance soil fertility, they may have damaging effects on soil biota when applied in excessive doses. In addition to its negative impacts on water, excessive manure application may therefore endanger crop health and the invertebrates and microorganisms which live in the soil (Domene et al., 2008).

6.6 Effect of manure and compost on nitrate levels in organic vegetables

In addition to negative environmental impacts, the overuse of manures and composts purchased from input markets may lead to severe problems in product quality and human health. Due to the high applications of organic composts and manures in Korean organic farming systems, high nitrate concentrations may occur in the edible parts of leaf and root vegetables, and in severe cases these may be higher than the amounts present in conventionally farmed products (Sohn, 1994, 2005; Sohn, 1999a,b).

Because of the high level of nitrate intake from vegetables (85% of daily intake) (European Food Safety Authority, 2008), Asian consumers ingest more nitrate than Westerners who are predominantly oriented towards meat-based foods. The reported levels of nitrate intake through vegetables from a few selected countries include 72.4% in Germany, 75% in the United States, 89.9% in Japan and 97.3% in Korea (Sohn, 1999a,b).

As vegetables predominate in the Korean diet, the average daily nitrate intake of a Korean consumer exceeds 219 mg which is the acceptable FAO/WHO daily intake (ADI) (Sohn, 2000a,b). This is often obtained through a high proportion of nitrate accumulating brassicas. Sohn's study (2009) aimed to deliver basic data on nitrate in certified organic Chinese cabbage, the vegetable most favoured by Koreans, in order to minimise the daily nitrate intake from this brassica. Sohn (2009) found that when a cabbage leaf was eaten whole, the intake of NO_3 from an inner leaf of small size was significantly lower than that from an outer leaf. The total NO_3 content in the smallest inner leaf weighing 3 g was 2.6 mg, while the total NO_3 content of the largest outer leaf with a weight of 132 g was 375 mg.

The chlorophyll content, which depended on the position of the leaf, was found to increase with the age of the leaf in both the midribs and leaf blades. In the case of the leaf blade, as the chlorophyll content increased, the nitrate content also tended to increase noticeably. Sohn (2000a,b) found that the further outward a leaf was situated, the higher the nitrate content. This result was consistent with reports by Sohn and Yoneyama (1996), Sohn (1999a,b) and Sohn and Alley (1997).

Detailed analysis showed that if the daily nitrate intake is to be reduced, it is advisable to process and eat the inner leaves of Chinese cabbages rather than the outer leaves.

As a general rule, it was recommended that one-third of Chinese cabbage leaves with high concentrations of nitrate be discarded before consumption. As a lower content of chlorophyll was shown to be associated with a lower nitrate content, it was also strongly recommended that the dark green leaves be discarded. It was considered essential to check the dried outer leaves of Chinese cabbages for their nitrate content before using them as the primary base for vegetable soups, which is a favourite local delicacy during the winter.

Organic vegetable growers in Asia, especially those involved in intensive vegetable cultivation, should try to avoid the problems of high nitrate concentration in leaves and roots in order to reduce high daily intake rates for consumers. Future advanced nutrient application strategies should therefore include a precise soil and plant diagnosis on nutrient availability in addition to the nutrient content of organic amendments.

7 Improving organic agriculture in Asia

7.1 Research needs

Until quite recently, both farmers and consumers in the Asian region often considered organic farming to be similar to the traditional farming methods practised by their ancestors (Willer and Lernoud, 2017) and little was known about modern scientifically based organic farming. Asia remains behind the advances made in European countries and in Northern America, where techniques have been developed by academic scientists and research institutions and have been widely extended by efficient advisory services. Institutional research in organic vegetable production in most tropical countries is either non-existent or in a nascent state.

This is still the case for most of the international agricultural research centres (CGIAR) with the exception of the World Vegetable Centre (AVRDC) located in Taiwan. Under its Director General Thomas Lumpkin (co-author of the foundational book on *Azolla* (Lumpkin and Plucknett, 1982)), AVRDC established an organic farming research programme in 2004 which focused primarily on organic vegetables (Juroszek et al., 2007). Unfortunately, this programme, which might have become the cornerstone of a scientifically based organic farming system in Asia, was only in existence for four years and despite yielding some promising results, was dropped by the next General Director when Lumpkin left to become Director General of CIMMYT in 2008.

7.2 Developing farmers' skills

Despite improvements in management, varieties and cultivars, organic agriculture yields are normally lower than those which can be achieved under controlled experimental conditions and in mainstream agriculture. A literature survey by Juroszek et al. (2008b) shows the lack of research on organic vegetable cultivation, especially in the tropics.

An individual farmer's crop management skills will have a pronounced impact even in highly productive environments (Aini et al., 2005; Juroszek et al., 2007, 2008a). Meanwhile, until the dearth of publicly accessible research literature on organic vegetable production is improved, the participatory approaches practised by networks of farmers, extension services and hands-on agriculture-oriented researchers may be able to overcome the problem by delivering proven strategies based on experience. Farmers who are 'organic by default' can benefit from information and training on improved science-based organic vegetable production techniques for increased productivity (Palada et al., 2006).

After weed control, alternative pest management options are the biggest challenge in Asian organic vegetable production. Currently, it appears to be easier for farmers to maintain or increase soil fertility and provide appropriate crop nutrition levels than to solve the plant protection problems outlined by Juroszek et al. (2008a).

7.3 ISOFAR World Organic Expo 2015

Each region or country has its own agricultural history and scientific tradition. South Korea's unique evolution from an agriculturally dominated society to a modern technically oriented economy and an urban society is described below. Its urban society is aware of the environmental problems for which mainstream agriculture is responsible and Korean politicians have recognised the advantages of organic farming, officially supporting numerous workshops, symposia and congresses which were organised by Sang Mok Sohn and other pioneers including external experts from Europe since the mid-1990s (Table 1).

In order to overcome the obstacles to modern scientifically based organic farming, the International Society of Organic Agriculture Research, ISOFAR (www.isofar.org), was launched in 2003 in Berlin at the Berlin-Brandenburg Academy for Science and Humanities at the birthplace of Alexander von Humboldt. Under the guidance of ISOFAR's founding president, Professor Dr. Ulrich Köpke, chair of Organic Agriculture at Bonn University, Germany, ISOFAR developed a close partnership and friendship with Professor Dr. Sang Mok Sohn, Dankook University, Cheonan, South Korea, who was an original board member of ISOFAR and became its second president in 2011. In the same year, the Third Scientific Conference of ISOFAR, embedded in the 17th IFOAM Organic World Congress of the International Federation of Organic Agriculture Movements (IFOAM), took place in Namyangju, Gyeonggi Province, South Korea. The Memorandum of Understanding to prepare the First ISOFAR World Organic Expo (WOE) was also signed in 2011 (Fig. 1).

Table 1 Scientific events on organic farming in South Korea

Date	Name of scientific event on organic agriculture
19–21 July 1995	2nd IFOAM/Asia-Pacific Regional Conference
15 November 2001	1st Gangjin International Organic Agriculture Symposium
12–15 October 2002	1st ARNOA International Conference on Development of Basic Standard for Organic Rice Cultivation
10–12 October 2003	2nd ARNOA International Conference on Development of Basic Standard for Organic Rice Cultivation
4–6 October 2004	3rd ARNOA International Conference on Development of Basic Standard for Organic Rice Cultivation
11–13 November 2004	6th IFOAM-Asia Scientific Conference
18 November 2004	2nd Gangjin International Organic Agriculture Symposium
24 July–16 August 2009	Uljin Environmental Friendly Agriculture Expo
14–15 March 2011	ISOFAR International Symposium on Asian Organic Agriculture
18 August – 11 October 2015	ISOFAR World Organic Expo in Goesan/Korea
7 October 2015	ISOFAR International Symposium on Organic Textile Industry, Organic Industry and Service Industry

Figure 1 Preparing ISOFAR's world organic expo (WOE). Memorandum of understanding: signing ceremony. Vice Governor of Chungbuk Province, Dr. G.G. Bahk (left) and ISOFAR President, Professor Dr. S.M. Sohn (right), at University Club, Bonn University, Germany, 21 December 2011.

Sang Mok Sohn proposed the WOE, which took place in Chungbuk Province, South Korea, in September 2015. With the slogan, 'Organic Life – Science meets public', and opening greetings from Mr. Ban Ki-Moon, the then UN Secretary-General, the WOE welcomed more than 1.1 million visitors during the three-week event. In addition to greenhouse and field demonstrations on an area of about 90 ha, the WOE included ten thematic halls, nine scientific symposia and seven seminars.

The WOE was the cornerstone for a joint ISOFAR publication on Organic Farming 3.0, later published in ISOFAR's international scientific journal *Organic Farming* (Rahmann et al., 2017). To date, ISOFAR's WOE represents the biggest international organic event ever held and has led to a wider recognition of organic agriculture, especially in the Asian region. The history of organic agriculture suggests that consumers and politicians have been the main drivers of its evolution in Asia where – as experienced in other continents – scientists have been slower to bring forward their contributions.

8 Conclusions and future trends

As in other parts of the world, social pressures and political decisions have been the main driving forces in the evolution of organic agriculture in Asia. As the East Asian diet, beyond the staple of rice, predominantly consists of a broad spectrum of fresh and processed (fermented) vegetables, the dietary culture could become a further driver of the future organic market through the demand of high-quality organic health food rich in desirable ingredients.

Two leading international institutes, IRRI and AVRDC, are located in the region, and have demonstrated that part of their work can be effective in advancing scientifically based organic agriculture in Asia. Decision-makers should advocate new research programmes in these institutions. These organic research approaches should interact with the region/site-specific national research programmes for farmers. Successful and proven research and extension service strategies should be tested for use in other regions of Asia.

9 References

Aini, Z., Sivapragasam, A., Vimala, P. and Mohamad Roff, M. N. (2005). *Organic Vegetable Cultivation in Malaysia*. Malaysian Agriculture Research and Development Institute (MARDI), Kuala Lumpur, Malaysia.

Becker, M. and Angulo, C. (2018). The evolution of lowland rice-based production systems in Asia: Historic trends, determinants of change, future perspectives. *Advances in Agronomy* 158 (*accepted, in press*).

Cagauan, A. G., Van Hove, C., Orden, E., Ramilo, N. and Brankaert, R. D. (1996). Preliminary results of a case study on integrated rice–fish–azolla–duck farming system in the Philippines (pp. 35–61). In: H. Hayakawa, M. Sasaki and K. Kimura (Eds), *Integrated Systems of Animal Production in the Asian Region*. Proceedings of a symposium held in conjunction with the 18th AAAP Animal Science Congress, Chiba, Japan, 13–18 October 1996. FAO, Rome, Italy, 111p.

Cagauan, A. G., Brankaert, R. D. and Van Hove, C. (2000). Integrating fish and azolla into rice–duck farming in Asia. Naga, *The ICLARM Quarterly* 23(1), 4–10.

Calhoun, F. G., Bigham, J. M. and Slater, B. K. (2002). Relationships among plant available phosphorus, fertilizer sales, and water quality in northwestern Ohio. *Journal of Environmental Quality* 31, 38–46.

Chang, J. (2018). Asia sector report. In: H. Willer and J. Lernoud (Eds), *The World of Organic Agriculture – Statistics and Emerging Trends 2018*. Research Institute of Organic Agriculture (FiBL), Frick and IFOAM – Organics International, Bonn, Germany, p. 188.

Danuwat, P. O. (2006). Organic agriculture in Southeast Asia. http://info.organic.org.tw/supergood/ezcatfiles/organic/img/img/1085/61701466.pdf

Davis, A. S. Hill, J. D., Chase, C. A., Johanns, A. M. and Lieman, M. (2012). Increasing cropping system diversity balances productivity, profitability and environmental health. *PLoS ONE* 7(10). doi:10.1371/journal.pone.0047149.

Domene, X., Ramírez, W., Mattana, S., Alcañiz, J. M. and Andrés, P. (2008). Ecological risk assessment of organic waste amendments using the species sensitivity distribution from a soil organisms test battery. *Environmental Pollution* 155(2), 227.

European Food Safety Authority (2008). Opinion of the scientific panel on contaminants in the food chain: Risk assessment on nitrate in vegetables. *The EFSA Journal* 689, 1–79.

FAO (2004). Rice and human nutrition. *Rice is Life: International Year of Rice 2004*. FAO, Rome, Italy.

FAO (2007). *FAO Statistical Yearbook*, 2005–2006, issue 1. http://www.fao.org/statistics/yearbook/vol_1_1/pdf/b03.pdf

Gutbrod, K. G. (1987). Effect of *Azolla* on irrigated rice in Brazil. Göttinger Beiträge zur Land- und Forstwirtschaft in den Tropen und Subtropen, Heft 36. Diss. agr. University of Goettingen, Germany, 165p.

Gutser, R., Ebertseder, T., Weber A., Schraml, M. and Schmidhalter, U. (2005). Short-term and residual availability of nitrogen after long-term application of organic fertilizer on arable land. *Journal of Plant Nutrition Science* 168, 439–46.

IPCC (1997). Intergovernmental Panel on Climate Change (IPCC), 1997. In: Houghton, J. T., et al. (Eds), *Greenhouse Gas Inventory Reference Manual*, Vol. 3. IPCC Technical Support Unit, London.

Joshi, R. C., Martin, E. C., Wada, T. and Sebastian, L. S. (2005). Role of golden apple snail in organic rice cultivation and weed management. In: U. Köpke, U. Niggli, D. Neuhoff, P. Cornish, W. Lockeretz and H. Willer (Eds), *Researching Sustainable Systems. Proceedings of the First Scientific Conference of the International Society of Organic Agriculture Research (ISOFAR)*, Adelaide, Australia, 21–23 September, pp. 112–15.

Juroszek, P. and Tsai, H.-H. (2008). Research needs in organic vegetable production systems in tropical countries with a focus on Asia. In: D. Neuhoff, N. Halberg, T. Alföldi, W. Lockeretz, A. Thommen, I. A. Rasmussen, J. Hermansen, M. Vaarst, L. Lueck, F. Caporali, H. H. Jensen, P. Migliorini and H. Willer (Eds), *Cultivating the Future Based on Science, Volume 1: Organic Crop Production*. Proceedings of the Second Scientific Conference of the International Society of Organic Agriculture Research (ISOFAR), Modena, Italy, 18–20 June 2008, pp. 635–638. Archived at http://orgprints.org/1180.

Juroszek, P., Ma, C.-H., Tsai, H.-H., Wu, D.-L. and Palada, M. C. (2007). Organic farming research at AVRDC – The world vegetable center: Developing systems for smallholder farmers in the tropics. In: E. Tielkes (Ed.), *Utilisation of Diversity and Land Use Systems: Sustainable and Organic Approaches to Meet Human Needs*. International Research on Food Security, Natural Resource Management and Rural Development, Tropentag, 9–11 October 2007, Witzenhausen. Book of Abstracts. Cuvillier Verlag, D-37075 Göttingen, Germany. Website: https://cuvillier.de. Web version: http://www.abstracts2007.tropentag.de/

Juroszek, P., Lumpkin, T. A., Easdown, W. and Tsai, H.-H. (2008a). Organic vegetable production and research needs in the tropics and subtropics. In: U. Köpke (Ed.), *Organic Agriculture in the Tropics and Subtropics*. ISOFAR, Verlag Dr. Köster, Berlin, Germany, 556p. ISBN:9783736923720 (E-book).

Juroszek, P., Lumpkin, T. A. and Palada, M. C. (2008b). Sustainable vegetable production systems. *Acta Horticulturae* 767, 133–50. doi:10.17660/ActaHortic.2008.767.13.

Köpke, U. (1989). Möglichkeiten und Grenzen der Produktionstechnik. Wie können Landwirte ökologiegerechte Landwirtschaftsstrukturen erhalten bzw. entwickeln – im Acker- und Pflanzenbau? In: N. Knauer (Hrsg.), *Ökologische Leistungen in der Landwirtschaft: Möglichkeiten und Perspektiven*. Agrarspektrum 15, 43–56.

Kücke, M. and Sohn, S. M. (2008). Soil fertility and soil cultivation in organic farming systems. *Proceedings of Asia Pacific Agriculture Forum*, 10–12 September 2008, Jeju National University, pp. 223–50.

Liu, C. C. and Liu, X. S. 1995. A brief introduction on controlling artificial biosphere in paddy field and its new cropping system (pp. 343–50). In: J.-J. Symoens and J.-C. Micha (Eds), *Proceedings of the Management of Integrated Freshwater Agro-piscicultural Ecosystems in Tropical Areas*, Brussels, 16–19 May, 587p.

Lumpkin, T. A. and Plucknett, D. L. (1982). *Azolla as a Green Manure: Use and Management in Crop Production*. Westview Tropical Agriculture Series No. 5. Westview Press, Boulder, CO, 230p.

Moeskopsa, B., Sukristiyonubowo, Buchan, D., Sleutel, S., Herawaty, L., Husen, E., Saraswati, R., Setyorini, D. and De Nevea, S. (2010). Soil microbial communities and activities under intensive organic and conventional vegetable farming in West Java, Indonesia. *Applied Soil Ecology* 45(2), 112–20.

Ong, K. W. (2016). Organic Asia 2015 (pp. 174–84). In: *The World of Organic Agriculture – Statistics & Emerging Trends 2016*. FiBL and IFOAM – Organics International, Bonn, Germany, 340p.

organicfacts.net (2016). https://www.organicfacts.net (26 November 2016).

Palada, M., Kalb, T. and Lumpkin, T. (2006). The role of AVRDC – the world vegetable center in enhancing and promoting vegetable production in the tropics. *HortScience*, 41(3), 556–65.

Qiao, Y. (2011). Organic agriculture development in China. In: H. Willer and L. Kilcher (Eds), *The World of Organic Agriculture, Statistics and Emerging Trends*. IFOAM & FiBL, Frick and Bonn.

Rahmann, G., Reza Ardekani, M., Barberi, P., Boehm, H., Canali, S., Chander, M., David, W., Dengel, L., Erisman, J. W., Galvis-Martinez, A. C., et al. (2017). Organic agriculture 3.0 is innovation with research. *Organic Agriculture*, 7(3), 169–97.

Reinfelder, H. (1992). Vergleichende Untersuchungen zur Prognose des Stickstoff-düngebedarfs verschiedener Böden Baden-Würtembergs. Dissertation, Hohenheimer Bodenkundliche Hefte, Heft 5. Institut für Bodenkunde und Standortslehre, Universität Hohenheim, Stuttgart.

Robertson, G. P., Paul, E. A. and Harwood, R. P. (2000). Greenhouse gases in intensive agriculture: Contribution of individual gases to radiative forcing of the atmosphere. *Science* 289(5486), 1922–5.

Scharpf, H. C. (1991). *Nutrient Influences on the Nitrate Content of Vegetables*. The Fertilizer Society, Peterborough, pp. 3–9.

Seo, J. H., Kim, D. H., Kim, C. G., Choi, S. H. and Koh, M. H. (1997). Variation of soil inorganic nitrogen and utilization efficiency of nitrogen in maize and oat at different level of nitrogen application. *RDA Journal of Agricultural Science* 39(1), 43–9.

Sohn, S. M. (1994). NO3-accumulationin edible parts of Chinese cabbage and radish cultivated by conventional and organic farming and its limit value for safe agricultural products. *10th International Organic Agriculture IFOAM Conference*. Lincoln University, New Zealand, 139p.

Sohn, S. M. (1999a). Chemical characteristics of rooted soil profile and nitrate content of cucumber by organic farming only applying organic fertilizer. *Proceedings of 2nd Workshop of Sustainable Development in Horticulture in Asia and Oceania*, pp. 1–17. Organized by Monbusho International Scientific Research Program: Cooperative Research by The University of Tokyo, Massey University and Seoul National University. Seoul National University, Suweon, Korea, 29 September 1999.

Sohn, S. M. (1999b). Strategy for soil fertility of organic farming by use of organic fertilization. *Proceeding of Symposium on Environmentally Sound Agriculture and Rural Development for 21st Century*. Organized by Korean Society of Organic Agriculture, 10 December 1999, National Agricultural Cooperation Agency, pp. 161–96.

Sohn, S. M. (2000a). Appropriate fertilization of organic fertilizer. *Agronomy and Horticulture* 165, 115–17.

Sohn, S. M. (2000b). Development of reduction technology for nitrate accumulation in vegetables. Ministry of Agriculture and Forestry, 315p.

Sohn, S. M. (2005). Leaching of mineral nitrogen, organic nitrogen and phosphorus in conventional and organic farming in Korean greenhouse. *15th IFOAM Organic World Congress*, 24 September 2005. Adelaide Convention Center, Adelaide, South Australia.

Sohn, S. M. (2009). Current research on organic agriculture in the Asia-Pacific region and worldwide. *Proceedings of APEC Symposium 'Go Organic'*, 19–21 August 2009, Bangkok, Thailand, pp. 312–25.

Sohn, S. M. (2016). Nitrate and chlorophyll contents in organically cultivated Chinese cabbages. *ISOFAR Scientific Conference*. Proceedings of Organic Farming Technology, 22 October 2016. Dankook University, Cheonan, South Korea, pp. 24–31.

Sohn, S. M. and Alley, M. M. (1997). NO3-N accumulation in vegetables and soil profiles due to excessive organic fertilizer application. *Agronomy Abstracts 1997, Annual Meetings of ASA/CSSA/SSSA at Anaheim, CA*, 26–31 October 1997.

Sohn, S. M. and Küecke, M. (2008). Leaching of mineral and organic nitrogen and phosphorus under Korean greenhouse conditions in conventional and organic farming. *Proceedings of 5th International Crop Science Congress*, 13–18 April 2008, Jeju/Korea.

Sohn, S. M. and Yoneyama, T. (1996). Nitrate in vegetables: Physiology, accumulation in plant and intake by human. *Agriculture & Horticulture* 71(11), 1179–82.

United Nations (UN) (2003). *Organic Fruit and Vegetables from the Tropics*. United Nations, New York and Geneva. http://www.unctad.org/en/docs/ditccom20032_en.pdf

Ventura, W. and Watanabe, I. (1993). Green manure production of *Azolla microphylla* and *Sesbania rostrata* and their long-term effects on rice yields and soil fertility. *Biology and Fertility of Soils* 15, 241–8.

Ventura, W., Mascarina, G. B., Furoc, R. E. and Watanabe, I. (1987). *Azolla* and *Sesbania* as bio fertilisers for lowland rice. *The Philippine Journal of Crop Science* 12, 61–9.

Willer, H. and Lernoud, J. (2017). *The World of Organic Agriculture: Statistics and Emerging Trends 2016.* Research Institute of Organic Agriculture (FiBL)/IFOAM-Organics International, Bonn, Germany.

Willer, H. and Lernoud, J. (2018). *The World of Organic Agriculture: Statistics and Emerging Trends 2017.* Research Institute of Organic Agriculture (FiBL)/IFOAM – Organics International, Bonn, Germany.

Xi, Y.-G. and Pei, Q. (2009). Energy evaluation of organic rice–duck mutualism system. *Ecological Engineering* 35, 1677–83.

Yusa, Y., Wada, T. and Takahashi, K. (2003). Apple snails in Japan: The problems, control strategies and possible benefit. Contribution presented at *the Korean-Japan Joint Conference on Applied Entomology and Zoology*, 28–31 May 2003, Busan, South Korea, 105p.

Zhang, L., Yang, Z. and Chen, G. (2007). Energy synthesis analysis of cropping-grazing systems in the Inner Mongolia Autonomous Region of China. *Energy Policy* 35, 3843–55.

Zhou, Z. (2003). China's market for organic fruits and vegetables: Current situation and prospects. *Proceedings of the Seminar on the Production and Exports of Organic Fruit and Vegetables in Asia*, Bangkok, Thailand, 3–5 November 2003. Organized by FAO, IFOAM, and EarthNet Foundation. http://www.fao.org/docrep/006/AD429E/ad429e17.htm#P618_6608

Chapter 21

Organic agriculture and agroecology in Latin America

Roberto Ugás, Universidad Nacional Agraria La Molina, Peru

1 Introduction
2 Principles
3 Organisations and movements
4 Production and business
5 Institutions and policies
6 Performance of organic systems
7 Summary and future trends
8 Acknowledgements
9 References

1 Introduction

Latin America and the Caribbean (LAC) have a rich and diverse history, incorporating indigenous cultures, ancient civilisations, European colonisation, African slavery, Chinese and Japanese immigration and global exchanges that make it complex and difficult to describe as a whole. There are 33 independent countries in the region and 13 dependencies or other territories, with a population of over 600 million people (Box 1). LAC is a region of high biodiversity, with dramatic landscapes ranging from the Sierra Madre in Mexico to the forests and volcanoes of Central America, tropical islands in the Caribbean, the extensive Andes mountain range, the dry deserts of Western South America and the grasslands of the Southern Cone, Patagonia and the Amazon. Brazil, Colombia, Mexico, Peru, Ecuador and Venezuela are generally considered to be among the world's ten most biodiverse countries as measured by species richness.

Agriculture most likely arose independently in Mesoamerica (today's Mexico and Guatemala) and in the Central Andes (particularly Peru and Bolivia) and the continent provided key crops for today's industry (corn, potatoes, tomatoes, chilli peppers, cacao, tobacco, beans and cotton). More recently, new products derived from the local biodiversity, such as avocado, quinoa, amaranth, maca, chia and stevia, are marketed worldwide, very often based on their 'superfood' health-promoting attributes.

http://dx.doi.org/10.19103/AS.2017.0029.23
Published by Burleigh Dodds Science Publishing Limited, 2019.

Promoting cultural and biological diversity is essential for organic agriculture and agroecology in the region. Around 400 million ha are controlled by 15 million family farms in LAC, 14 million of which are smallholders (Berdegué and Fuentealba, 2011). In 2010, 42 million indigenous people (7.8% of the total population) represented approximately 14% of the poor in the region (World Bank, 2015). Rural poverty in LAC is twice as high compared to urban areas, while extreme poverty is three times higher in rural areas (CEPAL, 2018).

This chapter provides an account of the development and current status of agroecology and organic agriculture in LAC. The development of these two movements is rooted in work on agrobiodiversity and traditional agriculture. It is fuelled by a reaction against industrial agriculture and the persistence of rural poverty, together with the opportunities provided by a growing market for organic products. The chapter highlights the main contributions made by LAC farmers, practitioners, researchers and value chain actors to the worldwide growth of agroecology and organic agriculture. These contributions include:

- The development of the concept of agroecology, grounded in rural development work with farmer's groups, NGOs and donors;
- The establishment of group certification schemes, which are now a standard way for millions of smallholders in the global South to access international markets;
- The growth of participatory guarantee systems for local organic markets, grounded in community interactions; and
- The emergence of a wide range of organic and biodiverse products from the region in world markets.

LAC is underrepresented in international scientific discussions. Much of the research on agroecology and organic farming is published in Spanish, very often in project reports, outreach materials and specialised journals. Perhaps the largest source of research on organic agriculture and agroecology in LAC comes from university theses, which generally do not find their way into scientific journals. This chapter reviews recent peer-reviewed research in English, analysing data gathered in LAC which, when translated into Spanish, will make it available to a larger audience. It may also help to identify research gaps and needs, as larger numbers of Latin Americans attend universities and research institutions in the region and worldwide. Finally, it may also facilitate connections with the research communities in Asia and Africa, promoting much needed South–South cooperation.

Box 1 Latin America and the Caribbean in figures: challenges for sustainability

LAC hosts 24% of the world's arable land, 33% of water resources and produces 11% of the value of world food production. Over 40% of the Earth's biodiversity is found within South America. Thirty-three countries spread over 21 million km^2 are home for close to 600 million people (80% urban), speaking four European languages (Spanish, Portuguese, English and French) and more than 500 indigenous languages. In the global trade, LAC accounts for 16% of the world's food and agriculture exports and 4% of its imports.

Published by Burleigh Dodds Science Publishing Limited, 2019.

	LAC	OECD
Population annual growth (%)	1.1	0.6
Human development index (HDI)	0.751	0.887
Life expectancy at birth (years)	75.2	80.3
Gross national income per capita ($)	14028	37916
Gross national income per capita ($ female/male)	10053/ 18091	28441/ 47684
Government expenditure on education (% GDP)	5.4	5.1
Mean years of schooling	8.3	11.9
Population with at least some secondary education (% age 25 or older)	58.1	85.5
Youth not in school or employment (% of people ages 15–24)	19.5	15
Employment in agriculture (% total population)	14.8	4.8
Agriculture, value added (% GDP)	5.5	1.6
Agriculture, value added per worker (constant 2010 $)	7140	17215
Homicide rate, number of unlawful deaths per 100000 people	21.6	3.3
Suicide rate, number of deaths (male) per 100000 people	9.9	17.4
Mobile phone subscriptions per 100 people	110.6	115.9
Confidence in judicial system (% answering 'yes')	32	51

- More than 50% of the region's food production comes from its 14 million smallholder farmers
- More than 50% of the productive land is controlled by 1% of the farms
- Rural population in poverty: 49%
- The poor spend 50–80% of their incomes on food
- Undernourishment: 5.5% (34.3 million people or 4.3% of the world's undernourished)
- Child malnutrition (stunting moderate or severe) under age five: 13%
- Child anaemia under age six, countries' range: 4–61.3%
- At least 200 million ha of degraded land, mostly as a result of unsustainable agriculture practices
- Share of freshwater withdrawal for agriculture: 72%
- Pesticide active ingredient use in arable land and permanent crops, t/1000 ha: 130.5
- Of the ten countries most impacted by extreme events (1992–2011), four are in LAC
- Crop and livestock production losses after medium- to large-scale climate-related disasters: floods (60%), drought (29%), storms (11%)
- Methane emissions from agricultural sector: 66%
- Three main sources of agricultural greenhouse gas emissions: enteric fermentation (58%), manure left on pasture (23%), synthetic fertilisers (6%)
- Food loss and waste, % total: 40% in production, 22% in post-harvest handling and storage
- Area of biotech crops: 75 million ha (42% of world total)

Sources include CGIAR, FAO, IFAD, WFP, ISAAA, Oxfam, Rabobank, UNDP, UNEP, World Bank and WRI.

2 Principles

Agroecology can be viewed as scientific discipline, agricultural practice and political or social movement (Wezel et al., 2009; Petersen, 2018), and has its roots in peasant agriculture, particularly traditional farming systems adapted to local conditions, managing harsh environments in a sustainable way, to meet subsistence needs (Altieri et al., 2012). The application of agroecology was introduced as a means to enhance the productivity of peasant farming systems in LAC (Altieri, 1999). When comparing agroecology to organic farming, many agroecologists suggest that organic agriculture simply follows standards set by certification agencies or governments and consists merely of a substitution of inputs, rather than a system redesign based on ecological processes. It is sometimes presented as an intermediate stage before a truly agroecological transition. On the other hand, many in the organic movement see agroecology as an intermediate stage towards organic cultivation, since agroecology is not always able to enforce limitations in the use of substances that are prohibited or restricted in organic regulations, thus jeopardising consumer trust. But this is not a black-and-white situation for farmers. As an example, Morris et al. (2013) analysed farmers' motivations for the conventional management (with pesticides) of subsistence food crops in El Salvador, in contrast to the adoption of organic management of their coffee destined for export. Bellon et al. (2011) consider that while organic agriculture has grown around key concepts such as farming systems and value chains, agroecology deals more with agroecosystems and food sovereignty, particularly from the perspective of smallholders.

Agroecology, in the most common Latin American understanding, has been presented as 'the most conspicuous example of ecological intensification for family agriculture in terms of technological and institutional development', while 'organic agriculture is one of the largest existing models of ecological intensification worldwide in terms of surface area, although not all forms of organic farming can be considered to be ecologically intensive or sustainable' (Tittonell, 2014).

The IAASTAD report for LAC (2010) states that the agroecological system is highly productive and sustainable, and creates a market niche for certified organic products, yet has been limited by a lack of government-wide support. There is also debate as to whether it can satisfy the increasing global demand for food. More recently, as agroecology has been endorsed by corporations or international organisations, some authors question whether 'agroecology will end up just being part of the toolbox of the agro industrial food system or will be strengthened as a mobilizing political alternative' (Giraldo and Rosset, 2016). Others question the feasibility of generalising an extensive organic farming scenario based on the techniques currently used by organic farmers, suggesting the need to intensify organic farming according to agroecological criteria. However, agroecological intensification cannot be prolonged indefinitely over time since it may be limited by the land available (González de Molina and Guzmán, 2017).

Although agroecology and organic agriculture are often used as synonyms, agroecology generally refers to a movement and has a political and ideological emphasis. The term 'organic' is used in the context of an industry, market or set of regulations, although it also refers to a movement or sector. Many organic farms are based on the substitution of chemicals for organic inputs without a fundamental system redesign (e.g. in many horticultural crops for export), while within the practice of agroecology farmers may use inputs not permitted in organic agriculture (e.g. when the boundaries between agroecology and smallholder farmers as a whole are not so clear). The agroecological and organic movements should also be viewed in the context of broader discussions in LAC, particularly related to the decrease in rural population, the widespread implementation of policies that favour export-oriented agribusiness and neglect smallholder farming for local markets, as well as climate change and biodiversity conservation. The concept of food empires has been used with regard to the systematic disassembling of the resource base on which peasant farming is grounded, the intensification of extraction processes and the takeover of strategic resources, such as land, water, genetic materials and market outlets (van der Ploeg, 2014). As an alternative model, the concept of the peasant food web has been proposed, not as a pseudonym for agroecology or organic farming but in opposition to the industrial food chain, stating that 'much (or most) of what peasants produce is de facto organic' while 'peasants make their decisions about synthetic fertilisers or pesticides for ethical, economic, environmental, or access reasons' (ETC Group, 2017). This last statement, in operational terms, may directly contradict regulations governing organic agriculture.

In spite of these differences, the practice of collaboration is widespread and should be strengthened in order to achieve greater impact. Examples of this are farmers markets across LAC, many of which require organic certification: agroecological farmers learn about the fulfilment and verification of organic standards while sharing their knowledge about ecological and social intensification, as has happened with participatory guarantee systems (PGS). Another example is partnerships advocating for improved policies, for instance Brazil's National Plan for Agroecology and Organic Production (CIAPO, 2013), the most comprehensive plan for organic agriculture approved in the global South, with implications ranging from guarantee systems to public procurement of food. At the international level, an example of this collaboration is the agreement between the International Federation of Organic Agriculture Movements (IFOAM) and the Latin American Scientific Society for Agroecology (SOCLA). IFOAM's 2011 position paper 'recognizes that smallholders have a fundamental role in the stewardship of biodiversity and regards organic agriculture based on the scientific discipline of agroecology as the most appropriate way to achieve ecological, agronomic and socio-economic intensification of smallholder agriculture' (2011).

3 Organisations and movements

In the second half of the twentieth century, thousands of local communities of practice emerged in LAC, linking smallholders with a new set of individuals, organisations, institutions, markets and regulations, and also with new ideologies and politics. Many of these initiatives were based on themes such as ethnobiology, seeds and genetic resources, conservation agriculture, participatory development of technologies and appropriate technology. Traditional agriculture, rural modernisation and globalisation are still central to current agroecological debates. From the 1980s, strong leadership of NGOs led to the use of local resources (rather than the purchase of external inputs), agroecology and organic agriculture, particularly in relation to traditional agrarian systems, local markets and food sovereignty, with active national movements in Mexico, Costa Rica, Peru, Argentina, Chile and Brazil, among others. Some of these NGOs moved into certification and export markets. Many had close contacts with organic and fair trade movements, as well as sufficient knowhow to help governments draft regulations and in negotiation with foreign governments (IFAD, 2003).

Coffee, the main organic product in the region, is dominated by smallholders, particularly in the humid tropical mountain regions of Central America and the Andean countries. They are typically organised in cooperatives, whose intensity and effectiveness of collective action varies greatly between and within countries. Cooperatives and other forms of farmer organisations, along with the establishment of group certification schemes, have been the main driver in the growth of organic agriculture for export in LAC. The performance of cooperatives may also be influenced by market conditions. In Costa Rica, Snider et al. (2017) found that low consumer demand, high certification costs and weak financial incentives encourage cooperatives to certify only a portion of their members, rewarding mainly progressive farmers in a way that ultimately undermines the sustainability of the cooperative as a whole.

At the national and regional level, local agroecological and organic movements are mainly facilitated by NGOs and farmer's organisations, some of which are supported by development projects. A notable exception is Argentina, where the local organic movement is steered by a coalition of small- and medium-sized businesses, NGOs and individuals. Early in the evolution of organic agriculture in the region, it was noted that producer associations played a major role in incorporating small farmers into organic production and recommended more support for farmer organisations to give them a stronger voice (IFAD, 2003). Today, associations of organic producers operate in most countries. Those in Bolivia (AOPEB) and Peru (ANPE) have a national spread, include several indigenous groups, and coordinate an organic guarantee system (PGS). From an analysis of Brazil, Bolivia and Peru, Hart et al. (2016) found that the contribution of producer movements to multifunctional landscapes depends on a strong focus on farm and landscape stewardship, and having the political support and capacity to engage meaningfully in multi-stakeholder processes.

Regional networks provide key opportunities for dialogue and knowledge exchange, help to coordinate with donors active in similar countries and improve advocacy at the international level. A network of NGO and peasant movements, the Agroecological Movement of Latin America (MAELA), is active in debates around agroecology and food sovereignty, and involved in the Civil Society Mechanism for the UN Committee on World Food Security. IFOAM has become a central link between organic movements and businesses worldwide, highlighting smallholder and LAC issues in international negotiations, including support for the declaration of 2014 as the International Year of Family Farming. Initially an informal network, the Latin American group of IFOAM was formally established in 2006. National organic movements organised the Organic World Congress in Rio de Janeiro, Brazil (1992), and in Mar del Plata, Argentina (1998). The Latin American Meetings of Ecological and Organic Agriculture (ELAO) started as a Central American initiative, bringing together farmers, NGOs and researchers, with the support of the Research Institute of Organic Agriculture (FiBL) in Switzerland. Other organisations active in the analysis and promotion of family farming, with links to agroecological movements, are REAF (MERCOSUR's Specialised Meeting on Family Farming, established in 2004) and Via Campesina.

A new development is FAO Agroecology, a knowledge hub that has organised regional symposia to guide technical support to governments and other stakeholders, emphasising the positive impact of the agroecological movement. Meanwhile, as agroecology gains momentum in international discussions, concern has been expressed over a reformist version that would co-opt agroecology into the Green Revolution, rather than a radical version that centres agroecology within a politically transformative peasant movement for food sovereignty (Holt-Giménez and Altieri, 2013). The more market-oriented organic sector, often involved in the export business, is particularly active in special committees in local chambers of commerce or similar structures, notably in Brazil, Argentina, Colombia and Mexico, as well as in commodity-specific organisations such as the coffee boards in most of tropical America. The agroecological and organic movements have acted jointly to support common objectives in major international meetings on biodiversity, climate change, access to genetic resources, food security, development goals and others (Box 2).

Box 2 Highlights of the development of research, education and extension in agroecology and organic agriculture in LAC

Key publications

The books on *Agroecology* by Altieri (1987) and by Gliessman (1990) explained the foundations of traditional agriculture and its role in rural development. Other key books include those by Hernández Xolocotzi (*Maize Genetic Resources*, 1951), Primavesi (*Ecological Soil Management*, 1979), Bunch (*People-oriented Agrarian Improvement*, 1982), Hart (*Agroecosystems*, 1985), Geilfus (*Participatory Development*, 1997), Bunch (*People-centred Agricultural Development*, 1982) and Benzig (*Organic Agriculture in the Andes*, 2001), among others. The magazines *LEISA* (in Spanish) and (in Portuguese), based in Peru and Brazil, became the largest printed regular source of information for practitioners dealing with low-external-input sustainable agriculture (LEISA). Latin American Scientific Society for Agroecology (SOCLA) collaborates with the Spanish peer-reviewed journal *Agroecología*, while the journal *Agroecology and Sustainable Food Systems*, edited by S. Gliessman, is publishing an increasing number of papers in English by Latin American researchers.

Farmer to Farmer Movement (Movimiento Campesino a campesino)

This movement was started in the late 1970s in Mexico and Central America as a reaction to the vertical approach of extension services based on technological transfer. Originally linked to development projects, it combined a recognition of agroecological techniques from traditional agriculture and the development of innovations based on direct farmer intervention and informal knowledge networks, aimed at reducing risk or increasing productivity. Centred on peasant pedagogy, it also has political ramifications with regard to cultural resistance and social change (Holt-Giménez, 2008). Elements of the farmer to farmer movement were implemented in several development projects and also government programmes in line with indigenous knowledge systems, as with the Andean peasant extension agents *kamayoc* and *yachachiq* (Quechua, Peru) and *yapuchiris* (Aymara, Bolivia).

Leading agroecological organisations

The Latin American Consortium for Agroecology and Development (CLADES), originally based in Chile, was a consortium of NGO that set the foundations for the understanding and promotion of agroecology in much of LAC; its distance education programme for the Andean countries, based in Peru, trained close to 3000 farmers, technicians and professional staff. SIMAS, the Mesoamerican Information Service on Sustainable Agriculture, was established in Nicaragua in

2003 and is an active channel for exchange of information in Central America. The SOCLA was established in 2007 and is the largest network of practitioners and researchers dealing with alternative agriculture and is organised in national chapters. Its massive Latin American Congress on Agroecology is organised every 3 years.

University education on agroecology, organic farming

In the early 1990s, universities in Spain organised the first graduate programme on agroecology, attended by many Latin Americans. Today, several LAC universities and research institutions offer academic training in agroecology and organic farming, but often lack regular adequate funding. SOCLA has sponsored PhD programmes on agroecology in Colombia and Nicaragua. A few universities operate certified organic experimental farms, such as La Molina (Peru) and Chapingo (Mexico). Researchers and practitioners from LAC are getting involved in the Technology Innovation Platform of IFOAM (Niggli et al., 2017).

Long-term studies and sustainability assessments

FiBL (Switzerland) implemented the Farming Systems Comparison in the Tropics Project, with long-term experiments to study conventional and organic cocoa production in monocrop and agroforestry systems in Bolivia. The Committee on Sustainability Assessment (COSA) works in 12 coffee and cacao producing countries (including six from Latin America) and has collected nearly 18000 farm and village-level surveys for the impact assessment of the application of voluntary sustainability standards. New collaborations have been proposed between the conservation research community and the sustainability standards community to develop common indicators and monitoring protocols, foster data sharing and synthesis, and link research and practice more effectively (Milder et al, 2014).

4 Production and business

4.1 Certified and non-certified organic agriculture

LAC has 13% of the world's certified organic land, accounting for 0.9% of its total agricultural land. In 2015 there were over 460000 certified (third-party and PGS) farmers in the region, managing 6.7 million ha of land. When wild collection and other non-agricultural areas are added, this totals 10.9 million ha. Nearly 80% of organic agricultural land is permanent pasture, while 12% is devoted to permanent crops. Cereals occupy almost 40% of the organic arable area, with over 70% of farmers cultivating quinoa and amaranth (Willer and Lernoud, 2017). Coffee (480000 ha) and cocoa (190000 ha), which often grow in the same farming system, are the top permanent crops, and account for more than 50% of the world's organic production. Bananas and avocados are the top organic tropical fruits, while Brazil nuts and honey are the main products from organic wild collection areas. Table 1 shows the leading Latin American countries and products per category; it should be noted, however, that Brazil might be underrepresented because detailed data from that country was not made available (Willer and Lernoud, 2017).

Table 1 Top organic countries and products in Latin America and the Caribbean, 2015

	Top countries, crops or products
Organic agricultural land	Argentina, Uruguay, Brazil, Mexico, Peru
Organic share of total agricultural land	Uruguay, French Guiana, Dominican Republic, Argentina, Peru
Organic wild collection areas	Mexico, Brazil, Bolivia, Argentina, Peru
Organic producers, number	Mexico, Peru, Paraguay, Dominican Republic, Bolivia
Land use	Grassland/grazing areas: 65%, Permanent crops: 12%, Arable crops: 5%, No details: 18%
Grassland/grazing areas	Argentina, Uruguay, Malvinas (Falklands)
Main crops, land	Coffee, cocoa beans, cereals and Andean grains, tropical and subtropical fruit, sugarcane, oilseeds, citrus fruit, temperate fruit
Coffee	Mexico, Peru, Honduras, Nicaragua, Bolivia
Cocoa beans	Dominican Republic, Peru, Panama, Ecuador, Bolivia
Cereals and Andean grains	Bolivia, Argentina, Peru, Ecuador, Mexico
Tropical and subtropical fruit	Mexico, Dominican Republic, Ecuador, Peru, Costa Rica
Sugarcane	Paraguay, Argentina
Oilseeds	Argentina, Paraguay, Bolivia, Mexico, Nicaragua
Citrus fruit	Mexico, Dominican Republic, Argentina, Paraguay, Chile
Temperate fruit	Argentina, Chile, Peru
Vegetables	Mexico, Argentina, Chile, Peru, Ecuador
Cotton	Peru, Brazil, Colombia
Olives	Argentina, Peru, Chile
Grapes	Chile, Argentina
Organic share of total agricultural land	Quinoa: 70%, cocoa: 11%, coffee: 9%, banana: 2.4%, sugarcane: 0.5%
Wild collection products, land	Beekeeping, nuts, palm heart, rose hips, medicinal and aromatic plants
Beehives, number	Brazil, Cuba, Mexico, Chile, Nicaragua
Aquaculture, land	Costa Rica, Ecuador

With information from Willer and Lernoud (2017).

Between 2001 and 2007, organic wild collection areas predominated, until 2008 when a rapid switch to organic agricultural land occurred (Fig. 1). Total organic agricultural land in LAC has decreased since its historical peak of 2009; however, the number of organic producers has steadily increased, to a record number in 2015. Although the overall decrease in certified land is to a large extent related to a reduction in certified grassland in Argentina, several factors are affecting growth and permanence in organic certification. These include coffee rust (*Hemileia vastatrix*) and its devastating effects throughout the continent (and governments distributing fungicides and fertilisers to coffee farmers); a

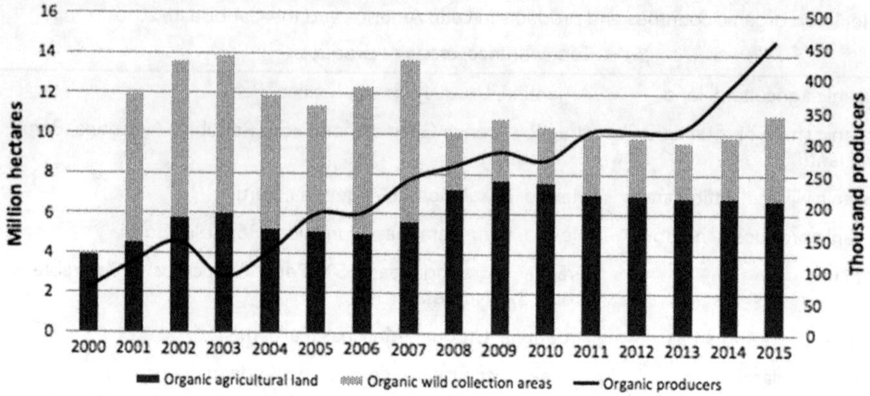

Figure 1 Development of organic agriculture in Latin America and the Caribbean: organic areas and producers, 2000–15. Based on Willer and Lernoud (2017) and additional data from the same authors.

reduction in premium prices for certain organic products; organic farmers switching to sustainability certification schemes allowing the use of some synthetic fertilisers and pesticides; and decreased support for local organic markets due to changes in government political orientation or reduced financial capacity.

A typical Latin American organic farmer grows coffee and cacao in a family farm of a few ha with varying degrees of agroforestry. Farmers generally belong to a cooperative or association for certification and marketing purposes. Research on organic agriculture in LAC has a strong focus on coffee, and lately also on coffee rust. The coffee rust epidemics in LAC can likely be traced back to the use of agrochemicals and a drastic reduction of shade, combining to steadily weaken the complex web of ecological interactions that kept the disease in check (Vandermeer et al., 2014). The issue of shade is complex and controversial, and there may be several factors involved. It is understood that heterogeneous agricultural landscapes that retain abundant tree cover contribute to biodiversity conservation (Harvey et al., 2008) and traditional ways of life, while shade has shown to improve coffee quality by promoting slower, more balanced and uniform ripening of berries (Muschler, 2001; Vaast et al., 2006). DaMatta (2004) points out that the ecophysiological constraints of the coffee crop make it obligatory to grow in shade, although higher yields may be obtained from unshaded plantations. López-Bravo et al. (2012) concluded that the advantage provided by shade in controlling coffee rust has to be balanced against the disadvantage of reduced yields in the short term. Rice (2008) found that the consumption and sale of all non-coffee products in smallholder coffee farms accounted for a fifth to a third of the total value realised from the agroforestry system, and that fuel wood and construction materials accounted for much of this value.

In an attempt to explain the causes of coffee rust, Avelino et al. (2015) posit that low profitability leads to suboptimal coffee management, resulting in plants succumbing to pests and diseases. Coupled with this, a reduction in diurnal thermal amplitudes, enhanced by weather conditions consistent with climate change, likely decrease the latency period of the disease. The impact of organic agricultural practices on coffee rust has not been studied in depth. Regarding brown eye spot, caused by *Cercospora coffeicola* (significant

Published by Burleigh Dodds Science Publishing Limited, 2019.

in Brazil), the highest values for incidence, severity and leaf fall were found in conventional coffee grown under full sun, whereas the lowest values occurred in organic crops grown under shade (Souza et al., 2015).

Quinoa is an ancient Andean grain native to the Central Andes that has become renowned as a gluten-free superfood because of its complete amino acid profile and other nutritional properties. The increase in organic land area given to quinoa in Bolivia and Peru is notable, a recent process in which high external demand was followed by very high prices and overproduction, leading to concern over the impact this would have on peasant communities, traditional farming systems and the environment (Jacobsen, 2011). Over a short time period, Peru increased its production of mechanised conventional quinoa in the irrigated coastal desert, while Bolivia and the Peruvian Altiplano peasants focused on organic quinoa. There inevitably followed a drop in prices. Winkel et al. (2016) analysed the dynamics of impoverished quinoa growers' communities in southern Bolivia that deliberately entered the global food market, pointing to landscape homogenisation, social inequity and increased dependence on external factors as potential sources of unsustainability. Bellemare et al. (2016), however, found a link between higher quinoa prices and the increased well-being of Peruvian Altiplano farmers and advocated support for organic differentiation in the marketplace.

4.2 Characterisation of organic production systems

Organic production in LAC is dominated by smallholders – typically cultivating up to 5 ha of land. The IAASTAD report (2010) considers three general agricultural systems in Latin America:

1. Traditional–indigenous, based on local/ancestral knowledge, very much tied to local conditions, including peasant systems.
2. Conventional, with a market-based approach, focused on intensive production practices, tending towards monoculture and the use of external inputs.
3. Agroecological, based on the combination of agroecology and traditional knowledge, favouring the use of organic inputs and the integration of natural processes.

Many grassroots organisations from the agroecological and organic sectors are involved in interacting with policy-makers and discussing budget allocation at local levels, so a thorough understanding of the economic characteristics of organic agriculture is essential. Berdegué and Fuentealba (2011) propose a 'simplification of the heterogeneity of smallholder agriculture' in Latin America, useful for policy-making, as follows:

- Asset-poor smallholders in territorial and regional contexts that are not conducive to economic growth and social development.
- Smallholder agriculture with some limitations of assets in territorial and regional contexts where there is a measure of economic growth and social development.
- Asset-rich smallholders in territorial and regional contexts that are highly conducive to economic growth and social development.

Coffee agroecosystems occur in landscapes where a gradient of farming systems has been identified. These range from rustic (where coffee bushes are inserted into the understorey of a natural forest); traditional polyculture or coffee garden (where coffee and other food crops are grown under an intact original canopy); commercial polyculture (where the overstorey is significantly reduced); shaded monoculture (with a sparse canopy over monoculture coffee); to unshaded monoculture or full sun (Moguel and Toledo, 1999). At one end of the scale, coffee is introduced as a non-timber product in natural forests, whilst at the other it, in effect, creates a biological desert (Barton et al., 2002). Certified organic production occurs along this gradient of agroecosystems.

Drawing upon African and Latin American cases, Bennett and Franzel (2013) categorise systems as traditional, organic-by-default, resource-conserving, organic agriculture and certified organic agriculture. Organic-by-default is defined by the lack of use of fertilisers due to logistic or economic reasons, while much of traditional agriculture, although not all, harnesses local ecological processes to enhance productivity while conserving the natural resource base. The same authors distinguish between organic-by-default subsistence farmers, transitional farmers and cash croppers, depending on their market orientation. Cáceres (2005) refers to non-certified organic farmers in Argentina as those that are producing organic products but are prevented from obtaining certification due to economic reasons.

Latin America is also experiencing a process of bifurcation similar to that of industrialised countries, referred to as the conventionalisation of organic agriculture: 'As agribusiness enters organics, a bi-polar production system emerges made up of larger conventional operations that mix input substitution strategies with monoculture production of high value crops targeted to indirect markets, while smaller farms employ artisanal practices to grow a variety of crops using more sustainable agronomic practices targeted to direct markets' (Constance et al., 2015). But both systems may also exchange information and experience, such as has been the case with *biol* (a liquid fertiliser based on the on-farm anaerobic fermentation of manure and other organic matters), the technique most widely used by organic smallholders for plant nutrition, today also practised by large organic and conventional holdings and found in the market for organic inputs. The most common examples of organic production systems in LAC are presented in Table 2.

4.3 Guarantee systems

As most LAC countries have national regulations, originally intended to promote organic exports, third-party certification bodies are the norm. Most countries have a local certification body, with the exception of Chile, El Salvador, Paraguay, Venezuela and most of the Caribbean islands. The 2013 Certification Directory listed 50 offices of foreign and local certification bodies in 13 countries in the region, but this number has very likely grown. NGOs and donor agencies were key players in the establishment of local certification bodies, after an initial period of foreign inspectors travelling all over the region. Later, these certification bodies became separate legal entities to fulfil accreditation requirements. Biolatina, based in Peru, resulted from the merger of certification bodies from Nicaragua, Bolivia, Colombia and Peru in 1998 and is one of the few examples of its kind in the developing world (Ugas, 2003). The contribution of organic certification bodies, both local and international, in terms of market development and capacity building is often underestimated. Group certification with Internal Control Systems (ICS), which started in Central America and Mexico, is the main methodology allowing the certification of smallholders worldwide, in association with farmers' cooperatives and NGOs. This has been a decisive innovation in rural LAC.

Published by Burleigh Dodds Science Publishing Limited, 2019.

Table 2 Examples of organic production systems in LAC

Production system	Main market	Examples
Custodian farmers, who actively maintain, adapt and disseminate agricultural biodiversity and related knowledge	Local	The most common organic farmer in some countries, generally non-certified organic. FAO's Programme on Globally Important Agricultural Heritage Systems has designated three sites in LAC (in Peru, Chile and Mexico).
Smallholders organised in cooperatives or other forms of collective action – mixed farming systems with a diversified portfolio	Local	Main characters of NGO or government projects for the development of local organic markets, often with PGS.
Smallholders organised in cooperatives or other forms of collective action – mixed farming systems with a main cash crop	Export	Coffee and cacao/Mexico to Bolivia; quinoa/Peru, Bolivia; banana/Dominican Republic, Ecuador, Peru; nuts, medicinal and cosmetic plants/ Amazonia.
Animal production		Organic meat for export/Brazil, Argentina, Uruguay; poultry and eggs for local markets; guinea pig in Andean countries.
Medium and large enterprises that may also produce conventional, monocultures, reliance on external inputs	Export	Vegetables/Mexico; apples/Argentina; orange/ Brazil; banana/Dominican Republic; blueberries/ Chile, Peru; wine/Chile, Argentina; mango/Brazil, Mexico, Peru.
Urban and periurban farmers mostly selling at farmers markets	Local	Vegetable growers, egg producers, small fruits and berries, herbs throughout the region.
Food processors, micro and small enterprises often selling at farmers markets and organic/natural shops	Local	Diversified processed products for farmers markets, boutique shops or natural products outlets.
Food processors, industrial	Local and export	Medium-sized companies or organic lines of conventional processors, selling brands locally or regionally.
Input manufacturers and distributors	Local	Liquid fertilisers from anaerobic fermentation, as a component of a diversified smallholder operation or as a commercial side product of animal husbandry; microbial preparations for widespread use/Brazil, Cuba; guano/Peru; commercial lines of organic inputs in all of LAC; imported organic inputs.
Seed producers	Local and export	Very few for the local markets (difficult implementation of organic regulation); organic seed production for export.
Wild collectors	Export and local	Brazil nut/Bolivia, Brazil, Peru; palm heart/Peru, Guyana; rose hip/Chile, Argentina; herbs and medicinal plants.
Biodynamic operations	Export	Close to 6000 ha. Main countries: Chile, Argentina, Brazil (ha), Paraguay, Argentina (farms), Brazil (processing).
Agrotourism	Local and foreign	In all countries but more frequent in Brazil, Argentina, Costa Rica.

Since the mid-1980s, and largely based on experience gained in Mexico and Central America, private certification bodies developed group certification systems for smallholder farmers in developing countries. Setting criteria for ICS (e.g. Augstburger et al., 2002), and a process of harmonisation led by IFOAM and its Accreditation Committee, was followed by the establishment of European Union guidelines for the evaluation of equivalence of organic producer group certification (IFOAM, 2012). This was essential to guarantee the supply of organic tropical and subtropical products in Northern markets, as well as, most importantly, being a socioeconomic innovation that allowed thousands of smallholder coffee and other crop growers in Latin America and worldwide to enter international markets. When managed well, ICSs are powerful tools to strengthen the capacity and function of smallholder organisations, with regard to production, market and advocacy.

From the outset of third-party certification in LAC, local groups complained the system was inappropriate for the reality of smallholders and local markets. The most common criticisms were that it was costly, involved intensive documentary requirements, lacked integration with social processes and was potentially under-supervised, with only one inspection each year. Many, from a more ideological side, criticised the imposition of foreign norms and control systems. The social movements and agroecological organisations in Southern Brazil developed what eventually became known as PGS, a component of social innovation partly described by Rover et al. (2017). A landmark meeting in Brazil (Lernoud and Fonseca, 2004) eventually triggered the expansion and adoption of PGSs in many developing countries. The 2017 IFOAM PGS database shows that 22726 farmers are involved in PGS initiatives in 13 countries in the region.

A PGS can provide guarantees to consumers, while avoiding the entry barriers of third-party certification. In a recent study (Home et al., 2017), farmers in seven countries were interviewed and the results analysed in the light of empowerment and social process theory. The conclusions suggested that participation in a PGS can empower farmers by basing their activities on long-lasting social processes and connection with their communities. Combining tradition and bottom-up collaboration within local-level social structures enables PGSs to overcome their challenges, ensuring their future as an alternative system to third-party certification. Three outcomes were identified: social cohesion, collective infrastructure and farmer opportunity. A major criticism of some PGS initiatives is that they are often started, not because there is a significant market, but because politically motivated actors wish to challenge the 'imposition' from the North and local competent authorities. A further concern is that in many cases there is a need for external (NGO) funding to maintain the system, as farmers rarely pay the full cost of its operation (Home et al., 2017). It is clear that market demand is a requisite for the existence of a PGS but national and subnational governments in LAC should also see PGS as a tool to provide public services, particularly technical assistance, and to empower farmer's groups and local organisations. Clark and Martínez (2016) analysed the role provincial governments in Ecuador can play through public certification, potentially broadening access to certification to marginalised producers, concluding that this may not be effective unless it is accompanied by other measures and policies favourable to small-scale producers.

PGSs and similar schemes have been the subject of intensive advocacy activities by national agroecological and organic movements and are currently approved for the local markets in several LAC countries. Some organisations have attempted to circumvent organic regulations and stringent procedures by developing their own mark and building trust in

Published by Burleigh Dodds Science Publishing Limited, 2019.

the marketplace through PGS and outreach. One notable example of this is the registered collective mark '*Frutos de la Tierra*' (Fruits of the Earth), owned by Peru's ANPE, the leading national association of (non-certified) ecological producers (Van den Eeckhout and Ugás, 2013). Governments also promote smallholder products through marks, like Chile's '*Manos Campesinas*' (Peasant Hands), although it is not exclusive to ecological producers.

4.4 Markets for organic products

4.4.1 Farmers markets

Farmers markets continue to be the main platform for organic products. Gaining authorisation from local governments to allow such events is one of the main advocacy activities for NGOs and farmers' organisations. These markets are showcases for biodiversity and tradition. Major cities have several regular weekly organic markets or farmers markets where organic products are offered, notably Sao Paulo, Lima, Mexico City and San José. Farmers markets are also a vital component in most PGSs and serve to motivate organic consumer groups. Increasingly, these organic markets are a space for exchange and dialogue between rural producers and a growing number of urban farmers throughout the region, as well as key players such as chefs, health-conscious opinion leaders and social media. Some have become tourist attractions and are actively promoted by local governments and communities. Originally, organic farmers markets were seen as serving the local elite, but this is rapidly changing as they appear in medium-sized cities and middle-class neighbourhoods. Loconto et al. (2016) analysed innovations in market institutions for organic and sustainable products, including examples in Andean countries.

4.4.2 Organic retail

In recent years LAC has witnessed a marked increase in the availability of organic products in supermarkets, specialised shops, delivery schemes and online sales, along with the growth of the middle class and some countries moving into medium income. In Lima, for example, organic stores have greater success in areas of the city with middle and high incomes as well as higher than average levels of education (Higuchi, 2015). There is naturally a downside. Growth has brought with it market confusion, unsubstantiated claims and fraud. Several LAC countries have established compulsory (although rarely fully enforced) national logos for organic products (Fig. 2). For fair trade and organic bananas in the Dominican Republic, supermarkets capture 28–48% of the retail value while farmers and their organisations receive 15–20% (Vagneron and Roquigny, 2011), although the local market for this type of banana is very small compared with exports. Current market trends, however, may be increasing the power of buyers and working to displace or disempower small organic producers. Strong producer associations and transnational movement ties have countered these trends with some success in the Dominican Republic (Raynolds, 2008).

4.4.3 Specialised or institutional purchases

There are good examples of public procurement of organic food for school meals and social programmes, as well as companies sourcing organic food for their catering. The restaurant industry could also become a specialised outreach mechanism for organic producers,

Figure 2 Examples of national logos for organic products in LAC.

showcasing their produce to local consumers. Most important may be the connection with the gastronomic movement to inspire social change in countries such as Peru, where there is a prominent 'cook–peasant alliance' (Fan, 2013), Mexico and Colombia.

4.4.4 Exports

Although data is missing for several countries, Brazil, Mexico and Peru are among the top exporters. Mexico, Peru and Ecuador are among the top five suppliers of organic products to the US market. Main exports consist of permanent crops (coffee, cocoa, fruit crops), cereals and sugarcane. Several organic exports are also certified for fair trade and participate in biotrade and sustainability initiatives. When combined (not split per product or per tariff heading), total organic exports are among the top agricultural exports in some countries. Other movements have shared common stakeholders, such as the indigenous groups participating in UCIRI in Mexico, and the key leadership groups that pioneered organic and fair trade exports in the 1980s and set the pace for the region. The Small Producers' Symbol (SPP) is a recent development.

Growth in international markets has promoted investment in organic produce, often by national and foreign companies already involved in conventional production. Examples are fruit juices (Brazil), pineapples (Costa Rica), mangos and avocados (Mexico), wines (Argentina, Chile), palm oil (Colombia), blueberries (Peru) and bananas (Dominican Republic, Ecuador, Peru), in larger fields or monocultures and a higher use of external inputs. Coffee, however, is the largest export in the region and is fundamentally a smallholder business. The trade show Biofach 2017 in Nuremberg was attended by 156 exhibitors from ten LAC countries, with strong participation from Peru, Mexico and Argentina. Entrepreneurship, stronger farmers' groups, support from national agencies for export promotion and key donors have been instrumental in increasing organic exports. Biofach America Latina in Sao Paulo, Brazil, together with Bio Brazil Fair are the largest international trade fairs for organic products in LAC. In 2016 Biofach America Latina featured 117 exhibitors and received 19 240 visitors from 31 countries.

5 Institutions and policies

5.1 National standards and regulations for organic agriculture

All Spanish-speaking countries, and Brazil, have fully implemented regulations, with the exception of Cuba, El Salvador and Venezuela (Willer and Lernoud, 2017). Bolivia, Brazil, Chile, Costa Rica, Ecuador, Mexico and Uruguay have all adopted or accept organic guarantee systems specific to smallholders selling in local markets. The competent authorities on organic production generally fall under the plant protection services and have a continental coordinating network (Interamerican Commission on Organic Agriculture), established in 2008 within the Interamerican Institute for Cooperation in Agriculture. Argentina was visionary in the early establishment of its national control system and became the first developing country in the European Union's Third Country List (1997). Costa Rica followed in 2003, whilst Chile is the first country to sign an equivalence agreement with the EU (in 2017). Also in 2017, Panama started a national public certification programme, supported by the Spanish government (Flores, in Willer and Lernoud, 2017), the first of its kind in the region.

5.2 National programmes for organic agriculture and agroecology

Over the last 20 years, several governments have implemented plans of various forms and with varying success rates, including Brazil's National Plan for Agroecology and Organic Production (PLANAPO) (CIAPO, 2013), Peru's National Plan for Organic Banana (PNBO), Costa Rica's National Program for Organic Agriculture (PNAO) and Cuba's 'Alternative Model' of agriculture in the 1990s, which allowed mass adoption of agroecological methods (Nelson et al., 2009). In its guidelines to public support for organic agriculture, IFOAM (Katto-Andrighetto et al., 2017) mentions several Latin American examples, although only a few are considered to have implemented and upscaled programmes successfully. This might be somewhat biased as many good examples from national, subnational and local governments remain unpublished or only in Spanish. Policies promoting agroecology in seven LAC countries have been documented recently (Sabourin, 2017), showing great potential for upscaling. A good promising example is the combination of a conditioned cash transfer programme for resource-poor Andean farmers in Peru with farming system support through agroecological technologies (Escobal and Ponce, 2015). Other policy instruments relevant to organic agriculture and agroecology, such as payment for environmental services or conservation of genetic resources, are still limited in the region. Most countries, however, have in the last decade passed legislation for public support for family farming, though many remain unregulated or without funding. Many of these policies have been advanced by the local movements (NGOs, producers' and consumers' organisations), with the result that their advocacy and pressure activities increasingly take a larger share of their time and resources. Detailed accounts of the development of local agroecological movements and, to some extent, the impact of their advocacy work, can be found in a special issue of the journal *ASFS* (2017).

Brazil provides perhaps the world's largest example of public procurement with a specific target towards organic food. In 2009, the National School Feeding Program (PNAE) set an objective of purchasing at least 30% of ingredients for school meals from local family farmers, using organic and agroecological processes where possible (Blanc and Kledal, 2012; Swensson, 2015). In 2015 the aim was to serve 42.6 million students. Even though these policies have been weakened, they continue to provide inspiration and influence over public decision-making in the provision of better food. In 2015 the city of Sao Paulo, the largest in LAC, passed a municipal law that made the purchase of organic food from family farmers compulsory for school meals, with the expectation that by the year 2026 organic food will be included in 100% of school meals (Lei n° 16.140/2015 – Alimentação Escolar).

6 Performance of organic systems

> An interaction network that results in autonomous pest control . . . like a spider web, achieving structure and resilience from the multiple interconnections of beetles, ants, fungi, scales, leaf miners . . . an ecosystem service emanating from ecological complexity.

The wonderful realisation mentioned above came after 10 years of research at Finca Irlanda, a 300-ha pioneer of organic and biodynamic coffee in Chiapas, Mexico (Vandermeer et al., 2010). Organic coffee is produced primarily by smallholders in mountainous areas with difficult access, scarce public services often of poor quality, utilising agricultural systems heavily dependent on decisions taken by institutions they are unable to influence. Poverty and exclusion are high in these areas, but organic coffee certification and markets have allowed a transition from a low input, extensive production system into an intensified system with higher yields and quality, catering to differentiated markets. However, performance of individual farms or grower groups varies to a great extent.

Research into the performance and impact of agroecology and organic production systems has taken place over a few decades. It is very often descriptive in nature, dealing with individual farms or case studies, and published mainly in Spanish (i.e. the *LEISA* magazine special edition on the measurement of the impact of agroecology, Gianella and Pinzás, 2016). The MESMIS programme (the Spanish acronym for Indicator-based Sustainability Assessment Framework) has been used since the late 1990s to conduct smallholder sustainability assessments through a long-term, participatory, interdisciplinary and multi-institutional approach (Astier et al., 2012). The early assessment of a large number of project reports and questionnaires in the developing world carried out by Pretty et al. (2006), to assess the adoption of sustainable agriculture, included a few cases from LAC and found a mean relative yield increase of 79% across the very wide variety of systems and crop types, as well as substantial benefits for poor households.

While many descriptive studies have documented the experience of farming communities using agroecological approaches, there is poorly documented quantitative evidence on the social and economic indicators of agroecology (D'Annolfo et al., 2017). After Hurricane Mitch in Nicaragua, field research was conducted by over 800 farmers, and agroecological plots on sustainable farms were found to have more topsoil, higher field moisture, more vegetation, less erosion and lower economic losses after the hurricane than control plots on conventional farms (Holt-Giménez, 2002). Many case studies already involve large numbers of farmers, but it is only in the last decade that research has moved into a more detailed investigation of biological, agronomic or socioeconomic parameters involving meta-studies, larger household surveys or data from longer-term experiments. This type of

Published by Burleigh Dodds Science Publishing Limited, 2019.

research is still carried out primarily by European or North American research institutions, sometimes involving Latin American authors. Tuck et al. (2014), while studying the effects of organic farming on biodiversity, found that organic farming increased species richness by about 30%, although he warned that current studies are heavily biased towards Europe and North America, while other regions with large areas of organic farming remain poorly investigated.

New research is being published based on data from LAC. Blackman and Rivera (2011) found 46 relevant studies to assess if sustainability certification (mostly in coffee and in LAC) creates financial incentives for producers to improve their environmental, social and economic performance, deciding that the methods used in 11 studies likely generated credible results. They concluded there is little evidence to support the hypotheses and that many studies do not ensure comparability between certified and non-certified farms. In a meta-analysis of 17 papers (two from LAC countries) where yields were analysed, D'Annolfo et al. (2017) found that yields were 16% greater for agroecological practices compared to conventional practices. However, results for farm profitability showed that despite the slight increase when adopting agroecological practices, this was not statistically significant. Bray and Neilson (2017) reviewed 51 studies (37 from LAC, of which 16 had organic certification), finding evidence of enhanced livelihood assets arising from certification under specific institutional and contextual settings, but a greater number of studies found either neutral or mixed impacts. Stark et al. (2018) characterised crop–livestock integration in Guadeloupe, Brazil and Cuba and quantified benefits by using the ecological network analysis approach. They found that a higher intensity of crop-livestock integration flows improves the efficiency of the system while a more complex and homogeneous nutrient flow network improves the resilience of the system.

The performance of organic systems in LAC, analysed with quantitative methods and often large samples, has increased in the last decade in peer-reviewed literature in English, as shown in Table 3 for organic coffee. This is highly relevant when analysing the upscaling potential of agroecology and organic agriculture, its performance and impact, in order to devise better policy instruments and advocate for change at government level and globally. Early assessments of organic production show that shifting to organic production leads to improved incomes for small farmers and higher net revenues relative to their previous situation, based on case studies on coffee, honey, cacao, banana, sugar cane and vegetables (IFAD, 2003). The literature refers to higher incomes when transitioning to certified organic production, generally for export. Evidence from the research reported in Table 3 does not fully support this, although the continuous growth in the number of organic producers in LAC in the last decade points to potential benefits for farmers to shift and/or remain as organic producers.

Income and labour, on the other hand, can be tricky to tackle from a research point of view. For sustainability certifications (not organic) Vellema et al. (2015) found that the time and effort required to attain the higher coffee income offered by certified production in Colombia means farmers have to cease other activities, the substitution effect cancelling out the income effect, such that there is no increase in total household income. In Bolivia, however, Armengot et al. (2016) compared the productivity and the return on labour (per working day) of four different cacao production systems – agroforestry and monocultures under organic and conventional management – during the first 5 years of an experiment. They found that cacao yields were 41% higher in monocultures, but the revenues derived

Table 3 Examples of performance studies (quantitative) on certified organic coffee agroecosystems in Latin America

Country	Households or farms[a]	Selected findings and reference
Colombia	224	Strong certification schemes can have significant effects on farm practice associated with environmental quality; organic cultivation has mixed effects on net returns, typically limiting overall economic benefits (Ibañez and Blackman, 2016).
Colombia	134	Productivity was found to be higher on organic farms compared to those farmed conventionally. High levels of production on organic farms are feasible if certain conditions are guaranteed (management of fertilisation, shade, pruning) (Bravo-Monroy et al., 2016).
Costa Rica	2603	Certified growers tend to be younger and certified farms tend to be larger; organic certification improves coffee growers' environmental performance (Blackman and Naranjo, 2012).
El Salvador	29	In a certified organic coffee cooperative, farmers used at least one chemical fertiliser (97%), herbicide (97%) and insecticide (93%) in their milpa agroecosystem (maize and beans); half of cooperative farmers spent more money on staple crop farming than the value of the crops produced (Morris et al., 2013).
Guatemala, El Salvador, Nicaragua, Mexico	469	Certifications alone (fair trade or organic) will not bring significant poverty alleviation to most coffee families: revenue contribution to household livelihoods was relatively small and households did not fare better in food security. Certifications are effective in supporting capacity building and networks that leverage global development funding (Méndez et al., 2010).
Mexico	107	Higher price and maintained yield suggest an augmented income stream for organic producers, while consumers believe organic does not deforest, no measurements are taken during inspection; the certification process will need to evolve to effectively address land use change (Jurjonas et al., 2016).
Mexico, Peru	1080	Yields rather than price premiums are most important for increasing net cash returns (Barham and Weber, 2012).
Nicaragua	327	Poverty is higher in organic and organic-fair trade households than in conventional households; due to constrained availability of family labour. Additional labour has to be hired which offsets saved input costs (Beuchelt and Zeller, 2011).
Nicaragua	278	Organic farms had low investment, low productivity strategies compared to C.A.F.E. Practice farms; the higher price paid for most certified coffee at least partially mitigates biodiversity/productivity trade-offs for the farmer, which could be an incentive to sustain otherwise less economically productive high biodiversity production systems (Haggar et al., 2017).
Nicaragua	292	Participation in organic/fair trade certified coffee markets did achieve a broader set of outcomes than merely accessing favourable prices, including building a more competitive value chain, more viable cooperatives and specific assets by some of the poorest farming households; significant variation due to pre-existing endowments was evident in the ability of households to build assets (Donovan and Poole, 2014).

Country	Households or farms[a]	Selected findings and reference
Nicaragua	96	Organic coffee farmers do fairly well in terms of reduced emissions and carbon sequestration compared to conventional farmers; still significant potential for GHG reductions and carbon sequestration within and outside the coffee plots; sustainable intensification of non-coffee activities (bean, maize, cattle) can improve ecosystem services (Rahn et al., 2014).
Nicaragua	233	Fair trade certified farmers have significantly higher yield compared to organic; even though organic coffee has fetched higher prices, this premium is compensated by the lower yield of organic coffee; no evidence of statistically significant impacts of either fair trade or organic certification on smallholder famers' total household income (Jena et al., 2017).
Peru	360	Organic fair trade farmers appeared with significantly higher risk acceptance, indicating improved entrepreneurial capacity; organic yield levels were slightly higher than their counterparts but no significant difference could be found (Ruben and Fort, 2012).

[a] All studies were based primarily on household surveys, except for Blackman and Naranjo (2012), based on observations from census.

from agroforestry by crops economically overcompensated for this difference, with a return on labour roughly twice as high in the agroforestry systems compared to the monocultures (but similar cacao yields and return on labour in conventional and organically managed agroforestry systems). A common critique of organic systems is precisely the additional labour often required. However, the concept of 'contributive justice' encourages the additional labour intensity of agroecological farming, whereby farmers develop skills and acquire valuable experiential knowledge, leading to the development of a number of capabilities and more bargaining power, facilitating self-determination (Timmermann and Félix, 2015). The high and rapid urbanisation of LAC is clearly affecting the availability of labour for commercial farms and may impede the growth of organic agriculture in certain circumstances.

While recent research generates mixed results for organic systems with regard to income and household revenue, the majority of studies tend to provide evidence of positive effects in key areas such as yield (particularly when transitioning from a situation closer to organic rather than from chemical conventional agriculture), environmental performance, reduction in the use of hazardous chemicals, improved nutrition derived from diversified systems and stronger farmer's organisations. Bray and Neilson (2017) report that certification is generally more likely to generate positive than negative impacts, although a considerable degree of uncertainty persists, and organic certification is more likely to support positive impact if associated with fair trade schemes. The authors also point out that stronger pre-existing producer institutions are more likely to result in benefits for individual households. The same authors recognise that existing studies rarely present either reliable baseline data or a realistic control group for comparison, making causation difficult to establish. Additionally, understanding the situation before the transition to organic is essential when analysing data. In a Colombian study, for instance, farmers who adopted organic practices were more educated, included married couples as decision-makers, had larger farms with technological tools to facilitate post-harvest tasks and were associated with a cooperative, achieving higher production levels than conventional coffee farmers (Bravo-Monroy et al., 2016). A study in Mesoamerica found that fair trade organisational norms combine with organic procedural norms to bring significant impacts in three areas: women's organisations have greater access to network benefits, women gain greater control over farm practices and women enjoy increased access to cash (Lyon et al., 2010).

Newer fields of research, such as the effects of organic management in LAC on climate change, life cycle assessments and food composition and human nutrition, are gaining more attention. Studying coffee agroforestry systems in Costa Rica, Häger (2012) found that total C-stocks were 43% higher on organic farms than on conventional farms, and that they differed in vegetation structure, but not in species diversity, concluding that organic farms in particular may contribute to GHG mitigation. Whilst studying pollination services and coffee suitability under climate change in LAC, Imbach et al. (2017) found that the maintenance of heterogeneous agricultural landscapes is a no-regret adaptation strategy, providing support for production systems that promote heterogeneity, such as organic agriculture. Research conducted on the life cycle assessment of organic orange juice imported from Brazil to Denmark found that the transport stage contributed 58% of its total greenhouse gas emissions; however, this was mainly attributable to the internal transport in Brazil and in Europe rather than the long-distance sea transport (Trydeman, in Halberg and Muller, 2013). In a major meta-analysis, Barański et al. (2014) found that organic crops, on average, have higher concentrations of antioxidants, lower concentrations of cadmium and a lower incidence of pesticide residues than non-organic comparators across regions and production seasons. Brazil was the fifth largest source of studies among all countries, but the remainder of LAC was unrepresented. Also in Brazil, an initial hypothesis that microbial diversity is reduced in pesticide-treated farms as compared to organic farms was not supported after analysing coffee rhizosphere with deep sequencing techniques (Caldwell et al., 2015). New research techniques may challenge some of the preconceptions in agroecology and organic agriculture, but will also foster a culture of continuous improvement based on evidence (Box 3).

Box 3 Evaluating the effects of agroecology on rural households in the Peruvian Andes

While studies evaluating the effects of certified organic agriculture have increased in the last decade, quantitative studies on non-certified organic agriculture are almost non-existent in the peer-reviewed literature. Expert consultations, key informant interviews and a household survey were carried out in three regions of the Peruvian Andes (Cajamarca, Huánuco and Cusco) during the dry season of 2014. Sampling criteria included households that complied with selection indicators to be considered either as agroecological or conventional, with a total sample of 451 households. Most agroecological households have been participating for 3 years or more in development projects that included a PGS and support for local marketing through the National Association of Ecological Producers (ANPE Peru) and several NGOs. Conventional farms, located in similar territories, have not participated in agroecological development projects and used at least one substance prohibited in organic agriculture.

Selected indicators (% of respondents)	Agroecological	Conventional
Persons per household	4.2	3.8
Distinguish an agroecological practice	99.1*	34.8*
Hire farm labour	61*	49
Grow new crops in the last season	58.4*	25.7
Exchange seeds with other farmers	24**	14.8
Use organic fertiliser	97.7*	57.8*
Erosion has increased	18.1	39.1*
Use sprinkler irrigation (powered by gravity)	30.3*	13
House water is clean	75.1*	59.6*
Health condition has improved	67	43.5
Income has barely improved	54.8*	48.3
Partnership is very important	30.8*	18.3*
Last year not concerned about food scarcity	23.5*	13.5
Access to government food programme	21.7*	7.8

* Difference is significant at the $P < 0.05$ level; ** difference is significant at the $P < 0.001$ level.

These are robust findings in terms of the differentiated effects of agroecological practices on rural livelihoods, particularly with regard to the natural and social capitals, as well as food security, but not so strong with regard to other capitals (human, physical and financial). Effects of agroecological practice on food security and income cannot be easily separated from participation in government-conditioned cash transfer and food programmes, or its interactions fully understood. In spite of the mixed results, there are strong policy implications (development of local markets, PGS as guarantee system as well as a tool to strengthen organisations and provide technical assistance, specific measures to upscale the participation of women) as well as a clear societal demand: the provision of basic public services of a good quality as requisites for sustainable production systems and rural businesses run by smallholders.

Source: Ugas et al. (2014)

7 Summary and future trends

Vibrant organic and agroecological territories, communities, organisations and businesses, often interacting successfully with governments at national and subnational levels, have provided evidence for the ecological and socioeconomic intensification of smallholder agriculture in LAC, as processes that improve livelihoods and resilience in a sustainable manner. Ample evidence from the region shows that favourable policies, capacity building, provision of good quality public services and infrastructure development are key elements promoting the transition of smallholders into more sustainable farming systems, including the trend towards organic value chains for local markets and exports. At the same time, however, outscaling and upscaling are urgently needed, particularly when confronted with rapid urbanisation and weak incentives to keep young people in rural areas. It is also the case that policies, in spite of political discourse, continue to favour industrial and export-oriented agriculture with stronger incentives than those for the massive sector of smallholder agriculture, the main provider of food.

Research, outreach and diversified opportunities for technical assistance (participatory, public and private) are required to put traditional agriculture, modern science and technological means to the service of a younger generation rooted in their territories and connected to the world; a rural generation very different from its predecessors. Similarly, researchers need to gather further evidence, involving larger data sets and stronger analytical tools, of the benefits and potential of organic agriculture and agroecology. These systems also need to avoid complacency and unjustified claims and strive to intensify ecological and socioeconomic processes to enhance the overall productivity of the farming system and increase income for the poor.

This cannot be achieved without collaboration between organic, agroecological and like-minded movements, sectors and industries, as proposed in the Organic 3.0 vision (Arbenz et al., 2017). If ideological, political or commercial differences prevent constructive interaction, common objectives and partnerships, they will favour the vision that separates man from nature, sees smallholder agriculture as a problem and poisons the planet and its inhabitants, failing to reduce inequalities. Ancient history and modern achievements in LAC offer abundant ground for wider support for organic agriculture and agroecology by consumers, market players and governments, based on stronger organisations and urban–rural connections.

8 Acknowledgements

P. Flores (IFOAM-OI, Peru), T. Gianella (LEISA, Peru) and G. Soto (ANAO, Costa Rica) provided valuable comments on sections of this chapter.

9 References

Altieri, M. 1987. *Agroecology: The Scientific Basis of Alternative Agriculture*. Westview Press, Boulder, CO.

Altieri, M. 1999. Applying agroecology to enhance the productivity of peasant farming systems in Latin America. *Environment, Development and Sustainability* 1(3–4): 197–217.

Altieri, M. 2015. Breve reseña sobre los orígenes y evolución de la agroecología en América Latina. *Agroecología* 10(2): 7–8.

Altieri, M., F. Funes-Monzote and P. Petersen. 2012. Agroecologically efficient agricultural systems for smallholder farmers: Contributions to food sovereignty. *Agronomy for Sustainable Development* 32: 1–13.

Arbenz, M., D. Gould and C. Stopes. 2017. ORGANIC 3.0 – the vision of the global organic movement and the need for scientific support. *Organic Agriculture* 7(3): 199–207.

Armengot, L., P. Barbieri, C. Andres, J. Milz and M. Schneider. 2016. Cacao agroforestry systems have higher return on labor compared to full-sun monocultures. *Agronomy for Sustainable Development* 36: 70.

ASFS. 2017. A brief history of agroecology in Spain and Latin America. *Agroecology and Sustainable Food Systems* 41(3–4): 229–30.

Astier, M., L. García-Barrios, Y. Galván-Miyoshi, C. E. González-Esquivel and O. R. Masera. 2012. Assessing the sustainability of small farmer natural resource management systems. A critical analysis of the MESMIS program (1995–2010). *Ecology and Society* 17(3): 25.

Augstburger, F., U. Eisenlohr, E. Rüegg and B. Wilhelm. 2002. *Manual for Quality Assurance: A Guideline for Internal Control Systems (ICS) in Smallholder Organisations*. Naturland Association, Graefelfing, Germany.

Avelino, J., M. Cristancho, S. Georgiou, P. Imbach, L. Aguilar, G. Bornemann, P. Läderach, F. Anzueto, A. Hruska and C. Morales. 2015. The coffee rust crises in Colombia and Central America (2008–2013): Impacts, plausible causes and proposed solutions. *Food Security* 7: 303–21.

Barański, M., D. Średnicka-Tober, N. Volakakis, C. Seal, R. Sanderson, G. Stewart, C. Benbrook, B. Biavati, E. Markellou, C. Giotis, J. Gromadzka-Ostrowska, E. Rembiałkowska, K. Skwarło-Sońta, R. Tahvonen, D. Janovská, U. Niggli, P. Nicot and C. Leifert. 2014. Higher antioxidant and lower cadmium concentrations and lower incidence of pesticide residues in organically grown crops: A systematic literature review and meta-analyses. *British Journal of Nutrition* 112(5): 794–811.

Barham, B. and J. Weber. 2012. The economic sustainability of certified coffee: Recent Evidence from Mexico and Peru. *World Development* 40(6): 1269–79.

Barton, D., J. Plaza and E. Contreras. 2002. Social dimensions of organic coffee production in Mexico: Lessons for eco-labeling initiatives. *Society and Natural Resources* 15: 5.

Bellemare, M., S. Gitter, A. Kasterine, E. Obregon and A. Zotz. 2016. *Trade in Quinoa: Impact on the Welfare of Peruvian Communities*. International Trade Centre, Geneva, Switzerland.

Bellon, S., C. Lamine, G. Ollivier and L. de Abreu. 2011. The relationships between organic farming and agroecology. *3rd ISOFAR Scientific Conference at the 17th IFOAM Organic World Congress*.

Bennett, M. and S. Franzel. 2013. Can organic and resource-conserving agriculture improve livelihoods? A synthesis. *International Journal of Agricultural Sustainability* 11(3): 193–215.

Berdegué, J. and R. Fuentealba. 2011. *Latin America: The State of Smallholders in Agriculture*. International Fund for Agricultural Development, Rome, Italy.

Beuchelt, T. and M. Zeller. 2011. Profits and poverty: Certification's troubled link for Nicaragua's organic and fairtrade coffee producers. *Ecological Economics* 70(7): 1316–24.

Blackman, A. and M. Naranjo. 2012. Does eco-certification have environmental benefits? Organic coffee in Costa Rica. *Ecological Economics* 83: 58–66.

Blackman, A. and J. Rivera. 2011. Producer-level benefits of sustainability certification. *Conservation Biology* 25: 1176–85.

Blanc, J. and P. Kledal. 2012. The Brazilian organic food sector: Prospects and constraints of facilitating the inclusion of smallholders. *Journal of Rural Studies* 28(1): 142–54.

Bravo-Monroy, L., S. Potts and J. Tzanopoulos. 2016. Drivers influencing farmer decisions for adopting organic or conventional coffee management practices. *Food Policy* 58: 49–61.

Bray, J. and J. Neilson. 2017. Reviewing the impacts of coffee certification programmes on smallholder livelihoods. *International Journal of Biodiversity Science, Ecosystem Services and Management* 13(1): 216–32.

Cáceres, D. 2005. Non-certified organic agriculture: An opportunity for resource-poor farmers? *Outlook on Agriculture* 34(3): 135–40.

Caldwell, A., L. Silva, C. da Silva and C. Ouverney. 2015. Prokaryotic diversity in the rhizosphere of organic, intensive, and transitional coffee farms in Brazil. *PLoS ONE* 10(6). doi:10.1371/journal.pone.0106355.

CIAPO. 2013. *Brasil Agroecológico – Plano Nacional de Agroecologia e Produção Orgânica – PLANAPO*. Câmara Interministerial de Agroecologia e Produção Orgânica, MDA, Brasilia, Brazil.

Clark, P. and L. Martínez. 2016. Local alternatives to private agricultural certification in Ecuador: Broadening access to 'new markets'? *Journal of Rural Studies* 45: 292–302.

Comisión Económica para América Latina y el Caribe (CEPAL). 2018. Panorama Social de América Latina, 2017 (LC/PUB.2018/1-P). Santiago, Chile.

Constance, D., J. Young Choi and D. Lara. 2015. Engaging the organic conventionalization debate. In: B. Freyer and J. Bingen (Eds), *Re-Thinking Organic Food and Farming in a Changing World, The International Library of Environmental, Agricultural and Food Ethics*, volume 22. Springer, Dordrecht, The Netherlands.

DaMatta, F. 2004. Ecophysiological constraints on the production of shaded and unshaded coffee: A review. *Field Crops Research* 86: 99–114.

D'Annolfo, R., B. Gemmill-Herren, B. Graeub and L. Garibaldi. 2017. A review of social and economic performance of agroecology. *International Journal of Agricultural Sustainability* 15(6): 632–44.

Donovan, J. and N. Poole. 2014. Changing asset endowments and smallholder participation in higher value markets: Evidence from certified coffee producers in Nicaragua. *Food Policy* 44: 1–13.

Escobal, J. and C. Ponce. 2015. Combining social protection with economic opportunities in rural Peru: Haku Wiñay. *Policy in Focus* 32: 22–5.

ETC Group. 2017. *Who Will Feed Us? The Industrial Food Chain vs. The Peasant Food Web*. ETC Group, USA, Canada and the Philippines, 65p.

Fan, J. 2013. Can ideas about food inspire real social change? The case of Peruvian gastronomy. *Gastronomica* 13(2): 29–40.

Gianella, T. and T. Pinzás (Eds). 2016. La medición del impacto de la agroecología. *LEISA revista de agroecología* 32(3).

Giraldo, O. and P. Rosset. 2016. La agroecología en una encrucijada: entre la institucionalidad y los movimientos sociales. *Guaju – Revista Brasileira de Desenvolvimento Territorial Sustentável* 2(1), 14–37.

Gliessman, S. 1990. *Agroecology: Researching the Ecological Basis for Sustainable Agriculture*. Springer, New York, NY.

Gómez, M., R. Schwentesius, J. Ortigoza and L. Gómez. 2010. Situation and challenges of the Mexican organic sector. *Revista Mexicana de Ciencias Agrícolas* 1(4): 593–608.

González de Molina, M. and G. Guzmán. 2017. Agroecology and ecological intensification: A discussion from a metabolic point of view. *Sustainability* 9: 86.

Halberg, N. and A. Muller (Eds). 2013. *Organic Agriculture and Sustainable Livelihoods*. Earthscan and Routledge, London, UK.

Häger, A. 2012. The effects of management and plant diversity on carbón storage in coffee agroforestry systems in Costa Rica. *Agroforestry Systems* 86: 159–74.

Haggar, J., G. Soto, F. Casanoves and E. de Melo. 2017. Environmental-economic benefits and trade-offs on sustainably certified coffee farms. *Ecological Indicators* 79: 330–7.

Hart, A., P. McMichael, J. Milder and S. Scherr. 2016. Multi-functional landscapes from the grassroots? The role of rural producer movements. *Agriculture and Human Values* 33: 305–22.

Harvey, C., O. Komar, R. Chazdon, B. Ferguson, B. Finegan, D. Griffith, M. Martínez-Ramos, H. Morales, R. Nigh, L. Soto-Pinto, M. van Breugel and M. Wishnie. 2008. Integrating agricultural landscapes with biodiversity conservation in the Mesoamerican hotspot. *Conservation Biology* 22(1): 8–15.

Higuchi, A. 2015. Characteristics of consumers of organic products and the increase in the supply of these products in metropolitan Lima, Peru. *Apuntes* XLII(77): 57–89.

Holt-Giménez, E. 2002. Measuring farmers' agroecological resistance after Hurricane Mitch in Nicaragua: A case study in participatory, sustainable land management impact monitoring. *Agriculture, Ecosystems and Environment* 93: 1–3.

Holt-Giménez, E. 2008. *Campesino a campesino: Voces de Latinoamérica – Movimiento Campesino para la Agricultura Sustentable*. SIMAS, Managua, Nicaragua.

Holt-Giménez, E. and M. Altieri. 2013. Agroecology, food sovereignty, and the new green revolution. *Agroecology and Sustainable Food Systems* 37(1): 90–102.

Home, R., H. Bouagnimbeck, R. Ugas, M. Arbenz and M. Stolze. 2017. Participatory guarantee systems: Organic certification to empower farmers and strengthen communities. *Agroecology and Sustainable Food Systems* 41(5): 526–45.

IAASTD. 2010. *International Assessment of Agricultural Knowledge, Science, and Technology for Development*. World Bank, Washington DC.

Ibañez, M. and A. Blackman. 2016. Is eco-certification a win–win for developing country agriculture? Organic coffee certification in Colombia. *World Development* 82: 14–27.

IFAD. 2003. *The Adoption of Organic Agriculture among Small Farmers in Latin America and the Caribbean – Thematic Evaluation*. International Fund for Agricultural Development, Rome, Italy.

IFOAM – Organics International. 2011. *The Role of Smallholders in Organic Agriculture*. International Federation of Organic Agriculture Movements, Bonn, Germany.

IFOAM. 2012. *Historical Background on Internal Control Systems for Group Certification*. IFOAM Head Office, Bonn, Germany.

Imbach, P., E. Fung, L. Hannah, C. Navarro-Racines, D. Roubik, T. Ricketts, C. Harvey, C. Donatti, P. Läderach, B. Locatelli and P. Roehrdanz. 2017. Coupling of pollination services and coffee suitability under climate change. *Proceedings of the National Academy of Sciences* 114(39): 10438–42.

Jacobsen, S. 2011. The situation for quinoa and its production in southern Bolivia: From economic success to environmental disaster. *Journal of Agronomy and Crop Science* 197(5): 390–9.

Jena, P., T. Stellmacher and U. Grote. 2017. Can coffee certification schemes increase incomes of smallholder farmers? Evidence from Jinotega, Nicaragua. *Environment, Development and Sustainability* 19(1): 45–66.

Jurjonas, M., K. Crossman, J. Solomon and W. Lopez. 2016. Potential links between certified organic coffee and deforestation in a protected area in Chiapas, Mexico. *World Development* 78: 13–21.

Katto-Andrighetto, J., D. Bowen, F. Varini and C. Kirchner. 2017. *Guidelines for Public Support to Organic Agriculture*. IFOAM – Organics International, Bonn, Germany.

Lernoud, A. and F. Fonseca, Eds 2004. *Proceedings of the Workshop on Alternatives on Certification for Organic Production*. IFOAM and MAELA, Torres, Brazil.

Loconto, A., A. Poisot and P. Santacoloma (Eds). 2016. *Innovative Markets for Sustainable Agriculture – How Innovations in Market Institutions Encourage Sustainable Agriculture in Developing Countries*. FAO/INRA, Rome, Italy.

López-Bravo, D., E. Virginio-Filho and J. Avelino. 2012. Shade is conducive to coffee rust as compared to full sun exposure under standardized fruit load conditions. *Crop Protection* 38: 21–9.

Lyon, S., J. Aranda and T. Mutersbaugh. 2010. Gender equity in fairtrade – organic coffee producer organizations: Cases from Mesoamerica. *Geoforum* 41: 93–103.

Méndez, E., C. Bacon, M. Olson, S. Petchers, D. Herrador, C. Carranza, L. Trujillo, C. Guadarrama-Zugasti and A. Cordón and A. Mendoza. 2010. Effects of fair trade and organic certifications on small-scale coffee farmer households in Central America and Mexico. *Renewable Agriculture and Food Systems* 25(3): 236–51.

Milder, J. C., M. Arbuthnot, A. Blackman, S. E. Brooks, D. Giovannucci, L. Gross, E. T. Kennedy, K. Komives, E. F. Lambin, A. Lee, D. Meyer, P. Newton, B. Phalan, G. Schroth, , B. Semroc, H. V. Rikxoort and M. Zrust. 2015. An agenda for assessing and improving conservation impacts of sustainability standards in tropical agriculture. *Conservation Biology* 29: 309–20.

Moguel, P. and V. Toledo. 1999. Biodiversity conservation in traditional coffee systems of Mexico. *Conservation Biology* 13(1): 11–21.

Morris, K., V. Mendez, S. Taylor and M. Olson. 2013. Conventional food plot management in an organic coffee cooperative: Explaining the paradox. *Agroecology and Sustainable Food Systems* 37(7): 762–87.

Muschler, R. 2001. Shade improves coffee quality in a sub-optimal coffee-zone of Costa Rica. *Agroforestry Systems* 51(2): 131–9.

Nelson, E., S. Scott, J. Cukier and A. Leiva. 2009. Institutionalizing agroecology: Successes and challenges in Cuba. *Agriculture and Human Values* 26: 233–43.

Niggli, U., C. Andres, H. Willer and B. Baker. 2017. *A Global Vision and Strategy for Organic Farming Research*. TIPI – Technology Innovation Platform of IFOAM – Organics International, % Research Institute of Organic Agriculture (FiBL), Frick, Switzerland.

Petersen, P. 2018. Agroecology and the restoration of organic metabolisms in agrifood systems. In: T. Marsden (Ed.), *The SAGE Handbook of Nature: Three Volume Set*. SAGE Publications Ltd.

Pretty, J., A. Noble, D. Bossio, J. Dixon, R. Hine, F. Penning and J. Morison. 2006. Resource-Conserving Agriculture Increases Yields in Developing Countries. *Environmental Science and Technology* 40(4): 1114–19.

Rahn, E, P. Läderach, M. Baca, C. Cressy, G. Schroth, D. Malin, H. van Rikxoort and J. Shriver. 2014. Climate change adaptation, mitigation and livelihood benefits in coffee production: Where are the synergies? *Mitigation and Adaptation Strategies for Global Change* 19:1119–37.

Raynolds, L. 2008. The organic agro-export boom in the Dominican Republic. Maintaining tradition or fostering transformation? *Latin American Research Review* 43(1): 161–84.

Rice, R. 2008. Agricultural intensification within agroforestry: The case of coffee and wood products. *Agriculture, Ecosystems and Environment* 128(4).

Rover, O., B. Corrado de Gennaro and L. Roselli. 2017. Social innovation and sustainable rural development: The case of a Brazilian agroecology network. *Sustainability* 9: 3.

Ruben, R. and R. Fort. 2012. The impact of fair trade certification for coffee farmers in Peru. *World Development* 40(3): 570–82.

Sabourin, E. (Coord.). 2017. *Políticas Públicas a favor de la Agroecología en América Latina y el Caribe*. Red Políticas Públicas y Desarrollo Territorial en América Latina- PP-AL. Brasilia, Brasil, 213p.

Snider, A., I. Gutiérrez, N. Sibelet and G. Faure. 2017. Small farmer cooperatives and voluntary coffee certifications: Rewarding progressive farmers of engendering widespread change in Costa Rica? *Food Policy* 69: 231–42.

Souza, A., L. Maffia, F. Silva, E. Mizubuti and H. Teixeira. 2015. A time series analysis of brown eye spot progress in conventional and organic coffee production systems. *Plant Pathology* 64:157–66.

Stark, F., E. González-García, L. Navegante, T. Miranda, R. Poccard-Chapuis, H. Archimède and. C. Moulin. 2018. Crop-livestock integration determines the agroecological performance of mixed farming systems in Latino-Caribbean farms. *Agronomy for Sustainable Development* 38: 4.

Swensson, L. 2015. *Institutional Procurement of Food from Smallholder Farmers. The Case of Brazil.* FAO, Rome, Italy.

Timmermann, C. and G. Félix. 2015. Agroecology as a vehicle for contributive justice. *Agriculture and Human Values* 32:523–38.

Tittonell, P. 2014. Ecological intensification of agriculture – sustainable by nature. *Current Opinion in Environmental Sustainability* 8:53–61.

Tuck, S., C. Winqvist, F. Mota, J. Ahnström, L. Turnbull, J. Bengtsson and A. McKenzie. 2014. Land-use intensity and the effects of organic farming on biodiversity: A hierarchical meta-analysis. *Journal of Applied Ecology* 51: 746–55.

Ugas, R. 2003. Cooperation for certification in the south. *The Organic Standard* 29: 8–9.

Ugas, R., S. Vargas, P. Condor and H. Van den Eeckhout. 2014. Agroecology and rural livelihoods: Desirable or real effects on food security among smallholders in the Peruvian Andes? Preliminary report of a household survey. AGROECO Project, UNALM and IDRC-CIFSRF, Lima, Peru.

USDA. 2018. Organic trade. https://www.ers.usda.gov/topics/natural-resources-environment/organic-agriculture/organic-trade/ (Last updated 19 January 2018).

Vaast, P., B. Bertrand, J. Perriot, B. Guyot and M. Génard. 2006. Fruit thinning and shade improve bean characteristics and beverage quality of coffee (*Coffea arabica* L.) under optimal conditions. *Journal of the Science of Food and Agriculture* 86(2): 197–204.

Vagneron, I. and S. Roquigny. 2011. Value distribution in conventional, organic and fair trade banana chains in the Dominican Republic. *Canadian Journal of Development Studies* 32(3): 324–38.

Van den Eeckhout, H. and R. Ugás. 2013. A collective mark to improve local value chains for Peru's smallholder organic products. In: *FAO-IFOAM Asia-Pacific Regional Symposium: Entrepreneurship and Innovation in Organic Farming*, 2–4 December 2013, Bangkok, Thailand.

van der Ploeg, J. D. 2014. Peasant-driven agricultural growth and food sovereignty. *Journal of Peasant Studies* 41(6): 999–1030.

Vandermeer, J., I. Perfecto and S. Philpott. 2010. Ecological complexity and pest control in organic coffee production: Uncovering an autonomous ecosystem service. *BioScience* 60: 527–37.

Vandermeer, J., D. Jackson and I. Perfecto. 2014. Qualitative dynamics of the coffee rust epidemic: Educating intuition with theoretical ecology. *BioScience* 64(3): 210–18.

Vellema, W., A. Buritica, C. Gonzalez and M. D'Haese. 2015. The effect of specialty coffee certification on household livelihood strategies and specialisation. *Food Policy* 57: 13–25.

Wezel, A., S. Bellon, T. Doré, C. Francis, D. Vallod and C. David. 2009. Agroecology as a science, a movement and a practice. A review. *Agronomy for Sustainable Development* 29(4): 503–15.

Willer, H. and J. Lernoud, eds 2017. *The World of Organic Agriculture. Statistics and Emerging Trends 2017*. Research Institute of Organic Agriculture (FiBL), Frick, and IFOAM – Organics International, Bonn.

Winkel, T., P. Bommel, M. Chevarría-Lazo, G. Cortes, C. Del Castillo, P. Gasselin, F. Léger, J. Nina-Laura, S. Rambal, M. Tichit, J. Tourrand, J. Joinville, A. Vassas-Toral, M. Vieira-Pak and R. Joffre. 2016. Panarchy of an indigenous agroecosystem in the globalized market: The quinoa production in the Bolivian Altiplano. *Global Environmental Change* 39: 195–204.

World Bank. 2015. *Indigenous Latin America in the Twenty-First Century*. World Bank, Washington DC.

Zamilpa, J., D. Ayala and R. Schwentesius. 2015. *Desafíos y prioridades de la agricultura orgánica en México, mirando a la Unión Europea*. Cámara de Diputados, México.

Index

'AAC Oravena' 20
AD. *see* Anaerobic digestion (AD)
AE. *see* Agronomic effects (AE)
Aerated steam treatment. *see* 'ThermoSeed™'
Africa, organic agriculture in
  ecological organic agriculture-initiative (EOA-I) 481
  overview 477–478
  status 479–481
  steps to improve
    government investment in agriculture 484–485
    information communication technology (ICT) 487–488
    irrigation agriculture and wetlands 483
    landownership laws and policies 486–487
    merits 482
    Network for Organic Agricultural Research in Africa (NOARA) 488
    organic soil amendments and bio-pesticides 484
    policies for value chain 486
    subsidization and certification 486
    varieties of major staple crops 482–483
African Organic Conference 480, 488
African Union Commission 481
Agricultural Marketing Service (AMS) 48–49
Agricultural practices and biodiversity 404–406
  crop rotations 404
  fertilisation 404–405
  plant protection 405–406
  tillage operations 406
Agricultural Transformation Agenda 485
Agroecological Movement of Latin America (MAELA) 514
Agroecology 290–291, 512–513, 515–516, 515–516, 525–526, 530–531
Agroecosystems 402
Agronomic effects (AE) 108
ALCA. *see* Attributional LCA (ALCA)
Alternaria radicina 36
'Alternative Model' 525
Aluminium sulphate 317
Amase® 41
AMF. *see* Arbuscular mycorrhizal fungi (AMF)
Ammonia volatilization 189–190
AMS. *see* Agricultural Marketing Service (AMS)
Amylase trypsin inhibitors (ATIs) 24
Anaerobic digestion (AD) 184, 188–189
Anecic group 259
Animal foods 442–443
ANPE Peru. *see* National Association of Ecological Producers (ANPE Peru)
AOPEB Bolivia 514
AOSA. *see* Association of Official Seed Analysts (AOSA)
'Apol-Humus' 220
Apoplastic pathway 253
Arbuscular mycorrhizal fungi (AMF) 93–94, 157, 215–216, 256, 262–263, 283, 407, 414
Area Under the Disease Progress Curve (AUDPC) 278
ARNOA. *see* Asian Research Network of Organic Agriculture (ARNOA)
Asia, organic agriculture in
  crops
    cereals 493–494
    permanent crops, shrimp, fish and textiles 494
  improvement
    farmers' skills 502–503
    ISOFAR WOE 2015 503–504
    research needs 502
  overview 493
  producers
    characteristics of agriculture 495
    governmental support and certification 494–495
  rice farming types
    integrated 498
    rice bran 497
    rice–duck 497–498
    rice–fish 498
    rice–snail 498
  rice production systems 495
    crop rotation and diversity 496
    green manure 496–497
  vegetable production 498
    local and domestic markets 499
    manure and compost effect on nitrate levels 501–502
    nutrient management 499–500
    organic manuring and conventional fertilisers 500–501
    pesticides 499
    tropical regions 499
Asian Research Network of Organic Agriculture (ARNOA) 495
Associated biodiversity 92
Association of Official Seed Analysts (AOSA) 45
AssureWel project 360
ATIs. *see* Amylase trypsin inhibitors (ATIs)
Attributional LCA (ALCA) 371–372
AUDPC. *see* Area Under the Disease Progress Curve (AUDPC)
Authorised pesticides
  commercial products 315–316

criteria 313–314
defining 314
natural and synthetic substances 313–314
developments in last 25 years
historically used substances 311
substances approved since 1991 311–312
traditional substances 311
process 312–313
decision 312–313
duration 313
implementation 313
request 312
substances exempt 314–315
AVRDC. *see* World Vegetable Center (AVRDC)
Azolla spp. 496–497
Azospirillum 213–214
Azotobacter 214

Balfour, Lady Eve 61, 346
Band steaming 329
Banker plant systems 306
Ban Ki-Moon 504
'Barnhouse' 11
Basin irrigation 249
Bees 409–410
Berlin-Brandenburg Academy for Science and Humanities 503
Best Practice Guideline for Agriculture & Value Chains 357
BGA. *see* Blue-green algae (BGA)
Biocontrol organisms 276
Biodiversity
at different spatial scales 416–417
impact of organic farming on selected functional groups
biological pest control 418–420
pollination 420–421
and land-use intensity 402
agricultural practices 404–406
in conventional and organic farming 403–404
organic farming impact on 406
arthropods 409–412
benefits for rare species 415
flora 407–409
habitat diversity on organic farms 416
soil biota 412–414
vertebrates 414–415
overview 401–402
'BioFeed Amin' 221
'BioFeed Quality' 222
Biogas digestates 223–224
Biological cropping methods 346
Biological nitrogen fixation (BNF) 111, 214
Biopores 387
Birds and mammals 414–415
Blue-green algae (BGA) 214–215
BNF. *see* Biological nitrogen fixation (BNF)
Bonn University 503
Bordeaux mixture 311
Border strip irrigation 249
Brassica juncea 91
Brush weeders 326, 327
Burgundy mixture 311
Butterflies 411–412

CA. *see* Conservation agriculture (CA)
CAADP. *see* Comprehensive Africa Agricultural Development Programme (CAADP)
Calcium hydroxide 315
California Certified Organic Farmers 346
Catch crops 116
CCPs. *see* Composite Cross Populations (CCPs)
CCs. *see* Composite crosses (CCs); Cover crops (CCs)
Chemical weathering 244
Chlorophyll 35
Civil Society Mechanism 514
CLADES. *see* Latin American Consortium for Agroecology and Development (CLADES)
CLAs. *see* Conjugated linoleic acids (CLAs)
CLCA. *see* Consequential life cycle assessment (CLCA)
Clostridium botulinum 448
CMF. *see* Conventional mineral fertilisation (CMF)
Codex Alimentarius 350, 501
COF. *see* Conventional farming systems (COF)
Coffee Development Authority (UCDA) 486
Committee on Sustainability Assessment (COSA) 516
Community-supported agriculture 494
Composite crosses (CCs) 283–287, 290–293
Composite Cross Populations (CCPs) 10–11, 25–26
Composted manure 188
Comprehensive Africa Agricultural Development Programme (CAADP) 488
Conjugated linoleic acids (CLAs) 445, 447
Consequential life cycle assessment (CLCA) 370, 372
Conservation agriculture (CA) 274
Conservation Technology Information Center (CTIC) 154
Conservation tillage
benefits
energy and labour efficiency 158
impact on climate change 158–159
soil fauna 156–157
soil physical properties 156
challenges
farmers' reluctance 163–164
pests and diseases 162
soil density, microclimate and nutrient supply 161–162

suitable farm machinery 162–163
weeds 160–161
yields 159–160
grass-clover ley termination with 165
overview 153–155
Conservation tillage techniques 165
Conventional farming systems (COF) 500
Conventional mineral fertilisation (CMF) 500
Copper 272
COSA. *see* Committee on Sustainability Assessment (COSA)
Council Directive 2014/150/EU 291
Cover crops (CCs)
carbon sequestration and greenhouse gas emissions 94–95
integration in no-till cultivation 95–96
nutrient cycling 89–90
overview 87–89
promoting biodiversity 92–94
soil structure and erosion control 89
weed, pest and disease control 90–92
CRISPR/CAS gene editing technique 3, 4
Crop breeding
breeding targets 13–14
initiatives and networks 11
issues for cultivars
biodiversity 9–12
competitiveness against weeds 8–9
nutritional quality 7
plant health 7–8
seed saving 6–7
social aspects 12
origins and aims 6
overview 3, 5–6
plant health 16–21
quality requirements 22–24
Crop propagation material 45
Crop rotation 260–261, 404, 496
and crop yields 120–122
designing
conversion period 123–124
distribution 124–125
length 124
sequence 124
spatial arrangements 125–126
measuring and modelling nutrient and humus balances 126–129
nutrient effects 109, 111–114, 116
overview 105–106
precrop effects 107–108
principles 106–107
on stockless farms 129
forage legume biomass management 131–132
legume-based 130–131
suppressing weeds, diseases and pests 117–120
Crop-specific precrop effects (CSPE) 108
Crotalaria juncea 91
CSPE. *see* Crop-specific precrop effects (CSPE)
CTIC. *see* Conservation Technology Information Center (CTIC)
Cyanobacteria 214–215
Cytoplasm 253

Dankook University 495
Darcy equation 247
Data acquisition 390
Data processing 390
Deep litters 181
Deep rooting systems 387
Demeter 346
Denitrification 194
Deoxynivalenol (DON) 162
Department of Input Support Services 486
Derris elliptica 311
Deutsche Bank 3
Diet and health
composition 437–438
determinants 438
effects of food composition 452
organic food
beneficial components and doses 443–445, 447
benefits 438–439
combined impact 450, 452
conventional and 440, 442–443
differences in components 440
harmful components and doses 447–449
social aspects 449–450
overview 435–436
psycho-social effects 453
Digital technologies 390–391, 394
Dilution effect 7
Disease-suppressive (DS) system 92
Disinfectants 310
DON. *see* Deoxynivalenol (DON)
Dormancy 45
'Dottenfelderhof' biodynamic farm 11
Double-layer ploughs 163
Drip irrigation 249
DS. *see* Disease-suppressive (DS) system
Dwarf bunt 18, 20
Dynamic management 283

Earthworms 157, 259, 412–413
'The East African Organic Products Standards' 486
EC Commission Regulation
540/2011 307
834/2007 47, 62, 360
889/2008 47, 63, 307, 313, 315, 360
2003/2003 227
2078/92 348
2371/2002 219
Eco-efficiency 369

Eco-functional intensification (EFI)
challenges
concept fuzziness 394
indeterministic system behaviour 392–393
limits to optimization 393
socioeconomic framework 393–394
upscaling and transferability 392
ecological regulation 388–389
material and energy flows 385–388
overview 383–385
technical developments 389–392
Ecological organic agriculture (EOA) 478
Ecological organic agriculture-initiative (EOA-I) 481, 482, 488
Ecological weed management (EWM) 335
Ecology principle 62–63
ECO-PB. *see* European Consortium for Organic Plant Breeding (ECO-PB)
Ecosystem diversity 402
Ecosystem services 61, 69, 76
ECX. *see* Ethiopia Commodity Exchange (ECX)
E-dressing. *see* 'E-ventus' technology
Effective Microorganisms (EM) 217
EFI. *see* Eco-functional intensification (EFI)
ELAO. *see* Latin American Meetings of Ecological and Organic Agriculture (ELAO)
Electron seed treatment. *see* 'E-ventus' technology
EM. *see* Effective Microorganisms (EM)
Embryogenesis 35
Endogeic earthworms 259
Environmental performance
efficiency, consistency and sufficiency 370–371
improvement 377–378
methods for assessing
food systems modeling 374–375, 377
life cycle assessment 371–372
multi-criteria assessment approaches 373
productivity analysis 373
proxy indicators and aggregations 377
notions 369–370
overview 365–368
Environmental Science Council 405
EOA. *see* Ecological organic agriculture (EOA)
EOA-I. *see* Ecological organic agriculture-initiative (EOA-I)
Epigeic earthworms 259
Eruca sativa 91
ET. *see* Evapotranspiration (ET)
Ethiopia Commodity Exchange (ECX) 487
Ethylene 311
EU. *see* European Union (EU)
EU-Core Organic consortium 287
EU organic legislation 314, 318–319
EU Regulation 356
European Academies 405
European Commission 9, 48, 51, 312–313, 356
European Consortium for Organic Plant Breeding (ECO-PB) 4
European Directive 2008/62/EC 9
European Union (EU) 5, 47, 51–52, 53, 211, 304, 307, 309, 312, 347–348, 349, 351, 356
Evaporation 251–252
Evapotranspiration (ET) 253
'E-ventus' technology 38
Evolutionary plant breeding 283–287
EWM. *see* Ecological weed management (EWM)
Expert Group for Technical Advice for Organic Production (EGTOP) 310, 311, 312–313, 316, 317, 319

F1-hybrids 5, 25
FAO. *see* Food and Agriculture Organisation (FAO)
FAO Agroecology 514
Farmers' declaration 480
Farmer to Farmer Movement. *see* Movimiento Campesino a campesino
Farm-saved seeds 49
Farmyard manure (FYM) 181, 465, 467–472
Federal Government of Nigeria 485
Federal Ministry of Agriculture & Rural Development (FMA&RD) 486
Feedlot manure 181
Ferric phosphate 311
FiBL. *see* Research Institute of Organic Agriculture (FiBL)
Finger weeders 326
Flood irrigation 249
Flora in grassland 408–409
FMA&RD. *see* Federal Ministry of Agriculture & Rural Development (FMA&RD)
FODMAPs 24
Food and Agriculture Organisation (FAO) 153–154, 274, 350, 357, 485
Forage legumes 116
Forschung und Züchtung Dottenfelderhof (FZD) 11
Free radical theory of ageing 443, 445
Furrow irrigation 249
Fusarium culmorum 17
Fusarium disease 16–17
Fusarium graminearum 17
Fusarium gramineum 162
FYM. *see* Farmyard manure (FYM)
FZD. *see* Forschung und Züchtung Dottenfelderhof (FZD)

Genetically modified Organisms (GMOs) 50, 308
Genetic diversity 402
Genetic engineering technologies 4
German Federal Ministry of Food and Agriculture 5

German Federal Plant Variety Office 8
German Fraunhofer Institut für Elektronenstrahl- und Plasmatechnik 38
GESS. *see* Growth Enhancement Support Scheme (GESS)
GHG. *see* Greenhouse gas (GHG) emissions
Global Programme for Food Security 481
Global warming potential (GWP) 367
Globodera pallida 91
Glutenin macropolymers (GMP) 23
GMOs. *see* Genetically modified Organisms (GMOs)
GMP. *see* Glutenin macropolymers (GMP)
Goëmar BM 86® 222
Golden Apple Snail (GAS). *see Pomacea canaliculata*
Grass–clover mixture precrop 120–121, 165
Grassland-based ruminant production 371, 375
Gravitational potential 247
Greenhouse gas (GHG) emissions 88–89
Green manure (GM). *see* Plant-based manures
Green Revolution 514
Growth Enhancement Support Scheme (GESS) 484, 485, 487
Guano 217–218
GWP. *see* Global warming potential (GWP)

Hedgerow management 275
Herbal decoctions 310
Herbicide-resistant crops 6
High potential PD 118
Home field advantage effect 7
Humic acids 224–225
'Humin Plus' 225
Humus balancing models 128

ICS. *see* Internal Control Systems (ICS)
ICT. *see* Information communication technology (ICT)
IFOAM. *see* International Federation for Organic Agriculture Movements (IFOAM)
IFOAM International Basic Standards for Organic Agriculture Production and Processing for Organic Agriculture 402
Indicator-based Sustainability Assessment Framework (MESMIS) 526
Information communication technology (ICT) 487–488
Inoculation biocontrol 306
Integrated organic approach 3
Integrated weed management (IWM)
  intra-row weed control 326–330
  overview 323–324, 326
  problematic 330–331
  in reduced- and no-till organic systems 331, 333
  and sustainable organic crop production 333–334
Interamerican Commission on Organic Agriculture 525
Interamerican Institute for Cooperation in Agriculture 525
Internal Control Systems (ICS) 520, 522
International Federation for Organic Agriculture Movements (IFOAM) 4, 47, 51, 55, 106, 133, 276, 293, 348, 356, 357, 358, 360, 385, 481, 486, 501, 503, 513, 514, 522
International Rice Research Institute (IRRI) 495, 505
International Rules for Seed Testing 45
International Seed Testing Association (ISTA) 45–46
International Society of Organic Agriculture Research (ISOFAR) 503
International Society of Organic Agriculture Research (ISOFAR) World Organic Expo (WOE) 2015 503–504
International Year of Family Farming 514
Inundation biocontrol 306
IRRI. *see* International Rice Research Institute (IRRI)
Irrigation systems 249
ISARA Lyon 167, 169
ISOFAR. *see* International Society of Organic Agriculture Research (ISOFAR)
ISOFAR WOE 2015. *see* International Society of Organic Agriculture Research (ISOFAR) World Organic Expo (WOE) 2015
ISTA. *see* International Seed Testing Association (ISTA)
IWM. *see* Integrated weed management (IWM)

'Jularo' 14

Kampmann, Wolfgang 11
Kelpak® 222
Keyserlingk Institute 11
'Klostergut Wiebrechtshausen' estate 11
Köpke, Ulrich 503
Korean Organic Farming system 500
'Kultursaat' 11

LAC. *see* Latin America and Caribbean (LAC)
LAI. *see* Leaf area index (LAI)
Lambda-cyhalothrin 312
Land equivalent ratio (LER) 392
Land reform in Nigeria 487
Land Use Act (1978) 487
Land-use intensity 402–404
Latin America, organic agriculture in
  institutions and policies
    national programmes 525–526
    national standards and regulations 525
  organisations and movements 513–515
  overview 509–511

performance 526–527, 529–531
principles 512–513
production and business
certified and non-certified 516–519
characterisation 519–520
guarantee systems 520, 522–523
markets for products 523–524
Latin America and Caribbean (LAC) 510, 515–516
see also Latin America, organic agriculture in
Latin American Congress on Agroecology 516
Latin American Consortium for Agroecology and Development (CLADES) 515
Latin American Meetings of Ecological and Organic Agriculture (ELAO) 514
Latin American Scientific Society for Agroecology (SOCLA) 513, 515, 516
LCA. *see* Life cycle assessment (LCA)
Leaf area index (LAI) 9, 259–260
Lecithin 315
Legume fatigue 119
Leguminous alley cropping systems 125, 128
LEISA. *see* Low-external-input sustainable agriculture (LEISA)
LER. *see* Land equivalent ratio (LER)
Lessivage 244
Life cycle assessment (LCA) 357, 367–369, 371–372
Lime sulphur 317
Liquid manure
application methods 190
storage 185–186
London–van der Waals forces 245
Long-term nitrogen availability
factors affecting residual effects 198
management criteria 198
processes 196–197
Loose smut 20
Louis Bolk Institute 11
Low-external-input sustainable agriculture (LEISA) 515
Low potential PD 118
Lumpkin, Thomas 502

MAELA. *see* Agroecological Movement of Latin America (MAELA)
Malabo Declaration on Agricultural Growth and Transformation 484
Mandaamin Institute 7
Manure management 217–218
composition and properties 181
nitrogen losses 183–184
field application methods
liquid 190
solid 190–191
timing 191–193
nitrogen turnover and availability
long-term 196–198
short-term 193–196
overview 179–181
plant-based manures 200–201
storage and treatment
anaerobic digestion 188–189
hygienic aspects 184–185
liquid 185–186
solid 186–188
utilization
phosphorus 198–199
potassium 199
sulphur 200
Matric potential 247
Maturation process 35
May, Robert M. 392
Mechanical weed control 14
Meloidogyne incognita 91
MESMIS. *see* Indicator-based Sustainability Assessment Framework (MESMIS)
Metaldehyde 311
Metarhizium anisopliae 162
Microbial biocontrol agents 311
Microbial organisms 413–414
Mineral fertilizers 128
Ministry of Agriculture and Cooperative in Thailand 494
Minjingu rock phosphate (MRP) 465, 467–472
Mixed cropping 125
Mouldboard ploughs 162–163
Movimiento Campesino a campesino 515
MRP. *see* Minjingu rock phosphate (MRP)
Müller–Rusch movement 346
'Mycosat' 216

National Annex 53–54
National Association of Ecological Producers (ANPE Peru) 514, 523, 530
National Organic Movements 474
National Organic Program (NOP) 48–49, 347
National Plan for Agroecology and Organic Production (PLANAPO) 513, 525
National Plan for Organic Banana (PNBO) 525
National Program for Organic Agriculture (PNAO) 525
National School Feeding Program (PNAE) 525
Natural biocontrol mechanisms 272
Nature & More 357
'Naturland' 11, 347
NBTs. *see* New breeding techniques (NBTs)
NCGS. *see* Non-celiac gluten sensitivity (NCGS)
Network for Organic Agricultural Research in Africa (NOARA) 488
New breeding techniques (NBTs) 4
N-fixing crops 179
Nitrate 448
Nitrite 448
Nitrogen
mineralization and immobilization 193–195
supply and demand for 69

Nitrous oxide (N_2O) 158–159
NOARA. *see* Network for Organic Agricultural Research in Africa (NOARA)
Non-celiac gluten sensitivity (NCGS) 24
Non-crop rotation strategies 119
Non-N-fixing crops 179–180
NOP. *see* National Organic Program (NOP)
No-tillage (NT) practices 94, 161–162, 163–164
 in France 169
 in Switzerland 168
Nutrient balances 126–127
Nutrient supply 259–260

Oats 20
On-farm breeding 7
On-land ploughs 163
Onsite breeding 11
Optimal trait values 13
Optimizing Subsidiary Crop Applications in Rotations (OSCAR) 277, 279, 281, 282, 283
'Organic 3.0' 289, 357–358, 368
Organic breeding initiatives and networks 11
Organic Farming 504
Organic fertilizers
 animal excrement 217–218
 animal origin
  chitin 219–220
  raw/processed animal residues 218–219
 biochar 227–228
 and biofertilizers
  mycorrhiza 215–216
  nitrogen-fixing microorganisms 212–215
  non-nitrogen-fixing bacteria 215
 composts 226–227
 microorganisms to improve effectiveness 216–217
 overview 211–212
 plant origin
  fermented products 223–224
  leonardites 224–225
  molasses 220–221
  mushroom culture wastes 223
  natural waste fibres 221
  peat 225
  plant-derived hydrolysates 221
  sapropel and sediments 225–226
  sawdust and wood chips 222–223
  seaweeds and seaweed products 221–222
  stillage and stillage extract 224
  wood ash 223
 untreated minerals and by-products 227
Organic Food Protection Act (1990) 347
'Organic Life – Science meets public' 504
'Organic lifestyle' 449–450
Organic products market
 exports 524
 farmers 523
 retail 523
 specialised/institutional purchases 523–524
'Organic Regulation' 310
Organic Rice Standards 495
Organics International 276, 293, 358, 481, 486
Organic standards
 aligning rules with sustainability assessments 356–359
 historic development and control systems 346–348
 overview 345–346
 production principles 348–349
 production regulation 349–351
  commitment to good practice 355
  defining outcomes 356
  prohibitions and restrictions 351, 354
  use of inputs 354–355
Organic Systems Plan 55
Organic World Congress 488, 514
OSCAR. *see* Optimizing Subsidiary Crop Applications in Rotations (OSCAR)
Osmotic potential 247

PA. *see* Productivity analysis (PA)
Participatory guarantee system (PGS) 479, 494, 514, 522
Passive capillary samplers (PCAPS) 500
PC. *see* Previous crop (PC)
PCAPS. *see* Passive capillary samplers (PCAPS)
PD. *see* Pests and diseases (PD)
Pea cover crop 168
Peruvian Andes 530–531
Pests and diseases (PD)
 categories 117–118
 control 125
 management
  agroecology-based ecological plant protection 289–292
  integrating diversity through evolutionary breeding 283–287
  overview 271–272
  plant protection principles 272–276
  soil regeneration and effects on potato health 276–283
'Peter Kunz' 11
PGPR. *see* Plant-growth-promoting rhizobacteria (PGPR)
PGS. *see* Participatory guarantee system (PGS)
Phosphonates 317
Physical weathering 244
Phytochemicals 452
PLANAPO. *see* National Plan for Agroecology and Organic Production (PLANAPO)
Planned biodiversity 92
Plant-based manures 200–201, 496–497
Plant foods 440, 442

Plant-growth-promoting rhizobacteria (PGPR) 212
Plant protection
animal origin substances 308
authorised substances availability 309
direct and indirect 303–304
efforts to replace copper fungicides
alternatives 317–318
minimisation of use 316–317
sources in agricultural production 318
European Union legislation 307
grey areas in legislation 309–310
invertebrate biocontrol agents 307
materials authorisation dynamics
commercial products 315–316
criteria 313–314
developments in last 25 years 310–312
process 312–313
substances exempt 314–315
microbial biocontrol agents 307–308
microbial origin substances 308
paraffin oil 308
permitted/prohibited substances 304
pesticides regulation 304–305
plant origin substances 308
practices
pests control 306
plant diseases control 305–306
weeds control 305
Plasmalemma 253
PNAE. *see* National School Feeding Program (PNAE)
PNAO. *see* National Program for Organic Agriculture (PNAO)
PNBO. *see* National Plan for Organic Banana (PNBO)
Podzolization 244
Polyphagous beetles 410–411
Pomacea canaliculata 498
Potassium hydrogen carbonate 317
Prairie Recommending Committee for Oat and Barley (PRCOB) 20
Pratylenchus penetrans 91
PRCOB. *see* Prairie Recommending Committee for Oat and Barley (PRCOB)
Presidential Technical Committee 487
Pressure potential 247
Previous crop (PC) 108, 127
Priming technique 42
Processed foods 443
Productivity analysis (PA) 373, 374
'Productivity and Growth in Organic Value Chains' (ProGrOV) project in East Africa
field study approach
experimental design and treatments 465
site description 464–465
legumes integration and farmyard manure
on kale performance 471–472
on maize performance 470–471
on tomato performance 469
overview 463–464
use of mulch 466–467
ProGrOV. *see* 'Productivity and Growth in Organic Value Chains' (ProGrOV) project in East Africa
Pseudomonas chlororaphis 41
Push-and-pull system 117
Pyrethroids deltamethrin 312

Ratoon cropping 125
RDA. *see* Rural Development Administration (RDA)
Recreation and aesthetics 126
Registration costs 309
Relay cropping 125
RePCro regenerative potato cropping system 278, 282–283, 292
Research Institute of Organic Agriculture (FiBL) 165, 168, 514, 516
Réseau Semences Paysannes 49
Resource Pool Diversity Hypothesis (RPDH) 334
Rhizobia 41, 213
Rhizobium 213
Rhizoctonia solani 92
Robotic weeding 328
Rock phosphate 128
Rodale Institute in Pennsylvania 331
Rove beetles 410
RPDH. *see* Resource Pool Diversity Hypothesis (RPDH)
Rural Development Administration (RDA) 495
Rural Development Programmes 348
Rust diseases 17
Ryania speciosa 311
Rye cover crop 169

SAFA. *see* Sustainability Assessment of Food and Agriculture systems (SAFA)
Sang Mok Sohn 503, 504
'Seasol' 222
Secondary metabolites 444–445
Sedimentation value 23
Seed-borne diseases 34
Seed filling phase 35
Seed production
application of biologicals 41–42
of arable crops 51
certification and testing
germination 45–46
health 46–47
contamination issues
accidental chemical 50
and availability of organic seeds 51–53
genetically modified seeds 50–51
policy 53–55

disease
and control of pathogens 37–38
natural products with biocidal activity 39–40
physical seed sanitation methods 38–39
maturity 35
overview 33–34
priming and pelleting 42–43
storage 43, 45
quality control issues 35
regulations 47–49
Seed Treatments for Organic Vegetable Production (STOVE) 38, 39
Selection with markers and advanced reproductive technologies (SMART) 3
Sequential cropping 125
Sewage sludge 67
Shallower rooting systems 387
Shallow on-land ploughing 167
Shaw, W. C. 324
Short-term nitrogen availability 193
management criteria 196
processes in soil 194–196
Silage maize 168
SIMAS Mesoamerican Information Service on Sustainable Agriculture 515–516
Slurry 217–218
Small Producers' Symbol (SPP) 524
SMART. *see* Selection with markers and advanced reproductive technologies (SMART)
SMART-Farm. *see* Sustainability Monitoring and Assessment Routine (SMART-Farm) Tool
SMS. *see* Spent mushroom substrates (SMS)
SOAAN. *see* Sustainable Organic Agriculture Action Network (SOAAN)
SOC. *see* Soil organic carbon (SOC)
SOCLA. *see* Latin American Scientific Society for Agroecology (SOCLA)
Software-based models 126
Soil Association standards 346
Soil biodiversity 262–263
Soil-borne diseases 118
Soil compaction 167
Soil disinfestation 276
Soil fertility and health
designing rotations 66
management and functions/ecosystem services 69, 76
organic matter maintenance 63, 65
overview 61–63
supply and demand for nitrogen 69
sustainable sources of nutrients 67–68
Soil formation 244–245
Soil genesis 244
Soil inversion techniques 162
Soil organic carbon (SOC) 156, 158, 159, 180
Soil organic matter (SOM) 256, 261
Soil porosity 245–246
Soil structure 167, 255–257
Solar energy 251
Solid manure 186–188
application methods 190–191
composting
advantages 188
ammonia losses 186
vs. anaerobic storage 186
crop nitrogen availability 187
effects on soil organic carbon 188
gaseous losses 187
storage 186
Solid phase residues 224
SOM. *see* Soil organic matter (SOM)
Species diversity 402
Spent mushroom substrates (SMS) 223
Spiders 411
SPP. *see* Small Producers' Symbol (SPP)
Sprinkler irrigation 249
SSNC. *see* Swedish Society for Nature Conservation (SSNC)
Standard-setting process 359
see also Organic standards
Starter-effect 199
Stegemann, Ernst 6
Steiner, Rudolf 6, 346
STOVE. *see* Seed Treatments for Organic Vegetable Production (STOVE)
Streptomyces sp. 92
Strip tillage 163
Stubble catch crop 125
Sub-irrigation 249
Subsoil drainage schemes 248–249
Sulphur 272
Sulphur-based potassium fertilizer 128
Sustainability Assessment of Food and Agriculture systems (SAFA) 357, 374
Sustainability Monitoring and Assessment Routine (SMART-Farm) Tool 374
Sustainable farming and consumption 358
Sustainable Organic Agriculture Action Network (SOAAN) 357, 360
Swedish Society for Nature Conservation (SSNC) 481
Swiss Agency for Development and Cooperation 481

'Talkunar' 20
Tanzania Certification Body (TANCERT) 486
Technology Innovation Platform of IFOAM–Organics International (TIPI) 368
'ThermoSeed™' 38
Tillage systems 258–259
TIPI. *see* Technology Innovation Platform of IFOAM–Organics International (TIPI)
Torsion weeders 326, 327
Transmembrane pathway 253

Transmissible spongiform encephalopathy (TSE) 218
Trichogramma brassicae 306
TSE. *see* Transmissible spongiform encephalopathy (TSE)

UCDA. *see* Coffee Development Authority (UCDA)
Uganda Organic Certification LTD (UGOCERT) 486
UN Committee on World Food Security 514
Undersown crop 125
United Nations 3, 6
United States Department of Agriculture (USDA) 48–49, 347
US Environmental Protection Agency 484

Value for cultivation and use (VCU) 21, 24
Vascular bundles 252
Vermicompost 226–227
Verticillium dahliae 91
Volatile fatty acids (VFAs) 183
von Humboldt, Alexander 503

Wageningen University & Research 39, 50
Water and weathering 244–245
Water management
  aspects affecting availability and use
    soil management 255–261
  developments
    managing soil limitations 262–263
  soil–plant–atmosphere continuum 250–253
  soil water balance 253–254
  use efficiency by crops 254–255
  water on earth and in atmosphere 243–249
Water movement 246–248
Water storage and availability 245–246
Water use efficiency (WUE) 254–255
Weed control 466–467
Weed flora in arable fields 407–408
Wheat 16–20
Winter barley 21
World Agriculture Report 5, 6
World Bank 5
World Vegetable Center (AVRDC) 495, 502, 505
'Worldwide Integrated Assessment of the Impacts of Systemic Pesticides on Biodiversity and Ecosystems' 405
WUE. *see* Water use efficiency (WUE)

Xanthomonas campestris 36

www.ingramcontent.com/pod-product-compliance
Lightning Source LLC
LaVergne TN
LVHW020628300926
849161LV00004B/645